POWER ELECTRONICS

ABOUT THE AUTHORS

Ned Mohan is a professor in the Department of Electrical Engineering at the University of Minnesota, where he holds the Oscar A. Schott Chair in Power Electronics. He has worked on several power electronics projects sponsored by the industry and the electric power utilities, including the Electric Power Research Institute. He has numerous publications and patents in this field.

Tore M. Undeland is a Professor in Power Electronics in the Faculty of Electrical Engineering and Computer Science at the Norwegian Institute of Technology. He is also Scientific Advisor to the Norwegian Electric Power Research Institute of Electricity Supply. He has been a visiting scientific worker in the Power Electronics Converter Department of ASEA in Vaasteras, Sweden, and at Siemens in Trondheim, Norway, and a visiting professor in the Department of Electrical Engineering at the University of Minnesota. He has worked on many industrial research and development projects in the power electronics field and has numerous publications.

William P. Robbins is a professor in the Department of Electrical Engineering at the University of Minnesota. Prior to joining the University of Minnesota, he was a research engineer at the Boeing Company. He has taught numerous courses in electronics and semiconductor device fabrication. His research interests are in ultrasonics, pest insect detection via ultrasonics, and micromechanical devices, and he has numerous publications in this field.

POWER ELECTRONICS

Converters, Applications, and Design

SECOND EDITION

NED MOHAN
Department of Electrical Engineering
University of Minnesota
Minneapolis, Minnesota

TORE M. UNDELAND
Faculty of Electrical Engineering and Computer Science
Norwegian Institute of Technology
Trondheim, Norway

WILLIAM P. ROBBINS
Department of Electrical Engineering
University of Minnesota
Minneapolis, Minnesota

JOHN WILEY & SONS, INC.
New York   Chichester   Brisbane   Toronto   Singapore

Acquisitions Editor	Steven M. Elliot
Developmental Editor	Sean M. Culhane
Marketing Manager	Susan Elbe
Senior Production Editor	Savoula Amanatidis
Text Designer	Lynn Rogan
Cover Designer	David Levy
Manufacturing Manager	Lori Bulwin
Illustration Coordinator	Jaime Perea

This book was typeset in Times Roman by The Clarinda Company, and printed and bound by Hamilton Printing Company. The cover was printed by NEBC.

PSpice is a registered trademark of MicroSim Corporation.
MATLAB is a registered trademark of The MathWorks, Inc.

Library of Congress Cataloging in Publication Data:

Mohan, Ned.
    Power electronics : converters, applications, and design / Ned
Mohan, Tore M. Undeland, William P. Robbins.—2nd ed.
      p.   cm.
    Includes bibliographical references and indexes.
    ISBN 0-471-58408-8 (cloth)
  1. Power electronics. 2. Electric current converters. 3. Power
semiconductors. I. Undeland, Tore M. II. Robbins, William P.
III. Title.
TK7881.15.M64 1995
621.317—dc20   94-21158
                                        CIP

Printed in the United States of America.

10 9

To Our Families . . .
Mary, Michael, and Tara
Mona, Hilde, and Arne
Joanne and Jon

PREFACE

SECOND EDITION

The first edition of this book was published in 1989. The basic intent of this edition remains the same; that is, as a cohesive presentation of power electronics fundamentals for applications and design in the power range of 500 kW or less, where a huge market exists and where the demand for power electronics engineers is likely to be. Based on the comments collected over a five-year period, we have made a number of substantial changes to the text. The key features are as follows:

- An introductory chapter has been added to provide a review of basic electrical and magnetic circuit concepts, making it easier to use this book in introductory power electronics courses.
- A chapter on computer simulation has been added that describes the role of computer simulations in power electronics. Examples and problems based on PSpice® and MATLAB® are included. However, we have organized the material in such a way that any other simulation package can be used instead or the simulations can be skipped altogether.
- Unlike the first edition, the diode rectifiers and the phase-controlled thyristor converters are covered in a complete and easy-to-follow manner. These two chapters now contain 56 problems.
- A new chapter on the design of inductors and transformers has been added that describes easy-to-understand concepts for step-by-step design procedures. This material will be extremely useful in introducing the design of magnetics into the curriculum.
- A new chapter on heat sinks has been added.

ORGANIZATION OF THE BOOK

This book is divided into seven parts. Part 1 presents an introduction to the field of power electronics, an overview of power semiconductor switches, a review of pertinent electric and magnetic circuit concepts, and a generic discussion of the role of computer simulations in power electronics.

Part 2 discusses the generic converter topologies that are used in most applications. The actual semiconductor devices (transistors, diodes, and so on) are assumed to be ideal, thus allowing us to focus on the converter topologies and their applications.

Part 3 discusses switch-mode dc and uninterruptible power supplies. Power supplies represent one of the major applications of power electronics.

Part 4 considers motor drives, which constitute another major applications area.

Part 5 includes several industrial and commercial applications in one chapter. Another chapter describes various high-power electric utility applications. The last chapter in this part of the book examines the harmonics and electromagnetic interference concerns and remedies for interfacing power electronic systems with the electric utilities.

Part 6 discusses the power semiconductor devices used in power electronic converters including diodes, bipolar junction thyristors, metal–oxide–semiconductor (MOS) field effect transistors, thyristors, gate turn-off thyristors, insulated gate bipolar transistors, and MOS-controlled thyristors.

Part 7 discusses the practical aspects of power electronic converter design including snubber circuits, drive circuits, circuit layout, and heat sinks. An extensive new chapter on the design of high-frequency inductors and transformers has been added.

PSPICE SIMULATIONS FOR TEACHING AND DESIGN

As a companion to this book, a large number of computer simulations are available directly from Minnesota Power Electronics Research and Education, P.O. Box 14503, Minneapolis, MN 55414 (Phone/Fax: 612-646-1447) to aid in teaching and in the design of power electronic systems. The simulation package comes complete with a diskette with 76 simulations of power electronic converters and systems using the classroom (evaluation) version of PSpice for IBM-PC-compatible computers, a 261-page detailed manual that describes each simulation and a number of associated exercises for home assignments and self-learning, a 5-page instruction set to illustrate PSpice usage using these simulations as examples, and two high-density diskettes containing a copy of the classroom (evaluation) version of PSpice. This package (for a cost of $395 plus a postage of $4 within North America and $25 outside) comes with a site license, which allows it to be copied for use at a single site within a company or at an educational institution in regular courses given to students for academic credits.

SOLUTIONS MANUAL

As with the first edition of this book, a solutions manual with completely worked-out solutions to all the problems is available from the publisher.

ACKNOWLEDGMENTS

We wish to thank all the instructors who have allowed us this opportunity to write the second edition of our book by adopting its first edition. Their comments have been most useful. We are grateful to Professors Peter Lauritzen of the University of Washington, Thomas Habetler of the Georgia Institute of Technology, Daniel Chen of the Virginia Institute of Technology, Alexander Emanuel of the Worcester Polytechnic Institute, F. P. Dawson of the University of Toronto, and Marian Kazimierczuk of the Wright State University for their helpful suggestions in the second edition manuscript. We express our sincere appreciation to the Wiley editorial staff, including Steven Elliot, Sean Culhane, Lucille Buonocore, and Savoula Amanatidis, for keeping us on schedule and for many spirited discussions.

Ned Mohan
Tore M. Undeland
William P. Robbins

CONTENTS

PART 1   INTRODUCTION   **1**

Chapter 1   Power Electronic Systems   **3**

1-1 Introduction   3
1-2 Power Electronics versus Linear Electronics   4
1-3 Scope and Applications   7
1-4 Classification of Power Processors and Converters   9
1-5 About the Text   12
1-6 Interdisciplinary Nature of Power Electronics   13
1-7 Convention of Symbols Used   14
    Problems   *14*
    References   *15*

Chapter 2   Overview of Power Semiconductor Switches   **16**

2-1 Introduction   16
2-2 Diodes   16
2-3 Thyristors   18
2-4 Desired Characteristics in Controllable Switches   20
2-5 Bipolar Junction Transistors and Monolithic Darlingtons   24
2-6 Metal–Oxide–Semiconductor Field Effect Transistors   25
2-7 Gate-Turn-Off Thyristors   26
2-8 Insulated Gate Bipolar Transistors   27
2-9 MOS-Controlled Thyristors   29
2-10 Comparison of Controllable Switches   29
2-11 Drive and Snubber Circuits   30
2-12 Justification for Using Idealized Device Characteristics   31
    Summary   *32*
    Problems   *32*
    References   *32*

Chapter 3   Review of Basic Electrical and Magnetic Circuit Concepts   **33**

3-1 Introduction   33
3-2 Electric Circuits   33
3-3 Magnetic Circuits   46
    Summary   *57*
    Problems   *58*
    References   *60*

Chapter 4   Computer Simulation of Power Electronic Converters and Systems   **61**

4-1 Introduction   61
4-2 Challenges in Computer Simulation   62
4-3 Simulation Process   62
4-4 Mechanics of Simulation   64
4-5 Solution Techniques for Time-Domain Analysis   65
4-6 Widely Used, Circuit-Oriented Simulators   69
4-7 Equation Solvers   72
Summary   74
Problems   74
References   75

PART 2   GENERIC POWER ELECTRONIC CIRCUITS   **77**

**Chapter 5   Line-Frequency Diode Rectifiers: Line-Frequency ac →
Uncontrolled dc**   **79**

5-1 Introduction   79
5-2 Basic Rectifier Concepts   80
5-3 Single-Phase Diode Bridge Rectifiers   82
5-4 Voltage-Doubler (Single-Phase) Rectifiers   100
5-5 Effect of Single-Phase Rectifiers on Neutral Currents in Three-Phase, Four-Wire Systems   101
5-6 Three-Phase, Full-Bridge Rectifiers   103
5-7 Comparison of Single-Phase and Three-Phase Rectifiers   112
5-8 Inrush Current and Overvoltages at Turn-On   112
5-9 Concerns and Remedies for Line-Current Harmonics and Low Power Factor   113
Summary   113
Problems   114
References   116
Appendix   117

**Chapter 6   Line-Frequency Phase-Controlled Rectifiers and
Inverters: Line-Frequency ac ↔ Controlled dc**   **121**

6-1 Introduction   121
6-2 Thyristor Circuits and Their Control   122
6-3 Single-Phase Converters   126
6-4 Three-Phase Converters   138
6-5 Other Three-Phase Converters   153
Summary   153
Problems   154
References   157
Appendix   158

Chapter 7   dc–dc Switch-Mode Converters   **161**

7-1 Introduction   161
7-2 Control of dc–dc Converters   162

7-3 Step-Down (Buck) Converter   164
7-4 Step-Up (Boost) Converter   172
7-5 Buck–Boost Converter   178
7-6 Cúk dc–dc Converter   184
7-7 Full Bridge dc–dc Converter   188
7-8 dc–dc Converter Comparison   195
    Summary   196
    Problems   197
    References   199

Chapter 8   Switch-Mode dc–ac Inverters: dc ↔ Sinusoidal ac   **200**

8-1 Introduction   200
8-2 Basic Concepts of Switch-Mode Inverters   202
8-3 Single-Phase Inverters   211
8-4 Three-Phase Inverters   225
8-5 Effect of Blanking Time on Output Voltage in PWM Inverters   236
8-6 Other Inverter Switching Schemes   239
8-7 Rectifier Mode of Operation   243
    Summary   244
    Problems   246
    References   248

Chapter 9   Resonant Converters: Zero-Voltage and/or Zero-Current Switchings   **249**

9-1 Introduction   249
9-2 Classification of Resonant Converters   252
9-3 Basic Resonant Circuit Concepts   253
9-4 Load-Resonant Converters   258
9-5 Resonant-Switch Converters   273
9-6 Zero-Voltage-Switching, Clamped-Voltage Topologies   280
9-7 Resonant-dc-Link Inverters with Zero-Voltage Switchings   287
9-8 High-Frequency-Link Integral-Half-Cycle Converters   289
    Summary   291
    Problems   291
    References   295

PART 3   POWER SUPPLY APPLICATIONS   **299**

Chapter 10   Switching dc Power Supplies   **301**

10-1 Introduction   301
10-2 Linear Power Supplies   301
10-3 Overview of Switching Power Supplies   302
10-4 dc–dc Converters with Electrical Isolation   304
10-5 Control of Switch-Mode dc Power Supplies   322
10-6 Power Supply Protection   341
10-7 Electrical Isolation in the Feedback Loop   344
10-8 Designing to Meet the Power Supply Specifications   346
    Summary   349

Problems   349
References   351

**Chapter 11   Power Conditioners and Uninterruptible Power
            Supplies**   **354**

11-1 Introduction   354
11-2 Power Line Disturbances   354
11-3 Power Conditioners   357
11-4 Uninterruptible Power Supplies (UPSs)   358
      Summary   363
      Problems   363
      References   364

PART 4   MOTOR DRIVE APPLICATIONS   **365**

Chapter 12   Introduction to Motor Drives   **367**

12-1 Introduction   367
12-2 Criteria for Selecting Drive Components   368
      Summary   375
      Problems   376
      References   376

Chapter 13   dc Motor Drives   **377**

13-1 Introduction   377
13-2 Equivalent Circuit of dc Motors   377
13-3 Permanent-Magnet dc Motors   380
13-4 dc Motors with a Separately Excited Field Winding   381
13-5 Effect of Armature Current Waveform   382
13-6 dc Servo Drives   383
13-7 Adjustable-Speed dc Drives   391
      Summary   396
      Problems   396
      References   398

Chapter 14   Induction Motor Drives   **399**

14-1   Introduction   399
14-2   Basic Principles of Induction Motor Operation   400
14-3   Induction Motor Characteristics at Rated (Line) Frequency
       and Rated Voltage   405
14-4   Speed Control by Varying Stator Frequency and Voltage   406
14-5   Impact of Nonsinusoidal Excitation on Induction Motors   415
14-6   Variable-Frequency Converter Classifications   418
14-7   Variable-Frequency PWM-VSI Drives   419
14-8   Variable-Frequency Square-Wave VSI Drives   425
14-9   Variable-Frequency CSI Drives   426
14-10 Comparison of Variable-Frequency Drives   427

14-11 Line-Frequency Variable-Voltage Drives   428
14-12 Reduced Voltage Starting ("Soft Start") of Induction Motors   430
14-13 Speed Control by Static Slip Power Recovery   431
    Summary   432
    Problems   433
    References   434

Chapter 15   Synchronous Motor Drives   **435**

15-1 Introduction   435
15-2 Basic Principles of Synchronous Motor Operation   435
15-3 Synchronous Servomotor Drives with Sinusoidal Waveforms   439
15-4 Synchronous Servomotor Drives with Trapezoidal Waveforms   440
15-5 Load-Commutated Inverter Drives   442
15-6 Cycloconverters   445
    Summary   445
    Problems   446
    References   447

PART 5   OTHER APPLICATIONS   **449**

Chapter 16   Residential and Industrial Applications   **451**

16-1 Introduction   451
16-2 Residential Applications   451
16-3 Industrial Applications   455
    Summary   459
    Problems   459
    References   459

Chapter 17   Electric Utility Applications   **460**

17-1 Introduction   460
17-2 High-voltage dc Transmission   460
17-3 Static var Compensators   471
17-4 Interconnection of Renewable Energy Sources and Energy Storage
     Systems to the Utility Grid   475
17-5 Active Filters   480
    Summary   480
    Problems   481
    References   482

**Chapter 18   Optimizing the Utility Interface with Power
              Electronic Systems**   **483**

18-1 Introduction   483
18-2 Generation of Current Harmonics   484
18-3 Current Harmonics and Power Factor   485
18-4 Harmonic Standards and Recommended Practices   485
18-5 Need for Improved Utility Interface   487

18-6 Improved Single-Phase Utility Interface   488
18-7 Improved Three-Phase Utility Interface   498
18-8 Electromagnetic Interference   500
     Summary   *502*
     Problems   *503*
     References   *503*

PART 6   SEMICONDUCTOR DEVICES   **505**

Chapter 19   Basic Semiconductor Physics   **507**

19-1 Introduction   507
19-2 Conduction Processes in Semiconductors   507
19-3 *pn* Junctions   513
19-4 Charge Control Description of *pn*-Junction Operation   518
19-5 Avalanche Breakdown   520
     Summary   *522*
     Problems   *522*
     References   *523*

Chapter 20   Power Diodes   **524**

20-1 Introduction   524
20-2 Basic Structure and $I-V$ Characteristics   524
20-3 Breakdown Voltage Considerations   526
20-4 On-State Losses   531
20-5 Switching Characteristics   535
20-6 Schottky Diodes   539
     Summary   *543*
     Problems   *543*
     References   *545*

Chapter 21   Bipolar Junction Transistors   **546**

21-1 Introduction   546
21-2 Vertical Power Transistor Structures   546
21-3 $I-V$ Characteristics   548
21-4 Physics of BJT Operation   550
21-5 Switching Characteristics   556
21-6 Breakdown Voltages   562
21-7 Second Breakdown   563
21-8 On-State Losses   565
21-9 Safe Operating Areas   567
     Summary   *568*
     Problems   *569*
     References   *570*

Chapter 22   Power MOSFETs   **571**

22-1 Introduction   571
22-2 Basic Structure   571

22-3 *I–V* Characteristics   574
22-4 Physics of Device Operation   576
22-5 Switching Characteristics   581
22-6 Operating Limitations and Safe Operating Areas   587
Summary   593
Problems   594
References   595

Chapter 23   Thyristors   **596**

23-1 Introduction   596
23-2 Basic Structure   596
23-3 *I–V* Characteristics   597
23-4 Physics of Device Operation   599
23-5 Switching Characteristics   603
23-6 Methods of Improving *di/dt* and *dv/dt* Ratings   608
Summary   610
Problems   611
References   612

Chapter 24   Gate Turn-Off Thyristors   **613**

24-1 Introduction   613
24-2 Basic Structure and *I–V* Characteristics   613
24-3 Physics of Turn-Off Operation   614
24-4 GTO Switching Characteristics   616
24-5 Overcurrent Protection of GTOs   623
Summary   624
Problems   624
References   625

Chapter 25   Insulated Gate Bipolar Transistors   **626**

25-1 Introduction   626
25-2 Basic Structure   626
25-3 *I–V* Characteristics   628
25-4 Physics of Device Operation   629
25-5 Latchup in IGBTs   631
25-6 Switching Characteristics   634
25-7 Device Limits and SOAs   637
Summary   639
Problems   639
References   640

Chapter 26   Emerging Devices and Circuits   **641**

26-1 Introduction   641
26-2 Power Junction Field Effect Transistors   641
26-3 Field-Controlled Thyristor   646
26-4 JFET-Based Devices versus Other Power Devices   648
26-5 MOS-Controlled Thyristors   649

26-6 Power Integrated Circuits   656
26-7 New Semiconductor Materials for Power Devices   661
    Summary   664
    Problems   665
    References   666

PART 7   PRACTICAL CONVERTER DESIGN CONSIDERATIONS   **667**

Chapter 27   Snubber Circuits   **669**

27-1 Function and Types of Snubber Circuits   669
27-2 Diode Snubbers   670
27-3 Snuber Circuits for Thyristors   678
27-4 Need for Snubbers with Transistors   680
27-5 Turn-Off Snubber   682
27-6 Overvoltage Snubber   686
27-7 Turn-On Snubber   688
27-8 Snubbers for Bridge Circuit Configurations   691
27-9 GTO Snubber Considerations   692
    Summary   693
    Problems   694
    References   695

Chapter 28   Gate and Base Drive Circuits   **696**

28-1 Preliminary Design Considerations   696
28-2 dc-Coupled Drive Circuits   697
28-3 Electrically Isolated Drive Circuits   703
28-4 Cascode-Connected Drive Circuits   710
28-5 Thyristor Drive Circuits   712
28-6 Power Device Protection in Drive Circuits   717
28-7 Circuit Layout Considerations   722
    Summary   728
    Problems   729
    References   729

Chapter 29   Component Temperature Control and Heat Sinks   **730**

29-1 Control of Semiconductor Device Temperatures   730
29-2 Heat Transfer by Conduction   731
29-3 Heat Sinks   737
29-4 Heat Transfer by Radiation and Convection   739
    Summary   742
    Problems   743
    References   743

Chapter 30   Design of Magnetic Components   **744**

30-1 Magnetic Materials and Cores   744
30-2 Copper Windings   752

30-3 Thermal Considerations   754
30-4 Analysis of a Specific Inductor Design   756
30-5 Inductor Design Procedures   760
30-6 Analysis of a Specific Transformer Design   767
30-7 Eddy Currents   771
30-8 Transformer Leakage Inductance   779
30-9 Transformer Design Procedure   780
30-10 Comparison of Transformer and Inductor Sizes   789
    Summary   *789*
    Problems   *790*
    References   *792*

Index   **793**

PART 1

INTRODUCTION

CHAPTER 1

POWER ELECTRONIC SYSTEMS

1-1 INTRODUCTION

In broad terms, the task of power electronics is to process and control the flow of electric energy by supplying voltages and currents in a form that is optimally suited for user loads. Figure 1-1 shows a power electronic system in a block diagram form. The power input to this power processor is usually (but not always) from the electric utility at a line frequency of 60 or 50 Hz, single phase or three phases. The phase angle between the input voltage and the current depends on the topology and the control of the power processor. The processed output (voltage, current, frequency, and the number of phases) is as desired by the load. If the power processor's output can be regarded as a voltage source, the output current and the phase angle relationship between the output voltage and the current depend on the load characteristic. Normally, a feedback controller compares the output of the power processor unit with a desired (or a reference) value, and the error between the two is minimized by the controller. The power flow through such systems may be reversible, thus interchanging the roles of the input and the output.

In recent years, the field of power electronics has experienced a large growth due to confluence of several factors. The controller in the block diagram of Fig. 1-1 consists of linear integrated circuits and/or digital signal processors. Revolutionary advances in microelectronics methods have led to the development of such controllers. Moreover, these advances in semiconductor fabrication technology have made it possible to significantly improve the voltage- and current-handling capabilities and the switching speeds of power semiconductor devices, which make up the power processor unit of Fig. 1-1. At the same time, the market for power electronics has significantly expanded. Electric utilities in the United States expect that by the year 2000 over 50% of the electrical load may be supplied through power electronic systems such as in Fig. 1-1. This growth in market may even be

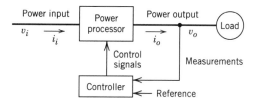

Figure 1-1 Block diagram of a power electronic system.

higher in other parts of the world where the cost of energy is significantly higher than that in the United States. Various applications of power electronics are considered in Section 1-3.

1-2   POWER ELECTRONICS VERSUS LINEAR ELECTRONICS

In any power conversion process such as that shown by the block diagram in Fig. 1-1, a small power loss and hence a high energy efficiency is important because of two reasons: the cost of the wasted energy and the difficulty in removing the heat generated due to dissipated energy. Other important considerations are reduction in size, weight, and cost.

The above objectives in most systems cannot be met by linear electronics where the semiconductor devices are operated in their linear (active) region and a line-frequency transformer is used for electrical isolation. As an example, consider the direct current (dc) power supply of Fig. 1-2a to provide a regulated output voltage V_o to a load. The utility input may be typically at 120 or 240 V and the output voltage may be, for example, 5 V. The output is required to be electrically isolated from the utility input. In the linear power supply, a line-frequency transformer is used to provide electrical isolation and for stepping down the line voltage. The rectifier converts the alternating current (ac) output of the transformer low-voltage winding into dc. The filter capacitor reduces the ripple in the dc voltage v_d. Figure 1-2b shows the v_d waveform, which depends on the utility voltage magnitude (normally in a $\pm$ 10% range around its nominal value). The transformer turns

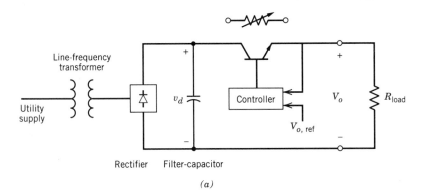

(a)

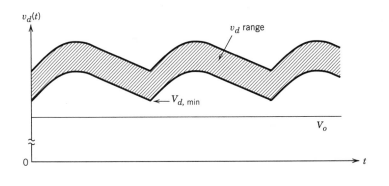

(b)

Figure 1-2   Linear dc power supply.

ratio must be chosen such that the minimum of the input voltage v_d is greater than the desired output V_o. For the range of the input voltage waveforms shown in Fig. 1-2b, the transistor is controlled to absorb the voltage difference between v_d and V_o, thus providing a regulated output. The transistor operates in its active region as an adjustable resistor, resulting in a low energy efficiency. The line-frequency transformer is relatively large and heavy.

In power electronics, the above voltage regulation and the electrical isolation are achieved, for example, by means of a circuit shown in Fig. 1-3a. In this system, the utility input is rectified into a dc voltage v_d, without a line-frequency transformer. By operating the transistor as a switch (in a switch mode, either fully *on* or fully *off*) at some high switching frequency f_s, for example at 300 kHz, the dc voltage v_d is converted into an ac voltage at the switching frequency. This allows a high-frequency transformer to be used for stepping down the voltage and for providing the electrical isolation. In order to simplify this circuit for analysis, we will begin with the dc voltage v_d as the dc input and omit the transformer, resulting in an equivalent circuit shown in Fig. 1-3b. Suffice it to

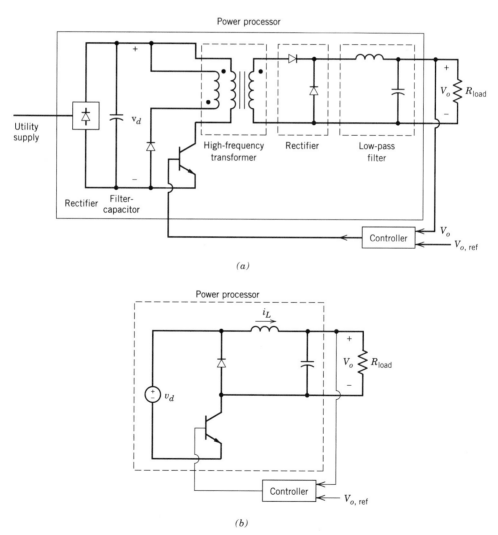

(a)

(b)

Figure 1-3 Switch-mode dc power supply.

say at this stage (this circuit is fully discussed in Chapters 7 and 10) that the transistor–diode combination can be represented by a hypothetical two-position switch shown in Fig. 1-4a (provided $i_L(t) > 0$). The switch is in position a during the interval t_{on} when the transistor is on and in position b when the transistor is off during t_{off}. As a consequence, v_{oi} equals V_d and zero during t_{on} and t_{off}, respectively, as shown in Fig. 1-4b. Let us define

$$v_{oi}(t) = V_{oi} + v_{ripple}(t) \qquad (1\text{-}1)$$

where V_{oi} is the average (dc) value of v_{oi}, and the instantaneous ripple voltage $v_{ripple}(t)$, which has a zero average value, is shown in Fig. 1-4c. The L–C elements form a low-pass filter that reduces the ripple in the output voltage and passes the average of the input voltage, so that

$$V_o = V_{oi} \qquad (1\text{-}2)$$

where V_o is the average output voltage. From the repetitive waveforms in Fig. 1-4b, it is easy to see that

$$V_o = \frac{1}{T_s} \int_0^{T_s} v_{oi}\, dt = \frac{t_{on}}{T_s} V_d \qquad (1\text{-}3)$$

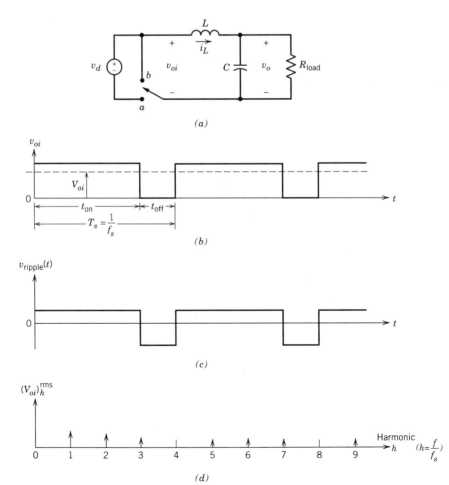

(a)

(b)

(c)

(d)

Figure 1-4   Equivalent circuit, waveforms, and frequency spectrum of the supply in Fig. 1-3.

As the input voltage v_d changes with time, Eq. 1-3 shows that it is possible to regulate V_o at its desired value by controlling the ratio t_{on}/T_s, which is called the duty ratio D of the transistor switch. Usually, T_s ($=1/f_s$) is kept constant and t_{on} is adjusted.

There are several characteristics worth noting. Since the transistor operates as a switch, fully on or fully off, the power loss is minimized. Of course, there is an energy loss each time the transistor switches from one state to the other state through its active region (discussed in Chapter 2). Therefore, the power loss due to switchings is linearly proportional to the switching frequency. This switching power loss is usually much lower than the power loss in linear regulated power supplies.

At high switching frequencies, the transformer and the filter components are very small in weight and size compared with line-frequency components. To elaborate on the role of high switching frequencies, the harmonic content in the waveform of v_{oi} is obtained by means of Fourier analysis (see Problem 1-3 and its further discussion in Chapter 3) and plotted in Fig. 1-4d. It shows that v_{oi} consists of an average (dc) value and of harmonic components that are at a multiple of the switching frequency f_s. If the switching frequency is high, these ac components can be easily eliminated by a small filter to yield the desired dc voltage. The selection of the switching frequency is dictated by the compromise between the switching power dissipation in the transistor, which increases with the switching frequency, and the cost of the transformer and filter, which decreases with the switching frequency. As transistors with higher switching speeds become available, the switching frequencies can be increased and the transformer and filter size reduced for the same switching power dissipation.

An important observation in the switch-mode circuit described above is that both the input and the output are dc, as in the linear regulated supply. The high switching frequencies are used to synthesize the output waveform, which in this example is dc. In many applications, the output is a low-frequency sine wave.

1-3 SCOPE AND APPLICATIONS

The expanded market demand for power electronics has been due to several factors discussed below (see references 1–3).

1. *Switch-mode (dc) power supplies and uninterruptible power supplies.* Advances in microelectronics fabrication technology have led to the development of computers, communication equipment, and consumer electronics, all of which require regulated dc power supplies and often uninterruptible power supplies.

2. *Energy conservation.* Increasing energy costs and the concern for the environment have combined to make energy conservation a priority. One such application of power electronics is in operating fluorescent lamps at high frequencies (e.g., above 20 kHz) for higher efficiency. Another opportunity for large energy conservation (see Problem 1-7) is in motor-driven pump and compressor systems [4]. In a conventional pump system shown in Fig. 1-5a, the pump operates at essentially a constant speed, and the pump flow rate is controlled by adjusting the position of the throttling valve. This procedure results in significant power loss across the valve at reduced flow rates where the power drawn from the utility remains essentially the same as at the full flow rate. This power loss is eliminated in the system of Fig. 1-5b, where an adjustable-speed motor drive adjusts the pump speed to a level appropriate to deliver the desired flow rate. As will be discussed in Chapter 14 (in combination with Chapter 8), motor speeds can be adjusted very efficiently using power electronics. Load-proportional, capacity-

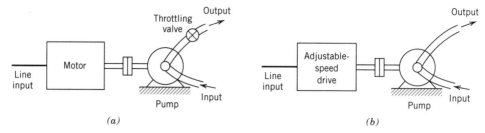

Figure 1-5   Energy conservation: (a) conventional drive, (b) adjustable-speed drive.

modulated heat pumps and air conditioners are examples of applying power electronics to achieve energy conservation.

3. *Process control and factory automation.* There is a growing demand for the
enhanced performance offered by adjustable-speed-driven pumps and compressors in process control. Robots in automated factories are powered by electric
servo (adjustable-speed and position) drives. It should be noted that the availability of process computers is a significant factor in making process control and
factory automation feasible.

4. *Transportation.* In many countries, electric trains have been in widespread use
for a long time. Now, there is also a possibility of using electric vehicles in large

TABLE 1-1 Power Electronic Applications

(a) *Residential*	(d) *Transportation*
Refrigeration and freezers	Traction control of electric vehicles
Space heating	Battery chargers for electric vehicles
Air conditioning	Electric locomotives
Cooking	Street cars, trolley buses
Lighting	Subways
Electronics (personal computers,	Automotive electronics including engine
other entertainment equipment)	controls
(b) *Commercial*	(e) *Utility systems*
Heating, ventilating, and air	High-voltage dc transmission (HVDC)
conditioning	Static var compensation (SVC)
Central refrigeration	Supplemental energy sources (wind,
Lighting	photovoltaic), fuel cells
Computers and office equipment	Energy storage systems
Uninterruptible power supplies	Induced-draft fans and boiler
(UPSs)	feedwater pumps
Elevators	(f) *Aerospace*
(c) *Industrial*	Space shuttle power supply systems
Pumps	Satellite power systems
Compressors	Aircraft power systems
Blowers and fans	(g) *Telecommunications*
Machine tools (robots)	Battery chargers
Arc furnaces, induction furnaces	Power supplies (dc and UPS)
Lighting	
Industrial lasers	
Induction heating	
Welding	

metropolitan areas to reduce smog and pollution. Electric vehicles would also require battery chargers that utilize power electronics.

5. *Electro-technical applications.* These include equipment for welding, electroplating, and induction heating.

6. *Utility-related applications.* One such application is in transmission of power over high-voltage dc (HVDC) lines. At the sending end of the transmission line, line-frequency voltages and currents are converted into dc. This dc is converted back into the line-frequency ac at the receiving end of the line. Power electronics is also beginning to play a significant role as electric utilities attempt to utilize the existing transmission network to a higher capacity [5]. Potentially, a large application is in the interconnection of photovoltaic and wind-electric systems to the utility grid.

Table 1-1 lists various applications that cover a wide power range from a few tens of watts to several hundreds of megawatts. As power semiconductor devices improve in performance and decline in cost, more systems will undoubtedly use power electronics.

1-4 CLASSIFICATION OF POWER PROCESSORS AND CONVERTERS

1-4-1 POWER PROCESSORS

For a systematic study of power electronics, it is useful to categorize the power processors, shown in the block diagram of Fig. 1-1, in terms of their input and output form or frequency. In most power electronic systems, the input is from the electric utility source. Depending on the application, the output to the load may have any of the following forms:

1. dc
   (a) regulated (constant) magnitude
   (b) adjustable magnitude

2. ac
   (a) constant frequency, adjustable magnitude
   (b) adjustable frequency and adjustable magnitude

The utility and the ac load, independent of each other, may be single phase or three phase. The power flow is generally from the utility input to the output load. There are exceptions, however. For example, in a photovoltaic system interfaced with the utility grid, the power flow is from the photovoltaics (a dc input source) to the ac utility (as the output load). In some systems the direction of power flow is reversible, depending on the operating conditions.

1-4-2 POWER CONVERTERS

The power processors of Fig. 1-1 usually consist of more than one power conversion stage (as shown in Fig. 1-6) where the operation of these stages is decoupled on an instantaneous basis by means of energy storage elements such as capacitors and inductors. Therefore, the instantaneous power input does not have to equal the instantaneous power output. We will refer to each power conversion stage as a converter. Thus, a converter is a basic module (building block) of power electronic systems. It utilizes power semicon-

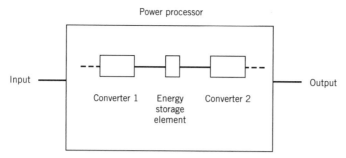

Figure 1-6   Power processor block diagram.

ductor devices controlled by signal electronics (integrated circuits) and possibly energy storage elements such as inductors and capacitors. Based on the form (frequency) on the two sides, converters can be divided into the following broad categories:

1. ac to dc
2. dc to ac
3. dc to dc
4. ac to ac

We will use *converter* as a generic term to refer to a single power conversion stage that may perform any of the functions listed above. To be more specific, in ac-to-dc and dc-to-ac conversion, *rectifier* refers to a converter when the average power flow is from the ac to the dc side. *Inverter* refers to the converter when the average power flow is from the dc to the ac side. In fact, the power flow through the converter may be reversible. In that case, as shown in Fig. 1-7, we refer to that converter in terms of its rectifier and inverter modes of operation.

As an example, consider that the power processor of Fig. 1-6 represents the block diagram of an adjustable-speed ac motor drive (described in Chapter 14). As shown in Fig. 1-8, it consists of two converters: converter 1 operating as a rectifier that converts line-frequency ac into dc and converter 2 operating as an inverter that converts dc into adjustable-magnitude, adjustable-frequency ac. The flow of power in the normal (dominant) mode of operation is from the utility input source to the output motor load. During regenerative braking, the power flow reverses direction (from the motor to the utility), in which case converter 2 operates as a rectifier and converter 1 operates as an inverter. As mentioned earlier, an energy storage capacitor in the dc link between the two converters decouples the operation of the two converters on an instantaneous basis. Further insight can be gained by classifying converters according to how the devices within the converter are switched. There are three possibilities:

1. *Line frequency (naturally commutated) converters,* where the utility line voltages present at one side of the converter facilitate the turn-off of the power semicon-

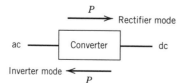

Figure 1-7   ac-to-dc converters.

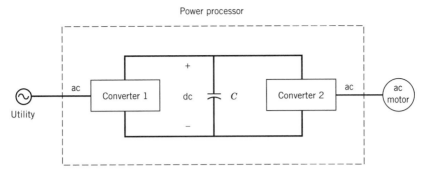

Figure 1-8 Block diagram of an ac motor drive.

ductor devices. Similarly, the devices are turned on, phase locked to the line-voltage waveform. Therefore, the devices switch on and off at the line frequency of 50 or 60 Hz.

2. *Switching (forced-commutated) converters,* where the controllable switches in the converter are turned on and off at frequencies that are high compared to the line frequency. In spite of the high switching frequency internal to the converter, the converter output may be either dc or at a frequency comparable to the line frequency. As a side note in a switching converter, if the input appears as a voltage source, then the output must appear as a current source, or vice versa.

3. *Resonant and quasi-resonant converters,* where the controllable switches turn on and/or turn off at zero voltage and/or zero current.

1-4-3 MATRIX CONVERTER AS A POWER PROCESSOR

In the above two sections, we discussed that most practical power processors utilize more than one converter whose instantaneous operation is decoupled by an energy storage element (an inductor or a capacitor). Theoretically, it is possible to replace the multiple conversion stages and the intermediate energy storage element by a single power conversion stage called the matrix converter. Such a converter uses a matrix of semiconductor bidirectional switches, with a switch connected between each input terminal to each output terminal, as shown in Fig. 1-9a for an arbitrary number of input and output phases. With this general arrangement of switches, the power flow through the converter can reverse. Because of the absence of any energy storage element, the instantaneous power input must be equal to the power output, assuming idealized zero-loss switches. However, the phase angle between the voltages and currents at the input can be controlled and does not have to be the same as at the output (i.e., the reactive power input does not have to equal the reactive power output). Also, the form and the frequency at the two sides are independent, for example, the input may be three-phase ac and the output dc, or both may be dc, or both may be ac.

However, there are certain requirements on the switches and restrictions on the converter operation: If the inputs appear as voltage sources as shown in Fig. 1-9a, then the outputs must appear as current sources or vice versa. If both sides, for example, were to appear as voltage sources, the switching actions will inevitably connect voltage sources of unequal magnitude directly across each other; an unacceptable condition. The switching functions in operating such a converter must ensure that the switches do not short-circuit the voltage sources and do not open-circuit the current sources. Otherwise, the converter will be destroyed.

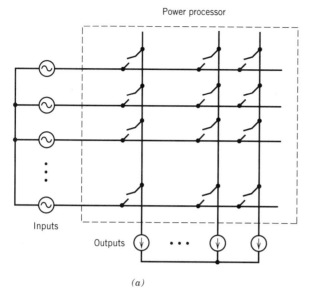

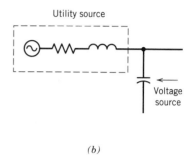

Figure 1-9 (*a*) Matrix converter. (*b*) Voltage source.

Through a voltage source, the current can change instantaneously, whereas the voltage across a current source can change instantaneously. If the input in Fig. 1-9*a* is a utility source, it is not an ideal voltage source due to its internal impedance corresponding to the transmission and distribution lines, transformers, etc., which are at the back of the utility outlet. To make it appear like a voltage source will require that we connect a small capacitance in parallel with it, as shown in Fig. 1-9*b* to overcome the effect of the internal impedance.

The switches in a matrix converter must be bidirectional, that is, they must be able to block voltages of either polarity and be able to conduct current in either direction. Such switches are not available and must be realized by a combination of the available unidirectional switches and diodes discussed in Chapter 2. There are also limits on the ratio of the magnitudes of the input and the output quantities.

In spite of numerous laboratory prototypes reported in research publications, the matrix converters so far have failed to show any significant advantage over conventional converters and hence have not found applications in practice. Therefore, we will not discuss them any further in this book.

1-5 ABOUT THE TEXT

The purpose of this book is to facilitate the study of practical and emerging power electronic converters made feasible by the new generation of power semiconductor devices. This book is divided into seven parts.

Part 1 of the book, which includes Chapter 1–4, presents an introduction, a brief review of basic concepts and devices, and computer simulations of power electronic systems. An overview of power semiconductor devices (discussed in detail in later parts of the book) and the justification for assuming them as ideal switches are presented in

Chapter 2. The basic electrical and magnetic concepts relevant to the discussion of power electronics are reviewed in Chapter 3. In Chapter 4, we briefly describe the role of computer simulations in the analysis and design of power electronic systems. Some of the simulation software packages suited for this purpose are also presented.

Part 2 (Chapters 5–9) describes power electronic converters in a generic manner. This way, the basic converter topologies used in more than one application can be described once, rather than repeating them each time a new application is encountered. This generic discussion is based on the assumption that the actual power semiconductor switches can be treated as ideal switches. Chapter 5 describes line-frequency diode rectifiers for ac-to-dc conversion. The ac-to-dc conversion using line-commutated (naturally commutated) thyristor converters operating in the rectifier and the inverter mode is discussed in Chapter 6. Switching converters for dc to dc, and dc to sinusoidal ac using controlled switches are described in Chapters 7 and 8, respectively. The discussion of resonant converters in a generic manner is presented in Chapter 9.

We decided to discuss ac-to-ac converters in the application-based chapters due to their application-specific nature. The matrix converters, which in principle can be ac-to-ac converters, were briefly described in Section 1-4-3. The static transfer switches are discussed in conjunction with the uninterruptible power supplies in Section 11-4-4. Converters where only the voltage magnitude needs to be controlled without any change in ac frequency are described in Section 14-12 for speed control of induction motors and in Section 17-3 for static var compensators (thyristor-controlled inductors and thyristor-switched capacitors). Cycloconverters for very large synchronous-motor drives are discussed in Section 15-6. High-frequency-link integral-half-cycle converters are discussed in Section 9-8. Integral-half-cycle controllers supplied by line-frequency voltages for heating-type applications are discussed in Section 16-3-3.

Part 3 (Chapters 10 and 11) deals with power supplies: switching dc power supplies (Chapter 10) and uninterruptible ac power supplies (Chapter 11). Part 4 describes motor drive applications in Chapters 12–15.

Other applications of power electronics are covered in Part 5, which includes residential and industrial applications (Chapter 16), electric utility applications (Chapter 17), and the utility interface of power electronic systems (Chapter 18).

Part 6 (Chapters 19–26) contains a qualitative description of the physical operating principles of semiconductor devices used as switches. Finally, Part 7 (Chapters 27–30) presents the practical design considerations of power electronic systems, including protection and gate-drive circuits, thermal management, and the design of magnetic components.

The reader is also urged to read the overview of the textbook presented in the Preface.

1-6   INTERDISCIPLINARY NATURE OF POWER ELECTRONICS

The discussion in this introductory chapter shows that the study of power electronics encompasses many fields within electrical engineering, as illustrated by Fig. 1-10. These include power systems, solid-state electronics, electrical machines, analog/digital control and signal processing, electromagnetic field calculations, and so on. Combining the knowledge of these diverse fields makes the study of power electronics challenging as well as interesting. There are many potential advances in all these fields that will improve the prospects for applying power electronics to new applications.

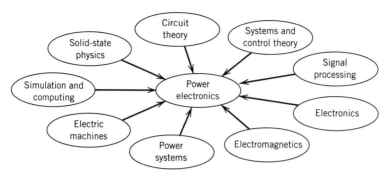

Figure 1-10   Interdisciplinary nature of power electronics.

1-7   CONVENTION OF SYMBOLS USED

In this textbook, for instantaneous values of variables such as voltage, current, and power that are functions of time, the symbols used are lowercase letters v, i, and p, respectively. We may or may not explicitly show that they are functions of time, for example, using v rather than $v(t)$. The uppercase symbols V and I refer to their values computed from their instantaneous waveforms. They generally refer to an average value in dc quantities and a root-mean-square (rms) value in ac quantities. If there is a possibility of confusion, the subscript avg or rms is added explicitly. The peak values are always indicated by the symbol "^" on top of the uppercase letters. The average power is always indicated by P.

PROBLEMS

1-1   In the power processor of Fig. 1-1, the energy efficiency is 95%. The output to the three-phase load is as follows: 200 V line-to-line (rms) sinusoidal voltages at 52 Hz and line current of 10 A at a power factor of 0.8 (lagging). The input to the power processor is a single-phase utility voltage of 230 V at 60 Hz. The input power is drawn at a unity power factor. Calculate the input current and the input power.

1-2   Consider a linear regulated dc power supply (Fig. 1-2a). The instantaneous input voltage corresponds to the lowest waveform in Fig. 1-2b, where $V_{d,\min} = 20$ V and $V_{d,\max} = 30$ V. Approximate this waveform by a triangular wave consisting of two linear segments between the above two values. Let $V_o = 15$ V and assume that the output load is constant. Calculate the energy efficiency in this part of the power supply due to losses in the transistor.

1-3   Consider a switch-mode dc power supply represented by the circuit in Fig. 1-4a. The input dc voltage $V_d = 20$ V and the switch duty ratio $D = 0.75$. Calculate the Fourier components of v_{oi} using the description of Fourier analysis in Chapter 3.

1-4   In Problem 1-3, the switching frequency $f_s = 300$ kHz and the resistive load draws 240 W. The filter components corresponding to Fig. 1-4a are $L = 1.3 \ \mu\text{H}$ and $C = 50 \ \mu\text{F}$. Calculate the attenuation in decibels of the ripple voltage in v_{oi} at various harmonic frequencies. (*Hint:* To calculate the load resistance, assume the output voltage to be a constant dc without any ripple.)

1-5   In Problem 1-4, assume the output voltage to be a pure dc $V_o = 15$ V. Calculate and draw the voltage and current associated with the filter inductor L, and the current through C. Using the capacitor current obtained above, estimate the peak-to-peak ripple in the voltage across C, which was initially assumed to be zero. (*Hint:* Note that under steady-state conditions, the average value of the current through C is zero.)

1-6 Considering only the switching frequency component in v_{oi} in Problems 1-3 and 1-4, calculate the peak-to-peak ripple in the output voltage across C. Compare the result with that obtained in Problem 1-5.

1-7 Reference 4 refers to a U.S. Department of Energy report that estimated that over 100 billion kWh/year can be saved in the United States by various energy conservation techniques applied to the pump-driven systems. Calculate (a) how many 1000-MW generating plants running constantly supply this wasted energy, which could be saved, and (b) the savings in dollars if the cost of electricity is 0.1 \$/kWh.

REFERENCES

1. B. K. Bose, "Power Electronics—A Technology Review," *Proceedings of the IEEE,* Vol. 80, No. 8, August 1992, pp. 1303–1334.
2. E. Ohno, "The Semiconductor Evolution in Japan—A Four Decade Long Maturity Thriving to an Indispensable Social Standing," *Proceedings of the International Power Electronics Conference (Tokyo),* 1990, Vol. 1, pp. 1–10.
3. M. Nishihara, "Power Electronics Diversity," *Proceedings of the International Power Electronics Conference (Tokyo),* 1990, Vol. 1, pp. 21–28.
4. N. Mohan and R. J. Ferraro, "Techniques for Energy Conservation in AC Motor Driven Systems," Electric Power Research Institute Final Report EM-2037, Project 1201-1213, September 1981.
5. N. G. Hingorani, "Flexible ac Transmission," *IEEE Spectrum,* April 1993, pp. 40–45.
6. N. Mohan, "Power Electronic Circuits: An Overview," *IEEE/IECON Conference Proceedings,* 1988, Vol. 3, pp. 522–527.

CHAPTER 2

OVERVIEW OF POWER SEMICONDUCTOR SWITCHES

2-1 INTRODUCTION

The increased power capabilities, ease of control, and reduced costs of modern power semiconductor devices compared to those of just a few years ago have made converters affordable in a large number of applications and have opened up a host of new converter topologies for power electronic applications. In order to clearly understand the feasibility of these new topologies and applications, it is essential that the characteristics of available power devices be put in perspective. To do this, a brief summary of the terminal characteristics and the voltage, current, and switching speed capabilities of currently available power devices are presented in this chapter.

If the power semiconductor devices can be considered as ideal switches, the analysis of converter topologies becomes much easier. This approach has the advantage that the details of device operation will not obscure the basic operation of the circuit. Therefore, the important converter characteristics can be more clearly understood. The summary of device characteristics will enable us to determine how much the device characteristics can be idealized.

Presently available power semiconductor devices can be classified into three groups according to their degree of controllability:

1. *Diodes*. On and off states controlled by the power circuit.
2. *Thyristors*. Latched on by a control signal but must be turned off by the power circuit.
3. *Controllable switches*. Turned on and off by control signals.

The controllable switch category includes several device types including bipolar junction transistors (BJTs), metal–oxide–semiconductor field effect transistors (MOSFETs), gate turn off (GTO) thyristors, and insulated gate bipolar transistors (IGBTs). There have been major advances in recent years in this category of devices.

2-2 DIODES

Figures 2-1a and 2-1b show the circuit symbol for the diode and its steady-state i–v characteristic. When the diode is forward biased, it begins to conduct with only a small

16

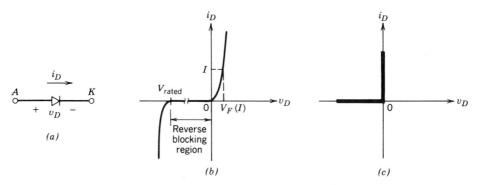

Figure 2-1   Diode: (a) symbol, (b) i–v characteristic, (c) idealized characteristic.

forward voltage across it, which is on the order of 1 V. When the diode is reverse biased, only a negligibly small leakage current flows through the device until the reverse breakdown voltage is reached. In normal operation, the reverse-bias voltage should not reach the breakdown rating.

In view of a very small leakage current in the blocking (reverse-bias) state and a small voltage in the conducting (forward-bias) state, as compared to the operating voltages and currents of the circuit in which the diode is used, the i–v characteristic for the diode can be idealized, as shown in Fig. 2-1c. This idealized characteristic can be used for analyzing the converter topology but should not be used for the actual converter design, when, for example, heat sink requirements for the device are being estimated.

At turn-on, the diode can be considered an ideal switch because it turns on rapidly compared to the transients in the power circuit. However, at turn-off, the diode current reverses for a reverse-recovery time t_{rr}, as is indicated in Fig. 2-2, before falling to zero. This reverse-recovery (negative) current is required to sweep out the excess carriers in the diode and allow it to block a negative polarity voltage. The reverse-recovery current can lead to overvoltages in inductive circuits. In most circuits, this reverse current does not affect the converter input/output characteristic and so the diode can also be considered as ideal during the turn-off transient.

Depending on the application requirements, various types of diodes are available:

1. *Schottky diodes.* These diodes are used where a low forward voltage drop (typically 0.3 V) is needed in very low output voltage circuits. These diodes are limited in their blocking voltage capabilities to 50–100 V.

2. *Fast-recovery diodes.* These are designed to be used in high-frequency circuits in combination with controllable switches where a small reverse-recovery time is needed. At power levels of several hundred volts and several hundred amperes, such diodes have t_{rr} ratings of less than a few microseconds.

3. *Line-frequency diodes.* The on-state voltage of these diodes is designed to be as low as possible and as a consequence have larger t_{rr}, which are acceptable for

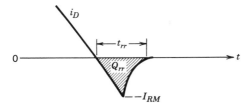

Figure 2-2   Diode turn-off.

line-frequency applications. These diodes are available with blocking voltage ratings of several kilovolts and current ratings of several kiloamperes. Moreover, they can be connected in series and parallel to satisfy any voltage and current requirement.

2-3   THYRISTORS

The circuit symbol for the thyristor and its i–v characteristic are shown in Figs. 2-3a and 2-3b. The main current flows from the anode (A) to the cathode (K). In its off-state, the thyristor can block a forward polarity voltage and not conduct, as is shown in Fig. 2-3b by the off-state portion of the i–v characteristic.

The thyristor can be triggered into the on state by applying a pulse of positive gate current for a short duration provided that the device is in its forward-blocking state. The resulting i–v relationship is shown by the on-state portion of the characteristics shown in Fig. 2-3b. The forward voltage drop in the on state is only a few volts (typically $1 - 3$ V depending on the device blocking voltage rating).

Once the device begins to conduct, it is latched on and the gate current can be removed. The thyristor cannot be turned off by the gate, and the thyristor conducts as a diode. Only when the anode current tries to go negative, under the influence of the circuit in which the thyristor is connected, does the thyristor turn off and the current go to zero. This allows the gate to regain control in order to turn the device on at some controllable time after it has again entered the forward-blocking state.

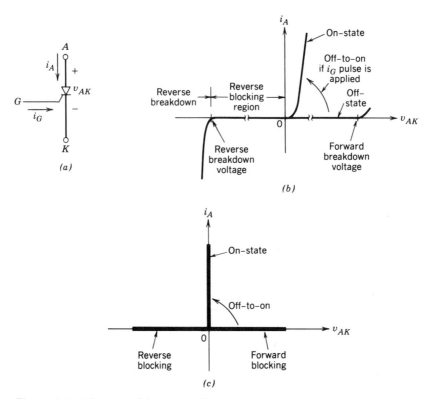

Figure 2-3   Thyristor: (a) symbol, (b) i–v characteristics, (c) idealized characteristics.

In reverse bias at voltages below the reverse breakdown voltage, only a negligibly small leakage current flows in the thyristor, as is shown in Fig. 2-3b. Usually the thyristor voltage ratings for forward- and reverse-blocking voltages are the same. The thyristor current ratings are specified in terms of maximum rms and average currents that it is capable of conducting.

Using the same arguments as for diodes, the thyristor can be represented by the idealized characteristics shown in Fig. 2-3c in analyzing converter topologies.

In an application such as the simple circuit shown in Fig. 2-4a, control can be exercised over the instant of current conduction during the positive half cycle of source voltage. When the thyristor current tries to reverse itself when the source voltage goes negative, the idealized thyristor would have its current become zero immediately after $t = \frac{1}{2}T$, as is shown in the waveform in Fig. 2-4b.

However, as specified in the thyristor data sheets and illustrated by the waveforms in Fig. 2-4c, the thyristor current reverses itself before becoming zero. The important parameter is not the time it takes for the current to become zero from its negative value, but rather the turn-off time interval t_q defined in Fig. 2-4c from the zero crossover of the current to the zero crossover of the voltage across the thyristor. During t_q a reverse voltage must be maintained across the thyristor, and only after this time is the device capable of blocking a forward voltage without going into its on state. If a forward voltage is applied to the thyristor before this interval has passed, the device may prematurely turn on, and damage to the device and/or circuit could result. Thyristor data sheets specify t_q with a specified reverse voltage applied during this interval as well as a specified rate of rise of voltage beyond this interval. This interval t_q is sometimes called the circuit-commutated recovery time of the thyristor.

Depending on the application requirements, various types of thyristors are available. In addition to voltage and current ratings, turn-off time t_q, and the forward voltage drop,

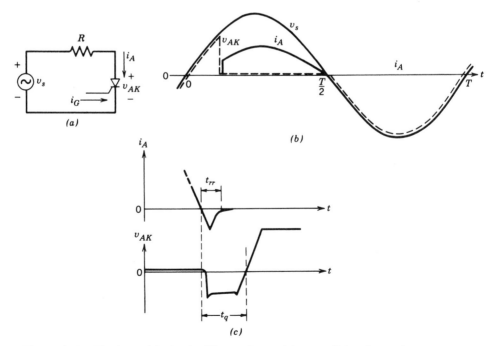

Figure 2-4   Thyristor: (a) circuit, (b) waveforms, (c) turn-off time interval t_q.

other characteristics that must be considered include the rate of rise of the current (*di/dt*) at turn-on and the rate of rise of voltage (*dv/dt*) at turn-off.

1. *Phase-control thyristors*. Sometimes termed *converter thyristors*, these are used primarily for rectifying line-frequency voltages and currents in applications such as phase-controlled rectifiers for dc and ac motor drives and in high-voltage dc power transmission. The main device requirements are large voltage and current-handling capabilities and a low on-state voltage drop. This type of thyristor has been produced in wafer diameters of up to 10 cm, where the average current is about 4000 A with blocking voltages of 5–7 kV. On-state voltages range from 1.5 V for 1000-V devices to 3.0 V for the 5–7-kV devices.

2. *Inverter-grade thyristors*. These are designed to have small turn-off times t_q in addition to low on-state voltages, although on-state voltages are larger in devices with shorter values of t_q. These devices are available with ratings up to 2500 V and 1500 A. Their turn-off times are usually in the range of a few microseconds to 100 μs depending on their blocking voltage ratings and on-state voltage drops.

3. *Light-activated thyristors*. These can be triggered on by a pulse of light guided by optical fibers to a special sensitive region of the thyristor. The light-activated triggering of the thyristor uses the ability of light of appropriate wavelengths to generate excess electron–hole pairs in the silicon. The primary use of these thyristors are in high-voltage applications such as high-voltage dc transmission where many thyristors are connected in series to make up a converter valve. The differing high potentials that each device sees with respect to ground poses significant difficulties in providing triggering pulses. Light-activated thyristors have been reported with ratings of 4 kV and 3 kA, on-state voltages of about 2 V, and light trigger power requirements of 5 mW.

Other variations of these thyristors are gate-assisted turn-off thyristors (GATTs), asymmetrical silicon-controlled rectifiers (ASCRs), and reverse-conducting thyristors (RCTs). These are utilized based on the application.

2-4   DESIRED CHARACTERISTICS IN CONTROLLABLE SWITCHES

As mentioned in the introduction, several types of semiconductor power devices including BJTs, MOSFETs, GTOs, and IGBTs can be turned on and off by control signals applied to the control terminal of the device. These devices we term *controllable switches* and are represented in a generic manner by the circuit symbol shown in Fig. 2-5. No current flows when the switch is off, and when it is on, current can flow in the direction of the arrow only. The ideal controllable switch has the following characteristics:

1. Block arbitrarily large forward and reverse voltages with zero current flow when off.

2. Conduct arbitrarily large currents with zero voltage drop when on.

Figure 2-5   Generic controllable switch.

3. Switch from on to off or vice versa instantaneously when triggered.

4. Vanishingly small power required from control source to trigger the switch.

Real devices, as we intuitively expect, do not have these ideal characteristics and hence will dissipate power when they are used in the numerous applications already mentioned. If they dissipate too much power, the devices can fail and, in doing so, not only will destroy themselves but also may damage the other system components.

Power dissipation in semiconductor power devices is fairly generic in nature; that is, the same basic factors governing power dissipation apply to all devices in the same manner. The converter designer must understand what these factors are and how to minimize the power dissipation in the devices.

In order to consider power dissipation in a semiconductor device, a controllable switch is connected in the simple circuit shown in Fig. 2-6a. This circuit models a very commonly encountered situation in power electronics; the current flowing through a

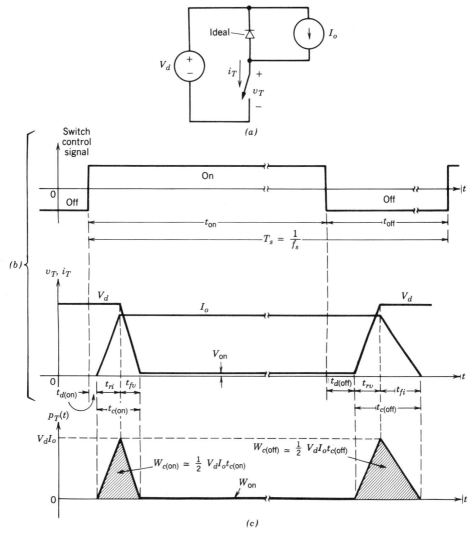

Figure 2-6 Generic-switch switching characteristics (linearized): (a) simplified clamped-inductive-switching circuit, (b) switch waveforms, (c) instantaneous switch power loss.

switch also must flow through some series inductance(s). This circuit is similar to the circuit of Fig. 1-3b, which was used to introduce switch-mode power electronic circuits. The dc current source approximates the current that would actually flow due to inductive energy storage. The diode is assumed to be ideal because our focus is on the switch characteristics, though in practice the diode reverse-recovery current can significantly affect the stresses on the switch.

When the switch is on, the entire current I_o flows through the switch and the diode is reverse biased. When the switch is turned off, I_o flows through the diode and a voltage equal to the input voltage V_d appears across the switch, assuming a zero voltage drop across the ideal diode. Figure 2-6b shows the waveforms for the current through the switch and the voltage across the switch when it is being operated at a repetition rate or switching frequency of $f_s = 1/T_s$, with T_s being the switching time period. The switching waveforms are represented by linear approximations to the actual waveforms in order to simplify the discussion.

When the switch has been off for a while, it is turned on by applying a positive control signal to the switch, as is shown in Fig. 2-6b. During the turn-on transition of this generic switch, the current buildup consists of a short delay time $t_{d(on)}$ followed by the current rise time t_{ri}. Only after the current I_o flows entirely through the switch can the diode become reverse biased and the switch voltage fall to a small on-state value of V_{on} with a voltage fall time of t_{fv}. The waveforms in Fig. 2-6b indicate that large values of switch voltage and current are present simultaneously during the turn-on crossover interval $t_{c(on)}$, where

$$t_{c(on)} = t_{ri} + t_{fv} \tag{2-1}$$

The energy dissipated in the device during this turn-on transition can be approximated from Fig. 2-6c as

$$W_{c(on)} = \tfrac{1}{2} V_d I_o t_{c(on)} \tag{2-2}$$

where it is recognized that no energy dissipation occurs during the turn-on delay interval $t_{d(on)}$.

Once the switch is fully on, the on-state voltage V_{on} will be on the order of a volt or so depending on the device, and it will be conducting a current I_o. The switch remains in conduction during the on interval t_{on}, which in general is much larger than the turn-on and turn-off transition times. The energy dissipation W_{on} in the switch during this on-state interval can be approximated as

$$W_{on} = V_{on} I_o t_{on} \tag{2-3}$$

where $t_{on} \gg t_{c(on)}, t_{c(off)}$.

In order to turn the switch off, a negative control signal is applied to the control terminal of the switch. During the turn-off transition period of the generic switch, the voltage build-up consists of a turn-off delay time $t_{d(off)}$ and a voltage rise time t_{rv}. Once the voltage reaches its final value of V_d (see Fig. 2-6a), the diode can become forward biased and begin to conduct current. The current in the switch falls to zero with a current fall time t_{fi} as the current I_o commutates from the switch to the diode. Large values of switch voltage and switch current occur simultaneously during the crossover interval $t_{c(off)}$, where

$$t_{c(off)} = t_{rv} + t_{fi} \tag{2-4}$$

The energy dissipated in the switch during this turn-off transition can be written, using Fig. 2-6c, as

$$W_{c(off)} = \tfrac{1}{2} V_d I_o t_{c(off)} \tag{2-5}$$

where any energy dissipation during the turn-off delay interval $t_{d(\text{off})}$ is ignored since it is small compared to $W_{c(\text{off})}$.

The instantaneous power dissipation $p_T(t) = v_T i_T$ plotted in Fig. 2-6c makes it clear that a large instantaneous power dissipation occurs in the switch during the turn-on and turn-off intervals. There are f_s such turn-on and turn-off transitions per second. Hence the average switching power loss P_s in the switch due to these transitions can be approximated from Eqs. 2-2 and 2-5 as

$$P_s = \tfrac{1}{2} V_d I_o f_s (t_{c(\text{on})} + t_{c(\text{off})}) \tag{2-6}$$

This is an important result because it shows that the switching power loss in a semiconductor switch varies linearly with the switching frequency and the switching times. Therefore, if devices with short switching times are available, it is possible to operate at high switching frequencies in order to reduce filtering requirements and at the same time keep the switching power loss in the device from being excessive.

The other major contribution to the power loss in the switch is the average power dissipated during the on-state P_{on}, which varies in proportion to the on-state voltage. From Eq. 2-3, P_{on} is given by

$$P_{\text{on}} = V_{\text{on}} I_o \frac{t_{\text{on}}}{T_s} \tag{2-7}$$

which shows that the on-stage voltage in a switch should be as small as possible.

The leakage current during the off state (switch open) of controllable switches is negligibly small, and therefore the power loss during the off state can be neglected in practice. Therefore, the total average power dissipation P_T in a switch equals the sum of P_s and P_{on}.

Form the considerations discussed in the preceding paragraphs, the following characteristics in a controllable switch are desirable:

1. Small leakage current in the off state.
2. Small on-state voltage V_{on} to minimize on-state power losses.
3. Short turn-on and turn-off times. This will permit the device to be used at high switching frequencies.
4. Large forward- and reverse-voltage-blocking capability. This will minimize the need for series connection of several devices, which complicates the control and protection of the switches. Moreover, most of the device types have a minimum on-state voltage regardless of their blocking voltage rating. A series connection of several such devices would lead to a higher total on-state voltage and hence higher conduction losses. In most (but not all) converter circuits, a diode is placed across the controllable switch to allow the current to flow in the reverse direction. In those circuits, controllable switches are not required to have any significant reverse-voltage-blocking capability.
5. High on-state current rating. In high-current applications, this would minimize the need to connect several devices in parallel, thereby avoiding the problem of current sharing.
6. Positive temperature coefficient of on-state resistance. This ensures that paralleled devices will share the total current equally.
7. Small control power required to switch the device. This will simplify the control circuit design.
8. Capability to withstand rated voltage and rated current simultaneously while switching. This will eliminate the need for external protection (snubber) circuits across the device.

9. Large dv/dt and di/dt ratings. This will minimize the need for external circuits otherwise needed to limit dv/dt and di/dt in the device so that it is not damaged.

We should note that the clamped-inductive-switching circuit of Fig. 2-6*a* results in higher switching power loss and puts higher stresses on the switch in comparison to the resistive-switching circuit shown in Problem 2-2 (Fig. P2-2).

We now will briefly consider the steady-state i–v characteristics and switching times of the commonly used semiconductor power devices that can be used as controllable switches. As mentioned previously, these devices include BJTs, MOSFETs, GTOs, and IGBTs. The details of the physical operation of these devices, their detailed switching characteristics, commonly used drive circuits, and needed snubber circuits are discussed in Chapters 19–28.

2-5   BIPOLAR JUNCTION TRANSISTORS AND MONOLITHIC DARLINGTONS

The circuit symbol for an NPN BJT is shown in Fig. 2-7*a*, and its steady-state i–v characteristics are shown in Fig. 2-7*b*. As shown in the i–v characteristics, a sufficiently large base current (dependent on the collector current) results in the device being fully on. This requires that the control circuit provide a base current that is sufficiently large so that

$$I_B > \frac{I_C}{h_{FE}} \tag{2-8}$$

where h_{FE} is the dc current gain of the device.

The on-state voltage $V_{CE(\text{sat})}$ of the power transistors is usually in the 1–2-V range, so that the conduction power loss in the BJT is quite small. The idealized i–v characteristics of the BJT operating as a switch are shown in Fig. 2-7*c*.

Bipolar junction transistors are current-controlled devices, and base current must be supplied continuously to keep them in the on state. The dc current gain h_{FE} is usually only 5–10 in high-power transistors, and so these devices are sometimes connected in a Darlington or triple Darlington configuration, as is shown in Fig. 2-8, to achieve a larger current gain. Some disadvantages accrue in this configuration including slightly higher overall $V_{CE(\text{sat})}$ values and slower switching speeds.

Whether in single units or made as a Darlington configuration on a single chip [a monolithic Darlington (MD)], BJTs have significant storage time during the turn-off transition. Typical switching times are in the range of a few hundred nanoseconds to a few microseconds.

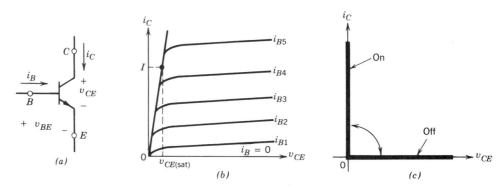

Figure 2-7   A BJT: (*a*) symbol, (*b*) *i*–*v* characteristics, (*c*) idealized characteristics.

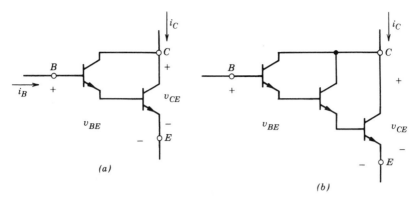

Figure 2-8 Darlington configurations: (a) Darlington, (b) triple Darlington.

Including MDs, BJTs are available in voltage ratings up to 1400 V and current ratings of a few hundred amperes. In spite of a negative temperature coefficient of on-state resistance, modern BJTs fabricated with good quality control can be paralleled provided that care is taken in the circuit layout and that some extra current margin is provided, that is, where theoretically four transistors in parallel would suffice based on equal current sharing, five may be used to tolerate a slight current imbalance.

2-6 METAL–OXIDE–SEMICONDUCTOR FIELD EFFECT TRANSISTORS

The circuit symbol of an n-channel MOSFET is shown in Fig. 2-9a. It is a voltage-controlled device, as is indicated by the i–v characteristics shown in Fig. 2-9b. The device is fully on and approximates a closed switch when the gate–source voltage is below the threshold value, $V_{GS(\text{th})}$. The idealized characteristics of the device operating as a switch are shown in Fig. 2-9c.

Metal–oxide–semiconductor field effect transistors require the continuous application of a gate–source voltage of appropriate magnitude in order to be in the on state. No gate current flows except during the transitions from on to off or vice versa when the gate capacitance is being charged or discharged. The switching times are very short, being in

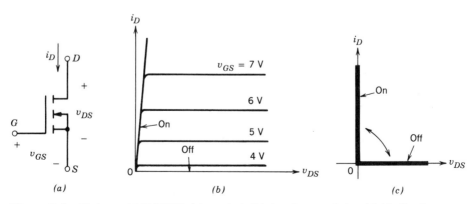

Figure 2-9 N-channel MOSFET: (a) symbol, (b) i–v characteristics, (c) idealized characteristics.

the range of a few tens of nanoseconds to a few hundred nanoseconds depending on the device type.

The on-state resistance $r_{DS(\text{on})}$ of the MOSFET between the drain and source increases rapidly with the device blocking voltage rating. On a per-unit area basis, the on-state resistance as a function of blocking voltage rating BV_{DSS} can be expressed as

$$r_{DS(\text{on})} = k \, BV_{DSS}^{2.5-2.7} \tag{2-9}$$

where k is a constant that depends on the device geometry. Because of this, only devices with small voltage ratings are available that have low on-state resistance and hence small conduction losses.

However, because of their fast switching speed, the switching losses can be small in accordance with Eq. 2-6. From a total power loss standpoint, 300–400-V MOSFETs compete with bipolar transistors only if the switching frequency is in excess of 30–100 kHz. However, no definite statement can be made about the crossover frequency because it depends on the operating voltages, with low voltages favoring the MOSFET.

Metal–oxide–semiconductor field effect transistors are available in voltage ratings in excess of 1000 V but with small current ratings and with up to 100 A at small voltage ratings. The maximum gate–source voltage is ±20 V, although MOSFETs that can be controlled by 5-V signals are available.

Because their on-state resistance has a positive temperature coefficient, MOSFETs are easily paralleled. This causes the device conducting the higher current to heat up and thus forces it to equitably share its current with the other MOSFETs in parallel.

2-7 GATE-TURN-OFF THYRISTORS

The circuit symbol for the GTO is shown in Fig. 2-10a and its steady-state i–v characteristic is shown in Fig. 2-10b.

Like the thyristor, the GTO can be turned on by a short-duration gate current pulse, and once in the on-state, the GTO may stay on without any further gate current. However, unlike the thyristor, the GTO can be turned off by applying a negative gate-cathode voltage, therefore causing a sufficiently large negative gate current to flow. This negative gate current need only flow for a few microseconds (during the turn-off time), but it must have a very large magnitude, typically as large as one-third the anode current being turned off. The GTOs can block negative voltages whose magnitude depends on the details of the

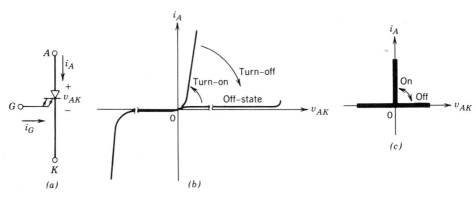

Figure 2-10 A GTO: (a) symbol, (b) i–v characteristics, (c) idealized characteristics.

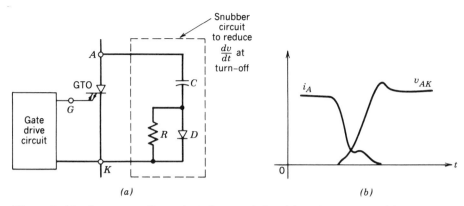

Figure 2-11 Gate turn-off transient characteristics: (*a*) snubber circuit, (*b*) GTO turn-off characteristic.

GTO design. Idealized characteristics of the device operating as a switch are shown in Fig. 2-10*c*.

Even though the GTO is a controllable switch in the same category as MOSFETs and BJTs, its turn-off switching transient is different from that shown in Fig. 2-6*b*. This is because presently available GTOs cannot be used for inductive turn-off such as is illustrated in Fig. 2-6 unless a snubber circuit is connected across the GTO (see Fig. 2-11*a*). This is a consequence of the fact that a large *dv/dt* that accompanies inductive turn-off cannot be tolerated by present-day GTOs. Therefore a circuit to reduce *dv/dt* at turn-off that consists of *R, C,* and *D*, as shown in Fig. 2-11*a*, must be used across the GTO. The resulting waveforms are shown in Fig. 2-11*b*, where *dv/dt* is significantly reduced compared to the *dv/dt* that would result without the turn-off snubber circuit. The details of designing a snubber circuit to shape the switching waveforms of GTOs are discussed in Chapter 27.

The on-state voltage (2–3 V) of a GTO is slightly higher than those of thyristors. The GTO switching speeds are in the range of a few microseconds to 25 μs. Because of their capability to handle large voltages (up to 4.5 kV) and large currents (up to a few kilo-amperes), the GTO is used when a switch is needed for high voltages and large currents in a switching frequency range of a few hundred hertz to 10 kHz.

2-8 INSULATED GATE BIPOLAR TRANSISTORS

The circuit symbol for an IGBT is shown in Fig. 2-12*a* and its *i–v* characteristics are shown in Fig. 2-12*b*. The IGBTs have some of the advantages of the MOSFET, the BJT, and the GTO combined. Similar to the MOSFET, the IGBT has a high impedance gate, which requires only a small amount of energy to switch the device. Like the BJT, the IGBT has a small on-state voltage even in devices with large blocking voltage ratings (for example, V_{on} is 2–3 V in a 1000-V device). Similar to the GTO, IGBTs can be designed to block negative voltages, as their idealized switch characteristics shown in Fig. 2-12*c* indicate.

Insulated gate bipolar transistors have turn-on and turn-off times on the order of 1 μ*s* and are available in module ratings as large as 1700 V and 1200 A. Voltage ratings of up to 2–3 kV are projected.

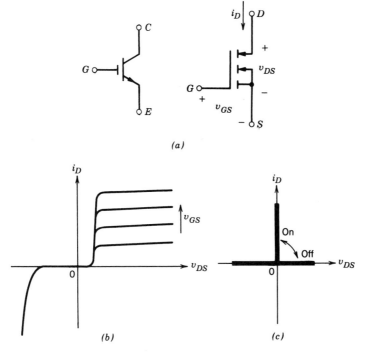

Figure 2-12   An IGBT: (a) symbol, (b) i–v characteristics, (c) idealized characteristics.

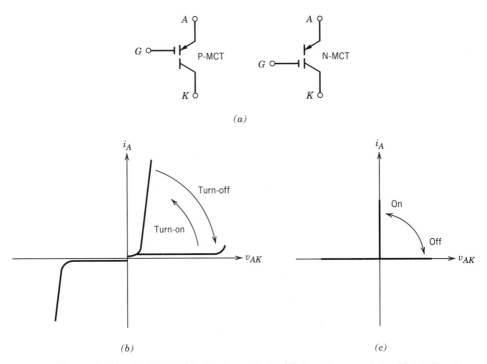

Figure 2-13   An MCT: (a) circuit symbols, (b) i–v characteristic, (c) idealized characteristics.

2-9   MOS-CONTROLLED THYRISTORS

The MOS-controlled thyristor (MCT) is a new device that has just appeared on the commercial market. Its circuit symbol is shown in Fig. 2-13a, and its i–v characteristic is shown in Fig. 2-13b. The two slightly different symbols for the MCT denote whether the device is a P-MCT or an N-MCT. The difference between the two arises from the different locations of the control terminals, a subject discussed in detail in Chapter 26.

From the i–v characteristic it is apparent that the MCT has many of the properties of a GTO, including a low voltage drop in the on state at relatively high currents and a latching characteristic (the MCT remains on even if the gate drive is removed). The MCT is a voltage-controlled device like the IGBT and the MOSFET, and approximately the same energy is required to switch an MCT as for a MOSFET or an IGBT.

The MCT has two principal advantages over the GTO, including much simpler drive requirements (no large negative gate current required for turn-off like the GTO) and faster switching speeds (turn-on and turn-off times of a few microseconds). The MCTs have smaller on-state voltage drops compared to IGBTs of similar ratings and are presently available in voltage ratings to 1500 V with current ratings of 50 A to a few hundred amperes. Devices with voltage ratings of 2500–3000 V have been demonstrated in prototypes and will be available soon. The current ratings of individual MCTs are significantly less than those of GTOs because individual MCTs cannot be made as large in cross-sectional area as a GTO due to their more complex structure.

2-10   COMPARISON OF CONTROLLABLE SWITCHES

Only a few definite statements can be made in comparing these devices since a number of properties must be considered simultaneously and because the devices are still evolving at a rapid pace. However, the qualitative observations given in Table 2-1 can be made.

It should be noted that in addition to the improvements in these devices, new devices are being investigated. The progress in semiconductor technology will undoubtedly lead to higher power ratings, faster switching speeds, and lower costs. A summary of power device capabilities is shown in Fig. 2-14.

On the other hand, the forced-commutated thyristor, which was once widely used in circuits for controllable switch applications, is no longer being used in new converter designs with the possible exception of power converters in multi-MVA ratings. This is a pertinent example of how the advances in semiconductor power devices have modified converter design.

Table 2-1 Relative Properties of Controllable Switches

Device	Power Capability	Switching Speed
BJT/MD	Medium	Medium
MOSFET	Low	Fast
GTO	High	Slow
IGBT	Medium	Medium
MCT	Medium	Medium

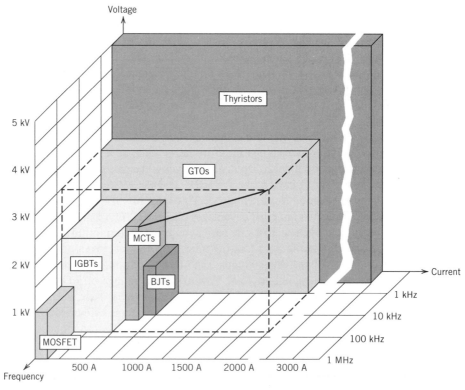

Figure 2-14 Summary of power semiconductor device capabilities. All devices except the MCT have a relatively mature technology, and only evolutionary improvements in the device capabilities are anticipated in the next few years. However, MCT technology is in a state of rapid expansion, and significant improvements in the device capabilities are possible, as indicated by the expansion arrow in the diagram.

2-11 DRIVE AND SNUBBER CIRCUITS

In a given controllable power semiconductor switch, its switching speeds and on-state losses depend on how it is controlled. Therefore, for a proper converter design, it is important to design the proper drive circuit for the base of a BJT or the gate of a MOSFET, GTO, or IGBT. The future trend is to integrate a large portion of the drive circuitry along with the power switch within the device package, with the intention that the logic signals, for example, from a microprocessor, can be used to control the switch directly. These topics are discussed in Chapters 20–26. In Chapters 5–18 where idealized switch characteristics are used in analyzing converter circuits, it is not necessary to consider these drive circuits.

Snubber circuits, which were mentioned briefly in conjunction with GTOs, are used to modify the switching waveforms of controllable switches. In general, snubbers can be divided into three categories:

1. Turn-on snubbers to minimize large overcurrents through the device at turn-on.
2. Turn-off snubbers to minimize large overvoltages across the device during turn-off.
3. Stress reduction snubbers that shape the device switching waveforms such that the voltage and current associated with a device are not high simultaneously.

In practice, some combination of snubbers mentioned before are used, depending on the type of device and converter topology. The snubber circuits are discussed in Chapter 27. Since ideal switches are assumed in the analysis of converters, snubber circuits are neglected in Chapters 5–18.

The future trend is to design devices that can withstand high voltage and current simultaneously during the short switching interval and thus minimize the stress reduction requirement. However, for a device with a given characteristic, an alternative to the use of snubbers is to alter the converter topology such that large voltages and currents do not occur at the same time. These converter topologies, called resonant converters, are discussed in Chapter 9.

2-12 JUSTIFICATION FOR USING IDEALIZED DEVICE CHARACTERISTICS

In designing a power electronic converter, it is extremely important to consider the available power semiconductor devices and their characteristics. The choice of devices depends on the application. Some of the device properties and how they influence the selection process are listed here:

1. On-state voltage or on-state resistance dictates the conduction losses in the device.
2. Switching times dictate the energy loss per transition and determine how high the operating frequency can be.
3. Voltage and current ratings determine the device power-handling capability.
4. The power required by the control circuit determines the ease of controlling the device.
5. The temperature coefficient of the device on-state resistance determines the ease of connecting them in parallel to handle large currents.
6. Device cost is a factor in its selection.

In designing a converter from the system viewpoint, the voltage and current requirements must be considered. Other important considerations include acceptable energy efficiency, the minimum switching frequency to reduce the filter and the equipment size, cost, and the like. Hence the device selection must ensure a proper match between the device capabilities and the requirements on the converter.

These observations help to justify the use of idealized device characteristics in analyzing converter topologies and their operation in various applications as follows:

1. Since the energy efficiency is usually desired to be high, the on-state voltage must be small compared to the operating voltages, and hence it can be ignored in analyzing converter characteristics.
2. The device switching times must be short compared to the period of the operating frequency, and thus the switchings can be assumed to be instantaneous.
3. Similarly, the other device properties can be idealized.

The assumption of idealized characteristics greatly simplifies the converter analysis with no significant loss of accuracy. However, in designing the converters, not only must the device properties be considered and compared, but the converter topologies must also be carefully compared based on the properties of the available devices and the intended application.

SUMMARY

Characteristics and capabilities of various power semiconductor devices are presented. A justification is provided for assuming ideal devices, unless stated explicitly, in Chapters 5–18. The benefits of this approach are the ease of analysis and a clear explanation of the converter characteristics, unobscured by the details of device operation.

PROBLEMS

2-1   The data sheets of a switching device specify the following switching times corresponding to the linearized characteristics shown in Fig. 2-6b for clamped-inductive switchings:

$$t_{ri} = 100 \text{ ns} \qquad t_{fv} = 50 \text{ ns} \qquad t_{rv} = 100 \text{ ns} \qquad t_{fi} = 200 \text{ ns}$$

Calculate and plot the switching power loss as a function of frequency in a range of 25–100 kHz, assuming $V_d = 300$ V and $I_o = 4$A in the circuit of Fig. 2-6a.

2-2   Consider the resistive-switching circuit shown in Fig. P2-2. $V_d = 300$ V, $f_s = 100$ kHz and $R = 75$ Ω, so that the on-state current is the same as in Problem 2-1. Assume the switch turn-on time to be the sum of t_{ri} and t_{fv} in Problem 2-1. Similarly, assume the turn-off time to be the sum of t_{rv} and t_{fi}.

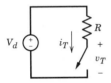

Figure P2-2

Assuming linear voltage- and current-switching characteristics, plot the switch voltage and current and the switching power loss as a function of time. Compare the average power loss with that in Problem 2-1.

REFERENCES

1. R. Sittig and P. Roggwiller (Eds.), *Semiconductor Devices for Power Conditioning*, Plenum, New York, 1982.
2. M. S. Adler, S. W. Westbrook, and A. J. Yerman, "Power Semiconductor Devices—An-Assessment," IEEE Industry Applications Society Conference Record, 1980, pp. 723–728.
3. David L. Blackburn, "Status and Trends in Power Semiconductor Devices," EPE '93, 5th European Conference on Power Electronics and Applications, Conference Record, 1993, Vol. 2, pp. 619–625.
4. B. Jayant Baliga, *Modern Power Devices*, John Wiley & Sons, Inc., New York, 1987.
5. User's Guide to MOS Controlled Thyristors, Harris Semiconductor, 1993.

CHAPTER 3

REVIEW OF BASIC ELECTRICAL AND MAGNETIC CIRCUIT CONCEPTS

3-1   INTRODUCTION

The purpose of this chapter is twofold: (1) to briefly review some of the basic definitions and concepts that are essential to the study of power electronics and (2) to introduce simplifying assumptions that allow easy evaluation of power electronic circuits.

3-2   ELECTRIC CIRCUITS

An attempt is made to use Institute of Electrical and Electronics Engineers (IEEE) standard letter and graphic symbols as much as possible. Moreover, the units used belong to the International System of Units (SI). The lowercase letters are used to represent instantaneous value of quantities that may vary as a function of time. The uppercase letters are used to represent either the average or the rms values. As an example, a voltage v_{oi} and its average value V_{oi} are shown in Fig. 1-4b. A value that is average or rms may be stated explicitly or it may be obvious from the context.

The positive direction of a current is shown explicitly by a current arrow in the circuit diagram. The voltage at any node is defined with respect to the circuit ground, for example, v_a is the voltage of node a with respect to ground. The symbol v_{ab} refers to the voltage of node a with respect to node b, where, $v_{ab} = v_a - v_b$.

3-2-1   DEFINITION OF STEADY STATE

In power electronic circuits, diodes and semiconductor switches are constantly changing their on or off status. Therefore the question arises: When is such a circuit in steady state? A steady-state condition is reached when the circuit waveforms repeat with a time period T that depends on the specific nature of that circuit.

3-2-2   AVERAGE POWER AND rms CURRENT

Consider the circuit of Fig. 3-1, where the instantaneous power flow from subcircuit 1 to subcircuit 2 is

$$p(t) = vi \tag{3-1}$$

Both v and i may vary as a function of time. If v and i waveforms repeat with a time period T in steady state, then the average power flow can be calculated as

$$P_{av} = \frac{1}{T} \int_0^T p(t)\, dt = \frac{1}{T} \int_0^T vi\, dt \tag{3-2}$$

Under the conditions stated earlier, if subcircuit 2 consists purely of a resistive load, then $v = Ri$ and in Eq. 3-2

$$P_{av} = R \frac{1}{T} \int_0^T i^2 dt \tag{3-3}$$

In terms of the rms value I of the current, the average power flow can be expressed as

$$P_{av} = RI^2 \tag{3-4}$$

A comparison of Eqs. 3-3 and 3-4 reveals that the rms value of the current is

$$I = \sqrt{\frac{1}{T} \int_0^T i^2 \cdot dt} \tag{3-5}$$

which shows the origin of the term root-mean-square.

If i is a constant dc current, then Eqs. 3-4 and 3-5 are still valid with the average and the rms values being equal.

3-2-3   STEADY-STATE ac WAVEFORMS WITH SINUSOIDAL VOLTAGES AND CURRENTS

Consider the ac circuit of Fig. 3-2a, with an inductive load under a steady-state operation, where

$$v = \sqrt{2}V \cos \omega t \qquad i = \sqrt{2}I \cos(\omega t - \phi) \tag{3-6}$$

and V and I are the rms values. The v and i waveforms are plotted as functions of ωt in Fig. 3-2b.

3-2-3-1   Phasor Representation

Since both v and i vary sinusoidally with time at the same frequency, they can be represented in a complex plane by means of the projection of the rotating phasors to the horizontal real axis, as shown in Fig. 3-2c. Conventionally, these phasors rotate in a

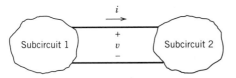

Figure 3-1   Instantaneous power flow.

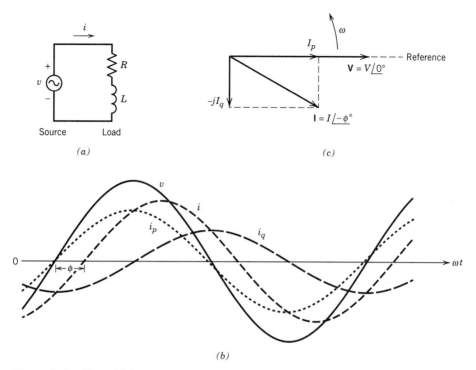

Figure 3-2 Sinusoidal steady state.

counterclockwise direction with an angular frequency ω, and their rms values (rather than their peak values) are used to represent their magnitudes:

$$\mathbf{V} = Ve^{jo} \quad \text{and} \quad \mathbf{I} = Ie^{-j\phi} \tag{3-7}$$

Considering Eq. 3-6, the phasor diagram in Fig. 3-2c corresponds to the time instant when v attains its positive-maximum value.

In Eq. 3-7 $\mathbf{V}$ and $\mathbf{I}$ are related by the complex load impedance $\mathbf{Z} = R + j\omega L = Ze^{j\phi}$ at the operating frequency ω in the following manner:

$$\mathbf{I} = \frac{\mathbf{V}}{\mathbf{Z}} = \frac{Ve^{jo}}{Ze^{j\phi}} = \frac{V}{Z}e^{-j\phi} = Ie^{-j\phi} \tag{3-8}$$

where $I = V/Z$.

3-2-3-2 Power, Reactive Power, and Power Factor

The complex power $\mathbf{S}$ is defined as

$$\mathbf{S} = \mathbf{V}\mathbf{I}^* = Ve^{j0} \cdot Ie^{j\phi} = VIe^{j\phi} = Se^{j\phi} \tag{3-9}$$

Therefore, the magnitude of the complex power, which is also called the *apparent power* and is expressed in the units volt-amperes, is

$$S = VI \tag{3-10}$$

The real average power P is

$$P = \text{Re}[\mathbf{S}] = VI \cos \phi \tag{3-11}$$

which is expressed as a product of V and the current component $I_p = I \cos \phi$, which is in phase with the voltage in the phasor diagram of Fig. 3-2c. The out-of-phase component is $I_q = I \sin \phi$. The in-phase current component $i_p(t)$ and the out-of-phase current component $i_q(t)$ can be expressed as

$$i_p(t) = \sqrt{2}I_p \cos \omega t = (\sqrt{2}I \cos \phi)\cos \omega t \qquad (3\text{-}12)$$

and

$$i_q(t) = \sqrt{2}I_q \sin \omega t = (\sqrt{2}I \sin \phi)\sin \omega t \qquad (3\text{-}13)$$

where $i(t) = i_p(t) + i_q(t)$. These two current components are plotted in Fig. 3-2b.

It should be noted that i_p and i_q result in instantaneous power flow components $p_1 = v \cdot i_p$ and $p_2 = v \cdot i_q$, where $p = p_1 + p_2$. Both p_1 and p_2 pulsate at 2ω, twice the source frequency ω. Here p_1 has an average value given by Eq. 3-11; the average value of p_2 is zero.

In the phasor diagram of Fig. 3-2c, only I_p ($=I \cos \phi$) is responsible for the power transfer, not I_q ($=I \sin \phi$). It is common to define a quantity called *reactive power* Q with the units of var (volt-ampere-reactive) using I_q. Defining the complex power $\mathbf{S} = P + jQ$ and using Eqs. 3-9 and 3-10,

$$Q = VI \sin \phi = VI_q = (S^2 - P^2)^{1/2} \qquad (3\text{-}14)$$

An inductive load shown in Fig. 3-2a has a positive value of ϕ, where the current lags the voltage. In accordance with Eq. 3-14, an inductive load draws positive vars, also called lagging vars. Conversely, a capacitive load draws negative vars, also called leading vars (in other words, it supplies positive vars to the electrical system).

The physical significance of S, P, and Q should be understood. The cost of most electrical equipment such as generators, transformers, and transmission lines increases with $S = VI$, since their electrical insulation level and magnetic core size depend on V and their conductor size depends on I. Power P has a physical significance since it represents the rate of useful work being performed plus the power losses. In most situations, it is desirable to have the reactive power Q be zero.

Based on the above discussion, another quantity called the *power factor* is defined, which is a measure of how effectively the load draws the real power:

$$\text{Power factor} = \frac{P}{S} = \frac{P}{VI} = \cos \phi \qquad (3\text{-}15)$$

which is dimensionless. Ideally, the power factor should be 1.0 (that is, Q should be zero) to draw power with a minimum current magnitude and hence minimize losses in the electrical equipment and possibly in the load.

■ *Example 3-1* An inductive load connected to a 120-V, 60-Hz ac source draws 1 kW at a power factor of 0.8. Calculate the capacitance required in parallel with the load in order to bring the combined power factor to 0.95 (lagging).

Solution

For the load:

$$P_L = 1000 \; W$$

$$S_L = \frac{1000}{0.8} = 1250 \text{ VA}$$

$$Q_L = \sqrt{S_L^2 - P_L^2} = 750 \text{ VA (lagging)}$$

Therefore, the complex power of the load is

$$\mathbf{S}_L = P_L + jQ_L$$
$$= 1000 + j750 \text{ VA}$$

The reactive power drawn by a capacitor is represented as $-jQ_C$ because the capacitor current leads the voltage by 90°. Therefore, the total complex power delivered from the source is

$$S = (P_L + jQ_L) - jQ_C$$
$$= P_L + j(Q_L - Q_C)$$

Since the combined power factor is 0.95 (lagging),

$$S = \sqrt{P_L^2 + (Q_L - Q_C)^2} = \frac{P_L}{0.95}$$

$$(Q_L - Q_C) = P_L \sqrt{\left(\frac{1}{0.95^2} - 1\right)} = 328.7 \text{ VA (lagging)}$$

and therefore,

$$Q_C = 750 - 328.7 = 421.3 \text{ VA (leading)}$$

Since

$$Q_C = \frac{V^2}{X_C} = \frac{V^2}{(1/\omega C)} = V^2 \omega C$$

$$C = \frac{421.3 \times 10^6}{2\pi \times 60 \times 120^2} = 77.6 \text{ }\mu F$$

∎

3-2-3-3 Three-Phase Circuits

During a balanced steady-state operating condition, it is possible to analyze three-phase circuits such as that in Fig. 3-3a on a per-phase basis. The positive phase sequence is commonly assumed to be a–b–c. Using rms values to represent the magnitudes,

$$\mathbf{I}_a = \frac{\mathbf{V}_a}{\mathbf{Z}} = \frac{Ve^{jo}}{Ze^{j\phi}} = \frac{V}{Z} e^{-j\phi} = Ie^{-j\phi}$$
$$\mathbf{I}_b = \mathbf{I}_a e^{-j2\pi/3} = Ie^{-j(\phi + 2\pi/3)}$$
$$\mathbf{I}_c = \mathbf{I}_a e^{j2\pi3} = Ie^{-j(\phi - 2\pi/3)}$$

(3-16)

where $I = V/Z$. Assuming $\mathbf{Z}$ to be an inductive impedance with a positive value of ϕ, the phase voltage and current phasors are shown in Fig. 3-3b.

It is possible to calculate the line-to-line voltages from the phase voltages, recognizing for example that $v_{ab} = v_a - v_b$. Figure 3-3c shows line-to-line voltage phasors where $\mathbf{V}_{ab} = V_{LL}e^{j\pi/6}$ leads $\mathbf{V}_a$ by 30° and the line-to-line rms voltage magnitude is

$$V_{LL} = \sqrt{3} \text{ V}$$

(3-17)

It is possible to calculate power on a per-phase basis as

$$S_{\text{phase}} = VI \quad \text{and} \quad P_{\text{phase}} = VI \cos \phi$$

(3-18)

Therefore, in a balanced system the total three-phase power can be expressed as

$$S_{\text{3-phase}} = 3S_{\text{phase}} = 3VI = \sqrt{3}V_{LL}I$$

(3-19)

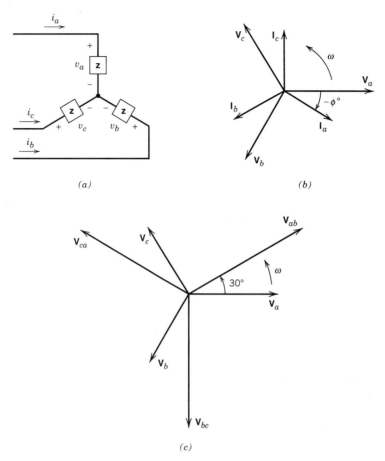

Figure 3-3   Three-phase circuit.

and

$$P_{3\text{-phase}} = 3P_{\text{phase}} = 3VI \cos \phi = \sqrt{3}V_{LL}I \cos \phi \qquad (3\text{-}20)$$

The above three-phase circuit operates on the same power factor as the per-phase power factor of $\cos \phi$.

It should be noted that even if a three-phase circuit is operating with nonsinusoidal voltages and currents, its total power can still be calculated on a per-phase basis provided the circuit is operating under a balanced, steady-state condition.

3-2-4   NONSINUSOIDAL WAVEFORMS IN STEADY STATE

In power electronic circuits, dc or low frequency ac waveforms are synthesized by using segments of an input waveform. The motor voltage produced by the power electronics inverter in an ac motor drive is shown in Fig. 3-4a. Often, the line current drawn from the utility by the power electronic equipment is highly distorted, as shown in Fig. 3-4b. In steady state, such waveforms repeat with a time period T and a frequency $f \, (=\omega/2\pi) = 1/T$. This repetition frequency is called the fundamental frequency, and it is usually designated by a subscript 1. In addition to a dominant component at the fundamental

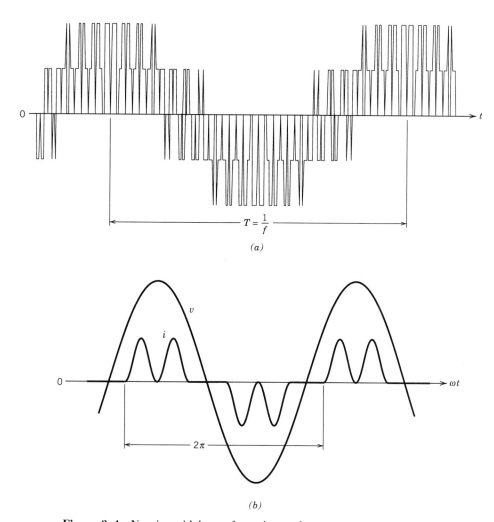

Figure 3-4 Nonsinusoidal waveforms in steady state.

frequency, the waveforms in Fig. 3-4 contain components at the unwanted frequencies that are harmonics (multiples) of the fundamental frequency. These components can be calculated by means of Fourier analysis.

3-2-4-1 Fourier Analysis of Repetitive Waveforms

In general, a nonsinusoidal waveform $f(t)$ repeating with an angular frequency ω can be expressed as

$$f(t) = F_0 + \sum_{h=1}^{\infty} f_h(t) = \tfrac{1}{2}a_0 + \sum_{h=1}^{\infty} \{a_h\cos(h\omega t) + b_h\sin(h\omega t)\} \tag{3-21}$$

where $F_0 = \tfrac{1}{2}a_0$ is the average value. In Eq. 3-21,

$$a_h = \frac{1}{\pi} \int_0^{2\pi} f(t)\cos(h\omega t) \, d(\omega t) \qquad h = 0, \cdot \cdot \cdot, \infty \tag{3-22}$$

and

$$b_h = \frac{1}{\pi} \int_0^{2\pi} f(t) \sin(h\omega t) \, d(\omega t) \qquad h = 1, \cdot \cdot \cdot, \infty \tag{3-23}$$

From Eqs. 3-21 and 3-22, the average value (noting that $\omega = 2\pi/T$)

$$F_0 = \tfrac{1}{2}a_0 = \frac{1}{2\pi} \int_0^{2\pi} f(t) \, d(\omega t) = \frac{1}{T} \int_0^T f(t) \, dt \tag{3-24}$$

In Eq. 3-21, each frequency component $[f_h(t) = a_h\cos(h\omega t) + b_h\sin(h\omega t)]$ can be represented as a phasor in terms of its rms value,

$$\mathbf{F}_h = F_h e^{j\phi_h} \tag{3-25}$$

where the rms magnitude

$$F_h = \frac{\sqrt{a_h^2 + b_h^2}}{\sqrt{2}} \tag{3-26}$$

and phase ϕ_h is given by

$$\tan(\phi_h) = \frac{(-b_h)}{a_h} \tag{3-27}$$

As shown later, the rms value of the function $f(t)$ can be expressed in terms of the rms values of its Fourier series components

$$F = \left(F_0^2 + \sum_{h=1}^{\infty} F_h^2 \right)^{1/2} \tag{3-28}$$

It should be noted that many ac waveforms such as that in Fig. 3-4 have a zero average value ($F_0 = 0$). Moreover, by use of the waveform symmetry it is often possible to simplify the calculations of a_h and b_h in Eqs. 3-22 and 3-23. Table 3-1 summarizes the types of symmetry, required conditions, and the expressions for a_h and b_h.

3-2-4-2 Line-Current Distortion

Figure 3-5 shows a line current i_s drawn from the utility by the power electronic equipment that deviates significantly from a sinusoidal waveform. This distorted current can also lead to distortion in the utility-supplied voltage. However, the distortion in the utility voltage is usually small. For the sake of significantly simplifying our analysis, we will assume the utility input voltage to be purely sinusoidal at the fundamental frequency (with $\omega_1 = \omega$ and $f_1 = f$) as

$$v_s = \sqrt{2} \, V_s \sin \omega_1 t \tag{3-29}$$

The input current in steady state is the sum of its Fourier (harmonic) components as (here it is assumed that there is no dc component in i_s)

$$i_s(t) = i_{s1}(t) + \sum_{h \neq 1} i_{sh}(t) \tag{3-30}$$

where i_{s1} is the fundamental (line-frequency f_1) component and i_{sh} is the component at the h harmonic frequency $f_h(=hf_1)$. These current components in Eq. 3-30 can be expressed as

$$i_s(t) = \sqrt{2}I_{s1}\sin(\omega_1 t - \phi_1) + \sum_{h \neq 1} \sqrt{2}I_{sh}\sin(\omega_h t - \phi_h) \tag{3-31}$$

Table 3-1 Use of Symmetry in Fourier Analysis

Symmetry	Condition Required	a_h and b_h
Even	$f(-t) = f(t)$	$b_h = 0 \qquad a_h = \dfrac{2}{\pi}\displaystyle\int_0^{\pi} f(t)\cos(h\omega t)\, d(\omega t)$
Odd	$f(-t) = -f(t)$	$a_h = 0 \qquad b_h = \dfrac{2}{\pi}\displaystyle\int_0^{\pi} f(t)\sin(h\omega t)\, d(\omega t)$
Half-wave	$f(t) = -f(t + \tfrac{1}{2}T)$	$a_h = b_h = 0$ for even h $a_h = \dfrac{2}{\pi}\displaystyle\int_0^{\pi} f(t)\cos(h\omega t)\, d(\omega t)$ for odd h $b_h = \dfrac{2}{\pi}\displaystyle\int_0^{\pi} f(t)\sin(h\omega t)\, d(\omega t)$ for odd h
Even quarter-wave	Even and half-wave	$b_h = 0$ for all h $a_h = \begin{cases} \dfrac{4}{\pi}\displaystyle\int_0^{\pi/2} f(t)\cos(h\omega t)\, d(\omega t) & \text{for odd } h \\[2mm] 0 & \text{for even } h \end{cases}$
Odd quarter-wave	Odd and half-wave	$a_h = 0$ for all h $b_h = \begin{cases} \dfrac{4}{\pi}\displaystyle\int_0^{\pi/2} f(t)\sin(h\omega t)\, d(\omega t) & \text{for odd } h \\[2mm] 0 & \text{for even } h \end{cases}$

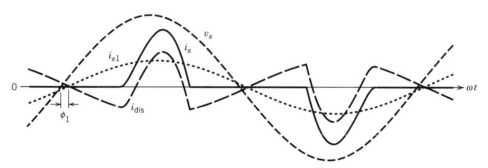

Figure 3-5 Line-current distortion.

where ϕ_1 is the phase angle between the assumed sinusoidal input voltage v_s and i_{s1} (a positive value of ϕ_1 means that the current i_{s1} lags the voltage). The rms value I_s of the line current can be calculated by applying the definition of rms given by Eq. 3-5 to the i_s waveform, as in the following equation (where $T_1 = 1/f_1 = 2\pi/\omega_1$):

$$I_S = \left(\frac{1}{T_1} \int_0^{T_1} i_s^2(t)\, dt \right)^{1/2} \tag{3-32}$$

Substituting for i_s from Eq. 3-30 into Eq. 3-32 and noting that the integrals of all the cross-product terms (i.e., the product of two different frequency components) are individually zero,

$$I_S = \left(I_{s1}^2 + \sum_{h \neq 1} I_{sh}^2 \right)^{1/2} \tag{3-33}$$

The amount of distortion in the voltage or current waveform (here in the input current) is quantified by means of an index called the *total harmonic distortion* (THD). The distortion component i_{dis} of the current from Eq. 3-30 is

$$i_{\text{dis}}(t) = i_s(t) - i_{s1}(t) = \sum_{h \neq 1} i_{sh}(t) \tag{3-34}$$

It is plotted in Fig. 3-5. In terms of the rms values,

$$I_{\text{dis}} = [I_s^2 - I_{s1}^2]^{1/2} = \left(\sum_{h \neq 1} I_{sh}^2\right)^{1/2} \tag{3-35}$$

The THD in the current is defined as

$$\%\text{THD}_i = 100 \times \frac{I_{\text{dis}}}{I_{s1}} \tag{3-36}$$

$$= 100 \times \frac{\sqrt{I_s^2 - I_{s1}^2}}{I_{s1}}$$

$$= 100 \times \sqrt{\sum_{h \neq 1} \left(\frac{I_{sh}}{I_{s1}}\right)^2}$$

where the subscript i indicates the THD in current. A similar index THD_v can be expressed by using voltage components in Eq. 3-36.

In many applications, it is important to know the peak value $I_{s,\text{peak}}$ of the i_s waveform in Fig. 3-5 as a ratio of the total rms current I_s. This ratio is defined as

$$\text{Crest factor} = \frac{I_{s,\text{peak}}}{I_s} \tag{3-37}$$

3-2-4-3   Power and Power Factor

Starting with the basic definition of average power, in Fig. 3-5

$$P = \frac{1}{T_1} \int_0^{T_1} p(t) \, dt = \frac{1}{T_1} \int_0^{T_1} v_s(t) i_s(t) \, dt \tag{3-38}$$

Using v_s from Eq. 3-29 and i_s from Eq. 3-31 and once again noting that the integrals of all cross-product terms are individually zero,

$$P = \frac{1}{T_1} \int_0^{T_1} \sqrt{2} V_s \sin \omega_1 t \cdot \sqrt{2} I_{s1} \sin(\omega_1 t - \phi_1) \, dt = V_s I_{s1} \cos \phi_1 \tag{3-39}$$

Note that the current components at harmonic frequencies do not contribute to the average (real) power drawn from the sinusoidal voltage source v_s. The apparent power S is the product of the rms voltage V_s and the rms current I_s (as in Eq. 3-10 for sinusoidal quantities),

$$S = V_s I_s \tag{3-40}$$

The power factor (PF) is the same as in Eq. 3-15 for sinusoidal quantities:

$$PF = \frac{P}{S} \tag{3-41}$$

Using Eqs. 3-39 through 3-41,

$$PF = \frac{V_s I_{s1} \cos \phi_1}{V_s I_s} = \frac{I_{s1}}{I_s} \cos \phi_1 \qquad (3\text{-}42)$$

The displacement power factor (DPF, which is the same as the power factor in linear circuits with sinusoidal voltages and currents) is defined as the cosine of the angle ϕ_1:

$$DPF = \cos \phi_1 \qquad (3\text{-}43)$$

Therefore, the power factor with a nonsinusoidal current is

$$PF = \frac{I_{s1}}{I_s} DPF \qquad (3\text{-}44)$$

From Eq. 3-35, we note that a large distortion in the current waveform will result in a small value of I_{s1}/I_s and hence a low power factor. In terms of Eqs. 3-36 and 3-44, the power factor can be expressed as

$$PF = \frac{1}{\sqrt{1 + THD_i^2}} DPF \qquad (3\text{-}45)$$

3-2-5 INDUCTOR AND CAPACITOR RESPONSE

As shown by phasors in Fig. 3-6 under a sinusoidal steady-state condition, the current lags the voltage by 90° in an inductor and leads the voltage by 90° in a capacitor. The voltages and currents are related by

$$\mathbf{I}_L = \frac{\mathbf{V}_L}{j\omega L} = \left(\frac{\mathbf{V}_L}{\omega L}\right) e^{-j\pi/2} \quad \text{in an inductor} \qquad (3\text{-}46)$$

and

$$\mathbf{I}_c = j\omega C \mathbf{V}_c = (\omega C \mathbf{V}_c) e^{j\pi/2} \quad \text{in a capacitor} \qquad (3\text{-}47)$$

In an inductor $L(di_L/dt) = v_L(t)$, and therefore,

$$i_L(t) = i_L(t_1) + \frac{1}{L} \int_{t_1}^{t} v_L \, d\xi \qquad t > t_1 \qquad (3\text{-}48)$$

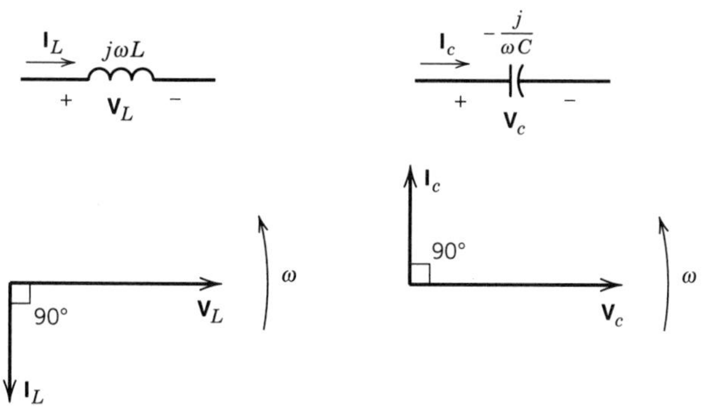

Figure 3-6 Phasor representation.

where ξ is the variable of integration and $i_L(t_1)$ is the inductor current at time t_1. Figure 3-7a shows the inductor current in response to a voltage pulse where initially at $t = t_1$, the inductor has a current of $i_L(t_1)$. It can be seen that even though the inductor voltage may jump instantaneously, the inductor current cannot change instantaneously.

Figure 3-7b shows the capacitor voltage in response to a current pulse, where $v_c(t_1)$ is the initial capacitor voltage at $t = t_1$. Since $C(dv_c/dt) = i_c$,

$$v_c(t) = v_c(t_1) + \frac{1}{C} \int_{t_1}^{t} i_c \, d\xi \qquad t > t_1 \tag{3-49}$$

where ξ is the variable of integration. A capacitor being an electrical dual of an inductor, the capacitor current can jump instantaneously but the capacitor voltage cannot change instantaneously.

3-2-5-1   Average V_L and I_c in Steady State

Now we will consider a concept that is frequently used in power electronics. Consider the circuits of Figs. 3-8a and 3-9a in steady state, though the circuit voltages and currents may not be sinusoidal or constant dc. A steady-state condition implies that the circuit voltage and current waveforms repeat with a time period T; that is,

$$v(t + T) = v(t) \quad \text{and} \quad i(t + T) = i(t) \tag{3-50}$$

In case of an inductor operating under a steady-state condition, substituting $t = t_1 + T$ in Eq. 3-48 and recognizing that $i_L(t_1 + T) = i_L(t_1)$ from Eq. 3-50 result in

$$\int_{t_1}^{t_1+T} v_L \, d\xi = 0$$

or

$$\frac{1}{T} \int_{t_1}^{t_1+T} v_L \, d\xi = 0 \tag{3-51}$$

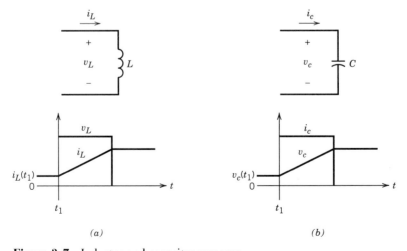

(a)   (b)

Figure 3-7 Inductor and capacitor response.

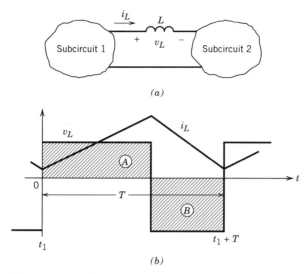

Figure 3-8 Inductor response in steady state.

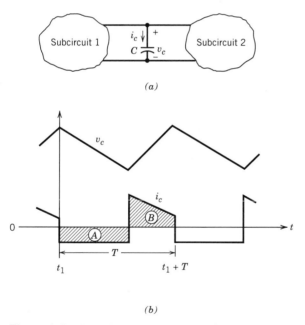

Figure 3-9 Capacitor response in steady state.

where ξ is the variable of integration. Equation 3-51 implies that in steady state, the average inductor voltage (averaged over one time period) must be zero. This is illustrated by the waveforms of Fig. 3-8b, where area $A = B$. As a physical explanation of this property, the integral of inductor voltage is equal to the change in the inductor flux linkage, and Eq. 3-51 implies that the net change of flux linking the inductor over one time period of repetition is zero, which is a necessary condition for steady-state operation.

In case of a capacitor operating under a steady-state condition, substituting $t = t_1 + T$ in Eq. 3-49 and recognizing that $v_c(t_1 + T) = v_c(t_1)$ from Eq. 3-50 result in

$$\int_{t_1}^{t_1+T} i_c \, d\xi = 0$$

or

$$\frac{1}{T} \int_{t_1}^{t_1+T} i_c \, d\xi = 0 \tag{3-52}$$

where ξ is the variable of integration. Equation 3-52 implies that in steady state, the average capacitor current (averaged over one time period) must be zero. This is illustrated by the waveforms of Fig. 3-9b, where area $A = B$. As a physical explanation of this property, the integral of capacitor current is equal to the change in charge of the capacitor, and Eq. 3-52 implies that the net change of charge on the capacitor over one time period of repetition is zero, which is a necessary condition for steady-state operation.

3-3 MAGNETIC CIRCUITS

Power electronic equipment often contains magnetic components such as inductors and transformers. Some of the basic definitions and concepts of magnetic circuits are reviewed here.

3-3-1 AMPERE'S LAW

A current-carrying conductor produces a magnetic field of intensity H whose SI unit is amperes per meter (A/m). According to Ampere's law (Fig. 3-10a), the line integral of the magnetic field intensity H (with units of A/m) equals the total (enclosed) current:

$$\oint H \, dl = \Sigma \, i \tag{3-53}$$

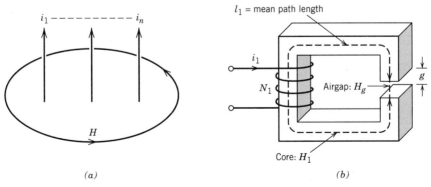

(a)   (b)

Figure 3-10 (a) General formulation of Ampere's law. (b) Specific example of Ampere's law in the case of a winding on a magnetic core with an airgap.

For most practical circuits, the above equation can be written as

$$\sum_{k} H_k l_k = \sum_{m} N_m i_m \tag{3-54}$$

For the circuit in Fig. 3-10b, Eq. 3-54 becomes $H_1 l_1 + H_g l_g = N_1 i_1$.

3-3-2 RIGHT-HAND RULE

The direction of the H-field produced by a current-carrying conductor is defined by the right-hand rule, in which a screw with a right-hand thread moves in the same direction as the current direction. The direction of the H-field is determined by the rotational direction of the screw, which is shown in Fig. 3-11a. This rule is applied in determining the direction of the H-field produced in the core of a coil such as shown in Fig. 3-11b. The direction of this H-field depends on the current direction as well as on how the coil is wound.

3-3-3 FLUX DENSITY OR B-FIELD

The H-field is related to the flux density B or the B-field by the property of the medium in which these fields exist:

$$B = \mu H \tag{3-55}$$

where B is in SI units of webers per square meter^2 (Wb/m^2) or tesla (T), where one tesla equals one weber per square meter, and μ is called the permeability of the medium in SI units of henrys per meter (H/m). The permeability μ of a medium is defined in terms of the permeability of free space or air, μ_0, and a relative permeability μ_r:

$$\mu = \mu_0 \mu_r \tag{3-56}$$

where $\mu_0 = 4\pi \times 10^{-7}$ H/m and μ_r may range from 1.0 for air or a nonmagnetic medium to several thousands for iron.

Equation 3-55 exhibits a linear relationship between the B- and H-fields provided μ stays constant, for example, in nonmagnetic materials or in the magnetic materials operating in a linear region, well below their saturation flux density B_s, as shown in Fig. 3-12. Beyond B_s, a magnetic material begins to saturate, as shown in Fig. 3-12, and the incremental permeability $\mu_\Delta = \Delta B / \Delta H$ can be much smaller than its permeability in the linear region.

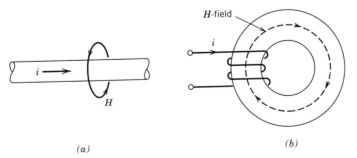

(a)

(b)

Figure 3-11 Determination of the magnetic field direction via the right-hand rule in (a) the general case and (b) a specific example of a current-carrying coil wound on a toroidal core.

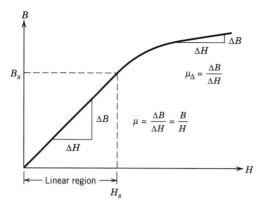

Figure 3-12 Relation between *B*- and *H*-fields.

3-3-4 CONTINUITY OF FLUX

The *B*-field or the flux density represents the density of flux lines per unit area. The magnetic flux ϕ crossing an area can be obtained by the surface integral of the *B*-field normal to that area,

$$\phi = \iint_A B \, dA \tag{3-57}$$

Since the magnetic flux lines form closed loops, the flux lines entering a closed surface area must equal those leaving it. This is called the *continuity of flux* and can be expressed as the closed surface integral of *B*, which is zero:

$$\phi = \iint_{A \text{ (closed surface)}} B \, dA = 0 \tag{3-58}$$

For example, in the magnetic circuit of Fig. 3-13, Eqs. 3-57 and 3-58 result in

$$B_1 A_1 + B_2 A_2 + B_3 A_3 = 0 \quad \text{or} \quad \phi_1 + \phi_2 + \phi_3 = 0$$

In general,

$$\sum_k \phi_k = 0 \tag{3-59}$$

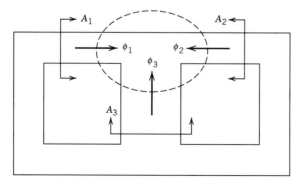

Figure 3-13 Continuity of flux.

3-3-5 MAGNETIC RELUCTANCE AND PERMEANCE

Ampere's law in the form of Eq. 3-54 and the continuity of flux given by Eq. 3-58 can be combined to define the reluctance of a magnetic circuit. In general, for a magnetic circuit of the type shown in Fig. 3-10*b*,

$$\sum_k H_k l_k = \sum_k H_k(\mu_k A_k)\frac{l_k}{\mu_k A_k} = \sum_k (B_k A_k)\frac{l_k}{\mu_k A_k} = \sum_k \phi_k \frac{l_k}{\mu_k A_k} = \phi \sum_k \frac{l_k}{\mu_k A_k}$$

where $\phi_k = \phi$ for each k by applying the continuity-of-flux Eq. 3-58. Therefore, from Eq. 3-54 and the equation above

$$\phi \sum_k \frac{l_k}{\mu_k A_k} = \sum_m N_m i_m \qquad (3\text{-}60)$$

For each section k, the term in the summation on the left-hand-side of Eq. 3-60 is defined as the magnetic reluctance in the path of the magnetic flux lines:

$$\mathcal{R}_k = \frac{l_k}{\mu_k A_k} \qquad (3\text{-}61)$$

and therefore,

$$\phi \sum_k \mathcal{R}_k = \sum_m N_m i_m \qquad (3\text{-}62)$$

For the simple magnetic structure shown in Fig. 3-14,

$$\phi \mathcal{R} = \mathrm{Ni} \qquad (3\text{-}63)$$

As seen from Eq. 3-61, the reluctance of a magnetic circuit depends on the property of the magnetic medium μ and its geometry l and A.

Knowing $\mathcal{R}_k$ and i_m in Eq. 3-62 for circuits similar to that in Fig. 3-10*b*, the flux ϕ can be calculated as

$$\phi = \frac{\displaystyle\sum_m N_m i_m}{\displaystyle\sum_k \mathcal{R}_k} \qquad (3\text{-}64)$$

Permeance of a magnetic circuit is defined as the reciprocal of its reluctance:

$$\mathcal{P} = \frac{1}{\mathcal{R}} \qquad (3\text{-}65)$$

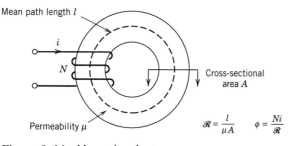

Figure 3-14 Magnetic reluctance.

Table 3-2 Electrical–Magnetic Analogy

Magnetic Circuit	Electric Circuit
mmf Ni	v
Flux ϕ	i
reluctance $\mathcal{R}$	R
permeability μ	$1/\rho$, where ρ = resistivity

3-3-6 MAGNETIC CIRCUIT ANALYSIS

To analyze magnetic circuits, it is often convenient to invoke an analogy between magnetic and electrical quantities, given in Table 3-2 which is valid on a quasi-static basis, that is at a given instant of time.

This leads to the analogy between magnetic and electrical circuit equations given in Table 3-3.

Figure 3-15*a* shows a magnetic circuit and Fig. 3-15*b* shows its electrical analog, which can be easily analyzed.

3-3-7 FARADAY'S VOLTAGE INDUCTION LAW

Consider a stationary coil with or without a magnetic core, as shown in Fig. 3-16*a*. By convention, the voltage polarity is chosen to be positive at the terminal where the positive current enters. This is using the same set of references as we do when we state Ohm's law, $v = Ri$. Having chosen the current direction, the positive flux direction is established by the right-hand rule, which is vertically up in Fig. 3-16*a* for a positive current. Then by Faraday's law, a time-varying flux linkage of the coil $N\phi$ is related to the induced voltage as

$$e = +\frac{d(N\phi)}{dt} = N\frac{d\phi}{dt} \tag{3-66}$$

where the positive voltage polarity and flux direction are as shown in Fig. 3-16*a*.

This induced voltage polarity can be confirmed by Lenz's law of electromagnetic induction. Let us treat the changing external flux ϕ_e as the cause and the induced voltage as the effect. The polarity of the induced voltage is such as to circulate a current (if the circuit is closed) to oppose the change of flux linkage. Therefore, in Fig. 3-16*b*, if ϕ_e is increasing with time in the direction indicated, then the polarity of the induced voltage can be obtained by hypothetically closing the circuit through a resistance R. The current should flow out of the top terminal (from the right-screw rule) in order to oppose the change in flux linkage of the coil by producing ϕ_i. This implies that the induced voltage would have to be positive at the top terminal of the coil with respect to the bottom terminal.

Table 3-3 Magnetic–Electrical Circuit Equation Analogy

Magnetic	Electrical (dc)
$\dfrac{Ni}{\phi} = \mathcal{R} = \dfrac{l}{\mu A}$	Ohm's law: $\dfrac{v}{i} = R = \dfrac{l}{A/\rho}$
$\phi \displaystyle\sum_k \mathcal{R}_k = \sum_m N_m i_m$	Kirchhoff's voltage law: $i \displaystyle\sum_k R_k = \sum_m v_m$
$\displaystyle\sum \phi_k = 0$	Kirchhoff's current law: $\displaystyle\sum_k i_k = 0$

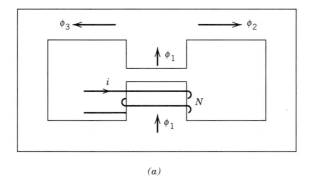

(a)

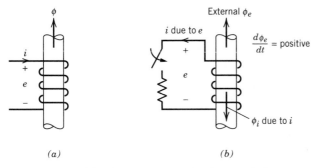

(b)

Figure 3-15   *(a)* Magnetic circuit. *(b)* An electrical analog.

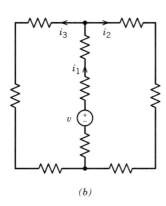

(a)   *(b)*

Figure 3-16   *(a)* Flux direction and voltage polarity.
(b) Lenz's law.

3-3-8   SELF-INDUCTANCE *L*

A coil such as that shown in Fig. 3-17 has a self-inductance or simply inductance *L*, which is defined as

$$L = \frac{N\phi}{i} \quad \text{or} \quad N\phi = Li \tag{3-67}$$

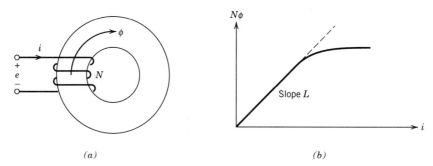

(a)   *(b)*

Figure 3-17   Self-inductance L.

where generally in the linear range of the core material, L is independent of i. Substituting for $N\phi$ from Eq. 3-67 into Eq. 3-66 yields

$$e = L\frac{di}{dt} + i\frac{dL}{dt} = L\frac{di}{dt} \quad \text{for a stationary coil} \tag{3-68}$$

In coils with ferromagnetic cores, $N\phi$ varies with i, as shown in Fig. 3-17b, and can be linearized as indicated by the dotted line. The slope of this linearized characteristic equals the linearized inductance L.

In Eq. 3-67, substitution for ϕ from Eq. 3-63 yields

$$L = \frac{N}{i}\frac{Ni}{\mathcal{R}} = \frac{N^2}{\mathcal{R}} \tag{3-69}$$

which shows that the coil inductance is a property of the magnetic circuit and is independent of i, provided magnetic saturation does not occur.

3-3-9   TRANSFORMERS

3-3-9-1   Transformers with Lossless Cores

A transformer consists of two or more coils that are magnetically coupled. Figure 3-18a shows a cross section of a conceptual transformer with two coils. We assume that the

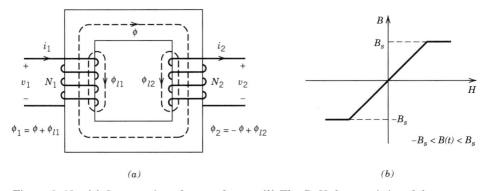

(a)   *(b)*

Figure 3-18   (a) Cross section of a transformer. (b) The B–H characteristics of the core.

transformer core has the *B–H* characteristic shown in Fig. 3-18b and that $B(t)$ is always less than B_s. The total flux ϕ_1 in coil 1 is given by

$$\phi_1 = \phi + \phi_{l1} \tag{3-70}$$

and the total flux ϕ_2 in coil 2 is given by

$$\phi_2 = -\phi + \phi_{l2} \tag{3-71}$$

In Eqs. 3-70 and 3-71 ϕ_{l1} and ϕ_{l2} are the leakage fluxes in coils 1 and 2, respectively, and are diagrammed in Fig. 3-18a. The flux ϕ in the core links the two coils and is given by

$$\phi = \frac{N_1 i_1 - N_2 i_2}{\mathcal{R}_c} = \frac{N_1 i_m}{\mathcal{R}_c} \tag{3-72}$$

where $\mathcal{R}_c$ is the reluctance of the core and i_m is the magnetizing current given by Eq. 3-72 as

$$i_m = i_1 - \frac{N_2 i_2}{N_1} \tag{3-73}$$

The leakage fluxes are given by

$$\phi_{l1} = \frac{N_1 i_1}{\mathcal{R}_{l1}} \tag{3-74}$$

and

$$\phi_{l2} = \frac{N_2 i_2}{\mathcal{R}_{l2}} \tag{3-75}$$

where $\mathcal{R}_{l1}$ and $\mathcal{R}_{l2}$ are the reluctances of the leakage-flux paths. Even in well-designed transformers, the leakage fluxes are a nonnegligible part of the total coil fluxes. This results in leakage reluctances that must be accounted for in any transformer description.

The voltages v_1 and v_2 at the terminals of the transformer are given by

$$v_1 = R_1 i_1 + N_1 \frac{d\phi_1}{dt} \tag{3-76}$$

and

$$v_2 = -R_2 i_2 - N_2 \frac{d\phi_2}{dt} \tag{3-77}$$

The resistances R_1 and R_2 account for the ohmic losses in the windings caused by the finite conductivity of the conductors. The negative signs in Eq. 3-77 are the result of setting the polarity of the voltage v_2 as positive where the current i_2 leaves the terminal of coil 2 (where the direction of ϕ_2 is consistent with the right-hand rule applied to coil 2). Using Eqs. 3-70, 3-72, and 3-74 in Eq. 3-76 to express the fluxes in terms of the currents i_1 and i_2 yields

$$v_1 = R_1 i_1 + \frac{N_1^2}{\mathcal{R}_{l1}} \frac{di_1}{dt} + \frac{N_1^2}{\mathcal{R}_c} \frac{di_m}{dt} \tag{3-78}$$

Similarly the voltage v_2 can be expressed as

$$v_2 = -R_2 i_2 - \frac{N_2^2}{\mathcal{R}_{l2}} \frac{di_2}{dt} + \frac{N_1 N_2}{\mathcal{R}_c} \frac{di_m}{dt} \tag{3-79}$$

The notation in the expression for v_1 (Eq. 3-78) can be simplified by defining the following quantities:

$$\text{the induced emf in coil 1 } e_1 = \frac{N_1^2}{\mathfrak{R}_c}\frac{di_m}{dt} = L_m\frac{di_m}{dt} \qquad (3\text{-}80a)$$

where,

$$\text{the magnetizing inductance } L_m = \frac{N_1^2}{\mathfrak{R}_c} \qquad (3\text{-}80b)$$

and,

$$\text{the leakage inductance for coil 1 } L_{l1} = \frac{N_1^2}{\mathfrak{R}_{l1}} \qquad (3\text{-}81)$$

Using these definitions in Eq. 3-78 yields

$$v_1 = i_1 R_1 + L_{l1}\frac{di_1}{dt} + L_m\frac{di_m}{dt} = R_1 i_1 + L_{l1}\frac{di_1}{dt} + e_1 \qquad (3\text{-}82)$$

If the third term on the right-hand side of Eq. 3-79 is multiplied by N_1/N_1 and we define

$$L_{l2} = \frac{N_2^2}{\mathfrak{R}_{l2}} \qquad (3\text{-}83)$$

then Eq. 3-79 can be expressed as

$$v_2 = -R_2 i_2 - L_{l2}\frac{di_2}{dt} + \frac{N_2}{N_1} e_1 = -R_2 i_2 - L_{l2}\frac{di_2}{dt} + e_2 \qquad (3\text{-}84)$$

where e_2 is the induced emf in coil 2.

Equations 3-82 and 3-84 form the basis of the transformer equivalent circuit shown in Fig. 3-19a.

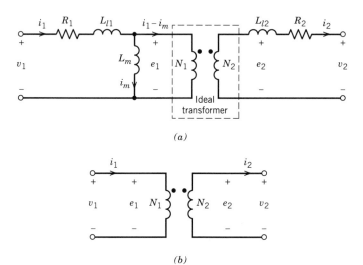

(a)

(b)

Figure 3-19 Equivalent circuit for (a) a physically realizable transformer wound on a lossless core and (b) an ideal transformer.

3-3-9-2 Ideal Transformers

Sometimes the equivalent circuit of a transformer can be simplified with the following idealizations:

1. $R_1 = R_2 = 0$ (i.e., windings made from perfect conductors).
2. $\mathscr{R}_c = 0$ (core permeability $\mu = \infty$) and hence $L_m = \infty$.
3. $\mathscr{R}_{l1} = \mathscr{R}_{l2} = \infty$ and thus $L_{l1} = L_{l2} = 0$ (leakage fluxes ϕ_{l1} and $\phi_{l2} = 0$).

If these idealizations can be made, then the transformer equivalent reduces to the ideal transformer equivalent shown in Fig. 3-19b. In the ideal-transformer approximations, Eqs. 3-82 and 3-84 reduce to

$$v_1 = e_1 \qquad v_2 = e_2 = \frac{N_2}{N_1} e_1 = \frac{N_2}{N_1} v_1 \quad \text{or} \quad \frac{v_1}{N_1} = \frac{v_2}{N_2} \tag{3-85}$$

Equation 3-72 can be rewritten as $\mathscr{R}_c\phi = N_1 i_1 - N_2 i_2 = 0$ when $\mathscr{R}_c = 0$. Thus we can express the currents as

$$N_2 i_2 = N_1 i_1 \quad \text{or} \quad \frac{i_1}{i_2} = \frac{N_2}{N_1} \tag{3-86}$$

Equations 3-85 and 3-86 are the mathematical description of an ideal transformer. Note that the equivalent circuit for the physically realizable transformer of Fig. 3-19a contains an ideal transformer.

3-3-9-3 Transformers with Cores Having Hysteresis

If the transformer is wound on a magnetic core having a B–H characteristic with hysteresis such as is shown in Fig. 3-20, then the time-varying flux in the core will dissipate power in the core. This dissipation or loss must be accounted for in the equivalent circuit of the transformer. According to Eq. 3-72, it is the magnetizing current i_m that generates the flux in the core, and in the magnetically lossless equivalent circuit of Fig. 3-19a, i_m flows through the magnetizing inductance L_m. In the presence of core losses, i_m still generates the flux in the core, so a convenient way to model the core losses in the transformer equivalent circuit is by a resistance either in series with L_m or in parallel with L_m. The standard practice is to include a resistance R_m in parallel with L_m, as is shown in Fig. 3-21a, which makes the core losses equal to $e_1{}^2/R_m$.

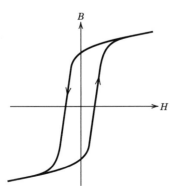

Figure 3-20 B–H characteristic of a transformer core having hysteresis and hence magnetic losses.

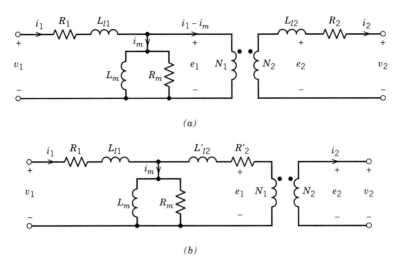

(a)

(b)

Figure 3-21 Equivalent circuit of a transformer including the effects of hysteresis loss. (*a*) Circuit components are on both sides (coil 1 and coil 2 sides) of the ideal transformer. (*b*) Components from the secondary (coil 2) side are reflected across the ideal transformer to the primary (coil 1) side.

3-3-9-4 Per-Unit Leakage Inductances

The total leakage inductance seen from one side (e.g., from the coil 1 side) can be written as

$$L_{l,\text{total}} = L_{l1} + L'_{l2} \tag{3-87}$$

where L'_{l2} is L_{l2} reflected across the ideal transformer in the equivalent circuit of Fig. 3-21*a* to the primary (coil 1 side) of the transformer. Here L'_{l2} can be expressed as

$$L'_{l2} = \left(\frac{N_1}{N_2}\right)^2 L_{l2} \tag{3-88}$$

In a similar fashion the winding resistance R_2 can be reflected across the ideal transformer to the coil 1 side of the equivalent circuit and can be written as

$$R'_2 = \left(\frac{N_1}{N_2}\right)^2 R_2 \tag{3-89}$$

The resulting equivalent circuit for the transformer with all the nonideal but still linear components on the primary (coil 1) side of the transformer is shown in Fig. 3-21*b*. For approximate analysis in transformers with ratings above a few kilo-volt-amperes, it is usually possible to ignore the winding resistances and to assume that the magnetic core is ideal (i.e., R_1, $R_2 = 0$ and R_m, $L_m = \infty$).

The total leakage inductance $L_{l,\text{total}} = L_{l1} + L'_{l2}$ is often specified in per unit or percentage of the transformer voltage and volt-ampere ratings. This is illustrated by means of an example.

■ *Example 3-2* A 110/220-V, 60-Hz, single-phase, 1-kVA transformer has a leakage reactance of 4%. Calculate its total leakage inductance referred to (a) the 110-V side and (b) the 220-V side.

Solution

Assuming the 110-V side to be side 1 and the 220-V side to be side 2, $N_1/N_2 = 0.5$. Let L_{lt1} be the total leakage inductance referred to side 1 and L_{lt2} referred to side 2.

(a) For side 1,

$$V_{1,\text{rated}} = 110 \text{ V}$$

$$I_{1,\text{rated}} = \frac{1000}{110} = 9.09 \text{ A}$$

$$Z_{1,\text{base}} = \frac{V_{1,\text{rated}}}{I_{1,\text{rated}}} = \frac{110}{9.09} = 12.1 \ \Omega$$

Therefore,

$$L_{lt1} = \frac{0.04 \times Z_{1,\text{base}}}{2\pi \times 60} = 1.28 \text{ mH}$$

(b) For side 2,

$$V_{2,\text{rated}} = 220 \text{ V}$$

$$I_{2,\text{rated}} = \frac{1000}{220} = 4.54 \text{ A}$$

$$Z_{2,\text{base}} = \frac{V_{2,\text{rated}}}{I_{2,\text{rated}}} = 48.4 \ \Omega$$

Therefore,

$$L_{lt2} = \frac{0.04 \times Z_{2,\text{base}}}{2\pi \times 60} = 5.15 \text{ mH}$$

Note that $L_{lt1} = (N_1/N_2)^2 \, L_{lt2}$.   ∎

SUMMARY

1. The steady state in a power electronic circuit is defined to be the condition when the circuit waveforms repeat with a time period T that depends on the specific nature of the circuit.

2. For sinusoidal voltages and currents in single-phase and three-phase circuits, the following quantities are defined: rms values, average power, reactive power or vars, and power factor.

3. Fourier analysis is reviewed to express nonsinusoidal waveforms in steady state in terms of their harmonic frequency components.

4. A distortion index called Total Harmonic Distortion (THD) is defined for nonsinusoidal voltages and currents.

5. Assuming sinusoidal voltages supplied by the utility, the displacement power factor (DPF) and the power factor (PF) are defined for nonlinear loads which draw nonsinusoidal currents.

6. In steady state in power electronic circuits, the average voltage across an inductor is shown to be zero. Similarly, the average current through a capacitor is zero.

7. The following fundamentals are reviewed: Ampere's law, right-hand rule, flux density or B-field, continuity of flux, and magnetic reluctance and permeance.

8. For magnetic circuit analysis, an analogy between the magnetic circuit and instantaneous electric circuit is invoked. Faraday's voltage induction law and the definition of self-inductance are reviewed.

9. Basic concepts relating to transformers are reviewed.

PROBLEMS

3-1 Using rms values to represent their magnitudes, the voltage across and the current into a load are as follows in phasor form:
$$\mathbf{V} = Ve^{j0^\circ} \quad \text{and} \quad \mathbf{I} = Ie^{-j\phi^\circ}$$
Show that the instantaneous power $p(t) = v(t) \cdot i(t)$ can be written as $p(t) = P + P \cos 2\omega t + Q \sin 2\omega t$, where average power $P = VI \cos \phi$ and reactive power $Q = VI \sin \phi$.

3-2 In Problem 3-1, $V = 120$ V and $\mathbf{I} = e^{-j30^\circ}$ A.

(a) Plot the following as a function of ωt:

 (i) v, i, and $p(t)$

 (ii) i_p as defined in Eq. 3-12 and $p_1 = v \cdot i_p$

 (iii) i_q as defined in Eq. 3-13 and, $p_2 = v \cdot i_q$

(b) Calculate the average power P.

(c) Calculate the peak value of p_2 in part (a) and Q from Eq. 3-14.

(d) Calculate the load power factor (PF). Is the load inductive or capacitive? Does the load draw positive vars?

3-3 For the waveforms in Fig. P3-3, calculate their average value and the rms values of the fundamental and the harmonic frequency components.

3-4 In the waveforms of Fig. P3-3 of Problem 3-3, $A = 10$ and $u = 20^\circ$ ($u_1 = u_2 = u/2$), where applicable. Calculate their total rms values as follows:

(a) By using the results of Problem 3-3 in Eq. 3-28.

(b) By using the definition of the rms value as given in Eq. 3-5.

3-5 Refer to Problem 3-4 and calculate the following:

(a) For each of the waveforms a–e, calculate the ratio of (i) the fundamental frequency component to the total rms value and (ii) the distortion component to the total rms value.

(b) For the waveforms f and g, calculate the ratio of the average value to the total rms value.

3-6 A sinusoidal voltage $v = \sqrt{2}V \sin \omega t$ is applied to a single-phase load. The current drawn by the load corresponds to one of the waveforms a–e of Fig. P3-3. The zero crossing of the current waveform *lags* the voltage waveform by $\omega t = \phi^\circ$. Using the results of Problems 3-3 and 3-4, calculate the average power drawn by the load, the displacement power factor (DPF), total harmonic distortion (THD), and the power factor (PF) for each of the current waveforms for the following numerical values: $V = 120$ V, $A = 10$ A, $\phi = 30^\circ$, and $u = 20^\circ$ ($u_1 = u_2 = u/2$), where applicable.

3-7 A balanced three-phase inductive load is supplied in steady state by a balanced three-phase voltage source with a phase voltage of 120 V rms. The load draws a total of 10 kW at a power factor of 0.85 (lagging). Calculate the rms value of the phase currents and the magnitude of the per-phase load impedance. Draw a phasor diagram showing all three voltages and currents.

3-8 An input voltage of a repetitive waveform is filtered and then applied across the load resistance, as shown in Fig. P3-8. Consider the system to be in steady state. It is given that $L = 5$ μH and $P_{\text{Load}} = 250$ W.

(a) Calculate the average output voltage V_o.

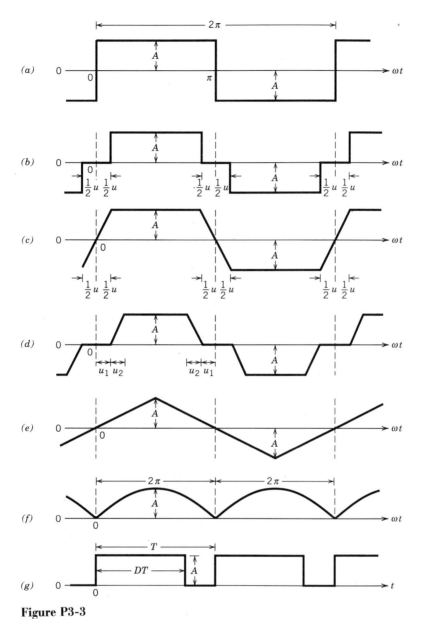

Figure P3-3

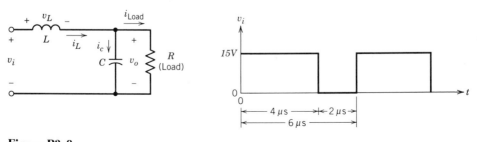

Figure P3-8

(b) Assume that $C{\to}\infty$ so that $v_o(t) \simeq V_o$. Calculate I_{Load} and the rms value of the capacitor current i_c.

(c) In part (b), plot v_L and i_L.

3-9 The repetitive waveforms for the current into, and the voltage across a load in Fig. P3-9 are shown by linear segments. Calculate the average power P into the load.

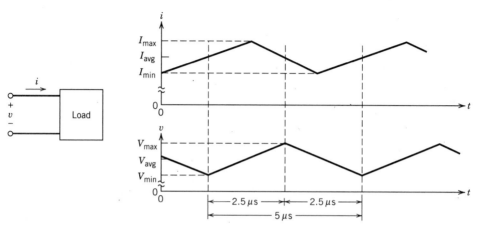

Figure P3-9

3-10 In Fig. P3-9, it is given that the load voltage maximum (minimum) is greater (lower) than V_{avg} by 1%. Similarly in the current, the fluctuation around its average value is $\pm 5\%$. Calculate the percent error if the average power is assumed to be $V_{\text{avg}}I_{\text{avg}}$, compared to its exact value.

3-11 A transformer is wound on a toroidal core. The primary winding is supplied with a square-wave voltage with a ± 50-V amplitude and a frequency of 100 kHz. Assuming a uniform flux density in the core, calculate the minimum number of primary winding turns required to keep the peak flux density in the core below 0.15 Wb/m^2 if the core cross-sectional area is 0.635 cm^2.

Plot the voltage and flux density waveforms in steady state as functions of time.

3-12 A toroidal core has distributed airgaps that make the relative permeability equal to 125. The cross-sectional area is 0.113 cm^2 and the mean path length is 3.12 cm. Calculate the number of turns required to obtain an inductance of 25 μH.

3-13 In Example 3-2, calculate the transformer voltage regulation in percent, if the input voltage is 110 V to the transformer which supplies its full-load kVA to a load at the following power factors:

(a) 1.0

(b) 0.8 (lagging)

Note that the transformer voltage regulation is defined as

$$\text{Percent regulation} = 100 \times \frac{V_{\text{out,No-load}} - V_{\text{out,full-load}}}{V_{\text{out,no-load}}}$$

3-14 Refer to Problem 3-11 and calculate the magnetizing inductance L_m if the mean path length equals 3.15 cm and the relative permeability $\mu_r = 2500$.

REFERENCES

1. H. P. Hsu, *Fourier Analysis,* Simon & Schuster, New York, 1967.
2. Any introductory textbook on electrical circuits and electromagnetic fields.

CHAPTER 4

COMPUTER SIMULATION OF POWER ELECTRONIC CONVERTERS AND SYSTEMS

4-1   INTRODUCTION

The purpose of this chapter is to briefly describe the role of computer simulations in the analysis and design of power electronics systems. We will discuss the simulation process and some of the simulation software packages suited for this application.

In power electronic systems such as shown by Fig. 1-1 and repeated as Fig. 4-1, converters for power processing consist of passive component, diodes, thyristors, and other solid-state switches. Therefore, the circuit topology changes as these switches open and close as a function of time under the guidance of the controller. Usually it is not possible, and often not desirable, to solve for the circuit states (voltages and currents) in a closed form as a function of time. However, by means of computer simulation, it is possible to model such circuits.

We will use computer simulations throughout this book as *optional* learning aid. Computer simulations are commonly used in research to analyze the behavior of new circuits, which leads to improved understanding of the circuit. In industry, they are used to shorten the overall design process, since it is usually easier to study the influence of a parameter on the system behavior in simulation, as compared to accomplishing the same in the laboratory on a hardware breadboard [1].

The simulations are used to calculate the circuit waveforms, the dynamic and steady-state performance of systems, and the voltage and current ratings of various components. Usually, there will be several iterations between various steps. As the confidence in the simulation develops, it may be possible to extend simulations to include power loss

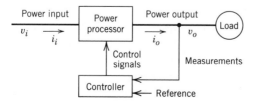

Figure 4-1 Power electronics system: a block diagram.

61

calculations. This allows a thermal design to ensure temperature rise within the system to acceptable levels.

However, it should be noted that in power electronics, much more so than in signal-level electronics, computer simulation and a proof-of-concept hardware prototype in the laboratory are complimentary to each other. That is, computer simulation should not be looked upon as a substitute for a hardware (breadboard) prototype.

4-2   CHALLENGES IN COMPUTER SIMULATION

At the outset, we need to realize that there are several factors that make simulation of power electronic systems very challenging:

1. Solid-state switches including diodes and thyristors present extreme nonlinearity during their transition from one state to the other. The simulation program ought to be able to represent this switching of states in an appropriate manner.

2. The simulation may take a long time. The time constants, or in other words the response time of various parts within the system, may differ by several orders of magnitude. For example, in a motor drive, the semiconductor switches have switching times of microseconds or less, whereas the mechanical time constant or the response time of the motor and the load may be of the order of seconds or even minutes. This requires simulation to proceed with a very small time step to have the resolution to represent the smallest time constants, for example switching, with accuracy. In the same simulation, the maximum simulation time is usually large and is dictated by the longest time constant.

3. Accurate models are not always available. This is specially true for power semi-conductor devices (even for simple power diodes) but is also the case for magnetic components such as inductors and transformers.

4. The controller in the block diagram of Fig. 4-1, which may be analog and/or digital, needs to be modeled along with the power converters.

5. Even if only the steady-state waveforms are of interest, the simulation time is usually long due to unknown values of the initial circuit states at the start of the simulation.

The challenges listed above dictate that we carefully evaluate the objective of the simulation. In general, it is not desirable to simulate all aspects of the system in detail (at least not initially, but it may be done as the last step). The reason is that the simulation time may be very long and the output at the end of the simulation may be overwhelming, thus obscuring the phenomena of interest. In this respect, the best simulation is the simplest possible simulation that meets the immediate objective. In other words, we must simplify the system to meet the simulation objectives. Some of the choices are discussed in the next section.

4-3   SIMULATION PROCESS

In power electronics, several types of analyses need to be carried out. For each type of analysis, there is an appropriate degree of simulation detail in which the circuit components and the controller should be represented. In the following sections, we will discuss these various types of analyses. It is important to note that at each step it may be desirable to verify simulation results by a hardware prototype in the laboratory.

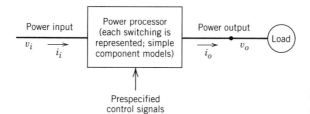

Figure 4-2 Open-loop, large-signal simulation.

4-3-1 OPEN-LOOP, LARGE-SIGNAL SIMULATION

In order to get a better understanding of the behavior of a new system, we often start by simulating the power processor with prespecified control signals, as shown in the block diagram of Fig. 4-2. The objective of this simulation is to obtain various voltage and current waveforms within the converters of the power processor to verify that the circuit behaves properly, as predicted by the analytical calculations. At the end, this step provides us with a choice of circuit topology and the component values.

This simulation includes each switch opening and closing, and the simulation is carried out over a large number of switching cycles to reach steady state. Most often, at this stage of simulation, no benefit is gained by including very detailed models of the circuit components. Therefore, the circuit components, especially the switching devices, should be represented by their simple (idealized) models. Because the design of the controller still remains to be carried out and the dynamic behavior of the system to changes in the operating conditions is not of interest at this early stage, the controller is not represented. Therefore, it is called an open-loop simulation.

4-3-2 SMALL-SIGNAL (LINEAR) MODEL AND CONTROLLER DESIGN

With a chosen circuit topology and the component values, we can develop a linear (small-signal) model (Fig. 4-3) of the power processor as a transfer function using the techniques described later in this book. The important item to note is that in such a model, the switches are represented by their averaged characteristics. Once we have a linearized model of the circuit, there are well-known methods from control theory for designing the controller to ensure stability and the dynamic response to disturbances or small changes (indicated by Δ in Fig. 4-3) in the input, the load, and the reference. There are specialized software packages available commercially that automate the controller design process [2].

4-3-3 CLOSED-LOOP, LARGE-SIGNAL SYSTEM BEHAVIOR

Once the controller has been designed, the system performance must be verified by combining the controller and the circuit under a close-loop operation, in response to large disturbances such as step changes in load and inputs. The block diagram is shown in Fig. 4-4. This large-signal simulation is carried out in time domain over a long time span that

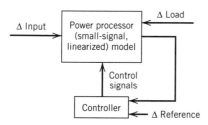

Figure 4-3 Small-signal (linear) model and controller design.

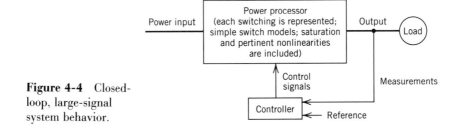

Figure 4-4   Closed-loop, large-signal system behavior.

includes many (thousands of) switching cycles. Therefore, the switching devices should be represented by their simple (idealized) models. However, saturation and other pertinent nonlinearilities and losses may be included. It is sufficient to represent the controller in such a simulation in as simple a manner as possible, rather than representing it in detail at a component level by operational amplifiers, comparators, and so on.

4-3-4 SWITCHING DETAILS

In the previous simulation step, the objective was to obtain the overall system behavior that is only marginally affected by semiconductor device nonidealities. Now the objective is to obtain overvoltages, power losses, and other component stresses due to the nonideal nature of switching devices and the stray inductances and capacitances within the power processor. This knowledge is necessary in the selection of component ratings, assessing the need for protection circuitry such as snubbers and the need to minimize stray inductances and capacitances. The block diagram is shown in Fig. 4-5.

   To obtain this information, only a few switching cycles need to be simulated with worst-case initial values of voltages and currents obtained in the previous simulations. Only a part of the circuit under investigation should be modeled in detail, rather than the entire circuit. Simulation over only a few cycles is needed to obtain the worst-case stresses because they repeat with each cycle. To this end, we need detailed and accurate models of switching devices.

4-4   MECHANICS OF SIMULATION [1]

Having established the various types of analyses that need to be carried out, the next step is to determine the best tools for the job. There are two basic choices: (1) circuit-oriented simulators and (2) equation solvers. These are now discussed in a generic manner.

4-4-1 CIRCUIT-ORIENTED SIMULATORS

Over the years, considerable effort has been put into developing software for circuit-oriented simulators. In these software packages, the user needs to supply the circuit

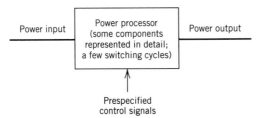

Figure 4-5   Switching details.

topology and the component values. The simulator internally generates the circuit equations that are totally transparent to the user. Depending on the simulator, the user may have the flexibility of selecting the details of the component models. Most simulators allow controllers to be specified by means of a transfer function or by models of components such as operational amplifiers, comparators, and so on.

4-4-2 EQUATION SOLVERS

An alternative to the use of circuit-oriented simulators is to describe the circuit and the controller by means of differential and algebraic equations. We must develop the equations for all possible states in which the circuit may operate. There may be many such states. Then, we must describe the logic that determines the circuit state and the corresponding set of differential equations based on the circuit conditions. These algebraic/differential equations can be solved by using a high-level language such as C or FORTRAN or by means of software packages specifically designed for this purpose that provide a choice of integration routines, graphical output, and so on.

4-4-3 COMPARISON OF CIRCUIT-ORIENTED SIMULATORS AND EQUATION SOLVERS

With circuit-oriented simulators, the initial setup time is small, and it is easy to make changes in the circuit topology and control. The focus remains on the circuit rather than on the mathematics of the solution. Many built-in models for the components and the controllers (analog and digital) are usually available. It is possible to segment the overall system into smaller modules or building blocks that can be individually tested and then brought together.

On the negative side, there is little control over the simulation process that can lead to long simulation times or, even worse, to numerical convergence or oscillation problems, causing the simulation to halt. Steps to overcome these difficulties are usually not apparent and may require trial and error.

Equation solvers, on the other hand, give total control over the simulation process, including the integration method to be used, time step of simulation, and so on. This results in a smaller simulation (execution) time. Being general-purpose tools, equation solvers can also be useful in applications other than power electronics simulation.

On the negative side, a long time is usually required for the initial setup because the user must develop all possible combinations of differential and algebraic equations. Even a minor change in the circuit topology and control may require just as much effort as the initial setup.

On balance, the circuit-oriented simulators are much easier and therefore are more widely used. The equation solvers tend to be used in specialized circumstances. The characteristics of some of the widely used software packages are discussed in Sections 4-6 and 4-7.

4-5 SOLUTION TECHNIQUES FOR TIME-DOMAIN ANALYSIS

Both circuit-oriented simulators and equation solvers must solve differential equations as a function of time. With either approach, the user should know the concepts fundamental to the solution of these equations.

In power electronics, the circuits are usually linear, but they change as a function of time due to the action of switches. A set of differential equations describes the system for each circuit state. In this section, we will discuss one numerical solution technique by means of a simple example that allows system variables to be calculated as a function of time.

4-5-1   LINEAR DIFFERENTIAL EQUATIONS

Figure 4-6a shows a simplified equivalent circuit to represent a switch-mode, regulated voltage supply of Figs. 1-3 and 1-4a in Chapter 1. In this example, change in the state of the switch in Fig. 1-3 affects the voltage v_{oi} in the circuit of Fig. 4-6a ($v_{oi} = V_d$ when the switch is on, otherwise $v_{oi} = 0$). Thus, the same circuit topology applies in both states of the switch. The inductor resistance r_L is included.

The waveform of the voltage v_{oi} is shown in Fig. 4-6b, where the switch duty ratio $D = t_{on}/T_s$ is dictated by the controller in the actual system, based on the operating conditions.

The equations are written in terms of the capacitor voltage v_c and the inductor current i_L, the so-called state variables, since they describe the state of the circuit. It is assumed that at time $t = 0$ at the beginning of the simulation, the initial inductor current $i_L(0)$ and the initial capacitor voltage $v_c(0)$ are known. By applying Kirchhoff's current law (KCL) and Kirchhoff's voltage law (KVL) in the circuit of Fig. 4-6a, we get the following two equations:

$$r_L i_L + L \frac{di_L}{dt} + v_c = v_{oi} \quad \text{(KVL)} \tag{4-1}$$

$$i_L - C \frac{dv_c}{dt} - \frac{v_c}{R} = 0 \quad \text{(KCL)} \tag{4-2}$$

By dividing both sides of Eq. 4-1 by L and Eq. 4-2 by C, we can express them in the usual state variable matrix form

$$\begin{bmatrix} \dfrac{di_L}{dt} \\ \dfrac{dv_c}{dt} \end{bmatrix} = \begin{bmatrix} -\dfrac{r_L}{L} & -\dfrac{1}{L} \\ \dfrac{1}{C} & -\dfrac{1}{CR} \end{bmatrix} \begin{bmatrix} i_L \\ v_c \end{bmatrix} + \begin{bmatrix} \dfrac{1}{L} \\ 0 \end{bmatrix} v_{oi}(t) \tag{4-3}$$

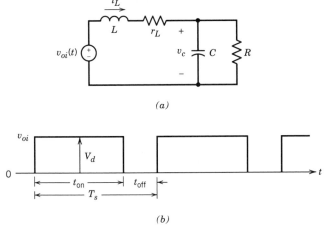

(a)

(b)

Figure 4-6   Simplified equivalent circuit of a switch-mode, regulated dc power supply (same as in Fig. 1-3).

The above equation can be written as

$$\frac{d\mathbf{x}(t)}{dt} = \mathbf{A}\mathbf{x}(t) + \mathbf{b}g(t) \tag{4-4}$$

where $\mathbf{x}(t)$ is a state variable vector and $g(t)$ is the single input:

$$\mathbf{x}(t) = \begin{bmatrix} i_L \\ v_c \end{bmatrix} \quad \text{and} \quad g(t) = v_{oi}$$

The matrix $\mathbf{A}$ and the vector $\mathbf{b}$ are

$$\mathbf{A} = \begin{bmatrix} -\dfrac{r_L}{L} & -\dfrac{1}{L} \\ \dfrac{1}{C} & -\dfrac{1}{CR} \end{bmatrix} \quad \text{and} \quad \mathbf{b} = \begin{bmatrix} \dfrac{1}{L} \\ 0 \end{bmatrix} \tag{4-5}$$

In general, the state transition matrix $\mathbf{A}$ and the input vector $\mathbf{b}$ may be functions of time and Eq. 4-4 may be written as

$$\frac{d\mathbf{x}(t)}{dt} = \mathbf{A}(t)\mathbf{x}(t) + \mathbf{b}(t)g(t) \tag{4-6}$$

With a time step of integration Δt, the solution of Eq. 4-6 at time t can be expressed in terms of the solution at $t - \Delta t$:

$$\mathbf{x}(t) = \mathbf{x}(t - \Delta t) + \int_{t-\Delta t}^{t} \left[\mathbf{A}(\zeta)\mathbf{x}(\zeta) + \mathbf{b}(\zeta)g(\zeta) \right] d\zeta \tag{4-7}$$

where ζ is a variable of integration.

4-5-2 TRAPEZOIDAL METHOD OF INTEGRATION

There are many elegant numerical methods for solving the integral in Eq. 4-7. However, we will discuss only one technique, called the trapezoidal method, which is used in two of the widely used circuit-solving programs called SPICE and EMTP.

This method uses an approximation of linear interpolation between the values at time $t - \Delta t$ and t, assuming that $\mathbf{x}(t)$ is known at time t. Since $\mathbf{x}(t)$ is what is being calculated in Eq. 4-7, this assumption of its a priori knowledge puts this method into a category of ''implicit'' methods. Figure 4-7 graphically illustrates this method of integration for a single variable $x(t)$, where A and b are scalars. As the name implies, the trapezoidal area, using the linear interpolation between the values at time $t - \Delta t$ and t, approximates the value of the integral. Applying this method to Eq. 4-7 yields

$$\mathbf{x}(t) = \mathbf{x}(t - \Delta t) + \tfrac{1}{2}\Delta t \left[\mathbf{A}(t - \Delta t)\mathbf{x}(t - \Delta t) + \mathbf{A}(t)\mathbf{x}(t) \right] + \tfrac{1}{2}\Delta t \left[\mathbf{b}(t - \Delta t)g(t - \Delta t) \right.$$

$$\left. + \mathbf{b}(t)g(t) \right] \tag{4-8}$$

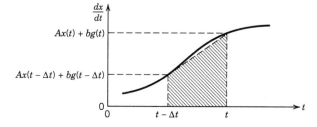

Figure 4-7
Trapezoidal method
of integration.

We should note that the above equation is an algebraic equation that in this case is also linear. Therefore, rearranging terms, we get

$$\left[\mathbf{I} - \tfrac{1}{2}\Delta t\, \mathbf{A}(t)\right]\mathbf{x}(t) = \left[\mathbf{I} + \tfrac{1}{2}\Delta t\, \mathbf{A}(t - \Delta t)\right]\mathbf{x}(t - \Delta t) + \tfrac{1}{2}\Delta t\left[\mathbf{b}(t - \Delta t)g(t - \Delta t)\right.$$
$$\left. + \mathbf{b}(t)g(t)\right] \tag{4-9}$$

Multiplying both sides of Eq. 4-9 by the inverse of $[\mathbf{I} - \tfrac{1}{2}\Delta t\, \mathbf{A}(t)]$ allows $\mathbf{x}(t)$ to be solved:

$$\mathbf{x}(t) = \left[\mathbf{I} - \tfrac{1}{2}\Delta t\, \mathbf{A}(t)\right]^{-1}\left\{\left[\mathbf{I} + \tfrac{1}{2}\Delta t \mathbf{A}(t - \Delta t)\right]\mathbf{x}(t - \Delta t) + \tfrac{1}{2}\Delta t\left[\mathbf{b}(t - \Delta t)g(t - \Delta t) + \mathbf{b}(t)g(t)\right]\right\} \tag{4-10}$$

Commonly in power electronic systems, $\mathbf{A}$ and $\mathbf{b}$ change when the circuit state changes due to a switching action. However, for any circuit state over an interval during which the state is not changing, the state transition matrix $\mathbf{A}$ and the vector $\mathbf{b}$ are independent of time. Therefore, $\mathbf{A}(t - \Delta t) = \mathbf{A}(t) = \mathbf{A}$ and $\mathbf{b}(t - \Delta t) = \mathbf{b}(t) = \mathbf{b}$. Use of this information in Eq. 4-10 results in

$$\mathbf{x}(t) = \mathbf{M}\mathbf{x}(t - \Delta t) + \mathbf{N}\left[g(t - \Delta t) + g(t)\right] \tag{4-11}$$

where

$$\mathbf{M} = \left[\mathbf{I} - \tfrac{1}{2}\Delta t\, \mathbf{A}\right]^{-1}\left[\mathbf{I} + \tfrac{1}{2}\Delta t\, \mathbf{A}\right] \tag{4-12}$$

and

$$\mathbf{N} = \left[\mathbf{I} - \tfrac{1}{2}\Delta t\, \mathbf{A}\right]^{-1}\left(\tfrac{1}{2}\Delta t\right)\mathbf{b} \tag{4-13}$$

need to be calculated only once for any circuit state, provided the time step Δt chosen for the numerical solution is kept constant.

An obvious question at this point is why solve the circuit state with a small time step Δt rather than choosing a Δt that takes the circuit from its previous switch state to its next switch state. In the solution of linear circuits, Δt must be chosen to be much smaller than the shortest time constant of interest in the circuit. However, in power electronics, Δt is often even smaller, dictated by the resolution with which the switching instants should be represented. We do not know a priori at what time the circuit will go to its next state since the values of the circuit variables themselves determine the instant of time when such a transition should take place. Another important point to note is that when the circuit state changes, the values of the state variables at the final time in the previous state are used as initial values at the beginning of the next state.

4-5-3   NONLINEAR DIFFERENTIAL EQUATIONS [1]

In power electronic systems, nonlinearities are introduced by component saturation (due to component values which depend on the associated currents and voltages) and limits imposed by the controller. An example is the output capacitance of a MOSFET which is a function of the voltage across it. In such systems, the differential equations can be written as (where $\mathbf{f}$ is a general non-linear function)

$$\dot{\mathbf{x}} = \mathbf{f}(\mathbf{x}(t),t) \tag{4-14}$$

The solution of the above equation can be written as

$$\mathbf{x}(t) = \mathbf{x}(t - \Delta t) + \int_{t - \Delta t}^{t} \mathbf{f}(\mathbf{x}(\zeta),\zeta)d\zeta \tag{4-15}$$

For example, applying the Trapezoidal rule to the integral in the equation above results in

$$\mathbf{x}(t) = \mathbf{x}(t - \Delta t) + \frac{\Delta t}{2} \{\mathbf{f}(\mathbf{x}(t),t) + \mathbf{f}(\mathbf{x}(t - \Delta t),t - \Delta t)\}. \qquad (4\text{-}16)$$

Equation 4-16 is nonlinear, and cannot be solved directly. This is because in the right side of Eq. 4-16, $\mathbf{f}(\mathbf{x}(t),t)$ depends on $\mathbf{x}(t)$. Such equations are solved by iterative procedures which converge to the solution within a reasonably small number of iterations. One of the commonly used solution techniques is the Newton-Raphson iterative procedure.

4-6 WIDELY USED, CIRCUIT-ORIENTED SIMULATORS

Several general-purpose, circuit-oriented simulators are available. These include SPICE, EMTP, SABER, and KREAN, to name a few. Two of these, SPICE and EMTP, are easily available and are widely used. Both have strengths and weaknesses. SPICE was developed for simulating integrated circuits, whereas EMTP was developed for power systems modeling. Because of their widespread popularity, we will describe both briefly in the following sections.

4-6-1 SPICE [3]

The abbreviation SPICE stands for Simulation Program with Integrated Circuit Emphasis. It was developed at the University of California, Berkeley. SPICE can handle nonlinearities and provides an automatic control on the time step of integration. There are several commercial versions of SPICE that operate on personal computers under several popular operating systems. One commercial version of SPICE is called PSpice [4].

In PSpice, many features are added to make it a multilevel simulator where the controllers can be represented by their behavior models, that is, by their input–output behavior, without resorting to a device-level simulation. There is an option for entering the input data by drawing the circuit schematic. In addition to its use in industry, PSpice has also become very popular in teaching undergraduate core courses in circuits and electronics. Therefore, many students are familiar with PSpice. One of the reasons for the popularity of PSpice is the availability and the capability to share its evaluation (classroom) version freely at no cost. This evaluation version is very powerful for power electronics simulations. For example, all simulations in reference 5 use only the evaluation version of PSpice. For these reasons, PSpice is used in this book in examples and in homework problems.

To illustrate how the information about a circuit is put into a circuit-oriented program in general and PSpice in particular, a very simple example is presented. We will consider the circuit of Fig. 1-3, redrawn in Fig. 4-8a, where the control signal for the switch under an open-loop operation is the waveform shown in Fig. 4-8b. Note that we explicitly include the representation of the diode and the switch, whereas we could have represented this circuit by the simple equivalent circuit of Fig. 4-6, which needs to be modified if the inductor current in this circuit becomes discontinuous. This shows the power of a circuit-oriented simulators that automatically takes into account the various circuit states without the user having to specify them. In the present simulation, the diode is represented by a simple built-in model within PSpice, and the switching device is represented by a simple voltage-controlled switch. In a circuit-oriented simulator like PSpice, detailed device models can be substituted if we wish to investigate switching details.

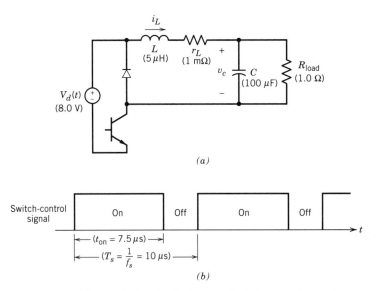

Figure 4-8   (a) Circuit for simulation. (b) Switch control waveform.

As a first step, we must assign node numbers as shown in the diagram of Fig. 4-9a, where one of the nodes has to be selected as a ground (0) node. The transistor in Fig. 4-8a is modeled by a voltage-controlled switch SW in the diagram of Fig. 4-9a whose state is determined by the voltage at its control terminals. In the on state with a control voltage greater then V_{on} (= 1 V default value), the switch has a small on-state resistance R_{on} (= 1 Ω default value). In the off state with a control voltage less than V_{off} (= 0 V default value), the switch is in its off state and is represented by a large resistance R_{off} (= 10^6 Ω default value). Of course, the default values are optional and the user can specify values that are more appropriate.

The listing of the input circuit file to PSpice is shown in Fig. 4-9b. The repetitive control voltage shown in Fig. 4-8b, which determines the state of the switch SW, is modeled by means of a voltage source called VCNTL within PSpice. There is a built-in model for diodes whose parameters such as the on-state resistance parameter R_s and the zero-bias junction capacitance C_{jo} can be changed; otherwise the default values are used by the program.

A sudden discontinuity in SPICE can result in the program proceeding with extremely small time steps and at worst may result in a problem of convergence, where the voltages at some node or nodes at some time step may fail to converge. If this were to happen, the simulation would stop with an error message. There are few definite rules to avoid the solution from failing to converge. Therefore, it is always better to avoid sudden discontinuities, such as by using an R–C "numerical snubber" across the diode in Fig. 4-9a to "soften" the discontinuity presented by the diode current suddenly going to zero. Similarly, the rise and fall times of VCNTL in Fig. 4-9a, represented by PULSE in Fig. 4-9b, are specified as 1 ns each rather than as zero.

The output waveforms from the simulation are shown in Fig. 4-10. These are produced by a graphical postprocessor (called Probe) within PSpice that is very easy to use.

4-6-2   EMTP SIMULATION PROGRAM [6]

Another widely used, general-purpose circuit simulation program is called EMTP (Electro-Magnetic Transients Program). Unlike SPICE, which has its origin in microelectron-

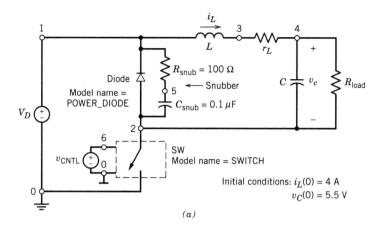

(a)

```
PSpice  Example
*
DIODE    2   1    POWER_DIODE
Rsnub    1   5    100.0
Csnub    5   2    0.1uF
*
SW       2   0   6   0    SWITCH
VCNTL    6   0    PULSE(0V,1V,0s,1ns,1ns,7.5us,10us)
*
L        1   3    5uH     IC=4A
rL       3   4    1m
C        4   2    100uF  IC=5.5V
RLOAD    4   2    1.0
*
VD       1   0    8.0V
*
.MODEL  POWER_DIODE  D(RS=0.01,CJO=10pF)
.MODEL  SWITCH  VSWITCH(RON=0.01)
.TRAN    10us   500.0us    0s   0.2us     uic
.PROBE
.END
```

(b)

Figure 4-9 PSpice simulation of circuit in Fig. 4-8.

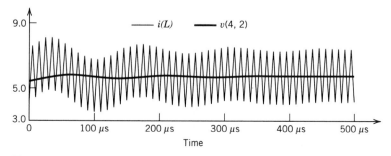

Figure 4-10 Results of PSpice simulation: i_L and v_c.

ics modeling, EMTP was originally developed for the electric power industry at the Bonneville Power Administration in Portland, Oregon. ATP (Alternative Transients Program) is a version of EMTP that is also available for personal computers under MS-DOS operating system [7]. Similar to SPICE, EMTP uses a trapezoidal rule of integration, but the time step of integration is kept constant.

Because of the built-in models for various power system components such as three-phase transmission lines, EMTP is a very powerful program for modeling power electronics applications in power systems.

Compared to SPICE, the switches in EMTP are treated quite differently. When a switch is closed, the row and column (in the network matrix) corresponding to the terminal nodes of the switch are coalesced together. There is a very powerful capability to represent analog and digital controllers that can be specified with almost the same ease as in a high-level language. The electrical network and the controller can pass values of various variables back and forth at each time step.

4-6-3 SUITABILITY OF PSpice AND EMTP

For power electronics simulation, both PSpice and EMTP are very useful. PSpice is better suited for use in power electronic courses for several reasons. Its evaluation version is available at no cost (in fact, its copying and sharing is welcomed and encouraged), and it is very easy to install on either an IBM-compatible or a Macintosh computer. It is user friendly with an easy-to-learn graphical postprocessing package for plotting of results. Because of the availability of semiconductor device models, it is also well suited for applications where such detailed representations are necessary. Perhaps in the near future, the models will improve to the point where power losses can be calculated accurately for a thermal design.

On the other hand, EMTP is better suited for simulating high-power electronics in power systems. It has the capability for representing controllers with the same ease as in a high-level language. The control over the time step Δt results in execution (run) times that are acceptable. For the reasons listed above, EMTP is very well suited for analyzing complex power electronic systems at a system level where it is adequate to represent switching devices by means of ideal switches and the controller by transfer functions and logical expressions.

A large number of power electronic exercises using the evaluation (classroom) version of PSpice [5] and EMTP [8] are available as aids in learning power electronics. They are also ideal for learning to use these software packages by examples.

4-7 EQUATION SOLVERS

If we choose an equation solver, then we must write the differential and algebraic equations to describe various circuit states and the logical expressions within the controller that determine the circuit state. Then, these differential/algebraic equations are simultaneously solved as a function of time.

In the most basic form, we can solve these equations by programming in any one of the higher level languages such as FORTRAN, C, or Pascal. It is also possible to access libraries in any of these languages, which consist of subroutines for specific applications, such as to carry out integration or for matrix inversion. However, it is far more convenient to use a package such as MATLAB [9] or a host of other packages where many of these convenience features are built in. Each of these packages use their own syntax and also excel in certain applications.

The program MATLAB can easily perform array and matrix manipulations, where, for example, $y = \mathbf{a} \cdot \mathbf{b}$ results in a value of y that equals cell-by-cell multiplication of two arrays $\mathbf{a}$ and $\mathbf{b}$. Similarly, to invert a matrix, all one needs to specify is $\mathbf{Y} = \text{inv}(\mathbf{X})$. Powerful plotting routines are built in. MATLAB is widely used in industry. Also, such programs are used in the teaching of undergraduate courses in control systems and signal processing. Therefore, the students are usually familiar with MATLAB prior to taking power electronics courses. If not, it is possible to learn their use quickly, especially by means of examples. For these reasons, MATLAB is utilized in this book for solution of some of the examples and the homework problems. SIMULINK is a powerful graphical pre-processor or user-interface to MATLAB which allows dynamic systems to be described in an easy block-diagram form.

As an example of MATLAB, the solution of the circuit in Fig. 4-8 using the trapezoidal method of integration is shown in Fig. 4-11. The circuit of Fig. 4-8 reduces to the equivalent circuit shown previously in Fig. 4-6 (provided $i_L(t) > 0$). As shown in Fig. 4-11a, the input voltage v_{oi} is generated in MATLAB by comparing a sawtooth waveform

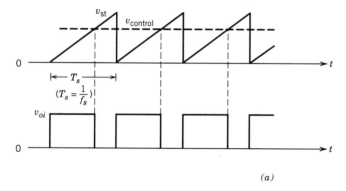

(a)

```
% Solution of the Circuit in Fig. 4-6 using Trapezoidal Method of Integration.
clc,clg,clear
% Input Data
Vd=8; L=5e-6; C=100e-6; rL=1e-3; R=1.0; fs=100e3; Vcontrol=0.75;
Ts=1/fs; tmax=50*Ts; deltat=Ts/50;
%
time= 0:deltat:tmax;
vst= time/Ts - fix(time/Ts);
voi= Vd * (Vcontrol > vst);
%
A=[-rL/L -1/L; 1/C -1/(C*R)];
b=[1/L 0]';
MN=inv(eye(2) - deltat/2 * A);
M=MN * (eye(2)+ deltat/2 * A);
N=MN * deltat/2 * b;
%
iL(1)=4.0; vC(1)=5.5;
timelength=length(time);
%
for k = 2:timelength
x = M * [iL(k-1) vC(k-1)]' + N * (voi(k) + voi(k-1));
iL(k) = x(1); vC(k) = x(2);
end
%
plot(time,iL,time,vC)
meta Example
```

(b)

Figure 4-11 MATLAB simulation of circuit in Fig. 4-6.

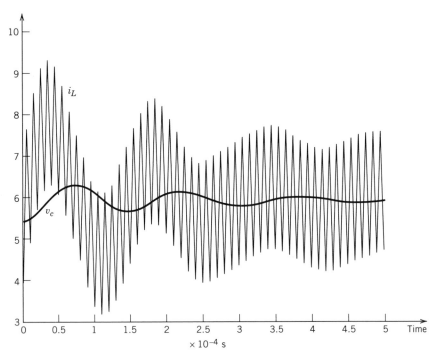

Figure 4-12   MATLAB simulation results.

v_{st} at the switching frequency f_s with a dc control voltage $v_{control}$. When the control voltage is greater than v_{st}, $v_{oi} = V_d$; otherwise it is zero. The MATLAB listing is shown in Fig. 4-11b. The output waveforms are shown in Fig. 4-12. These waveforms differ slightly from the PSpice simulation results shown in Fig. 4-10 due to the on-state voltage drop across the diode in PSpice.

SUMMARY

Modeling and computer simulations play an important role in the analysis, design, and education of power electronic systems. Because of the challenges involved in such simulations, it is important to simplify the system being simulated to be consistent with the simulation objectives. Over the years, several simulation packages have been developed. It is necessary to carefully evaluate the advantages and shortcomings of each package prior to selecting one for a given set of objectives.

PROBLEMS

4-1 Generate a triangular waveform with a peak of ± 1 V at 100 kHz using MATLAB.

4-2 Using the PSpice listing in Fig. 4-9b, obtain the switch current and the diode voltage waveforms.

4-3 Using the MATLAB listing in Fig. 4-11b, obtain the inductor current and the capacitor voltage waveforms similar to those in Fig. 4-12.

4-4 In the PSpice simulation of Fig. 4-9, make R_{Load} = 10 Ω. Evaluate the effects of the following changes on the simulation results:

    (a) Remove the R-C snubber across the diode.

    (b) In the pulse waveform of the control voltage VCNTL, make the rise and fall times zero.

    (c) Remove the R-C snubber across the diode *simultaneously* with the following changes: i) make the diode model to be as follows:
.MODEL POWER_DIODE D(IS=3e-15, RS=0.1, CJO=10PF)
and, ii) add the following Options statement:
.OPTIONS ABSTOL=1N, VNTOL=1M, RELTOL=0.015

    (d) In part c, make the rise and fall times of the control voltage VCNTL zero.

4-5 To bring the results of the PSpice simulation of Fig. 4-9 closer to the MATLAB simulation of Fig. 4-11, insert an ideal dc voltage source of 0.7 V in series with the diode in Fig. 4-9a to compensate for the on-state voltage drop of the diode. Compare the simulation results with those from MATLAB in Fig. 4-12.

4-6 In the PSpice simulation of Fig. 4–9a, idealize the switch-diode combination, which allows replacing it by a pulse input voltage waveform (between 0 and 8 V) as the input v_{oi}. Compare the PSpice simulation results with the MATLAB simulation results of Fig. 4-12.

4-7 Repeat the MATLAB simulation of Fig. 4-11 by using the built-in integration routine ODE45 in MATLAB.

4-8 In the PSpice simulation of Fig. 4-8, change R_{Load} to be 10 Ω which causes the inductor current i_L to become discontinuous (i.e., it becomes zero for a finite interval during each switching cycle). Obtain i_L and v_c waveforms.

4-9 Repeat problem 4-8 (with R_{Load} = 10 Ω) using MATLAB, recognizing that i_L becomes discontinuous at this low output power level.

4-10 Since the output capacitor C is usually large in circuits similar to that in Fig. 4-8a, the capacitor voltage changes slowly. Therefore, rewrite differential equations in terms of i_L and v_c by making the following assumptions: i) calculate $i_L(t)$ based on $v_c(t - \Delta t)$ and, ii) use the calculated value of $i_L(t)$ in the previous step to calculate $v_c(t)$. Simulate using MATLAB and compare results with those in Fig. 4-12.

4-11 In the circuit of Fig. 4-8a, ignore the input voltage source and the transistor switch. Let $i_L(o)$ = 4 A, and $v_c(o)$ = 5.5 V. Assuming the diode to be ideal, simulate this circuit in MATLAB using the trapezoidal rule of integration. Plot the inductor voltage v_L.

REFERENCES

1. N. Mohan, W. P. Robbins, T. M. Undeland, R. Nilssen and O. Mo, "Simulation of Power-Electronics and Motion Control Systems– An Overview," Proceedings of the IEEE, Vol. 82, No. 8, Aug. 1994, pp. 1287–1302.

2. B. C. Kuo and D. C. Hanselman, MATLAB Tools for Control System Analysis and Design, Prentice Hall, Englewood (NJ), 1994.

3. L. W. Nagel, "SPICE2 A Computer Program to Simulate Semiconductor Circuits," Memorandum No. ERL-M520, University of California, Berkeley, 1975.

4. PSpice, MicroSim Corporation, 20 Fairbanks, Irvine, CA 92718.

5. "Power Electronics: Computer Simulation, Analysis, and Education Using Evaluation Version of PSpice," on diskette with a manual, Minnesota Power Electronics, P.O. Box 14503, Minneapolis, MN 55414. (See page viii.)

6. W. S. Meyer and T. H. Liu, "EMTP Rule Book," Bonneville Power Administration, Portland, OR 97208.

7. ATP version of EMTP, Canadian/American EMTP User Group, The Fontaine, Unit 6B, 1220 N.E., 17th Avenue, Portland, OR 97232.
8. ''Computer Exercises for Power Electronics Education using EMTP,'' University of Minnesota Media Distribution, Box 734 Mayo Building, 420 Delaware Street, Minneapolis, MN 55455.
9. MATLAB, The Math Works Inc., 24 Prime Park Way, Natick, MA 01760.

PART 2

GENERIC POWER ELECTRONIC CONVERTERS

CHAPTER 5

LINE-FREQUENCY DIODE RECTIFIERS: LINE-FREQUENCY ac → UNCONTROLLED dc

5-1 INTRODUCTION

In most power electronic applications, the power input is in the form of a 50- or 60-Hz sine wave ac voltage provided by the electric utility, that is first converted to a dc voltage. Increasingly, the trend is to use the inexpensive rectifiers with diodes to convert the input ac into dc in an uncontrolled manner, using rectifiers with diodes, as illustrated by the block diagram of Fig. 5-1. In such diode rectifiers, the power flow can only be from the utility ac side to the dc side. A majority of the power electronics applications such as switching dc power supplies, ac motor drives, dc servo drives, and so on, use such uncontrolled rectifiers. The role of a diode rectifier in an ac motor drive was discussed by means of Fig. 1-8 in Chapter 1. In most of these applications, the rectifiers are supplied directly from the utility source without a 60-Hz transformer. The avoidance of this costly and bulky 60-Hz transformer is important in most modern power electronic systems.

The dc output voltage of a rectifier should be as ripple free as possible. Therefore, a large capacitor is connected as a filter on the dc side. As will be shown in this chapter, this capacitor gets charged to a value close to the peak of the ac input voltage. As a consequence, the current through the recitifier is very large near the peak of the 60-Hz ac input voltage and it does not flow continuously; that is, it becomes zero for finite durations during each half-cycle of the line frequency. These rectifiers draw highly distorted current from the utility. Now and even more so in the future, harmonic standards and guidelines will limit the amount of current distortion allowed into the utility, and the simple diode rectifiers may not be allowed. Circuits to achieve a nearly sinusoidal current rectification at a unity power factor for many applications are discussed in Chapter 18.

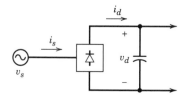

Figure 5-1 Block diagram of a rectifier.

Rectifiers with single-phase and three-phase inputs are discussed in this chapter. As discussed in Chapter 2, the diodes are assumed to be ideal in the analysis of rectifiers. In a similar manner, the electromagnetic interference (EMI) filter at the ac input to the rectifiers is ignored, since it does not influence the basic operation of the rectifier. Electromagnetic interference and EMI filters are discussed in Chapter 18.

5-2   BASIC RECTIFIER CONCEPTS

Rectification of ac voltages and currents is accomplished by means of diodes. Several simple circuits are considered to illustrate the basic concepts.

5-2-1   PURE RESISTIVE LOAD

Consider the circuit of Fig. 5-2a, with a sinusoidal voltage source v_s. The waveforms in Fig. 5-2b show that both the load voltage v_d and the current i have an average (dc) component. Because of the large ripple in v_d and i, this circuit is of little practical significance.

5-2-2   INDUCTIVE LOAD

Let us consider the load to be inductive, with an inductor in series with a resistor, as shown in Fig. 5-3a. Prior to $t = 0$, the voltage v_s is negative and the current in the circuit is zero. Subsequent to $t = 0$, the diode becomes forward biased and a current begins to flow. Then, the diode can be replaced by a short, as shown in the equivalent circuit of Fig. 5-3e. The current in this circuit is governed by the following differential equation:

$$v_s = Ri + L\frac{di}{dt} \tag{5-1}$$

where the voltage across the inductor $v_L = L\,di/dt$. The resulting voltages and current are shown in Figs. 5-3b and c. Until t_1, $v_s > v_R$ (hence $v_L = v_s - v_R$ is positive), the current

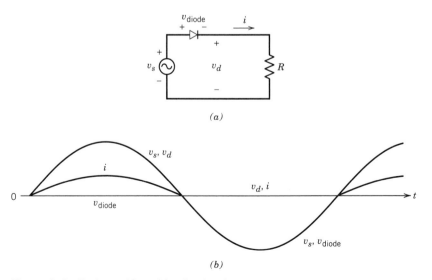

Figure 5-2   Basic rectifier with a load resistance.

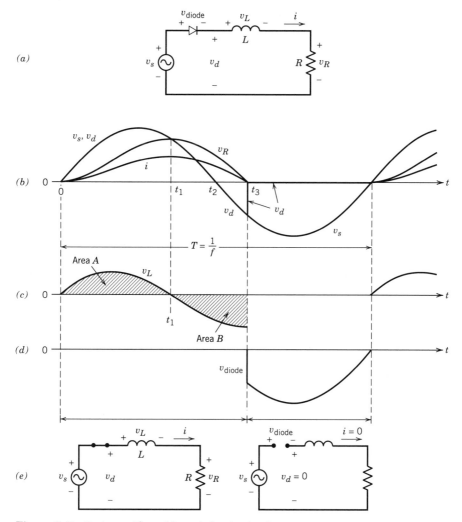

Figure 5-3 Basic rectifier with an inductive load.

builds up, and the inductor stored energy increases. Beyond t_1, v_L becomes negative, and the current begins to decrease. After t_2, the input voltage v_s becomes negative but the current is still positive and the diode must conduct because of the inductor stored energy.

The instant t_3, when the current goes to zero and the diode stops conducting, can be obtained as follows (also discussed in Section 3-2-5-1): The inductor equation $v_L = L \, di/dt$ can be rearranged as

$$\frac{1}{L} v_L \, dt = di \tag{5-2}$$

Integrating both sides of the above equation between zero and t_3 and recognizing that $i(0)$ and $i(t_3)$ are both zero give

$$\frac{1}{L} \int_0^{t_3} v_L \, dt = \int_{i(0)}^{i(t_3)} di = i(t_3) - i(0) = 0 \tag{5-3}$$

From the above equation, we can observe that

$$\int_0^{t_3} v_L \, dt = 0 \tag{5-4}$$

A graphical interpretation of the above equation is as follows: Equation 5-4 can be written as

$$\int_0^{t_1} v_L \, dt + \int_{t_1}^{t_3} v_L \, dt = 0 \tag{5-5}$$

which in terms of the volt-second areas A and B of Fig. 5-3c is

$$\text{Area } A - \text{Area } B = 0 \tag{5-6}$$

Therefore, the current goes to zero at t_3 when area $A = B$ in Fig. 5-3c.

Beyond t_3, the voltages across both R and L are zero and a reverse polarity voltage ($= -v_s$) appears across the diode, as shown in Fig. 5-3d. These waveforms repeat with the time period $T = 1/f$.

The load voltage v_d becomes negative during the interval from t_2 to t_3. Therefore, in comparison to the case of purely resistive load of Fig. 5-2a, the average load voltage is less.

5-2-3   LOAD WITH AN INTERNAL dc VOLTAGE

Next, we will consider the circuit of Fig. 5-4a where the load consists of an inductor L and a dc voltage E_d. The diode begins to conduct at t_1 when v_s exceeds E_d. The current reaches its peak at t_2 (when v_s is again equal to E_d) and decays to zero at t_3, with t_3 determined by the requirement that the volt-second area A be equal to area B in the plot of v_L shown in Fig. 5-4c. The voltage across the diode is shown in Fig. 5-4d.

5-3   SINGLE-PHASE DIODE BRIDGE RECTIFIERS

A commonly used single-phase diode bridge rectifier is shown in Fig. 5-5. A large filter capacitor is connected on the dc side. The utility supply is modeled as a sinusoidal voltage source v_s in series with its internal impedance, which in practice is primarily inductive. Therefore, it is represented by L_s. To improve the line-current waveform, an inductor may be added in series on the ac side, which in effect will increase the value of L_s. The objective of this chapter is to thoroughly analyze the operation of this circuit. Although the circuit appears simple, the procedure to obtain the associated voltage and current waveforms in a closed form is quite tedious. Therefore, we will simulate this circuit using PSpice and MATLAB. However, we will next analyze many simpler and hypothetical circuits in order to gain insight into the operation of the circuit in Fig. 5-5.

5-3-1   IDEALIZED CIRCUIT WITH $L_S = 0$

As a first approximation to the circuit of Fig. 5-5, we will assume L_s to be zero and replace the dc side of the rectifier by a resistance R or a constant dc current source I_d, as shown in Figs. 5-6a and b, respectively. It should be noted in the circuit of Fig. 5-6a that although it is very unlikely that a pure resistive load will be supplied through a diode rectifier, this circuit models power-factor-corrected rectifiers discussed in Chapter 18.

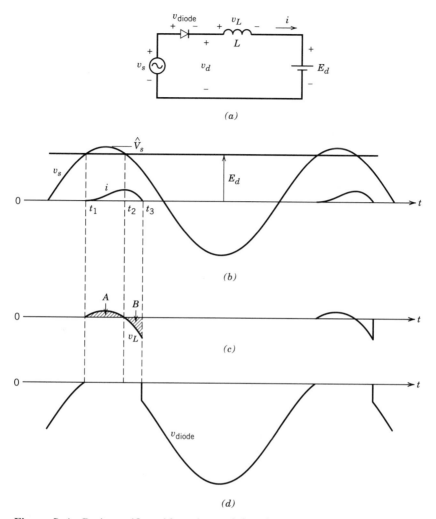

(a)

(b)

(c)

(d)

Figure 5-4   Basic rectifier with an internal dc voltage.

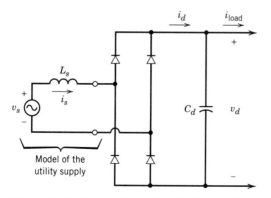

Figure 5-5   Single-phase diode bridge rectifier.

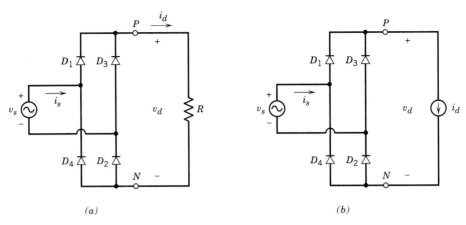

(a)   (b)

Figure 5-6   Idealized diode bridge rectifiers with $L_s = 0$.

Similarly, the representation of the load by a constant dc current in the circuit of Fig. 5-6b is an approximation to a situation where a large inductor may be connected in series at the dc output of the rectifier for filtering in Fig. 5-5. This is commonly done in phase-controlled thyristor converters, discussed in Chapter 6.

The circuits in Fig. 5-6 are redrawn in Fig. 5-7, which shows that this circuit consists of two groups of diodes: the top group with diodes 1 and 3 and the bottom group with diodes 2 and 4. With $L_s = 0$, it is easy to see the operation of each group of diodes. The current i_d flows continuously through one diode of the top group and one diode of the bottom group.

In the top group, the cathodes of the two diodes are at a common potential. Therefore, the diode with its anode at the highest potential will conduct i_d. That is, when v_s is positive, diode 1 will conduct i_d and v_s will appear as a reverse-bias voltage across diode 3. When v_s goes negative, the current i_d shifts (commutates) instantaneously to diode 3 since $L_s = 0$. A reverse-bias voltage appears across diode 1.

In the bottom group, the anodes of the two diodes are at a common potential. Therefore, the diode with its cathode at the lowest potential will conduct i_d. That is, when v_s is positive, diode 2 will carry i_d and v_s will appear as a reverse-bias voltage across diode 4. When v_s goes negative, the current i_d instantaneously commutates to diode 4 and a reverse-bias voltage appears across diode 2.

The voltage and current waveforms in the circuits of Fig. 5-6 are shown in Figs. 5-8a and b. There are several items worth noting. In both circuits, when v_s is positive, diodes

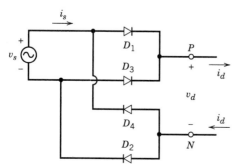

Figure 5-7   Redrawn rectifiers of Fig. 5-6.

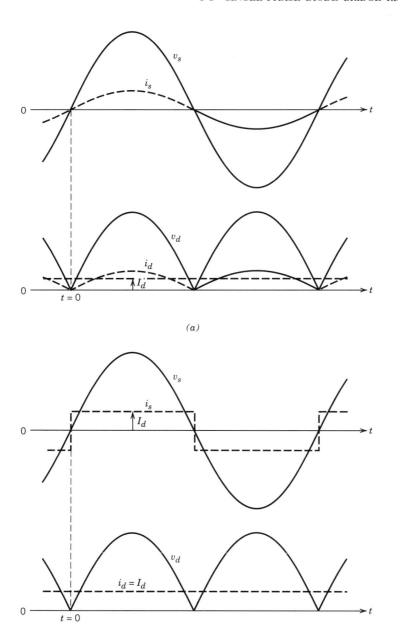

Figure 5-8   Waveforms in the rectifiers of (a) Fig. 5-6a and (b) Fig. 5-6b.

1 and 2 conduct and $v_d = v_s$ and $i_s = i_d$. When v_s goes negative, diodes 3 and 4 conduct and, therefore, $v_d = -v_s$ and $i_s = -i_d$. Therefore, at any time, the dc-side output voltage of the diode rectifier can be expressed as

$$v_d(t) = |v_s| \tag{5-7}$$

Similarly, the ac-side current can be expressed as

$$i_s = \begin{cases} i_d & \text{if } v_s > 0 \\ -i_d & \text{if } v_s < 0 \end{cases} \tag{5-8}$$

and the transition between the two values is instantaneous due to the assumption of zero L_s.

The average value V_{do} (where the subscript o stands for the idealized case with $L_s = 0$) of the dc output voltage in both circuits can be obtained by assigning an arbitrary time origin $t = 0$ in Fig. 5-8 and then integrating $v_s = \sqrt{2}V_s \sin \omega t$ over one-half time period (where $\omega = 2\pi f$ and $\omega T/2 = \pi$):

$$V_{do} = \frac{1}{(T/2)} \int_0^{T/2} \sqrt{2}V_s \sin \omega t \, dt = \frac{1}{\omega T/2}(\sqrt{2}V_s \cos \omega t)\Big|_{T/2}^0 = \frac{2}{\pi}\sqrt{2}V_s \quad (5\text{-}9)$$

Therefore,

$$V_{do} = \frac{2}{\pi}\sqrt{2}V_s = 0.9V_s \quad (5\text{-}10)$$

where V_s is the rms value of the input voltage.

With $i_d(t) = I_d$, v_s and i_s waveforms are shown in Fig. 5-9a along with the fundamental-frequency component i_{s1}. Applying the basic definition of the rms value to the i_s waveform in this idealized case yields

$$I_s = I_d \quad (5\text{-}11)$$

By Fourier analysis of i_s, the fundamental and the harmonic components have the following rms values in this idealized case:

$$I_{s1} = \frac{2}{\pi}\sqrt{2}I_d = 0.9I_d \quad (5\text{-}12)$$

$$I_{sh} = \begin{cases} 0 & \text{for even values of h and} \\ I_{s1}/h & \text{for odd values of h} \end{cases} \quad (5\text{-}13)$$

The harmonic components in i_s are shown in Fig. 5-9b. The total harmonic distortion can be calculated by Eq. 3-36 to be

$$THD = 48.43\% \quad (5\text{-}14)$$

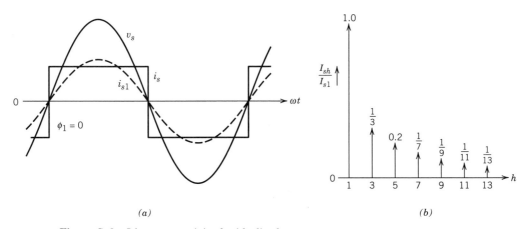

(a)   *(b)*

Figure 5-9   Line current i_s in the idealized case.

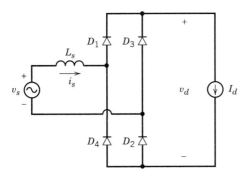

Figure 5-10   Single-phase rectifier with L_s.

By visual inspection of i_s waveform in Fig. 5-9a, it is apparent that i_{s1} is in phase with the v_s waveform. Therefore from Fig. 5-9a,

$$DPF = 1.0 \tag{5-15}$$

and

$$PF = DPF \frac{I_{s1}}{I_s} = 0.9 \tag{5-16}$$

5-3-2   EFFECT OF L_S ON CURRENT COMMUTATION

Next, we will look at the effect of a finite ac-side inductance L_s on the circuit operation. We will assume that the dc side can be represented by a constant dc current I_d shown in Fig. 5-10. Due to a finite L_s, the transition of the ac-side current i_s from a value of $+I_d$ to $-I_d$ (or vice versa) will not be instantaneous. The finite time interval required for such a transition is called the current commutation time, or the commutation interval u, and this process where the current conduction shifts from one diode (or a set of diodes) to the other is called the current commutation process.

In order to understand this process fully, let us first consider a simple hypothetical circuit of Fig. 5-11a with two diodes supplied by a sinusoidal voltage source $v_s = \sqrt{2}V_s \sin \omega t$. The output is represented by a constant dc current source I_d. For comparison purposes, Fig. 5-11b shows v_s, v_d, and i_s waveforms with $L_s = 0$.

Prior to time $\omega t = 0$, the voltage v_s is negative and the current I_d is circulating through D_2 with $v_d = 0$ and $i_s = 0$. When v_s becomes positive at $\omega t = 0$, a forward-bias voltage appears across D_1 and it begins to conduct. With a finite L_s, the buildup of i_s can be obtained from the circuit redrawn as in Fig. 5-12a (valid only for $0 < i_s < I_d$). Since D_2 is conducting, it provides a short-circuit (with $v_d = 0$, assuming an ideal diode) path through which i_s can build up. The two mesh currents shown are I_d and i_s. In terms of

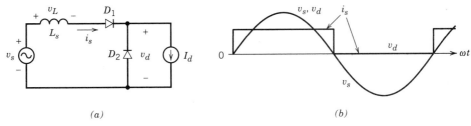

(a)   (b)

Figure 5-11   Basic circuit to illustrate current commutation. Waveforms assume $L_s = 0$.

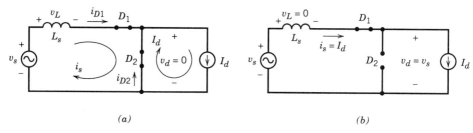

(a)   (b)

Figure 5-12 (a) Circuit during the commutation. (b) Circuit after the current commutation is completed.

these mesh currents, the diode current $i_{D2} = I_d - i_s$. Therefore, as i_s builds up to a value I_d during the commutation interval $\omega t = u$, i_{D2} is positive and D_2 conducts in the circuit of Fig. 5-12a. The current i_s cannot exceed I_d since it will result in a negative value of i_{D2} that is not possible. As a consequence, the diode D_2 stops conducting at $\omega t = u$, resulting in the circuit shown in Fig. 5-12b. The waveforms are plotted in Fig. 5-13 as a function of ωt.

It is clear from the above introduction that the current i_s through the inductor starts with a value of zero at the beginning of the commutation interval and ends up with a value of I_d at the end. Therefore, to obtain the length of the commutation interval u, we should consider the inductor equation. During the commutation interval, the input ac voltage appears as a current commutation voltage across the inductor in Fig. 5-12a:

$$v_L = \sqrt{2}V_s \sin \omega t = L_s \frac{di_s}{dt} \qquad 0 < \omega t < u \qquad (5\text{-}17)$$

The right side of the above equation can be written as $\omega L_s di_s/d(\omega t)$. Therefore,

$$\sqrt{2}V_s \sin \omega t \, d(\omega t) = \omega L_s \, di_s \qquad (5\text{-}18)$$

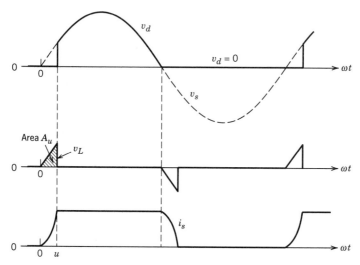

Figure 5-13 Waveforms in the basic circuit of Fig. 5-11. Note that a large value of L_s is used to clearly show the commutation interval.

Integrating both sides of Eq. 5-18 and recognizing that i_s goes from zero to I_d during the commutation interval from zero to u, we get

$$\int_0^u \sqrt{2}V_s \sin \omega t \, d(\omega t) = \omega L_s \int_0^{I_d} di_s = \omega L_s I_d \qquad (5\text{-}19)$$

In Eq. 5-19, the left side is the integral of the inductor voltage v_L during the commutation interval. The above voltage integral is the same as the volt-radian area A_u in Fig. 5-13:

$$A_u = \int_0^u \sqrt{2}V_s \sin \omega t \, d(\omega t) = \sqrt{2}V_s(1 - \cos u) \qquad (5\text{-}20)$$

Combining Eqs. 5-19 and 5-20 yields

$$A_u = \sqrt{2}V_s(1 - \cos u) = \omega L_s I_d \qquad (5\text{-}21)$$

The important observation from Eq. 5-21 is that the integral of the commutation voltage over the commutation interval can always be calculated by the product of ω, L_s, and the *change* in the current through L_s during commutation. From Eq. 5-21,

$$\cos u = 1 - \frac{\omega L_s I_d}{\sqrt{2}V_s} \qquad (5\text{-}22)$$

Equation 5-22 confirms that if $L_s = 0$, $\cos u = 1$ and the current commutation will be instantaneous with $u = 0$. For a given frequency ω, the commutation interval u increases with L_s and I_d and decreases with increasing voltage V_s.

The finite commutation interval reduces the average value of the output voltage. In Fig. 5-11b with $L_s = 0$, the average value V_{do} of v_d is

$$V_{do} = \frac{1}{2\pi} \int_0^\pi \sqrt{2}V_s \sin \omega t \, d(\omega t) = \frac{2\sqrt{2}}{2\pi} V_s = 0.45V_s \qquad (5\text{-}23)$$

With a finite L_s and hence a nonzero u in Fig. 5-13, $v_d = 0$ during the interval u. Therefore,

$$V_d = \frac{1}{2\pi} \int_u^\pi \sqrt{2}V_s \sin \omega t \, d(\omega t) \qquad (5\text{-}24)$$

which can be written as

$$V_d = \frac{1}{2\pi} \int_0^\pi \sqrt{2}V_s \sin \omega t \, d(\omega t) - \frac{1}{2\pi} \int_0^u \sqrt{2}V_s \sin \omega t \, d(\omega t) \qquad (5\text{-}25)$$

Substituting Eqs. 5-23 and 5-19 into Eq. 5-25 yields

$$V_d = 0.45V_s - \frac{\text{area } A_u}{2\pi} = 0.45V_s - \frac{\omega L_s}{2\pi} I_d \qquad (5\text{-}26)$$

where the reduction in the average output voltage by ΔV_d from V_{do} is

$$\Delta V_d = \frac{\text{area } A_u}{2\pi} = \frac{\omega L_s}{2\pi} I_d \qquad (5\text{-}27)$$

We will now extend this analysis to the circuit of Fig. 5-10, redrawn in Fig. 5-14a. The waveforms are shown in Fig. 5-14b. Once again, we need to consider the current

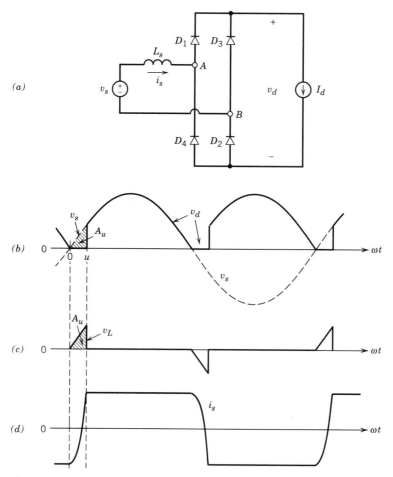

Figure 5-14   (a) Single-phase diode rectifier with L_s. (b) Waveforms.

commutation process. Prior to $\omega t = 0$ in Fig. 5-14, diodes 3 and 4 are conducting I_d (as in the circuit of Fig. 5-6b with $L_s = 0$) and $i_s = -I_d$. The circuit of Fig. 5-14a is carefully redrawn in Fig. 5-15 to show the current commutation process during $0 < \omega t < u$. Subsequent to $t = 0$, v_s becomes positive and diodes 1 and 2 become forward biased because of the short-circuit path provided by the conducting diodes 3 and 4. The three mesh currents are shown in Fig. 5-15, where the two commutation currents i_u are equal, based on the assumption of identical diodes. All four diodes conduct during the commutation interval, and therefore, $v_d = 0$. In terms of these mesh currents, we can express diode currents and the line current i_s during the commutation interval as

$$i_{D1} = i_{D2} = i_u \qquad i_{D3} = i_{D4} = I_d - i_u \tag{5-28}$$

and

$$i_s = -I_d + 2i_u \tag{5-29}$$

where i_u builds up from zero at the beginning to I_d at the end of the commutation interval. Therefore, at $\omega t = u$, $i_{D1} = i_{D2} = I_d$ and $i_s = I_d$. During this commutation of current from diodes 3 and 4 to diodes 1 and 2, the current through inductor L_s changes from $-I_d$ to I_d. Following the analysis previously carried out on the hypothetical circuit of Fig. 5-11a,

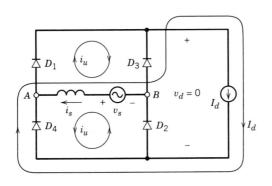

Figure 5-15 Redrawn circuit of Fig. 5-14a during current commutation.

volt-radian area A_u in the waveforms of Fig. 5-14b and c can be written from Eqs. 5-19 to 5-21 as

$$\text{Volt-radian area } A_u = \int_0^u \sqrt{2}V_s \sin \omega t \, d(\omega t) = \omega L_s \int_{-I_d}^{I_d} di_s = 2\omega L_s I_d \quad (5\text{-}30)$$

where the lower limit of integration now is $i_s(0) = -I_d$. Therefore,

$$A_u = \sqrt{2}V_s(1 - \cos u) = 2\omega L_s I_d \quad (5\text{-}31)$$

and

$$\cos u = 1 - \frac{2\omega L_s}{\sqrt{2}V_s} I_d \quad (5\text{-}32)$$

A similar commutation takes place one-half cycle later when i_s goes from I_d to $-I_d$.

In this circuit, the average value of v_d in the idealized case (with $L_s = 0$) was calculated in Eq. 5-10 as $V_{do} = 0.9V_s$. Therefore in the presence of L_s, the average value V_d can be calculated, following the procedure outlined previously by Eqs. 5-23 through 5-26. Alternatively, we can calculate V_d by inspecting Fig. 5-14b, where compared to the idealized case, the area A_u is "lost" every half-cycle from the integral of voltage v_d. Therefore,

$$V_d = V_{do} - \frac{\text{area } A_u}{\pi} = 0.9V_s - \frac{2\omega L_s I_d}{\pi} \quad (5\text{-}33)$$

5-3-3 CONSTANT dc-SIDE VOLTAGE $v_d(t) = V_d$

Next, we will consider the circuit shown in Fig. 5-16a, where the assumption is that the dc-side voltage is constant. It is an approximation to the circuit of Fig. 5-5 with a large value of C. Another assumption here is that the circuit conditions are such that the current i_d is zero during the zero crossing of v_s, as shown by the waveforms in Fig. 5-16c. Under these conditions, the equivalent circuit is drawn in Fig. 5-16b. Consider the waveforms in Fig. 5-16c. When v_s exceeds V_d at θ_b, diodes 1 and 2 begin to conduct. The current reaches its peak at θ_p, beyond which v_L becomes negative. The current becomes zero at θ_f when the volt-second areas A and B become equal and negative of each other. The current remains zero until $\pi + \theta_b$. With a given value of V_d, the average value I_d of the dc current can be calculated by the following procedure:

1. The angle θ_b can be calculated from the equation

$$V_d = \sqrt{2}V_s \sin \theta_b \quad (5\text{-}34)$$

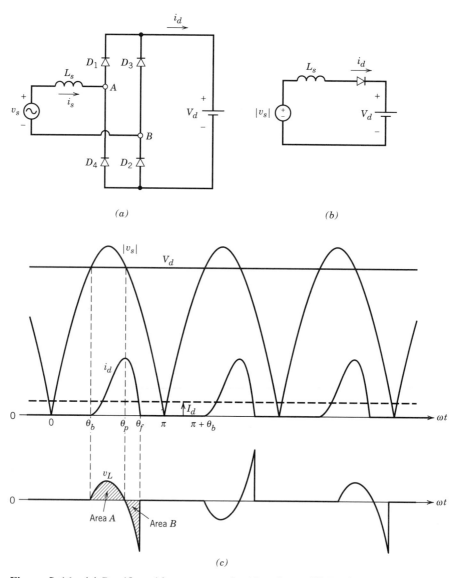

Figure 5-16   (*a*) Rectifier with a constant dc-side voltage. (*b*) Equivalent circuit. (*c*) Waveforms.

2. As shown in Fig. 5-16*c*, the inductor voltage starts at zero at θ_b and becomes zero at θ_p prior to becoming negative. From symmetry in Fig. 5-16*c*,

$$\theta_p = \pi - \theta_b \tag{5-35}$$

3. When the current is flowing, the inductor voltage v_L is given by

$$v_L = L_s \frac{di_d}{dt} = \sqrt{2} V_s \sin(\omega t) - V_d \tag{5-36}$$

and its integral with respect to ωt can be written as

$$\omega L_s \int_{\theta_b}^{\theta} di_d = \int_{\theta_b}^{\theta} (\sqrt{2} V_s \sin \omega t - V_d) \, d(\omega t) \tag{5-37}$$

where $\theta > \theta_b$. Recognizing that i_d at θ_b is zero, Eq. 5-37 results in

$$i_d(\theta) = \frac{1}{\omega L_s} \int_{\theta_b}^{\theta} (\sqrt{2}V_s \sin \omega t - V_d)\, d(\omega t) \qquad (5\text{-}38)$$

4. The angle θ_f at which i_d goes to zero can be obtained from Eq. 5-38 as

$$0 = \int_{\theta_b}^{\theta_f} (\sqrt{2}V_s \sin \omega t - V_d)\, d(\omega t) \qquad (5\text{-}39)$$

It corresponds to area $A = B$ in Fig. 5-16c.

5. The average value I_d of the dc current can be obtained by integrating $i_d(\theta)$ from θ_b to θ_f and then dividing by π:

$$I_d = \frac{\displaystyle\int_{\theta_b}^{\theta_f} i_d(\theta)\, d\theta}{\pi} \qquad (5\text{-}40)$$

It is intuitively obvious that for given circuit parameters, I_d will depend on the value of V_d and vice versa. To present the relationship between the two in a general manner, we will normalize V_d by V_{do} and I_d by $I_{\text{short circuit}}$, where

$$I_{\text{short circuit}} = \frac{V_s}{\omega L_s} \qquad (5\text{-}41)$$

is the rms current that will flow if the ac voltage source v_s was short circuited through L_s. Following the above procedure results in the plot shown in Fig. 5-17, where the current reaches zero as V_d approaches the peak value of the ac input voltage.

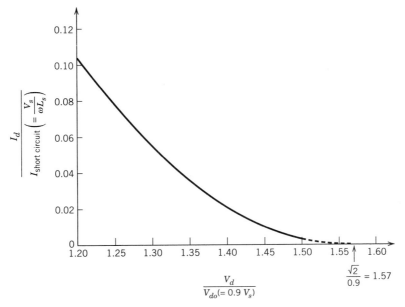

Figure 5-17 Normalized I_d versus V_d in the rectifier of Fig. 5-16a with a constant dc-side voltage.

5-3-3-1   Rectifier Characteristic

The approximation of a constant dc voltage $v_d(t) = V_d$ may be reasonable if the capacitance in the practical circuit of Fig. 5-5 is large. That is, if the load were replaced by an equivalent resistance R_{load} as in Fig. 5-20 later, the time constant $C_d R_{load}$ is much larger than the line-frequency cycle time, resulting in a very small ripple in v_d. This approximation allows us to present the characteristics of the rectifier in a general fashion. In Figs. 5-18 and 5-19, various quantities are plotted as a function of the dc current I_d, normalized with respect to the short-circuit current $I_{short\ circuit}$ (given by Eq. 5-41) in order to combine the effects of L_s and frequency ω in the same plots. (See Problem 5-17 for justification that

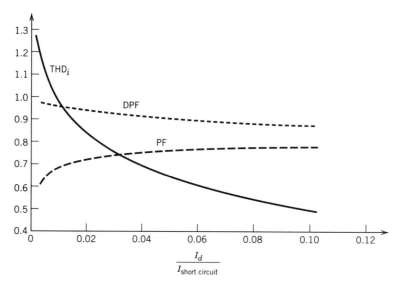

Figure 5-18   Total harmonic distortion, DPF, and PF in the rectifier of Fig. 5-16a with a constant dc-side voltage.

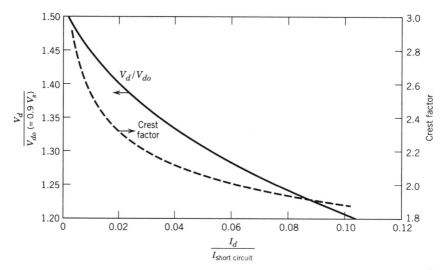

Figure 5-19   Normalized V_d and the crest factor in the rectifier of Fig. 5-16a with a constant dc-side voltage.

allows such a generalization.) For a given value of I_d, increasing L_s results in a smaller $I_{\text{short circuit}}$ and hence a larger $I_d/I_{\text{short circuit}}$. Therefore, Figs. 5-18 and 5-19 show that increasing L_s results in improved i_s waveform with a lower THD, a better power factor, and a lower (improved) crest factor.

5-3-4 PRACTICAL DIODE BRIDGE RECTIFIERS

Having considered the simplified circuits in the previous sections, we are now ready to consider the practical circuit of Fig. 5-5, which is redrawn in Fig. 5-20. The load is represented by an equivalent resistance R_{load}. In this circuit, there will be some ripple in the capacitor voltage, and therefore, it must be analyzed differently than the circuit of Fig. 5-16a. A circuit such as shown in Fig. 5-20 can be easily analyzed by a circuit simulation program such as PSpice. However, for educational purposes we will first analytically calculate the circuit waveforms.

5-3-4-1 Analytical Calculations under a Highly Discontinuous Current

The circuit operating conditions are assumed to result in a highly discontinuous i_d, similar to the waveform in Fig. 5-16c, where i_d goes to zero prior to the zero crossing of v_s every half-cycle. Then the equivalent circuit of Fig. 5-21 can be used to calculate the voltages and currents in Fig. 5-20. If the above condition is not met, the current commutation discussed in the earlier sections must be included, which makes analytical calculations difficult.

In order to describe the system in Fig. 5-21, the state variables chosen are the inductor current i_d and the capacitor voltage v_d. During each half-cycle of line frequency, there are two distinct intervals, similar to those shown earlier in Fig. 5-16c (where $t_b = \theta_b/\omega$ and $t_f = \theta_f/\omega$):

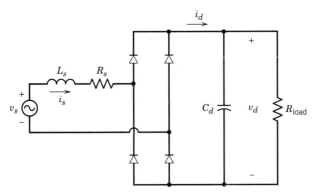

Figure 5-20 Practical diode–bridge rectifier with a filter capacitor.

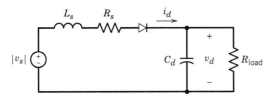

Figure 5-21 Equivalent circuit of Fig. 5-20.

(a) $t_b < t < t_f$. When the current is flowing during $t_b < t < t_f$, where t_b is the beginning of conduction and t_f is the final conduction time, the following equations describe the circuit every half-cycle of line frequency:

$$|v_s| = R_s i_d + L_s \frac{di_d}{dt} + v_d \quad \text{(using KVL)} \tag{5-42}$$

and

$$i_d = C_d \frac{dv_d}{dt} + \frac{v_d}{R_{\text{load}}} \quad \text{(using KCL)} \tag{5-43}$$

where KVL and KCL are the Kirchhoff voltage and current laws. Rearranging the above equations in the state variable form during $t_b < t < t_f$ yields

$$\begin{bmatrix} \dfrac{di_d}{dt} \\ \dfrac{dv_d}{dt} \end{bmatrix} = \begin{bmatrix} -\dfrac{R_s}{L_s} & -\dfrac{1}{L_s} \\ \dfrac{1}{C_d} & -\dfrac{1}{C_d R_{\text{load}}} \end{bmatrix} \begin{bmatrix} i_d \\ v_d \end{bmatrix} + \begin{bmatrix} \dfrac{1}{L_s} \\ 0 \end{bmatrix} |v_s| \tag{5-44}$$

This circuit is very similar to that in Fig. 4-6 in the simulation chapter. The state variable vector $\mathbf{x}$ consists of i_d and v_d. The state transition matrix is

$$\mathbf{A} = \begin{bmatrix} -\dfrac{R_s}{L_s} & -\dfrac{1}{L_s} \\ \dfrac{1}{C_d} & -\dfrac{1}{C_d R_{\text{load}}} \end{bmatrix} \tag{5-45}$$

and

$$\mathbf{b} = \begin{bmatrix} \dfrac{1}{L_s} \\ 0 \end{bmatrix} \tag{5-46}$$

Using Eqs. 4-11 through 4-13 for the trapezoidal rule of integration yields

$$\mathbf{x}(t) = \mathbf{M}\mathbf{x}(t - \Delta t) + \mathbf{N}[|v_s(t)| - |v_s(t - \Delta t)|] \tag{5-47}$$

where

$$\mathbf{M} = \left[\mathbf{I} - \frac{\Delta t}{2}\mathbf{A}\right]^{-1} \cdot \left[\mathbf{I} + \frac{\Delta t}{2}\mathbf{A}\right]$$
$$\mathbf{N} = \left[\mathbf{I} - \frac{\Delta t}{2}\mathbf{A}\right]^{-1} \frac{\Delta t}{2}\mathbf{b} \tag{5-48}$$

(b) $t_f < t < t_b + \frac{1}{2}T$. During the interval $t_f < t < t_b + \frac{1}{2}T$, when the diode bridge is not conducting,

$$i_d = 0 \tag{5-49}$$

and

$$\frac{dv_d}{dt} = -\frac{1}{C_d R_{\text{load}}} v_d \tag{5-50}$$

The solution to Eq. 5-50 can be expressed as

$$v_d(t) = v_d(t_f)e^{-(t - t_f)/(C_d R_{\text{load}})} \tag{5-51}$$

In the solution of Eqs. 5-47 and 5-51, we need the time instant t_b at which the current conduction starts. Since it is not known prior to the solution, we will use an estimated value. An exact value of t_b will result in the beginning of current conduction exactly one-half cycle later. Therefore, we will use this condition to check for the accuracy of our choice of t_b and slowly increment t_b from its initially chosen value until we reach the exact value within a small tolerance.

■ **Example 5-1**   Simulate the full-bridge rectifier of Fig. 5-20 using MATLAB with the following parameters: $V_s = 120$ V at 60 Hz, $L_s = 1$ mH, $R_s = 1$ mΩ, $C_d = 1000$ μF, and $R_{load} = 20$ Ω. Assume the diodes to be ideal and choose a time step $\Delta t = 25$ μs.

Solution   The MATLAB program listing is included in the Appendix at the end of this chapter, and the results with the correct initial value of $v_d(t_b)$ are shown in Fig. 5-22.

■

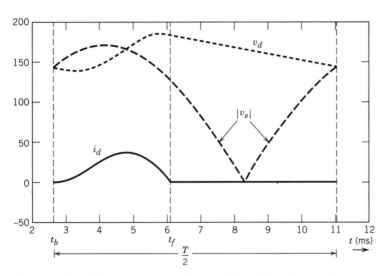

Figure 5-22   Waveforms in the circuit of Fig. 5-20, obtained in Example 5-1.

5-3-4-2   Circuit Simulation for General Operating Conditions

The discussion in the previous section is limited to a highly discontinuous current where the current commutation does not have to be considered. In general, the above condition may or may not hold, and the equations and their analytical solution become fairly complicated even for a simple circuit. Therefore, in general, it is preferable to use a circuit-oriented simulator as illustrated by Example 5-2.

■ **Example 5-2**   Simulate the circuit of Fig. 5-20 using PSpice with the same parameter values as in Example 5-1. Perform a Fourier analysis on the input current and the output dc voltage.

Solution   The PSpice network with node numbers and the input data file are included in the Appendix at the end of this chapter. The results are plotted in Fig. 5-23, where i_{s1} (the fundamental-frequency component of the input current i_s) has an rms value of 10.86 A that lags v_s by an angle $\phi_1 = 10°$. The harmonic components in the line current are listed in the PSpice output file included in the Appendix. Also from the output listing, the average values are $V_d = 158.45$ V and $I_d = 7.93$ A.

■

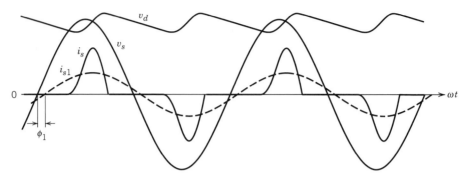

Figure 5-23   Waveforms in the circuit of Fig. 5-20, obtained in Example 5-2.

5-3-4-3   Line-Current Distortion

Figure 5-23 of Example 5-2 shows that the line current i_s at the input to the diode bridge rectifier deviates significantly from a sinusoidal waveform. This distorted current can also lead to distortion in the line voltage. We can quantify this distortion based on the theory discussed in Chapter 3. The fundamental and the third harmonics of i_s are shown in Fig. 5-24, along with the distortion component.

■ *Example 5-3*   In the rectifiers of Examples 5-1 and 5-2, calculate THD_i and the crest factor in the input current, the DPF, the PF, the average output voltage V_d, and $I_d/I_{\text{short circuit}}$.

Solution   Based on the Fourier analysis using PSpice in Example 5-2, $\text{THD}_i = 88.8\%$. From the same analysis, $I_{s1} = 10.86$ A. Using Eq. 3-36, the rms value $I_s = 14.52$ A. In Fig. 5-24, $I_{s,\text{peak}} = 34.7$ A. Therefore, from the definition in Chapter 3 (Eq. 3-37), the crest factor is 2.39. Since $\phi_1 = -10°$ by the Fourier analysis, DPF = 0.985 (lagging).

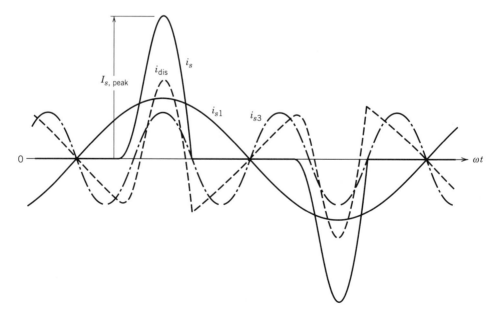

Figure 5-24   Distorted line current in the rectifier of Fig. 5-20.

The power factor is 0.736. In this example, $I_d = 7.93$ A. From Eq. 5-41, $I_{short\ circuit} = 318.3$ A. Therefore, $I_d/I_{short\ circuit} = 0.025$. The average output voltage $V_d = 158.45$ V ($V_d/V_{do} = 1.467$). ∎

■ **Example 5-4** In Example 5-3, $I_d/I_{short\ circuit}$ was calculated to be 0.025. For the same value of $I_d/I_{short\ circuit}$, use the results in Figs. 5-18 and 5-19 (obtained by assuming $C_{pu} \rightarrow \infty$) to calculate THD_i, DPF, PF, the crest factor, and V_d (nomalized). Compare the results with those in Example 5-3.

Solution Using Figs. 5-18 and 5-19, the following values are obtained: $THD_i =$ 79%, the crest factor is 2.25, DPF = 0.935, PF = 0.735, and $V_d/V_{do} = 1.384$. Before comparing with the results in Example 5-3, we should note that the power in these two cases is not the same, since a finite value of the filter capacitance in Example 5-3 results in a higher value of V_d and hence a higher power. In spite of this difference, the results are close, and the plots in Figs. 5-18 and 5-19 can be used for approximate calculations. ∎

5-3-4-4 Line-Voltage Distortion

Distorted currents drawn by loads such as the diode bridge rectifiers can result in distortion in the utility voltage waveform. For example, consider the circuit of Fig. 5-20 which is redrawn in Fig. 5-25. Here L_{s1} represents the internal impedance of the utility source and L_{s2} may be intentionally added as a part of the power electronics equipment. A resistance R_s is included that also can be used to represent the diode resistances.

The voltage across other equipment at the point of common coupling (PCC) is

$$v_{PCC} = v_s - L_{s1} \frac{di_s}{dt} \tag{5-52}$$

where v_s is assumed to be sinusoidal.

Expressing i_s in Eq. 5-52 in terms of its fundamental and harmonic components yields

$$v_{PCC} = \left(v_s - L_{s1} \frac{di_{s1}}{dt} \right) - L_{s1} \sum_{h \neq 1} \frac{di_{sh}}{dt} \tag{5-53}$$

where

$$(v_{PCC})_1 = v_s - L_{s1} \frac{di_{s1}}{dt} \tag{5-54}$$

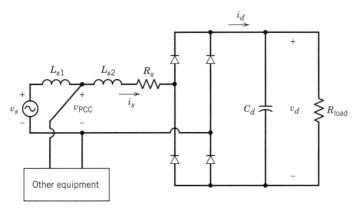

Figure 5-25 Line-voltage notching and distortion.

and the voltage distortion component due to the current harmonics is

$$(v_{PCC})_{dis} = -L_{s1} \sum_{h \neq 1} \frac{di_{sh}}{dt} \tag{5-55}$$

■ *Example 5-5*   Using the parameters in Example 5-1 and splitting L_s, for example, such that $L_{s1} = L_{s2} = 0.5$ mH, obtain the voltage waveform at the point of common coupling in the circuit of Fig. 5-25.

Solution   Using PSpice, the voltage and the current waveforms are shown in Fig. 5-26. The total harmonic distortion THD_v in the voltage at the point of common coupling is computed to be approximately 5.7%.

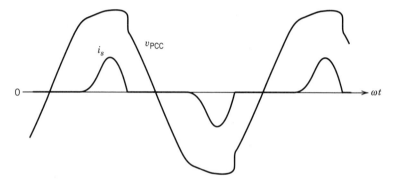

Figure 5-26   Voltage waveform at the point of common coupling in the circuit of Fig. 5-25.

5-4   VOLTAGE DOUBLER (SINGLE-PHASE) RECTIFIERS

In many applications, the input line-voltage magnitude may be insufficient to meet the dc output voltage requirement. More importantly, the equipment may be required to operate with a line voltage of 115 V as well as 230 V. Therefore, a voltage-doubler rectifier, as shown in Fig. 5-27, may be used to avoid a voltage stepup transformer.

When the switch in Fig. 5-27 is in the 230-V position with a line voltage of 230 V, the circuit is similar to the full-bridge rectifier circuit discussed earlier. With the switch in the 115-V position and the line voltage of 115 V, each capacitor gets charged to approximately the peak of the ac input voltage, and therefore, V_d (which is the sum of

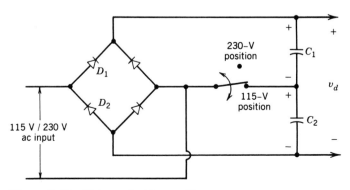

Figure 5-27   Voltage-doubler rectifier.

voltages across C_1 and C_2) is approximately the same as in the 230-V operation. The capacitor C_1 is charged through the diode D_1 during the positive half-cycle of the input ac voltage, and C_2 is charged through D_2 during the negative half-cycle. Therefore, in this mode the circuit operates as a voltage-doubler rectifier.

5-5   EFFECT OF SINGLE-PHASE RECTIFIERS ON NEUTRAL CURRENTS IN THREE-PHASE, FOUR-WIRE SYSTEMS

Often, large commercial and office buildings are supplied by a three-phase utility source. However, internally the voltage distribution and the load are primarily single phase, between one of the three line voltages and the neutral, as shown in a simplified manner in Fig. 5-28. An attempt is made to load all three phases equally. In case of linear loads, if all three phases are identically loaded, then the neural current i_n will be zero. In the following section, we will discuss the impact of nonlinear loads on the neutral current.

We will discuss the impact of single-phase diode rectifiers on i_n by first assuming that these are identical in each phase. Considering phase a, we can write i_a in terms of its fundamental and odd harmonics (even harmonics are zero) whose rms values I_{s1} and I_{sh} are the same in all three phases:

$$i_a = i_{a1} + \sum_{h=2k+1}^{\infty} i_{ah} \tag{5-56}$$

$$= \sqrt{2}I_{s1}\sin(\omega_1 t - \phi_1) + \sum_{h=2k+1}^{\infty} \sqrt{2}I_{sh}\sin(\omega_h t - \phi_h)$$

where $k = 1, 2, 3, \ldots$.

Assuming a balanced three-phase utility supply and identical loads, the currents in phases b and c are shifted by 120° and 240°, respectively, at the fundamental line frequency. Therefore,

$$i_b = \sqrt{2}I_{s1}\sin(\omega_1 t - \phi_1 - 120°) + \sum_{h=2k+1}^{\infty} \sqrt{2}I_{sh}\sin(\omega_h t - \phi_h - 120°h) \tag{5-57}$$

and

$$i_c = \sqrt{2}I_{s1}\sin(\omega_1 t - \phi_1 - 240°) + \sum_{h=2k+1}^{\infty} \sqrt{2}I_{sh}\sin(\omega_h t - \phi_h - 240°h) \tag{5-58}$$

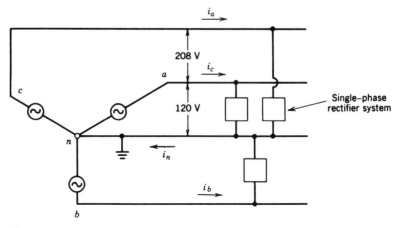

Figure 5-28   Three-phase, four-wire system.

These phase currents add up on the neutral wire, and therefore applying Kirchhoff's current law at node n

$$i_n = i_a + i_b + i_c \qquad (5\text{-}59)$$

Using Eqs. 5-56 through 5-58 into Eq. 5-59, we note that all the nontriplen harmonics and the fundamental frequency components add up to zero. The triplen harmonics add up to three times their individual values. Therefore,

$$i_n = 3 \sum_{h=3(2k-1)}^{\infty} \sqrt{2} I_{sh} \sin(\omega_h t - \phi_h) \qquad (5\text{-}60)$$

In terms of the rms values,

$$I_n = 3 \left(\sum_{h=3(2k-1)}^{\infty} I_{sh}^2 \right)^{1/2} \qquad (5\text{-}61)$$

As seen from the results in Section 5-3-4-3, the third harmonic dominates all other harmonic components. Therefore, in Eq. 5-61,

$$I_n \simeq 3 I_{s3} \qquad (5\text{-}62)$$

which means that the rms value of the neural current is approximately three times the third-harmonic rms current in the line conductors. Since the third-harmonic line current can be a significant percentage of the fundamental-frequency current, the neutral-wire current can be quite large. This realization has led to changes in the electrical wiring codes, which now stipulate that the neutral conductor should be able to carry at least as much current as the line conductors. In fact, if the line currents are highly discontinuous, the neutral current can be as large as (see Problem 5-21)

$$I_n = \sqrt{3} I_{\text{line}} \qquad (5\text{-}63)$$

■ *Example 5-6*   Assume that each nonlinear load in Fig. 5-28 can be represented by the single-phase load of Example 5-1. Obtain the neutral-wire current waveform and its rms value for the same per-phase voltage and the ac-side impedance as in Example 5-1.

Solution   Using PSpice, the neutral-wire current is plotted in Fig. 5-29 and its rms value is calculated to be approximately 25 A. It is almost $\sqrt{3}$ as large as the line current of 14.52 A (rms). ■

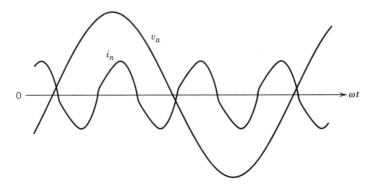

Figure 5-29   Neutral-wire current i_n.

5-6   THREE-PHASE, FULL-BRIDGE RECTIFIERS

In industrial applications where three-phase ac voltages are available, it is preferable to use three-phase rectifier circuits, compared to single-phase rectifiers, because of their lower ripple content in the waveforms and a higher power-handling capability. The three-phase, six-pulse, full-bridge diode rectifier shown in Fig. 5-30 is a commonly used circuit arrangement. A filter capacitor is connected at the dc side of the rectifier.

Similar to the analysis of single-phase, full-bridge rectifiers, we will begin with simplified circuits prior to the discussion of the circuit in Fig. 5-30.

5-6-1   IDEALIZED CIRCUIT WITH $L_S = 0$

In the circuit of Fig. 5-31a, the ac-side inductance L_s is assumed to be zero and the dc side is replaced by a constant dc current I_d. We will see later that replacing the dc current I_d by a load resistance R_{load} makes little difference in the circuit operation.

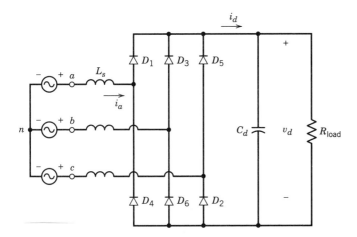

Figure 5-30   Three-phase, full-bridge rectifier.

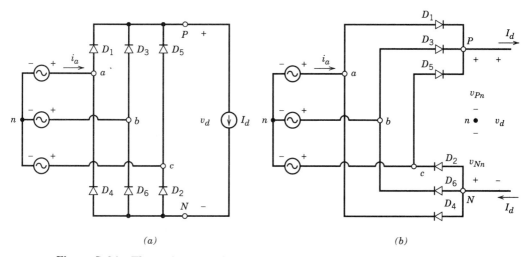

(a)   (b)

Figure 5-31   Three-phase rectifier with a constant dc current.

The rectifier in Fig. 5-31a can be redrawn as in Fig. 5-31b. With $L_s = 0$, the current I_d flows through one diode from the top group and one from the bottom group. Similar to the discussion in the case of single-phase rectifiers, in the top group, the diode with its anode at the highest potential will conduct and the other two become reversed biased. In the bottom group, the diode with its cathode at the lowest potential will conduct and the other two become reverse biased.

The voltage waveforms in the circuits of Fig. 5-31 are shown in Fig. 5-32a, where v_{Pn} is the voltage at the point P with respect to the ac voltage neutral point n. Similarly, v_{Nn} is the voltage at the negative dc terminal N. Since I_d flows continuously, at any time, v_{Pn} and v_{Nn} can be obtained in terms of one of the ac input voltages v_{an}, v_{bn}, and v_{cn}. Applying KVL in the circuit of Fig. 5-31 on an instantaneous basis, the dc-side voltage is

$$v_d = v_{Pn} - v_{Nn} \qquad (5\text{-}64)$$

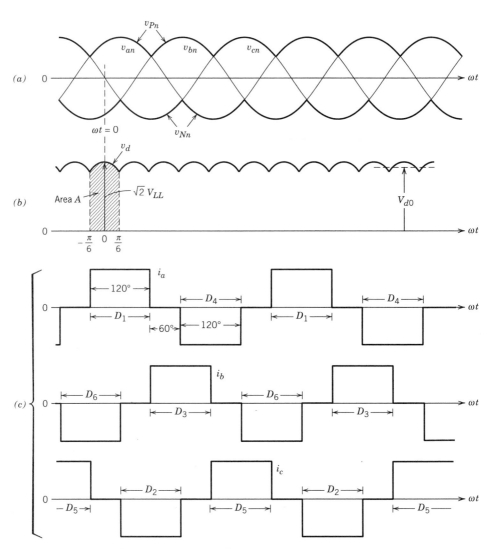

Figure 5-32   Waveforms in the circuit of Fig. 5-31.

The instantaneous waveform of v_d consists of six segments per cycle of line frequency. Hence, this rectifier is often termed a six-pulse rectifier. Each segment belongs to one of the six line-to-line voltage combinations, as shown in Fig. 5-32b. Each diode conducts for 120°. Considering the phase a current waveform in Fig. 5-32c,

$$i_a = \begin{cases} I_d & \text{when diode 1 is conducting} \\ -I_d & \text{when diode 4 is conducting} \\ 0 & \text{when neither diode 1 or 4 is conducting} \end{cases} \qquad (5\text{-}65)$$

The commutation of current from one diode to the next is instantaneous, based on the assumption of $L_s = 0$. The diodes are numbered in Fig. 5-31 in such a way that they conduct in the sequence 1, 2, 3, . . . ,. Next, we will compute the average value of the output dc voltage and rms values of the line currents, where the subscript o is added due to the assumption of $L_s = 0$.

To obtain the average value of the output dc voltage, it is sufficient to consider only one of the six segments and obtain its average over a 60° or $\pi/3$-rad interval. Arbitrarily, the time origin $t = 0$ is chosen in Fig. 5-32a when the line-to-line voltage v_{ab} is at its maximum. Therefore,

$$v_d = v_{ab} = \sqrt{2}V_{LL}\cos \omega t \qquad -\tfrac{1}{6}\pi < \omega t < \tfrac{1}{6}\pi \qquad (5\text{-}66)$$

where V_{LL} is the rms value of line-to-line voltages.

By integrating v_{ab}, the volt-second area A is given by

$$A = \int_{-\pi/6}^{\pi/6} \sqrt{2}V_{LL}\cos \omega t \, d(\omega t) = \sqrt{2}V_{LL} \qquad (5\text{-}67)$$

and therefore dividing A by the $\pi/3$ interval yields

$$V_{do} = \frac{1}{\pi/3} \int_{-\pi/6}^{\pi/6} \sqrt{2}V_{LL}\cos \omega t \, d(\omega t) = \frac{3}{\pi} \sqrt{2}V_{LL} = 1.35V_{LL} \qquad (5\text{-}68)$$

One of the phase voltages and the corresponding phase current (labeled v_s and i_s) are redrawn in Fig. 5-33a. Using the definition of rms current in the phase current waveform of Fig. 5-33a, the rms value of the line current i_s in this idealized case is

$$I_s = \sqrt{\tfrac{2}{3}}I_d = 0.816I_d \qquad (5\text{-}69)$$

By means of Fourier analysis of i_s in this idealized case, the fundamental-frequency component i_{s1} shown in Fig. 5-33a has an rms value

$$I_{s1} = \frac{1}{\pi} \sqrt{6}I_d = 0.78I_d \qquad (5\text{-}70)$$

The harmonic components I_{sh} can be expressed in terms of the fundamental-frequency component as

$$I_{sh} = \frac{I_{s1}}{h} \qquad (5\text{-}71)$$

where $h = 5, 7, 11, 13, \ldots$. The even and triplen harmonics are zero, as shown in Fig. 5-33b. Since i_{s1} is in phase with its utility phase voltage,

$$\text{DPF} = 1.0 \qquad (5\text{-}72)$$

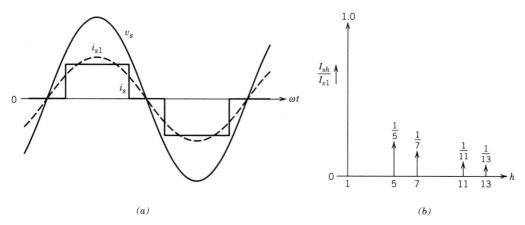

(a)   *(b)*

Figure 5-33 Line current in a three-phase rectifier in the idealized case with $L_s = 0$ and a constant dc current.

and therefore,

$$\text{PF} = \frac{3}{\pi} = 0.955 \tag{5-73}$$

The voltage waveforms will be identical if the load on the dc side is represented by a resistance R_{load} instead of a current source I_d. The phase currents will also flow during identical intervals, as in Fig. 5-32. The only difference will be that the current waveforms will not have a flat top, as in Fig. 5-32.

5-6-2 EFFECT OF L_S ON CURRENT COMMUTATION

Next, we will include L_s on the ac side and represent the dc side by a current source $i_d = I_d$, as shown in Fig. 5-34. Now the current commutations will not be instantaneous. We will look at only one of the current commutations because all others are identical in a balanced circuit. Consider the commutation of current from diode 5 to diode 1, beginning at t or $\omega t = 0$ (the time origin is chosen arbitrarily). Prior to this, the current i_d is flowing through diodes 5 and 6. Figure 5-35a shows the subcircuit pertinent to this current commutation.

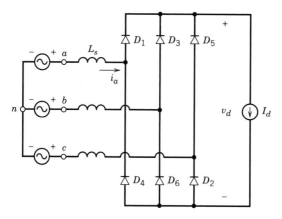

Figure 5-34 Three-phase rectifier with a finite L_s and a constant dc current.

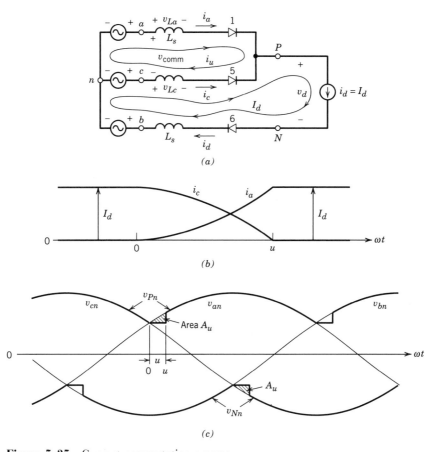

Figure 5-35 Current commutation process.

The current commutation only involves phases a and c, and the commutation voltage responsible is $v_{comm} = v_{an} - v_{cn}$. The two mesh currents i_u and I_d are labeled in Fig. 5-35a. The commutation current i_u flows due to a short-circuit path provided by the conducting diode 5. In terms of the mesh currents, the phase currents are

$$i_a = i_u$$

and

$$i_c = I_d - i_u \tag{5-74}$$

These are plotted in Fig. 5-35b, where i_u builds up from zero to I_d at the end of the commutation interval $\omega t_u = u$. In the circuit of Fig. 5-35a,

$$v_{La} = L_s \frac{di_a}{dt} = L_s di_u/dt \tag{5-75}$$

and

$$v_{Lc} = L_s \frac{di_c}{dt} = -L_s \frac{di_u}{dt} \tag{5-76}$$

noting that $i_c = I_d - i_u$ and therefore $di_c/dt = d(I_d - i_u)/dt = - di_u/dt$. Applying KVL in the upper loop in the circuit of Fig. 5-35a and using the above equations yield

$$v_{\text{comm}} = v_{an} - v_{cn} = v_{La} - v_{Lc} = 2L_s \frac{di_u}{dt} \qquad (5\text{-}77)$$

Therefore from the above equation,

$$L_s \frac{di_u}{dt} = \frac{v_{an} - v_{cn}}{2} \qquad (5\text{-}78)$$

The commutation interval u can be obtained by multiplying both sides of Eq. 5-78 by ω and integrating:

$$\omega L_s \int_0^{I_d} di_u = \int_0^u \frac{v_{an} - v_{cn}}{2} \, d(\omega t) \qquad (5\text{-}79)$$

where the time origin is assumed to be at the beginning of the current commutation. With this choice of time origin, we can express the line-to-line voltage $(v_{an} - v_{cn})$ as

$$v_{an} - v_{cn} = \sqrt{2} V_{LL} \sin \omega t \qquad (5\text{-}80)$$

Using Eq. 5-80 in Eq. 5-79 yields

$$\omega L_s \int_0^{I_d} di_u = \omega L_s I_d = \frac{\sqrt{2} V_{LL}(1 - \cos u)}{2} \qquad (5\text{-}81)$$

or

$$\cos u = 1 - \frac{2\omega L_s I_d}{\sqrt{2} V_{LL}} \qquad (5\text{-}82)$$

If the current commutation was instantaneous due to zero L_s, then the voltage v_{Pn} will be equal to v_{an} beginning with $\omega t = 0$, as in Fig. 5-35c. However, with a finite L_s, during $0 < \omega t < \omega t_u$ in Fig. 5-35c

$$v_{Pn} = v_{an} - L_s \frac{di_u}{dt} = \frac{v_{an} + v_{cn}}{2} \quad \text{(using Eq. 5-78)} \qquad (5\text{-}83)$$

where the voltage across $L_s[=L_s(di_u/dt)]$ is the drop in the voltage v_{Pn} during the commutation interval shown in Fig. 5-35c. The integral of this voltage drop is the area A_u, which according to Eq. 5-81 is

$$A_u = \omega L_s I_d \qquad (5\text{-}84)$$

This area is ''lost'' every $60°(\pi/3 \text{ rad})$ interval, as shown in Fig. 5-35c. Therefore, the average dc voltage output is reduced from its V_{do} value, and the voltage drop due to commutation is

$$\Delta V_d = \frac{\omega L_s I_d}{\pi/3} = \frac{3}{\pi} \omega L_s I_d \qquad (5\text{-}85)$$

Therefore, the average dc voltage in the presence of a finite commutation interval is

$$V_d = V_{do} - \Delta V_d = 1.35 V_{LL} - \frac{3}{\pi} \omega L_s I_d \qquad (5\text{-}86)$$

where V_{do} is the average voltage with an instantaneous commutation due to $L_s = 0$, given by Eq. 5-68.

5-6-3   A CONSTANT dc-SIDE VOLTAGE $v_d(t) = V_d$

Next, we will consider the circuit shown in Fig. 5-36a, where the assumption is that the dc-side voltage is a constant dc. It is an approximation to the circuit of Fig. 5-30 with a large value of C. In order to simplify our analysis, we will make an assumption that the current i_d on the dc side of the rectifier flows discontinuously, and therefore only two diodes—one from the top group and one from the bottom group—conduct at any given time. This assumption allows the equivalent circuit shown in Fig. 5-36b, where the input voltage is made up of the line-to-line voltage segments shown in Fig. 5-36c. The diode D_P corresponds to one of the diodes D_1, D_3, and D_5 from the top group. Similarly, the diode D_N corresponds to one of the diodes D_2, D_4, and D_6. Similar to the analysis in Section 5-3-3 for a single-phase input, the resulting phase current waveform is shown in Fig. 5-36c.

5-6-3-1   Distortion in the Line-Current Waveforms

It is useful to know the power factor, total harmonic distortion, and the dc output voltage in the practical circuit of Fig. 5-30. However, to present these results in a generalized manner requires the assumption of a constant dc output voltage, as made in this section.

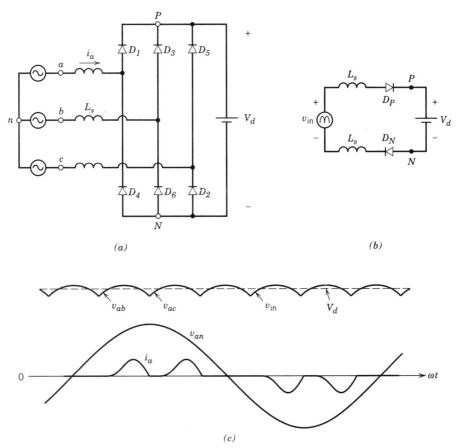

Figure 5-36   (a) Three-phase rectifier with a finite L_s and a constant dc voltage. (b) Equivalent circuit. (c) Waveforms.

The dc-side current I_d is normalized by the per-phase short-circuit current. This per-phase short-circuit current can be obtained in terms of the line-to-line input ac voltages:

$$I_{\text{short circuit}} = \frac{V_{LL}/\sqrt{3}}{\omega_1 L_s} \tag{5-87}$$

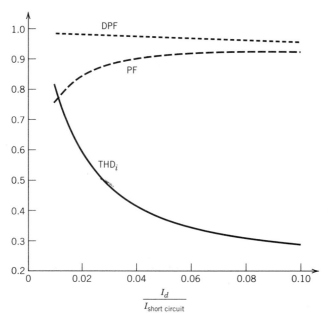

Figure 5-37 Total harmonic distortion, DPF, and PF in the rectifier of Fig. 5-36 with a constant dc voltage.

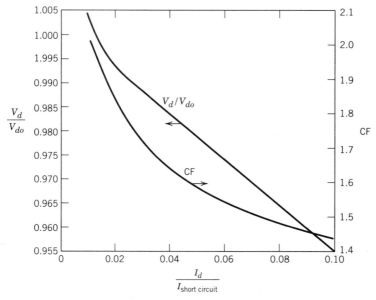

Figure 5-38 Normalized V_d and crest factor in the rectifier of Fig. 5-36 with a constant dc voltage.

Figures 5-37 and 5-38 show the plots of PF, DPF, THD, crest factor (CF), and V_d/V_{do} as a function of I_d normalized by $I_{\text{short circuit}}$. (See Problem 5-17 for justification that allows such a generalization.)

5-6-4 PRACTICAL THREE-PHASE DIODE BRIDGE RECTIFIERS

In three-phase rectifiers, even relatively small values of L_s result in a continuously flowing i_d, making the analysis by means of differential equations quite complicated. Therefore, the practical circuit shown in Fig. 5-30 is generally simulated using software such as PSpice.

■ *Example 5-7*   The three-phase diode rectifier circuit of Fig. 5-30 is supplying approximately 2.2 kW load with $V_{LL} = 208$ V at 60 Hz, $L_s = 1$ mH, and $C_d = 1100$ μF. The load can be represented by an equivalent resistance of 35.0 Ω. Obtain the circuit waveforms by means of PSpice simulation.

Solution   The PSpice network with node numbers and the input data file are included in Appendix at the end of this chapter. The voltage and current waveforms are shown in Fig. 5-39.

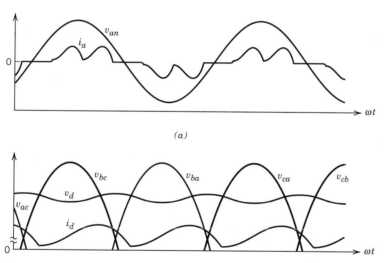

(a)

Figure 5-39   Waveforms in the rectifier of Fig. 5-30, obtained in Example 5-7.

The average dc voltage is 278.0 V, and it has a peak-to-peak ripple of 4.2 V, which in percentage of average dc voltage is 1.5%. The input current has a THD of 54.9%, the DPF is 0.97 (leading), and the PF is 0.85. The average dc current is 7.94 A.   ■

■ *Example 5-8*   In Example 5-7, the dc side has a filter capacitor of 1100 μF with an average value of 278.0 V and a small, superimposed ripple voltage. The results in Figs. 5-37 and 5-38 are obtained by assuming a dc-side voltage of a constant dc value. The objective of this example is to illustrate the effect of assuming a constant dc voltage of 278.0 V (same average value as in Example 5-7).

Solution   With $V_d = 278.0$ V, $V_d/V_{do} = 0.9907$. From Fig. 5-38, this corresponds to $I_d/I_{\text{short circuit}} = 0.025$. This value in Fig. 5-37 approximately corresponds to THD =

50%, DPF = 0.98, and PF = 0.87. All these values are very close to those calculated in Example 5-7 with a dc-side filter capacitor of 1100 μF. ∎

5-7 COMPARISON OF SINGLE-PHASE AND THREE-PHASE RECTIFIERS

Comparison of the line-current waveforms in Figs. 5-23 and 5-39*a* shows that the line current in a single-phase rectifier contains significantly more distortion compared to a three-phase rectifier. This results in a much poorer power factor in a single-phase rectifier compared to a three-phase rectifier. This is confirmed by comparing Figs. 5-18 and 5-19 for single-phase rectifiers with the results for three-phase rectifiers in Figs. 5-37 and 5-38. The displacement power factor (cos ϕ_1) is high in both rectifiers.

Comparison of the i_d waveforms in Figs. 5-22 and 5-39*b* shows that the ripple in the dc current is smaller in a three-phase rectifier in comparison to a single-phase rectifier. The ripple current, which flows through the filter capacitor, dictates the capacitance and the current-handling capability required of the filter capacitor. Therefore, in some applications, the filter capacitance required may be much smaller in a three-phase rectifier compared to a single-phase rectifier.

In a three-phase rectifier, the maximum regulation in the dc voltage V_d from no-load to a full-load condition will generally be less than 5%, as seen from Fig. 5-38. This regulation is often much larger in single-phase rectifiers.

Based on the foregoing discussion and the fact that the use of single-phase rectifiers in three-phase, four-wire systems introduces large currents in the neutral (even in a balanced system), it is always preferable to use a three-phase rectifier over a single-phase rectifier.

5-8 INRUSH CURRENT AND OVERVOLTAGES AT TURN-ON

In the previous sections, we have considered only the steady-state rectifier operation. However, considerable overvoltages and large inrush currents can result at turn-on if the ac voltage is suddenly applied to the circuit by means of a contactor.

For the worst-case analysis, the filter capacitor is assumed to be initially completely discharged. Furthermore, it is assumed that at turn-on ($\omega t = 0$), the ac source input is at its peak value ($\sqrt{2}V_{LL}$ in the three-phase circuit). Therefore, the theoretical maximum voltage across the capacitor due to this series *L–C* connection approaches

$$V_{d,\max} = 2\sqrt{2}V_s \quad \text{(single-phase)} \tag{5-88}$$

$$V_{d,\max} = 2\sqrt{2}V_{LL} \quad \text{(three-phase)} \tag{5-89}$$

Normally, the load across the filter capacitor is a voltage-sensitive electronic circuit such as a switch-mode inverter in an ac motor drive, and this large overvoltage can cause serious damage both to the dc capacitor and the electronic load. Moreover, large inrush currents at turn-on may destroy the diodes in the rectifier. It will also result in a momentary voltage drop at the point of common coupling.

To overcome these problems, one possible solution is to use a current-limiting resistor on the dc side in series between the rectifier output and the filter capacitor. This resistor is shorted out, either by a mechanical contactor or by a thyristor after a few cycles subsequent to turn-on, in order to avoid power dissipation and a substantial loss of efficiency due to its presence. An alternative circuit is presented in reference 3.

5-9   CONCERNS AND REMEDIES FOR LINE-CURRENT HARMONICS AND LOW POWER FACTOR

Typical ac current waveforms in single-phase and three-phase diode rectifier circuits are far from a sinusoid. The power factor is also very poor because of the harmonic contents in the line current.

As power electronic systems proliferate, ac-to-dc rectifiers are playing an increasingly important role. A large number of systems injecting harmonic currents into the utility grid can have significant impact on the quality of the ac voltage waveform (i.e., it will become distorted), thus causing problems with other sensitive loads connected to the same supply. Moreover, these harmonic currents cause additional harmonic losses in the utility system and may excite electrical resonances, leading to large overvoltages. Another problem caused by harmonics in the line current is to overload the circuit wiring. For example, a 120-V, 1.7-kW unity power factor load will draw only a current of 14 A and therefore can easily be supplied by a 15-A service. However, a 1.7-kW rectifier load with a power factor of 0.6 will draw a current of 23.6 A, in excess of 15 A, thus tripping the circuit breaker.

Standards for harmonics and the remedies for a poor line-current waveform and the input power factor are important concerns of power electronic systems. These are discussed in detail in Chapter 18.

SUMMARY

1. Line-frequency diode rectifiers are used to convert 50- or 60-Hz ac input into a dc voltage in an uncontrolled manner. A large filter capacitor is connected across the rectifier output since in most power electronic applications, a low ripple in the output dc voltage V_d is desirable.

2. Based on simplifying assumptions, analytical expressions are derived from the commonly used full-bridge rectifier topologies with single-phase and three-phase inputs.

3. In practical circuits where analytical expressions will be unnecessarily complicated, the simulation methods are presented to obtain the rectifier voltage and current waveforms.

4. Various rectifier characteristics such as the total harmonic distortion in the input current, the displacement power factor, and the power factor are presented in a generalized manner for both the one-phase and three-phase rectifiers.

5. In diode rectifiers with small L_s or L_d (where L_d is the inductance between the rectifier dc output and the filter capacitor), the current i_d and i_s are highly discontinuous, and as a consequence, the rms value of the input current I_s becomes large, and the power is drawn from the utility source at a very poor power factor.

6. In case of a single-phase ac input, voltage-doubler rectifiers can be used to approximately double the output dc voltage magnitude, compared to a full-bridge rectifier. These are sometimes used in low-power equipment, which may be required to operate from dual voltages of 115 and 230 V.

7. Effect of single-phase rectifiers on the neutral-wire current in three-phase four-wire systems is analyzed.

8. Comparison of single-phase and three-phase diode rectifiers shows that the three-phase rectifiers are preferable in most respects.

9. Both single-phase and three-phase diode rectifiers inject large amounts of harmonic currents into the utility system. As the power electronic systems proliferate, remedies

for the poor input current waveform would have to be implemented. These topics are discussed in Chapter 18.

PROBLEMS

BASIC CONCEPTS

5-1 In the basic circuit of Fig. 5-3a, V_s = 120 V at 60 Hz, L = 10 mH, and R = 5 Ω. Calculate and plot the current i along with v_s.

5-2 In the basic circuit of Fig. 5-4a, V_s = 120 V at 60 Hz, L = 10 mH, and V_d = 150 V. Calculate and plot the current i along with v_s.

5-3 The voltage v across a load and the current i into the positive-polarity terminal are as follows (where ω_1 and ω_3 are not equal):

$$v(t) = V_d + \sqrt{2}V_1\cos(\omega_1 t) + \sqrt{2}V_1\sin(\omega_1 t) + \sqrt{2}V_3\cos(\omega_3 t) \quad \text{V}$$

$$i(t) = I_d + \sqrt{2}I_1\cos(\omega_1 t) + \sqrt{2}I_3\cos(\omega_3 t - \phi_3) \quad \text{A}$$

Calculate the following:

(a) The average power P supplied to the load

(b) The rms value of $v(t)$ and $i(t)$

(c) The power factor at which the load is operating

SINGLE-PHASE RECTIFIERS

5-4 In the single-phase diode rectifier circuit shown in Fig. 5-6b with zero L_s and a constant dc current I_d = 10 A, calculate the average power supplied to the load:

(a) If v_s is a sinusoidal voltage with V_s = 120 V at 60 Hz

(b) If v_s has the pulse waveform shown in Fig. P5-4

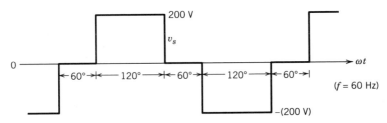

Figure P5-4

5-5 Consider the basic commutation circuit of Fig. 5-11a with I_d = 10 A.

(a) With V_s = 120 V at 60 Hz and L_s = 0, calculate V_d and the average power P_d.

(b) With V_s = 120 V at 60 Hz and L_s = 5 mH, calculate u, V_d, and P_d.

(c) Here v_s has a 60-Hz square waveform with an amplitude of 200 V, and L_s = 5mH. Plot the i_s waveform and calculate u, V_d, and P_d.

(d) Repeat part (c) if v_s has the pulse waveform shown in Fig. P5-4.

5-6 In the simplified single-phase rectifier circuit shown in Fig. 5-6b with L_s = 0 and a constant dc current I_d, obtain the average and the rms values of the current through each diode as a ratio of I_d.

5-7 In the single-phase rectifier circuit of Fig. 5-20, assume the ac-side impedance to be negligible. Instead, an inductance L_d is placed between the rectifier output and the filter capacitor. Derive the minimum value of L_d in terms of V_s, ω, and I_d that will result in a continuous i_d assuming that the ripple in v_d is negligible.

5-8 In the single-phase rectifier circuit shown in Fig. 5-14a, V_s = 120 V at 60 Hz, L_s = 1 mH, and I_d = 10 A. Calculate u, V_d, and P_d. What is the percentage voltage drop in V_d due to L_s?

5-9 Repeat Problem 5-8.

(a) If v_s has a 60-Hz square waveform with an amplitude of 200 V

(b) If v_s has the pulse waveform shown in Fig. P5-4

5-10 In the single-phase rectifier circuit of Fig. 5-16a, L_s = 1 mH and V_d = 160 V. The input voltage v_s has the pulse waveform shown in Fig. P5-4. Plot i_s and i_d waveforms. (*Hint*: i_s and i_d flow discontinuously.)

5-11 In the single-phase rectifier of Fig. 5-16a, V_s = 120 V at 60 Hz, L_s = 1 mH, and V_d = 150 V. Calculate the waveform for i_d shown in Fig. 5-16c and indicate the values of θ_b, θ_f, and $I_{d,\text{peak}}$. Also calculate the average value I_d.

5-12 Using the MATLAB program listing in the Appendix at the end of this chapter, calculate V_d and P_d in Example 5-1. Compare the results with the plot in Fig. 5-22.

5-13 The single-phase rectifier circuit of Example 5-2 with R_s = 0.4 Ω is supplying a load of 1 kW. Modify the basic PSpice input listed in Appendix at the end of this chapter for Example 5-2 to obtain the plot of the v_d waveform, its average value V_d, and its peak-to-peak ripple if the load is represented as:

(a) Absorbing a constant instantaneous power $p_d(t)$ = 1 kW. (*Hint*: Represent the load by a voltage-dependent current source, for example, using the statement GDC 5 6 VALUE={1000.0/V(5,6)}.)

(b) A constant equivalent resistance that absorbs 1 kW based on V_d in part (a)

(c) A dc current source I_d that absorbs 1 kW based on V_d in part (a)

Compare the peak-to-peak ripple in the dc output voltage for these three types of load representations.

5-14 In the single-phase rectifier circuit of Fig. 5-6b with i_d = I_d, obtain the THD, DPF, PF, and CF.

5-15 Using the MATLAB program in Problem 5-12, calculate the THD, DPF, PF, and CF.

5-16 In the single-phase rectifier circuit of Fig. 5-20, V_s = 120 V at 60 Hz, L_s = 2 mH, R_s = 0.4 Ω, and the instantaneous load power $p_d(t)$ = 1 kW. Using PSpice, evaluate the effect of the dc-side filter capacitance by plotting the THD, DPF, PF, and $\Delta V_{d(\text{peak}-\text{peak})}$ for the following values of C_d: 200, 500, 1000, and 1500 μF.

5-17 The generalized results for the THD, DPF, and PF are presented in Figs. 5-18 and 5-19 for single-phase rectifiers and in Figs. 5-37 and 5-38 for three-phase rectifiers. Show that assuming the dc side of the rectifier to be represented by a purely dc voltage source allows us to present the results in a generalized manner as a function of $I_d/I_{\text{short circuit}}$.

5-18 Calculate the voltage distortion at the point of common coupling in the circuit of Fig. 5-25. Here V_s = 120 V at 60 Hz, L_{s1} = L_{s2} = 1 mH, and the dc side of the rectifier is represented by a dc current source of 10 A.

SINGLE-PHASE VOLTAGE-DOUBLER AND MIDPOINT RECTIFIERS

5-19 Consider the voltage-doubler circuit of Fig. 5-27, where V_s = 120 V at 60 Hz, L_s = 1 mH, C_1 = C_2 = 1000 μF, and the load is represented by a dc current source of 10 A. Use PSpice.

(a) Obtain v_{c1}, v_{c2}, and v_d waveforms.

(b) Obtain $\Delta V_{d(\text{peak}-\text{peak})}$ as a ratio of V_d.

(c) Compare with the result in part (b) if a single-phase, full-bridge rectifier is used with V_s = 240 V, L_s = 1 mH, C_d = 500 μF, and a load of 10 A.

5-20 A midpoint rectifier is shown in Fig. P5-20, where we assume the transformer to be ideal and the dc-side load to be represented by a current source. Calculate the volt-ampere rating of the transformer as a ratio of the average power supplied to the load.

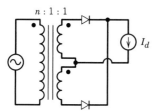

$n : 1 : 1$

I_d

Figure P5-20 Midpoint rectifier.

THREE-PHASE, FOUR-WIRE SYSTEMS, NEUTRAL CURRENT

5-21 In the three-phase, four-wire system of Fig. 5-28, all single-phase rectifier loads are identical and the conditions are such that each line current flows for less than 60° during each half-cycle of the line-to-neutral voltage. Show that in terms of their rms values $I_n = \sqrt{3} I_{\text{line}}$.

5-22 Write a PSpice input circuit file and execute it to obtain the results of Example 5-6.

THREE-PHASE RECTIFIERS

5-23 In the simplified three-phase rectifier circuit of Fig. 5-31a, obtain the average and the rms values of current through each diode as a ratio of the dc-side current I_d.

5-24 For simplification in the three-phase rectifier circuit of Fig. 5-35a, assume the commutation voltages to be increasing linearly rather than sinusoidally.

   (a) Obtain the expression for u, following a derivation similar to that of Eq. 5-82.

   (b) For $V_{LL} = 208$ V at 60 Hz, $L_s = 2$ mH, and $I_d = 10$ A, compare the results from the expression in part (a) and Eq. 5-82.

5-25 Using PSpice in Example 5-7, evaluate the effect of the filter capacitance on $\Delta V_{d(\text{peak}-\text{peak})}$, THD, DPF, and PF for the following values of C_d: 220, 550, 1100, 1500, and 2200 μF.

5-26 In the three-phase rectifier circuit of Fig. 5-30, assume the ac-side inductance L_s to be negligible. Instead, an inductance L_d is placed between the rectifier output and the filter capacitor. Derive the minimum value of L_d in terms of V_{LL}, ω, and I_d that will result in a continuous i_d, assuming that the ripple in v_d is negligible.

5-27 Using Fourier analysis, prove Eqs. 5-69 through 5-73 for three-phase rectifiers.

5-28 Using PSpice, this problem is intended to compare the performance of single-phase rectifiers with three-phase rectifiers in terms of the THD, DPF, PF, and $\Delta V_{d(\text{peak}-\text{peak})}$ while supplying the same load. In the circuits of Figs. 5-20 and 5-30, $V_s = 120$ V and $V_{LL} = 208$ V, respectively, at 60 Hz. Assume $L_s = 1$ mH and $R_s = 0.2$ Ω. The instantaneous load is constant at 5 kW, as examplified in Problem 5-13(a). The filter capacitor C_d has a value of 1100 μF in the single-phase rectifier. Choose its value in the three-phase rectifier to provide the same average energy storage as in the single-phase case.

5-29 Evaluate the effect of unbalanced voltages on the current waveforms in a three-phase rectifier. In the system of Example 5-7, assume $V_{an} = 110$ V and $V_{bn} = V_{cn} = 120$ V. Using PSpice, obtain the input current waveforms and their harmonic components.

INRUSH CURRENTS

5-30 In the single-phase rectifier of Example 5-2, obtain the maximum inrush current and the corresponding instant of switching with initial capacitor voltage equal to zero.

5-31 In the three-phase rectifier of Example 5-7, obtain the maximum inrush current and the corresponding instant of switching with initial capacitor voltage equal to zero.

REFERENCES

1. P. M. Camp, "Input Current Analysis of Motor Drives with Rectifier Converters," *IEEE-IAS Conference Record*, 1985, pp. 672–675.

2. B. Brakus, "100 Amp Switched Mode Charging Rectifier for Three-Phase Mains," *IEEE-Intelec,* 1984, pp. 72–78.

3. T. M. Undeland and N. Mohan, "Overmodulation and Loss Considerations in High Frequency Modulated Transistorized Induction Motor Drives," *IEEE Transactions on Power Electronics,* Vol. 3, No. 4, October 1988, pp. 447–452.

4. M. Grotzbach and B. Draxler "Line Side Behavior of Uncontrolled Rectifier Bridges with Capacitive DC Smoothing," paper presented at the European Power Electronics Conference (EPE), Aachen, 1989, pp. 761–764.

5. W. F. Ray, "The Effect of Supply Reactance on Regulation and Power Factor for an Uncontrolled 3-Phase Bridge Rectifier with a Capacitive Load," IEE Conference Publication, No. 234, 1984, pp. 111–114.

6. W. F. Ray, R. M. Davis, and I. D. Weatherhog, "The Three-Phase Bridge Rectifier with a Capacitive Load," IEE Conference Publication, No. 291, 1988, pp. 153–156.

7. R. Gretsch, "Harmonic Distortion of the Mains Voltage by Switched-Mode Power Supplies—Assessment of the Future Development and Possible Mitigation Measures," European Power Electronics Conference (EPE), Aachen, 1989, pp. 1255–1260.

APPENDIX

(PSpice examples are adapted from "Power Electronics: Computer Simulation, Analysis and Education Using Evaluation Version of PSpice," Minnesota Power Electronics, P.O. Box 14503, Minneapolis, MN 55414.)

MATLAB PROGRAM LISTING FOR EXAMPLE 5-1

```
% Single-Phase, Diode-Rectifier Bridge
clc,clg,clear
% Data
ls=1e-3; rs=0.001; cd=1000e-6; rload=20; deltat=25e-6;
freq=60; thalf=1/(2*freq); ampl=170; w=2*pi*freq;
% Matrix A, see Eq. 5-45
A=[-rs/ls -1/ls; 1/cd -1/(cd*rload)];
% Vector b, see Eq. 5-46
b=[1/ls; 0];
%
M=inv(eye(2) - deltat/2 * A)*(eye(2) + deltat/2 * A); % see Eq. 5-48
N=deltat/2 * inv(eye(2) - deltat/2 * A) * b; % see Eq. 5-48
%
for alfa0=55:0.5:75
alfa0
% Initial Conditions
vc0=ampl*sin(alfa0*pi/180);
il0=0;k=1;time(1)=alfa0/(360*freq);
il(1)=il0;vc(1)=vc0;vs(1)=vc0;
x=[il(1) vc(1)]';
%
while il(k) >= 0
 k=k+1;
  time(k)=time(k-1) + deltat;
   y=M*x + N*(ampl*sin(w*time(k)) + ampl*sin(w*time(k-1))); % see Eq. 5-47
    il(k)=y(1);
   vc(k)=y(2);
  vs(k)=ampl*sin(w*time(k));
 x=y;
end
%
time1=time(k);
il1=0;
vc1=vc(k);
```

```
%
while vc(k) > ampl*abs(sin(w*time(k)))
 k=k+1;
  time(k)=time(k-1) + deltat;
   vc(k)= vc1*exp(-(time(k)-time1)/(cd*rload)); % see Eq. 5-51
  vs(k)=ampl*abs(sin(w*time(k)));
 il(k)=0;
end
if(abs(time(k) - thalf -time(1)) <= 2*deltat), break, end
end
plot(time(1:k),il(1:k),time(1:k),vs(1:k),time(1:k),vc(1:k))
```

PSPICE INPUT CIRCUIT FILE FOR EXAMPLE 5-2

```
* Single-Phase, Diode-Bridge Rectifier
LS        1    2    1mH
RS        2    3    1m
*
rdc       4    5    1u
RLOAD     5    6    20.0
CD        5    6    1000uF IC=160V
*
XD1       3    4    DIODE_WITH_SNUB
XD3       0    4    DIODE_WITH_SNUB
XD2       6    0    DIODE_WITH_SNUB
XD4       6    3    DIODE_WITH_SNUB
*
VS        1    0    SIN(0 170V 60.0 0 0 0)
*
.TRAN     50us  50ms  0s    50us   UIC
.PROBE
.FOUR     60.0 v(1) i(LS)    i(rdc)    v(5,6)

.SUBCKT DIODE_WITH_SNUB 101 102
* Power Electronics: Simulation, Analysis  Education.....by N. Mohan.
DX        101 102    POWER_DIODE
RSNUB     102 103    1000.0
CSNUB     103 101    0.1uF
.MODEL    POWER_DIODE  D(RS=0.01, CJO=100pF)
.ENDS

.END
```

PSPICE OUTPUT OF EXAMPLE 5-2

FOURIER COMPONENTS OF TRANSIENT RESPONSE V(1)

HARMONIC NO	FREQUENCY (HZ)	FOURIER COMPONENT	NORMALIZED COMPONENT	PHASE (DEG)	NORMALIZED PHASE (DEG)
1	6.000E+01	1.700E+02	1.000E+00	-1.266E-04	0.000E+00

FOURIER COMPONENTS OF TRANSIENT RESPONSE I(LS)

HARMONIC NO	FREQUENCY (HZ)	FOURIER COMPONENT	NORMALIZED COMPONENT	PHASE (DEG)	NORMALIZED PHASE (DEG)
1	6.000E+01	1.536E+01	1.000E+00	-1.003E+01	0.000E+00
2	1.200E+02	6.405E-02	4.171E-03	-9.138E+01	-8.135E+01
3	1.800E+02	1.174E+01	7.648E-01	1.489E+02	1.589E+02
4	2.400E+02	4.198E-02	2.734E-03	8.531E+01	9.534E+01
5	3.000E+02	6.487E+00	4.224E-01	-5.632E+01	-4.629E+01
6	3.600E+02	1.585E-02	1.032E-03	-1.028E+02	-9.275E+01
7	4.200E+02	2.207E+00	1.438E-01	8.052E+01	9.055E+01
8	4.800E+02	2.778E-03	1.809E-04	-8.191E+01	-7.187E+01
9	5.400E+02	1.032E+00	6.724E-02	1.535E+02	1.636E+02

TOTAL HARMONIC DISTORTION = 8.879830E+01 PERCENT

```
FOURIER COMPONENTS OF TRANSIENT RESPONSE I(rdc)
 DC COMPONENT = 7.931217E+00


FOURIER COMPONENTS OF TRANSIENT RESPONSE V(5, 6)
 DC COMPONENT = 1.584512E+02
```

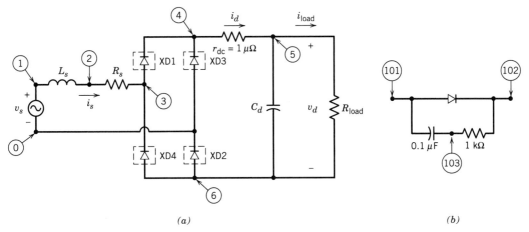

(a)

(b)

Figure 5A-1 (a) PSpice Input Circuit for Example 5-2, (b) Subcircuit Diode__with__Snub.

PSPICE INPUT CIRCUIT FILE FOR EXAMPLE 5-7

```
* Three-Phase, Diode-Bridge Rectifier
LSA       1    11   1mH
LSB       2    21   1mH
LSC       3    31   1mH
RSA       11   12   1m
RSB       21   22   1m
RSC       31   32   1m
*
LD        4    5    1uH
RD        5    6    1u
RLOAD     6    7    35.0
CD        6    7    1100uF   IC=276V
*
XD1       12   4    DIODE_WITH_SNUB
XD3       22   4    DIODE_WITH_SNUB
XD5       32   4    DIODE_WITH_SNUB
XD4       7    12   DIODE_WITH_SNUB
XD6       7    22   DIODE_WITH_SNUB
XD2       7    32   DIODE_WITH_SNUB
*
VSA       1    0    SIN(0 170 60.0 0 0 0)
VSB       2    0    SIN(0 170 60.0 0 0 -120)
VSC       3    0    SIN(0 170 60.0 0 0 -240)
*
.TRAN     50us  100ms   0s   50us   UIC
.PROBE
.FOUR     60.0   i(LSA) v(6,7) i(LD)
```

```
.SUBCKT  DIODE_WITH_SNUB  101  102
DX     101  102  POWER_DIODE
RSNUB 102  103  1000.0
CSNUB 103  101  0.1uF
.MODEL     POWER_DIODE D(RS=0.01, CJO=100pF)
.ENDS

.END
```

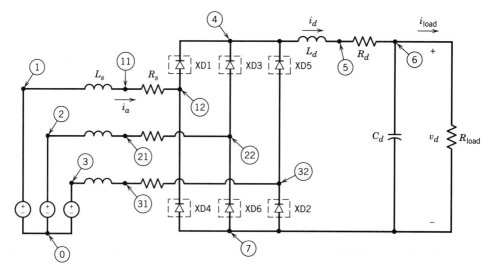

Figure 5A-2 PSpice Input Circuit for Example 5-7.

CHAPTER 6

LINE-FREQUENCY PHASE-CONTROLLED RECTIFIERS AND INVERTERS: LINE-FREQUENCY ac ↔ CONTROLLED dc

6-1   INTRODUCTION

In Chapter 5 we discussed the line-frequency diode rectifiers that are increasingly being used at the front end of the switch-mode power electronic systems to convert line frequency ac input to an uncontrolled dc output voltage.

However, in some applications such as battery chargers and a class of dc- and ac-motor drives it is necessary for the dc voltage to be controllable. The ac to controlled-dc conversion is accomplished in line-frequency phase-controlled converters by means of thyristors. In the past, these converters were used in a large number of applications for controlling the flow of electric power. Owing to the increasing availability of better controllable switches in high voltage and current ratings, new use of these thyristor converters nowadays is primarily in three-phase, high-power applications. This is particularly true in applications, most of them at high power levels, where it is necessary or desirable to be able to control the power flow in both directions between the ac and the dc sides. Examples of such applications are converters in high-voltage dc power transmission (Chapter 17) and some dc motor and ac motor drives with regenerative capabilities (Chapters 13–15).

As the name of these converters implies, the line-frequency voltages are present on their ac side. In these converters, the instant at which a thyristor begins or ceases to conduct depends on the line-frequency ac voltage waveforms and the control inputs. Moreover, the transfer or commutation of current from one device to the next occurs naturally because of the presence of these ac voltages.

It should be noted that the uncontrollable, line-frequency diode rectifiers of Chapter 5 are a subset of the controlled converters discussed in this chapter. The reasons for discussing the diode rectifiers separately in Chapter 5 have to do with their increasing importance and the way in which the dc current i_d flows through them. In Chapter 5, it was pointed out that in many applications of line-frequency diode rectifiers, the current i_d through them does not flow continuously due to the capacitor filter for smoothing the

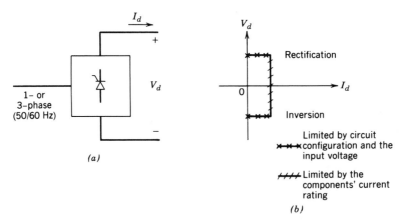

Figure 6-1   Line-frequency controlled converter.

dc-side voltage. Therefore, the emphasis in Chapter 5 was on the discontinuous-current-conduction mode. However, in most applications of the controlled converters of this chapter, the dc-side current i_d flows continuously, and hence the emphasis is on the continuous-current-conduction mode, although the discontinuous-current-conduction mode is also briefly discussed.

A fully controlled converter is shown in Fig. 6-1a in block diagram form. For given ac line voltages, the *average* dc-side voltage can be controlled from a positive maximum to a negative minimum value in a continuous manner. The converter dc current I_d (or i_d on an instantaneous basis) cannot change direction, as will be explained later. Therefore, a converter of this type can operate in only two quadrants (of the V_d-I_d plane), as shown in Fig. 6-1b. Here, the positive values of V_d and I_d imply *rectification* where the power flow is from the ac to the dc side. In an *inverter* mode, V_d becomes negative (but I_d stays positive) and the power is transferred from the dc to the ac side. The inverter mode of operation on a sustained basis is possible only if a source of power, such as batteries, is present on the dc side.

In some applications, such as in reversible-speed dc motor drives with regenerative braking, the converter must be capable of operating in all four quadrants. This is accomplished by connecting two two-quadrant converters (described above) in antiparallel or back to back, as discussed in Chapter 13.

In analyzing the converters in this chapter, the thyristors are assumed to be ideal, except for the consideration of the thyristor turn-off time t_q, which was described in Chapter 2.

6-2   THYRISTOR CIRCUITS AND THEIR CONTROL

For given ac input voltages, the magnitude of the average output voltage in thyristor converters can be controlled by delaying the instants at which the thyristors are allowed to start conduction. This is illustrated by the simple circuits of Fig. 6-2.

6-2-1   BASIC THYRISTOR CIRCUITS

In Fig. 6-2a, a thyristor connects the line-frequency source v_s to a load resistance. In the positive half-cycle of v_s, the current is zero until $\omega t = \alpha$, at which time the thyristor is supplied a positive gate pulse of a short duration. With the thyristor

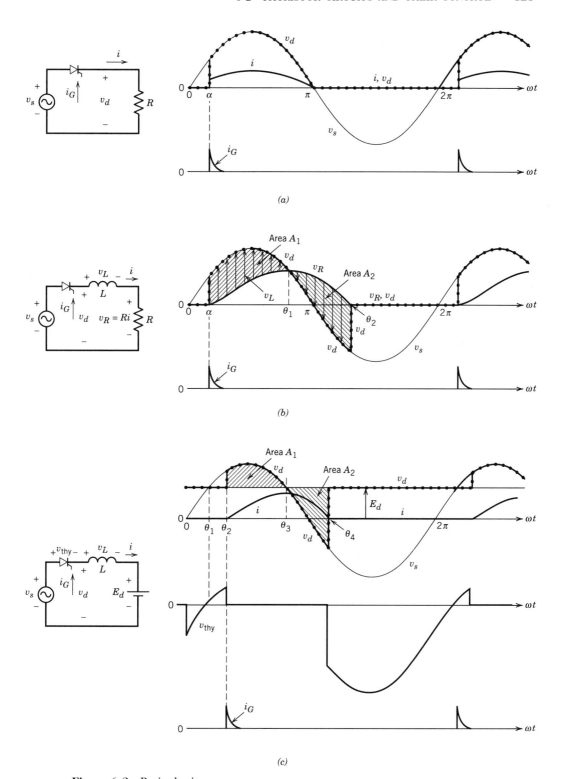

Figure 6-2   Basic thyristor converters.

conducting, $v_d = v_s$. For the rest of the positive half-cycle, the current waveform follows the ac voltage waveform and becomes zero at $\omega t = \pi$. Then the thyristor blocks the current from flowing during the negative half-cycle of v_s. The current stays zero until $\omega t = 2\pi + \alpha$, at which time another short-duration gate pulse is applied and the next cycle of the waveform begins. By adjusting α, the average value of the load voltage v_d can be controlled.

In Fig. 6-2b, the load consists of both R and L. Initially, the current is zero. The thyristor conduction is delayed until $\omega t = \alpha$. Once the thyristor is fired or gated on at $\omega t = \alpha$ during the positive half-cycle of v_s when the voltage across the thyristor is positive, the current begins to flow and $v_d = v_s$. The voltage across the inductor can be written as

$$v_L(t) = L\frac{di}{dt} = v_s - v_R \tag{6-1}$$

where $v_R = Ri$. In Fig. 6-2b, v_R (which is proportional to the current) is plotted and v_L is shown as the difference between v_s and v_R. During α to θ_1, v_L is positive and the current increases, since

$$i(\omega t) = \frac{1}{\omega L}\int_{\alpha}^{\omega t} v_L(\zeta)\,d\zeta \tag{6-2}$$

where ζ is a variable of integration. Beyond $\omega t = \theta_1$, v_L becomes negative and the current (as well as v_R) begins to decline. The instant at which the current becomes zero and stays zero due to the thyristor is dictated by Eq. 6-2. Graphically in Fig. 6-2b, $\omega t = \theta_2$ is the instant at which area A_1 equals area A_2 and the current becomes zero. These areas represent the time integral of v_L, which must be zero over one cycle of repetition in steady state, as explained in Section 3-2-5-1. It should be noted that the current continues to flow for a while even after v_s has become negative, as discussed in Section 5-2-2. The reason for this has to do with the stored energy in the inductor, a part of which is supplied to R and the other part is absorbed by v_s when it becomes negative.

In Fig. 6-2c, the load consists of an inductor and a dc voltage E_d. Here, with the current initially zero, the thyristor is reverse biased until $\omega t = \theta_1$, as shown in Fig. 6-2c. Therefore, it cannot conduct until $\omega t = \theta_1$. The thyristor conduction is further delayed until θ_2, when a positive gate pulse is applied. With the current flowing

$$v_L(t) = L\frac{di}{dt} = v_s - E_d \tag{6-3}$$

In terms of ωt,

$$i(\omega t) = \frac{1}{\omega L}\int_{\theta_2}^{\omega t} [v_s(\zeta) - E_d]\,d\zeta \tag{6-4}$$

where ζ is an arbitrary variable of integration. The current peaks at θ_3 where $v_d = E_d$. The current goes to zero at $\omega t = \theta_4$, at which instant area A_1 equals area A_2, and the time integral of the inductor voltage over one time period of repetition becomes zero.

6-2-2 THYRISTOR GATE TRIGGERING

By controlling the instant at which the thyristor is gated on, the average current in the circuits of Fig. 6-2 can be controlled in a continuous manner from zero to a maximum value. The same is true for the power supplied by the ac source.

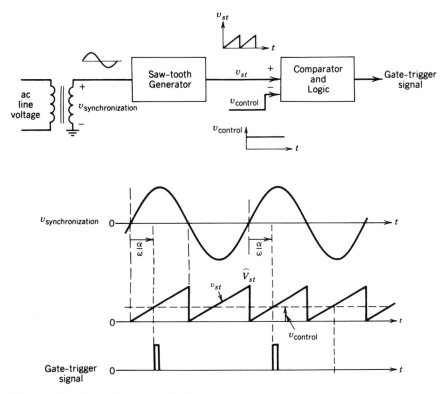

Figure 6-3 Gate trigger control circuit.

Versatile integrated circuits, such as the TCA780, are available to provide delayed gate trigger signals to the thyristors. A simplified block diagram of a gate trigger control circuit is shown in Fig. 6-3. Here, a sawtooth waveform (synchronized to the ac input) is compared with the control signal $v_{control}$, and the delay angle α with respect to the positive zero crossing of the ac line voltage is obtained in terms of $v_{control}$ and the peak of the sawtooth waveform $\hat{V}_{st}$:

$$\alpha° = 180° \frac{v_{control}}{\hat{V}_{st}}$$

(6-5)

Another gate trigger signal can easily be obtained, delayed with respect to the negative zero crossing of the ac line voltage.

6-2-3 PRACTICAL THYRISTOR CONVERTERS

Full-bridge converters for single- and three-phase utility inputs are shown in Fig. 6-4. The dc-side inductance may be a part of the load, for example, in dc motor drives. Prior to the analysis of the full-bridge converters in Fig. 6-4, it will be helpful to analyze some simpler and possibly hypothetical circuits. This simplification is achieved by assuming ac-side inductance to be zero and the dc side current to be purely dc. Next, the effect of L_s on the converter waveforms will be analyzed. Finally, the effect of the ripple in i_d (as well as a discontinuous i_d) will be included. These converters will also be analyzed for their inverter mode of operation.

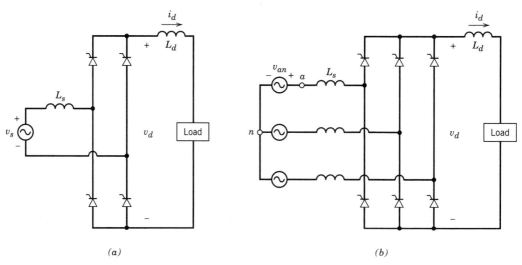

Figure 6-4   Practical thyristor converters.

6-3   SINGLE-PHASE CONVERTERS

6-3-1   IDEALIZED CIRCUIT WITH $L_S = 0$ AND $i_d(t) = I_d$

The practical circuit of Fig. 6-4a, with the assumption of $L_s = 0$ and a purely dc current $i_d(t) = I_d$, is shown in Fig. 6-5a. It can be redrawn as in Fig. 6-5b. The current I_d flows through one thyristor of the top group (thyristors 1 and 3) and one thyristor of the bottom group (thyristors 2 and 4). If the gate currents to the thyristors were continuously applied, the thyristors in Fig. 6-5 would behave as diodes and their operation would be similar to that described in Section 5-3-1 of the previous chapter. The voltage and current waveforms under these conditions are shown in Fig. 6-6a.

The instant of natural conduction for a thyristor refers to the instant at which the thyristor would begin to conduct if its gate current were continuously applied (or as if it was a diode). Therefore, in Fig. 6-6a, the instant of natural conduction is $\omega t = 0$ for thyristors 1 and 2 and $\omega t = \pi$ for thyristors 3 and 4.

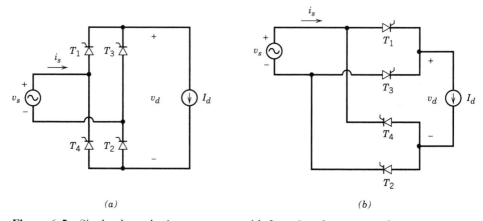

Figure 6-5   Single-phase thyristor converter with $L_s = 0$ and a constant dc current.

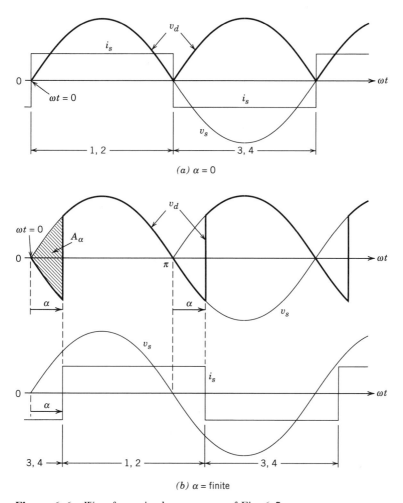

(a) $\alpha = 0$

(b) α = finite

Figure 6-6   Waveforms in the converter of Fig. 6-5.

Next, consider the effect of applying gate current pulses that are delayed by an angle α (called the delay angle or firing angle) with respect to the instant of natural conduction. Now prior to $\omega t = 0$, the current is flowing through thyristors 3 and 4, and $v_d = -v_s$. As shown in Fig. 6-6b, the voltage across thyristor 1 becomes forward biased beyond $\omega t = 0$, but it cannot conduct until $\omega t = \alpha$ when a gate current pulse is applied. The situation is identical for thyristor 2. As a consequence of this finite delay angle α (note that in Fig. 6-6a, $\alpha = 0$), v_d becomes negative during the interval from 0 to α.

At $\omega t = \alpha$, the commutation of current from thyristors 3 and 4 to thyristors 1 and 2 is instantaneous due to the assumption of $L_s = 0$. When thyristors 1 and 2 are conducting, $v_d = v_s$. Thyristors 1 and 2 conduct until $\pi + \alpha$ when thyristors 3 and 4 are triggered, delayed by the angle α with respect to their instant of natural conduction ($\omega t = \pi$). A similar commutation of current takes place from thyristors 1 and 2 to thyristors 3 and 4.

Comparing the effect of the delay angle α on the v_d waveform in Fig. 6-6b with that in Fig. 6-6a shows that the average value V_d of the dc voltage can be controlled by the delay angle. The expression for V_d can be obtained as

$$V_{d\alpha} = \frac{1}{\pi} \int_{\alpha}^{\pi + \alpha} \sqrt{2} V_s \sin \omega t \; d(\omega t) = \frac{2\sqrt{2}}{\pi} V_s \cos \alpha = 0.9 V_s \cos \alpha \qquad (6\text{-}6)$$

Let V_{do} be the average value of the dc voltage in Fig. 6-6a with $\alpha = 0$ (as in Chapter 5) and $L_s = 0$, where

$$V_{do} = \frac{1}{\pi} \int_0^\pi \sqrt{2} V_s \sin \omega t \, d(\omega t) = \frac{2\sqrt{2}}{\pi} V_s = 0.9 V_s \qquad (6\text{-}7)$$

Then, the drop in the average value due to α is

$$\Delta V_{d\alpha} = V_{do} - V_{d\alpha} = 0.9 V_s (1 - \cos \alpha) \qquad (6\text{-}8)$$

This "lossless" voltage drop in V_d is equal to the volt-radian area A_α shown in Fig. 6-6b divided by π.

The variation of V_d as a function of α is shown in Fig. 6-7, which shows that the average dc voltage becomes negative beyond $\alpha = 90°$. This region is called the inverter mode of operation and is discussed later in a separate section.

The average power through the converter can be calculated as

$$P = \frac{1}{T} \int_0^T p(t) \, dt = \frac{1}{T} \int_0^T v_d i_d \, dt \qquad (6\text{-}9)$$

With a constant dc current ($i_d = I_d$),

$$P = I_d \left(\frac{1}{T} \int_0^T v_d \, dt \right) = I_d V_d = 0.9 V_s I_d \cos \alpha \qquad (6\text{-}10)$$

6-3-1-1 dc-Side Voltage

As seen from the waveform of v_d in Fig. 6-6, the dc-side voltage has a dc (average) component $V_{d\alpha}$ ($=0.9 V_s \cos \alpha$ from Eq. 6-6). In addition, v_d has an ac ripple that repeats at twice the line frequency. The magnitude of harmonic components in v_d for various values of α can be calculated by means of Fourier analysis.

6-3-1-2 Line Current i_s

The input line current i_s in Fig. 6-6a is a square wave with an amplitude of I_d. The entire waveform in Fig. 6-6b is phase shifted by the delay angle α with respect to the input

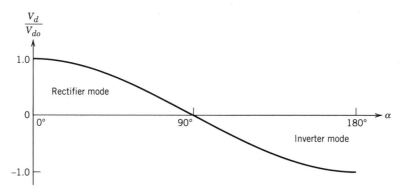

Figure 6-7   Normalized V_d as a function of α.

voltage v_s waveform. The current i_s in Fig. 6-8a can be expressed in terms of its Fourier components as

$$i_s(\omega t) = \sqrt{2}I_{s1}\sin(\omega t - \alpha) + \sqrt{2}I_{s3}\sin[3(\omega t - \alpha)] + \sqrt{2}I_{s5}\sin[5(\omega t - \alpha)] + \cdots$$

(6-11)

where only odd harmonics h are present. The rms value of its fundamental-frequency component i_{s1}, plotted in Fig. 6-8a, is

$$I_{s1} = \frac{2}{\pi}\sqrt{2}I_d = 0.9\,I_d$$

(6-12)

From Fourier analysis, the harmonics of i_s can be expressed as

$$I_{sh} = \frac{I_{s1}}{h}$$

(6-13)

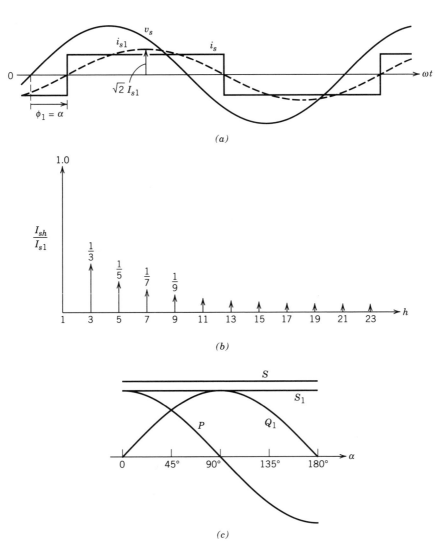

(a)

(b)

(c)

Figure 6-8 The ac-side quantities in the converter of Fig. 6-5.

which are plotted in Fig. 6-8b. By applying the basic definition of rms from Eq. 3-5 to the i_s waveform, the rms value I_s can be shown to be equal to the dc current:

$$I_s = I_d \qquad (6\text{-}14)$$

From Eqs. 6-12 and 6-14, the total harmonic distortion can be calculated as

$$\%\text{THD} = 100 \times \frac{\sqrt{I_s^2 - I_{s1}^2}}{I_{s1}} = 48.43\% \qquad (6\text{-}15)$$

6-3-1-3 Power, Power Factor, and Reactive Volt-Amperes

Looking at the waveform of i_{s1} in Fig. 6-8a, it is obvious that in the circuit of Fig. 6-5,

$$\text{DPF} = \cos \phi_1 = \cos \alpha \qquad (6\text{-}16)$$

and from Eqs. 6-12, 6-14, and 6-16,

$$PF = \frac{I_{s1}}{I_s} \text{DPF} = 0.9 \cos \alpha \qquad (6\text{-}17)$$

Based on the ac-side quantities, the power into the converter is

$$P = V_s I_{s1} \cos \phi_1 \qquad (6\text{-}18)$$

Using Eqs. 6-12 and 6-16 in 6-18 yields

$$P = 0.9 V_s I_d \cos \alpha \qquad (6\text{-}19a)$$

which is identical to the average power given by Eq. 6-10, directly calculated based on the dc-side quantities. The fundamental-frequency current results in fundamental reactive volt-amperes,

$$Q_1 = V_s I_{s1} \sin \phi_1 = 0.9 V_s I_d \sin \alpha \qquad (6\text{-}19b)$$

and the fundamental-frequency apparent power S_1, where

$$S_1 = V_s I_{s1} = (P^2 + Q_1^2)^{1/2} \qquad (6\text{-}19c)$$

With a constant I_d, P, S ($=V_s I_s$), Q_1, and S_1 as functions of α are plotted in Fig. 6-8c.

6-3-2 EFFECT OF L_S

Next, we will include the ac-side inductance in Fig. 6-9, which generally cannot be ignored in practical thyristor converters. Now for a given delay angle α, the current commutation takes a finite commutation interval u, as shown in Fig. 6-10a. In principle,

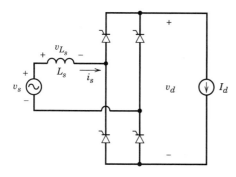

Figure 6-9 Single-phase thyristor converter with a finite L_s and a constant dc current.

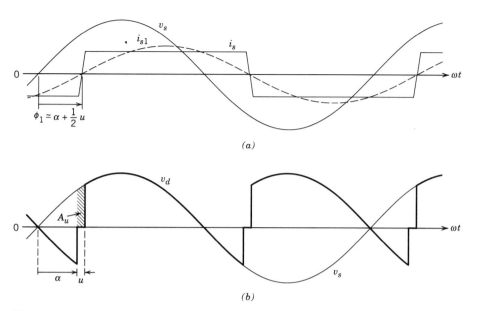

Figure 6-10 Waveforms in the converter of Fig. 6-9.

this commutation process is similar to that in diode bridge rectifiers discussed in Chapter 5. During the commutation interval, all four thyristors conduct, and therefore, $v_d = 0$ and the voltage $v_{L_s} = v_s$ in Fig. 6-9:

$$v_s = v_{L_s} = L_s \frac{di_s}{dt} \tag{6-20}$$

Multiplying both sides by $d(\omega t)$ and integrating over the commutation interval yield

$$\int_\alpha^{\alpha+u} \sqrt{2}V_s \sin \omega t \, d(\omega t) = \omega L_s \int_{-I_d}^{I_d} (di_s) = 2\omega L_s I_d \tag{6-21}$$

The left side of Eq. 6-21 is the area A_u (in volt-radians) in Fig. 6-10b:

$$A_u = \int_\alpha^{\alpha+u} \sqrt{2}V_s \sin \omega t \, d(\omega t) \tag{6-22}$$

Carrying out the integration in Eq. 6-22 and combining with Eq. 6-21 give

$$A_u = \sqrt{2}V_s[\cos \alpha - \cos(\alpha + u)] = 2\omega L_s I_d \tag{6-23}$$

and

$$\cos(\alpha + u) = \cos \alpha - \frac{2\omega L_s I_d}{\sqrt{2}V_s} \tag{6-24}$$

For $\alpha = 0$, Eq. 6-24 is identical to Eq. 5-32 for diode rectifiers. The effect of α on u is shown in Example 6-1.

Comparing the waveforms for v_d in Figs. 6-6b and 6-10b, we note that L_s results in an additional voltage drop ΔV_{du}, proportional to the volt-radian area A_u:

$$\Delta V_{du} = \frac{A_u}{\pi} = \frac{2\omega L_s I_d}{\pi} \tag{6-25}$$

Using Eqs. 6-6 and 6-25 yields

$$V_d = 0.9V_s\cos\alpha - \frac{2}{\pi}\omega L_s I_d \tag{6-26}$$

■ *Example 6-1*   In the converter circuit of Fig. 6-8a, L_s is 5% with the rated voltage of 230 V at 60 Hz and the rated volt-amperes of 5 kVA. Calculate the commutation angle u and V_d/V_{do} with the rate input voltage, power of 3 kW, and $\alpha = 30°$.

Solution

The rated current is

$$I_\text{rated} = \frac{5000}{230} = 21.74 \text{ A}$$

The base impedance is

$$Z_\text{base} = \frac{V_\text{rated}}{I_\text{rated}} = 10.58 \text{ }\Omega$$

Therefore,

$$L_s = \frac{0.05Z_\text{base}}{\omega} = 1.4 \text{ mH}$$

The average power through the converter can be calculated using Eq. 6-26:

$$P_d = V_d I_d = 0.9V_s I_d\cos\alpha - \frac{2}{\pi}\omega L_s I_d^2 = 3 \text{ kW}$$

Using the given values in the above equation gives

$$I_d^2 - 533.53I_d + 8928.6 = 0$$

Therefore,

$$I_d = 17.3 \text{ A}$$

Using this value of I_d in Eqs. 6-24 and 6-26 results in

$$u = 5.9° \quad \text{and} \quad V_d = 173.5 \text{ V}$$

6-3-2-1   Input Line Current i_s

The input current i_s in Fig. 6-10a has an essentially (not exactly) a trapezoidal waveform. With this assumption, the angle ϕ_1 is approximately equal to $\alpha + \frac{1}{2}u$. Therefore,

$$DPF \simeq \cos(\alpha + \tfrac{1}{2}u) \tag{6-27}$$

The rms value of the fundamental-frequency current component can be obtained by equating power on the ac and the dc sides:

$$V_s I_{s1}\text{ }DPF = V_d I_d \tag{6-28}$$

Using Eqs. 6-26 through 6-28 yields

$$I_{s1} \simeq \frac{0.9V_s I_d\cos\alpha - (2/\pi)\text{ }\omega L_s I_d^2}{V_s\cos(\alpha + u/2)} \tag{6-29}$$

The rms value of the line current can be calculated by applying the definition of rms to its trapezoidal waveform. Combining this with the above equations allows the power factor and the total harmonic distortion in the line current to be calculated.

6-3-3 PRACTICAL THYRISTOR CONVERTERS

The circuit of Fig. 6-4a is redrawn in Fig. 6-11a, where the load is represented by a dc voltage source E_d in series with L_d, which may be a part of the load; otherwise it is externally added. A small resistance r_d is also included. Such a representation applies to battery chargers, discussed in Chapter 11, and dc motor drives, discussed in Chapter 13. The waveforms are shown in Fig. 6-11b for $\alpha = 45°$ and a continuously flowing i_d. There is a finite commutation interval u due to L_s. Also due to L_s and the ripple in i_d, the v_d waveform differs from the instantaneous $|v_s(t)|$ waveform by the voltage drop across L_s. It is reasonable to express the average value of v_d in terms of Eq. 6-26 if i_d is flowing continuously:

$$V_d \simeq 0.9 V_s \cos \alpha - \frac{2}{\pi} \omega L_s I_{d,\min} \qquad (6\text{-}30)$$

where $I_{d,\min}$ is the minimum value of i_d that occurs at $\omega t \simeq \alpha$.

To obtain the average value I_d of the dc current in the circuit of Fig. 6-11a,

$$v_d = r_d i_d + L_d \frac{di_d}{dt} + E_d \qquad (6\text{-}31)$$

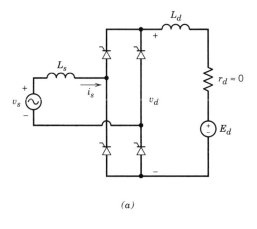

(a)

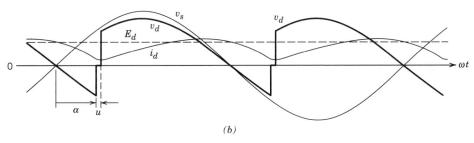

(b)

Figure 6-11 (a) A practical thyristor converter. (b) Waveforms.

Integrating both sides of Eq. 6-31 over one time period T and dividing by T, we get the average voltages:

$$\frac{1}{T}\int_0^T v_d \, dt = \frac{r_d}{T}\int_0^T i_d \, dt + \frac{L_d}{T}\int_{I_d(0)}^{I_d(T)} di_d + E_d \qquad (6\text{-}32)$$

In the steady state the waveforms repeat with the time period T, and hence, $I_d(0) = I_d(T)$. Therefore, the average voltage accross L_d in the steady state in Eq. 6-32 is zero. In terms of average values, Eq. 6-32 can be written as

$$V_d = r_d I_d + E_d \qquad (6\text{-}33)$$

In accordance with Eq. 6-30 we can control the average dc voltage V_d by means of α and thus control I_d and the power delivered to the load. The ac-side current waveforms associated with practical converters are analyzed by means of a computer simulation.

■ *Example 6-2*   The single-phase thyristor converter of Fig. 6-11a is supplied by a 240-V, 60-Hz source. Assume $L_s = 1.4$ mH and the delay angle $\alpha = 45°$. The load can be represented by $L_d = 9$ mH and $E_d = 145$ V. Using PSpice, obtain the v_d and i_s waveforms and calculate I_{s1}, I_s, DPF, PF, and %THD.

Solution

The PSpice circuit and the input data file, where the thyristors are represented by a subcircuit called SCR, are included in the Appendix at the end of this chapter. The thyristor in this subcircuit is modeled by means of a voltage-controlled switch whose current is monitored by means of a 0-V source. The voltage-controlled switch is in its on state if the gate pulse is present and/or if the current is flowing through it. The details of this subcircuit are explained in reference 1. The call to this subcircuit includes the time delay in gate pulse (TDLY) with respect to the instant of natural condition and the initial gate voltage (ICGATE) at the start of the simulation.

The dc-side waveforms of Fig. 6-11b shown earlier were obtained for the above circuit conditions. The input current i_s and i_{s1} waveforms are shown in Fig. 6-12, and the calculated values are as follows: $I_{s1} = 59.68$ A, $I_s = 60.1$ A, DPF = 0.576, PF = 0.572, and THD = 12.3%.

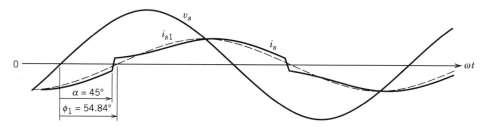

Figure 6-12   Waveforms in Example 6-2 for the circuit of Fig. 6-11a.

6-3-3-1   Discontinuous-Current Conduction

At light loads with low values of I_d, the i_d waveform becomes discontinuous. For example, beyond a certain value of E_d in Example 6-2 with a delay angle of 45°, i_d becomes discontinuous. Figure 6-13 shows the waveforms for $E_d = 180$ V, where the effect of the snubber network on the waveforms is ignored. A larger value of E_d will result in a smaller average value I_d of the dc current.

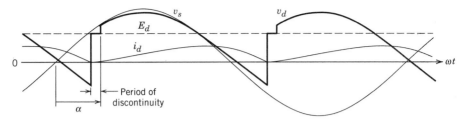

Figure 6-13 Waveforms in a discontinuous-current-conduction mode.

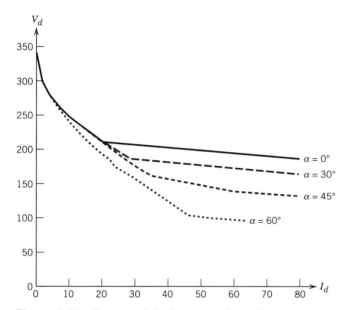

Figure 6-14 V_d versus I_d in the single-phase thyristor converter of Fig. 6-11a.

Using the values given in Example 6-2 for the converter of Fig. 6-11a, the relation between E_d $(=V_d)$ and I_d is plotted in Fig. 6-14 for several values of the delay angle α. This figure shows that at a constant α, if I_d falls below a threshold that depends on α, V_d increases sharply. In order to keep V_d constant, the delay angle will have to be increased at low values of I_d.

6-3-4 INVERTER MODE OF OPERATION

It was mentioned in Section 6-1 that the thyristor converters can also operate in an inverter mode, where V_d has a negative value, as shown in Fig. 6-1b, and hence the power flows from the dc side to the ac side. The easiest way to understand the inverter mode of operation is to assume that the dc side of the converter can be replaced by a current source of a constant amplitude I_d, as shown in Fig. 6-15a. For a delay angle α greater than 90° but less than 180°, the voltage and current waveforms are shown in Fig. 6-15b. The average value of v_d is negative, given by Eq. 6-26, where 90° < α < 180°. Therefore, the average power P_d $(=V_dI_d)$ is negative, that is, it flows from the dc to the ac side. On the ac side, $P_{ac} = V_sI_{s1}\cos\phi_1$ is also negative because $\phi_1 > 90°$.

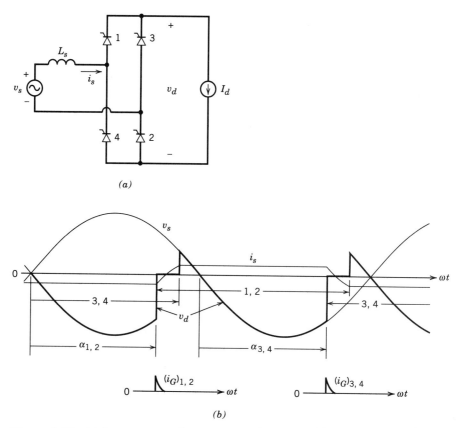

Figure 6-15   (a) Inverter, assuming a constant dc current. (b) Waveforms.

There are several points worth noting here. This inverter mode of operation is possible since there is a source of energy on the dc side. On the ac side, the ac voltage source facilitates the commutation of current from one pair of thyristors to another. The power flows into this ac source.

Generally, the dc current source is not a realistic dc-side representation of systems where such a mode of operation may be encountered. Figure 6-16a shows a voltage source E_d on the dc side that may represent a battery, a photovoltaic source, or a dc voltage produced by a wind-electric system. It may also be encountered in a four-quadrant dc motor supplied by a back-to-back connected thyristor converter.

An assumption of a very large value of L_d allows us to assume i_d to be a constant dc, and hence the waveforms of Fig. 6-15b also apply to the circuit of Fig. 6-16a. Since the average voltage across L_d is zero,

$$E_d = V_d = V_{do}\cos \alpha - \frac{2}{\pi} \omega L_s I_d \qquad (6\text{-}34)$$

The equation is exact if the current is constant at I_d; otherwise, a value of i_d at $\omega t = \alpha$ should be used in Eq. 6-34 instead of I_d. Figure 6-16b shows that for a given value of α, for example, α_1, the intersection of the dc source voltage $E_d = E_{d1}$, and the converter characteristic at α_1 determines the dc current I_{d1} and hence the power flow P_{d1}.

During the inverter mode, the voltage waveform across one of the thyristors is shown in Fig. 6-17. An extinction angle γ is defined to be

$$\gamma = 180° - (\alpha + u) \qquad (6\text{-}35)$$

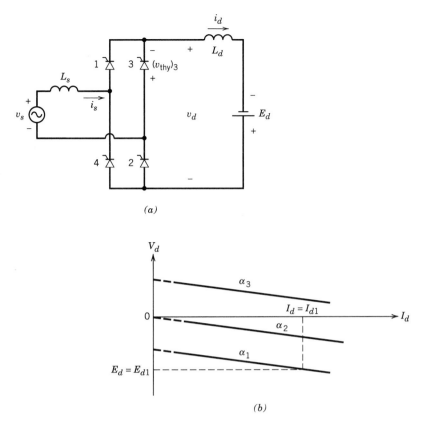

(a)

(b)

Figure 6-16 *(a)* Thyristor inverter with a dc voltage source. *(b)* $\mathbf{V_d}$ versus I_d.

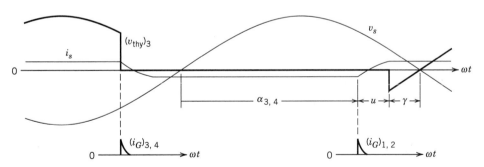

Figure 6-17 Voltage across a thyristor in the inverter mode.

during which the voltage across the thyristor is negative and beyond which it becomes positive. As discussed in Section 2-3 dealing with thyristors, the extinction time interval $t_\gamma = \gamma/\omega$ should be greater than the thyristor turn-off time t_q. Otherwise, the thyristor will prematurely begin to conduct, resulting in the failure of current to commutate from one thyristor pair to the other, an abnormal operation that can result in large destructive currents.

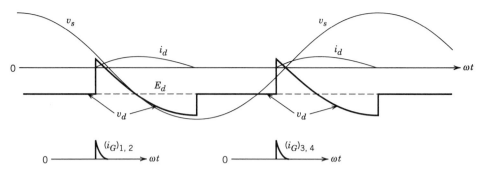

Figure 6-18   Waveforms at inverter start-up.

6-3-4-1   Inverter Start-up

For startup of the inverter in Fig. 6-16a, the delay angle α is initially made sufficiently large (e.g., 165°) so that i_d is discontinuous, as shown in Fig. 6-18. Then, α is decreased by the controller such that the desired I_d and P_d are obtained.

6-3-5   ac VOLTAGE WAVEFORM (LINE NOTCHING AND DISTORTION)

Thyristor converters result in line noise. Two main reasons for line noise are line notching and voltage distortion. Both of these topics are discussed in Section 6-4-5 for three-phase converters. A similar analysis can be carried out for single-phase converters.

6-4   THREE-PHASE CONVERTERS

6-4-1   IDEALIZED CIRCUIT WITH $L_S = 0$ AND $i_d(t) = I_d$

The practical circuit of Fig. 6-4b with the assumption of $L_s = 0$ and a purely dc current $i_d(t) = I_d$ is shown in Fig. 6-19a. It can be redrawn as in Fig. 6-19b. The current i_d flows

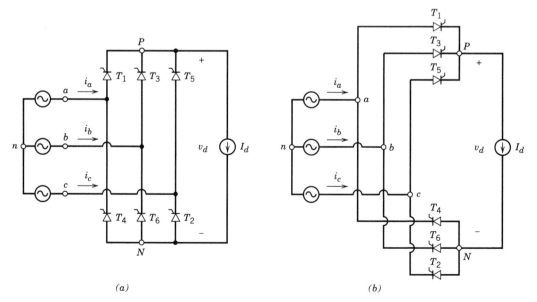

(a)   (b)

Figure 6-19   Three-phase thyristor converter with $L_s = 0$ and a constant dc current.

through one of the thyristors of the top group (thyristors 1, 3, and 5) and one of the bottom group (2, 4, and 6). If the gate currents were continuously applied, the thyristors in Fig. 6-19 would behave as diodes and their operation would be similar to that described in the previous chapter. Under these conditions ($\alpha = 0$ and $L_s = 0$), the voltages and the current in phase a are shown in Fig. 6-20a. The average dc voltage V_{do} is as in Eq. 5-68:

$$V_{do} = \frac{3\sqrt{2}}{\pi} V_{LL} = 1.35 V_{LL} \qquad (6\text{-}36)$$

Using the same definition as in Section 6-3-1, instants of natural conduction for the various thyristors are shown in Fig. 6-20a by 1, 2, ... The effect of the firing or delay angle α on the converter waveforms are shown in Figs. 6-20b through d. Focusing on the commutation of current from thyristor 5 to 1, we see that thyristor 5 keeps on conducting until $\omega t = \alpha$, at which instant the current commutates instantaneously to thyristor 1 due to zero L_s. The current in phase a is shown in Fig. 6-20c. Similar delay by an angle α takes place in the conduction of other thyristors. The line-to-line ac voltages and the dc output voltage v_d ($= v_{Pn} - v_{Nn}$) are shown in Fig. 6-20d.

The expression for the average dc voltage can be obtained from the waveforms in Figs. 6-20b and d. The volt-second area A_α (every 60°) results in the reduction in the average dc voltage with a delay angle α compared to V_{do} in Fig. 6-20a. Therefore,

$$V_{d\alpha} = V_{do} - \frac{A_\alpha}{\pi/3} \qquad (6\text{-}37)$$

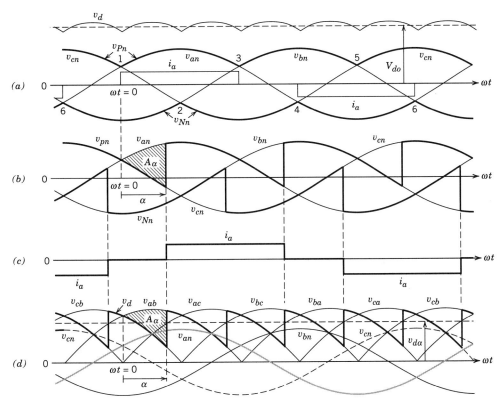

Figure 6-20 Waveforms in the converter of Fig. 6-19.

From Fig. 6-20b, the volt-radian area A_α is the integral of $v_{an} - v_{cn}$ ($=v_{ac}$). This can be confirmed by Fig. 6-20d, where A_α is the integral of $v_{ab} - v_{cb}$ ($=v_{ac}$). With the time origin chosen in Fig. 6-20,

$$v_{ac} = \sqrt{2}V_{LL}\sin \omega t \qquad (6\text{-}38)$$

Therefore,

$$A_\alpha = \int_0^\alpha \sqrt{2}V_{LL}\sin \omega t \, d(\omega t) = \sqrt{2}V_{LL}(1 - \cos \alpha) \qquad (6\text{-}39)$$

Substituting A_α in Eq. 6-37 and using Eq. 6-36 for V_{do} yield

$$V_{d\alpha} = \frac{3\sqrt{2}}{\pi} V_{LL}\cos \alpha = 1.35V_{LL}\cos \alpha = V_{do}\cos \alpha \qquad (6\text{-}40)$$

The above procedure to obtain $v_{d\alpha}$ is straightforward when $\alpha < 60°$. For $\alpha > 60°$ we get the same result but an alternate derivation may be easier (see ref. 2).

Equation 6-40 shows that $V_{d\alpha}$ is independent of the current magnitude I_d so long as i_d flows continuously (and $L_s = 0$). The control of V_d as a function of α is similar to the single-phase case shown by Fig. 6-7. The dc voltage waveform for various values of α is shown in Fig. 6-21. The average power is

$$P = V_d I_d = 1.35V_{LL}I_d\cos \alpha \qquad (6\text{-}41)$$

6-4-1-1   dc-Side Voltage

Each of the dc-side voltage waveforms shown in Fig. 6-21 consists of a dc (average) component $V_{d\alpha}$ ($=1.35V_{LL}\cos \alpha$ given by Eq. 6-40). As seen from Fig. 6-21, the ac ripple in v_d repeats at six times the line frequency. The harmonic components can be obtained by means of Fourier analysis.

6-4-1-2   Input Line Currents i_a, i_b, and i_c

The input currents i_a, i_b, and i_c have rectangular waveforms with an amplitude I_d. The waveform of i_a is phase shifted by the delay angle α in Fig. 6-22a with respect to its waveform in Fig. 6-20a with $\alpha = 0$. It can be expressed in terms of its Fourier components (with ωt defined to be zero at the positive zero crossing of v_{an}) as

$$
\begin{aligned}
i_a(\omega t) = {} & \sqrt{2}I_{s1}\sin(\omega t - \alpha) - \sqrt{2}I_{s5}\sin[5(\omega t - \alpha)] - \sqrt{2}I_{s7}\sin[7(\omega t - \alpha)] \\
& + \sqrt{2}I_{s11}\sin[11(\omega t - \alpha)] + \sqrt{2}I_{s13}\sin[13(\omega t - \alpha)] \\
& - \sqrt{2}I_{s17}\sin[17(\omega t - \alpha)] - \sqrt{2}I_{s19}\sin[19(\omega t - \alpha)] \cdots
\end{aligned} \qquad (6\text{-}42)
$$

where only the nontriplen odd harmonics h are present and

$$h = 6n \pm 1 \qquad (n = 1, 2, \ldots) \qquad (6\text{-}43)$$

The rms value of the fundamental-frequency component is

$$I_{s1} = 0.78I_d \qquad (6\text{-}44)$$

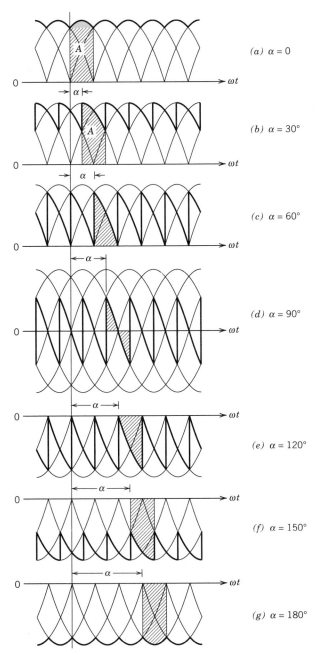

Figure 6-21 The dc-side voltage waveforms as a function of α where $V_{d\alpha} = A/(\pi/3)$. (From ref. 2 with permission.)

and the rms values of the harmonic components are inversely proportional to their harmonic order,

$$I_{sh} = \frac{I_{s1}}{h} \quad \text{where } h = 6n \pm 1 \tag{6-45}$$

as plotted in Fig. 6-22b.

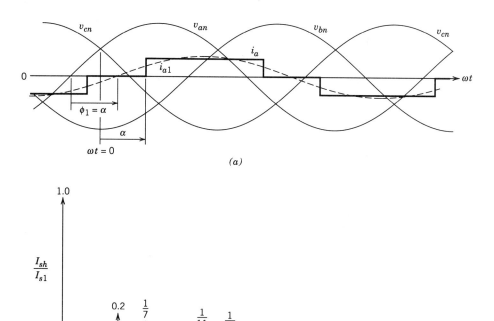

Figure 6-22   Line current in a three-phase thyristor converter of Fig. 6-19.

From the i_a waveform of Fig. 6-22, the total rms value of the phase current can be calculated as

$$I_s = \sqrt{\tfrac{2}{3}}I_d = 0.816 I_d \tag{6-46}$$

Therefore, with $i_d = I_d$ and $L_s = 0$, from Eqs. 6-45 and 6-46

$$\frac{I_{s1}}{I_s} = \frac{3}{\pi} = 0.955 \tag{6-47a}$$

and therefore, in i_s

$$\text{THD} = 31.08\% \tag{6-47b}$$

6-4-1-3 Power, Power Factor, and Reactive Volt-Amperes

With $L_s = 0$, $\phi_1 = \alpha$, as shown in Fig. 6-22a, and

$$\text{DPF} = \cos \phi_1 = \cos \alpha \tag{6-48}$$

Using Eq. 3-44 for the power factor, Eqs. 6-47 and 6-48 yield

$$\text{PF} = \frac{3}{\pi} \cos \alpha \tag{6-49}$$

The current waveforms along with the phasor representation of its fundamental-frequency component are shown in Fig. 6-23 for various values of α. Similar to Eqs. 6-19a through

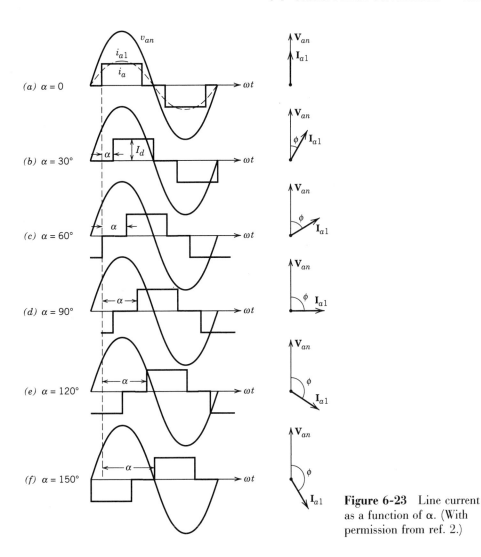

(a) $\alpha = 0$

(b) $\alpha = 30°$

(c) $\alpha = 60°$

(d) $\alpha = 90°$

(e) $\alpha = 120°$

(f) $\alpha = 150°$

Figure 6-23 Line current as a function of α. (With permission from ref. 2.)

6-19c for the single-phase converters, the equations can be written for P, Q_1 and S_1, along with a plot similar to that in Fig. 6-8c, for the three-phase converters.

6-4-2 EFFECT OF L_S

Next, we will include the ac-side inductance L_s in Fig. 6-24, which cannot be ignored in practical thyristor converters. In fact, the German VDE standards require that this inductance must be a minimum of 5%, that is,

$$\omega L_s \geq 0.05 \frac{V_{LL}/\sqrt{3}}{I_{s1}} \tag{6-50}$$

Now for a given delay angle α, the current commutation takes a finite commutation interval u. Consider the situation where thyristors 5 and 6 have been conducting previously, and at $\omega t = \alpha$ the current begins to commutate from thyristor 5 to 1. Only the thyristors involved in current conduction are drawn in Fig. 6-25a. The instant when v_{an} becomes more positive than v_{cn} (the instant of natural conduction for thyristor 1) is chosen as the time origin $\omega t = 0$ in Fig. 6-25b.

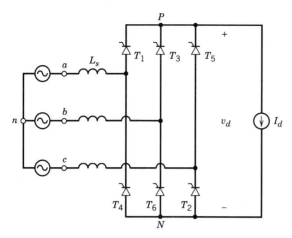

Figure 6-24 Three-phase converter with L_s and a constant dc current.

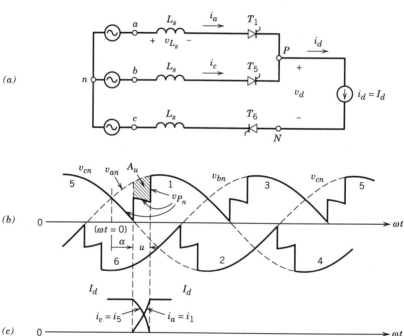

Figure 6-25 Commutation in the presence of L_s.

During the current commutation interval u, thyristors 1 and 5 conduct simultaneously and the phase voltages v_{an} and v_{cn} are shorted together through L_s in each phase. The current i_a builds up from 0 to I_d, whereas i_c decreases from I_d to zero, at which instant the current commutation from 5 to 1 is completed. Currents i_5 and i_1 through thyristors 5 and 1 are drawn in Fig. 6-25c. The complete i_a waveform is shown in Fig. 6-26.

In Fig. 6-25a during the commutation interval $\alpha < \omega t < \alpha + u$

$$v_{Pn} = v_{an} - v_{L_s} \tag{6-51}$$

where

$$v_{L_s} = L_s \frac{di_a}{dt} \tag{6-52}$$

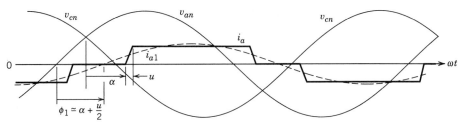

Figure 6-26 Line current in the presence of L_s.

The reduction in volt-radian area due to the commutation interval (in Fig. 6-25*b*) is

$$A_u = \int_\alpha^{\alpha+u} v_{L_s} \, d(\omega t) \tag{6-53}$$

Using Eq. 6-52 in Eq. 6-53 and recognizing that i_a changes from zero to I_d in the interval $\omega t = \alpha$ to $\omega t = \alpha + u$ give

$$A_u = \omega L_s \int_0^{I_d} di_a = \omega L_s I_d \tag{6-54}$$

Therefore, the average dc output voltage is reduced from $V_{d\alpha}$ (given by Eq. 6-40) by $A_u/(\pi/3)$:

$$V_d = \frac{3\sqrt{2}}{\pi} V_{LL}\cos \alpha - \frac{3\omega L_s}{\pi} I_d \tag{6-55}$$

In the foregoing derivation, the following should be noted: During the current commutation, phases a and c are shorted together. Therefore, during commutation

$$v_{Pn} = v_{an} - L_s \frac{di_a}{dt} \tag{6-56}$$

Also

$$v_{Pn} = v_{cn} - L_s \frac{di_c}{dt} \tag{6-57}$$

Therefore, from Eqs. 6-56 and 6-57

$$v_{Pn} \text{ (during commutation)} = \frac{v_{an} + v_{cn}}{2} - \frac{L_s}{2}\left(\frac{di_a}{dt} + \frac{di_c}{dt}\right) \tag{6-58}$$

Since $I_d \,(=i_a + i_c)$ is assumed to be constant during the commutation interval,

$$\frac{di_a}{dt} = -\frac{di_c}{dt} \tag{6-59}$$

Therefore, Eq. 6-58 reduces to

$$v_{Pn} = \tfrac{1}{2}(v_{an} + v_{cn}) \tag{6-60}$$

The v_{Pn} waveform during the commutation interval is shown in Fig. 6-25b. The importance of L_s in the inverter operation is discussed in the next section.

Even though an explicit expression for the commutation interval u is not needed to calculate V_d (see Eq. 6-55), it is required to ensure reliable operation in the inverter mode. Thus, this is the appropriate place to calculate this interval u. Combining Eqs. 6-56 and 6-60 in the circuit of Fig. 6-25a yields

$$L_s \frac{di_a}{dt} = \frac{v_{an}}{2} - \frac{v_{cn}}{2} = \frac{v_{ac}}{2} \tag{6-61}$$

With the time origin chosen in Fig. 6-25b, $v_{ac} = \sqrt{2}V_{LL}\sin \omega t$. Therefore

$$\frac{di_a}{d(\omega t)} = \sqrt{2} \frac{V_{LL}\sin \omega t}{2\omega L_s}$$

Its integration between $\omega t = \alpha$ and $\omega t = \alpha + u$, recognizing that during this interval i_a changes from zero to I_d, results in

$$\int_0^{I_d} di_a = \sqrt{2}\frac{V_{LL}}{2\omega L_s} \int_\alpha^{\alpha+u} \sin \omega t \, d(\omega t)$$

or

$$\cos(\alpha + u) = \cos \alpha - \frac{2\omega L_s}{\sqrt{2}\, V_{LL}} I_d \tag{6-62}$$

Thus, knowing α and I_d, the commutation interval u can be calculated.

6-4-2-1 Input Line Current i_s

Similar to the analysis for the single-phase converters in Section 6-3-2-1, the i_a waveform in Fig. 6-26 can be approximated to be trapezoidal. With this approximation,

$$\text{DPF} \simeq \cos\left(\alpha + \tfrac{1}{2}u\right) \tag{6-63}$$

Another expression for the displacement power factor, by equating ac-side and dc-side powers, is given by Eq. 6-64 [2]. Its derivation is left as an exercise (see Problem 6-11):

$$\text{DPF} \simeq \tfrac{1}{2}[\cos \alpha + \cos(\alpha + u)] \tag{6-64}$$

The ac-side inductance reduces the magnitudes of the harmonic currents. Figures 6-27a through d show the effects of L_s (and hence of u) on various harmonics for various values of α, where I_d is a constant dc. The harmonic currents are normalized by I_1 with $L_s = 0$, which is given by Eq. 6-44. Normally, the dc-side current is not a constant dc. Typical and idealized harmonics are shown in Table 6-1.

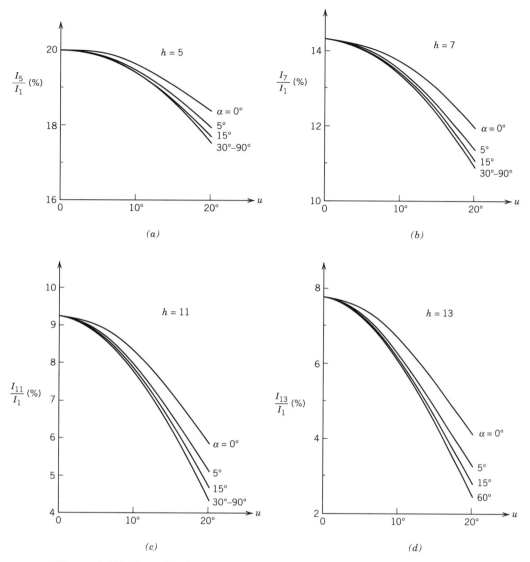

Figure 6-27 Normalized harmonic currents in the presence of L_s. (With permission from ref. 2).

Table 6-1 Typical and Idealized Harmonics

	h	5	7	11	13	17	19	23	25
Typical	I_h/I_1	0.17	0.10	0.04	0.03	0.02	0.01	0.01	0.01
Idealized	I_h/I_1	0.20	0.14	0.09	0.07	0.06	0.05	0.04	0.04

6-4-3 PRACTICAL CONVERTER

The circuit of a converter used in practice was shown in Fig. 6-4b. It is redrawn in Fig. 6-28, where the load is represented by a dc voltage source E_d and L_d is finite. A small resistance r_d is also included. Such practical converters are analyzed by means of a computer simulation in the following example.

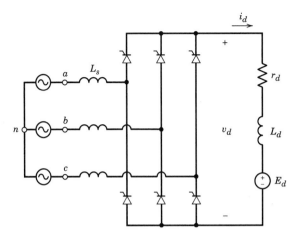

Figure 6-28 A practical thyristor converter.

■ *Example 6-3*   A three-phase thyristor converter is supplied by a 480-V (line-to-line), 60-Hz source. Its internal inductance L_{s1} = 0.2 mH. The converter has a series inductance L_{s2} = 1.0 mH. The load is represented as shown in Fig. 6-28, with L_d = 5 mH, r_d = 0, and E_d = 600 V and the delay angle α = 20°. Using PSpice simulation, obtain v_s, i_s, and v_d waveforms and calculate I_{s1}, I_s, DPF, PF, and %THD.

Solution   The PSpice circuit and the input data file are included in the Appendix at the end of this chapter. The waveforms are shown in Fig. 6-29. The calculated results are as follows: I_{s1} = 22.0 A, I_s = 22.94 A, DPF = 0.928, PF = 0.89, and THD = 29.24%.

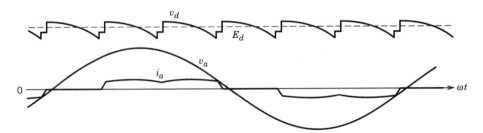

Figure 6-29   Waveforms in the converter of Fig. 6-28.   ■

6-4-3-1 Discontinuous-Current Conduction

As in a single-phase converter, the dc-side current i_d becomes discontinuous in the circuit of Fig. 6-28 below a certain average value for a given α. The waveforms are shown in Fig. 6-30. A larger value of E_d will result in a smaller I_d. In order to regulate V_d, α will have to be increased at lower values of I_d.

6-4-4 INVERTER MODE OF OPERATION

Once again, to understand the inverter mode of operation, we will assume that the dc side of the converter can be represented by a current source of a constant amplitude I_d, as shown in Fig. 6-31. For a delay angle α greater than 90° but less than 180°, the voltage and current waveforms are shown in Fig. 6-32a. The average value of V_d is negative according to Eq. 6-55. On the ac side, the negative power implies that the phase angle ϕ_1 between v_s and i_{s1} is greater than 90°, as shown in Fig. 6-32b.

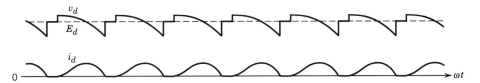

Figure 6-30 Waveforms in a discontinuous-current-conduction mode.

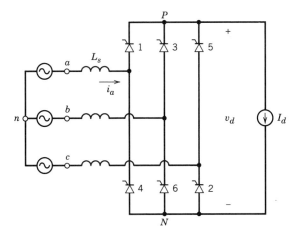

Figure 6-31 Inverter with a constant dc current.

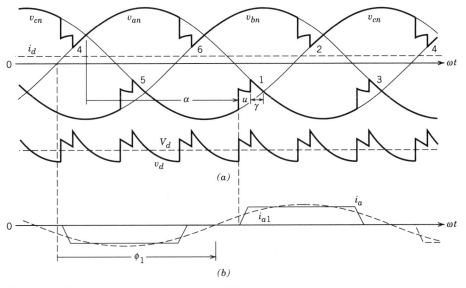

Figure 6-32 Waveforms in the inverter of Fig. 6-31.

In a practical circuit shown in Fig. 6-33a, the operating point for a given E_d and α can be obtained from the characteristics shown in Fig. 6-33b.

Similar to the discussion in connection with single-phase converters, the extinction angle γ ($=180° - \alpha - u$) must be greater than the thyristor turn-off interval ωt_q in the waveforms of Fig. 6-34, where v_5 is the voltage across thyristor 5.

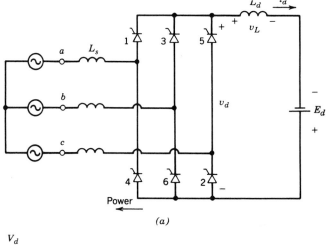

(a)

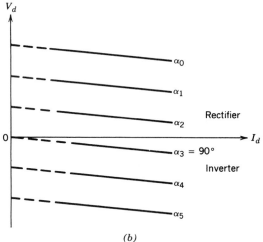

(b)

Figure 6-33 (a) Thyristor inverter with a dc voltage source. (b) V_d versus I_d.

6-4-4-1 Inverter Start-up

As discussed in Section 6-3-4-1 for start up of a single-phase inverter, the delay angle α in the three-phase inverter of Fig. 6-33a is initially made sufficiently large (e.g., 165°) so that i_d is discontinuous. Then, α is decreased by the controller such that the desired I_d and P_d are obtained.

6-4-5 ac VOLTAGE WAVEFORM (LINE NOTCHING AND DISTORTION)

Figure 6-35a shows a practical arrangement with $L_s = L_{s1} + L_{s2}$, where L_{s1} is the per-phase internal inductance of the ac source and L_{s2} is the inductance associated with the converter. The junction of L_{s1} and L_2 is also the point of common coupling where other loads may be connected, as shown in Fig. 6-35a. A thyristor converter results in line noise. Two main reasons for line noise are line notching and waveform distortion, which are considered in the following sections.

6-4-5-1 Line Notching

In the converter of Fig. 6-35a, there are six communications per line-frequency cycle. The converter voltage waveforms are shown in Fig. 6-35b. During each commutation, two out

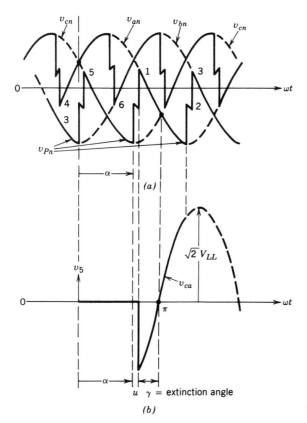

Figure 6-34 Voltage across a thyristor in the inverter mode.

of three phase voltages are shorted together by the converter thyristors through L_s in each phase. Consider a line-to-line voltage, for example, V_{AB}, at the converter internal terminals, as shown in Fig. 6-35c. It is short-circuited twice per cycle, resulting in deep notches. There are four other notches where either phase A or phase B (but not both) is involved in commutation. Ringing due to stray capacitances C_s and snubbers in Fig. 6-35a have been omitted for clarity.

The area for each of the deep notches in Fig. 6-35c, where both phases A and B are shorted, would be twice that of A_u in Fig. 6-25b. Therefore, from Eq. 6-54

$$\text{Deep notch area } A_n = 2\omega L_s I_d \quad [\text{volt radians}] \tag{6-65}$$

The notch width u can be calculated from Eq. 6-62. Assuming that u is small, as an approximation, we may write

$$\text{Deep notch area depth} \simeq \sqrt{2}V_{LL}\sin\alpha \quad (\alpha = \text{delay angle}) \tag{6-66}$$

Therefore, from Eqs. 6-65 and 6-66, considering one of the deep notches yields

$$\text{Notch width } u = \frac{\text{notch area}}{\text{notch depth}} \simeq \frac{2\omega L_s I_d}{\sqrt{2}V_{LL}\sin\alpha} \quad \text{rad} \tag{6-67}$$

Otherwise, Eq. 6-62 can be used to calculate u.

The shallow notches in Fig. 6-35c have the same width u as is calculated for the deep notches in Eq. 6-67. However, the depth and the area for each of the shallow notches is one-half compared with those of the deep notches.

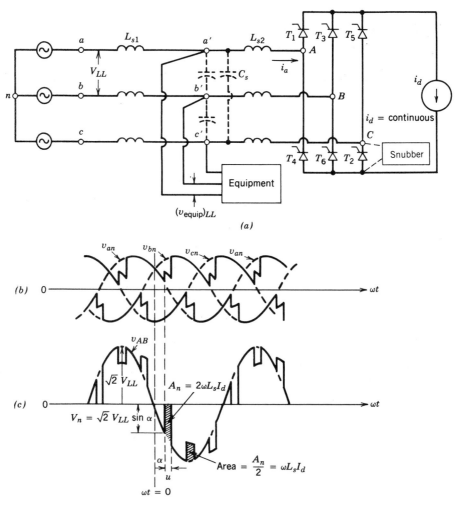

Figure 6-35   Line notching in other equipment voltage: (a) circuit, (b) phase voltages, (c) line-to-line voltage v_{AB}.

In practice, the line notching at the point of common coupling is of concern. The notches in the line-to-line voltage $v_{a'b'}$ have the same width u, as shown in Fig. 6-35c and given by Eq. 6-67. However, the notch depths and the notch areas are a factor ρ times the corresponding depth and areas, respectively, in Fig. 6-35c, where

$$\rho = \frac{L_{s1}}{L_{s1} + L_{s2}} \tag{6-68}$$

Therefore, for a given ac system (i.e., given L_{s1}), a higher value of L_{s2} will result in smaller notches at the point of common coupling. The German VDE standards recommend that ωL_{s2} must be a minimum of 5%, that is, in Fig. 6-35a

$$\omega L_{s2} I_{a1} \geq 0.05 \frac{V_{LL}}{\sqrt{3}} \tag{6-69}$$

In Table 6-2, guidelines suggested by IEEE standard 519-1981 for the line noise at the point of common coupling are given. This standard also suggests that the converter equipment should be capable of performing satisfactorily on supply systems containing

Table 6-2 Line Notching and Distortion Limits for 460-V Systems

Class	Line Notch Depth ρ(%)	Line Notch Area (V·μs)	Voltage Total Harmonic Distortion (%)
Special applications	10	16,400	3
General system	20	22,800	5
Dedicated system	50	36,500	10

line notches of 250 μs width (5.4 electrical degrees) and a notch depth of 70% of the rated maximum line voltage.

Due to the stray capacitances (or if any filter capacitor is used at the input) and snubbers across thyristors, there would be ringing at the end of each commutation interval. These transient voltages can overload the transient suppressors within the equipment.

6-4-5-2   Voltage Distortion

The voltage distortion at the point of common coupling can be calculated by means of phase quantities by knowing the harmonic components I_h of the converter input current and the ac source inductance (L_{s1}):

$$\text{Voltage \%THD} = \frac{\left[\sum_{h \neq 1}(I_h \times \omega L_{s1})^2\right]^{1/2}}{V_{\text{phase(fundamental)}}} \times 100 \qquad (6\text{-}70)$$

The recommended limits for this system are given in Table 6-2.

It should be noted that the total harmonic distortion in the voltage at the point of common coupling can also be calculated by means of the notches in its line-to-line voltage waveform as follows: The total rms value of the harmonic components (other than the fundamental) can be approximately obtained by root-mean-squaring the six notches per cycle in the line-to-line voltage waveform; moreover the fundamental-frequency component of the line-to-line voltage at the point of common coupling can be approximated as V_{LL} (see Problem 6-21).

6-5   OTHER THREE-PHASE CONVERTERS

Only the three-phase, full-bridge, six-pulse converters have been analyzed in detail in the preceding sections. There are several other types of converters: 12-pulse and higher pulse number bridge converters, the 6-pulse bridge converter with a star-connected transformer, the 6-pulse bridge converter with an interphase transformer, half-controlled bridge rectifiers, and so on. The choice of converter topology depends on the application. For example, 12-pulse bridge converters are used for high-voltage dc transmission, as discussed in Chapter 17. A detailed description of these converters is presented in reference 2.

SUMMARY

1. Line-frequency controlled rectifiers and inverters are used for controlled transfer of power between the line-frequency ac and the adjustable-magnitude dc. By controlling the delay angle of thyristors in these converters, a smooth transition can be made from

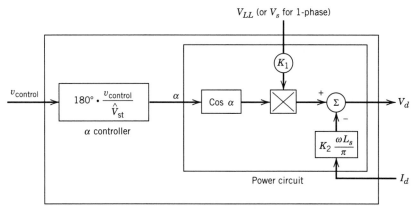

Single-phase full-bridge: $K_1 = 0.9$, $K_2 = 2$
Three-phase full-bridge: $K_1 = 1.35$, $K_2 = 3$

Figure 6-36   Summary of thyristor converter output voltage with a dc current I_d.

the rectification mode to the inversion mode or vice versa. The dc-side voltage can reverse polarity but the dc-side current remains unidirectional.

2. Because of the increasing importance of diode rectifiers, discussed in Chapter 5, the phase-controlled converters are mostly used at high power levels.

3. Phase-controlled converters inject large harmonics into the utility system. At small values of V_d (compared to its maximum possible value), these operate at a very poor power factor as well as a poor displacement power factor. Additionally, these converters produce notches in the line-voltage waveform.

4. The relationship between the control input and the average converter output in the steady state can be summarized as shown in Fig. 6-36. In a dynamic sense, the output of the converter does not respond instantaneously to the change in the control input. This delay, which is a fraction of the line-frequency cycle, is larger in a single-phase converter than in a three-phase converter.

PROBLEMS

BASIC CONCEPTS

6-1   In the circuit of Fig. P6-1, v_{s1} and v_{s2} have an rms value of 120 V at 60 Hz, and the two are 180° out of phase. Assume $L_s = 5$ mH and $I_d = 10$ A is a dc current. For the following two values of the delay angle α, obtain v_{s1}, i_{s1}, and v_d waveforms. Calculate the average value V_d and the commutation interval u at (a) 45° and (b) 135°.

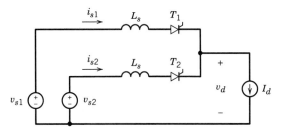

Figure P6-1

6-2 In the circuit of Fig. P6-2, the balanced three-phase voltages v_a, v_b, and v_c have an rms value of 120 V at 60 Hz. Assume $L_s = 5$ mH and $I_d = 10$ A is a dc current. For the following two values of the delay angle α, obtain v_a, i_a, and v_d waveforms. Calculate the average value V_d and the commutation interval u at (a) 45° and (b) 135°.

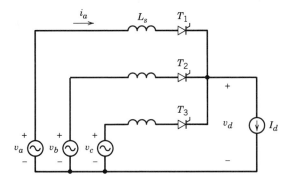

Figure P6-2

6-3 In the single-phase converter of Fig. 6-5, the input voltage has a square waveform with amplitude of 200 V at a frequency of 60 Hz. Assume $I_d = 10$ A. Obtain an analytical expression for V_d in terms of V_s, I_d, and α. Obtain the v_d waveform and its average value V_d for α equal to 45° and 135°.

6-4 In the single-phase converter of Fig. 6-9, the input voltage has a square waveform with amplitude of 200 V at a frequency of 60 Hz. Assume $L_s = 3$ mH and $I_d = 10$ A.

(a) Obtain analytical expressions for u and V_d in terms of V_s, L_s, ω, I_d, and α. Why is u independent of α, unlike in Eq. 6-24.

(b) Obtain the i_s and v_d waveforms and calculate the commutation interval u and V_d for the following values of the delay angle α: 45° and 135°.

SINGLE-PHASE CONVERTERS

6-5 Consider the single-phase, half-controlled converter shown in Fig. P6-5, where v_s is sinusoidal.

(a) Draw v_s, i_s, and v_d waveforms and identify the devices conducting for various intervals for the following values of α: 45°, 90°, and 135°.

(b) Calculate DPF, PF, and %THD for $V_d = \frac{1}{2}V_{do}$, where V_{do} is the dc output at $\alpha = 0$.

(c) Repeat part (b) for a full-bridge converter.

(d) Compare results in parts (b) and (c).

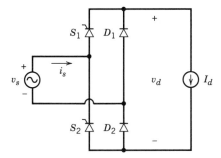

Figure P6-5

6-6 In terms of V_s and I_d in the single-phase converter of Fig. 6-5a, compute the peak inverse voltage and the average and the rms values of the current through each thyristor.

6-7 The single-phase converter of Fig. 6-9 is supplying a dc load of 1 kW. A 1.5-kVA-isolation transformer with a source-side voltage rating of 120 V at 60 Hz is used. It has a total leakage reactance of 8% based on its ratings. The ac source voltage of nominally 115 V is in the range of −10% and +5%. Assume L_d is large enough to allow the assumption of $i_d = I_d$.

Calculate the minimum transformer turns ratio if the dc load voltage is to be regulated at a constant value of 100 V. What is the value of α when $V_s = 115$ V + 5%.

6-8   Equation 6-26 can be expressed in terms of the equivalent circuit shown in Fig. P6-8, where R_u is a ''lossless'' resistor to represent the voltage drop due to I_d. Express this equivalent circuit in terms of the extinction angle γ rather than the delay angle α to represent the inverter mode of operation.

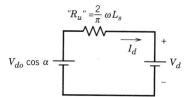

Figure P6-8

6-9   In the single-phase inverter of Fig. 6-16a, $V_s = 120$ V at 60 Hz, $L_s = 1.2$ mH, $L_d = 20$ mH, $E_d = 88$ V, and the delay angle $\alpha = 135°$. Using PSpice, obtain v_s, i_s, v_d, and i_d waveforms in steady state.

6-10 In the inverter of Problem 6-9, vary the delay angle α from a value of 165° down to 120° and plot I_d versus α. Obtain the delay angle α_b below which i_d becomes continuous. How does the slope of the characteristic in this range depend on L_s?

THREE-PHASE CONVERTERS

6-11 In the three-phase converter of Fig. 6-24, derive the expression for the displacement power factor given by Eq. 6-64.

6-12 In the three-phase converter of Fig. 6-24, $V_{LL} = 460$ V at 60 Hz and $L_s = 25$ μH. Calculate the commutation angle u if $V_d = 525$ V and $P_d = 500$ kW.

6-13 In terms of V_{LL} and I_d in the three-phase converter of Fig. 6-19a, compute the peak inverse voltage and the average and the rms values of the current through each thyristor.

6-14 In the three-phase converter of Fig. 6-28, derive the expression for the minimum dc current I_{dB} that results in a continuous-current conduction for given V_{LL}, ω, L_d, and $\alpha = 30°$. Assume that L_s and r_d are negligible and E_d is a dc voltage.

6-15 Consider the three-phase, half-controlled converter shown in Fig. P6-15. Calculate the value of the delay angle α for which $V_d = 0.5 V_{do}$. Draw v_d waveform and identify the devices that conduct during various intervals. Obtain the DPF, PF, and %THD in the input line current and compare results with a full-bridge converter operating at $V_d = 0.5 V_{do}$. Assume $L_s = 0$.

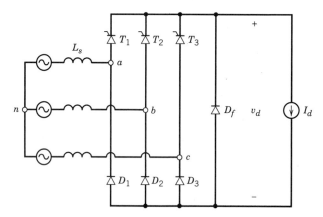

Figure P6-15

6-16 Repeat Problem 6-15 by assuming that diode D_f is not present in the converter of Fig. P6-15.

6-17 The three-phase converter of Fig. 6-24 is supplying a dc load of 12 kW. A Y–Y connected isolation transformer has a per-phase rating of 5 kVA and an ac source-side voltage rating of 120 V at 60 Hz. It has a total per-phase leakage reactance of 8% based on its ratings. The ac source voltage of nominally 208 V (line to line) is in the range of −10% and +5%. Assume L_d is large enough to allow the assumption of $i_d = I_d$.

Calculate the minimum transformer turns ratio if the dc load voltage is to be regulated at a constant value of 300 V. What is the value of α when $V_{LL} = 208$ V +5%.

6-18 In the three-phase inverter of Fig. 6-33a, $V_{LL} = 460$ V at 60 Hz, $E = 550$ V, and $L_s = 0.5$ mH. Assume L_d is very large, to yield $i_d(t) = I_d$. Calculate α and γ if the power flow is 55 kW.

6-19 In typical applications, L_s is finite. Moreover, i_d is not a pure dc current. Table 6-1 lists typical and idealized values of ac-side current harmonics in a six-pulse, full-bridge controlled converter as functions of its fundamental current component.

Calculate the ratio I_1/I and the THD in the current for typical as well as idealized harmonics.

6-20 In the three-phase converter of Fig. 6-19, assume that the input ac voltages and the dc current I_d remain constant. Plot the locus of the reactive volt-amperes due to the fundamental-frequency component of the line current versus the real power for various values of the delay angle α.

6-21 In the circuit of Fig. 6-35a, L_{s1} corresponds to the leakage inductance of a 60-Hz transformer with the following ratings: three-phase kVA rating of 500 kVA, line-to-line voltage of 480 V, and an impedance of 6%. Assume L_{s2} is due to a 200-ft-long cable, with a per-phase inductance of 0.1 μH/ft. The ac input voltage is 460 V line to line and the dc side of the rectifier is delivering 25 kW at a voltage of 525 V.

Calculate the notch width in microseconds and the line notch depth ρ in percentage at the point of common coupling. Also, calculate the area for a deep line notch at the point of common coupling in volt-microseconds and compare the answers with the recommended limits in Table 6-2.

6-22 Repeat Problem 6-21 if a 480-V 1 : 1 transformer is also used at the input to the rectifier, which has a leakage impedance of 3%. The three-phase rating of the transformer equals 40 kVA.

6-23 Calculate the THD in the voltage at the point of common coupling in Problems 6-21 and 6-22.

6-24 Using the typical harmonics in the input current given in Table 6-1, obtain the THD in the voltage at the point of common coupling in Problem 6-21.

6-25 With the parameters for the converter in Example 6-3, use the PSpice listing of the Appendix at the end of this chapter to evaluate the voltage distortion at the point of common coupling.

REFERENCES

1. N. Mohan, "Power Electronics: Computer Simulation, Analysis, and Education Using the Evaluation Version of PSpice," Minnesota Power Electronics Research and Education, P.O. Box 14503, Minneapolis, MN 55414.

2. E. W. Kimbark, *Direct Current Transmission,* Vol. 1, Wiley-Interscience, New York, 1971.

3. B. M. Bird and K. G. King, *An Introduction to Power Electronics,* Wiley, New York, 1983.

4. Institute of Electrical and Electronics Engineers, "IEEE Guide for Harmonic Control and Reactive Compensation of State Power Converters," ANSI/IEEE Standard 519-1981, IEEE, New York.

5. D. A. Jarc and R. G. Schieman, "Power Line Considerations for Variable Frequency Drives," *IEEE Transactions on Industry Applications,* Vol. IAS, No. 5, September/October 1985, pp. 1099–1105.

6. Institute of Electrical and Electronics Engineers, "IEEE Standard Practice and Requirements for General Purpose Thyristor DC Drives," IEEE Standard 597-1983, IEEE, New York.

7. M. Grotzbach, W. Frankenberg, "Injected Currents of Controlled AC/DC Converters for Harmonic Analysis in Industrial Power Plants," *Proceedings of the IEEE International Conference on Harmonics in Power Systems,* September 1992, Atlanta, GA, pp. 107–113.

8. N. G. Hingorani, J. L. Hays and R. E. Crosbie, "Dynamic Simulation of HVDC Transmission Systems on Digital Computers," *Proceedings of the IEEE,* Vol. 113, No. 5, May 1966, pp. 793–802.

APPENDIX

(PSpice examples are adapted from ''Power Electronics: Computer Simulation, Analysis and Education Using Evaluation Version of PSpice,'' Minnesota Power Electronics, P.O. Box 14503, Minneapolis, MN 55414.)

PSPICE INPUT CIRCUIT FILE FOR EXAMPLE 6-2

```
* Single-Phase, Thyristor-Bridge Rectifier
.PARAM PERIOD = {1/60}, ALFA= 45.0, PULSE_WIDTH=0.5ms
.PARAM HALF_PERIOD = {1/120}
*
LS1    1   2   0.1mH   IC = 10A
LS2    2   3   1.3mH   IC = 10A
LD     4   5   9mH
VD     5   6   145V
*
XTHY1  3   4   SCR   PARAMS:   TDLY=0               ICGATE=2V
XTHY3  0   4   SCR   PARAMS:   TDLY={HALF_PERIOD}   ICGATE=0V
XTHY2  6   0   SCR   PARAMS:   TDLY=0               ICGATE=2V
XTHY4  6   3   SCR   PARAMS:   TDLY={HALF_PERIOD}   ICGATE=0V
*
VS     1   0   SIN(0 340V 60 0 0 {ALFA})
*
.TRAN    50us    100ms    0    50us    UIC
.PROBE
.FOUR 60.0 v(1) i(ls1) i(ld)

.SUBCKT SCR 101 103 PARAMS: TDLY=1ms ICGATE=0V
* Power Electronics: Simulation, Analysis  Education.....by N. Mohan.
SW       101 102   53 0 SWITCH
VSENSE   102 103   0V
RSNUB    101 104   200
CSNUB    104 103   1uF
*
VGATE    51  0    PULSE(0 1V {TDLY} 0 0 {PULSE_WIDTH} {PERIOD})
RGATE    51  0    1MEG
EGATE    52  0    TABLE {I(VSENSE)+V(51)} = (0.0,0.0) (0.1,1.0) (1.0,1.0)
RSER     52  53   1
CSER     53  0    1uF  IC={ICGATE}
*
.MODEL   SWITCH VSWITCH ( RON=0.01 )
.ENDS

.END
```

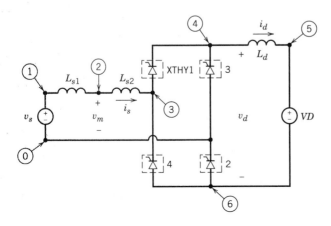

Figure 6A-1 Spice Input Circuit for Example 6-2.

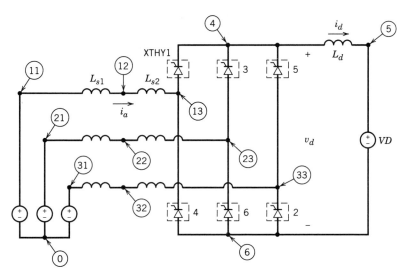

Figure 6A-2   PSpice Input Circuit for Example 6-3.

PSPICE INPUT CIRCUIT FILE FOR EXAMPLE 6-3

```
Example 6-3
* Three-Phase, Thyristor-Bridge Rectifier
.PARAM PERIOD= {1/60}, DEG120= {1/(3*60)}
.PARAM ALFA= 20.0, PULSE_WIDTH=0.5ms
*
LS1A      11    12    0.2mH      IC=45A
LS2A      12    13    1.0mH      IC=45A
LS1B      21    22    0.2mH      IC=-45A
LS2B      22    23    1.0mH      IC=-45A
LS1C      31    32    0.2mH
LS2C      32    33    1.0mH
*
LD        4     5     5mH        IC=45A
VD        5     6     600.0V
*
XTHY1     13    4     SCR     PARAMS: TDLY=0              ICGATE=2V
XTHY3     23    4     SCR     PARAMS: TDLY={DEG120}       ICGATE=0V
XTHY5     33    4     SCR     PARAMS: TDLY={2*DEG120}     ICGATE=0V
XTHY2     6     33    SCR     PARAMS: TDLY={DEG120/2}     ICGATE=0V
XTHY4     6     13    SCR     PARAMS: TDLY={3*DEG120/2}   ICGATE=0V
XTHY6     6     23    SCR     PARAMS: TDLY={5*DEG120/2}   ICGATE=2V
*
VSA       11    0     SIN(0 391.9V 60 0 0 {30+ALFA})
VSB       21    0     SIN(0 391.9V 60 0 0 {-90+ALFA})
VSC       31    0     SIN(0 391.9V 60 0 0 {-210+ALFA})
*
.TRAN     50us      50ms      0s      50us      UIC
.PROBE
.FOUR     60.0   v(11)  i(LS1A)  i(LD)
```

```
.SUBCKT SCR 101 103 PARAMS: TDLY=1ms ICGATE=0V
* Power Electronics: Simulation, Analysis  Education.....by N. Mohan.
SW        101 102   53 0 SWITCH
VSENSE    102 103   0V
RSNUB     101 104   200
CSNUB     104 103   1uF
*
VGATE     51  0     PULSE(0 1V {TDLY} 0 0 {PULSE_WIDTH} {PERIOD})
RGATE     51  0     1MEG
EGATE     52  0     TABLE {I(VSENSE)+V(51)} = (0.0,0.0) (0.1,1.0) (1.0,1.0)
RSER      52  53 1
CSER      53  0     1uF  IC={ICGATE}
*
.MODEL    SWITCH VSWITCH ( RON=0.01 )
.ENDS

.END
```

CHAPTER 7

dc–dc SWITCH-MODE CONVERTERS

7-1 INTRODUCTION

The dc–dc converters are widely used in regulated switch-mode dc power supplies and in dc motor drive applications. As shown in Fig. 7-1, often the input to these converters is an unregulated dc voltage, which is obtained by rectifying the line voltage, and therefore it will fluctuate due to changes in the line-voltage magnitude. Switch-mode dc-to-dc converters are used to convert the unregulated dc input into a controlled dc output at a desired voltage level.

Looking ahead to the application of these converters, we find that these converters are very often used with an electrical isolation transformer in the switch-mode dc power supplies and almost always without an isolation transformer in case of dc motor drives. Therefore, to discuss these circuits in a generic manner, only the nonisolated converters are considered in this chapter, since the electrical isolation is an added modification.

The following dc–dc converters are discussed in this chapter:

1. Step-down (buck) converter
2. Step-up (boost) converter
3. Step-down/step-up (buck–boost) converter
4. Cúk converter
5. Full-bridge converter

Of these five converters, only the step-down and the step-up are the basic converter topologies. Both the buck–boost and the Cúk converters are combinations of the two basic topologies. The full-bridge converter is derived from the step-down converter.

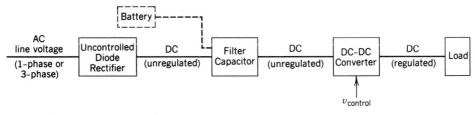

Figure 7-1 A dc–dc converter system.

The converters listed are discussed in detail in this chapter. Their variations, as they apply to specific applications, are described in the chapters dealing with switch-mode dc power supplies and dc motor drives.

In this chapter, the converters are analyzed in steady state. The switches are treated as being ideal, and the losses in the inductive and the capacitive elements are neglected. Such losses can limit the operational capacity of some of these converters and are discussed separately.

The dc input voltage to the converters is assumed to have zero internal impedance. It could be a battery source; however, in most cases, the input is a diode rectified ac line voltage (as is discussed in Chapter 5) with a large filter capacitance, as shown in Fig. 7-1 to provide a low internal impedance and a low-ripple dc voltage source.

In the output stage of the converter, a small filter is treated as an integral part of the dc-to-dc converter. The output is assumed to supply a load that can be represented by an equivalent resistance, as is usually the case in switch-mode dc power supplies. A dc motor load (the other application of these converters) can be represented by a dc voltage in series with the motor winding resistance and inductance.

7-2 CONTROL OF dc–dc CONVERTERS

In dc–dc converters, the average dc output voltage must be controlled to equal a desired level, though the input voltage and the output load may fluctuate. Switch-mode dc–dc converters utilize one or more switches to transform dc from one level to another. In a dc–dc converter with a given input voltage, the average output voltage is controlled by controlling the switch on and off durations (t_{on} and t_{off}). To illustrate the switch-mode conversion concept, consider a basic dc–dc converter shown in Fig. 7-2a. The average value V_o of the output voltage v_o in Fig. 7-2b depends on t_{on} and t_{off}. One of the methods for controlling the output voltage employs switching at a constant frequency (hence, a constant switching time period $T_s = t_{on} + t_{off}$) and adjusting the on duration of the switch to control the average output voltage. In this method, called *pulse-width modulation* (PWM) switching, the switch duty ratio D, which is defined as the ratio of the on duration to the switching time period, is varied.

The other control method is more general, where both the switching frequency (and hence the time period) and the on duration of the switch are varied. This method is used only in dc–dc converters utilizing force-commutated thyristors and therefore will not be discussed in this book. Variation in the switching frequency makes it difficult to filter the ripple components in the input and the output waveforms of the converter.

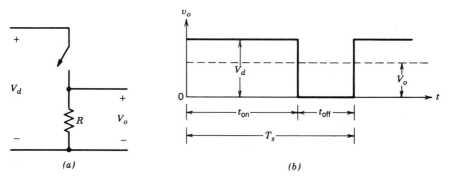

Figure 7-2 Switch-mode dc–dc conversion.

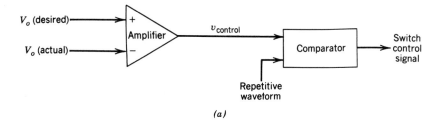

(a)

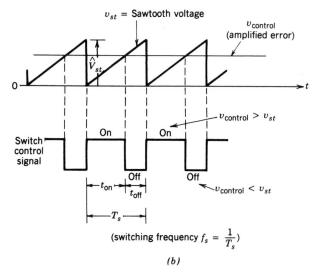

(b)

Figure 7-3 Pulse-width modulator: *(a)* block diagram; *(b)* comparator signals.

In the PWM switching at a constant switching frequency, the switch control signal, which controls the state (on or off) of the switch, is generated by comparing a signal-level control voltage $v_{control}$ with a repetitive waveform as shown in Figs. 7-3a and 7-3b. The control voltage signal generally is obtained by amplifying the error, or the difference between the actual output voltage and its desired value. The frequency of the repetitive waveform with a constant peak, which is shown to be a sawtooth, establishes the switching frequency. This frequency is kept constant in a PWM control and is chosen to be in a few kilohertz to a few hundred kilohertz range. When the amplified error signal, which varies very slowly with time relative to the switching frequency, is greater than the sawtooth waveform, the switch control signal becomes high, causing the switch to turn on. Otherwise, the switch is off. In terms of $v_{control}$ and the peak of the sawtooth waveform $\hat{V}_{st}$ in Fig. 7-3, the switch duty ratio can be expressed as

$$D = \frac{t_{on}}{T_s} = \frac{v_{control}}{\hat{V}_{st}} \qquad (7\text{-}1)$$

The dc–dc converters can have two distinct modes of operation: (1) continuous current conduction and (2) discontinuous current conduction. In practice, a converter may operate in both modes, which have significantly different characteristics. Therefore, a converter and its control should be designed based on both modes of operation.

7-3   STEP-DOWN (BUCK) CONVERTER

As the name implies, a step-down converter produces a lower average output voltage than the dc input voltage V_d. Its main application is in regulated dc power supplies and dc motor speed control.

Conceptually, the basic circuit of Fig. 7-2a constitutes a step-down converter for a purely resistive load. Assuming an ideal switch, a constant instantaneous input voltage V_d, and a purely resistive load, the instantaneous output voltage waveform is shown in Fig. 7-2b as a function of the switch position. The average output voltage can be calculated in terms of the switch duty ratio:

$$V_o = \frac{1}{T_s} \int_0^{T_s} v_o(t)\, dt = \frac{1}{T_s} \left(\int_0^{t_{on}} V_d\, dt + \int_{t_{on}}^{T_s} 0\, dt \right) = \frac{t_{on}}{T_s} V_d = D V_d \qquad (7\text{-}2)$$

Substituting for D in Eq. 7-2 from Eq. 7-1 yields

$$V_o = \frac{V_d}{\hat{V}_{st}} v_{control} = k v_{control}$$

where

$$k = \frac{V_d}{\hat{V}_{st}} = \text{constant}$$

By varying the duty ratio t_{on}/T_s of the switch, V_o can be controlled. Another important observation is that the average output voltage V_o varies linearly with the control voltage, as is the case in linear amplifiers. In actual applications, the foregoing circuit has two drawbacks: (1) In practice the load would be inductive. Even with a resistive load, there would always be certain associated stray inductance. This means that the switch would have to absorb (or dissipate) the inductive energy and therefore it may be destroyed. (2) The output voltage fluctuates between zero and V_d, which is not acceptable in most applications. The problem of stored inductive energy is overcome by using a diode as shown in Fig. 7-4a. The output voltage fluctuations are very much diminished by using a low-pass filter, consisting of an inductor and a capacitor. Figure 7-4b shows the waveform of the input v_{oi} to the low-pass filter (same as the output voltage in Fig. 7-2b without a low-pass filter), which consists of a dc component V_o, and the harmonics at the switching frequency f_s and its multiples, as shown in Fig. 7-4b. The low-pass filter characteristic with the damping provided by the load resistor R is shown in Fig. 7-4c. The corner frequency f_c of this low-pass filter is selected to be much lower than the switching frequency, thus essentially eliminating the switching frequency ripple in the output voltage.

During the interval when the switch is on, the diode in Fig. 7-4a becomes reverse biased and the input provides energy to the load as well as to the inductor. During the interval when the switch is off, the inductor current flows through the diode, transferring some of its stored energy to the load.

In the steady-state analysis presented here, the filter capacitor at the output is assumed to be very large, as is normally the case in applications requiring a nearly constant instantaneous output voltage $v_o(t) \simeq V_o$. The ripple in the capacitor voltage (output voltage) is calculated later.

From Fig. 7-4a we observe that in a step-down converter, the average inductor current is equal to the average output current I_o, since the average capacitor current in steady state is zero (as discussed in Chapter 3, Section 3-2-5-1).

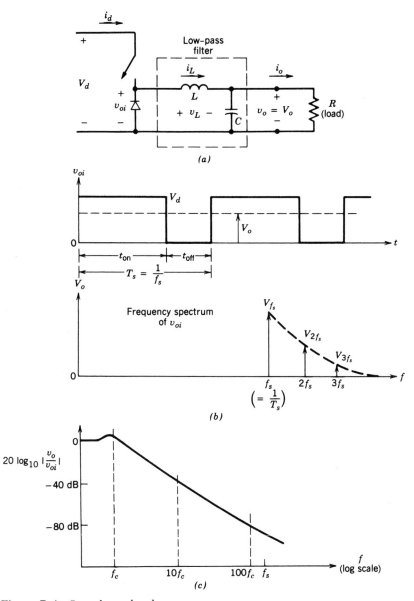

Figure 7-4 Step-down dc–dc converter.

7-3-1 CONTINUOUS-CONDUCTION MODE

Figure 7-5 shows the waveforms for the continuous-conduction mode of operation where the inductor current flows continuously [$i_L(t) > 0$]. When the switch is on for a time duration t_{on}, the switch conducts the inductor current and the diode becomes reverse biased. This results in a positive voltage $v_L = V_d - V_o$ across the inductor in Fig. 7-5a. This voltage causes a linear increase in the inductor current i_L. When the switch is turned off, because of the inductive energy storage, i_L continues to flow. This current now flows through the diode, and $v_L = -V_o$ in Fig. 7-5b.

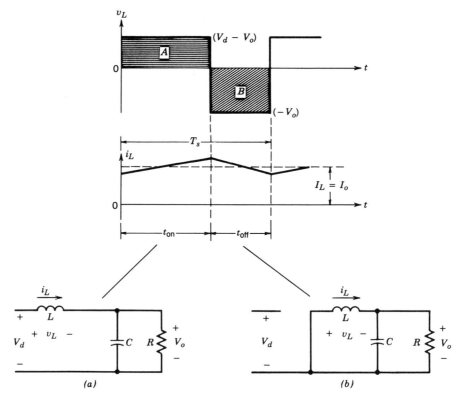

Figure 7-5   Step-down converter circuit states (assuming i_L flows continuously): (a) switch on; (b) switch off.

Since in steady-state operation the waveform must repeat from one time period to the next, the integral of the inductor voltage v_L over one time period must be zero, as discussed in Chapter 3 (Eq. 3-51), where $T_s = t_{on} + t_{off}$:

$$\int_0^{T_s} v_L \, dt = \int_0^{t_{on}} v_L \, dt + \int_{t_{on}}^{T_s} v_L \, dt = 0$$

In Fig. 7-5, the foregoing equation implies that the areas A and B must be equal. Therefore,

$$(V_d - V_o)t_{on} = V_o(T_s - t_{on})$$

or

$$\frac{V_o}{V_d} = \frac{t_{on}}{T_s} = D \quad \text{(duty ratio)} \tag{7-3}$$

Therefore, in this mode, the voltage output varies linearly with the duty ratio of the switch for a given input voltage. It does not depend on any other circuit parameter. The foregoing equation can also be derived by simply averaging the voltage v_{oi} in Fig. 7-4b and recognizing that the average voltage across the inductor in steady-state operation is zero:

$$\frac{V_d t_{on} + 0 \cdot t_{off}}{T_s} = V_o$$

or

$$\frac{V_o}{V_d} = \frac{t_{on}}{T_s} = D$$

Neglecting power losses associated with all the circuit elements, the input power P_d equals the output power P_o:

$$P_d = P_o$$

Therefore,

$$V_d I_d = V_o I_o$$

and

$$\frac{I_o}{I_d} = \frac{V_d}{V_o} = \frac{1}{D} \tag{7-4}$$

Therefore, in the continuous-conduction mode, the step-down converter is equivalent to a dc transformer where the turns ratio of this equivalent transformer can be continuously controlled electronically in a range of 0–1 by controlling the duty ratio of the switch.

We observe that even though the average input current I_d follows the transformer relationship, the instantaneous input current waveform jumps from a peak value to zero every time the switch is turned off. An appropriate filter at the input may be required to eliminate the undesirable effects of the current harmonics.

7-3-2 BOUNDARY BETWEEN CONTINUOUS AND DISCONTINUOUS CONDUCTION

In this section, we will develop equations that show the influence of various circuit parameters on the conduction mode of the inductor current (continuous or discontinuous). At the edge of the continuous-current-conduction mode, Fig. 7-6a shows the waveforms for v_L and i_L. Being at the boundary between the continuous and the discontinuous mode, by definition, the inductor current i_L goes to zero at the end of the off period.

At this boundary, the average inductor current, where the subscript B refers to the boundary, is

$$I_{LB} = \frac{1}{2} i_{L,peak} = \frac{t_{on}}{2L} (V_d - V_o) = \frac{DT_s}{2L} (V_d - V_o) = I_{oB} \tag{7-5}$$

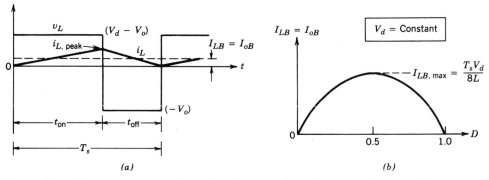

Figure 7-6 Current at the boundary of continuous–discontinuous conduction: (a) current waveform; (b) I_{LB} versus D keeping V_d constant.

Therefore, during an operating condition (with a given set of values for T_s, V_d, V_o, L, and D), if the average output current (and, hence, the average inductor current) becomes less than I_{LB} given by Eq. 7-5, then i_L will become discontinuous.

7-3-3   DISCONTINUOUS-CONDUCTION MODE

Depending on the application of these converters, either the input voltage V_d or the output voltage V_o remains constant during the converter operation. Both of these types of operation are discussed below.

7-3-3-1   Discontinuous-Conduction Mode with Constant V_d

In an application such as a dc motor speed control, V_d remains essentially constant and V_o is controlled by adjusting the converter duty ratio D.

Since $V_o = DV_d$, the average inductor current at the edge of the continuous-conduction mode from Eq. 7-5 is

$$I_{LB} = \frac{T_s V_d}{2L} D(1 - D) \tag{7-6}$$

Using this equation, we find that Fig. 7-6b shows the plot of I_{LB} as a function of the duty ratio D, keeping V_d and all other parameters constant. It shows that the output current required for a continuous-conduction mode is maximum at $D = 0.5$:

$$I_{LB,\text{max}} = \frac{T_s V_d}{8L} \tag{7-7}$$

From Eqs. 7-6 and 7-7

$$I_{LB} = 4I_{LB,\text{max}} D(1 - D) \tag{7-8}$$

Next the voltage ratio V_o/V_d will be calculated in the discontinuous mode. Let us assume that initially the converter is operating at the edge of continuous conduction, as in Fig. 7-6a, for given values of T, L, V_d, and D. If these parameters are kept constant and the output load power is decreased (i.e., the load resistance goes up), then the average inductor current will decrease. As is shown in Fig. 7-7, this dictates a higher value of V_o than before and results in a discontinuous inductor current.

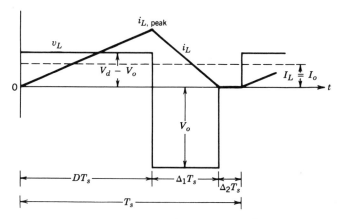

Figure 7-7   Discontinuous conduction in step-down converter.

During the interval $\Delta_2 T_s$ where the inductor current is zero, the power to the load resistance is supplied by the filter capacitor alone. The inductor voltage v_L during this interval is zero. Again, equating the integral of the inductor voltage over one time period to zero yields

$$(V_d - V_o) DT_s + (-V_o)\Delta_1 T_s = 0 \tag{7-9}$$

$$\therefore \frac{V_o}{V_d} = \frac{D}{D + \Delta_1} \cdot \tag{7-10}$$

where $D + \Delta_1 < 1.0$. From Fig. 7-7,

$$i_{L,\text{peak}} = \frac{V_o}{L}\Delta_1 T_s \tag{7-11}$$

Therefore,

$$I_o = i_{L,\text{peak}} \frac{D + \Delta_1}{2} \tag{7-12}$$

$$= \frac{V_o T_s}{2L}(D + \Delta_1)\,\Delta_1 \quad \text{(using Eq. 7-11)} \tag{7-13}$$

$$= \frac{V_d T_s}{2L}D\Delta_1 \quad \text{(using Eq. 7-10)} \tag{7-14}$$

$$= 4I_{LB,\text{max}}D\Delta_1 \quad \text{(using Eq. 7-7)} \tag{7-15}$$

$$\therefore \Delta_1 = \frac{I_o}{4I_{LB,\text{max}}D} \tag{7-16}$$

From Eqs. 7-10 and 7-16

$$\frac{V_o}{V_d} = \frac{D^2}{D^2 + \frac{1}{4}\,(I_o/I_{LB,\text{max}})} \tag{7-17}$$

Figure 7-8 shows the step-down converter characteristic in both modes of operation for a constant V_d. The voltage ratio (V_o/V_d) is plotted as a function of $I_o/I_{LB,\text{max}}$ for various values of duty ratio using Eqs. 7-3 and 7-17. The boundary between the continuous and the discontinuous mode, shown by the dashed curve, is established by Eq. 7-3 and 7-8.

7-3-3-2 Discontinuous-Conduction Mode with Constant V_o

In applications such as regulated dc power supplies, V_d may fluctuate but V_o is kept constant by adjusting the duty ratio D.

Since $V_d = V_o/D$, the average inductor current at the edge of the continuous-conduction mode from Eq. 7-5 is

$$I_{LB} = \frac{T_s V_o}{2L}(1 - D) \tag{7-18}$$

Equation 7-18 shows that if V_o is kept constant, the maximum value of I_{LB} occurs at $D = 0$:

$$I_{LB,\text{max}} = \frac{T_s V_o}{2L} \tag{7-19}$$

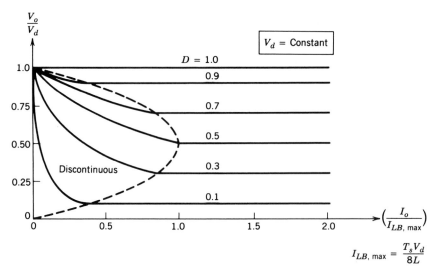

Figure 7-8   Step-down converter characteristics keeping V_d constant.

It should be noted that the operation corresponding to $D = 0$ and a finite V_o is, of course, hypothetical because it would require V_d to be infinite.

From Eqs. 7-18 and 7-19

$$I_{LB} = (1 - D)I_{LB,\max} \tag{7-20}$$

For the converter operation where V_o is kept constant, it will be useful to obtain the required duty ratio D as a function of $I_o/I_{LB,\max}$. Using Eqs. 7-10 and 7-13 (which are valid in the discontinuous-conduction mode whether V_o or V_d is kept constant) along with Eq. 7-19 for the case where V_o is kept constant yields

$$D = \frac{V_o}{V_d}\left(\frac{I_o/I_{LB,\max}}{1 - V_o/V_d}\right)^{1/2} \tag{7-21}$$

The duty ratio D as a function of $I_o/I_{LB,\max}$ is plotted in Fig. 7-9 for various values of V_d/V_o, keeping V_o constant. The boundary between the continuous and the discontinuous mode of operation is obtained by using Eq. 7-20.

7-3-4   OUTPUT VOLTAGE RIPPLE

In the previous analysis, the output capacitor is assumed to be so large as to yield $v_o(t) = V_o$. However, the ripple in the output voltage with a practical value of capacitance can be calculated by considering the waveforms shown in Fig. 7-10 for a continuous-conduction mode of operation. Assuming that all of the ripple component in i_L flows through the capacitor and its average component flows through the load resistor, the shaded area in Fig. 7-10 represents an additional charge ΔQ. Therefore, the peak-to-peak voltage ripple ΔV_o can be written as

$$\Delta V_o = \frac{\Delta Q}{C} = \frac{1}{C}\frac{1}{2}\frac{\Delta I_L}{2}\frac{T_s}{2}$$

From Fig. 7-5 during t_{off}

$$\Delta I_L = \frac{V_o}{L}(1 - D)T_s \tag{7-22}$$

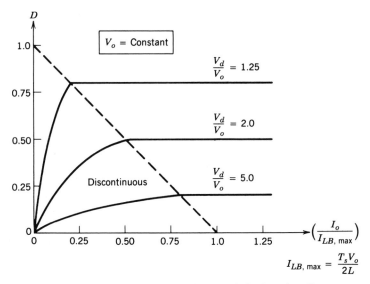

Figure 7-9 Step-down converter characteristics keeping V_o constant.

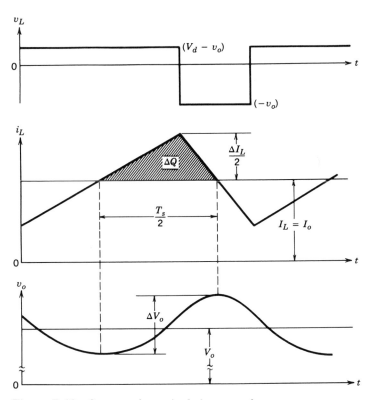

Figure 7-10 Output voltage ripple in a step-down converter.

Therefore, substituting ΔI_L from Eq. 7-22 into the previous equation gives

$$\Delta V_o = \frac{T_s}{8C} \frac{V_o}{L}(1 - D)T_s \tag{7-23}$$

$$\therefore \frac{\Delta V_o}{V_o} = \frac{1}{8} \frac{T_s^2(1 - D)}{LC} = \frac{\pi^2}{2}(1 - D)\left(\frac{f_c}{f_s}\right)^2 \tag{7-24}$$

where switching frequency $f_s = 1/T_s$ and

$$f_c = \frac{1}{2\pi\sqrt{LC}} \tag{7-25}$$

Equation 7-24 shows that the voltage ripple can be minimized by selecting a corner frequency f_c of the low-pass filter at the output such that $f_c \ll f_s$. Also, the ripple is independent of the output load power, so long as the converter operates in the continuous-conduction mode. A similar analysis can be performed for the discontinuous-conduction mode.

We should note that in switch-mode dc power supplies, the percentage ripple in the output voltage is usually specified to be less than, for instance, 1%. Therefore, the analysis in the previous sections assuming $v_o(t) = V_o$ is valid. It should be noted that the output ripple in Eq. 7-24 is consistent with the discussion of the low-pass filter characteristic in Fig. 7-4c.

7-4   STEP-UP (BOOST) CONVERTER

Figure 7-11 shows a step-up converter. Its main application is in regulated dc power supplies and the regenerative braking of dc motors. As the name implies, the output voltage is always greater than the input voltage. When the switch is on, the diode is reversed biased, thus isolating the output stage. The input supplies energy to the inductor. When the switch is off, the output stage receives energy from the inductor as well as from the input. In the steady-state analysis presented here, the output filter capacitor is assumed to be very large to ensure a constant output voltage $v_o(t) \simeq V_o$.

7-4-1   CONTINUOUS-CONDUCTION MODE

Figure 7-12 shows the steady-state waveforms for this mode of conduction where the inductor current flows continuously [$i_L(t) > 0$].

Since in steady state the time integral of the inductor voltage over one time period must be zero,

$$V_d t_{on} + (V_d - V_o)t_{off} = 0$$

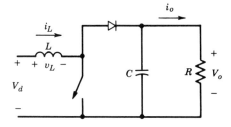

Figure 7-11   Step-up dc–dc converter.

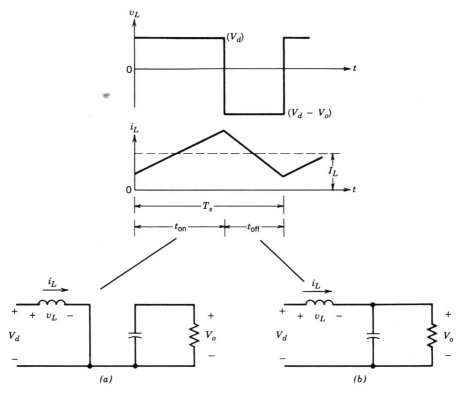

Figure 7-12   Continuous-conduction mode: (a) switch on; (b) switch off.

Dividing both sides by T_s and rearranging terms yield

$$\frac{V_o}{V_d} = \frac{T_s}{t_{\text{off}}} = \frac{1}{1 - D} \tag{7-26}$$

Assuming a lossless circuit, $P_d = P_o$,

$$\therefore V_d I_d = V_o I_o$$

and

$$\frac{I_o}{I_d} = (1 - D) \tag{7-27}$$

7-4-2   BOUNDARY BETWEEN CONTINUOUS AND DISCONTINUOUS CONDUCTION

Figure 7-13a shows the waveforms at the edge of continuous conduction. By definition, in this mode i_L goes to zero at the end of the off interval. The average value of the inductor current at this boundary is

$$I_{LB} = \frac{1}{2} i_{L,\text{peak}} \quad (\text{Fig. 7-13}a)$$

$$= \frac{1}{2} \frac{V_d}{L} t_{\text{on}}$$

$$= \frac{T_s V_o}{2L} D(1 - D) \quad (\text{using Eq. 7-26}) \tag{7-28}$$

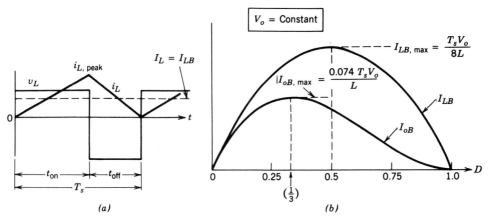

Figure 7-13 Step-up dc–dc converter at the boundary of continuous–discontinuous conduction.

Recognizing that in a step-up converter the inductor current and the input current are the same ($i_d = i_L$) and using Eq. 7-27 and 7-28, we find that the average output current at the edge of continuous conduction is

$$I_{oB} = \frac{T_s V_o}{2L} D(1 - D)^2 \tag{7-29}$$

Most applications in which a step-up converter is used require that V_o be kept constant. Therefore, with V_o constant, I_{oB} are plotted in Fig. 7-13b as a function of duty ratio D. Keeping V_o constant and varying the duty ratio imply that the input voltage is varying.

Figure 7-13b shows that I_{LB} reaches a maximum value at $D = 0.5$:

$$I_{LB,\text{max}} = \frac{T_s V_o}{8L} \tag{7-30}$$

Also, I_{oB} has its maximum at $D = \frac{1}{3} = 0.333$:

$$I_{oB,\text{max}} = \frac{2}{27} \frac{T_s V_o}{L} = 0.074 \frac{T_s V_o}{L} \tag{7-31}$$

In terms of their maximum values, I_{LB} and I_{oB} can be expressed as

$$I_{LB} = 4D(1 - D)I_{LB,\text{max}} \tag{7-32}$$

and

$$I_{oB} = \frac{27}{4} D(1 - D)^2 I_{oB,\text{max}} \tag{7-33}$$

Figure 7-13b shows that for a given D, with constant V_o, if the average load current drops below I_{oB} (and, hence, the average inductor current below I_{LB}), the current conduction will become discontinuous.

7-4-3 DISCONTINUOUS-CONDUCTION MODE

To understand the discontinuous-current-conduction mode, we would assume that as the output load power decreases, V_d and D remain constant (even though, in practice, D

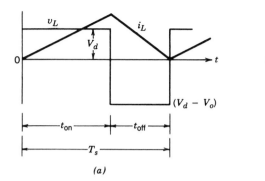 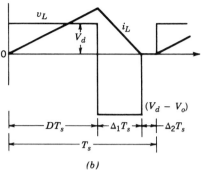

Figure 7-14 Step-up converter waveforms: (a) at the boundary of continuous–discontinuous conduction; (b) at discontinuous conduction.

would vary in order to keep V_o constant). Figure 7-14 compares the waveforms at the boundary of continuous conduction and discontinuous conduction, assuming that V_d and D are constant.

In Fig. 7-14b, the discontinuous current conduction occurs due to decreased $P_o (=P_d)$ and, hence, a lower $I_L (=I_d)$, since V_d is constant. Since $i_{L,\text{peak}}$ is the same in both modes in Fig. 7-14, a lower value of I_L (and, hence a discontinuous i_L) is possible only if V_o goes up in Fig. 7-14b.

If we equate the integral of the inductor voltage over one time period to zero,

$$V_d DT_s + (V_d - V_o)\Delta_1 T_s = 0$$

$$\therefore \frac{V_o}{V_d} = \frac{\Delta_1 + D}{\Delta_1} \tag{7-34}$$

and

$$\frac{I_o}{I_d} = \frac{\Delta_1}{\Delta_1 + D} \quad (\text{since } P_d = P_o) \tag{7-35}$$

From Fig. 7-14b, the average input current, which is also equal to the inductor current, is

$$I_d = \frac{V_d}{2L} DT_s(D + \Delta_1) \tag{7-36}$$

Using Eq. 7-35 in the foregoing equation yields

$$I_o = \left(\frac{T_s V_d}{2L}\right) D\Delta_1 \tag{7-37}$$

In practice, since V_o is held constant and D varies in response to the variation in V_d, it is more useful to obtain the required duty ratio D as a function of load current for various values of V_o/V_d. By using Eqs. 7-34, 7-37, and 7-31, we determine that

$$D = \left[\frac{4}{27} \frac{V_o}{V_d} \left(\frac{V_o}{V_d} - 1\right) \frac{I_o}{I_{oB,\text{max}}}\right]^{1/2} \tag{7-38}$$

In Fig. 7-15, D is plotted as a function of $I_o/I_{oB,\text{max}}$ for various values of V_d/V_o. The boundary between continuous and discontinuous conduction is shown by the dashed curve.

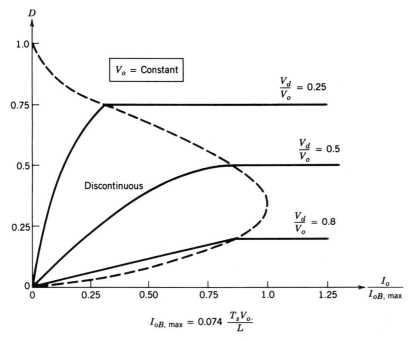

$$I_{oB,\,max} = 0.074\,\frac{T_s V_o}{L}$$

Figure 7-15   Step-up converter characteristics keeping V_o constant.

In the discontinuous mode, if V_o is not controlled during each switching time period, at least

$$\frac{L}{2}\,i_{L,\text{peak}}^2 = \frac{(V_d D T_s)^2}{2L}\quad\text{W-s}$$

are transferred from the input to the output capacitor and to the load. If the load is not able to absorb this energy, the capacitor voltage V_o would increase until an energy balance is established. If the load becomes very light, the increase in V_o may cause a capacitor breakdown or a dangerously high voltage to occur.

■ *Example 7-1*   In a step-up converter, the duty ratio is adjusted to regulate the output voltage V_o at 48 V. The input voltage varies in a wide range from 12 to 36 V. The maximum power output is 120 W. For stability reasons, it is required that the converter always operate in a discontinuous-current-conduction mode. The switching frequency is 50 kHz.

Assuming ideal components and C as very large, calculate the maximum value of L that can be used.

Solution   In this converter, $V_o = 48$ V, $T_s = 20\,\mu s$, and $I_{o,\text{max}} = 120$ W/48 V $= 2.5$ A. To find the maximum value of L that keeps the current conduction discontinuous, we will assume that at the extreme operating condition, the inductor current is at the edge of continuous conduction.

For the given range of V_d (12–36 V), D is in a range of 0.75–0.25 (corresponding to the current conduction bordering on being continuous). For this range of D, from Fig. 7-13*b*, I_{oB} has the smallest value at $D = 0.75$.

Therefore, by substituting $D = 0.75$ in Eq. 7-29 for I_{oB} and equating it to $I_{o,\max}$ of 2.5 A, we can calculate

$$L = \frac{20 \times 10^{-6} \times 48}{2 \times 2.5} 0.75(1 - 0.75)^2$$

$$= 9 \ \mu H$$

Therefore, if $L = 9 \ \mu H$ is used, the converter operation will be at the edge of continuous conduction with $V_d = 12$ V and $P_o = 120$ W. Otherwise, the conduction will be discontinuous. To further ensure a discontinuous-conduction mode, a smaller than 9 μH inductance may be used.   ∎

7-4-4 EFFECT OF PARASITIC ELEMENTS

The parasitic elements in a step-up converter are due to the losses associated with the inductor, the capacitor, the switch, and the diode. Figure 7-16 qualitatively shows the effect of these parasitics on the voltage transfer ratio. Unlike the ideal characteristic, in practice, V_o/V_d declines as the duty ratio approaches unity. Because of very poor switch utilization at high values of duty ratio (as discussed in Section 7-8), the curves in this range are shown as dashed. These parasitic elements have been ignored in the simplified analysis presented here; however, these can be incorporated into circuit simulation programs on computers for designing such converters.

7-4-5 OUTPUT VOLTAGE RIPPLE

The peak-to-peak ripple in the output voltage can be calculated by considering the waveforms shown in Fig. 7-17 for a continuous mode of operation. Assuming that all the ripple current component of the diode current i_D flows through the capacitor and its average value flows through the load resistor, the shaded area in Fig. 7-17 represents charge ΔQ. Therefore, the peak–peak voltage ripple is given by

$$\Delta V_o = \frac{\Delta Q}{C} = \frac{I_o D T_s}{C} \quad \text{(assuming a constant output current)}$$

$$= \frac{V_o}{R} \frac{D T_s}{C} \tag{7-39}$$

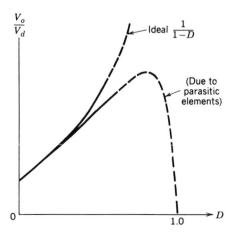

Figure 7-16 Effect of parasitic elements on voltage conversion ratio (step-up converter).

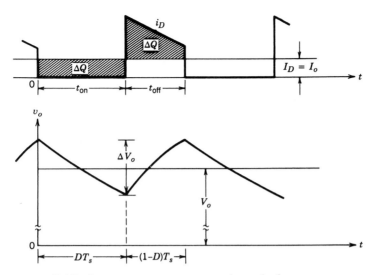

Figure 7-17   Step-up converter output voltage ripple.

$$\therefore \frac{\Delta V_o}{V_o} = \frac{DT_s}{RC}$$

$$= D\frac{T_s}{\tau} \quad \text{(where } \tau = RC \text{ time constant)} \tag{7-40}$$

A similar analysis can be performed for the discontinuous mode of conduction.

7-5   BUCK–BOOST CONVERTER

The main application of a step-down/step-up or buck–boost converter is in regulated dc power supplies, where a negative-polarity output may be desired with respect to the common terminal of the input voltage, and the output voltage can be either higher or lower than the input voltage.

A buck–boost converter can be obtained by the cascade connection of the two basic converters: the step-down converter and the step-up converter. In steady state, the output-to-input voltage conversion ratio is the product of the conversion ratios of the two converters in cascade (assuming that switches in both converters have the same duty ratio):

$$\frac{V_o}{V_d} = D\frac{1}{1 - D} \quad \text{(from Eqs. 7-3 and 7-26)} \tag{7-41}$$

This allows the output voltage to be higher or lower than the input voltage, based on the duty ratio D.

The cascade connection of the step-down and the step-up converters can be combined into the single buck–boost converter shown in Fig. 7-18. When the switch is closed, the input provides energy to the inductor and the diode is reverse biased. When the switch is open, the energy stored in the inductor is transferred to the output. No energy is supplied by the input during this interval. In the steady-state analysis presented here, the output capacitor is assumed to be very large, which results in a constant output voltage $v_o(t) \simeq V_o$.

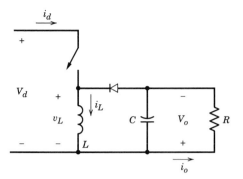

Figure 7-18 Buck–boost converter.

7-5-1 CONTINUOUS-CONDUCTION MODE

Figure 7-19 shows the waveforms for the continuous-conduction mode where the inductor current flows continuously.

Equating the integral of the inductor voltage over one time period to zero yields

$$V_d D T_s + (-V_o)(1 - D)T_s = 0$$

$$\therefore \frac{V_o}{V_d} = \frac{D}{1 - D} \tag{7-42}$$

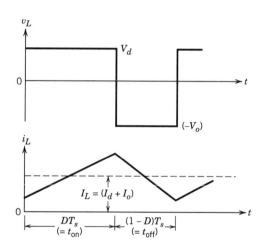

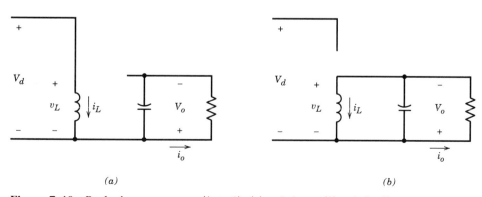

(a)   (b)

Figure 7-19 Buck–boost converter $(i_L > 0)$: (a) switch on; (b) switch off.

and

$$\frac{I_o}{I_d} = \frac{1 - D}{D} \quad \text{(assuming } P_d = P_o\text{)} \tag{7-43}$$

Equation 7-42 implies that depending on the duty ratio, the output voltage can be either higher or lower than the input.

7-5-2   BOUNDARY BETWEEN CONTINUOUS AND DISCONTINUOUS CONDUCTION

Figure 7-20a shows the waveforms at the edge of continuous conduction. By definition, in this mode i_L goes to zero at the end of the off interval.

From Fig. 7-20a,

$$I_{LB} = \tfrac{1}{2} i_{L,\text{peak}}$$

$$= \frac{T_s V_d}{2L} D \tag{7-44}$$

From Fig. 7-18,

$$I_o = I_L - I_d \tag{7-45}$$

(since the average capacitor current is zero).

By using Eqs. 7-42 through 7-45, we can obtain the average inductor current and the output current at the border of continuous conduction in terms of V_o,

$$I_{LB} = \frac{T_s V_o}{2L} (1 - D) \tag{7-46}$$

and

$$I_{oB} = \frac{T_s V_o}{2L} (1 - D)^2 \tag{7-47}$$

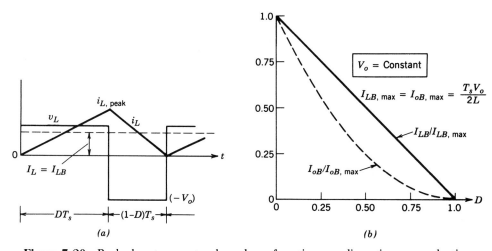

Figure 7-20   Buck–boost converter: boundary of continuous–discontinuous conduction.

Most applications in which a buck–boost converter may be used require that V_o be kept constant, though V_d (and, hence, D) may vary. Our inspection of Eqs. 7-46 and 7-47 shows that both I_{LB} and I_{oB} result in their maximum values at $D = 0$:

$$I_{LB,\text{max}} = \frac{T_sV_o}{2L} \tag{7-48}$$

and

$$I_{oB,\text{max}} = \frac{T_sV_o}{2L} \tag{7-49}$$

Using Eqs. 7-46 through 7-49 gives

$$I_{LB} = I_{LB,\text{max}}(1 - D) \tag{7-50}$$

and

$$I_{oB} = I_{oB,\text{max}}(1 - D)^2 \tag{7-51}$$

Figure 7-20*b* shows I_{LB} and I_{oB} as a function of D, keeping $V_o = $ const.

7-5-3 DISCONTINUOUS-CONDUCTION MODE

Figure 7-21 shows the waveforms with a discontinuous i_L. If we equate the integral of the inductor voltage over one time period to zero,

$$V_dDT_s + (-V_o)\Delta_1T_s = 0$$
$$\therefore \frac{V_o}{V_d} = \frac{D}{\Delta_1} \tag{7-52}$$

and

$$\frac{I_o}{I_d} = \frac{\Delta_1}{D} \quad (\text{since } P_d = P_o) \tag{7-53}$$

From Fig. 7-21

$$I_L = \frac{V_d}{2L} DT_s(D + \Delta_1) \tag{7-54}$$

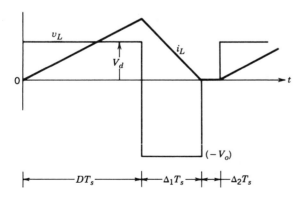

Figure 7-21 Buck–boost converter waveforms in a discontinuous-conduction mode.

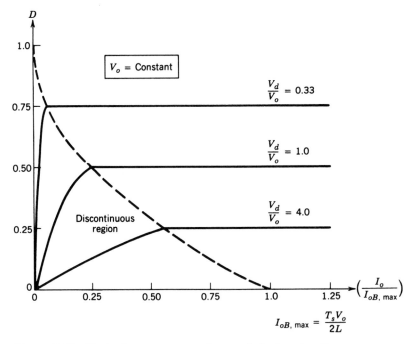

Figure 7-22 Buck–boost converter characteristics keeping V_o constant.

Since V_o is kept constant, it is useful to obtain D as a function of the output load current I_o for various values of V_o/V_d. Using the equations derived earlier, we find that

$$D = \frac{V_o}{V_d} \sqrt{\frac{I_o}{I_{oB,\text{max}}}} \tag{7-55}$$

Figure 7-22 shows the plot of D as a function of $I_o/I_{oB,\text{max}}$ for various values of V_d/V_o. The boundary between the continuous and the discontinuous mode is shown by the dashed curve.

■ *Example 7-2* In a buck–boost converter operating at 20 kHz, $L = 0.05$ mH. The output capacitor C is sufficiently large and $V_d = 15$ V. The output is to be regulated at 10 V and the converter is supplying a load of 10 W. Calculate the duty ratio D.

Solution In this example, $I_o = 10/10 = 1$ A. Initially, the mode of conduction is not known. If the current is assumed to be at the edge of continuous conduction, from Eq. 7-42

$$\frac{D}{1 - D} = \frac{10}{15}$$
$$D = 0.4 \quad \text{(initial estimate)}$$

From Eq. 7-49

$$I_{oB,\text{max}} = \frac{0.05 \times 10}{2 \times 0.05} = 5 \text{ A}$$

Using $D = 0.4$ and $I_{oB,\text{max}} = 5$ A in Eq. 7-51, we find that

$$I_{oB} = 5(1 - 0.4)^2$$
$$= 1.8 \text{ A}$$

Since the output current $I_o = 1$ A is less than I_{oB}, the current conduction is discontinuous.

Therefore, using Eq. 7-55,

$$\text{Duty ratio } D = \frac{10}{15}\sqrt{\frac{1.0}{5.0}}$$

$$= 0.3 \quad \text{(discontinuous conduction)} \qquad \blacksquare$$

7-5-4 EFFECT OF PARASITIC ELEMENTS

Analogous to the step-up converter, the parasitic elements have significant impact on the voltage conversion ratio and the stability of the feedback-regulated buck–boost converter. As an example, Fig. 7-23 qualitatively shows the effect of these parasitic elements. The curves are shown as dashed because of the very poor switch utilization, making very high duty ratios impractical. The effect of these parasitic elements can be modeled in the circuit simulation programs for designing such converters.

7-5-5 OUTPUT VOLTAGE RIPPLE

The ripple in the output voltage can be calculated by considering the waveform shown in Fig. 7-24 for a continuous mode of operation. Assuming that all the ripple current component of i_D flows through the capacitor and its average value flows through the load resistor, the shaded area in Fig. 7-24 represents charge ΔQ. Therefore, the peak-to-peak voltage ripple is calculated as

$$\Delta V_o = \frac{\Delta Q}{C} = \frac{I_o DT_s}{C} \quad \text{(assuming a constant output current)} \qquad (7\text{-}56)$$

$$= \frac{V_o}{R}\frac{DT_s}{C}$$

$$\frac{\Delta V_o}{V_o} = \frac{DT_s}{RC} \qquad (7\text{-}57)$$

$$= D\frac{T_s}{\tau}$$

where $\tau = RC$ time constant.

A similar analysis can be performed for the discontinuous mode of operation.

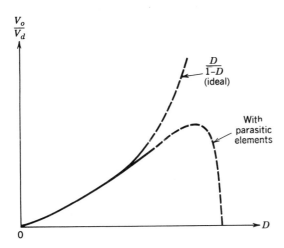

Figure 7-23 Effect of parasitic elements on the voltage conversion ratio in a buck–boost converter.

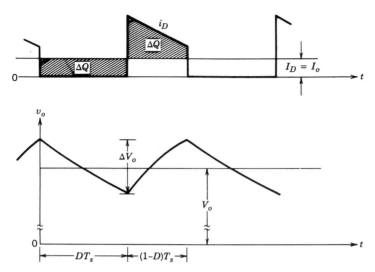

Figure 7-24   Output voltage ripple in a buck–boost converter.

7-6   CÚK dc–dc CONVERTER

Named after its inventor, the Cúk converter is shown in Fig. 7-25. This converter is obtained by using the duality principle on the circuit of a buck–boost converter, discussed in the previous section. Similar to the buck–boost converter, the Cúk converter provides a negative-polarity regulated output voltage with respect to the common terminal of the input voltage. Here, the capacitor C_1 acts as the primary means of storing and transferring energy from the input to the output.

In steady state, the average inductor voltages V_{L1} and V_{L2} are zero. Therefore, by inspection of Fig. 7-25,

$$V_{C1} = V_d + V_o \tag{7-58}$$

Therefore, V_{C1} is larger than both V_d and V_o. Assuming C_1 to be sufficiently large, in steady state the variation in v_{C1} from its average value V_{C1} can be assumed to be negligibly small (i.e., $v_{C1} \simeq V_{C1}$), even though it stores and transfers energy from the input to the output.

When the switch is off, the inductor currents i_{L1} and i_{L2} flow through the diode. The circuit is shown in Fig. 7-26a. Capacitor C_1 is charged through the diode by energy from both the input and L_1. Current i_{L1} decreases, because V_{C1} is larger than V_d. Energy stored in L_2 feeds the output. Therefore, i_{L2} also decreases.

When the switch is on, V_{C1} reverse biases the diode. The inductor currents i_{L1} and i_{L2} flow through the switch, as shown in Fig. 7-26b. Since $V_{C1} > V_0$, C_1 discharges through

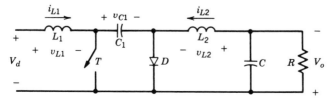

Figure 7-25   Cúk converter.

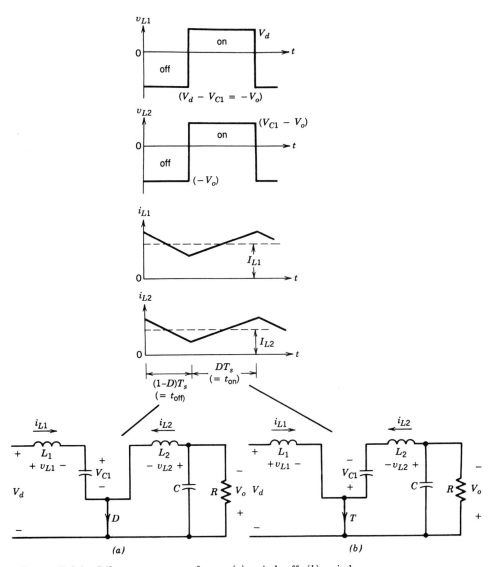

Figure 7-26 Cúk converter waveforms: (a) switch off; (b) switch on.

the switch, transferring energy to the output and L_2. Therefore, i_{L2} increases. The input feeds energy to L_1 causing i_{L1} to increase.

The inductor currents i_{L1} and i_{L2} are assumed to be continuous. The voltage and current expressions in steady state can be obtained in two different ways.

If we assume the capacitor voltage V_{C1} to be constant, then equating the integral of the voltages across L_1 and L_2 over one time period to zero yields

$$L_1: \quad V_d DT_s + (V_d - V_{C1})(1 - D)T_s = 0$$

$$\therefore V_{C1} = \frac{1}{1 - D} V_d \tag{7-59}$$

$$L_2: \quad (V_{C1} - V_o)DT_s + (-V_o)(1 - D)T_s = 0$$

$$\therefore V_{C1} = \frac{1}{D} V_o \tag{7-60}$$

From Eqs. 7-59 and 7-60

$$\frac{V_o}{V_d} = \frac{D}{1-D} \tag{7-61}$$

Assuming $P_d = P_o$ gives

$$\frac{I_o}{I_d} = \frac{1-D}{D} \tag{7-62}$$

where $I_{L1} = I_d$ and $I_{L2} = I_o$.

There is another way to obtain these expressions. Assume that the inductor currents i_{L1} and i_{L2} are essentially ripple free (i.e., $i_{L1} = I_{L1}$ and $i_{L2} = I_{L2}$). When the switch is off, the charge delivered to C_1 equals $I_{L1}(1 - D)T_s$. When the switch is on, the capacitor discharges by an amount $I_{L2}DT_s$. Since in steady state the net change of charge associated with C_1 over one time period must be zero,

$$I_{L1}(1 - D)T_s = I_{L2}DT_s \tag{7-63}$$

$$\therefore \frac{I_{L2}}{I_{L1}} = \frac{I_o}{I_d} = \frac{1-D}{D} \tag{7-64}$$

and

$$\frac{V_o}{V_d} = \frac{D}{1-D} \quad (\text{since } P_o = P_d) \tag{7-65}$$

Both methods of analysis yield identical results. The average input and output relations are similar to that of a buck–boost converter.

In practical circuits, the assumption of a nearly constant V_{C1} is reasonably valid. An advantage of this circuit is that both the input current and the current feeding the output stage are reasonably ripple free (unlike the buck–boost converter where both these currents are highly discontinuous). It is possible to simultaneously eliminate the ripples in i_{L1} and i_{L2} completely, leading to lower external filtering requirements. A significant disadvantage is the requirement of a capacitor C_1 with a large ripple-current-carrying capability.

Detailed analysis of this converter has been adequately reported in the technical literature and will not be undertaken here.

■ *Example 7-3* In a Cúk converter operating at 50 kHz, $L_1 = L_2 = 1$ mH and $C_1 = 5$ μF. The output capacitor is sufficiently large to yield an essentially constant output voltage. Here $V_d = 10$ V and the output V_o is regulated to be constant at 5 V. It is supplying 5 W to a load. Assume ideal components.

Calculate the percentage errors in assuming a constant voltage across C_1 or in assuming constant currents i_{L1} and i_{L2}.

Solution

(a) If the voltage across C_1 is assumed to be constant, from Eq. 7-58

$$v_{C1} = V_{C1} = 10 + 5 = 15 \text{ V}$$

Initially, we will assume the current conduction to be continuous. Therefore, from Eq. 7-61

$$\frac{D}{1-D} = \frac{5}{10}$$

$$\therefore D = 0.333$$

Therefore, from Fig. 7-26 during the off interval

$$\Delta i_{L1} = \frac{V_{C1} - V_d}{L_1} (1 - D)T_s$$

$$= \frac{(15 - 10)}{10^{-3}} (1 - 0.333) \times 20 \times 10^{-6}$$

$$= 0.067 \ A$$

and

$$\Delta i_{L2} = \frac{V_o}{L_2} (1 - D)T_s$$

$$= \frac{5}{10^{-3}} (1 - 0.333) \times 20 \times 10^{-6}$$

$$= 0.067 \ A$$

Note that Δi_{L1} and Δi_{L2} would be equal (since $V_{C1} - V_d = V_o$), provided that $L_1 = L_2$.

At an output load of 5 W, using Eq. 7-62,

$$I_o = 1 \ A \quad \text{and} \quad I_d = 0.5 \text{A}$$

Since $\Delta i_{L1} < I_d \ (=I_{L1})$ and $\Delta i_{L2} < I_o \ (=I_{L2})$, the mode of operation is continuous, as assumed earlier.

Therefore, the percentage errors in assuming constant I_{L1} and I_{L2} would be

$$\frac{\Delta i_{L1}}{I_{L1}} = \frac{0.067 \times 100}{0.5} = 13.4\%$$

and

$$\frac{\Delta i_{L2}}{I_{L2}} = \frac{0.067 \times 100}{1.0} = 6.7\%$$

(b) If i_{L1} and i_{L2} are assumed to be constant, from Fig. 7-26 during the off interval

$$\Delta V_{C1} = \frac{1}{C} \int_0^{(1-D)T_s} i_{L1} \, dt$$

Assuming

$$i_{L1} = I_{L1} = 0.5 \ A$$

yields

$$\Delta V_{C1} = \frac{1}{5 \times 10^{-6}} \times 0.5 \times (1 - 0.333) 20 \times 10^{-6}$$

$$= 1.33 \ V$$

Therefore, the percentage error in assuming a constant voltage across C_1 is

$$\frac{\Delta V_{C1}}{V_{C1}} = \frac{1.33 \times 100}{15} = 8.87\%$$

This example shows that as a first approximation, to illustrate the principle of operation, it is reasonable to assume either a constant v_{C1} or constant i_{L1} and i_{L2} in this problem. ∎

7-7 FULL-BRIDGE dc–dc CONVERTER

There are three distinct applications of the full-bridge switch-mode converters shown in Fig. 7-27:

- dc motor drives
- dc-to-ac (sine-wave) conversion in single-phase uninterruptible ac power supplies
- dc-to-ac (high intermediate frequency) conversion in switch-mode transformer-isolated dc power supplies

Even though the full-bridge topology remains the same in each of these three applications, the type of control depends on the application. However, the full-bridge converter, as used in dc motor drives, is covered in this "generic" chapter because it provides a good basis for understanding the switch-mode dc-to-ac (sine-wave) converters of Chapter 8.

In the full-bridge converter shown in Fig. 7-27, the input is a fixed-magnitude dc voltage V_d. The output of the converter in this chapter is a dc voltage V_o, which can be controlled in magnitude as well as polarity. Similarly, the magnitude and the direction of the output current i_o can be controlled. Therefore, a full-bridge converter such as that shown in Fig. 7-27 can operate in all four quadrants of the i_o–v_o plane, and the power flow through the converter can be in either direction.

In a converter topology such as that of the full-bridge converter shown in Fig. 7-27, where diodes are connected in antiparallel with the switches, a distinction must be made between the on state versus the conducting state of a switch. Because of the diodes in antiparallel with the switches, when a switch is turned on, it may or may not conduct a current, depending on the direction of the output current i_o. If the switch conducts a current, then it is in a conducting state. No such distinction is required when the switch is turned off.

The full-bridge converter consists of two legs, A and B. Each leg consists of two switches and their antiparallel diodes. The two switches in each leg are switched in such

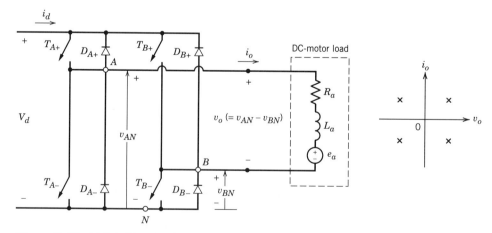

Figure 7-27 Full-bridge dc–dc converter.

a way that when one of them is in its off state, the other switch is on. Therefore, the two switches are never off simultaneously. In practice, they are both off for a short time interval, known as the blanking time, to avoid short circuiting of the dc input. This blanking time is neglected in this chapter since we are assuming the switches to be ideal, capable of turning off instantaneously.

We should note that if the converter switches in each leg are switched in such a way that both the switches in a leg are not off simultaneously, then the output current i_o in Fig. 7-27 will flow continuously. Therefore, the output voltage is solely dictated by the status of the switches. For example, consider leg A in Fig. 7-27. The output voltage v_{AN}, with respect to the negative dc bus N, is dictated by the switch states as follows: When T_{A+} is on, the output current will flow through T_{A+} if i_o is positive or it will flow through D_{A+} if i_o is negative. In either case, T_{A+} being on ensures that point A in Fig. 7-27 is at the same potential as the positive terminal of the dc input, and therefore

$$v_{AN} = V_d \quad \text{(if } T_{A+} \text{ is on and } T_{A-} \text{ is off)} \tag{7-66a}$$

Similarly, when T_{A-} is on, a negative i_o will flow through T_{A-} (since D_{A+} is reverse biased) and a positive i_o will flow through D_{A-}. Therefore

$$v_{AN} = 0 \quad \text{(if } T_{A-} \text{ is on and } T_{A+} \text{ is off)} \tag{7-66b}$$

Equations 7-66a and 7-66b show that v_{AN} depends only on the switch status and is independent of the direction of i_o. Therefore, the output voltage of the converter leg A, averaged over one switching frequency time period T_s, depends only on the input voltage V_d and the duty ratio of T_{A+}:

$$V_{AN} = \frac{V_d t_{\text{on}} + 0 \cdot t_{\text{off}}}{T_s} = V_d \cdot \text{duty ratio of } T_{A+} \tag{7-67}$$

where t_{on} and t_{off} are the on and off intervals of T_{A+}, respectively.

Similar arguments apply to the converter leg B, and V_{BN} depends on V_d and the duty ratio of the switch T_{B+}:

$$V_{BN} = V_d \cdot \text{duty ratio of } T_{B+} \tag{7-68}$$

independent of the direction of i_o. Therefore, the converter output $V_o\,(=V_{AN} - V_{BN})$ can be controlled by controlling the switch duty ratios and is independent of the magnitude and the direction of i_o.

(It should be noted that it is possible to control the output voltage of a converter leg by turning both switches off simultaneously for some time interval. However, this scheme would make the output voltage dependent on the direction of i_o in Fig. 7-27. This is obviously undesirable, since it would introduce nonlinearity in the relationship between the control voltage and the average output voltage. Therefore, such a scheme will not be considered with a load, such as a dc motor load, shown in Fig. 7-27.)

In the single-switch converters discussed previously, the polarity of the output voltage is unidirectional, and therefore, the converter switch is pulse-width modulated by comparing a switching-frequency sawtooth waveform with the control voltage v_{control}. In contrast, the output voltage of the full-bridge converter is reversible in polarity and therefore, a switching-frequency triangular waveform is used for PWM of the converter switches. Two such PWM switching strategies are described below:

1. *PWM with bipolar voltage switching*, where (T_{A+}, T_{B-}) and (T_{A-}, T_{B+}) are treated as two switch pairs; switches in each pair are turned on and off simultaneously.

2. *PWM with unipolar voltage switching* is also referred to as the *double-PWM switching*. Here the switches in each inverter leg are controlled independently of the other leg.

As we mentioned earlier, the output current through these PWM full-bridge dc–dc converters, while supplying dc loads of the type shown in Fig. 7-27, does not become discontinuous at low values of I_o, unlike the single-switch converters discussed in the previous sections.

In the full-bridge converter of Fig. 7-27, the input current i_d changes direction instantaneously. Therefore, it is important that the input to this converter be a dc voltage source with a low internal impedance. In practice, the large filter-capacitor shown in the block diagram of Fig. 7-1 provides this low-impedance path to i_d.

7-7-1   PWM WITH BIPOLAR VOLTAGE SWITCHING

In this type of voltage switching, switches (T_{A+}, T_{B-}) and (T_{B+}, T_{A-}) are treated as two switch pairs (two switches in a pair are simultaneously turned on and off). One of the two switch pairs is always on.

The switching signals are generated by comparing a switching-frequency triangular waveform (v_{tri}) with the control voltage $v_{control}$. When $v_{control} > v_{tri}$, T_{A+} and T_{B-} are turned on. Otherwise, T_{A-} and T_{B+} are turned on. The switch duty ratios can be obtained from the waveforms in Fig. 7-28a as follows by arbitrarily choosing a time origin as shown in the figure:

$$v_{tri} = \hat{V}_{tri} \frac{t}{T_s/4} \qquad 0 < t < \tfrac{1}{4} T_s \tag{7-69}$$

At $t = t_1$ in Fig. 7-28a, $v_{tri} = v_{control}$. Therefore, from Eq. 7-69

$$t_1 = \frac{v_{control}}{\hat{V}_{tri}} \frac{T_s}{4} \tag{7-70}$$

By studying Fig. 7-28, we find that the on duration t_{on} of switch pair 1 (T_{A+}, T_{B-}) is

$$t_{on} = 2t_1 + \tfrac{1}{2} T_s \tag{7-71}$$

Therefore, their duty ratio from Eq. 7-71 is

$$D_1 = \frac{t_{on}}{T_s} = \frac{1}{2} \left(1 + \frac{v_{control}}{\hat{V}_{tri}} \right) \qquad (T_{A+}, T_{B-}) \tag{7-72}$$

Therefore, the duty ratio D_2 of the switch pair 2 (T_{B+}, T_{A-}) is

$$D_2 = 1 - D_1 \qquad (T_{B+}, T_{A-}) \tag{7-73}$$

By using the foregoing duty ratios, we can obtain V_{AN} and V_{BN} in Fig. 7-28 from Eqs. 7-67 and 7-68, respectively. Therefore,

$$V_o = V_{AN} - V_{BN} = D_1 V_d - D_2 V_d = (2D_1 - 1)V_d \tag{7-74}$$

Substituting D_1 from Eq. 7-72 into Eq. 7-74 yields

$$V_o = \frac{V_d}{\hat{V}_{tri}} v_{control} = k v_{control} \tag{7-75}$$

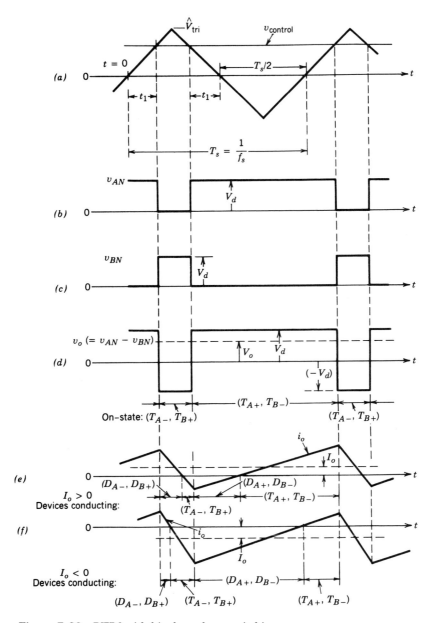

Figure 7-28 PWM with bipolar voltage switching.

where $k = V_d/\hat{V}_{tri}$ = const. This equation shows that in this switch-mode converter, similar to the single-switch converters discussed previously, the average output voltage varies linearly with the input control signal, similar to a linear amplifier. We will learn in Chapter 8 that a finite blanking time has to be used between turning off one switch pair and turning on the other switch pair. This blanking time introduces a slight nonlinearity in the relationship between $v_{control}$ and V_o.

The waveform for the output voltage v_o in Fig. 7-28d shows that the voltage jumps between $+V_d$ and $-V_d$. This is the reason why this switching strategy is referred to as the *bipolar voltage-switching PWM.*

We should also note that the duty ratio D_1 in Eq. 7-72 can vary between 0 and 1, depending on the magnitude and the polarity of $v_{control}$. Therefore V_o can be continuously varied in a range from $-V_d$ to V_d. Here, the output voltage of the converter is independent of the output current i_o, since the blanking time has been neglected.

The average output current I_o can be either positive or negative. For small values of I_o, i_o during a cycle can be both positive and negative; this is shown in Fig. 7-28e for $I_o > 0$, where the average power flow is from V_d to V_o, and in Fig. 7-28f for $I_o < 0$, where the average power flow is from V_o to V_d.

7-7-2 PWM WITH UNIPOLAR VOLTAGE SWITCHING

An inspection of Fig. 7-27 shows that regardless of the direction of i_o, $v_o = 0$ if T_{A+} and T_{B+} are both on. Similarly, $v_o = 0$ if T_{A-} and T_{B-} are both on. This property can be exploited to improve the output voltage waveform.

In Fig. 7-29, a triangular waveform is compared with the control voltage $v_{control}$ and $-v_{control}$ for determining the switching signals for leg A and leg B, respectively. A comparison of $v_{control}$ with v_{tri} controls leg A switches, whereas leg B switches are controlled by comparing $-v_{control}$ with v_{tri} in the following manner:

$$T_{A+} \text{ on: } \quad \text{if } v_{control} > v_{tri} \tag{7-76}$$

and

$$T_{B+} \text{ on: } \quad \text{if } -v_{control} > v_{tri} \tag{7-77}$$

Output voltages of each leg and v_o are shown in Fig. 7-29. By examining Fig. 7-29 and comparing it to Fig. 7-28, it can be seen that the duty ratio D_1 of the switch T_{A+} is given by Eq. 7-72 of the previous switching strategy. Similarly, the duty ratio D_2 of the switch T_{B+} is given by Eq. 7-73. That is,

$$D_1 = \frac{1}{2}\left(\frac{v_{control}}{\hat{V}_{tri}} + 1\right) \qquad T_{A+} \tag{7-78}$$

and

$$D_2 = 1 - D_1 \qquad T_{B+} \tag{7-79}$$

Therefore, from Eq. 7-74, which is also valid in this case,

$$V_o = (2D_1 - 1)V_d = \frac{V_d}{\hat{V}_{tri}} v_{control} \tag{7-80}$$

Therefore, the average output voltage V_o in this switching scheme is the same as in the bipolar voltage-switching scheme and varies linearly with $v_{control}$.

Figures 7-29e and 7-29f show the current waveforms and the devices conducting for $I_o > 0$ and $I_o < 0$, respectively, where V_o is positive in both cases.

If the switching frequencies of the switches are the same in these two PWM strategies, then the unipolar voltage switching results in a better output voltage waveform and in a better frequency response, since the "effective" switching frequency of the output voltage waveform is doubled and the ripple is reduced. This is illustrated by means of the following example.

■ *Example 7-4*   In a full-bridge dc–dc converter, the input V_d is constant and the output voltage is controlled by varying the duty ratio. Calculate the rms value of the ripple V_r in the output voltage as a function of the average V_o for

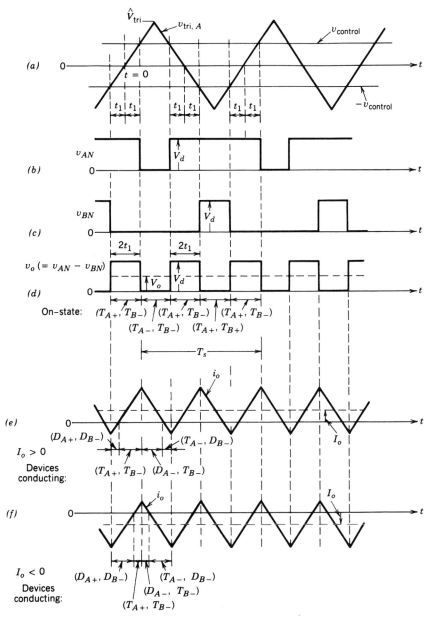

Figure 7-29 PWM with unipolar voltage switching.

(a) PWM with bipolar voltage switching and

(b) PWM with unipolar voltage switching.

Solution

(a) Using PWM with bipolar voltage switching, the waveform of the output voltage v_o is shown in Fig. 7-28d. For such a waveform, independent of the value of $v_{control}/\hat{V}_{tri}$, the root-mean-square (rms) value of the output voltage is

$$V_{o,rms} = V_d \qquad (7\text{-}81)$$

and the average value V_o is given by Eq. 7-74.

By using the definition of an rms value and Eqs. 7-74 and 7-81, we can calculate the ripple component V_r in the output voltage as

$$V_{r,\text{rms}} = \sqrt{V_{o,\text{rms}}^2 - V_o^2} = V_d\sqrt{1 - (2D_1 - 1)^2} = 2V_d\sqrt{D_1 - D_1^2} \quad (7\text{-}82)$$

As D_1 varies from 0 to 1, V_o varies from $-V_d$ to V_d. A plot of $V_{r,\text{rms}}$ as a function of V_o is shown in Fig. 7-30 by the solid curve.

(b) Using PWM with unipolar voltage switching, the waveform of the output voltage v_o is shown in Fig. 7-29d. To find the rms value of the output voltage, the interval t_1 in Fig. 7-29 can be written as

$$t_1 = \frac{v_{\text{control}}}{\hat{V}_{\text{tri}}}\frac{T_s}{4} \quad \text{for } v_{\text{control}} > 0 \quad (7\text{-}83)$$

By examining the v_o waveform in Fig. 7-29d for $v_{\text{control}} > 0$, the rms voltage can be obtained as

$$\begin{aligned}
V_{o,\text{rms}} &= \sqrt{\frac{4t_1 V_d^2}{T_s}} \\
&= \sqrt{\frac{v_{\text{control}}}{\hat{V}_{\text{tri}}}} V_d \\
&= \sqrt{(2D_1 - 1)}V_d \quad \text{(using Eq. 7-80) } (7\text{-}84)
\end{aligned}$$

Therefore, using Eqs. 7-80 and 7-84

$$V_{r,\text{rms}} = \sqrt{V_{o,\text{rms}}^2 - V_o^2} = \sqrt{6D_1 - 4D_1^2 - 2V_d} \quad (7\text{-}85)$$

where $v_{\text{control}} > 0$ and $0.5 < D_1 < 1$. As $v_{\text{control}}/\hat{V}_{\text{tri}}$ varies from 0 to 1, D_1 varies from 0.5 to 1.0 and the plot of $V_{r,\text{rms}}$ as a function of V_o is shown by the dashed curve in Fig. 7-30. Similarly, the curve corresponding to $v_{\text{control}}/\hat{V}_{\text{tri}}$ in the range from -1.0 to 0 can be obtained.

Figure 7-30 shows that with the same switching frequency, PWM with unipolar voltage switching results in a lower rms ripple component in the output voltage. ∎

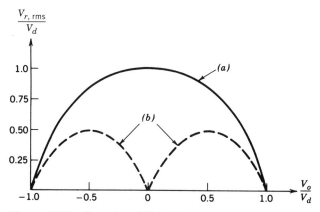

Figure 7-30 $V_{r,\text{rms}}$ in a full-bridge converter using PWM: (a) with bipolar voltage switching; (b) with unipolar voltage switching.

7-8 dc–dc CONVERTER COMPARISON

The step-down, step-up, buck–boost, and Cúk converters in their basic forms are capable of transferring energy only in one direction. This is a consequence of their capability to produce only unidirectional voltage and unidirectional current. A full-bridge converter is capable of a bidirectional power flow, where both V_o and I_o can be reversed independent of one another. This capability to operate in four quadrants of the V_o–I_o plane allows a full-bridge converter to be used as a dc-to-ac inverter. In dc-to-ac inverters, which are described in Chapter 8, the full-bridge converter operates in all four quandrants during each cycle of the ac output.

To evaluate how well the switch is utilized in the previously discussed converter circuits, we make the following assumptions:

1. The average current is at its rated (designed maximum) value I_o. The ripple in the inductor current is negligible; therefore $i_L(t) = I_L$. This condition implies a continuous-conduction mode for all converters.

2. The output voltage v_o is at its rated (designed maximum) value V_o. The ripple in v_o is assumed to be negligible; therefore $v_o(t) = \hat{V}_o$.

3. The input voltage V_d is allowed to vary. Therefore, the switch duty ratio must be controlled to hold V_o constant.

With the foregoing steady-state operating conditions, the switch peak voltage rating V_T and the peak current rating I_T are calculated. The switch power rating is calculated as $P_T = V_T I_T$. The switch utilization is expressed as P_o/P_T, where $P_o = V_o I_o$ is the rated output power.

In Fig. 7-31, the switch utilization factor P_o/P_T is plotted for the previously considered converters. This shows that in the step-down and the step-up converters, if the input and the output voltages are of the same order of magnitude, then the switch utilization is very good. In the buck–boost and the Cúk converter, the switch is poorly utilized. The maximum switch utilization of 0.25 is realized at $D = 0.5$, which corresponds to $V_o = V_d$.

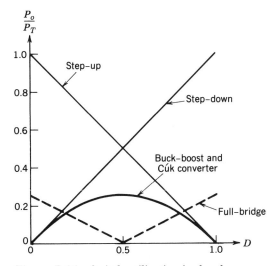

Figure 7-31 Switch utilization in dc–dc converters.

In the nonisolated full-bridge converter, the switch utilization factor is plotted as a function of the duty ratio of one of the switches (e.g., switch T_{A+} in Fig. 7-27). Here, the overall switch utilization is also poor. It is maximum at $V_o = -V_d$ and $V_o = V_d$, respectively.

In conclusion, for these *nonisolated* dc–dc converters, if at all possible, it is preferable to use either the step-down or the step-up converter from the switch utilization consideration. If both higher as well as lower output voltages compared to the input are necessary or a negative polarity output compared to the input is desired, then the buck–boost or the Cúk converter should be used. Similarly, the nonisolated full-bridge converter should only be used if four-quadrant operation is required.

SUMMARY

In this chapter various nonisolated converter topologies for dc-to-dc conversion are discussed. Except for the full-bridge converter, all others operate in a single quadrant of the output voltage–current plane, thereby allowing power to flow only in one direction.

In any converter circuit operating in steady state, a capacitor can be represented by its instantaneous voltage as an equivalent voltage source. Similarly, an inductor can be represented by its instantaneous current as an equivalent current source. Following this procedure, various converters discussed in this chapter can be drawn as shown in Fig. 7-32.

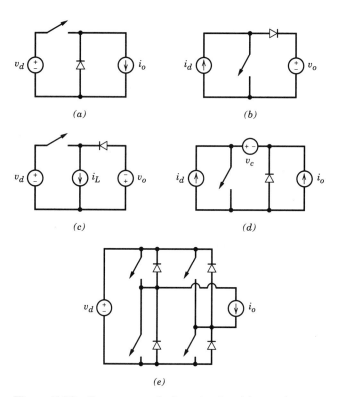

Figure 7-32   Converter equivalent circuits: (*a*) step-down; (*b*) step-up; (*c*) step-down/step-up; (*d*) Cúk; (*e*) full-bridge.

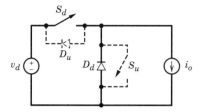

Figure 7-33   Reversible power flow with reversible direction of the output current i_o.

In all converters, the switching action does not cause a jump (or discontinuity) in the value of the voltage source or in the current source. In step-down (including full-bridge) and step-up converters, the energy transfer is between a voltage and a current source, as seen from the switch-diode combination. In step-down/step-up and Cúk converters, the energy transfer is between two similar types of sources but they are separated by a source of the other type (for example, two voltage sources are separated by a current source in the step-down/step-up converter).

It is possible to achieve a reversible flow of power by adding an additional diode and a switch (shown by dashed lines) as in Fig. 7-33. This converter with a positive value of i_o, and with S_d and D_d operating, resembles a step-down converter where the flow of power is from the voltage source to the equivalent current source. On the other hand, with a negative value of i_o, and with S_u and D_u operating, it resembles a step-up connector where the flow of power is from the equivalent current source to the voltage source.

PROBLEMS

STEP-DOWN CONVERTERS

7-1 In a step-down converter, consider all components to be ideal. Let $v_o \approx V_o$ be held constant at 5 V by controlling the switch duty ratio D. Calculate the minimum inductance L required to keep the converter operation in a continuous-conduction mode under all conditions if V_d is 10–40 V, $P_o \geq$ 5 W, and f_s = 50 kHz.

7-2 Consider all components to be ideal. Assume V_o = 5 V, f_s = 20 kHz, L = 1 mH, and C = 470 μF. Calculate ΔV_o (peak–peak) if V_d = 12.6 V, and I_o = 200 mA.

7-3 In Problem 7-2, calculate the rms value of the ripple current through L and, hence, through C.

7-4 Derive an expression for ΔV_o (peak–peak) in a discontinuous-conduction mode in terms of the circuit parameters.

7-5 In Problem 7-2, calculate ΔV_o (peak–peak) if I_o (instead of being 200 mA) is equal to $\frac{1}{2} I_{oB}$.

7-6 In the PSpice simulation of a step-down converter discussed in Chapter 4 (Figs. 4-8 through 4-10), decrease the load so that R_{load} = 80 Ω. Obtain the inductor current and the capacitor voltage waveforms in the discontinuous-conduction mode.

STEP-UP CONVERTERS

7-7 In a step-up converter, consider all components to be ideal. Let V_d be 8–16 V, V_o = 24 V (regulated), f_s = 20 kHz, and C = 470 μF. Calculate L_{min} that will keep the converter operating in a continuous-conduction mode if $P_o \geq$ 5 W.

7-8 In a step-up converter, V_d = 12 V, V_o = 24 V, I_o = 0.5 A, L = 150 μH, C = 470 μF, and f_s = 20 kHz. Calculate ΔV_o (peak–peak).

7-9 In Problem 7-8, calculate the rms value of the ripple in the diode current (which also flows through the capacitor).

7-10 Derive an expression for ΔV_o (peak–peak) in a discontinuous-conduction mode in terms of the circuit parameters.

7-11 In Problem 7-8, calculate ΔV_o peak–peak if I_o (instead of being 0.5 A) is equal to $\frac{1}{2} I_{oB}$.

BUCK–BOOST CONVERTERS

7-12 In a buck–boost converter, consider all components to be ideal. Let V_d be 8–40 V, $V_o = 15$ V (regulated), $f_s = 20$ kHz, and $C = 470$ µF. Calculate L_{min} that will keep the converter operating in a continuous-conduction mode if $P_o \geq 2$ W.

7-13 In a buck–boost converter, $V_d = 12$ V, $V_o = 15$ V, $I_o = 250$ mA, $L = 150$ µH, $C = 470$ µF, and $f_s = 20$ kHz. Calculate ΔV_o (peak–peak).

7-14 Calculate the rms value of the ripple current in Problem 7-13 through the diode and, hence, through the capacitor.

7-15 Derive an expression for ΔV_o (peak–peak) in a discontinuous-conduction mode in terms of the circuit parameters.

7-16 In Problem 7-13, calculate ΔV_o (peak–peak), if I_o (instead of being 250 mA) is equal to $\frac{1}{2} I_{oB}$.

CÚK CONVERTER

7-17 In the circuit of Example 7-3, calculate the rms current flowing through the capacitor C_1.

FULL-BRIDGE dc–dc CONVERTERS

7-18 In a full-bridge dc–dc converter using PWM bipolar voltage switching, $v_{control} = 0.5\ \hat{V}_{tri}$. Obtain V_o and I_d in terms of given V_d and I_O. By Fourier analysis, calculate the amplitudes of the switching-frequency harmonics in v_o and i_d. Assume that $i_O(t) \simeq I_O$.

7-19 Repeat Problem 7-18 for a PWM unipolar voltage-switching scheme.

7-20 Plot instantaneous power output $p_o(t)$ and the average power P_o, corresponding to i_o in Figs. 7-28e and 7-28f.

7-21 Repeat Problem 7-20 for Figs. 7-29e and 7-29f.

7-22 In a full-bridge dc–dc converter using PWM bipolar voltage switching, analytically obtain the value of (V_o/V_d) which results in the maximum (peak–peak) ripple in the output current i_o. Calculate this ripple in terms of V_d, L_a, and f_s. Assume that R_a is negligible.

7-23 Repeat Problem 7-22 for a PWM unipolar voltage-switching scheme.

7-24 Using PSpice, simulate the full-bridge dc–dc converter shown in Fig. P7-24 using (a) PWM bipolar voltage switching and (b) PWM unipolar voltage switching.

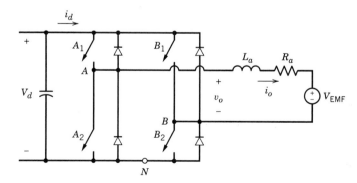

Nominal values: $V_d = 200$ V
$V_{EMF} = 79.5$ V
$R_a = 0.37\ \Omega$
$L_a = 1.5$ mH
I_o (avg) = 10 A
$f_s = 20$ kHz
duty-ratio D1 of T_{A1} and $T_{B2} = 0.708$
($\therefore v_{control} = 0.416$ V with $\hat{V}_{tri} = 1.0$ V)

Figure P7-24 From reference 5 of Chapter 4, "Power Electronics: Computer Simulation, Analysis and Education Using PSpice (evaluation, classroom version)," on diskette with a manual, Minnesota Power Electronics, P.O. Box 14503, Minneapolis, MN 55414.

REFERENCES

1. R. P. Severns and E. Bloom, *Modern DC-to-DC Switchmode Power Converter Circuits,* Van Nostrand Reinhold Company, New York, 1985.
2. G. Chryssis, *High Frequency Switching Power Supplies: Theory and Design,* McGraw-Hill, New York, 1984.
3. R. E. Tarter, *Principles of Solid-State Power Conversion,* Sams and Co., Indianapolis, IN, 1985.
4. R. D. Middlebrook and S. Cúk, *Advances in Switched-Mode Power Conversion,* Vols. I and II, TESLAco, Pasadena, CA, 1981.

CHAPTER 8

SWITCH-MODE dc–ac INVERTERS: dc ↔ SINUSOIDAL ac

8-1   INTRODUCTION

Switch-mode dc-to-ac inverters are used in ac motor drives and uninterruptible ac power supplies where the objective is to produce a sinusoidal ac output whose magnitude and frequency can both be controlled. As an example, consider an ac motor drive, shown in Fig. 8-1 in a block diagram form. The dc voltage is obtained by rectifying and filtering the line voltage, most often by the diode rectifier circuits discussed in Chapter 5. In an ac motor load, as will be discussed in Chapters 14 and 15, the voltage at its terminals is desired to be sinusoidal and adjustable in its magnitude and frequency. This is accomplished by means of the switch-mode dc-to-ac inverter of Fig. 8-1, which accepts a dc voltage as the input and produces the desired ac voltage input.

To be precise, the switch-mode inverter in Fig. 8-1 is a converter through which the power flow is reversible. However, most of the time the power flow is from the dc side to the motor on the ac side, requiring an inverter mode of operation. Therefore, these switch-mode converters are often referred to as switch-mode inverters.

To slow down the ac motor in Fig. 8-1, the kinetic energy associated with the inertia of the motor and its load is recovered and the ac motor acts as a generator. During the so-called braking of the motor, the power flows from the ac side to the dc side of the switch-mode converter and it operates in a rectifier mode. The energy recovered during the braking of the ac motor can be dissipated in a resistor, which can be switched in

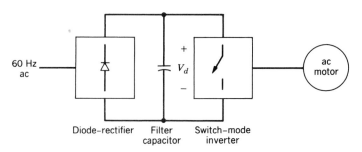

Figure 8-1   Switch-mode inverter in ac motor drive.

parallel with the dc bus capacitor for this purpose in Fig. 8-1. However, in applications where this braking is performed frequently, a better alternative is regenerative braking where the energy recovered from the motor load inertia is fed back to the utility grid, as shown in the system of Fig. 8-2. This requires that the converter connecting the drive to the utility grid be a two-quadrant converter with a reversible dc current, which can operate as a rectifier during the motoring mode of the ac motor and as an inverter during the braking of the motor. Such a reversible-current two-quadrant converter can be realized by two back-to-back connected line-frequency thyristor converters of the type discussed in Chapter 6 or by means of a switch-mode converter as shown in Fig. 8-2. There are other reasons for using such a switch-mode rectifier (called a rectifier because, most of the time, the power flows from the ac line input to the dc bus) to interface the drive with the utility system. A detailed discussion of switch-mode rectifiers is deferred to Chapter 18, which deals with issues regarding the interfacing of power electronics equipment with the utility grid.

In this chapter, we will discuss inverters with single-phase and three-phase ac outputs. The input to switch-mode inverters will be assumed to be a dc voltage source, as was assumed in the block diagrams of Fig. 8-1 and 8-2. Such inverters are referred to as voltage source inverters (VSIs). The other types of inverters, now used only for very high power ac motor drives, are the current source inverters (CSIs), where the dc input to the inverter is a dc current source. Because of their limited applications, the CSIs are not discussed in this chapter, and their discussion is deferred to ac motor drives Chapters 14 and 15.

The VSIs can be further divided into the following three general categories:

1. *Pulse-width-modulated inverters.* In these inverters, the input dc voltage is essentially constant in magnitude, such as in the circuit of Fig. 8-1, where a diode rectifier is used to rectify the line voltage. Therefore, the inverter must control the magnitude and the frequency of the ac output voltages. This is achieved by PWM of the inverter switches and hence such inverters are called PWM inverters. There are various schemes to pulse-width modulate the inverter switches in order to shape the output ac voltages to be as close to a sine wave as possible. Out of these various PWM schemes, a scheme called the sinusoidal PWM will be discussed in detail, and some of the other PWM techniques will be described in a separate section at the end of this chapter.

2. *Square-wave inverters.* In these inverters, the input dc voltage is controlled in order to control the magnitude of the output ac voltage, and therefore the inverter has to control only the frequency of the output voltage. The output ac voltage has a waveform similar to a square wave, and hence these inverters are called square-wave inverters.

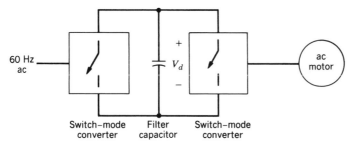

Figure 8-2 Switch-mode converters for motoring and regenerative braking in ac motor drive.

3. *Single-phase inverters with voltage cancellation.* In case of inverters with single-phase output, it is possible to control the magnitude and the frequency of the inverter output voltage, even though the input to the inverter is a constant dc voltage and the inverter switches are not pulse-width modulated (and hence the output voltage waveshape is like a square wave). Therefore, these inverters combine the characteristics of the previous two inverters. It should be noted that the voltage cancellation technique works only with single-phase inverters and not with three-phase inverters.

8-2   BASIC CONCEPTS OF SWITCH-MODE INVERTERS

In this section, we will consider the requirements on the switch-mode inverters. For simplicity, let us consider a single-phase inverter, which is shown in block diagram form in Fig. 8-3a, where the output voltage of the inverter is filtered so that v_o can be assumed to be sinusoidal. Since the inverter supplies an inductive load such as an ac motor, i_o will lag v_o, as shown in Fig. 8-3b. The output waveforms of Fig. 8-3b show that during interval 1, v_o and i_o are both positive, whereas during interval 3, v_o and i_o are both negative. Therefore, during intervals 1 and 3, the instantaneous power flow p_o ($= v_o i_o$) is from the dc side to the ac side, corresponding to an inverter mode of operation. In contrast, v_o and i_o are of opposite signs during intervals 2 and 4, and therefore p_o flows from the ac side to the dc side of the inverter, corresponding to a rectifier mode of operation. Therefore, the switch-mode inverter of Fig. 8-3a must be capable of operating in all four quadrants of the i_o–v_o plane, as shown in Fig. 8-3c during each cycle of the ac

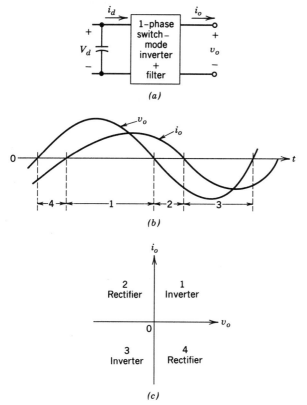

Figure 8-3   Single-phase switch-mode inverter.

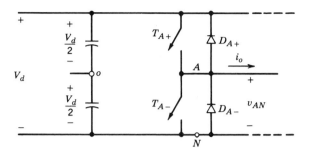

Figure 8-4 One-leg switch-mode inverter.

output. Such a four-quadrant inverter was first introduced in Chapter 7, where it was shown that in a full-bridge converter of Fig. 7-27, i_o is reversible and v_o can be of either polarity independent of the direction of i_o. Therefore, the full-bridge converter of Fig. 7-27 meets the switch-mode inverter requirements. Only one of the two legs of the full-bridge converter, for example leg A, is shown in Fig. 8-4. All the dc-to-ac inverter topologies described in this chapter are derived from the one-leg converter of Fig. 8-4. For ease of explanation, it will be assumed that in the inverter of Fig. 8-4, the midpoint "o" of the dc input voltage is available, although in most inverters it is not needed and also not available.

To understand the dc-to-ac inverter characteristics of the one-leg inverter of Fig. 8-4, we will first assume that the input dc voltage V_d is constant and that the inverter switches are pulse-width modulated to shape and control the output voltage. Later on, it will be shown that the square-wave switching is a special case of the PWM switching scheme.

8-2-1   PULSE-WIDTH-MODULATED SWITCHING SCHEME

We discussed the PWM of full-bridge dc–dc converters in Chapter 7. There, a control signal $v_{control}$ (constant or slowly varying in time) was compared with a repetitive switching-frequency triangular waveform in order to generate the switching signals. Controlling the switch duty ratios in this way allowed the average dc voltage output to be controlled.

In inverter circuits, the PWM is a bit more complex, since as mentioned earlier, we would like the inverter output to be sinusoidal with magnitude and frequency controllable. In order to produce a sinusoidal output voltage waveform at a desired frequency, a sinusoidal control signal at the desired frequency is compared with a triangular waveform, as shown in Fig. 8-5a. The frequency of the triangular waveform establishes the inverter switching frequency and is generally kept constant along with its amplitude $\hat{V}_{tri}$.

Before discussing the PWM behavior, it is necessary to define a few terms. The triangular waveform v_{tri} in Fig. 8-5a is at a switching frequency f_s, which establishes the frequency with which the inverter switches are switched (f_s is also called the carrier frequency). The control signal $v_{control}$ is used to modulate the switch duty ratio and has a frequency f_1, which is the desired fundamental frequency of the inverter voltage output (f_1 is also called the modulating frequency), recognizing that the inverter output voltage will not be a perfect sine wave and will contain voltage components at harmonic frequencies of f_1. The amplitude modulation ratio m_a is defined as

$$m_a = \frac{\hat{V}_{control}}{\hat{V}_{tri}}$$

(8-1)

where $\hat{V}_{control}$ is the peak amplitude of the control signal. The amplitude $\hat{V}_{tri}$ of the triangular signal is generally kept constant.

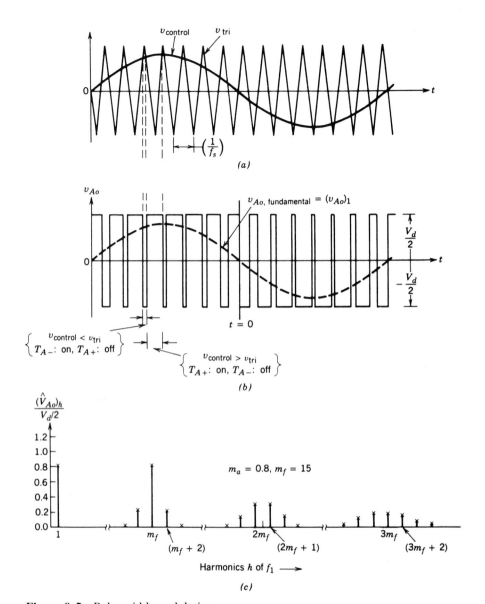

Figure 8-5   Pulse-width modulation.

The frequency modulation ratio m_f is defined as

$$m_f = \frac{f_s}{f_1} \tag{8-2}$$

In the inverter of Fig. 8-4b, the switches T_{A+} and T_{A-} are controlled based on the comparison of $v_{control}$ and v_{tri}, and the following output voltage results, independent of the direction of i_o:

$$v_{control} > v_{tri}, \qquad T_{A+} \text{ is on,} \qquad v_{Ao} = \tfrac{1}{2}V_d \tag{8-3}$$

or

$$v_{control} < v_{tri}, \qquad T_{A-} \text{ is on,} \qquad v_{Ao} = -\tfrac{1}{2}V_d$$

Since the two switches are never off simultaneously, the output voltage v_{Ao} fluctuates between two values ($\frac{1}{2}V_d$ and $-\frac{1}{2}V_d$). Voltage v_{Ao} and its fundamental frequency component (dashed curve) are shown in Fig. 8-5b, which are drawn for $m_f = 15$ and $m_a = 0.8$.

The harmonic spectrum of v_{Ao} under the conditions indicated in Figs. 8-5a and 8-5b is shown in Fig. 8-5c, where the normalized harmonic voltages $(\hat{V}_{Ao})_h/\frac{1}{2}V_d$ having significant amplitudes are plotted. This plot (for $m_a \leq 1.0$) shows three items of importance:

1. The peak amplitude of the fundamental-frequency component $(\hat{V}_{Ao})_1$ is m_a times $\frac{1}{2}V_d$. This can be explained by first considering a constant v_{control} as shown in Fig. 8-6a. This results in an output waveform v_{Ao}. From the discussion of Chapter 7 regarding the PWM in a full-bridge dc–dc converter, it can be noted that the average output voltage (or more specifically, the output voltage averaged over one switching time period $T_s = 1/f_s$) V_{Ao} depends on the ratio of v_{control} to $\hat{V}_{\text{tri}}$ for a given V_d:

$$V_{Ao} = \frac{v_{\text{control}}}{\hat{V}_{\text{tri}}}\frac{V_d}{2} \qquad v_{\text{control}} \leq \hat{V}_{\text{tri}} \tag{8-4}$$

Let us assume (though this assumption is not necessary) that v_{control} varies very little during a switching time period, that is, m_f is large, as shown in Fig. 8-6b. Therefore, assuming v_{control} to be constant over a switching time period, Eq. 8-4 indicates how the "instantaneous average" value of v_{Ao} (averaged over one switching time period T_s) varies from one switching time period to the next. This "instantaneous average" is the same as the fundamental-frequency component of v_{Ao}.

The foregoing argument shows why v_{control} is chosen to be sinusoidal to provide a sinusoidal output voltage with fewer harmonics. Let the control voltage vary sinusoidally at the frequency $f_1 = \omega_1/2\pi$, which is the desired (or the fundamental) frequency of the inverter output:

$$v_{\text{control}} = \hat{V}_{\text{control}} \sin \omega_1 t$$

where

$$\hat{V}_{\text{control}} \leq \hat{V}_{\text{tri}} \tag{8-5}$$

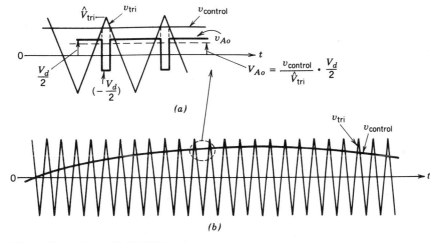

Figure 8-6 Sinusoidal PWM.

Using Eqs. 8-4 and 8-5 and the foregoing arguments, which show that the fundamental-frequency component $(v_{Ao})_1$ varies sinusoidally and in phase with $v_{control}$ as a function of time, results in

$$(v_{Ao})_1 = \frac{\hat{V}_{control}}{\hat{V}_{tri}} \sin \omega_1 t \frac{V_d}{2}$$

$$= m_a \sin \omega_1 t \frac{V_d}{2} \quad \text{for } m_a \leq 1.0 \qquad (8\text{-}6)$$

Therefore,

$$(\hat{V}_{Ao})_1 = m_a \frac{V_d}{2} \qquad m_a \leq 1.0 \qquad (8\text{-}7)$$

which shows that in a sinusoidal PWM, the amplitude of the fundamental-frequency component of the output voltage varies linearly with m_a (provided $m_a \leq 1.0$). Therefore, the range of m_a from 0 to 1 is referred to as the linear range.

2. The harmonics in the inverter output voltage waveform appear as sidebands, centered around the switching frequency and its multiples, that is, around harmonics m_f, $2m_f$, $3m_f$, and so on. This general pattern holds true for all values of m_a in the range 0–1.

For a frequency modulation ratio $m_f \leq 9$ (which is always the case, except in very high power ratings), the harmonic amplitudes are almost independent of m_f, though m_f defines the frequencies at which they occur. Theoretically, the frequencies at which voltage harmonics occur can be indicated as

$$f_h = (jm_f \pm k)f_1$$

that is, the harmonic order h corresponds to the kth sideband of j times the frequency modulation ratio m_f:

$$h = j(m_f) \pm k \qquad (8\text{-}8)$$

where the fundamental frequency corresponds to $h = 1$. For odd values of j, the harmonics exist only for even values of k. For even values of j, the harmonics exist only for odd values of k.

In Table 8-1, the normalized harmonics $(\hat{V}_{Ao})_h / \frac{1}{2}V_d$ are tabulated as a function of the amplitude modulation ratio m_a, assuming $m_f \geq 9$. Only those with significant amplitudes up to $j = 4$ in Eq. 8-8 are shown. A detailed discussion is provided in reference 6.

It will be useful later on to recognize that in the inverter circuit of Fig. 8-4

$$v_{AN} = v_{Ao} + \frac{1}{2}V_d \qquad (8\text{-}9)$$

Therefore, the harmonic voltage components in v_{AN} and v_{Ao} are the same:

$$(\hat{V}_{AN})_h = (\hat{V}_{Ao})_h \qquad (8\text{-}10)$$

Table 8-1 shows that Eq. 8-7 is followed almost exactly and the amplitude of the fundamental component in the output voltage varies linearly with m_a.

3. The harmonic m_f should be an odd integer. Choosing m_f as an odd integer results in an odd symmetry $[f(-t) = -f(t)]$ as well as a half-wave symmetry $[f(t) = -f(t + \frac{1}{2}T_1)]$ with the time origin shown in Fig. 8-5b, which is plotted for $m_f = 15$. Therefore, only odd harmonics are present and the even harmonics disappear from the waveform of v_{Ao}. Moreover, only the coefficients of the sine series in the Fourier analysis are finite; those for the cosine series are zero. The harmonic spectrum is plotted in Fig. 8-5c.

Table 8-1 Generalized Harmonics of v_{Ao} for a Large m_f.

h \\ m_a	0.2	0.4	0.6	0.8	1.0
1 Fundamental	0.2	0.4	0.6	0.8	1.0
m_f	1.242	1.15	1.006	0.818	0.601
$m_f \pm 2$	0.016	0.061	0.131	0.220	0.318
$m_f \pm 4$					0.018
$2m_f \pm 1$	0.190	0.326	0.370	0.314	0.181
$2m_f \pm 3$		0.024	0.071	0.139	0.212
$2m_f \pm 5$				0.013	0.033
$3m_f$	0.335	0.123	0.083	0.171	0.113
$3m_f \pm 2$	0.044	0.139	0.203	0.176	0.062
$3m_f \pm 4$			0.012	0.047	0.104	0.157
$3m_f \pm 6$				0.016	0.044
$4m_f \pm 1$	0.163	0.157	0.008	0.105	0.068
$4m_f \pm 3$	0.012	0.070	0.132	0.115	0.009
$4m_f \pm 5$			0.034	0.084	0.119
$4m_f \pm 7$				0.017	0.050

Note: $(\hat{V}_{Ao})_h / \frac{1}{2}V_d \, [= (\hat{V}_{AN})_h / \frac{1}{2}V_d]$ is tabulated as a function of m_a.

■ *Example 8-1* In the circuit of Fig. 8-4, $V_d = 300$ V, $m_a = 0.8$, $m_f = 39$, and the fundamental frequency is 47 Hz. Calculate the rms values of the fundamental-frequency voltage and some of the dominant harmonics in v_{Ao} using Table 8-1.

Solution From Table 8-1, the rms voltage at any value of h is given as

$$(V_{Ao})_h = \frac{1}{\sqrt{2}} \frac{V_d}{2} \frac{(\hat{V}_{Ao})_h}{V_d/2}$$

$$= 106.07 \frac{(\hat{V}_{Ao})_h}{V_d/2} \tag{8-11}$$

Therefore, from Table 8-1 the rms voltages are as follows:

Fundamental: $(V_{Ao})_1 = 106.07 \times 0.8 = 84.86$ V   at 47 Hz
$(V_{Ao})_{37} = 106.07 \times 0.22 = 23.33$ V   at 1739 Hz
$(V_{Ao})_{39} = 106.07 \times 0.818 = 86.76$ V   at 1833 Hz
$(V_{Ao})_{41} = 106.07 \times 0.22 = 23.33$ V   at 1927 Hz
$(V_{Ao})_{77} = 106.07 \times 0.314 = 33.31$ V   at 3619 Hz
$(V_{Ao})_{79} = 106.07 \times 0.314 = 33.31$ V   at 3713 Hz
etc.   ■

Now we discuss the selection of the switching frequency and the frequency modulation ratio m_f. Because of the relative ease in filtering harmonic voltages at high frequencies, it is desirable to use as high a switching frequency as possible, except for one significant drawback: Switching losses in the inverter switches increase proportionally with the switching frequency f_s. Therefore, in most applications, the switching frequency is selected to be either less than 6 kHz or greater than 20 kHz to be above the audible range. If the optimum switching frequency (based on the overall system performance) turns out to be somewhere in the 6–20-kHz range, then the disadvantages of increasing it to 20 kHz are often outweighed by the advantage of no audible noise with f_s of 20 kHz or greater. Therefore, in 50- or 60-Hz type applications, such as ac motor drives (where the fundamental frequency of the inverter output may be required to be as high as 200 Hz), the frequency modulation ratio m_f may be 9 or even less for switching frequencies of less than 2 kHz. On the other hand, m_f will be larger than 100 for switching frequencies higher than 20 kHz. The desirable relationships between the triangular waveform signal and the control voltage signal are dictated by how large m_f is. In the discussion here, $m_f = 21$ is treated as the borderline between large and small, though its selection is somewhat arbitrary. Here, it is assumed that the amplitude modulation ratio m_a is less than 1.

8-2-1-1 Small m_f ($m_f \leq 21$)

1. *Synchronous PWM.* For small values of m_f, the triangular waveform signal and the control signal should be synchronized to each other (synchronous PWM) as shown in Fig. 8-5a. This synchronous PWM requires that m_f be an integer. The reason for using the synchronous PWM is that the asynchronous PWM (where m_f is not an integer) results in subharmonics (of the fundamental frequency) that are very undesirable in most applications. This implies that the triangular waveform frequency varies with the desired inverter frequency (e.g., if the inverter output frequency and hence the frequency of v_{control} is 65.42 Hz and $m_f = 15$, the triangular wave frequency should be exactly $15 \times 65.42 = 981.3$ Hz).

2. *m_f should be an odd integer.* As discussed previously, m_f should be an odd integer except in single-phase inverters with PWM unipolar voltage switching, to be discussed in Section 8-3-2-2.

8-2-1-2 Large m_f ($m_f > 21$)

The amplitudes of subharmonics due to asynchronous PWM are small at large values of m_f. Therefore, at large values of m_f, the asynchronous PWM can be used where the frequency of the triangular waveform is kept constant, whereas the frequency of v_{control} varies, resulting in noninteger values of m_f (so long as they are large). However, if the inverter is supplying a load such as an ac motor, the subharmonics at zero or close to zero frequency, even though small in amplitude, will result in large currents that will be highly undesirable. Therefore, the asynchronous PWM should be avoided.

8-2-1-3 Overmodulation ($m_a > 1.0$)

In the previous discussion, it was assumed that $m_a \leq 1.0$, corresponding to a sinusoidal PWM in the linear range. Therefore, the amplitude of the fundamental-frequency voltage varies linearly with m_a, as derived in Eq. 8-7. In this range of $m_a \leq 1.0$, PWM pushes the harmonics into a high-frequency range around the switching frequency and its multiples. In spite of this desirable feature of a sinusoidal PWM in the linear range, one of the drawbacks is that the maximum available amplitude of the fundamental-frequency component is not as high as we wish. This is a natural consequence of the notches in the output voltage waveform of Fig. 8-5b.

To increase further the amplitude of the fundamental-frequency component in the output voltage, m_a is increased beyond 1.0, resulting in what is called *overmodulation*. Overmodulation causes the output voltage to contain many more harmonics in the sidebands as compared with the linear range (with $m_a \leq 1.0$), as shown in Fig. 8-7. The harmonics with dominant amplitudes in the linear range may not be dominant during

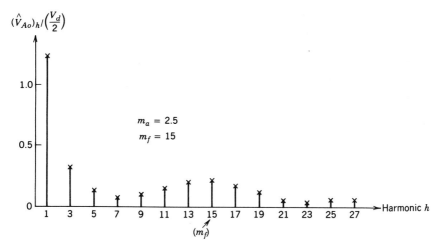

Figure 8-7   Harmonics due to overmodulation; drawn for $m_a = 2.5$ and $m_f = 15$.

overmodulation. More significantly, with overmodulation, the amplitude of the fundamental-frequency component does not vary linearly with the amplitude modulation ratio m_a. Figure 8-8 shows the normalized peak amplitude of the fundamental-frequency component $(\hat{V}_{Ao})_1/\frac{1}{2}V_d$ as a function of the amplitude modulation ratio m_a. Even at reasonably large values of m_f, $(\hat{V}_{Ao})_1/\frac{1}{2}V_d$ depends on m_f in the overmodulation region. This is contrary to the linear range $(m_a \leq 1.0)$ where $(\hat{V}_{Ao})_1/\frac{1}{2}V_d$ varies linearly with m_a, almost independent of m_f (provided $m_f > 9$).

With overmodulation regardless of the value of m_f, it is recommended that a synchronous PWM operation be used, thus meeting the requirements indicated previously for a small value of m_f.

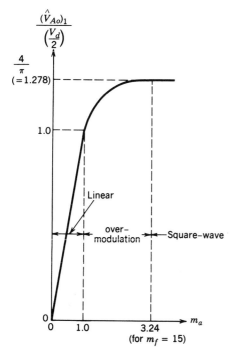

Figure 8-8   Voltage control by varying m_a.

As will be described in Chapter 11, the overmodulation region is avoided in uninterruptible power supplies because of a stringent requirement on minimizing the distortion in the output voltage. In induction motor drives described in Chapter 14, overmodulation is normally used.

For sufficiently large values of m_a, the inverter voltage waveform degenerates from a pulse-width-modulated waveform into a square wave, which is discussed in detail in the next section. From Fig. 8-8 and the discussion of square-wave switching to be presented in the next section, it can be concluded that in the overmodulation region with $m_a > 1$

$$\frac{V_d}{2} < (\hat{V}_{Ao})_1 < \frac{4}{\pi}\frac{V_d}{2} \tag{8-12}$$

8-2-2   SQUARE-WAVE SWITCHING SCHEME

In the square-wave switching scheme, each switch of the inverter leg of Fig. 8-4 is on for one half-cycle (180°) of the desired output frequency. This results in an output voltage waveform as shown in Fig. 8-9a. From Fourier analysis, the peak values of the fundamental-frequency and harmonic components in the inverter output waveform can be obtained for a given input V_d as

$$(\hat{V}_{Ao})_1 = \frac{4}{\pi}\frac{V_d}{2} = 1.273\left(\frac{V_d}{2}\right) \tag{8-13}$$

and

$$(\hat{V}_{Ao})_h = \frac{(\hat{V}_{Ao})_1}{h} \tag{8-14}$$

where the harmonic order h takes on only odd values, as shown in Fig. 8-9b. It should be noted that the square-wave switching is also a special case of the sinusoidal PWM switching when m_a becomes so large that the control voltage waveform intersects with the triangular waveform in Fig. 8-5a only at the zero crossing of v_{control}. Therefore, the output voltage is independent of m_a in the square-wave region, as shown in Fig. 8-8.

One of the advantages of the square-wave operation is that each inverter switch changes its state only twice per cycle, which is important at very high power levels where the solid-state switches generally have slower turn-on and turn-off speeds. One of the serious disadvantages of square-wave switching is that the inverter is not capable of regulating the output voltage magnitude. Therefore, the dc input voltage V_d to the inverter must be adjusted in order to control the magnitude of the inverter output voltage.

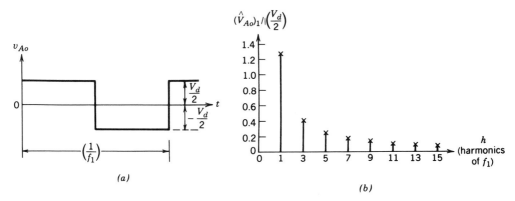

Figure 8-9   Square-wave switching.

8-3 SINGLE-PHASE INVERTERS

8-3-1 HALF-BRIDGE INVERTERS (SINGLE PHASE)

Figure 8-10 shows the half-bridge inverter. Here, two equal capacitors are connected in series across the dc input and their junction is at a midpotential, with a voltage $\frac{1}{2}V_d$ across each capacitor. Sufficiently large capacitances should be used such that it is reasonable to assume that the potential at point o remains essentially constant with respect to the negative dc bus N. Therefore, this circuit configuration is identical to the basic one-leg inverter discussed in detail earlier, and $v_o = v_{Ao}$.

Assuming PWM switching, we find that the output voltage waveform will be exactly as in Fig. 8-5b. It should be noted that regardless of the switch states, the current between the two capacitors C_+ and C_- (which have equal and very large values) divides equally. When T_+ is on, either T_+ or D_+ conducts depending on the direction of the output current, and i_o splits equally between the two capacitors. Similarly, when the switch T_- is in its on state, either T_- or D_- conducts depending on the direction of i_o, and i_o splits equally between the two capacitors. Therefore, the capacitors C_+ and C_- are "effectively" connected in parallel in the path of i_o. This also explains why the junction o in Fig. 8-10 stays at midpotential.

Since i_o must flow through the parallel combination of C_+ and C_-, i_o in steady state cannot have a dc component. Therefore, these capacitors act as dc blocking capacitors, thus eliminating the problem of transformer saturation from the primary side, if a transformer is used at the output to provide electrical isolation. Since the current in the primary winding of such a transformer would not be forced to zero with each switching, the transformer leakage inductance energy does not present a problem to the switches.

In a half-bridge inverter, the peak voltage and current ratings of the switches are as follows:

$$V_T = V_d \tag{8-15}$$

and

$$I_T = i_{o,\text{peak}} \tag{8-16}$$

8-3-2 FULL-BRIDGE INVERTERS (SINGLE PHASE)

A full-bridge inverter is shown in Fig. 8-11. This inverter consists of two one-leg inverters of the type discussed in Section 8-2 and is preferred over other arrangements in higher power ratings. With the same dc input voltage, the maximum output voltage of the

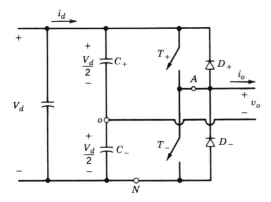

Figure 8-10 Half-bridge inverter.

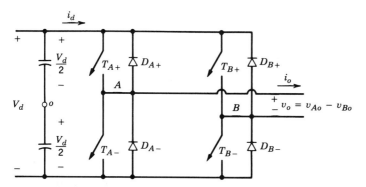

Figure 8-11   Single-phase full-bridge inverter.

full-bridge inverter is twice that of the half-bridge inverter. This implies that for the same power, the output current and the switch currents are one-half of those for a half-bridge inverter. At high power levels, this is a distinct advantage, since it requires less paralleling of devices.

8-3-2-1   PWM with Bipolar Voltage Switching

This PWM scheme was first discussed in connection with the full-bridge dc–dc converters in Chapter 7. Here, the diagonally opposite switches (T_{A+}, T_{B-}) and (T_{A-}, T_{B+}) from the two legs in Fig. 8-11 are switched as switch pairs 1 and 2, respectively. With this type of PWM switching, the output voltage waveform of leg A is identical to the output of the basic one-leg inverter in Section 8-2, which is determined in the same manner by comparison of v_{control} and v_{tri} in Fig. 8-12a. The output of inverter leg B is negative of

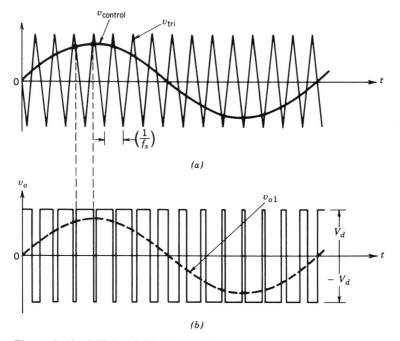

Figure 8-12   PWM with bipolar voltage switching.

the leg A output; for example, when T_{A+} is on and v_{Ao} is equal to $+\frac{1}{2}V_d$, T_{B-} is also on and $v_{Bo} = -\frac{1}{2}V_d$. Therefore

$$v_{Bo}(t) = -v_{Ao}(t) \qquad (8\text{-}17)$$

and

$$v_o(t) = v_{Ao}(t) - v_{Bo}(t) = 2v_{Ao}(t) \qquad (8\text{-}18)$$

The v_o waveform is shown in Fig. 8-12*b*. The analysis carried out in Section 8-2 for the basic one-leg inverter completely applies to this type of PWM switching. Therefore, the peak of the fundamental-frequency component in the output voltage ($\hat{V}_{o1}$) can be obtained from Eqs. 8-7, 8-12, and 8-18 as

$$\hat{V}_{o1} = m_a V_d \qquad (m_a \leq 1.0) \qquad (8\text{-}19)$$

and

$$V_d < \hat{V}_{o1} < \frac{4}{\pi}V_d \qquad (m_a > 1.0) \qquad (8\text{-}20)$$

In Fig. 8-12*b*, we observe that the output voltage v_o switches between $-V_d$ and $+V_d$ voltage levels. That is the reason why this type of switching is called a PWM with bipolar voltage switching. The amplitudes of harmonics in the output voltage can be obtained by using Table 8-1, as illustrated by the following example.

■ *Example 8-2* In the full-bridge converter circuit of Fig. 8-11, $V_d = 300$ V, $m_a = 0.8$, $m_f = 39$, and the fundamental frequency is 47 Hz. Calculate the rms values of the fundamental-frequency voltage and some of the dominant harmonics in the output voltage v_o if a PWM bipolar voltage-switching scheme is used.

Solution From Eq. 8-18, the harmonics in v_o can be obtained by multiplying the harmonics in Table 8-1 and Example 8-1 by a factor of 2. Therefore from Eq. 8-11, the rms voltage at any harmonic h is given as

$$(V_o)_h = \frac{1}{\sqrt{2}} \cdot 2 \cdot \frac{V_d}{2} \frac{(\hat{V}_{Ao})_h}{V_d/2} = \frac{V_d}{\sqrt{2}} \frac{(\hat{V}_{Ao})_h}{V_d/2}$$

$$= 212.13 \frac{(\hat{V}_{Ao})_h}{V_d/2} \qquad (8\text{-}21)$$

Therefore, the rms voltages are as follows:

Fundamental: $V_{o1} = 212.13 \times 0.8 = 169.7$ V   at 47 Hz
$(V_o)_{37} = 212.13 \times 0.22 = 46.67$ V   at 1739 Hz
$(V_o)_{39} = 212.13 \times 0.818 = 173.52$ V   at 1833 Hz
$(V_o)_{41} = 212.13 \times 0.22 = 46.67$ V   at 1927 Hz
$(V_o)_{77} = 212.13 \times 0.314 = 66.60$ V   at 3619 Hz
$(V_o)_{79} = 212.13 \times 0.314 = 66.60$ V   at 3713 Hz
etc.   ■

dc-Side Current i_d. It is informative to look at the dc-side current i_d in the PWM biopolar voltage-switching scheme.

For simplicity, fictitious *L-C* high-frequency filters will be used at the dc side as well as at the ac side, as shown in Fig. 8-13. The switching frequency is assumed to be very high, approaching infinity. Therefore, to filter out the high-switching-frequency compo-

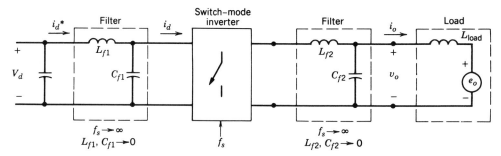

Figure 8-13 Inverter with "fictitious" filters.

nents in v_o and i_d, the filter components L and C required in both ac- and dc-side filters approach zero. This implies that the energy stored in the filters is negligible. Since the converter itself has no energy storage elements, the instantaneous power input must equal the instantaneous power output.

Having made these assumptions, v_o in Fig. 8-13 is a pure sine wave at the fundamental output frequency ω_1,

$$v_{o1} = v_o = \sqrt{2}V_o\sin \omega_1 t \tag{8-22}$$

If the load is as shown in Fig. 8-13, where e_o is a sine wave at frequency ω_1, then the output current would also be sinusoidal and would lag v_o for an inductive load such as an ac motor:

$$i_o = \sqrt{2}I_o\sin(\omega_1 t - \phi) \tag{8-23}$$

where ϕ is the angle by which i_o lags v_o.

On the dc side, the L-C filter will filter the high-switching-frequency components in i_d, and i_d^* would only consist of the low-frequency and dc components.

Assuming that no energy is stored in the filters,

$$V_d i_d^*(t) = v_o(t)i_o(t) = \sqrt{2}V_o\sin \omega_1 t\sqrt{2}I_o\sin(\omega_1 t - \phi) \tag{8-24}$$

Therefore

$$i_d^*(t) = \frac{V_oI_o}{V_d}\cos \phi - \frac{V_oI_o}{V_d}\cos(2\omega_1 t - \phi) = I_d + i_{d2} \tag{8-25}$$

$$= I_d - \sqrt{2}I_{d2}\cos(2\omega_1 t - \phi) \tag{8-26}$$

where

$$I_d = \frac{V_oI_o}{V_d}\cos \phi \tag{8-27}$$

and

$$I_{d2} = \frac{1}{\sqrt{2}}\frac{V_oI_o}{V_d} \tag{8-28}$$

Equation 8-26 for i_d^* shows that it consists of a dc component I_d, which is responsible for the power transfer from V_d on the dc side of the inverter to the ac side. Also, i_d^* contains a sinusoidal component at twice the fundamental frequency. The inverter input current i_d consists of i_d^* and the high-frequency components due to inverter switchings, as shown in Fig. 8-14.

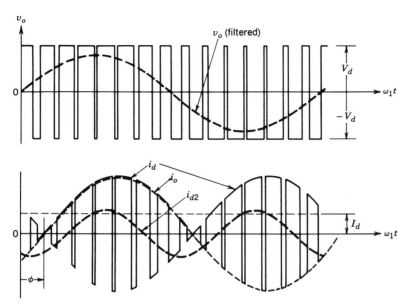

Figure 8-14 The dc-side current in a single-phase inverter with PWM bipolar voltage switching.

In practical systems, the previous assumption of a constant dc voltage as the input to the inverter is not entirely valid. Normally, this dc voltage is obtained by rectifying the ac utility line voltage. A large capacitor is used across the rectifier output terminals to filter the dc voltage. The ripple in the capacitor voltage, which is also the dc input voltage to the inverter, is due to two reasons: (1) The rectification of the line voltage to produce dc does not result in a pure dc as discussed in Chapters 5 and 6, dealing with the line-frequency rectifiers. (2) As shown earlier by Eq. 8-26, the current drawn by a single-phase inverter from the dc side is not a constant dc but has a second harmonic component (of the fundamental frequency at the inverter output) in addition to the high-switching-frequency components. The second harmonic current component results in a ripple in the capacitor voltage, although the voltage ripple due to the high switching frequencies is essentially negligible.

8-3-2-2 PWM with Unipolar Voltage Switching

In PWM with unipolar voltage switching, the switches in the two legs of the full-bridge inverter of Fig. 8-11 are not switched simultaneously, as in the previous PWM scheme. Here, the legs A and B of the full-bridge inverter are controlled separately by comparing v_{tri} with v_{control} and $-v_{\text{control}}$, respectively. As shown in Fig. 8-15a, the comparison of v_{control} with the triangular waveform results in the following logic signals to control the switches in leg A:

$$v_{\text{control}} > v_{\text{tri}}: \quad T_{A+} \text{ on} \quad \text{and} \quad v_{AN} = V_d \qquad (8\text{-}29)$$
$$v_{\text{control}} < v_{\text{tri}}: \quad T_{A-} \text{ on} \quad \text{and} \quad v_{AN} = 0$$

The output voltage of inverter leg A with respect to the negative dc bus N is shown in Fig. 8-15b. For controlling the leg B switches, $-v_{\text{control}}$ is compared with the same triangular waveform, which yields the following:

$$(-v_{\text{control}}) > v_{\text{tri}}: \quad T_{B+} \text{ on} \quad \text{and} \quad v_{BN} = V_d \qquad (8\text{-}30)$$
$$(-v_{\text{control}}) < v_{\text{tri}}: \quad T_{B-} \text{ on} \quad \text{and} \quad v_{BN} = 0$$

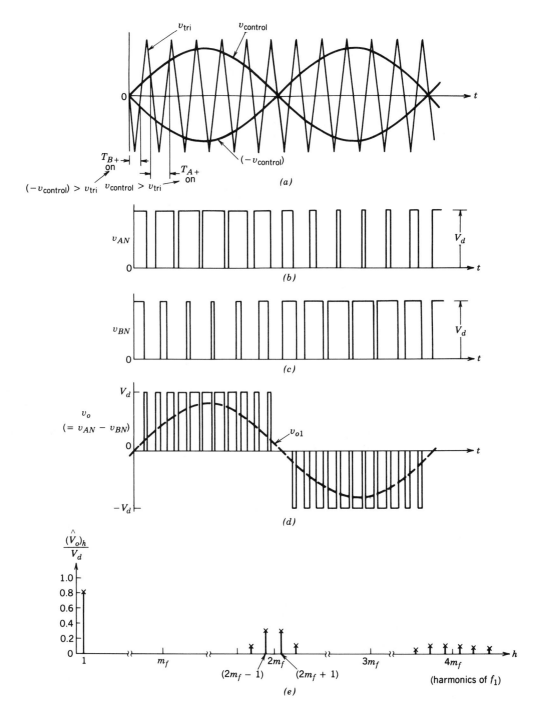

Figure 8-15 PWM with unipolar voltage switching (single phase).

Because of the feedback diodes in antiparallel with the switches, the foregoing voltages given by Eqs. 8-29 and 8-30 are independent of the direction of the output current i_o.

The waveforms of Fig. 8-15 show that there are four combinations of switch on-states and the corresponding voltage levels:

1. T_{A+}, T_{B-} on: $v_{AN} = V_d$, $v_{BN} = 0$; $v_o = V_d$
2. T_{A-}, T_{B+} on: $v_{AN} = 0$, $v_{BN} = V_d$; $v_o = -V_d$
3. T_{A+}, T_{B+} on: $v_{AN} = V_d$, $v_{BN} = V_d$; $v_o = 0$
4. T_{A-}, T_{B-} on: $v_{AN} = 0$, $v_{BN} = 0$; $v_o = 0$

$$(8\text{-}31)$$

We notice that when both the upper switches are on, the output voltage is zero. The output current circulates in a loop through T_{A+} and D_{B+} or D_{A+} and T_{B+} depending on the direction of i_o. During this interval, the input current i_d is zero. A similar condition occurs when both bottom switches T_{A-} and T_{B-} are on.

In this type of PWM scheme, when a switching occurs, the output voltage changes between zero and $+V_d$ or between zero and $-V_d$ voltage levels. For this reason, this type of PWM scheme is called PWM with a unipolar voltage switching, as opposed to the PWM with bipolar (between $+V_d$ and $-V_d$) voltage-switching scheme described earlier. This scheme has the advantage of "effectively" doubling the switching frequency as far as the output harmonics are concerned, compared to the bipolar voltage-switching scheme. Also, the voltage jumps in the output voltage at each switching are reduced to V_d, as compared to $2V_d$ in the previous scheme.

The advantage of "effectively" doubling the switching frequency appears in the harmonic spectrum of the output voltage waveform, where the lowest harmonics (in the idealized circuit) appear as sidebands of twice the switching frequency. It is easy to understand this if we choose the frequency modulation ratio m_f to be even (m_f should be odd for PWM with bipolar voltage switching) in a single-phase inverter. The voltage waveforms v_{AN} and v_{BN} are displaced by $180°$ of the fundamental frequency f_1 with respect to each other. Therefore, the harmonic components at the switching frequency in v_{AN} and v_{BN} have the same phase ($\phi_{AN} - \phi_{BN} = 180° \cdot m_f = 0°$, since the waveforms are $180°$ displaced and m_f is assumed to be even). This results in the cancellation of the harmonic component at the switching frequency in the output voltage $v_o = v_{AN} - v_{BN}$. In addition, the sidebands of the switching-frequency harmonics disappear. In a similar manner, the other dominant harmonic at twice the switching frequency cancels out, while its sidebands do not. Here also

$$\hat{V}_{o1} = m_a V_d \qquad (m_a \leq 1.0) \qquad (8\text{-}32)$$

and

$$V_d < \hat{V}_{o1} < \frac{4}{\pi} V_d \qquad (m_a > 1.0) \qquad (8\text{-}33)$$

■ *Example 8-3* In Example 8-2, suppose that a PWM with unipolar voltage-switching scheme is used, with $m_f = 38$. Calculate the rms values of the fundamental-frequency voltage and some of the dominant harmonics in the output voltage.

Solution Based on the discussion of unipolar voltage switching, the harmonic order h can be written as

$$h = j(2m_f) \pm k \qquad (8\text{-}34)$$

where the harmonics exist as sidebands around $2m_f$ and the multiples of $2m_f$. Since h is odd, k in Eq. 8-34 attains only odd values. From Example 8-2

$$(V_o)_h = 212.13 \frac{(V_{Ao})_h}{V_d/2} \qquad (8\text{-}35)$$

Using Eq. 8-35 and Table 8-1, we find that the rms voltages are as follows:

At fundamental or 47 Hz: $V_{o1} = 0.8 \times 212.13 = 169.7 \ V$

At $h = 2m_f - 1 = 75 \ or \ 3525 \ Hz$: $(V_o)_{75} = 0.314 \times 212.13 = 66.60 \ V$

At $h = 2m_f + 1 = 77 \ or \ 3619 \ Hz$: $(V_o)_{77} = 0.314 \times 212.13 = 66.60 \ V$

etc.

Comparison of the unipolar voltage switching with the bipolar voltage switching of Example 8-2 shows that, in both cases, the fundamental-frequency voltages are the same for equal m_a. However, with unipolar voltage switching, the dominant harmonic voltages centered around m_f disappear, thus resulting in a significantly lower harmonic content. ∎

dc-Side Current i_d.   Under conditions similar to those in the circuit of Fig. 8-13 for the PWM with bipolar voltage switching, Fig. 8-16 shows the dc-side current i_d for the PWM unipolar voltage-switching scheme, where $m_f = 14$ (instead of $m_f = 15$ for the bipolar voltage switching).

By comparing Figs. 8-14 and 8-16, it is clear that using PWM with unipolar voltage switching results in a smaller ripple in the current on the dc side of the inverter.

8-3-2-3   Square-Wave Operation

The full-bridge inverter can also be operated in a square-wave mode. Both types of PWM discussed earlier degenerate into the same square-wave mode of operation, where the switches (T_{A+}, T_{B-}) and (T_{B+}, T_{A-}) are operated as two pairs with a duty ratio of 0.5.

As is the case in the square-wave mode of operation, the output voltage magnitude given below is regulated by controlling the input dc voltage:

$$\hat{V}_{o1} = \frac{4}{\pi} V_d \tag{8-36}$$

8-3-2-4   Output Control by Voltage Cancellation

This type of control is feasible only in a single-phase, full-bridge inverter circuit. It is based on the combination of square-wave switching and PWM with a unipolar voltage switching. In the circuit of Fig. 8-17a, the switches in the two inverter legs are controlled separately (similar to PWM unipolar voltage switching). But all switches have a duty raio of 0.5, similar to a square-wave control. This results in waveforms for v_{AN} and v_{BN} shown in Fig. 8-17b, where the waveform overlap angle α can be controlled. During this overlap interval, the output voltage is zero as a consequence of either both top switches or both

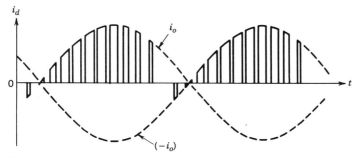

Figure 8-16   The dc-side current in a single-phase inverter with PWM unipolar voltage switching.

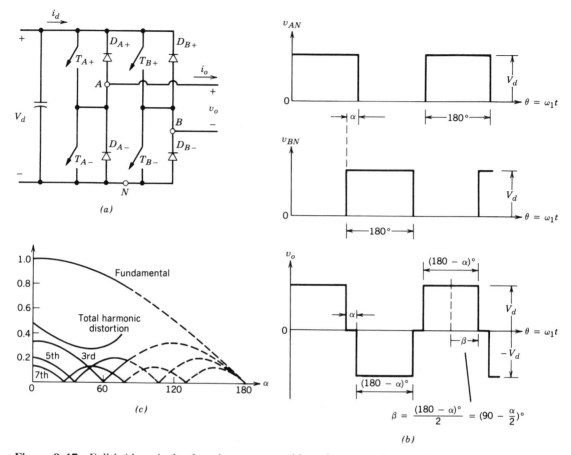

Figure 8-17 Full-bridge, single-phase inverter control by voltage cancellation: (a) power circuit; (b) waveforms; (c) normalized fundamental and harmonic voltage output and total harmonic distortion as a function of α.

bottom switches being on. With $\alpha = 0$, the output waveform is similar to a square-wave inverter with the maximum possible fundamental output magnitude.

It is easier to derive the fundamental and the harmonic frequency components of the output voltage in terms of $\beta = 90° - \frac{1}{2}\alpha$, as is shown in Fig. 8-17b:

$$(\hat{V}_o)_h = \frac{2}{\pi} \int_{-\pi/2}^{\pi/2} v_o \cos(h\theta) \, d\theta$$

$$= \frac{2}{\pi} \int_{-\beta}^{\beta} V_d \cos(h\theta) \, d\theta$$

$$\therefore (\hat{V}_o)_h = \frac{4}{\pi h} V_d \sin(h\beta) \qquad (8\text{-}37)$$

where $\beta = 90° - \frac{1}{2}\alpha$ and h is an odd integer.

Figure 8-17c shows the variation in the fundamental-frequency component as well as the harmonic voltages as a function of α. These are normalized with respect to the fundamental-frequency component for the square-wave ($\alpha = 0$) operation. The total harmonic distortion, which is the ratio of the rms value of the harmonic distortion to the rms value of the fundamental-frequency component, is also plotted as a function of α. Because of a large distortion, the curves are shown as dashed for large values of α.

8-3-2-5   Switch Utilization in Full-Bridge Inverters

Similar to a half-bridge inverter, if a transformer is utilized at the output of a full-bridge inverter, the transformer leakage inductance does not present a problem to the switches.

Independent of the type of control and the switching scheme used, the peak switch voltage and current ratings required in a full-bridge inverter are as follows:

$$V_T = V_d \tag{8-38}$$

and

$$I_T = i_{o,\text{peak}} \tag{8-39}$$

8-3-2-6   Ripple in the Single-Phase Inverter Output

The *ripple* in a repetitive waveform refers to the difference between the instantaneous values of the waveform and its fundamental-frequency component.

Figure 8-18a shows a single-phase switch-mode inverter. It is assumed to be supplying an induction motor load, which is shown by means of a simplified equivalent circuit with a counter electromotive force (emf) e_o. Since $e_o(t)$ is sinusoidal, only the sinusoidal (fundamental-frequency) components of the inverter output voltage and current are responsible for the real power transfer to the load.

We can separate the fundamental-frequency and the ripple components in v_o and i_o by applying the principle of superposition to the linear circuit of Fig. 8-18a. Let $v_o = v_{o1} + v_{\text{ripple}}$ and $i_o = i_{o1} + i_{\text{ripple}}$. Figures 8-18b, c show the circuits at the fundamental frequency and at the ripple frequency, respectively, where the ripple frequency component consists of sub-components at various harmonic frequencies.

Therefore, in a phasor form (with the fundamental-frequency components designated by subscript 1) as shown in Fig. 8-18d,

$$\mathbf{V}_{o1} = \mathbf{E}_o + \mathbf{V}_{L1} = \mathbf{E}_o + j\omega_1 L \mathbf{I}_{o1} \tag{8-40}$$

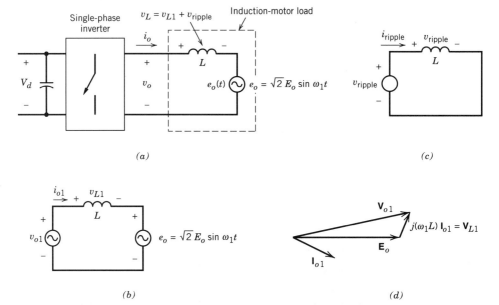

Figure 8-18   Single-phase inverter: (a) circuit; (b) fundamental-frequency components; (c) ripple frequency components; (d) fundamental-frequency phasor diagram.

Since the superposition principle is valid here, all the ripple in v_o appears across L, where

$$v_{\text{ripple}}(t) = v_o - v_{o1} \qquad (8\text{-}41)$$

The output current ripple can be calculated as

$$i_{\text{ripple}}(t) = \frac{1}{L} \int_0^t v_{\text{ripple}}(\zeta)\, d\zeta + k \qquad (8\text{-}42)$$

where k is a constant and ζ is a variable of integration.

With a properly selected time origin $t = 0$, the constant k in Eq. 8-42 will be zero. Therefore, Eqs. 8-41 and 8-42 show that the current ripple is independent of the power being transferred to the load.

As an example, Fig. 8-19a shows the ripple current for a square-wave inverter output. Figure 8-19b shows the ripple current in a PWM bipolar voltage switching. In drawing Figs. 8-19a and 8-19b, the fundamental-frequency components in the inverter output voltages are kept equal in magnitude (this requires a higher value of V_d in the PWM inverter). The PWM inverter results in a substantially smaller peak ripple current compared to the square-wave inverter. This shows the advantage of pushing the harmonics in the inverter output voltage to as high frequencies as feasible, thereby reducing the losses in the load by reducing the output current harmonics. This is achieved by using higher inverter switching frequencies, which would result in more frequent switchngs and hence higher switching losses in the inverter. Therefore, from the viewpoint of the overall system energy efficiency, a compromise must be made in selecting the inverter switching frequency.

8-3-3 PUSH–PULL INVERTERS

Figure 8-20 shows a push–pull inverter circuit. It requires a transformer with a center-tapped primary. We will initially assume that the output current i_o flows continuously. With this assumption, when the switch T_1 is on (and T_2 is off), T_1 would conduct for a positive value of i_o, and D_1 would conduct for a negative value of i_o. Therefore, regardless of the direction of i_o, $v_o = V_d/n$, where n is the transformer turns ratio between the primary half and the secondary windings, as shown in Fig. 8-20. Similarly, when T_2 is on (and T_1 is off), $v_o = -V_d/n$. A push–pull inverter can be operated in a PWM or a square-wave mode and the waveforms are identical to those in Figs. 8-5 and 8-12 for half-bridge and full-bridge inverters. The output voltage in Fig. 8-20 equals

$$\hat{V}_{o1} = m_a \frac{V_d}{n} \qquad (m_a \leq 1.0) \qquad (8\text{-}43)$$

and

$$\frac{V_d}{n} < \hat{V}_{o1} < \frac{4}{\pi}\frac{V_d}{n} \qquad (m_a > 1.0) \qquad (8\text{-}44)$$

In a push–pull inverter, the peak switch voltage and current ratings are

$$V_T = 2V_d \qquad I_T = i_{o,\text{peak}}/n \qquad (8\text{-}45)$$

The main advantage of the push–pull circuit is that no more than one switch in series conducts at any instant of time. This can be important if the dc input to the converter is from a low-voltage source, such as a battery, where the voltage drops across more than one switch in series would result in a significant reduction in energy efficiency. Also, the

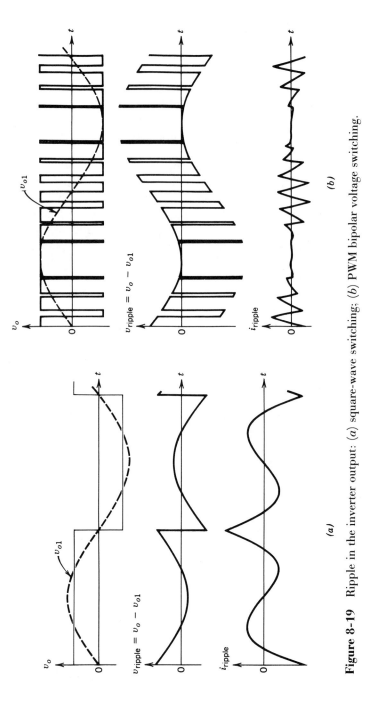

Figure 8-19 Ripple in the inverter output: (*a*) square-wave switching; (*b*) PWM bipolar voltage switching.

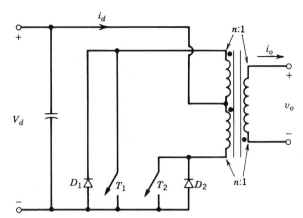

Figure 8-20 Push–pull inverter (single phase).

control drives for the two switches have a common ground. It is, however, difficult to avoid the dc saturation of the transformer in a push–pull inverter.

The output current, which is the secondary current of the transformer, is a slowly varying current at the fundamental output frequency. It can be assumed to be a constant during a switching interval. When a switching occurs, the current shifts from one half to the other half of the primary winding. This requires very good magnetic coupling between these two half-windings in order to reduce the energy associated with the leakage inductance of the two primary windings. This energy will be dissipated in the switches or in snubber circuits used to protect the switches. This is a general phenomenon associated with all converters (or inverters) with isolation where the current in one of the windings is forced to go to zero with every switching. This phenomenon is very important in the design of such converters.

In a pulse-width-modulated push–pull inverter for producing sinusoidal output (unlike those used in switch-mode dc power supplies), the transformer must be designed for the fundamental output frequency. The number of turns will therefore be high compared to a transformer designed to operate at the switching frequency in a switch-mode dc power supply. This will result in a high transformer leakage inductance, which is proportional to the square of the number of turns, provided all other dimensions are kept constant. This makes it difficult to operate a sine-wave-modulated PWM push–pull inverter at switching frequencies higher than approximately 1 kHz.

8-3-4 SWITCH UTILIZATION IN SINGLE-PHASE INVERTERS

Since the intent in this section is to compare the utilization of switches in various single-phase inverters, the circuit conditions are idealized. We will assume that $V_{d,\max}$ is the highest value of the input voltage, which establishes the switch voltage ratings. In the PWM mode, the input remains constant at $V_{d,\max}$. In the square-wave mode, the input voltage is decreased below $V_{d,\max}$ to decrease the output voltage from its maximum value. Regardless of the PWM or the square-wave mode of operation, we assume that there is enough inductance associated with the output load to yield a purely sinusoidal current (an idealized condition indeed for a square-wave output) with an rms value of $I_{o,\max}$ at the maximum load.

If the output current is assumed to be purely sinusoidal, the inverter rms volt-ampere output at the fundamental frequency equals $V_{o1}I_{o,\max}$ at the maximum rated output, where the subscript 1 designates the fundamental-frequency component of the inverter output.

With V_T and I_T as the peak voltage and current ratings of a switch, the combined utilization of all the switches in an inverter can be defined as

$$\text{Switch utilization ratio} = \frac{V_{o1}I_{o,\max}}{qV_TI_T} \qquad (8\text{-}46)$$

where q is the number of switches in an inverter.

To compare the utilization of switches in various single-phase inverters, we will initially compare them for a square-wave mode of operation at the maximum rated output. (The maximum switch utilization occurs at $V_d = V_{d,\max}$.)

Push–Pull Inverter

$$V_T = 2V_{d,\max} \qquad I_T = \sqrt{2}\frac{I_{o,\max}}{n} \qquad V_{o1,\max} = \frac{4}{\pi\sqrt{2}}\frac{V_{d,\max}}{n} \qquad q = 2 \quad (8\text{-}47)$$

$$(n = \text{turns ratio, Fig. 8-20})$$

$$\therefore \text{Maximum switch utilization ratio} = \frac{1}{2\pi} \simeq 0.16 \qquad (8\text{-}48)$$

Half-Bridge Inverter

$$V_T = V_{d,\max} \qquad I_T = \sqrt{2}I_{o,\max} \qquad V_{o1,\max} = \frac{4}{\pi\sqrt{2}}\frac{V_{d,\max}}{2} \qquad q = 2 \quad (8\text{-}49)$$

$$\therefore \text{Maximum switch utilization ratio} = \frac{1}{2\pi} \simeq 0.16 \qquad (8\text{-}50)$$

Full-Bridge Inverter

$$V_T = V_{d,\max} \qquad I_T = \sqrt{2}I_{o,\max} = \frac{4}{\pi\sqrt{2}}V_{d,\max} \qquad q = 4 \qquad (8\text{-}51)$$

$$\therefore \text{Maximum switch utilization ratio} = \frac{1}{2\pi} \simeq 0.16 \qquad (8\text{-}52)$$

This shows that in each inverter, the switch utilization is the same with

$$\text{Maximum switch utilization ratio} = \frac{1}{2\pi} \simeq 0.16 \qquad (8\text{-}53)$$

In practice, the switch utilization ratio would be much smaller than 0.16 for the following reasons: (1) switch ratings are chosen conservatively to provide safety margins; (2) in determining the switch current rating in a PWM inverter, one would have to take into account the variations in the input dc voltage available; and (3) the ripple in the output current would influence the switch current rating. Moreover, the inverter may be required to supply a short-term overload. Thus, the switch utilization ratio, in practice, would be substantially less than the 0.16 calculated.

At the lower output volt-amperes compared to the maximum rated output, the switch utilization decreases linearly. It should be noted that using a PWM switching with $m_a \le 1.0$, this ratio would be smaller by a factor of $(\pi/4)m_a$ as compared to the square-wave switching:

$$\text{Maximum switch utilization ratio} = \frac{1}{2\pi}\frac{\pi}{4}m_a = \frac{1}{8}m_a \qquad (8\text{-}54)$$

$$(\text{PWM, } m_a \le 1.0)$$

Therefore, the theoretical maximum switch utilization ratio in a PWM switching is only 0.125 at $m_a = 1$, as compared with 0.16 in a square-wave inverter.

■ *Example 8-4* In a single-phase full-bridge PWM inverter, V_d varies in a range of 295–325 V. The output voltage is required to be constant at 200 V (rms), and the maximum load current (assumed to be sinusoidal) is 10 A (rms). Calculate the combined switch utilization ratio (under these idealized conditions, not accounting for any overcurrent capabilities).

Solution In this inverter

$$V_T = V_{d,\max} = 325\ V$$
$$I_T = \sqrt{2}I_o = \sqrt{2} \times 10 = 14.14$$
$$q = \text{no. of switches} = 4$$

The maximum output volt-ampere (fundamental frequency) is

$$V_{o1}I_{o,\max} = 200 \times 10 = 2000\ VA \tag{8-55}$$

Therefore, from Eq. 8-46

$$\text{Switch utilization ratio} = \frac{V_{o1}I_{o,\max}}{qV_TI_T} = \frac{2000}{4 \times 325 \times 14.14} \approx 0.11$$

■

8-4 THREE-PHASE INVERTERS

In applications such as uninterruptible ac power supplies and ac motor drives, three-phase inverters are commonly used to supply three-phase loads. It is possible to supply a three-phase load by means of three separate single-phase inverters, where each inverter produces an output displaced by 120° (of the fundamental frequency) with respect to each other. Though this arrangement may be preferable under certain conditions, it requires either a three-phase output transformer or separate access to each of the three phases of the load. In practice, such access is generally not available. Moreover, it requires 12 switches.

The most frequently used three-phase inverter circuit consists of three legs, one for each phase, as shown in Fig. 8-21. Each inverter leg is similar to the one used for describing the basic one-leg inverter in Section 8-2. Therefore, the output of each leg, for example v_{AN} (with respect to the negative dc bus), depends only on V_d and the switch

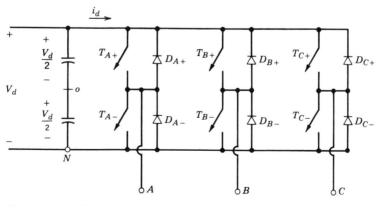

Figure 8-21 Three-phase inverter.

status; the output voltage is independent of the output load current since one of the two switches in a leg is always on at any instant. Here, we again ignore the blanking time required in practical circuits by assuming the switches to be ideal. Therefore, the inverter output voltage is independent of the direction of the load current.

8-4-1   PWM IN THREE-PHASE VOLTAGE SOURCE INVERTERS

Similar to the single-phase inverters, the objective in pulse-width-modulated three-phase inverters is to shape and control the three-phase output voltages in magnitude and frequency with an essentially constant input voltage V_d. To obtain balanced three-phase output voltages in a three-phase PWM inverter, the same triangular voltage waveform is compared with three sinusoidal control voltages that are 120° out of phase, as shown in Fig. 8-22a (which is drawn for $m_f = 15$).

It should also be noted from Fig. 8-22b that an identical amount of average dc component is present in the output voltages v_{AN} and v_{BN}, which are measured with respect to the negative dc bus. These dc components are canceled out in the line-to-line voltages, for example in v_{AB} shown in Fig. 8-22b. This is similar to what happens in a single-phase full-bridge inverter utilizing a PWM switching.

In the three-phase inverters, only the harmonics in the line-to-line voltages are of concern. The harmonics in the output of any one of the legs, for example v_{AN} in Fig. 8-22b, are identical to the harmonics in v_{Ao} in Fig. 8-5, where only the odd harmonics exist as sidebands, centered around m_f and its multiples, provided m_f is odd. Only considering the harmonic at m_f (the same applies to its odd multiples), the phase difference between the m_f harmonic in v_{AN} and v_{BN} is $(120\ m_f)°$. This phase difference will be equivalent to zero (a multiple of 360°) if m_f is odd and a multiple of 3. As a consequence, the harmonic at m_f is suppressed in the line-to-line voltage v_{AB}. The same argument applies in the suppression of harmonics at the odd multiples of m_f if m_f is chosen to be an odd multiple of 3 (where the reason for choosing m_f to be an odd multiple of 3 is to keep m_f odd and, hence, eliminate even harmonics). Thus, some of the dominant harmonics in the one-leg inverter can be eliminated from the line-to-line voltage of a three-phase inverter.

PWM considerations are summarized as follows:

1. *For low values of m_f,* to eliminate the even harmonics, a synchronized PWM should be used and m_f should be an odd integer. Moreover, m_f should be a multiple of 3 to cancel out the most dominant harmonics in the line-to-line voltage.

2. *For large values of m_f,* the comments in Section 8-2-1-2 for a single-phase PWM apply.

3. *During overmodulation ($m_a > 1.0$),* regardless of the value of m_f, the conditions pertinent to a small m_f should be observed.

8-4-1-1   Linear Modulation ($m_a \leq 1.0$)

In the linear region ($m_a \leq 1.0$), the fundamental-frequency component in the output voltage varies linearly with the amplitude modulation ratio m_a. From Figs. 8-5b and 8-22b, the peak value of the fundamental-frequency component in one of the inverter legs is

$$(\hat{V}_{AN})_1 = m_a \frac{V_d}{2} \tag{8-56}$$

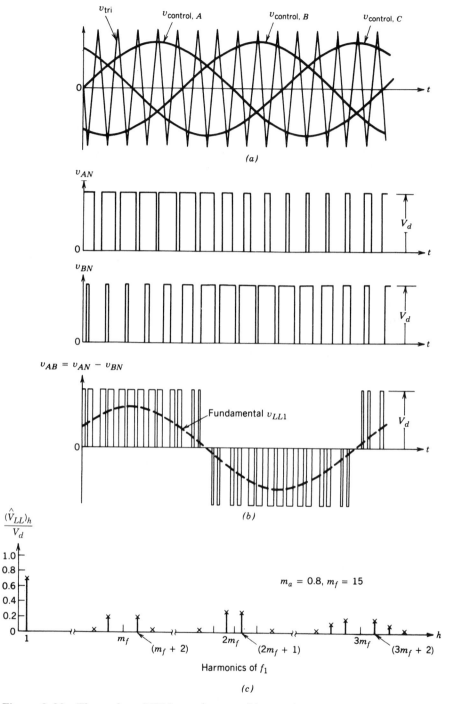

Figure 8-22 Three-phase PWM waveforms and harmonic spectrum.

Therefore, the line-to-line rms voltage at the fundamental frequency, due to 120° phase displacement between phase voltages, can be written as

$$V_{LL_1} \atop \text{(line–line, rms)} = \frac{\sqrt{3}}{\sqrt{2}} (\hat{V}_{AN})_1$$

$$= \frac{\sqrt{3}}{2\sqrt{2}} m_a V_d$$

$$\simeq 0.612 m_a V_d \qquad (m_a \leq 1.0) \qquad (8\text{-}57)$$

The harmonic components of the line-to-line output voltages can be calculated in a similar manner from Table 8-1, recognizing that some of the harmonics are canceled out in the line-to-line voltages. These rms harmonic voltages are listed in Table 8-2.

8-4-1-2 Overmodulation ($m_a > 1.0$)

In PWM overmodulation, the peak of the control voltages are allowed to exceed the peak of the triangular waveform. Unlike the linear region, in this mode of operation the fundamental-frequency voltage magnitude does not increase proportionally with m_a. This is shown in Fig. 8-23, where the rms value of the fundamental-frequency line-to-line voltage V_{LL1} is plotted as a function of m_a. Similar to a single-phase PWM, for sufficiently large values of m_a, the PWM degenerates into a square-wave inverter waveform. This results in the maximum value of V_{LL_1} equal to $0.78V_d$ as explained in the next section.

In the overmodulation region compared to the region with $m_a \leq 1.0$, more sideband harmonics appear centered around the frequencies of harmonics m_f and its multiples. However, the dominant harmonics may not have as large an amplitude as with $m_a \leq 1.0$. Therefore, the power loss in the load due to the harmonic frequencies may not be as high in the overmodulation region as the presence of additional sideband harmonics would suggest. Depending on the nature of the load and on the switching frequency, the losses due to these harmonics in overmodulation may be even less than those in the linear region of the PWM.

Table 8-2 Generalized Harmonics of v_{LL} for a Large and Odd m_f That Is a Multiple of 3.

h \ m_a	0.2	0.4	0.6	0.8	1.0
1	0.122	0.245	0.367	0.490	0.612
$m_f \pm 2$	0.010	0.037	0.080	0.135	0.195
$m_f \pm 4$				0.005	0.011
$2m_f \pm 1$	0.116	0.200	0.227	0.192	0.111
$2m_f \pm 5$				0.008	0.020
$3m_f \pm 2$	0.027	0.085	0.124	0.108	0.038
$3m_f \pm 4$		0.007	0.029	0.064	0.096
$4m_f \pm 1$	0.100	0.096	0.005	0.064	0.042
$4m_f \pm 5$			0.021	0.051	0.073
$4m_f \pm 7$				0.010	0.030

Note: $(V_{LL})_h/V_d$ are tabulated as a function of m_a where $(V_{LL})_h$ are the rms values of the harmonic voltages.

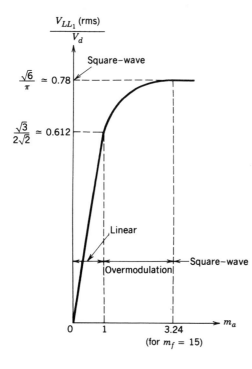

Figure 8-23 Three-phase inverter; $V_{LL_1}(\text{rms})/V_d$ as a function of m_a.

8-4-2 SQUARE-WAVE OPERATION IN THREE-PHASE INVERTERS

If the input dc voltage V_d is controllable, the inverter in Fig. 8-24a can be operated in a square-wave mode. Also, for sufficiently large values of m_a, PWM degenerates into square-wave operation and the voltage waveforms are shown in Fig. 8-24b. Here, each switch is on for 180° (i.e., its duty ratio is 50%). Therefore, at any instant of time, three switches are on.

In the square-wave mode of operation, the inverter itself cannot control the magnitude of the output ac voltages. Therefore, the dc input voltage must be controlled in order to control the output in magnitude. Here, the fundamental-frequency line-to-line rms voltage component in the output can be obtained from Eq. 8-13 for the basic one-leg inverter operating in a square-wave mode:

$$\begin{aligned} V_{LL_1} \atop (\text{rms}) &= \frac{\sqrt{3}}{\sqrt{2}} \frac{4}{\pi} \frac{V_d}{2} \\ &= \frac{\sqrt{6}}{\pi} V_d \\ &\simeq 0.78 V_d \end{aligned} \tag{8-58}$$

The line-to-line output voltage waveform does not depend on the load and contains harmonics ($6n \pm 1; n = 1, 2, \ldots$), whose amplitudes decrease inversely proportional to their harmonic order, as shown in Fig. 8-24c:

$$V_{LL_h} = \frac{0.78}{h} V_d \tag{8-59}$$

where

$$h = 6n \pm 1 \qquad (n = 1, 2, 3, \ldots)$$

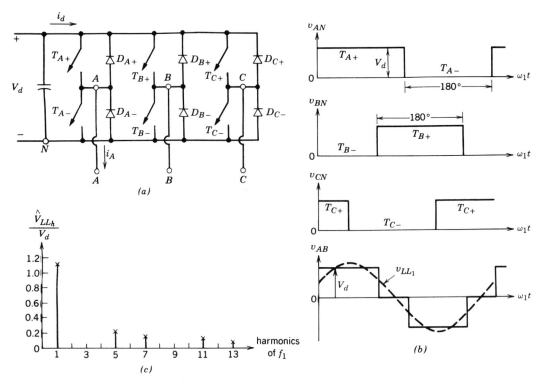

Figure 8-24   Square-wave inverter (three phase).

It should be noted that it is *not* possible to control the output magnitude in a three-phase, square-wave inverter by means of voltage cancellation as described in Section 8-3-2-4.

8-4-3   SWITCH UTILIZATION IN THREE-PHASE INVERTERS

We will assume that $V_{d,\max}$ is the maximum input voltage that remains constant during PWM and is decreased below this level to control the output voltage magnitude in a square-wave mode. We will also assume that there is sufficient inductance associated with the load to yield a pure sinusoidal output current with an rms value of $I_{o,\max}$ (both in the PWM and the square-wave mode) at maximum loading. Therefore, each switch would have the following peak ratings:

$$V_T = V_{d,\max} \tag{8-60}$$

and

$$I_T = \sqrt{2}I_{o,\max} \tag{8-61}$$

If V_{LL_1} is the rms value of the fundamental-frequency line-to-line voltage component, the three-phase output volt-amperes (rms) at the fundamental frequency at the rated output is

$$(\text{VA})_{\text{3-phase}} = \sqrt{3}V_{LL_1}I_{o,\max} \tag{8-62}$$

Therefore, the total switch utilization ratio of all six switches combined is

$$\text{Switch utilization ratio} = \frac{(VA)_{3\text{-phase}}}{6V_T I_T}$$

$$= \frac{\sqrt{3} V_{LL_1} I_{o,\max}}{6 V_{d,\max} \sqrt{2} I_{o,\max}}$$

$$= \frac{1}{2\sqrt{6}} \frac{V_{LL_1}}{V_{d,\max}} \qquad (8\text{-}63)$$

In the PWM linear region ($m_a \leq 1.0$) using Eq. 8-57 and noting that the maximum switch utilization occurs at $V_d = V_{d,\max}$,

$$\begin{array}{c}\text{Maximum switch utilization ratio}\\ \text{(PWM)}\end{array} = \frac{1}{2\sqrt{6}} \frac{\sqrt{3}}{2\sqrt{2}} m_a$$

$$= \tfrac{1}{8} m_a \qquad (m_a \leq 1.0)$$

$$(8\text{-}64)$$

In the square-wave mode, this ratio is $1/2\pi \simeq 0.16$ compared to a maximum of 0.125 for a PWM linear region with $m_a = 1.0$.

In practice, the same derating in the switch utilization ratio applies as discussed in Section 8-3-4 for single-phase inverters.

Comparings Eqs. 8-54 and 8-64, we observe that the maximum switch utilization ratio is the same in a three-phase, three-leg inverter as in a single-phase inverter. In other words, using the switches with identical ratings, a three-phase inverter with 50% increase in the number of switches results in a 50% increase in the output volt-ampere, compared to a single-phase inverter.

8-4-4 RIPPLE IN THE INVERTER OUTPUT

Figure 8-25a shows a three-phase, three-leg, voltage source, switch-mode inverter in a block diagram form. It is assumed to be supplying a three-phase ac motor load. Each phase of the load is shown by means of its simplified equivalent circuit with respect to the load neutral n. The induced back-emfs $e_A(t)$, $e_B(t)$, and $e_C(t)$ are assumed to be sinusoidal.

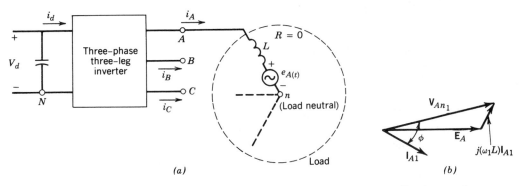

Figure 8-25 Three-phase inverter: (a) circuit diagram; (b) phasor diagram (fundamental frequency).

Under balanced operating conditions, it is possible to express the inverter phase output voltages v_{AN}, and so on (with respect to the load neutral n), in terms of the inverter output voltages with respect to the negative dc bus N:

$$v_{kn} = v_{kN} - v_{nN} \qquad (k = A, B, C) \qquad (8\text{-}65)$$

Each phase voltage can be written as

$$v_{kn} = L\frac{di_k}{dt} + e_{kn} \qquad (k = A, B, C) \qquad (8\text{-}66)$$

In a three-phase, three-wire load

$$i_A + i_B + i_C = 0 \qquad (8\text{-}67a)$$

and

$$\frac{d}{dt}(i_A + i_B + i_C) = 0 \qquad (8\text{-}67b)$$

Similarly, under balanced operating conditions, the three back-emfs are a balanced three-phase set of voltages, and therefore

$$e_A + e_B + e_C = 0 \qquad (8\text{-}68)$$

From the foregoing equations, the following condition for the inverter voltages can be written:

$$v_{An} + v_{Bn} + v_{Cn} = 0 \qquad (8\text{-}69)$$

Using Eqs. 8-65 through 8-69,

$$v_{nN} = \tfrac{1}{3}(v_{AN} + v_{BN} + v_{CN}) \qquad (8\text{-}70)$$

Substituting v_{nN} from Eq. 8-70 into Eq. 8-65, we can write the phase-to-neutral voltage for phase A as

$$v_{An} = \tfrac{2}{3}v_{AN} - \tfrac{1}{3}(v_{BN} + v_{CN}) \qquad (8\text{-}71)$$

Similar equations can be written for phase B and C voltages.

Similar to the discussion in Section 8-3-2-6 for the ripple in the single-phase inverter output, only the fundamental-frequency components of the phase voltage v_{An_1} and the output current i_{A1} are responsible for the real power transformer since the back-emf $e_A(t)$ is assumed to be sinusoidal and the load resistance is neglected. Therefore, in a phasor form as shown in Fig. 8-25b

$$\mathbf{V}_{An_1} = \mathbf{E}_A + j\omega_1 L\mathbf{I}_{A1} \qquad (8\text{-}72)$$

By using the principle of superposition, all the ripple in v_{An} appears across the load inductance L. Using Eq. 8-71, the waveform for the phase-to-load-neutral voltage V_{An} is shown in Figs. 8-26a and 8-26b for square-wave and PWM operations, respectively. Both inverters have identical magnitudes of the fundamental-frequency voltage component V_{An_1}, which requires a higher V_d in the PWM operation. The voltage ripple v_{ripple} ($=v_{An} - v_{An_1}$) is the ripple in the phase-to-neutral voltage. Assuming identical loads in these two cases, the output current ripple is obtained by using Eq. 8-42 and plotted in Fig. 8-26. This current ripple is independent of the power being transferred, that is, the current ripple would be the same so long as for a given load inductance L, the ripple in the inverter output voltage remains constant in magnitude and frequency. This comparison indicates that for large values of m_f, the current ripple in the PWM inverter will be significantly lower compared to a square-wave inverter.

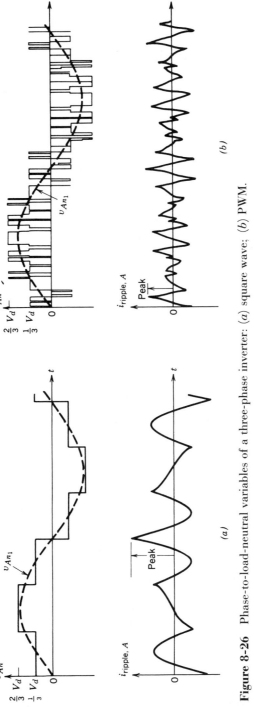

Figure 8-26 Phase-to-load-neutral variables of a three-phase inverter: (*a*) square wave; (*b*) PWM.

8-4-5 dc-SIDE CURRENT i_d

Similar to the treatment of a single-phase inverter, we now look at the voltage and current waveforms associated with the dc side of a pulse-width-modulated, three-phase inverter. The input voltage V_d is assumed to be dc without any ripple. If the switching frequency in Fig. 8-25a is assumed to approach infinity, then similar to Fig. 8-13, a fictitious filter with negligible energy storage can be inserted on the ac side and the current at the inverter output will be sinusoidal with no ripple. Because of the assumption of no energy storage in the fictitious ac-side filter, the instantaneous ac power output can be expressed in terms fundamental-frequency output voltages and currents. Similarly, on the dc side, a fictitious filter with no energy storage, as shown in Fig. 8-13, can be assumed. Then, the high-switching-frequency components in i_d are filtered. Now equating the instantaneous power input to the instantaneous power output, we get

$$V_d i_d^* = v_{An_1}(t)i_A(t) + v_{Bn_1}(t)i_B(t) + v_{Cn_1}(t)i_C(t) \qquad (8\text{-}73)$$

In a balanced steady-state operation, the three phase quantities are displaced by 120° with respect to each other. Assuming that ϕ is the phase angle by which a phase current lags the inverter phase voltage and $\sqrt{2}V_o$ and $\sqrt{2}I_o$ are the amplitudes of the phase voltages and currents, respectively, yields

$$i_d^* = \frac{2V_o I_o}{V_d}[\cos \omega_1 t \cos(\omega_1 t - \phi) + \cos(\omega_1 t - 120°)\cos(\omega_1 t - 120° - \phi)$$

$$+\cos(\omega_1 t + 120°) \cos(\omega_1 t + 120° - \phi)]$$

$$= \frac{3V_o I_o}{V_d}\cos \phi = I_d \quad \text{(a dc quantity)} \qquad (8\text{-}74)$$

The foregoing analysis shows that i_d^* is a dc quantity, unlike in the single-phase inverter, where i_d^* contained a component at twice the output frequency. However, i_d consists of high-frequency switching components as shown in Fig. 8-27, in addition to i_d^*. These high-frequency components, due to their high frequencies, would have a negligible effect on the capacitor voltage V_d.

8-4-6 CONDUCTION OF SWITCHES IN THREE-PHASE INVERTERS

We discussed earlier that the output voltage does not depend on the load. However, the duration of each switch conduction is dependent on the power factor of the load.

8-4-6-1 Square-Wave Operation

Here, each switch is in its on state for 180°. To determine the switch conduction interval, a load with a fundamental-frequency displacement angle of 30° (lagging) is assumed (as

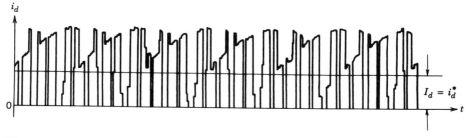

Figure 8-27 Input dc current in a three-phase inverter.

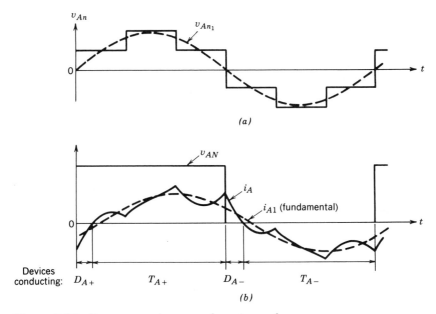

Figure 8-28 Square-wave inverter: phase A waveforms.

an example). The waveforms are shown in Fig. 8-28 for one of the three phases. The phase-to-neutral voltages V_{An} and V_{An_1} are shown in Fig. 8-28a. In Fig. 8-28b, V_{AN} (with respect to the negative dc bus), i_A, and its fundamental component i_{A_1} are plotted. Even though the switches T_{A+} and T_{A-} are in their on state for 180°, due to the lagging power factor of the load, their actual conduction intervals are smaller than 180°. It is easy to interpret that as the power factor (lagging) of the load decreases, the diode conduction intervals will increase and the switch conduction intervals will decrease. On the other hand, with a purely resistive load, theoretically the feedback diodes would not conduct at all.

8-4-6-2 PWM Operation

The voltage and current waveforms associated with a PWM inverter are shown in Fig. 8-29. Here, as an example, the load displacement power factor angle is assumed to be 30° (lagging). Also, the output current is assumed to be a perfect sinusoid. In Figs. 8-29a through 8-29c, the phase to the negative dc bus voltages and the phase current (v_{An}, i_A, etc.) are plotted for approximately one-fourth of the fundamental-frequency cycle.

By looking at the devices conducting in Figs. 8-29a through 8-29c, we notice that there are intervals during which the phase currents i_A, i_B, and i_C flow through only the devices connected to the positive dc bus (i.e., three out of T_{A+}, D_{A+}, T_{B+}, D_{B+}, T_{C+}, and D_{C+}). This implies that during these intervals, all three phases of the load are short-circuited and there is no power input from the dc bus (i.e., $i_d = 0$), as shown in Fig. 8-30a. Similarly, there are intervals during which all conducting devices are connected to the negative dc bus resulting in the circuit of Fig. 8-30b.

The output voltage magnitude is controlled by controlling the duration of these short-circuit intervals. Such intervals of three-phase short circuit do not exist in a square-wave mode of operation. Therefore, the output voltage magnitude in an inverter operating in a square-wave mode must be controlled by controlling the input voltage V_d.

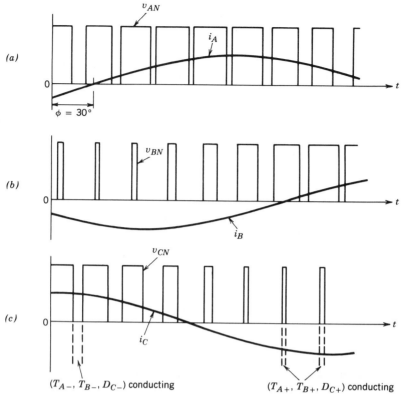

Figure 8-29   PWM inverter waveforms: load power factor angle = 30° (lag).

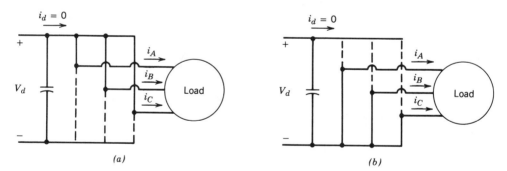

Figure 8-30   Short-circuit states in a three-phase PWM inverter.

8-5   EFFECT OF BLANKING TIME ON VOLTAGE IN PWM INVERTERS

The effect of blanking time on the output voltage is described by means of one leg of a single-phase or a three-phase full-bridge inverter, as shown in Fig. 8-31a. In the previous discussion, the switches were assumed to be ideal, which allowed the status of the two switches in an inverter leg to change simultaneously from on to off and vice versa. Concentrating on one switching time period, $v_{control}$ is a constant dc voltage, as explained in Fig. 8-6; its comparison with a triangular waveform v_{tri} determines the switching

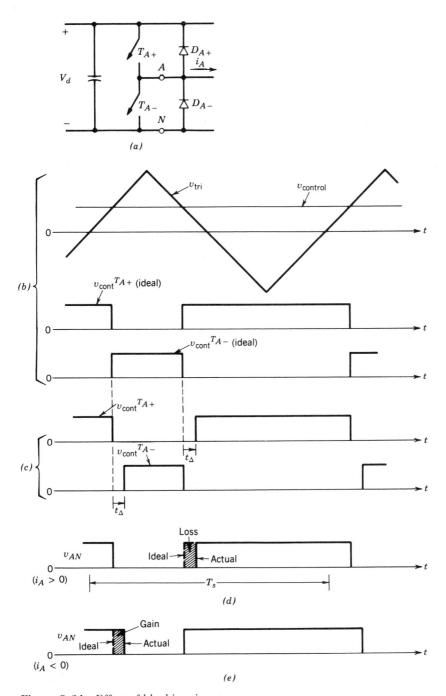

Figure 8-31   Effect of blanking time t_Δ.

instants and the switch control signals v_{cont} (ideal) as shown in Fig. 8-31*b*, assuming ideal switches.

In practice, because of the finite turn-off and turn-on times associated with any type of switch, a switch is turned off at the switching time instant determined in Fig. 8-31*b*. However, the turn-on of the other switch in that inverter leg is delayed by a blanking time t_Δ, which is conservatively chosen to avoid a "shoot through" or cross-conduction current through the leg. This blanking time is chosen to be just a few microseconds for fast switching devices like MOSFETs and larger for slower switching devices. The switch control signals for the two switches in the presence of a blanking time are shown in Fig. 8-31c.

Since both the switches are off during the blanking time, v_{AN} during that interval depends on the direction of i_A, as shown in Fig. 8-31d for $i_A > 0$, and in Fig. 8-31e for $i_A < 0$. The ideal waveforms (without the blanking time) are shown as dotted. Comparing the ideal waveform of v_{AN} without the blanking time to the actual waveform with the blanking time, the difference between the ideal and the actual output voltage is

$$v_\epsilon = (v_{AN})_{\text{ideal}} - (v_{AN})_{\text{actual}}$$

By averaging v_ϵ over one time period of the switching frequency, we can obtain the change (defined as a drop if positive) in the output voltage due to t_Δ:

$$\Delta V_{AN} = \begin{cases} +\dfrac{t_\Delta}{T_s}V_d & i_A > 0 \\[2ex] -\dfrac{t_\Delta}{T_s}V_d & i_A < 0 \end{cases} \tag{8-75}$$

Equation 8-75 shows that ΔV_{AN} does not depend on the magnitude of current but its polarity depends on the current direction. Moreover, ΔV_{AN} is proportional to the blanking time t_Δ and the switching frequency $f_s\ (=1/T_s)$, which suggests that at higher switching frequencies, faster switching devices that allow t_Δ to be small should be used.

Applying the same analysis to the leg B of the single-phase inverter of Fig. 8-32a and recognizing that $i_A = -i_B$, we determine that

$$\Delta V_{BN} = \begin{cases} -\dfrac{t_\Delta}{T_s}V_d & i_A > 0 \\[2ex] +\dfrac{t_\Delta}{T_s}V_d & i_A < 0 \end{cases} \tag{8-76}$$

Since $v_o = v_{AN} - v_{BN}$ and $i_o = i_A$, the instantaneous average value of the voltage difference, that is, the average value during T_s of the idealized waveform minus the actual waveform, is

$$\Delta V_o = \begin{cases} \Delta V_{AN} - \Delta V_{BN} = +\dfrac{2t_\Delta}{T_s}V_d & i_o > 0 \\[2ex] -\dfrac{2t_\Delta}{T_s}V_d & i_o < 0 \end{cases} \tag{8-77}$$

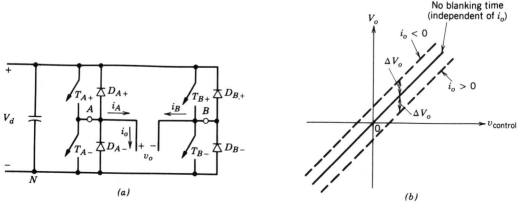

Figure 8-32   Effect of t_Δ on V_o, where ΔV_o is defined as a voltage drop if positive.

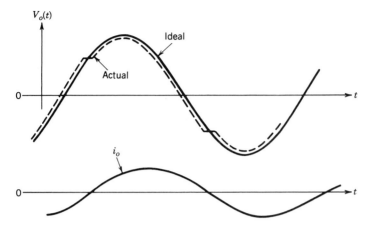

Figure 8-33 Effect of t_Δ on the sinusoidal output.

A plot of the instantaneous average value V_o as a function of $v_{control}$ is shown in Fig. 8-32b, with and without the blanking time.

If the full-bridge converter in Fig. 8-32a is pulse-width modulated for a dc-to-dc conversion as discussed in Chapter 7, then $v_{control}$ is a constant dc voltage in steady state. The plot of Fig. 8-32b is useful in determining the effect of blanking time in applications of such converters for dc motor drives, as will be discussed in Chapter 13.

For a sinusoidal $v_{control}$ in a single-phase full-bridge PWM inverter, the instantaneous average output $V_o(t)$ is shown in Fig. 8-33 for a load current i_o which is assumed to be sinusoidal and lagging behind $V_o(t)$. The distortion in $V_o(t)$ at the current zero crossings results in low order harmonics such as third, fifth, seventh, and so on, of the fundamental frequency in the inverter output. Similar distortions occur in the line-to-line voltages at the output of a three-phase PWM inverter, where the low-order harmonics are of the order $6m \pm 1$ ($m = 1, 2, 3, \ldots$) of the fundamental frequency. The effect of these distortions due to the blanking time is further discussed in Chapter 11, dealing with uninterruptible power supplies; Chapter 13, dealing with dc motor drives; and Chapter 15, dealing with synchronous motor drives.

8-6 OTHER INVERTER SWITCHING SCHEMES

In the previous sections, two commonly used inverter switching schemes, sinusoidal PWM and square wave, have been analyzed in detail. In this section, some other PWM schemes are briefly discussed. A detailed discussion of these techniques is presented in the references cited at the end of the chapter. A comprehensive review of various PWM techniques is presented in reference 14.

8-6-1 SQUARE-WAVE PULSE SWITCHING

Here, each phase voltage output is essentially square wave except for a few notches (or pulses) to control the fundamental amplitude. These notches are introduced without any regard to the harmonic content in the output, and therefore this type of scheme is no longer employed except in some thyristor inverters. A serious drawback of the foregoing techniques is that no attention is paid to the output harmonic content, which can become unacceptable. The advantage is in their simplicity and a small number of switchings required (which is significant in high-power thyristor inverters).

8-6-2 PROGRAMMED HARMONIC ELIMINATION SWITCHING

This technique combines the square-wave switching and PWM to control the fundamental output voltage as well as to eliminate the designated harmonics from the output.

The voltage v_{Ao}, of an inverter leg, normalized by $\frac{1}{2}V_d$ is plotted in Fig. 8-34a, where six notches are introduced in the otherwise square-wave output, to control the magnitude of the fundamental voltage and to eliminate fifth and seventh harmonics. On a half-cycle basis, each notch provides one degree of freedom, that is, having three notches per half-cycle provides control of fundamental and elimination of two harmonics (in this case fifth and seventh).

Figure 8-34a shows that the output waveform has odd half-wave symmetry (sometimes it is referred to as odd quarter-wave symmetry). Therefore, only odd harmonics

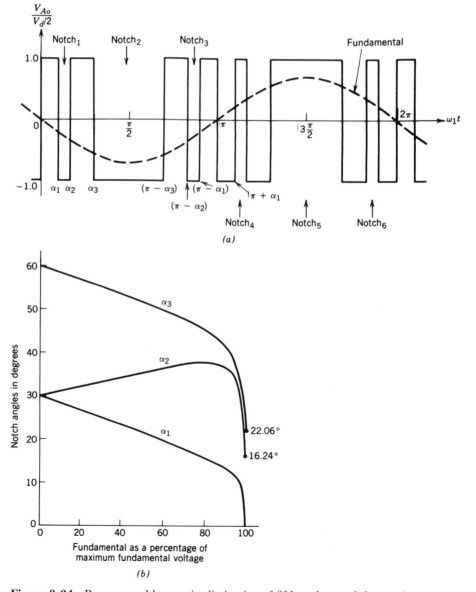

(a)

(b)

Figure 8-34   Programmed harmonic elimination of fifth and seventh harmonics.

(coefficients of sine series) will be present. Since in a three-phase inverter (consisting of three such inverter legs), the third harmonic and its multiples are canceled out in the output, these harmonics need not be eliminated from the output of the inverter leg by means of waveform notching.

A careful examination shows that the switching frequency of a switch in Fig. 8-34a is seven times the switching frequency associated with a square-wave operation.

In a square-wave operation, the fundamental-frequency voltage component is

$$\frac{(\hat{V}_{Ao})_1}{V_d/2} = \frac{4}{\pi} = 1.273 \quad \text{(Eq. 8-13 repeated)}$$

Because of the notches to eliminate fifth and seventh harmonics, the maximum available fundamental amplitude is reduced. It can be shown that

$$\frac{(\hat{V}_{Ao})_{1,\max}}{V_d/2} = 1.188 \tag{8-78}$$

The required values of α_1, α_2, and α_3 are plotted in Fig. 8-34b as a function of the normalized fundamental in the output voltage (see references 8 and 9 for details).

To allow control over the fundamental output and to eliminate the fifth-, seventh-, eleventh-, and the thirteenth-order harmonics, five notches per half-cycle would be needed. In that case, each switch would have 11 times the switching frequency compared with a square-wave operation.

With the help of very large scale integrated (VLSI) circuits and microcontrollers, this programmed harmonic elimination scheme can be implemented. Without making the switching frequency (and therefore the switching losses) very high, it allows the undesirable lower order harmonics to be eliminated. The higher order harmonics can be filtered by a small filter, if necessary. However, before deciding on this technique, it should be compared with a sinusoidal PWM technique with a low m_f to evaluate which one is better. It should be noted that the distortions due to the blanking time, as discussed in Section 8-5, occur here as well.

8-6-3 CURRENT-REGULATED (CURRENT-MODE) MODULATION

In applications such as dc and ac motor servo drives, it is the motor current (supplied by the switch-mode converter or inverter) that needs to be controlled, even though a VSI is often used. In this regard, it is similar to the switching dc power supplies of Chapter 10, where the output-stage current can be controlled in order to regulate the output voltage.

There are various ways to obtain the switching signals for the inverter switches in order to control the inverter output current. Two such methods are described.

8-6-3-1 Tolerance Band Control

This is illustrated by Fig. 8-35 for a sinusoidal reference current $i_A{}^*$, where the actual phase current i_A is compared with the tolerance band around the reference current associated with that phase. If the actual current in Fig. 8-35a tries to go beyond the upper tolerance band, T_{A-} is turned on (i.e., T_{A+} is turned off). The opposite switching occurs if the actual current tries to go below the lower tolerance band. Similar actions take place in the other two phases. This control is shown in a block diagram form in Fig. 8-35b.

The switching frequency depends on how fast the current changes from the upper limit to the lower limit and vice versa. This, in turn, depends on V_d, the load back-emf, and the load inductance. Moreover, the switching frequency does not remain constant but varies along the current waveform.

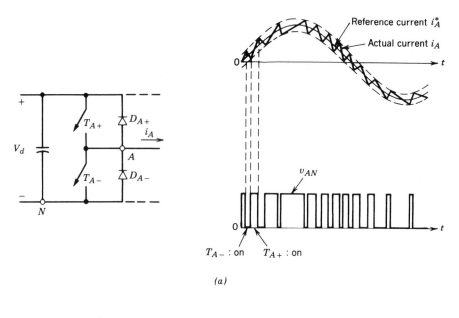

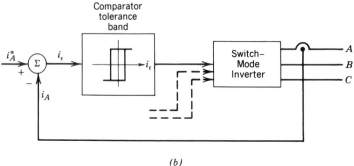

(b)

Figure 8-35 Tolerance band current control.

8-6-3-2 Fixed-Frequency Control

The fixed-frequency current control is shown in a block diagram form in Fig. 8-36. The error between the reference and the actual current is amplified or fed through a proportional integral (PI) controller. The output $v_{control}$ of the amplifier is compared with a fixed-frequency (switching frequency f_s) triangular waveform v_{tri}. A positive error ($i_A^* - i_A$) and, hence, a positive $v_{control}$ result in a larger inverter output voltage, thus bringing

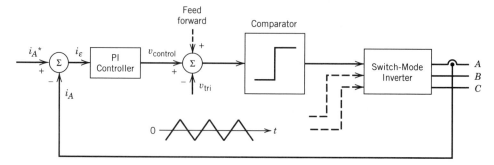

Figure 8-36 Fixed-frequency current control.

i_A to its reference value. Similar action takes place in the other two phases. Often, the load voltage (derived from the model of the load) is used as a compensating feed forward signal, shown dashed in Fig. 8-36.

It should be noted that many sophisticated switching techniques that minimize the total number of combined switchings in all three phases are discussed in the literature.

8-6-4 SWITCHING SCHEME INCORPORATING HARMONIC NEUTRALIZATION BY MODULATION AND TRANSFORMER CONNECTIONS

In some applications such as three-phase uninterruptible ac power supplies, it is usually required to have isolation transformers at the output. In such applications, the presence of output transformers is utilized in eliminating certain harmonics. In addition, the programmed harmonic elimination technique can be used to control the fundamental output and to eliminate (or reduce) a few more harmonics.

This arrangement is discussed in connection with the uninterruptible ac power supplies in Chapter 11.

8-7 RECTIFIER MODE OF OPERATION

As we discussed in the introduction in Section 8-1, these switch-mode converters can make a smooth transition from the inverter mode to the rectifier mode. The rectifier mode of operation results, for example, during braking (slowing down) of induction motors supplied through a switch-mode converter. This mode of operation is briefly discussed in this section. The switch-mode rectifiers, used for interfacing power electronics equipment with the utility grid, operate on the same basic principle and are discussed in detail in Chapter 19.

The rectifier mode of operation is discussed only for the three-phase converters; the same principle applies to single-phase converters. Assuming a balanced steady-state operating condition, a three-phase converter is discussed on a per-phase basis.

As an example, consider the three-phase system shown in Fig. 8-25a, which is redrawn in Fig. 8-37a. Consider only the fundamental frequency (where the subscript 1 is omitted), neglecting the switching-frequency harmonics. In Fig. 8-37b, a motoring mode of operation is shown where the converter voltage V_{An} applied to the motor leads E_A by an angle δ. The active (real) component $(I_A)_p$ of I_A is in phase with E_A, and therefore the converter is operating in an inverter mode.

The phase angle (as well as the magnitude) of the ac voltage produced by the converter can be controlled. If the converter voltage V_{An} is now made to lag E_A by the same angle δ as before (keeping V_{An} constant), the phasor diagram in Fig. 8-37c shows that the active component $(I_A)_p$ of I_A is now 180° out of phase with E_A, resulting in a rectifier mode of operation where the power flows from the motor to the dc side of the converter.

In fact, V_{An} can be controlled both in magnitude (within limits) and phase, thus allowing a control over the current magnitude and the power level, for example during the ac motor braking. Assuming that E_A cannot change instantaneously, Fig. 8-37d shows the locus of the V_{An} phasor, which would keep the magnitude of the current constant.

The waveforms of Fig. 8-22 can be used for explaining how to control V_{An} in magnitude, as well as in phase, with a given (fixed) dc voltage V_d. It is obvious that by controlling the amplitude of the sinusoidal reference waveform $v_{control,A}$, V_{An} can be varied. Similarly, by shifting the phase of $v_{control,A}$ with respect to E_A, the phase angle of

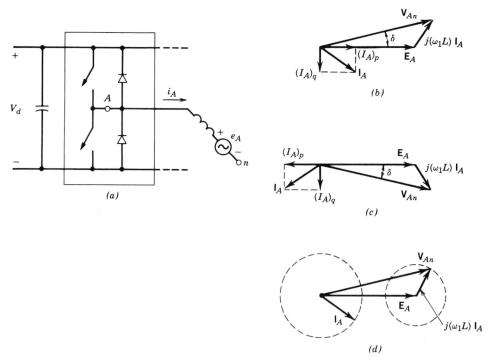

Figure 8-37   Operation modes: (*a*) circuit; (*b*) inverter mode; (*c*) rectifier mode; (*d*) constant I_A.

$\mathbf{V}_{An}$ can be varied. For a balanced operation, the control voltages for phases B and C are equal in magnitude, but ±120° displaced with respect to the control voltage of phase A.

Switch-mode rectifiers, where the rectifier is the primary mode of operation, are further discussed in Chapter 19, which deals with circuits for interfacing power electronics equipment with the utility grid.

SUMMARY

1. Switch-mode, voltage source dc-to-ac inverters are described that accept dc voltage source as input and produce either single-phase or three-phase sinusoidal output voltages at a low frequency relative to the switching frequency (current source inverters are described in Chapter 14).

2. These dc-to-ac inverters can make a smooth transition into the rectification mode, where the flow of power reverses to be from the ac side to the dc side. This occurs, for example, during braking of an induction motor supplied through such an inverter.

3. The sinusoidal PWM switching scheme allows control of the magnitude and the frequency of the output voltage. Therefore, the input to the PWM inverters is an uncontrolled, essentially constant dc voltage source. This switching scheme results in harmonic voltages in the range of the switching frequency and higher, which can be easily filtered out.

4. The square-wave switching scheme controls only the frequency of the inverter output. Therefore, the output magnitude must be controlled by controlling the magnitude of the input dc voltage source. The square-wave output voltage contains low-order harmonics. A variation of the square-wave switching scheme, called the voltage

cancellation technique, can be used to control both the frequency and the magnitude of the single-phase (but not three-phase) inverter output.

5. As a consequence of the harmonics in the inverter output voltage, the ripple in the output current does not depend on the level of power transfer at the fundamental frequency; instead the ripple depends inversely on the load inductance, which is more effective at higher frequencies.

6. In practice, if a switch turns off in an inverter leg, the turn-on of the other switch is delayed by a blanking time, which introduces low-order harmonics in the inverter output.

7. There are many other switching schemes in addition to the sinusoidal PWM. For example, the programmed harmonic elimination switching technique can be easily implemented with the help of VLSI circuits to eliminate specific harmonics from the inverter output.

8. The current-regulated (current-mode) modulation allows the inverter output current(s) to be controlled directly by comparing the measured actual current with the reference current and using the error to control the inverter switches. As will be discussed in Chapters 13–15 this technique is extensively used for dc and ac servo drives. The current-mode control is also used in dc-to-dc converters, as discussed in Chapter 10, dealing with switching dc power supplies.

9. The relationship between the control input and the full-bridge inverter output magnitude can be summarized as shown in Fig. 8-38a, assuming a sinusoidal PWM in the linear range of $m_a \leq 1.0$. For a square-wave switching, the inverter does not control the magnitude of the inverter output, and the relationship between the dc input voltage and the output magnitude is summarized in Fig. 8-38b.

10. These converters can be used for interfacing power electronics equipment with the utility source. As discussed in Chapter 18, because rectification is the primary mode of operation, these are called switch-mode ac-to-dc rectifiers.

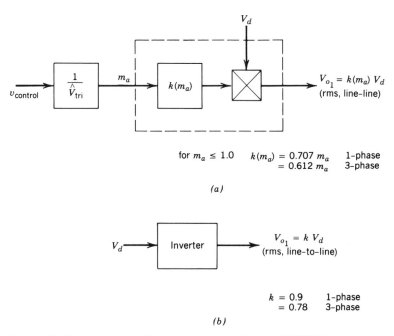

Figure 8-38 Summary of inverter output voltages: (a) PWM operation ($m_a \leq 1$); (b) square-wave operation.

PROBLEMS

SINGLE PHASE

8-1 In a single-phase full-bridge PWM inverter, the input dc voltage varies in a range of 295–325 V. Because of the low distortion required in the output v_o, $m_a \leq 1.0$.

(a) What is the highest V_{o1} that can be obtained and stamped on its nameplate as its voltage rating?

(b) Its nameplate volt-ampere rating is specified as 2000 VA, that is, $V_{o1,\max} I_{o,\max} = 2000$ VA, where i_o is assumed to be sinusoidal. Calculate the combined switch utilization ratio when the inverter is supplying its rated volt-amperes.

8-2 Consider the problem of ripple in the output current of a single-phase full-bridge inverter. Assume $V_{o1} = 220$ V at a frequency of 47 Hz and the type of load is as shown in Fig. 8-18a with $L = 100$ mH.

    If the inverter is operating in a square-wave mode, calculate the peak value of the ripple current.

8-3 Repeat Problem 8-2 with the inverter operating in a sinusoidal PWM mode, with $m_f = 21$ and $m_a = 0.8$. Assume a bipolar voltage switching.

8-4 Repeat Problem 8-2 but assume that the output voltage is controlled by voltage cancellation and V_d has the same value as required in the PWM inverter of Problem 8-3.

8-5 Calculate and compare the peak values of the ripple currents in Problems 8-2 through 8-4.

8-6 Using MATLAB, verify the results given in Table 8-1.

THREE-PHASE

8-7 Consider the problem of ripple in the output current of a three-phase square-wave inverter. Assume $(V_{LL})_1 = 200$ V at a frequency of 52 Hz and the type of load is as shown in Fig. 8-25a with $L = 100$ mH. Calculate the peak ripple current defined in Fig. 8-26a.

8-8 Repeat Problem 8-7 if the inverter of Problem 8-7 is operating in a synchronous PWM mode with $m_f = 39$ and $m_a = 0.8$. Calculate the peak ripple current defined in Fig. 8-26b.

8-9 Obtain an expression for the Fourier components in the waveform of Fig. 8-34a for programmed harmonic elimination of the fifth- and seventh-order harmonics. Show that for $\alpha_1 = 0$, $\alpha_2 = 16.24°$, and $\alpha_3 = 22.06°$, the fifth and the seventh harmonics are eliminated and the fundamental-frequency output of the inverter has a maximum amplitude given by Eq. 8-78.

8-10 In the three-phase, square-wave inverter of Fig. 8-24a, consider the load to be balanced and purely resistive with a load-neutral n. Draw the steady-state v_{An}, i_A, $i_{D_{A+}}$, and i_d waveforms, where $i_{D_{A+}}$ is the current through D_{A+}.

8-11 Repeat Problem 8-10 by assuming that the load is purely inductive, where the load resistance, though finite, can be neglected.

8-12 Consider only one inverter leg as shown in Fig. 8-4, where the output current lags $(v_{Ao})_1$ by an angle ϕ, as shown in Fig. P8-12a, and o is the fictitious midpoint of the dc input. Because of the blanking time t_Δ, the instantaneous error voltage v_ϵ is plotted in Fig. P8-12b, where

$$v_\epsilon = (v_{Ao})_{\text{ideal}} - (V_{Ao})_{\text{actual}}$$

Each v_ϵ pulse, either positive or negative, has an amplitude of V_d and a duration of t_Δ. In order to calculate the low-order harmonics of the fundamental frequency in the output voltage due to blanking time, these pulses can be replaced by an equivalent rectangular pulse (shown dashed in Fig. P8-12b) of amplitude K whose volt-second area per half-cycle equals that of v_ϵ pulses.

    Derive the following expression for the harmonics of the fundamental frequency in v_{Ao} introduced by the blanking time:

$$(\hat{V}_{Ao})_h = \frac{4}{\pi h} V_d t_\Delta f_s \qquad (h = 1, 3, 5, \ldots)$$

where f_s is the switching frequency.

8-13 Using PSpice, simulate the inverter of Fig. P8-13.

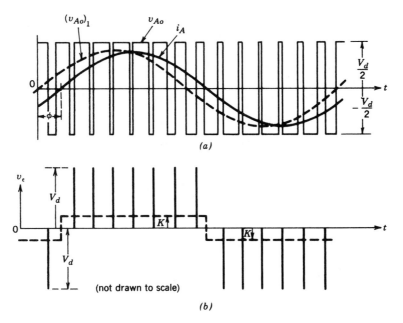

(a)

(b)

Figure P8-12

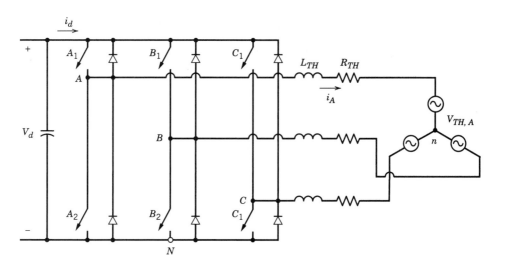

Nominal Values: $V_d = 313.97$V, $f_1 = 47.619$ Hz
$\hat{V}_{tri} = 1.0$ V, $m_a = 0.95$
$R_{TH} = 2\Omega$, $L_{TH} = 10$ mH
$v_{control,A} = 0.95 \cos (2\pi f_1 t - 90°)$V
$(\mathbf{V}_{th,A})_1 = 74.76 \ \underline{/-12.36°} \ V_{(rms)}$
$f_s = 1$ kHz, $\mathbf{I}_{A1} = 10 \ \underline{/-30°} \ A_{(rms)}$

Figure P8-13   From reference 5 of Chapter 4, "Power Electronics: Computer Simulation, Analysis and Education Using PSpice (evaluation, classroom version)," on a diskette with a manual, Minnesota Power Electronics, P.O. Box 14503, Minneapolis, MN 55414.

REFERENCES

1. K. Thorborg, *Power Electronics,* Prentice Hall International (U.K.) Ltd, London 1988.
2. A. B. Plunkett, "A Current-controlled PWM Transistor Inverter Drive," IEEE/IAS 1979 Annual Meeting, pp. 785–792.
3. T. Kenjo and S. Nagamori, *Permanent Magnet and Brushless DC Motors,* Clarendon, Oxford, 1985.
4. H. Akagi, A. Nabae, and S. Atoh, "Control Strategy of Active Filters Using Multiple Voltage-Source PWM Converters," *IEEE Transactions on Industry Applications,* Vol. IA-22, No. 3 May/June 1986, pp. 460–465.
5. T. Kato, "Precise PWM Waveform Analysis of Inverter for Selected Harmonic Elimination," 1986 IEEE/IAS Annual Meeting, pp. 611–616.
6. J. W. A. Wilson and J. A. Yeamans, "Intrinsic Harmonics of Idealized Inverter PWM Systems," 1976 IEEE/IAS Annual Meeting, pp. 967–973.
7. Y. Murai, T. Watanabe, and H. Iwasaki, "Waveform Distortion and Correction Circuit for PWM Inverters with Switching Lag-Times," 1985 IEEE/IAS Annual Meeting, pp. 436–441.
8. H. Patel and R. G. Hoft, "Generalized Techniques of Harmonic Elimination and Voltage Control in Thyristor Inverters: Part I—Harmonic Elimination," *IEEE Transactions on Industry Applications,* Vol. IA-9, No. 3, May/June 1973.
9. H. Patel and R. G. Hoft, "Generalized Techniques of Harmonic Elimination and Voltage Control in Thyristor Inverters: Part II—Voltage Control Techniques," *IEEE Transactions on Industry Applications,* Vol. IA-10, No. 5, September/October 1974.
10. I. J. Pitel, S. N. Talukdar, and P. Wood, "Characterization of Programmed-Waveform Pulsewidth Modulation," *IEEE Transactions on Industry Applications,* Vol. IA-16, No. 5, September/October 1980.
11. M. Boost and P. D. Ziogas, "State-of-the-Art PWM Techniques: A Critical Evaluation," IEEE Power Electronics Specialists Conference, 1986, pp. 425–433.
12. J. Rosa, "The Harmonic Spectrum of D.C. Link Currents in Inverters," *Proceedings of the Fourth International PCI Conference on Power Conversion,* Intertec Communications, Oxnard, CA, pp. 38–52.
13. EPRI Report, "AC/DC Power Converter for Batteries and Fuel Cells," Project 841-1, Final Report, September 1981, EPRI, Palo Alto, CA.
14. J. Holtz, "Pulsewidth Modulation—A Survey," IEEE Transactions on Industrial Electronics, Vol. 39, No. 5, Dec. 1992, pp. 410–420.

CHAPTER 9

RESONANT CONVERTERS: ZERO-VOLTAGE AND/OR ZERO-CURRENT SWITCHINGS

9-1 INTRODUCTION

In all the pulse-width-modulated dc-to-dc and dc-to-ac converter topologies discussed in Chapters 7 and 8, the controllable switches are operated in a switch mode where they are required to turn on and turn off the entire load current during each switching. In this switch-mode operation, as explained further in Section 9-1-1, the switches are subjected to high switching stresses and high switching power loss that increases linearly with the switching frequency of the PWM. Another significant drawback of the switch-mode operation is the EMI produced due to large di/dt and dv/dt caused by a switch-mode operation.

These shortcomings of switch-mode converters are exacerbated if the switching frequency is increased in order to reduce the converter size and weight and hence to increase the power density. Therefore, to realize high switching frequencies in converters, the aforementioned shortcomings are minimized if each switch in a converter changes its status (from on to off or vice versa) when the voltage across it and/or the current through it is zero at the switching instant. The converter topologies and the switching strategies, which result in zero-voltage and/or zero-current switchings, are discussed in this chapter. Since most of these topologies (but not all) require some form of LC resonance, these are broadly classified as "resonant converters."

9-1-1 SWITCH-MODE INDUCTIVE CURRENT SWITCHING

This topic was briefly reviewed in Chapter 2. To illustrate further the problems associated with switch-mode operation, consider one of the legs of a full-bridge dc–dc converter or a dc-to-ac inverter (single phase or three phase), as shown in Fig. 9-1. The output current can be in either direction and can be assumed to have a constant magnitude I_o due to the load inductance, during the very brief switching interval. The linearized voltage and current waveforms, for example, for the lower switch T_- are shown in Fig. 9-2a.

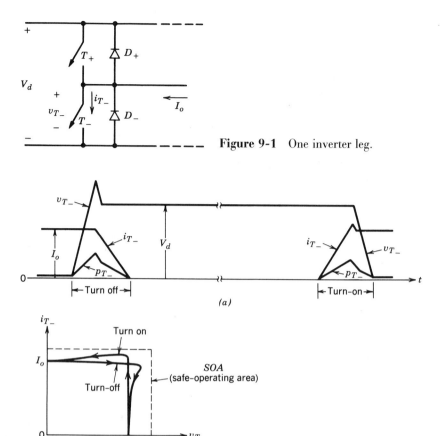

Figure 9-1 One inverter leg.

Figure 9-2 Switch-mode inductive current switchings.

Initially, I_o is assumed to be flowing through T_-. If a control signal is applied to turn T_- off, the switch voltage v_{T_-} increases to V_d (it overshoots V_d due to stray inductances), and then the switch current i_{T_-} decays to zero. After the turn-off of T_-, I_o flows through D_+. The power loss P_{T_-} ($=v_{T_-} \cdot i_{T_-}$) in the switch during turn-off is shown in Fig. 9-2a.

Now consider the turn-on of T_-. Prior to the turn-on of T_-, I_o is flowing through D_+. When the switch control signal is applied to turn T_- on, i_{T_-} increases to I_o plus the peak-reverse-recovery current of the diode D_+, as shown in Fig. 9-2a. Subsequently, the diode recovers and the switch voltage v_{T_-} and i_{T_-} results in a switching power loss in T_- during turn-on.

The average value of the switching loss P_{T_-}, being proportional to the switching frequency, limits how high the switching frequency can be pushed without significantly degrading the system efficiency. With the availability of fast switches (with the switching times as low as a few tens of nanoseconds), the present limit seems to be up to approximately 500 kHz with a reasonable energy efficiency.

Another significant disadvantage of the switch-mode operation is that it results in large di/dt and dv/dt due to fast switching transitions required to keep the switching losses in the switch as low as possible. Diodes with poor reverse-recovery characteristics significantly add to this phenomenon, which produces EMI.

Switch-mode inductive current switching results in switching loci in the $v_T - i_T$ plane, as shown in Fig. 9-2b. Because a large switch voltage and a large switch current occur simultaneously, the switch must be capable of withstanding high switching stresses, with a safe operating area (SOA), as shown by the dashed lines. This requirement to be able to withstand such large stresses results in undesirable design compromises in other characteristics of the power semiconductor devices.

9-1-2 ZERO-VOLTAGE AND ZERO-CURRENT SWITCHINGS

Switching frequencies in the megahertz range, even tens of megahertz, are being contemplated to reduce the size and the weight of transformers and filter components and, hence, to reduce the cost as well as the size and the weight of power electronics converters. Realistically, the switching frequencies can be increased to such high values only if the problems of switch stresses, switching losses, and the EMI associated with the switch-mode converters can be overcome.

The switch stresses, as discussed in later chapters in this book, can be reduced by connecting simple dissipative snubber circuits (consisting of diodes and passive components) in series and parallel with the switches in the switch-mode converters. Such snubber circuits are shown in Fig. 9-3a, and the switching loci that result in reduced switch stresses are shown in Fig. 9-3b. However, these dissipative snubbers shift the switching power loss from the switch to the snubber circuit and therefore do not provide a reduction in the overall switching power loss.

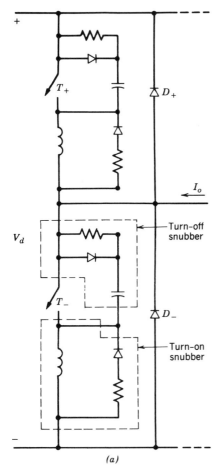

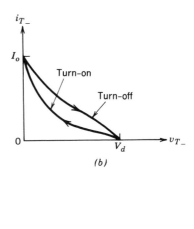

Figure 9-3 Dissipative snubbers: (*a*) snubber circuits; (*b*) switching loci with snubbers.

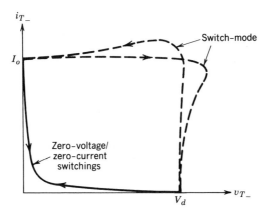

Figure 9-4   Zero-voltage-/ zero-current-switching loci.

In contrast to dissipative snubbers in switch-mode converters, the combination of proper converter topologies and switching strategies can overcome the problems of switching stresses, switching power losses, and the EMI by turning on and turning off each of the converter switches when either the switch voltage or the switch current is zero. Ideally, both the switch voltage and current should be zero when the switching transition occurs.

As a brief introduction, once again consider the one-leg inverter of Fig. 9-1. If both the turn-on and turn-off switchings occur under a zero-voltage and/or a zero-current condition, then the switching loci are shown in Fig. 9-4, where the switching loci in the switch mode are shown (by dashed curves) for comparison purposes. Such switching loci, without dissipative snubbers, reduce switch stresses, switching power losses, and the EMI.

9-2   CLASSIFICATION OF RESONANT CONVERTERS

The resonant converters are defined here as the combination of converter topologies and switching strategies that result in zero-voltage and/or zero-current switchings. One way to categories these converters is as follows:

1. Load-resonant converters
2. Resonant-switch converters
3. Resonant-dc-link converters
4. High-frequency-link integral-half-cycle converters

These classifications are explained further.

9-2-1   LOAD-RESONANT CONVERTERS

These converters consist of an LC resonant tank circuit. Oscillating voltage and current, due to LC resonance in the tank are applied to the load, and the converter switches can be switched at zero voltage and/or zero current. Either a series LC or a parallel LC circuit can be used. In these converter circuits, the power flow to the load is controlled by the resonant tank impedance, which in turn is controlled by the switching frequency f_s in comparison to the resonant frequency f_0 of the tank. These dc-to-dc and dc-to-ac converters can be subclassified as follows:

1. Voltage-source series-resonant converters
   (a) Series-loaded resonant (SLR) converters
   (b) Parallel-loaded resonant (PLR) converters
   (c) Hybrid-resonant converters

2. Current-source parallel-resonant converters

3. Class E and subclass E resonant converters

9-2-2 RESONANT-SWITCH CONVERTERS

In certain switch-mode converter topologies, an *LC* resonance can be utilized primarily to shape the switch voltage and current to provide zero-voltage and/or zero-current switchings. In such resonant-switch converters, during one switching-frequency time period, there are resonant as well as nonresonant operating intervals. Therefore, these converters in the literature have also been termed quasi-resonant converters. They can be subclassified as follows:

1. Resonant-switch dc–dc converters
   (a) Zero-current-switching (ZCS) converters
   (b) Zero-voltage-switching (ZVS) converters

2. Zero-voltage-switching, clamped-voltage (ZVS-CV) converters, which are also referred to as pseudo-resonant converters and resonant-transition converters, respectively in references 34 and 31.

9-2-3 RESONANT-dc-LINK CONVERTERS

In the conventional switch-mode PWM dc-to-ac inverters, the input V_d to the inverter is a fixed-magnitude dc, and the sinusoidal output (single phase or three phase) is obtained by switch-mode PWM switchings. However, in the resonant-dc-link converters, the input voltage is made to oscillate around V_d by means of an *LC* resonance so that the input voltage remains zero for a finite duration during which the status of the inverter switches can be changed, thus resulting in zero-voltage switchings.

9-2-4 HIGH-FREQUENCY-LINK INTEGRAL-HALF-CYCLE CONVERTERS

If the input to a single-phase or three-phase inverter is a high-frequency sinusoidal ac, then by using bidirectional switches it is possible to synthesize a low-frequency ac of adjustable magnitude and frequency or an adjustable-magnitude dc, where the switches are turned on and off at the zero crossings of the input voltage.

9-3 BASIC RESONANT CIRCUIT CONCEPTS

Some basic configurations encountered in the resonant converters discussed in this chapter are analyzed in a generic fashion. Appropriate assumptions are made to keep the analysis simple.

The initial conditions are indicated by uppercase letters, subscript 0, and square brackets, for example, $[V_{c0}]$ and $[I_{c0}]$.

9-3-1   SERIES-RESONANT CIRCUITS

9-3-1-1   Undamped Series-Resonant Circuit

Figure 9-5a shows an undamped series-resonant circuit where the input voltage is V_d at time t_0. The initial conditions are I_{L0} and V_{c0}. With the inductor current i_L and the capacitor voltage v_c as the state variables, the circuit equations are

$$L_r \frac{di_L}{dt} + v_c = V_d \tag{9-1}$$

and

$$C_r \frac{dv_c}{dt} = i_L \tag{9-2}$$

The solution of this set of equations for $t \geq t_0$ is as follows:

$$i_L(t) = I_{L0}\cos \omega_0(t - t_0) + \frac{V_d - V_{c0}}{Z_o}\sin \omega_0(t - t_0) \tag{9-3}$$

and

$$v_c(t) = V_d - (V_d - V_{c0})\cos \omega_0(t - t_0) + Z_0 I_{L0}\sin \omega_0(t - t_0) \tag{9-4}$$

where

$$\text{Angular resonance frequency} = \omega_0 = 2\pi f_0 = \frac{1}{\sqrt{L_r C_r}} \tag{9-5}$$

and

$$\text{Characteristic impedance} = Z_0 = \sqrt{\frac{L_r}{C_r}} \quad \Omega \tag{9-6}$$

To plot normalized v_c and i_L, the following base quantities are chosen:

$$V_{\text{base}} = V_d \tag{9-7}$$

and

$$I_{\text{base}} = \frac{V_d}{Z_0} \tag{9-8}$$

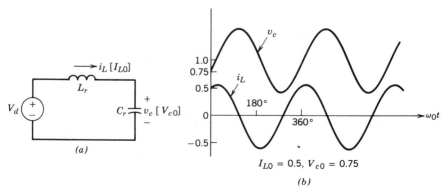

Figure 9-5   Undamped series-resonant circuit; i_L and v_c are normalized: (a) circuit; (b) waveforms with $I_{L0} = 0.5$, $V_{c0} = 0.75$.

As an example, the normalized i_L and v_c are plotted in Fig. 9-5b for $I_{L0} = 0.5$ and $V_{c0} = 0.75$.

9-3-1-2 Series-Resonant Circuit with a Capacitor-Parallel Load

Figure 9-6a shows a series-resonant circuit, where the capacitor is in parallel with a current I_o, which represents the load. In this circuit, V_d and I_o are dc quantities. The initial conditions are I_{L0} and V_{c0} at the initial time t_0. Therefore

$$v_c = V_d - L_r \frac{di_L}{dt} \tag{9-9}$$

and

$$i_L - i_c = I_o \tag{9-10}$$

By differentiating Eq. 9-9

$$i_c = C_r \frac{dv_c}{dt} = -L_r C_r \frac{d^2 i_L}{dt^2} \tag{9-11}$$

Substituting i_c from Eq. 9-11 into Eq. 9-10 yields

$$\frac{d^2 i_L}{dt^2} + \omega_0^2 i_L = \omega_0^2 I_o \tag{9-12}$$

where ω_0 is the same as in Eq. 9-5. Solution of these equations for $t \geq t_0$ is as follows:

$$i_L(t) = I_o + (I_{L0} - I_o)\cos \omega_0(t - t_0) + \frac{V_d - V_{c0}}{Z_0}\sin \omega_0(t - t_0) \tag{9-13}$$

and

$$v_c(t) = V_d - (V_d - V_{c0})\cos \omega_0(t - t_0) + Z_0(I_{L0} - I_o)\sin \omega_0(t - t_0) \tag{9-14}$$

where ω_0 is the angular resonant frequency as defined in Eq. 9-5 and Z_0 is the characteristic impedance defined in Eq. 9-6.

In a special case with $V_{c0} = 0$ and $I_{L0} = I_o$,

$$i_L(t) = I_o + \frac{V_d}{Z_0}\sin \omega_0(t - t_0) \tag{9-15}$$

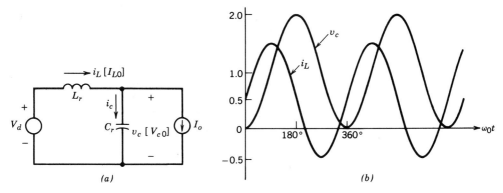

Figure 9-6 Series-resonant circuit with capacitor-parallel load (i_L and v_c are normalized): (a) circuit; (b) $\mathbf{V}_{c0} = 0$, $I_{L0} = I_o = 0.5$.

and

$$v_c(t) = V_d[1 - \cos \omega_0(t - t_0)] \tag{9-16}$$

For this special case, Fig. 9-6b shows the plot of i_L and v_c, which are normalized by using Eqs. 9-7 and 9-8, respectively, and $I_{L0} = I_o = 0.5$ per unit.

9-3-1-3 Frequency Characteristics of a Series-Resonant Circuit

It is informative to obtain the frequency characteristics of the series-resonant circuit of Fig. 9-7a. The resonance frequency ω_0 and the characteristic impedance Z_0 are defined by Eqs. 9-5 and 9-6, respectively. In the presence of a load resistance R, another quantity called the quality factor Q is defined as

$$Q = \frac{\omega_0 L_r}{R} = \frac{1}{\omega_0 C_r R} = \frac{Z_0}{R} \tag{9-17}$$

Figure 9-7b shows the magnitude Z_s of the circuit impedance as a function of frequency with Q as a parameter, keeping R constant. It shows that Z_s is a pure resistance equal to R at $\omega_s = \omega_0$ and is very sensitive to frequency deviation from ω_0 at higher values of Q.

Figure 9-7c shows the current phase angle θ ($= \theta_i - \theta_v$) as a function of frequency. The current leads at frequencies below ω_0 ($\omega_s < \omega_0$), where the capacitor impedance dominates over inductor impedance. At frequencies above ω_0 ($\omega_s > \omega_0$), the inductor impedance dominates over the capacitor impedance and the current lags the voltage, with the current phase angle θ approaching $-90°$.

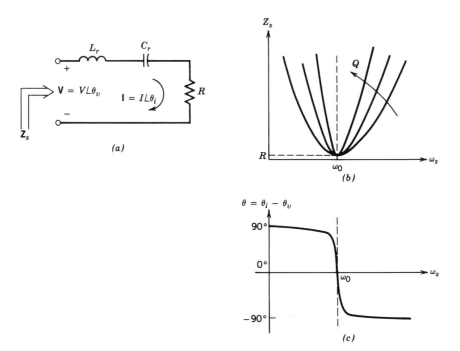

Figure 9-7 Frequency characteristics of a series-resonant circuit.

9-3-2 PARALLEL-RESONANT CIRCUITS

9-3-2-1 Undamped Parallel-Resonant Circuit

Figure 9-8a shows an undamped parallel-resonant circuit supplied by a dc current I_d. The initial conditions at time $t = t_0$ are I_{L0} and V_{c0}. With the inductor current i_L and the capacitor voltage v_c as the state variables, the circuit equations are

$$i_L + C_r \frac{dv_c}{dt} = I_d \qquad (9\text{-}18)$$

and

$$v_c = L_r \frac{di_L}{dt} \qquad (9\text{-}19)$$

The solution of the foregoing sets of equations for $t \geq t_0$ is as follows:

$$i_L(t) = I_d + (I_{L0} - I_d)\cos \omega_0(t - t_0) + \frac{V_{c0}}{Z_0}\sin \omega_0(t - t_0) \qquad (9\text{-}20)$$

and

$$v_c(t) = Z_0(I_d - I_{L0})\sin \omega_0(t - t_0) + V_{c0}\cos \omega_0(t - t_0) \qquad (9\text{-}21)$$

where

$$\omega_0 = \frac{1}{\sqrt{L_r C_r}} \qquad (9\text{-}22)$$

and

$$Z_0 = \sqrt{\frac{L_r}{C_r}} \qquad (9\text{-}23)$$

9-3-2-2 Frequency Characteristics of Parallel-Resonant Circuit

It is informative to obtain the frequency characteristics of the parallel-resonant circuit of Fig. 9-9a. The resonance frequency ω_0 and Z_0 are as defined by Eqs. 9-22 and 9-23, respectively. In the presence of a load resistor R, another quantity called the quality factor Q is defined, where

$$Q = \omega_0 R C_r = \frac{R}{\omega_0 L_r} = \frac{R}{Z_0} \qquad (9\text{-}24)$$

Figure 9-9b shows the magnitude Z_p of the circuit impedance as a function of frequency with Q as a parameter, keeping R constant.

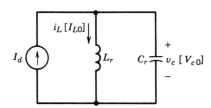

Figure 9-8 Undamped parallel-resonant circuit.

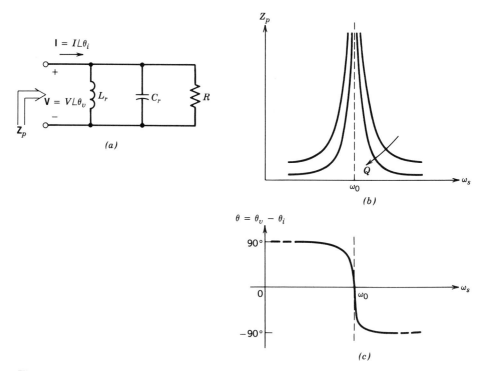

Figure 9-9   Frequency characteristics of a parallel-resonant circuit.

Figure 9-9c shows the voltage phase angle θ ($=\theta_v - \theta_i$) as a function of frequency. The voltage leads the current at frequencies below ω_0 ($\omega_s < \omega_0$), where the inductor impedance is lower than the capacitor impedance, and hence the inductor current dominates. At frequencies above ω_0 ($\omega_s > \omega_0$), the capacitor impedance is lower and the voltage lags the current, with the voltage phase angle θ approaching $-90°$.

9-4   LOAD-RESONANT CONVERTERS

In these resonant converters, an LC tank is used that results in oscillating load voltage and current and thus provides zero-voltage and/or zero-current switchings. Each circuit in this category is analyzed with a load that is most practical for the converter topology being considered. Only the steady-state operation is considered.

9-4-1   SERIES-LOADED RESONANT dc–dc CONVERTERS

A half-bridge configuration of the SLR converter is shown in Fig. 9-10a. The waveforms and the operating principles are the same for the full-bridge configurations. A transformer can be included to provide the output voltage of a desired magnitude as well as the electrical isolation between the input and the output.

The series-resonant tank is formed by L_r and C_r, and the current through the resonant tank circuit is full-wave rectified at the output, and $|i_L|$ feeds the output stage. Therefore, as the name suggests, the output load appears in series with the resonant tank.

The filter capacitor C_f at the output is usually very large, and therefore the output voltage across the capacitor can be assumed to be a dc voltage without any ripple. The

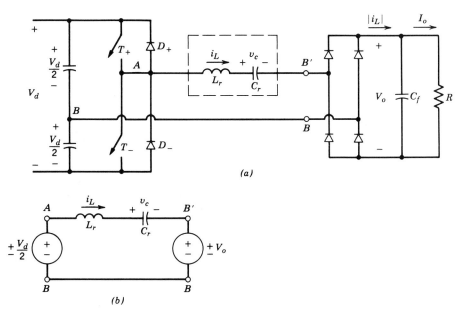

Figure 9-10 SLR dc–dc converter: (a) half-bridge; (b) equivalent circuit.

resistive power loss in the resonant circuit is assumed to be negligible, which greatly simplifies the analysis. The output voltage V_o is reflected across the rectifier input as $v_{B'B}$, where $v_{B'B} = V_o$ if i_L is positive and $v_{B'B} = -V_o$ if i_L is negative.

When i_L is positive, it flows through T_+ if it is on; otherwise it flows through the diode D_-. Similarly, when i_L is negative, it flows through T_- if it is on; otherwise it flows through the diode D_+. Therefore, in the circuit of Fig. 9-10a,

For $i_L > 0$

$$T_+ \text{ conducting:} \quad v_{AB} = +\tfrac{1}{2}V_d \quad v_{AB'} = +\tfrac{1}{2}V_d - V_o \quad (9\text{-}25)$$

$$D_- \text{ conducting:} \quad v_{AB} = -\tfrac{1}{2}V_d \quad v_{AB'} = -\tfrac{1}{2}V_d - V_o \quad (9\text{-}26)$$

For $i_L < 0$

$$T_- \text{ conducting:} \quad v_{AB} = -\tfrac{1}{2}V_d \quad v_{AB'} = -\tfrac{1}{2}V_d + V_o \quad (9\text{-}27)$$

$$D_+ \text{ conducting:} \quad v_{AB} = +\tfrac{1}{2}V_d \quad v_{AB'} = +\tfrac{1}{2}V_d + V_o \quad (9\text{-}28)$$

The foregoing equations show that the voltage applied across the tank ($v_{AB'}$) depends on which device is conducting and on the direction of i_L. The conditions described by Eqs. 7-25 through 9-28 can be represented by an equivalent circuit of Fig. 9-10b. The solution for the circuit of Fig. 9-5a is applied to the equivalent circuit of Fig. 9-10b for each interval, based on the initial conditions and the voltages V_{AB} and $V_{B'B}$, which appear as dc voltages for a given interval.

In the steady-state symmetrical operation, both the switches are operated identically. Similarly, the two diodes operate identically. Therefore, it is sufficient to analyze only one half-cycle of operation, since the other half is symmetrical. It can be shown that in the SLR converter of Fig. 9-10a, the output voltage V_o cannot exceed the input voltage $\tfrac{1}{2}V_d$, that is, $V_o \le \tfrac{1}{2}V_d$.

The switching frequency f_s ($=\omega_s/2\pi$), with which the circuit waveforms repeat, can be controlled to be less than or greater than the resonance frequency f_0 ($=\omega_0/2\pi$) if the converter consists of self-controlled switches. There are three possible modes of operation based on the ratio of switching frequency ω_s to the resonance frequency ω_0, which determines if i_L flows continuously or discontinuously.

9-4-1-1   Discontinuous-Conduction Mode with $\omega_s < \frac{1}{2}\omega_0$

By using Eqs. 9-3 and 9-4, Fig. 9-11 shows the circuit waveforms in steady state where, at $\omega_0 t_0$, switch T_+ is turned on and the inductor current builds up from its zero value. The capacitor voltage builds up from its initial negative value $V_{c0} = -2V_o$. Figure 9-11 also shows the circuits during various intervals with corresponding v_{AB} and $v_{B'B}$.

At $\omega_0 t_1$, 180° subsequent to $\omega_0 t_0$, the inductor current reverses and now must flow through D_+ since the other switch T_- is not yet turned on. After another 180° subsequent to $\omega_0 t_1$ with a smaller peak current in this half-cycle, the current goes to zero and remains zero as no switches are on. A symmetrical operation requires that v_c during the discontinuous interval $\omega_0(t_3 - t_2)$ be negative of V_{c0}, that is, equal to $2V_o$. During this interval, the capacitor voltage equal to $2V_o$ is less than $\frac{1}{2}V_d + V_o$ (since $V_o \leq \frac{1}{2}V_d$); therefore the current becomes discontinuous. At $\omega_0 t_3$, the next switch T_- is turned on and the next half-cycle ensues.

Because of the discontinuous interval in Fig. 9-11, one half-cycle of the operating frequency exceeds 360° of the resonance frequency f_0, and therefore in this mode of operation, $\omega_s < \frac{1}{2}\omega_0$. The average of the rectified inductor current $|i_L|$ equals the output dc current I_o, which is supplied to the load at a voltage of V_o.

Note that in this mode of operation, the switches turn off naturally at zero current and at zero voltage, since the inductor current goes through zero. The switches turn on at zero current but not at zero voltage. Also the diodes turn on at zero current and turn off naturally at zero current. Since the switches turn off naturally in this mode of operation, it is possible to use thyristors in low-switching-frequency applications.

The disadvantage of this mode is the relatively large peak current in the circuit and, therefore, higher conduction losses, compared with the continuous-conduction mode.

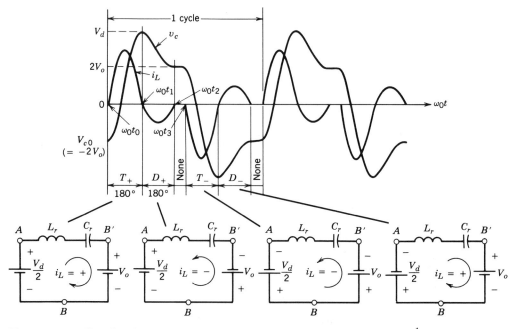

Figure 9-11   SLR dc–dc converter; discontinuous-conduction mode with $\omega_s < \frac{1}{2}\omega_0$.

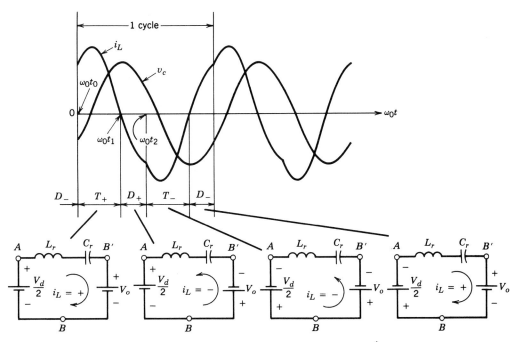

Figure 9-12 SLR dc–dc converter; continuous-conduction mode with $\frac{1}{2}\omega_0 < \omega_s < \omega_0$.

9-4-1-2 Continuous-Conduction Mode with $\frac{1}{2}\omega_0 < \omega_s < \omega_0$

The waveforms are shown in Fig. 9-12 where T_+ turns on at $\omega_0 t_0$, with a finite value of the inductor current and at a preconduction switch voltage of V_d. Here T_+ conducts for less than 180°. At $\omega_0 t_1$, i_L reverses and flows through D_+, thus naturally turning off T_+. At $\omega_0 t_2$, T_- is turned on and i_L transfers from D_+ to T_-. In this mode, D_+ conducts for less than 180° because T_- is switched on early, compared with the discontinuous-conduction mode.

In this mode of operation, the switches turn on at a finite current and at a finite voltage, thus resulting in a turn-on switching loss. Moreover, the freewheeling diodes must have good reverse-recovery characteristics to avoid large reverse current spikes flowing through the switches, for example, at $\omega_0 t_2$ through D_+ and T_-, and to minimize the diode turn-off losses. However, the turn-off switches occurs naturally at zero current and at zero voltage as the inductor current through them goes to zero and reverses through the freewheeling diodes. Therefore, it is possible to use thyristors as switches in low-switching-frequency applications.

9-4-1-3 Continuous-Conduction Mode with $\omega_s > \omega_0$

Compared with the previous continous-conduction mode, where the switches turn off naturally but turn on at a finite current, the switches in this mode with $\omega_s > \omega_0$ are forced to turn off a finite current, but they are turned on at zero current and zero voltage.

Figure 9-13 shows the circuit waveforms where T_+ starts conduction at $\omega_0 t_0$ at zero current when the inductor current reverses in direction. At $\omega_0 t_0$, before the half-cycle of the current oscillation ends, T_+ is forced to turn off, thus forcing the positive i_L to flow through D_-. Because of the large negative dc voltage applied across the LC tank ($v_{AB'} = -\frac{1}{2}V_d - V_o$), the current through the diode goes to zero quickly (note that its frequency

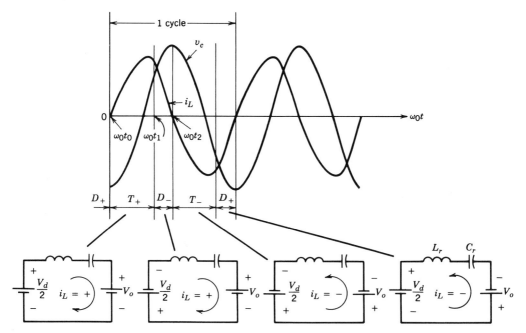

Figure 9-13 SLR dc–dc converter; continuous-conduction mode with $\omega_s > \omega_0$.

of oscillation ω_0 does not change) at $\omega_0 t_2$. Here T_- is gated on as soon as D_- begins to conduct so that it can conduct when i_L reverses. The combined conduction interval for T_+ and D_- is equal to one half-cycle of operation at the switching frequency of ω_s. This half-cycle is less than 180° of the resonance frequency ω_0, thus resulting in $\omega_s > \omega_0$.

There are several advantages in operating at $\omega_s > \omega_0$. Unlike the continuous-conduction mode with ω_s less than ω_0, the switches turn on at a zero current and zero voltage; thus, the freewheeling diodes do not need to have very fast reverse-recovery characteristics. A significant disadvantage would appear to be that the switches need to force turn off near the peak of i_L, thus causing a large turn-off switching loss. However, since the switches turn on not only at zero current but also at zero voltage (note that prior to turn-on of T_-, the freewheeling diode D_- across it is conducting), it is possible to use lossless snubber capacitors C_s in parallel with the switches, as shown in Fig. 9-14, which act as lossless turn-off snubbers for the switches.

Operation above the resonance frequency requires that the controllable switches be used.

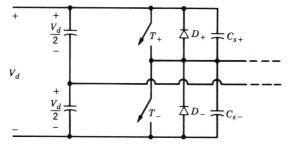

Figure 9-14 Lossless snubbers in an SLR converter at $\omega_s > \omega_0$.

9-4-1-4 Steady-State Operating Characteristics

It is useful to know the relationship of the peak and the average values of the circuit voltages and currents to the operating conditions (V_d, V_o, I_o, ω_0, etc.). The voltages, currents, and switching angular frequency ω_s are normalized by the following base quantities:

$$V_{\text{base}} = \tfrac{1}{2}V_d \qquad (9\text{-}29)$$

$$I_{\text{base}} = \frac{\tfrac{1}{2}V_d}{Z_0} \qquad (9\text{-}30)$$

$$\omega_{\text{base}} = \omega_0 \qquad (9\text{-}31)$$

Figure 9-15 shows normalized I_o versus ω_0 for two values of V_o. This figure shows that a SLR dc–dc converter in the discontinuous-conduction mode (corresponding to $\omega_s < 0.5$) operates as a current source, that is, I_o stays constant even though the load resistance and hence V_o may change. Because of this property, this converter exhibits an inherent overload protection capability in the discontinuous-conduction mode.

It should be noted that in Fig. 9-10a, I_o is the average value of the full-wave-rectified inductor current $|i_L|$, where the ripple in $|i_L|$ is assumed to flow through the output filter capacitor and its average value I_o flows through the output load resistance. In this converter, the peak value of the inductor current (which also is the peak value of the current through the switches) and the peak voltage across the capacitor C_r can be several times higher than I_o and V_d, respectively (see the problems at the end of the chapter). This aspect must be considered in comparing this converter with other converter topologies.

9-4-1-5 Control of SLR dc–dc Converters

As shown in Section 9-3-1-3 dealing with the frequency characteristics of series-resonant circuits, the resonant-tank impedance depends on the frequency of operation. Therefore, for a given applied input voltage V_d and a load resistance, V_o can be regulated by controlling the switching frequency f_s. This is shown in block diagram form in Fig. 9-16,

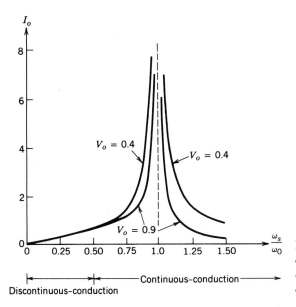

Figure 9-15 Steady-state characteristics of an SLR dc–dc converter; all parameters are normalized.

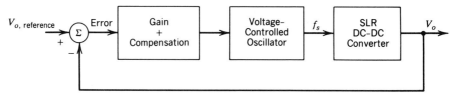

Figure 9-16 Control of SLR dc–dc converter.

where the error between the sensed output voltage and the reference voltage determines the output frequency f_s of the voltage-controlled oscillator, which in turn controls the two switches.

The variable frequency control described before is not optimum because of the complexity of its analysis and the design of EMI filters. As discussed in reference 7, a constant frequency control can be implemented in a full-bridge version of the SLR converter, where the switches in each leg of the converter operate at the 50% duty ratio at a constant frequency of $\omega_s > \omega_0$, but the phase delay between the output of the two converter legs is controlled. Such a control restricts the load to be in a limited range, beyond which the zero-voltage-/zero-current-switching characteristics of the converter do not hold.

It should be noted that the SLR converter can be used where the output is not a rectified dc; for example, SLR inverters are used for induction heating applications, where the load appears as a resistance rather than a dc voltage V_o.

9-4-2 PARALLEL-LOADED RESONANT dc–dc CONVERTERS

These converters are similar to the SLR converters in terms of operating with a series-resonant LC tank circuit. However, unlike the SLR converters, where the output stage or the load appears in series with the resonant tank, here the output stage is connected in parallel with the resonant-tank capacitor C_r, as shown in Fig. 9-17a. The isolation transformer is omitted for simplicity.

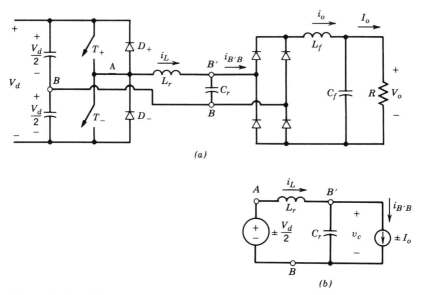

Figure 9-17 PLR dc–dc converter: (a) half-bridge; (b) equivalent circuit.

The PLR converters differ from the SLR converters in many important respects, for example, (1) PLR converters appear as a voltage source and hence, are better suited for multiple outlets; (2) unlike the SLR converters, the PLR converters do not possess inherent short-circuit protection capability, which obviously is a drawback; and (3) PLR converters can step up as well as step down the voltage, unlike the SLR converters, which can operate only as a step-down converter (not counting the transformer turns ratio).

In the following sections, only the modes in which a PLR converter is likely to operate are discussed. The discussion on the other modes can be found in the literature.

The voltage across the resonant-tank capacitor C_r is rectified, filtered, and then supplied to the load. To develop an equivalent circuit, the current through the output filter inductor in Fig. 9-17a can be assumed to be a ripple-free dc current I_o during a switching-frequency time period. This is a reasonable assumption, based on a high switching frequency and a sufficiently large value of the filter inductor. The voltage across the resonant tank depends on the devices conducting as follows:

$$T_+ \text{ or } D_+: \qquad v_{AB} = +\tfrac{1}{2}V_d \qquad\qquad (9\text{-}32)$$

and

$$T_- \text{ or } D_-: \qquad v_{AB} = -\tfrac{1}{2}V_d \qquad\qquad (9\text{-}33)$$

Based on the previous discussion, an equivalent circuit of Fig. 9-17b can be obtained where the input voltage to the tank (v_{AB}) is equal in magnitude to $\tfrac{1}{2}V_d$ but its polarity depends on which switch is turned on (T_+ or T_-). The current $i_{B'B}$, defined in Fig. 9-17a, equals I_o in magnitude, but its direction depends on the polarity of the voltage v_c across C_r at the input to the bridge rectifier.

The equivalent circuit of Fig. 9-17b is identical to that discussed in Section 9-3-1-2. Therefore, Eqs. 9-13 and 9-14 can be applied with the appropriate v_{AB} and $i_{B'B}$ and the initial conditions.

Unlike SLR converters, a PLR dc–dc converter can operate in a large number of combinations consisting of the states of i_L and v_c. However, only three modes are considered in the following sections.

9-4-2-1 Discontinuous Mode of Operation

In this mode of operation, both i_L and v_c remain zero simultaneously for some length of time. The steady-state waveforms for this mode of operation are plotted in Fig. 9-18, based on Eqs. 9-13 and 9-14. During steady-state operation, initially both i_L and v_c are zero and T_+ is turned on at $\omega_0 t_0$. So long as $|i_L| < I_o$, the output current circulates through the rectifier bridge, which appears as a short circuit across C_r and keeps its voltage at zero, as shown in Fig. 9-18. At $\omega_0 t_1$, i_L exceeds I_o and the difference $i_L - I_o$ flows through C_r, and v_c increases. Due to LC resonance, i_L reverses at $\omega_0 t_2$ and flows through D_+, since T_- is not turned on until some time later. During the interval $\omega_0(t_3 - t_1)$, i_L and v_c can be calculated from Eqs. 9-13 and 9-14 using $i_{L0} = I_o$ and $v_{c0} = 0$ as the initial conditions at time $\omega_0 t_1$. If the gate/base drive of T_+ is removed prior to $\omega_0 t_3$, i_L can no longer flow after $\omega_0 t_3$ and stays at zero. With $i_L = 0$, i_o flows through C_r, and v_c decays linearly to zero during the interval $\omega_0 t_3$ to $\omega_0 t_4$.

In this discontinuous mode of operation, both v_c and i_L stay at zero for an interval that can be varied in order to control the output voltage. Beyond this discontinuous interval, T_- is gated on at $\omega_0 t_5$ and the next half-cycle ensues with identical initial conditions of zero i_L and v_c as for the first half-cycle.

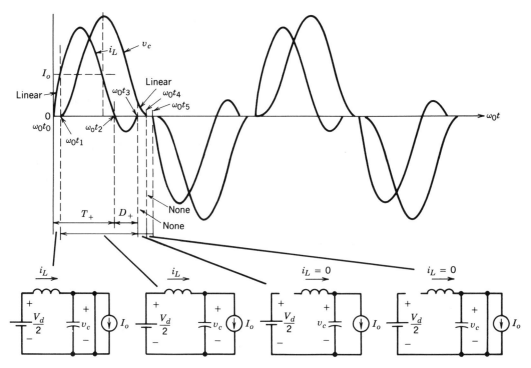

Figure 9-18   PLR dc–dc converter in a discontinuous mode.

Clearly, the foregoing operation corresponds to ω_s in a range from zero to approximately $\frac{1}{2}\omega_0$. Also, there are no turn-on or turn-off stresses on the switches or the diodes.

9-4-2-2   Continuous Mode of Operation Below ω_0

At switching frequencies higher than those in the discontinuous mode but less than ω_0, both v_c and i_L becomes continuous. The waveforms are shown in Fig. 9-19, where a switch turns on at a finite i_L and the current commutates from the diode connected in antiparallel with the other switch. This results in turn-on losses in the switches, and the diodes must have good reverse-recovery characteristics. However, there are no turn-off losses in the switches since the current through them commutates naturally when i_L reverses in direction.

9-4-2-3   Continuous Mode of Operation Above ω_0

This mode with continuous v_c and i_L occurs at $\omega_s > \omega_0$. The circuit waveforms are shown in Fig. 9-20. Here, the turn-on losses in the switches are eliminated since the switches turn on naturally when i_L, initially flowing through the diodes, reverses. However, this operating mode results in the turn-off losses in the switches, since a switch is forced to turn off, thus transferring its current to the diode connected in antiparallel with the other switch.

Similar to the SLR converter operating in a continuous-conduction mode with $\omega_s > \omega_0$, the switches here turn on at zero voltage, thus at the switching instant the snubber capacitor in parallel has no stored energy. Therefore, it is possible to eliminate the turn-off

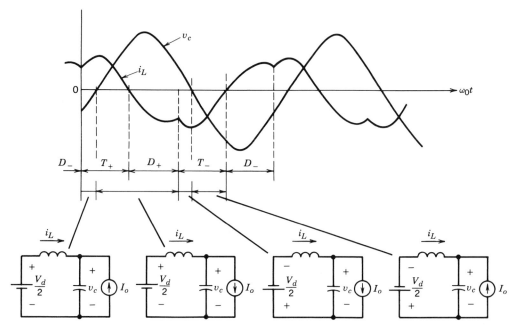

Figure 9-19 PLR dc–dc converter in a continuous mode with $\omega_s < \omega_0$.

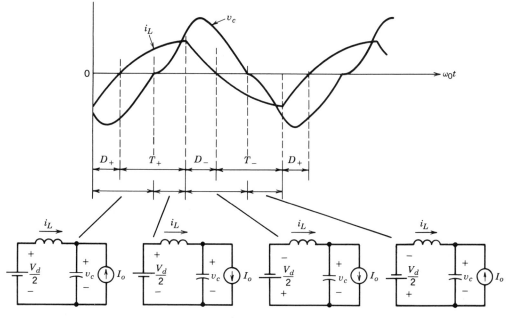

Figure 9-20 PLR dc–dc converter in a continuous mode with $\omega_s > \omega_0$.

losses by connecting a lossless snubber consisting of a capacitor (with no series resistor) in parallel with each switch, as in an SLR converter in Fig. 9-14.

9-4-2-4 Steady-State Operating Characteristics

The steady-state operating characteristics of the PLR dc–dc converters are shown in Fig. 9-21 for two values of I_o, where the variables are normalized by using the base quantities

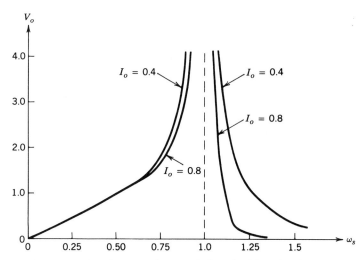

Figure 9-21   Steady-state characteristics of a PLR dc–dc converter. All quantities are normalized.

defined in Eqs. 9-29 through 9-31. Figure 9-21 shows the following important properties of the PLR converter:

- In the discontinuous mode of operation with $\omega_s < \frac{1}{2}\omega_0$, this converter exhibits a good voltage source characteristic and V_o remains independent of I_o. This property is useful in designing a converter with multiple outputs.
- Also in the frequency range $\omega_s < \frac{1}{2}\omega_0$, the output varies linearly with ω_s, thus simplifying the output regulation.
- It is also possible to operate in the high-frequency range $\omega_s > \omega_0$, and the maximum change required in the operating frequency is less than 50% to compensate for the output loading for a normalized output voltage of 1.0.
- It is possible to step up or step down the output voltage, that is, V_o can be less than or greater than 1.0.

In this converter, the peak inductor current (which is also the peak current through the switches) and the peak capacitor voltage can be several times higher than I_o and V_d, respectively (see the problems).

The converter characteristics shown in Fig. 9-21 suggest than an effective way to regulate the output is by controlling the frequency of operation ω_s.

9-4-3   HYBRID-RESONANT dc–dc CONVERTER

This topology consists of a series-resonant circuit as shown in Fig. 9-22 but the load is connected in parallel with only part of the capacitance, for example, one-third of the total capacitance, and the other two-thirds of the capacitance appears in series. The purpose of this topology is to benefit from the advantageous properties of both the SLR and the PLR converters, namely that an SLR converter offers an inherent current limiting under short-circuit conditions and a PLR converter acts as a voltage source, and thus regulating its voltage at no load with a high-Q resonant tank is not a problem. These converters can be analyzed based on the discussion presented in the previous two sections. These are analyzed in detail in reference 15.

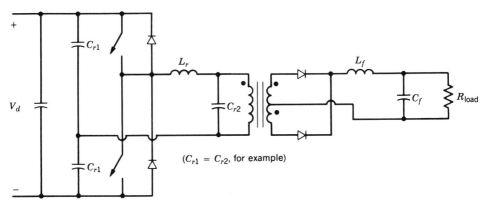

Figure 9-22 Hybrid-resonant dc–dc converter.

9-4-4 CURRENT-SOURCE, PARALLEL-RESONANT dc-TO-ac INVERTERS FOR INDUCTION HEATING

The basic principle of such an inverter is illustrated by means of the circuit of Fig. 9-23a, where a square-wave current source is applied to a parallel-resonant load. The induction coil and the load (RL combination) are modeled by means of a parallel combination of equivalent R_{load} and L_r, rather than a series RL. The capacitor C_r is added to resonate with L_r, rather than a series RL. The capacitor C_r is added to resonate with L_r in parallel. It is assumed that the harmonic impedance of the parallel-resonant load at the harmonic frequencies of the input current source is negligibly small, thus resulting in an essentially sinusoidal voltage v_o. Therefore, the analysis of Section 9-3-2-2 applies.

When the fundamental frequency ω_s of the source current i_o equals the natural resonance frequency $\omega_0 = (1/\sqrt{L_r C_r})$, the circuit phasor diagram is shown in Fig. 9-23b, where the fundamental-frequency component $\mathbf{V}_{o1}$ of the resulting voltage is in phase with the fundamental-frequency component $\mathbf{I}_{o1}$ of the input current.

Since the square-wave input current in practice is supplied by a thyristor inverter, the resonant load must supply the capacitive vars to the inverter. This implies that the load voltage $\mathbf{V}_{o1}$ should lag the input current $\mathbf{I}_{o1}$, which is possible only at a frequency $\omega_s > \omega_0$, as shown in Fig. 9-23c.

A current-source inverter consisting of thyristors is shown in Fig. 9-24a. To avoid a large di/dt (during current commutation) through the inverter thyristors, a small inductance L_c in series with the resonant load is purposely introduced. The inverter output

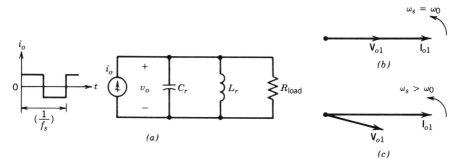

Figure 9-23 Basic circuit for current-source, parallel-resonant converter for induction heating: (a) basic circuit; (b) phasor diagram at $\omega_s = \omega_0$; (c) phasor diagram at $\omega_s > \omega_0$.

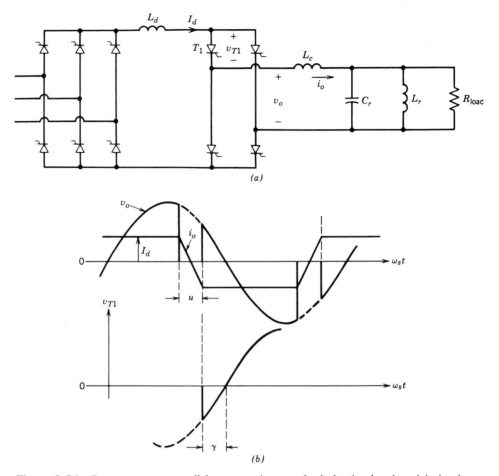

Figure 9-24   Current-source, parallel-resonant inverter for induction heating: (*a*) circuit;
(*b*) waveforms.

current i_o therefore deviates from its idealized square-wave shape and becomes trapezoidal, as shown in Fig. 9-24*b*.

The voltage across one of the thyristors T_1 shows that after it stops conducting a reverse voltage appears across it for a time interval equal to γ/ω_s; subsequently it is required to block a forward-polarity voltage. Therefore, γ/ω_s should be sufficiently larger than the specified turn-off time t_q of the thyristor that is being used.

One of the techniques to control the power output of this inverter is by controlling its switching frequency. As the switching frequency f_s is increased further above the natural resonance frequency f_0, the power output decreases if I_d is held constant by means of a controlled dc supply. Another obvious technique to control the power output is to control I_d, keeping the switching frequency of the inverter constant.

9-4-4-1   Start-up

In case of a current-fed parallel-resonant inverter, the load must be in resonance with C_r prior to the inverter operation in Fig. 9-24*a*. This is accomplished by means of a precharged capacitor dumping its charge onto the parallel-resonant load circuit, thus establishing oscillating load voltages and currents. Shortly after that, the inverter operation is initiated.

9-4-5 CLASS E CONVERTERS

In class E converters, the load is supplied through the sharply tuned series-resonant circuit shown in Fig. 9-25a. This results in an essentially sinusoidal current i_o. The input to the converter is through a sufficiently large inductor to allow the assumption that in steady state, the input to the converter is a dc current source I_d, as shown in Fig. 9-25a, where the current magnitude depends on the power output. The waveforms are shown in Fig. 9-25b for an optimum mode, which is discussed later on. When the switch is on, $I_d + i_o$ flows through the switch, as shown in Fig. 9-25c. When the switch is turned off, because of the capacitor C_1, the voltage across the switch builds up slowly, thus allowing a zero-voltage turn-off of the switch. With the switch off, the oscillating circuit is as shown in Fig. 9-25d, where the voltage across capacitor C_1 builds up, reaches its peak, and eventually comes back to zero, at which instant the switch is turned back on.

A class E converter operates at a switching f_s, which is slightly higher than the resonant frequency $f_0 = 1/(2\pi\sqrt{L_rC_r})$. During the interval when the switch is off, the

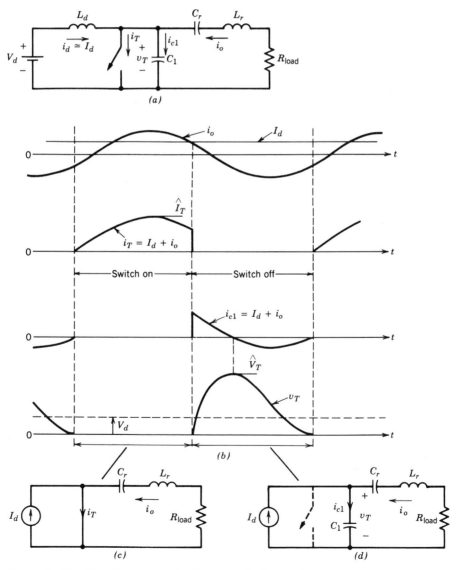

Figure 9-25 Class E converter (optimum mode, $D = 0.5$).

input supplies power to the circuit since v_T is positive, as shown in Fig. 9-25b. For a high-quality factor of the series $L_r C_r R$ circuit ($Q \geq 7$), which results in an essentially sinusoidal load current i_o, only a slight variation in f_s is needed to vary the output voltage. As f_s increases (where $f_s > f_0$), i_o and therefore v_R decrease.

Another observation that can be made is as follows: The average value of v_T equals V_d. If i_o is assumed to be purely sinusoidal, the average voltage across the load resistance R is zero. The average voltage across L_r is also zero in steady state. Therefore, C_r blocks the dc voltage V_d in addition to providing a resonant circuit.

The operation of a class E converter can be categorized in optimum and suboptimum modes. The circuit and the waveforms shown in Fig. 9-25 belong to the optimum mode of operation where the switch voltage returns to zero with a zero slope ($i_{c1} = 0$) and there is no need for a diode in antiparallel with the switch. This mode of operation requires that the load resistance R be equal to an optimum value R_{opt}. The switch duty ratio $D = 0.5$ results in a maximum power capability or, in other words, the maximum switch utilization ratio, where the switch utilization ratio is defined as the ratio of the output power P_o to the product of the peak switch voltage and the peak switch current. It is shown in the literature that the peak switch current is approximately $3I_d$ and the peak switch voltage is approximately $3.5V_d$.

The nonoptimum mode of operation occurs if $R < R_{\text{opt}}$. Here, the switch voltage reaches zero with a negative slope [$dv_T/dt < 0$, and hence, $i_{c1} = C_1(dv_T/dt) < 0$]. A diode is connected in antiparallel with the switch as shown in Fig. 9-26a to allow this

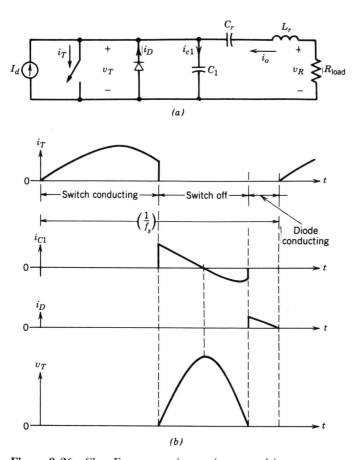

(a)

(b)

Figure 9-26 Class E converter (nonoptimum mode).

current to flow while keeping the switch voltage at zero (at one diode drop). The waveforms are shown in Fig. 9-26b, where the switch is turned on as soon as the diode starts to conduct. In a circuit with a high input voltage, it is important to reduce the peak switch voltage $\hat{V}_T$. It can be shown that for a smaller switch duty ratio, $\hat{V}_T$ decreases but the peak switch current $\hat{I}_T$ goes up.

The advantage of a class E converter is the elimination of switching losses and the reduction in EMI. Also, it is a single-switch topology and produces a sinusoidal output current. Significant disadvantages are high peak voltage and current associated with the switch and large voltages and currents through the resonant LC elements. For the resistive load shown in Fig. 9-26a (the optimum mode without an antiparallel diode with the switch is very restrictive), class E converters have been considered for high-frequency electronic lamp ballasts.

It is possible to obtain a dc–dc voltage conversion by rectifying the output current. Since the output load may vary over a large range, an impedance matching network is required between the output of the class E converter and the output rectification stage to ensure a lossless switching operation of the class E converter. Various suboptimum class E topologies are described in reference 22.

9-5   RESONANT-SWITCH CONVERTERS

Historically, prior to the availability of controllable switches with appreciable voltage- and current-handling capability, the switch-mode converters consisted of thyristors (currently, thyristors in switch-mode converters are used only at very high power levels). Such converters had topologies and control schemes similar to those described in Chapters 7 and 8 for switch-mode dc–dc converters and dc-to-ac inverters. Each thyristor in such a converter required a current commutation circuit, which consisted of an LC resonant circuit plus other auxillary thyristors and diodes, which turned the main thyristor off by forcing the current through it to go to zero. Because of the complexity and substantial losses in the commutation circuits, thyristors were replaced by controllable switches, as their power-handling capability improved.

A need to increase switching frequencies and to reduce EMI led to augmenting the controllable switches in certain basic switch-mode converter topologies of Chapters 7 and 8 by a simple LC resonant circuit, thereby shaping the switch voltage and current in order to yield zero-voltage and/or zero-current switchings. Such converters are termed resonant-switch converters. Often, the diode needed for the resonant-switch circuit operation is the same as that in the original switch-mode converter topology. Similarly, inductors (such as the transformer leakage inductance) and the capacitors (such as the output capacitance of the semiconductor switch), which appear as undesirable parasitics in switch-mode topologies, can be utilized to provide the resonant inductor and the capacitor needed for the resonant-switch circuit.

The output in some of these circuits is controlled by controlling the operating frequency; in others a constant-frequency square-wave or PWM control can be used with some additional constraints to provide zero-voltage and/or zero-current switchings.

A majority of such converters can be divided into three switching categories:

1. Zero-current-switching (ZCS) topology where the switch turns on and turns off at zero current. The peak resonant current flows through the switch but the peak switch voltage remains the same as in its switch-mode counterpart. Such a topology is shown in Fig. 9-27a for a step-down dc–dc converter.

2. Zero-voltage-switching (ZVS) topology where the switch turns on and turns off at zero voltage. The peak resonant voltage appears across the switch, but the peak

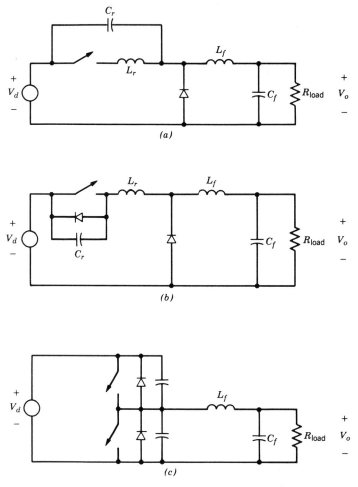

Figure 9-27   Resonant-switch converters: (a) ZCS dc–dc converter (step-down); (b) ZVS dc–dc converter (step-down); (c) ZVS-CV dc–dc converter (step-down).

switch current remains the same as in its switch-mode counterpart. Such a topology is shown in Fig. 9-27b for a step-down dc–dc converter.

3. Zero-voltage-switching, clamped-voltage (ZVS-CV) topology where the switch turns on and off at zero voltage as in category 2 above. However, a converter of this topology consists of at least one converter leg made up of two such switches. The peak switch voltage remains the same as in its switch-mode counterpart, but the peak switch current is generally higher. Such a converter topology is shown in Fig. 9-27c for a step-down dc–dc converter.

In the following sections, the operating principles of the converters belonging to all three switching categories are discussed.

9-5-1   ZCS RESONANT-SWITCH CONVERTERS

In such converters, the current produced by *LC* resonance flows through the switch, thus causing it to turn on and off at zero current. This can be easily explained in the step-down dc–dc converter of Fig. 9-28a, which has been modified as shown in Fig. 9-28b by the

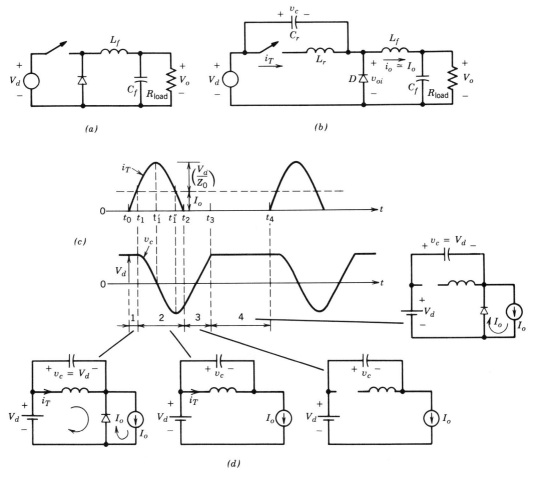

Figure 9-28 ZCS resonant-switch dc–dc converter.

addition of L_r and C_r. The filter inductor L_f is sufficiently large such that the current i_o can be assumed to be a current of constant magnitude I_o in Fig. 9-28*b*. The circuit waveforms in steady state are shown in Fig. 9-28*c* and the subcircuits are shown in Fig. 9-28*d*.

Prior to turning the switch on, the output current I_o freewheels through the diode D, and the voltage v_c across C_r equals V_d. At t_0, the switch is turned on at zero current. So long as i_T is less than I_o, D keeps on conducting and v_c stays at V_d. Therefore, i_T rises linearly, and at t_1, i_T equals I_o, which causes D to stop conducting. Now, L_r and C_r form a parallel-resonant circuit and the analysis of Section 9-3-2-1 applies. Use of Eq. 9-20 shows that at t_1', i_T peaks at $V_d/Z_0 + I_o$ and v_c reaches zero. The negative peak of v_c occurs at t_1'' when $i_T = I_o$. At t_2, i_T reaches zero and cannot reverse its direction. Thus the switch T is naturally turned off. Beyond t_2, the gate pulse from T is removed. Now I_o flows through C_r and v_c rises linearly to V_d at t_3, at which point the diode D turns on and v_c stays at V_d. After an interval during which i_T is zero and $v_c = V_d$, the gate pulse to T is again applied at t_4 to turn it on, and the next cycle ensues.

It is clear from the waveforms of Fig. 9-28*c* that the forward switch voltage is limited to V_d. The instantaneous voltage $v_{oi} = V_d - v_c$ across the output diode, as defined as Fig. 9-28*b*, is plotted in Fig. 9-29. By controlling the switch-off time interval $t_4 - t_3$, or in other words the switching frequency of operation, the average value of v_{oi} and, hence, the

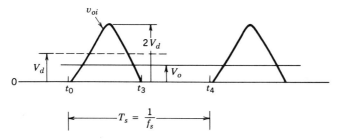

Figure 9-29 v_{oi} waveform in a ZCS resonant-switch dc–dc converter.

average power supplied to the output stage can be controlled. This in turn regulates the output voltage V_o for a given load current I_o.

From the waveforms of Fig. 9-28c, it can be seen that if $I_o > V_d/Z_0$, i_T will not come back to zero naturally and the switch will have to be forced off, thus resulting in turn-off losses.

Zero-current switching can also be obtained by connecting C_r in parallel with D as shown in Fig. 9-30a. As discussed previously, i_o can be assumed to be a current of constant magnitude I_o in Fig. 9-30a during a high-frequency resonant cycle.

Initially both the capacitor voltage (across C_r) and the inductor current (through L_r) are assumed to be zero and the load current I_o freewheels through the diode D. The converter operation can be divided into the following intervals for which the converter waveforms as well as the corresponding circuit states are shown in Fig. 9-30b and 9-30c:

1. *Time interval 1 (between* t_0 *and* t_1*).* At time t_0, the switch is turned on. Because of I_o flowing through the diode it appears as a short circuit and the entire input voltage V_d appears across L_r. Therefore, the switch current builds up linearly until it becomes equal to I_o at time t_1. Beyond this time, the diode turns off and the voltage clamp across C_r is removed.

2. *Time interval 2 (between* t_1 *and* t_2*).* Beyond t_1, $i_T > I_o$ and their difference ($i_T - I_o$) flows through C_r. At t_1', i_T peaks and $v_c = V_d$. At time t_1'', the switch current drops from its peak value to I_o and the capacitor voltage reaches $2V_d$. The switch current eventually drops to zero at t_2 and cannot reverse through the switch (if a BJT or a MOSFET is used as a switch, then a diode in series with it must be used to block a negative voltage and to prevent the flow of reverse current through the switch). Thus, the switch current is commutated off naturally and the gate/base drive from the switch should be removed at this point.

3. *Time interval (between* t_2 *and* t_3*).* Beyond the time t_2 with the switch off, the capacitor C_r discharges into the output load and the capacitor voltage linearly drops to zero at t_3.

4. *Time interval 4 (between* t_3 *and* t_4*).* Beyond t_3, the load current just freewheels through the diode until a time t_4, when the switch is turned on and the next switching cycle begins. This time interval is controlled to adjust the output voltage.

Under a steady-state operating condition, the average voltage across the filter inductor is zero; therefore the voltage across C_r averaged over one switching cycle equals the output voltage V_o. By controlling the freewheeling time interval 4 (i.e., by controlling the switching frequency), the output voltage V_o can be regulated.

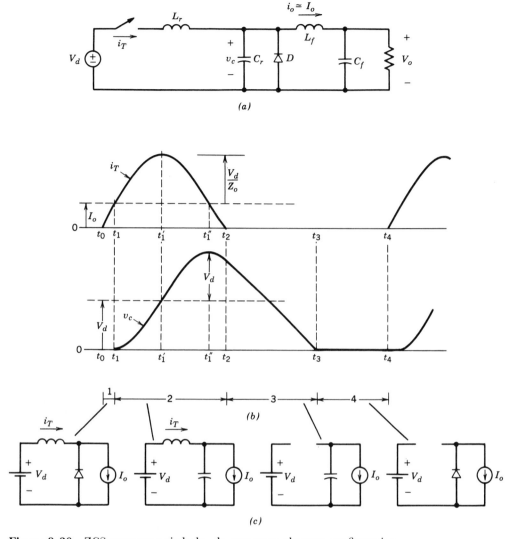

Figure 9-30 ZCS resonant-switch dc–dc converter; alternate configuration.

From the waveforms of Fig. 9-30b, the following circuit properties can be observed:

- The L_r and C_r together determine the natural resonance frequency $\omega_0 = 1/(2\pi \sqrt{L_r C_r})$, which can be made to be large (in the megahertz range) by proper selection of L_r and C_r. Both the switch turn-on and turn-off occur at zero current, thus reducing the switching losses. It should be noted that at turn-on, the voltage across the switch equals V_d. This results in losses, as discussed in Section 9-5-3.

- The load current I_o must be less than a maximum value of V_d/Z_0, which depends on the circuit parameters. Otherwise, the switch would have to turn off a finite amount of current.

- At a given switching frequency of operation, V_o declines with increasing load. Therefore, the switching frequency ω_s must be increased to regulate V_o. The opposite is true if the load decreases.

- By placing a diode in the antiparallel across the switch in Fig. 9-30a, the inductor current is allowed to reverse. This permits excess energy in the resonant circuit at light loads to be transferred back to the voltage source V_d. This significantly reduces the dependence of V_o on the output load.

Since the switching losses are minimized and the EMI is reduced, very high switching frequencies can be attained. One drawback of such a converter is that the switch peak current rating required is significantly higher than the load current. This also implies that the conduction losses in the switch would be higher compared with its switch-mode counterpart. In references 25 and 26, it has been shown that this principle can be applied to various other single-switch dc–dc converter topologies.

9-5-2   ZVS RESONANT-SWITCH CONVERTERS

In these converters, the resonant capacitor produces a zero voltage across the switch, at which instant the switch can be turned on or off. Such a step-down dc–dc converter circuit is shown in Fig. 9-31a, where a diode D_r is connected in antiparallel with the switch. As discussed previously, the output current i_o can be assumed to be a current of constant magnitude I_o in Fig. 9-31a during a high-frequency resonant cycle.

Initially, the switch is conducting I_o and therefore, $I_{L0} = I_o$ and $V_{c0} = 0$. The converter operation can be divided into the following intervals for which the converter waveforms as well as the corresponding circuit states are shown in Fig. 9-31b and 9-31c, respectively:

1. *Time interval 1 (between* t_0 *and* t_1*).* At time t_0, the switch is turned off. Because of C_r, the voltage across the switch builds up slowly but linearly from zero *to* V_d at t_1. This results in a zero-voltage turn-off of the switch.

2. *Time interval 2 (between* t_1 *and* t_2*).* Beyond t_1, since $v_c > V_d$, the diode D becomes forward biased, C_r and L_r resonate, and the analysis of Section 9-3-1-1 applies. At t_1', i_L goes through zero and v_c reaches its peak of $V_d + Z_0 I_o$. At t_1'', $v_c = V_d$ and $i_L = -I_o$. At t_2, the capacitor voltage reaches zero and cannot reverse its polarity because the diode D_r begins to conduct.

   Note that the load current I_o should be sufficiently large so that $Z_0 I_o > V_d$. Otherwise, the switch voltage will not come back to zero naturally and the switch will have to be turned on at a nonzero voltage, resulting in turn-on losses (the energy stored in C_r will dissipate in the switch).

3. *Time interval 3 (between* t_2 *and* t_3*).* Beyond t_2, the capacitor voltage is clamped to zero by D_r, which conducts the negative i_L. The gate drive to the switch is applied once its antiparallel diode begins to conduct. Now i_L increases linearly and goes through zero at time t_2', at which instant i_L begins to flow through the switch. Therefore, the switch turns on at a zero voltage and zero current. Here i_L increases linearly to I_o at t_3.

4. *Time interval 4 (between* t_3 *and* t_4*).* Once i_L reaches I_o at t_3, the freewheeling diode D turns off. Because a small negative slope is associated with di/dt through the diode at turn-off, there are no diode reverse-recovery problems like the ones encountered in the switch mode. The switch conducts I_o as long as it is kept on until t_4. The interval t_4-t_3 can be controlled. At t_4, the switch is turned off and the next cycle ensues.

It is clear from the waveforms of Fig. 9-31b that the switch current is limited to I_o. The voltage v_{oi} across the output diode as defined in Fig. 9-31a, is plotted in Fig. 9-32.

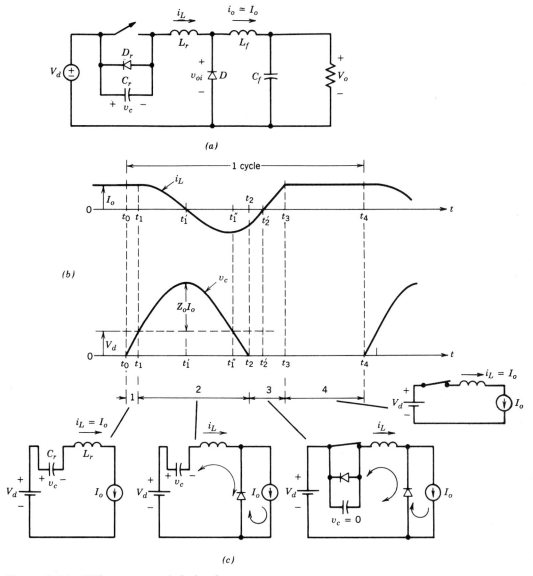

Figure 9-31 ZVS resonant-switch dc–dc converter.

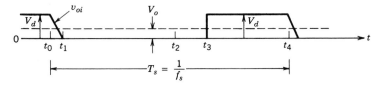

Figure 9-32 The v_{oi} waveform in a ZVS resonant-switch dc–dc converter.

By controlling the on interval t_4-t_3 of the switch, the average value of v_{oi} and, hence, the average power supplied to the output stage can be controlled. This in turn regulates the output voltage V_o for a given load current I_o.

This zero-voltage-switching approach can also be applied to various other single-switch dc–dc converter topologies, as described in reference 27.

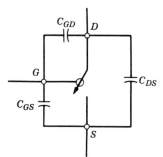

Figure 9-33 Switch internal capacitances.

9-5-3 COMPARISON OF ZCS AND ZVS TOPOLOGIES

Both of these techniques require a variable-frequency control to regulate the output voltage.

In the ZCS, the switch is required to conduct a peak current that is higher than the load current I_o by an amount V_d/Z_0. For natural turn-off of the switch at zero current, the load current I_o must not exceed V_d/Z_0. Therefore, there is a limit on how low the load resistance can become. By placing a diode in antiparallel with the switch, the output voltage can be made insensitive to the load variations.

In the ZVS topology discussed here, the switch is required to withstand a forward voltage that is higher than V_d by an amount Z_0I_o. For zero-voltage (lossless) turn-on of the switch, the load current I_o must be greater than V_d/Z_0. Therefore, if the output load current I_o varies in a wide range, then the foregoing two conditions result in a very large voltage rating of the switch (see Problem 9-13). Therefore, this technique is limited to an essentially constant load application. To overcome this limitation, a zero-voltage-switching multiresonant technique is described in reference 29.

In general, ZVS is preferable over ZCS at high switching frequencies. The reason has to do with the internal capacitances of the switch, as shown in Fig. 9-33. When the switch turns on at zero current but at a finite voltage, the charge on the internal capacitances is dissipated in the switch. As discussed in reference 30, this loss becomes significant at very high switching frequencies. However, no such loss occurs if the switch turns on at a zero voltage.

9-6 ZERO-VOLTAGE-SWITCHING, CLAMPED-VOLTAGE TOPOLOGIES

In the literature, these topologies have been referred to as the pseudo-resonant- and resonant-transition topologies. In these topologies, the switches turn on and turn off at zero voltage. But unlike the ZVS topology discussed in Section 9-5-2, the peak voltage of a switch is clamped at the input dc voltage. Such converters consist of at least one converter leg having two switches.

9-6-1 ZVS-CV dc–dc CONVERTERS

The basic principle is shown by means of the dc–dc step-down converter shown in Fig. 9-34a, consisting of two switches. The filter inductor L_f is very small compared with the normal switch-mode topology so that i_L becomes positive as well as negative during each

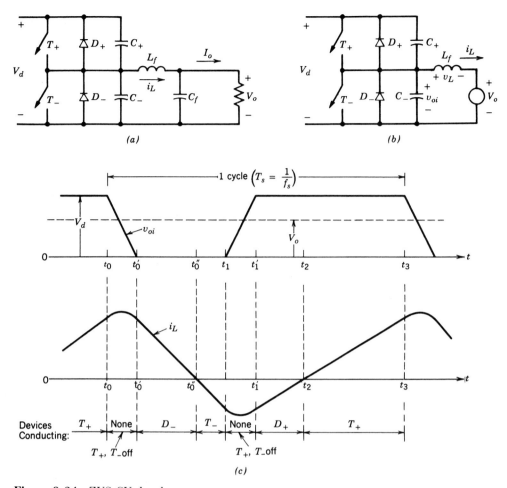

Figure 9-34 ZVS–CV dc–dc converter.

cycle of operation. Assuming C_f to be large, the filter capacitor and the load can be replaced by a dc voltage V_o in steady state as shown in the equivalent circuit of Fig. 9-34b. The waveforms are shown in Fig. 9-34c.

Initially, T_+ is conducting a positive i_L and v_L ($=V_d - V_o$) is positive. At time t_0, T_+ is turned off at zero voltage because of C_+ in Fig. 9-34a; the voltage across T_+ builds up slowly compared with its switching time. With T_- off and T_+ just off, the subcircuit is as shown in Fig. 9-35a. It can be redrawn as in Fig. 9-35b, where the initial voltage vV_d on C_- is shown explicitly by means of a voltage source. Since $C_+ = C_- = \frac{1}{2}C$, the Thévenin equivalent of the circuit on the left results in the circuit of Fig. 9-35c. Since C is very small, the resonant frequency $f_0 = 1/(2\pi\sqrt{L_fC})$ is much larger than the switching frequency of the converter. Moreover, $Z_0 = \sqrt{L_f/C}$ in this circuit is very large, resulting in a small variation in i_L during the time interval shown in Fig. 9-35d. At t_0', v_{oi}, the voltage across C_-, reaches zero, beyond which time this subcircuit has to be modified since v_{oi} cannot become negative because of the presence of D_- in the original circuit. During the time interval t_0'–t_0, the magnitude of dv/dt across both the capacitors is the same; therefore, $\frac{1}{2}i_L$ flows through each of the capacitors during this interval since $C_+ = C_-$.

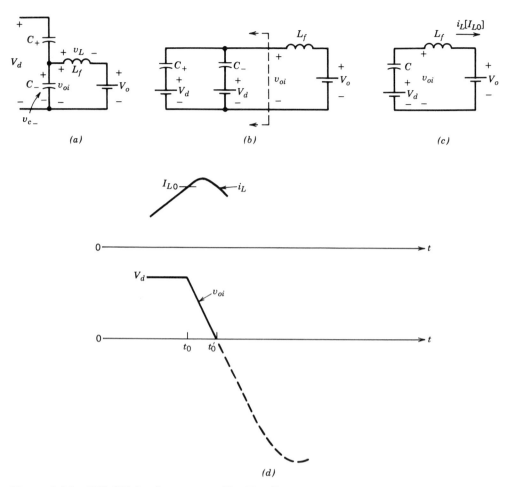

Figure 9-35   ZVS-CV dc–dc converter; T_+, T_- off.

The preceding discussion can be simplified if i_L during this interval is assumed to be essentially constant. This allows the assumption that v_{oi} changes linearly, as shown in Fig. 9-34c during the blanking-time interval where both the switches are off.

Beyond t_0', i_L decreases linearly, since it flows through D_-, and therefore, $v_L = -V_o$. Once D_- begins to conduct, T_- is gated on. At t_0'', i_L reverses direction and flows through T_-.

At t_1, T_- is turned off at zero voltage and after a capacitor charging–discharging interval $t_1'-t_1$, similar to t_0' to t_0, the negative i_L flows through D_+. Since v_L ($=V_d - V_o$) is positive beyond t_1', i_L increases. Here T_+ is gated on at zero voltage as soon as D_+ begins to conduct. At t_2, i_L becomes positive and flows through T_+.

At t_3, T_+ is turned off at zero voltage, thus completing a cycle with a time period $T_s = (t_3-t_0)$.

As shown by the waveforms in Fig. 9-34c, the switch voltage peak in this topology is clamped at V_d.

An important observation is that for the zero-voltage turn-off of a switch, a capacitor is connected directly across the switch. Therefore, the switch must be turned on only at zero voltage; otherwise the energy stored in the capacitor will be dissipated in the switch. Therefore, the diode in antiparallel with the switch must conduct prior to the closing of the switch. This requires that in the circuit of Fig. 9-34a, i_L must flow in both directions during each cycle to satisfy this requirement for both the switches.

In such a circuit, the output voltage can be regulated by means of a constant-switching-frequency PVM control. Assuming the blanking-time intervals $t_0'-t_0$ and $t_1'-t_1$, during which the resonant transition occurs, to be much smaller than the time period T_s of the switching frequency, the voltage v_{oi} in Fig. 9-34c is of a rectangular waveform. Since the average voltage across L_f is zero, the average value of v_{oi} equals V_o. Therefore, $V_o = DV_d$, where D is the duty ratio of the switch T_+ and DT_s is the time interval during which either T_+ or D_+ is conducting. The average value of i_L equals the output current I_o.

If a constant-frequency PWM control is used to regulate V_o, then L_f needs to be chosen such that even under the minimum value of V_d and the highest load (i.e., the minimum load resistance), i_L reaches a value less than zero.

The advantage of this ZVS-CV is that the switch voltages are clamped to V_d. The disadvantage is that because of the higher ripple in i_L, the switches need to carry higher peak currents as compared to the switch mode of operation.

9-6-2 ZVS-CV dc-TO-ac INVERTERS

It should be noted that the dc–dc converter discussed in Section 9-6-1 is capable of a two-quadrant operation where i_o can reverse. Therefore, such a converter can be modified as shown in Fig. 9-36a, which results in a half-bridge square-wave dc-to-ac inverter to supply an inductive load. The resulting waveforms with equal switch duty ratios are shown in Fig. 9-36b, and the switching losses are eliminated, since the switches turn on and turn off at zero voltage. The load current must lag the voltage (i.e., the load must be inductive like a motor load) for the switchings to occur at a zero voltage.

It is possible to operate the inverter of Fig. 9-36a in a current-regulated mode, similar to that discussed in Chapter 8. However, to achieve zero-voltage switchings, both switches must conduct every switching cycle, and therefore i_o must flow in each direction during every switching cycle. The waveforms for square-wave and PWM modes are shown in Figs. 9-36b and 9-36c, respectively. This concept can be extended to a three-phase inverter as shown in Fig. 9-37.

9-6-3 ZVS-CV dc–dc CONVERTER WITH VOLTAGE CANCELLATION

The ZVS-CV technique can be extended to the single-phase dc-to-ac inverter with voltage cancellation. The switch-mode circuit, which was discussed in detail in Chapter 8, is shown in Fig. 9-38a, and the resulting waveforms are shown in Fig. 9-38b, where both switches in each leg operate at a 50% duty ratio but the phase delay ϕ between the outputs of the two legs is controlled in order to control the output v_{AB} of the full bridge. The output voltage of the full bridge is stepped down through an isolation transformer and then rectified to yield an overall dc–dc converter.

The switch-mode circuit of Fig. 9-38a can be modified to provide ZVC-CV by adding L_A, C_{A+}, C_{A-} to leg A and L_B, C_{B+}, C_{B-} to leg B, as shown in Fig. 9-39a. For simplicity, the transformer is replaced by its magnetizing inductance L_m and its leakage inductance is neglected. The output stage is represented by the output current I_o. The resulting waveforms are shown in Fig. 9-39c, where the idealized switch-mode waveforms of Fig. 9-38 are repeated in Fig. 9-39b for comparison. Proper selection of inductance and capacitance values and a proper switching strategy result in a ZVS-CV switching, as discussed in reference 34.

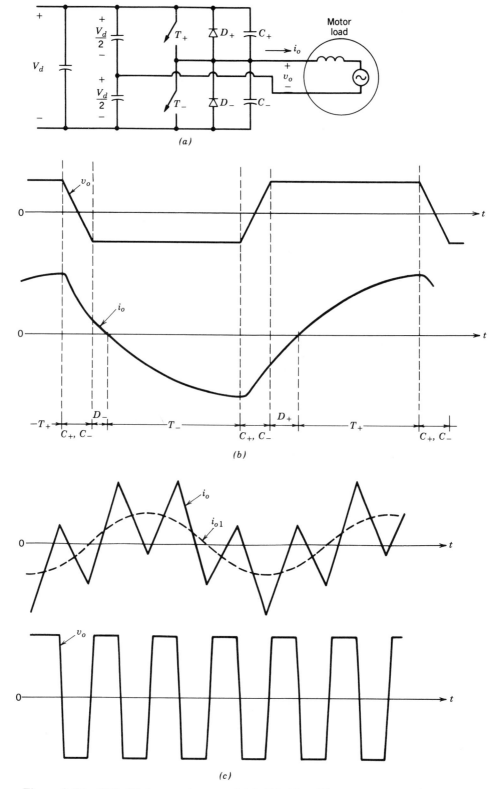

Figure 9-36 ZVS-CV dc-to-ac inverter: (*a*) half-bridge; (*b*) square-wave mode; (*c*) current-regulated mode.

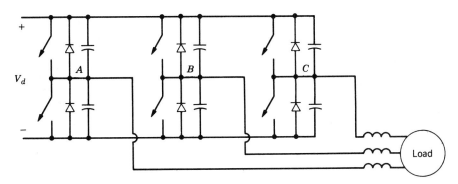

Figure 9-37   Three-phase, ZVS-CV dc-to-ac inverter.

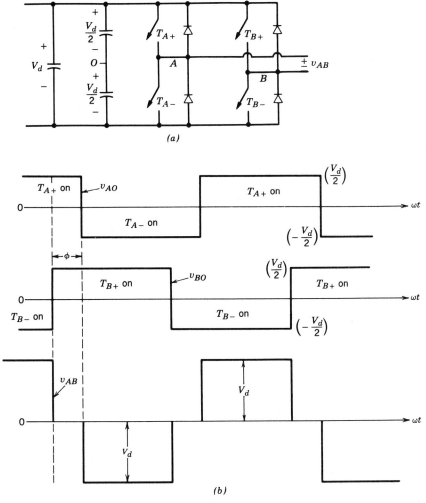

Figure 9-38   Voltage control by voltage cancellation: conventional switch-mode converter.

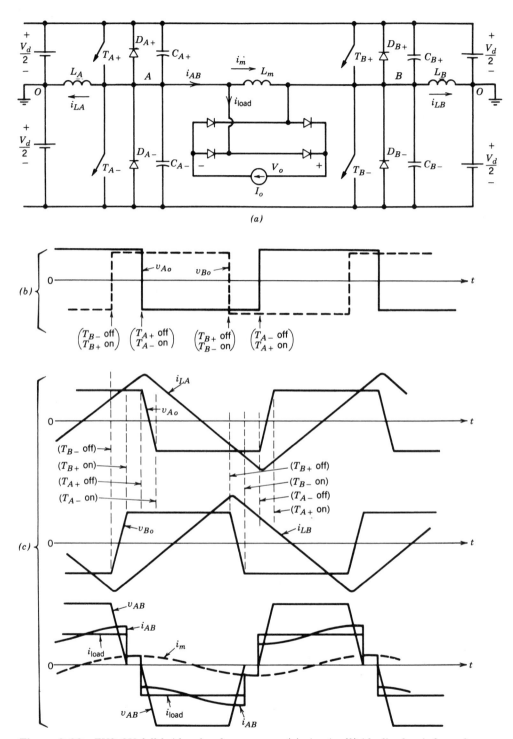

Figure 9-39 ZVS-CV full-bridge dc–dc converter: (a) circuit; (b) idealized switch-mode waveforms; (c) ZVS-CV waveforms.

9-7   RESONANT-dc-LINK INVERTERS WITH ZERO-VOLTAGE SWITCHINGS

In the conventional switch-mode PWM inverters of the type discussed in Chapter 8, the input is a dc voltage. To avoid the switching losses in the inverter, a new topology has been recently proposed in reference 35, where a resonant circuit is introduced in between the dc input voltage and the PWM inverter. As a result, the input voltage to the inverter in the basic configuration oscillates between zero and slightly greater than twice the dc input voltage. The inverter switches are turned on and turned off at zero voltage.

The basic concept is illustrated by means of the circuit of Fig. 9-40a. The resonant circuit consists of L_r, C_r, and a switch with an antiparallel diode. The load of

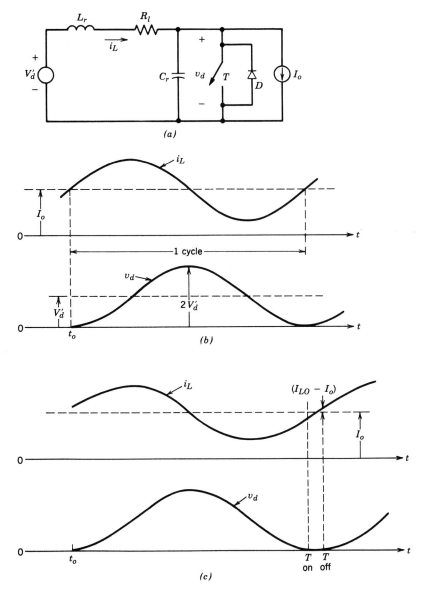

Figure 9-40   Resonant-dc-link inverter, basic concept: (a) basic circuit; (b) lossless $R_l = 0$; (c) losses are present.

the circuit is represented by a current I_o, which represents, for example, the current being supplied by the inverter to a motor load. Because of the internal inductance of the load, it is reasonable to assume I_o to be constant in magnitude during a resonant frequency cycle.

As a first step, R_l is assumed to be zero. Initially, the switch is closed and the difference of i_L and I_o flows through the diode–switch combination. Current i_L builds up linearly. At time t_0, with $i_L = I_{L0}$, the switch is turned off at zero voltage. The equations for the resonant circuit are as follows for $t > t_0$:

$$i_L(t) = I_o + \left[\frac{V_d'}{\omega_0 L_r}\sin \omega_0(t - t_0) + (I_{L0} - I_o)\cos \dot{\omega}_0(t - t_0) \right] \quad (9\text{-}34)$$

and

$$v_d(t) = V_d' + [\omega_0 L_r(I_{L0} - I_o)\sin \omega_0 t - V_d' \cos \omega_0 t] \quad (9\text{-}35)$$

where

$$\omega_0 = \frac{1}{\sqrt{L_r C_r}} \quad (9\text{-}36)$$

The waveforms in Fig. 9-40*b* for $I_{L0} = I_o$ show that v_d returns to zero and i_L returns to I_o after one resonant cycle from the switch opening. Therefore, in this idealized circuit without any losses, the switch T and diode D can be removed once the oscillations start.

In the basic circuit of Fig. 9-40*a*, R_l represents the losses. In order for the zero-voltage turn-on and turn-off of the switch to occur, v_d must return to zero. In the presence of losses in R_l, I_{L0} must be greater than I_o at the instant the switch is turned off. The waveforms are shown in Fig. 9-40*c*. If the switch is kept on too long and I_{L0} is much larger than I_o, then v_d will peak at a value significantly larger than $2V_d'$. Therefore, $I_{L0} - I_o$ must be controlled by controlling the time interval during which the switch remains closed.

The foregoing concept can be applied to the three-phase PWM inverter of Fig. 9-41. The resonant switch T and D of Fig. 9-40*a* are not needed since their function can be fulfilled by any of two switches comprising an inverter leg. The switches in any of the three inverter legs can be turned on and turned off at zero voltage when v_d reaches zero.

Further modifications to clamp the peak voltage across the switches to less than twice the input dc voltage have been discussed in the literature.

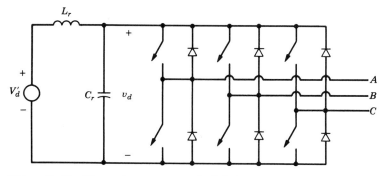

Figure 9-41   Three-phase resonant-dc-link inverter.

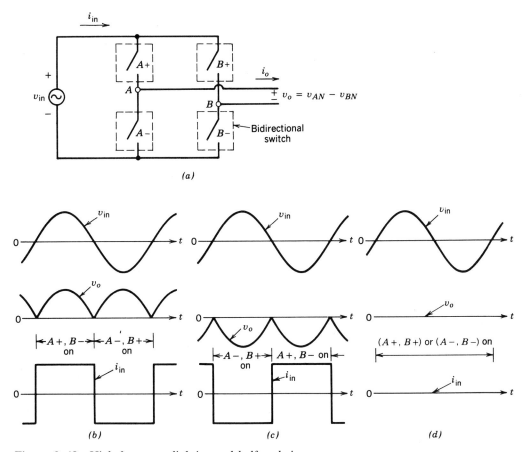

Figure 9-42   High-frequency-link integral-half-cycle inverter.

9-8   HIGH-FREQUENCY-LINK INTEGRAL-HALF-CYCLE CONVERTERS

Unlike the resonant-dc-link converters where the input to the single-phase or three-phase converter oscillates between zero and a value higher than the average input dc voltage, in the high-frequency-link converters the input to the single-phase or three-phase converter is a single-phase, high-frequency sinusoidal ac, as shown in Fig. 9-42a. As discussed in reference 38, by turning the inverter switches on or off when the input voltage passes through zero, the switching losses can be minimized.

Figure 9-42a shows a single-phase converter of this type with a high-frequency sinusoidal input voltage v_{in}. The output is synthesized to be a low-frequency ac, for example, to supply a motor load. This requires that all four switches be bidirectional. Each bidirectional switch in Fig. 9-42a can be obtained by connecting two unidirectional switches with reverse blocking capability in antiparallel.

To describe the operating principle, the output load current is assumed to be a constant I_o during a cycle of high-frequency ac input. Here I_o can be positive or negative. For either direction of I_o, v_{AB} can consist of two positive half-cycles, two negative half-cycles, or zero (or any combination of these three options). These three options and

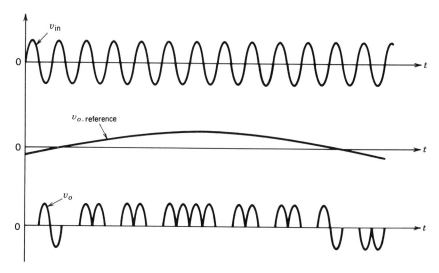

Figure 9-43 Synthesis of low-frequency ac output.

the corresponding i_{in} are shown in Fig. 9-42b through 9-42d for a positive I_o, for example. This control over v_o to be positive, negative, or zero during each high-frequency half-cycle allows a low-frequency output to be synthesized to be of the desirable frequency and magnitude, as shown in Fig. 9-43. This control is discussed in detail in reference 38. Since the low-frequency output consists of an integral number of half-cycles of the high-frequency input, these converters are labeled high-frequency-link integral-half-cycle converters.

This concept can also be extended to deliver three-phase ac output by the circuit of Fig. 9-44. It should be noted that in both the single-phase and three-phase converters, a parallel-resonant filter of the type shown in Fig. 9-44 must be used. It is tuned to be parallel resonant at the input voltage frequency f_{in}. Therefore, it does not draw any current from the high-frequency ac input. However, the capacitor C_f provides a low-impedance path to all other frequency components in i_{in} so that they do not have to be supplied by v_{in} through the stray inductance L_{stray}.

The low-frequency output may in fact be dc in the circuit of Fig. 9-42a. Also, the power can flow in either direction in these converters.

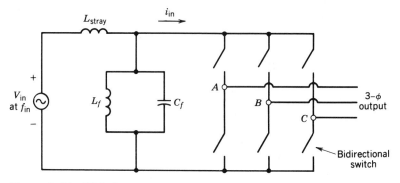

Figure 9-44 High-frequency ac to low-frequency three-phase ac converter.

These high-frequency-link converters are a form of cycloconverters, whereby the power is transferred between two ac systems operating at two different frequencies without an intermediate dc link. Unlike the phase-controlled, line-frequency cycloconverters using thyristors, which are described in Chapter 15, the bidirectional controllable switches in these high-frequency-link converters are turned on or off when the input high-frequency ac passes through zero.

SUMMARY

In this chapter, various techniques are discussed that can either eliminate or diminish the stresses and the switching losses in the semiconductor devices. The following converters are described:

1. Load-resonant converters
   (a) Series-loaded resonant (SLR) dc–dc converters
   (b) Parallel-loaded resonant (PLR) dc–dc converters
   (c) Hybrid-resonant, dc–dc converters
   (d) Current-source, parallel-resonant dc-to-ac inverters for induction heating
   (e) Class E converters
2. Resonant-switch converters
   (a) Zero-current-switching (ZCS) converters
   (b) Zero-voltage-switching (ZVS) converters
   (c) Zero-voltage-switching, clamped-voltage (ZVS-CV) converters
       (i) ZVS-CV dc–dc converters
       (ii) ZVS-CV dc-to-ac inverters
       (iii) ZVS-CV dc–dc converters with zero voltage cancellation
3. Resonant-dc-link inverters with zero-voltage switchings
4. High-frequency-link integral-half-cycle converters

An overview of these converters is provided in reference 41.

PROBLEMS

SLR dc–dc CONVERTERS

9-1 The SLR dc–dc converter of Fig. 9-10a is operating in a discontinuous-conduction mode with $\omega_s < 0.5\omega_0$. In the waveforms of Fig. 9-11 (with $t_0 = 0$), the initial conditions in terms of normalized quantities are always as follows: $V_{c0} = -2V_o$ and $I_{L0} = 0$. Show that in terms of normalized quantities, $V_{c,\text{peak}} = 2$ and $I_{L,\text{peak}} = 1 + V_o$.

9-2 Design an SLR dc–dc converter of Fig. 9-10a with an isolation transformer of turns-ratio $n : 1$, where $V_d = 155$ V, and the operating frequency $f_s = 100$ kHz. The output is at 5 V and 20 A.

(a) The foregoing converter is to operate in a discontinous-conduction mode with $\omega_s < 0.5\omega_0$. The normalized output voltage V_o is chosen to be 0.9 and the normalized frequency to be 0.45. Using the curves of Fig. 9-15, obtain turns ratio n, L_r, and C_r.

(b) Obtain the numerical value for the sum of peak energies stored in L_r and C_r:

$$S = \tfrac{1}{2}L_r I_{L,\text{peak}}^2 + \tfrac{1}{2}C_r V_{c,\text{peak}}^2$$

9-3 Repeat Problem 9-2 if the SLR converter is designed to operate in a continuous-conduction mode at below the resonant frequency.

    (a) Choose normalized output voltage as 0.9 and normalized output current as 1.4. Use the design curves of Fig. 9-15. Obtain n, L_r, and C_r.

    (b) Obtain S as defined in Problem 9-2(b) by means of the design curves of Fig. P9-3.

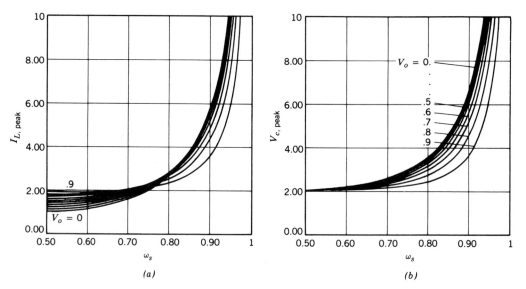

Figure P9-3   SLR dc–dc converter characteristics; all quantities are normalized. (Source: Ramesh Oruganti, Ph.D. Dissertation, VPI, 1987.)

9-4 Repeat Problem 9-3 for an operation above the resonant frequency ($\omega_s > \omega_0$) but with normalized output voltage of 0.9 and normalized output current of 0.4. Use the design curves of Fig. 9-15 and Fig. P9-4.

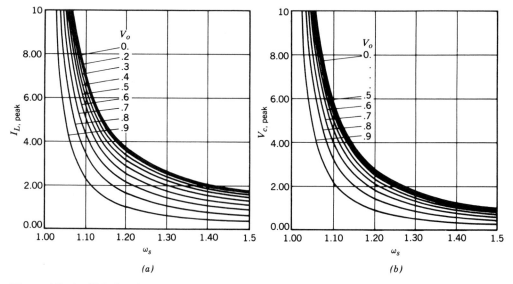

Figure P9-4   SLR dc–dc converter characteristics; all quantities are normalized. (Source: Ramesh Oruganti, Ph.D. Dissertation, VPI, 1987.)

9-5   Compare the values of S in Problems 9-2 through 9-4.

PLR dc–dc CONVERTERS

9-6   The PLR dc–dc converter of Fig. 9-17a with an isolation transformer of turns ratio n: 1 is operating in a discontinuous mode and the voltage and current waveforms are shown in Fig. 9-18.
     Show that in a discontinuous mode,

$$V_{c,\text{peak}} = V_d$$

and

$$I_{L,\text{peak}} = \frac{I_o}{n} + \omega_0 c_r \frac{V_d}{2}$$

9-7   Design a half-bridge, transformer-isolated PLR dc–dc converter. The input dc voltage $V_d = 155$ V and the operating frequency $f_s = 300$ kHz. The output is at 5 V and 20 A. Obtain the transformer turns ratio n, L_r, and C_r, assuming a discontinuous mode of operation, normalized operating frequency of 0.45, normalized $C_r = 1.2$ per unit, and normalized $L_r = 0.833$ per unit, where

$$(C_r)_{\text{base}} = \frac{I_o/n}{\omega_0 V_d/2} \quad \text{and} \quad (L_r)_{\text{base}} = \frac{V_d/2}{\omega_0 I_o/n}$$

Using the design curves of Fig. 9-21, and Problem 9-6, obtain the peak values of v_c and i_L. Calculate S, which was defined in Problem 9-2.

9-8   Design the converter of Problem 9-7, assuming a continuous mode below the resonant frequency. Let the normalized operating frequency be 0.8 and the values of normalized C_r and L_r as in Problem 9-7. Normalized I_o is 0.8.

(a) Calculate n, L_r, and C_r.

(b) Using the design curves of Figs. 9-21 and P9-8, obtain the peak values of v_c and i_L. Calculate S, as defined in Problem 9-2.

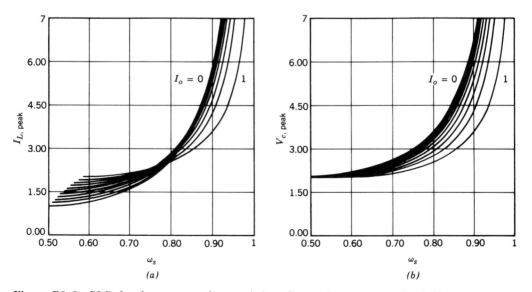

Figure P9-8   PLR dc–dc converter characteristics; all quantities are normalized. (Source: Ramesh Oruganti, Ph.D. Dissertation, VPI, 1987.)

9-9   Repeat Problem 9-8 for operation above the resonant frequency with the normalized frequency of 1.1. Use the design curves of Fig. 9-21 and P9-9.

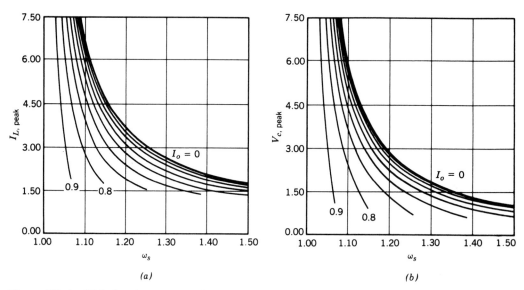

Figure P9-9   PLR dc–dc converter characteristics; all quantities are normalized. (Source: Ramesh Oruganti, Ph.D. Dissertation, VPI, 1987.)

9-10   Compare the values of S calculated in Problems 9-7 through 9-9.

ZCS RESONANT-SWITCH CONVERTERS

9-11   In the ZCS resonant-switch circuit of Fig. 9-30a, $f_0 = 1$ MHz, $Z_0 = 10\ \Omega$, $P_{load} = 10$ W, $V_d = 15$ V, and $V_o = 10$ V. Assume L_2 to be quite large and all components to be ideal.
   Calculate the i_L and v_c waveforms as a function of time. Sketch the waveforms for i_L and v_c and label the important transition points. Also label the peak values of I_L and v_c and the time instants at which they occur.

9-12   Repeat Problem 9-11 assuming a diode is connected in antiparallel with the switch in Fig. 9-30a.

9-13   Using PSpice, simulate the zero-current-switching, quasi-resonant boost converter shown in Fig. P9-13. Obtain v_c, i_L, and i_{diode} waveforms.

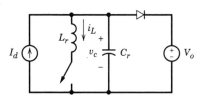

Nominal values $I_d = 26.667$ A, $V_o = 450$ V
    $L_r = 5.37\,\mu$H, $C_r = 117.9$ nF
    $f_s = 100$ kHz

Figure P9-13   Source: reference 5 of Chapter 4. "Power Electronics: Computer Simulation, Analysis, and Education Using PSpice (evaluation, classroom version)," on a diskette with a manual, Minnesota Power Electronics, P.O. Box 14503, Minneapolis, MN 55414.

ZVS RESONANT-SWITCH CONVERTERS

9-14   In the ZVS resonant-switch dc–dc converter of Fig. 9-31a, $V_d = 40$ V; I_o varies in the range 4–20 A. Calculate the theoretical minimum value of the switch voltage rating.

REFERENCES

SERIES-LOADED RESONANT dc–dc CONVERTERS

1. R. Oruganti, "State-Plane Analysis of Resonant Converters," Ph.D. Dissertation, Virginia Polytechnic Institute, 1987; available from University Microfilms International, Ann Arbor, MI.
2. R. J. King and T. A. Stuart, "A Normalized Model for the Half-Bridge Series Resonant Converters," *IEEE Transactions on Aerospace and Electronics Systems,* Vol. AES-17, No. 2, March 1981, pp. 190–198.
3. R. J. King and T. A. Stuart, "Modelling the Full-Bridge Series-Resonant Power Converter," *IEEE Transactions on Aerospace and Electronic Systems,* Vol. AES-18, No. 4, July 1982, pp. 449–459.
4. R. J. King and T. A. Stuart, "Inherent Over-Load Protection for the Series-Resonant Converters," *IEEE Transactions on Aeropace and Electronics Systems,* Vol. AES-19, No. 6, November 1983, pp. 820–830.
5. R. Oruganti and F. C. Lee, "Resonant Power Processors: Part 1—State Plane Analysis," *IEEE-IAS Annual Meeting Conference Records,* 1984, pp. 860–867.
6. A. F. Witulski and R. W. Erickson, "Design of the Series Resonant Converter for Minimum Component Stress," *IEEE Transactions on Aerospace and Electronic Systems,* VOl. AES-22, No. 4, July 1986, pp. 356–363.
7. J. G. Hayes, N. Mohan, and C. P. Henze, "Zero-Voltage Switching in a Digitally Controlled Resonant DC–DC Power Converter," *Proceedings of the 1988 IEEE Applied Power Electronics Conference,* pp. 360–367.

PARALLEL-LOADED RESONANT dc–dc CONVERTERS

8. N. Mapham, "An SCR Inverter with Good Regulation and Sine-Wave Output," *IEEE Transactions on Industry and General Applications,* Vol. IGA-3, March/April 1967, pp. 176–187.
9. V. T. Ranganathan, P. D. Ziogas, and V. R. Stefanovic, "A Regulated DC–DC Voltage Source Converter Using a High Frequency Link," *IEEE Transactions on Industry Applications,* Vol. IA-18, No. 3, May/June 1982, pp. 279–287.
10. R. Oruganti, "State-Plane Analysis of Resonant Converters," Ph.D. Dissertation, Virginia Polytechnic Institute, 1987.
11. M. C. W. Lindmark, "Switch-Mode Power Supply," U.S. Patent 4,097,773, June 27, 1978.
12. I. J. Pitel, "Phase-modulated Resonant Power Conversion Techniques for High Frequency Inverters," *IEEE-IAS Annual Meeting Proceedings,* 1985.
13. Y. G. Kang and A. K. Upadhyay, "Analysis and Design of a Half-Bridge Parallel Resonant Converter, *1987 IEEE Power Electronics Specialists Conference,* 1987, pp. 231–243.
14. F. S. Tasi, P. Materu, and F. C. Lee, "Constant-Frequency, Clamped-Mode Resonant Converters," *1987 IEEE Power Electronics Specialists Conference,* 1987, pp. 557–566.

HYBRID-RESONANT dc–dc CONVERTERS

15. D. V. Jones, "A New Resonant-Converter Topology," *Proceeding of the 1987 High Frequency Power Conversion Conference,* 1987, pp. 48–52.

CURRENT SOURCE PARALLEL-RESONANT dc-TO-ac INVERTERS FOR INDUCTION HEATING

16. K. Thorborg, *Power Electronics,* Prentice-Hall International (U.K.) Ltd, London, 1988.

CLASS E CONVERTERS

17. N. O. Sokal and A. D. Sokal, "Class-E, A New Class of High Efficiency Tuned Single-Ended Switching Power Amplifiers," *IEEE Journal of Solid State Circuits,* Vol. SC-10, June 1975, pp. 168–176.
18. F. H. Raab, "Idealized Operation of Class-E Tuned Power Amplifier," *IEEE Transactions on Circuits and Systems,* VOl. CAS-24, No. 12, December 1977, pp. 725–735.

19. K. Löhn, "On the Overall Efficiency of the Class-E Power Converters," *1986 IEEE Power Electronics Specialists Conference,* pp. 351–358.

20. M. Kazimierczuk and K. Puczko, "Control Circuit for Class-E Resonant DC/DC Converter," *Proceedings of the National Aerospace Electronics Conference 1987,* Vol. 2, 1987 pp. 416–423.

21. G. Lutteke and H. C. Raets, "220 V Mains 500 kHz Class-E Converter, *1985 IEEE Power Electronics Specialists Conference,* 1985, pp. 127–135.

22. H. Omori, T. Iwai, et al., "Comparative Studies between Regenerative and Non-Regenerative Topologies of Single-Ended Resonant Inverters," *Proceedings of the 1987 High Frequency Power Conversion Conference.*

ZCS AND ZVS RESONANT-SWITCH CONVERTERS

23. P. Vinciarelli, "Forward Converter Switching at Zero Current," U.S. Patent 4,415,959, Nov. 1983.

24. R. Oruganti, "State-Plane Analysis of Resonant Converters," Ph.D. Dissertation, Virginia Polytechnic Institute, 1987.

25. K. H. Liu and F. C. Lee, "Resonant Switches—A Unified Approach to Improve Performances of Switching Converters," *IEEE INTELEC Conference Record,* 1984, pp. 344–351.

26. K. H. Liu, R. Oruganti, and F. C. Lee, "Resonant Switches—Topologies and Characteristics," *1986 IEEE Power Electronics Specialists Conference,* 1986, pp. 106–116.

27. K. H. Liu and F. C. Lee, "Zero Voltage Switches and Quasi-Resonant DC–DC Converters," *1987 IEEE Power Electronics Specialists Conference,* 1986, pp. 58–70.

28. K. D. T. Ngo, "Generalization of Resonant Switches and Quasi-Resonant DC–DC Converters," *1987 IEEE Power Electronics Specialists Conference,* 1987, pp. 395–403.

29. W. A. Tabisz and F. C. Lee, "Zero-Voltage Switching Multi-Resonant Technique—A Novel Approach To Improve Performance of High-Frequency Quasi-Resonant Converters," *IEEE PESC Record,* 1988.

30. M. F. Schlecht and L. F. Casey, "Comparison of the Square-Wave and Quasi-Resonant Topologies," Second Annual Applied Power Electronics Conference, San Diego, CA, 1987, pp. 124–134.

ZERO-VOLTAGE-SWITCHING CLAMPED-VOLTAGE CONVERTERS

31. C. P. Henze, H. C. Martin, and D. W. Parsley, "Zero-Voltage Switching in High Frequency Power Converters Using Pulse Width Modulation," *Proceedings of the 1988 IEEE Applied Power Electronics Conference.*

32. R. Goldfarb, "A New Non-Dissipative Load-Line Shaping Technique Eliminates Switching Stress in Bridge Converters," *Proceedings of Powercon 8,* 1981, pp. D-4-1–D-4-6.

33. T. M. Undeland, "Snubbers for Pulse Width Modulated Bridge Converters with Power Transistors or GTOs," *1983 International Power Electronics Conference,* Tokyo, Japan, pp. 313–323.

34. O. D. Patterson and D. M. Divan, "Pseudo-Resonant Full-Bridge DC/DC Converter," *1987 IEEE Power Electronics Specialists Conference,* 1987, pp. 424–430.

RESONANT-dc-LINK CONVERTERS

35. D. M. Divan, "The Resonant DC Link Converter—A New Concept in Static Power Conversion," *1986 IEEE-IAS Annual Meeting Record,* 1986, pp. 648–656.

36. M. Kheraluwala and D. M. Divan, "Delta Modulation Strategies for Resonant Link Inverters," *1987 IEEE Power Electronics Specialists Conference,* 1987, pp. 271–278.

37. K. S. Rajashekara et al., "Resonant DC Link Inverter-Fed AC Machines Control," *1987 IEEE Power Electronics Specialists Conference,* 1987, pp. 491–496.

HIGH-FREQUENCY-LINK CONVERTERS

38. P. K. Sood, T. A. Lipo, and I. G. Hansen, "A Versatile Power Converter for High Frequency Link Systems," *1987 IEEE Applied Power Electronics Confrence,* 1987, pp. 249–256.

39. L. Gyugyi and F. Cibulka, "The High-Frequency Base Converter—A New Approach to Static High Frequency Conversion," *IEEE Transactions on Industry Applications,* Vol. IA-15, No. 4, July/August 1979, pp. 420–429.

40. P. M. Espelage and B. K. Bose, "High Frequency Link Power Conversion," *1975 IEEE-IAS Annual Meeting Record,* 1975, pp. 802–808.

SUMMARY

41. N. Mohan, "Power Electronic Circuits: An Overview," *1988 IEEE Industrial Electronics Society Conference,* 1988, pp. 522–527.

PART 3

POWER SUPPLY APPLICATIONS

CHAPTER 10

SWITCHING dc POWER SUPPLIES

10-1 INTRODUCTION

Regulated dc power supplies are needed for most analog and digital electronic systems. Most power supplies are designed to meet some or all of the following requirements:

- **Regulated output.** The output voltage must be held constant within a specified tolerance for changes within a specified range in the input voltage and the output loading.
- **Isolation.** The output may be required to be electrically isolated from the input.
- **Multiple outputs.** There may be multiple outputs (positive and negative) that may differ in their voltage and current ratings. Such outputs may be isolated from each other.

In addition to these requirements, common goals are to reduce power supply size and weight and improve their efficiency. Traditionally, linear power supplies have been used. However, advances in the semiconductor technology have lead to switching power supplies, which are smaller and much more efficient compared to linear power supplies. The cost comparison between linear and switching supplies depends on the power rating.

10-2 LINEAR POWER SUPPLIES

To appreciate the advantages of the switching supplies, it is desirable first to consider the linear power supplies. Figure 10-1a shows the schematic of a linear supply. In order to provide electrical isolation between the input and the output and to deliver the output in the desired voltage range, a 60-Hz transformer is needed. A transistor is connected in series that operates in its active region.

Comparing V_o with a reference voltage V_{ref}, the control circuit in Fig. 10-1a adjusts the transistor base current such that V_o ($=v_d - v_{CE}$) equals $V_{o,ref}$. The transistor in a linear supply acts as an adjustable resistor where the voltage difference $v_d - V_o$ between the input and the desired output voltage appears across the transistor and causes power losses in it. For a given range of 60 Hz ac input voltage, the rectified and filtered output $v_d(t)$ may be as shown in Fig. 10-1b. To minimize the transistor power losses, the transformer turns ratio should be carefully selected such that $V_{d,min}$ in Fig. 10-1b is greater than V_o but does not exceed V_o by a large margin.

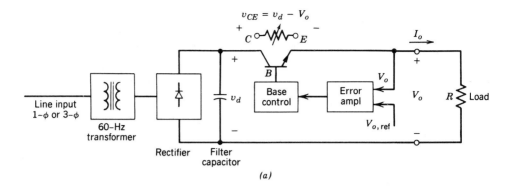

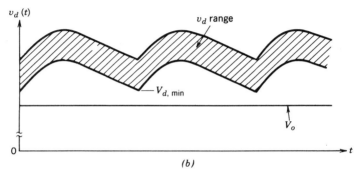

Figure 10-1   Linear power supply: (*a*) schematic; (*b*) selection of transformer turns ratio so that $V_{d,\min} > V_o$ by a small margin.

The preceding discussion points out two major shortcomings of a linear power supply:

1. A low-frequency (60-Hz) transformer is required. Such transformers are larger in size and weight compared to high-frequency transformers.
2. The transistor operates in its active region, incurring a significant amount of power loss. Therefore, the overall efficiencies of linear power supplies are usually in a range of 30–60%.

On the positive side, these supplies utilize simple circuitry and therefore may cost less in small power ratings (<25 W). Also, these supplies do not produce large EMI with other equipment.

10-3   OVERVIEW OF SWITCHING POWER SUPPLIES

As opposed to linear power supplies, in switching power supplies, the transformation of dc voltage from one level to another is accomplished by using dc-to-dc converter circuits (or those derived from them), which were discussed in Chapters 7 and 9. These circuits employ solid-state devices (transistors, MOSFETs, etc.), which operate as a switch: either completely off or completely on. Since the power devices are not required to operate in their active region, this mode of operation results in a lower power dissipation. Increased switching speeds, higher voltage and current ratings, and a relatively lower cost of these devices are the factors that have contributed to the emergence of switching power supplies.

Figure 10-2 shows a switching supply with electrical isolation in a simplified block diagram form. The input ac voltage is rectified into an unregulated dc voltage by means of a diode rectifier of the type discussed in Chapter 5. It should be noted that an EMI filter, as discussed in Chapter 18, is used at the input to prevent the conducted EMI. The dc–dc converter block in Fig. 10-2 converts the input dc voltage from one level to another dc level. This is accomplished by high-frequency switching, which produces high-frequency ac across the isolation transformer. The secondary output of the transformer is rectified and filtered to produce V_o. The output of the dc supply in Fig. 10-2 is regulated by means of a feedback control that employs a PWM controller as discussed in Chapter 7, where the control voltage is compared with a sawtooth waveform at the switching frequency. The electrical isolation in the feedback loop is provided either through an isolation transformer as shown or through an optocoupler.

In many applications, multiple outputs (both positive and negative) are required. These outputs may be required to be electrically isolated from each other, depending on the application. Figure 10-3 shows the block diagram of a switching supply where only one output V_{o1} is regulated and the other two are unregulated. If V_{o2} and/or V_{o3} needs to be regulated, then linear regulator(s) can be used to regulate the other output(s).

Two major advantages of switching power supplies over linear power supplies are now apparent. These are as follows:

- The switching elements (power transistors or MOSFETs) operate as a switch: either completely off or completely on. By avoiding their operation in their active region, a significant reduction in power losses is achieved. This results in a higher energy efficiency in a 70–90% range. Moreover, a transistor operating in on/off mode has a much larger power-handling capability compared to its linear mode.

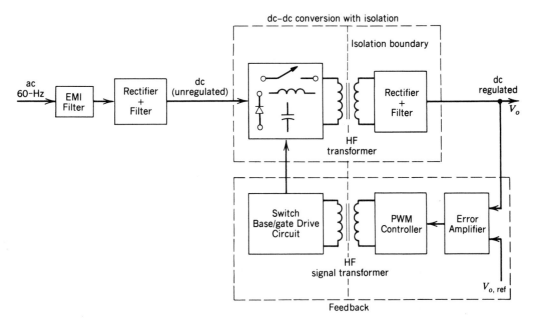

Figure 10-2   Schematic of a switch-mode dc power supply.

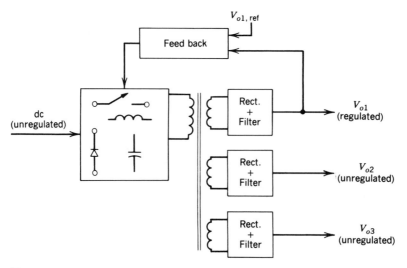

Figure 10-3   Multiple outputs.

- Since a high-frequency isolation transformer is used (as compared to a 50- or 60-Hz transformer in a linear power supply), the size and weight of switching supplies can be significantly reduced.

On the negative side, switching supplies are more complex, and proper measures must be taken to prevent EMI due to high-frequency switchings.

The above-mentioned advantages of switching supplies (over linear supplies) outweigh their shortcomings above a certain power rating. The power rating where this breakover occurs is steadily decreasing with time due to advances in semiconductor technology.

Switching dc power supplies, in general, utilize modifications of the following two classes of converter topologies:

1. Switch-mode dc–dc converters, discussed in Chapter 7, where the switches operate in a switch mode.
2. Resonant converters, discussed in Chapter 9, which utilize zero-voltage and/or zero-current switchings.

In this chapter, the switch-mode converter topologies are used to described the operation of switching power supplies. Many of the basic principles discussed in this chapter also apply to the switching power supplies with resonant converters.

10-4   dc–dc CONVERTERS WITH ELECTRICAL ISOLATION

10-4-1   INTRODUCTION TO dc–dc CONVERTERS WITH ISOLATION

As seen by the block diagram of Fig. 10-2, the electrical isolation in switching dc power supplies is provided by a high-frequency isolation transformer. Figure 10-4a shows a typical transformer core characteristic in terms of its B–H (hysteresis) loop. Here B_m is the maximum flux density beyond which saturation occurs and B_r is the remnant flux density.

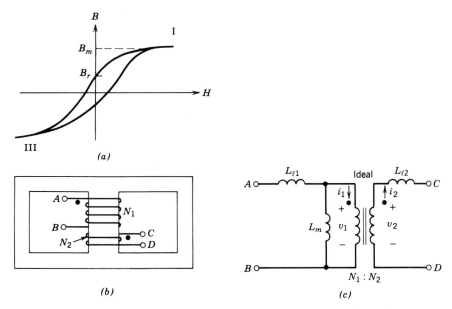

Figure 10-4 Transformer representation: (*a*) typical B–H loop of transformer core; (*b*) two-winding transformer; (*c*) equivalent circuit.

Various types of dc–dc converters (with isolation) can be divided into two basic categories, based on the way they utilize the transformer core:

1. Unidirectional core excitation where only the positive part (quadrant 1) of the B–H loop is used
2. Bidirectional core excitation where both the positive (quadrant 1) and the negative (quadrant 3) parts of the B–H loop are utilized alternatively

10-4-1-1 Unidirectional Core Excitation

Some of the dc–dc converters (without isolation) discussed in Chapter 7 can be modified to provide electrical isolation by means of unidirectional core excitation. Two such modifications are as follows:

- Flyback converter (derived from buck–boost converter)
- Forward converter (derived from step-down converter)

The output voltage of these converters is regulated by means of the PWM switching scheme discussed in Chapter 7.

10-4-1-2 Bidirectional Core Excitation

To provide electrical isolation by means of bidirectional core excitation, the single-phase switch-mode inverter topologies of Chapter 8 can be used to produce a square-wave ac at the input of the high-frequency isolation transformer in Fig. 10-2. We will discuss the following inverter topologies, which can constitute a switching dc power supply:

- Push–pull
- Half bridge
- Full bridge

As in Chapters 7 and 8, for analyzing the following circuits, the switches are treated as being ideal and the power losses in the inductive, capacitive, and transformer elements are neglected. Some of these losses limit the operational capabilities of these circuits and are discussed separately.

All of the circuits are analyzed under a steady-state operating condition, and the filter capacitor at the output is assumed to be so large (as in Chapter 7) as to allow the assumption that the output voltage $v_o(t) \simeq V_o$ (i.e., essentially a pure dc). The analysis is presented only for the continuous-conduction mode, and the analysis of the discontinuous-conduction mode is left as an exercise.

10-4-1-3   Isolation Transformer Representation

A high-frequency transformer is required to provide electrical isolation. Neglecting the losses in the transformer of Fig. 10-4b, an approximate equivalent circuit for a two-winding transformer is redrawn in Fig. 10-4c, where $N_1 : N_2$ is the transformer winding turns ratio, L_m is the magnetizing inductance referred to the primary side, and L_{l1} and L_{l2} are the leakage inductances. In the ideal transformer, $v_1/v_2 = N_1/N_2$ and $N_1 i_1 = N_2 i_2$.

In a switch-mode dc–dc converter, it is desirable to minimize the leakage inductances L_{l1} and L_{l2} by providing a tight magnetic coupling between the two windings. The energy associated with the leakage inductances has to be absorbed by the switching elements and their snubber circuits, thus clearly indicating a need to minimize the leakage inductances. Similarly, in a switch-mode dc–dc converter, it is desirable to make the magnetizing inductance L_m in Fig. 10-4c as high as possible to minimize the magnetizing current i_m that flows through the switches and thus increases their current ratings.

It is important to consider the effect of the transformer leakage inductances in switch selection and snubber design. However, these inductances have a minor effect on the converter voltage transfer characteristics and therefore have been neglected in the converter analysis to follow.

In one of the converter topologies to be discussed, called the flyback converter, the transformer is in fact intended to be a two-winding inductor, which has dual functions of providing energy storage as in an inductor and electrical isolation as in a transformer. Therefore, the previous comments to make L_m high do not apply to this topology. However, the simplified transformer equivalent circuit can still be used for analysis purposes.

The transformer design considerations in resonant power supplies are different than the ones discussed before for switch-mode power supplies. There, the leakage inductances and/or the magnetizing inductance may in fact be utilized to provide zero-voltage and/or zero-current switchings.

10-4-1-4   Control of dc–dc Converters with Isolation

In the single-switch topologies like the flyback and the forward converters, the output voltage V_o for a given input V_d is controlled by PWM in a manner similar to that used for their nonisolated counterparts discussed in Chapter 7.

In the push–pull, half-bridge, and full-bridge dc–dc converters, where the converter output is rectified to produce a dc output, the dc output voltage V_o is controlled by using the PWM scheme shown in Fig. 10-5, which controls the interval Δ during which all the switches are off simultaneously. This is unlike the PWM schemes used in Chapter 7 to control full-bridge dc–dc converters and in Chapter 8 to control single-phase dc-to-ac inverters.

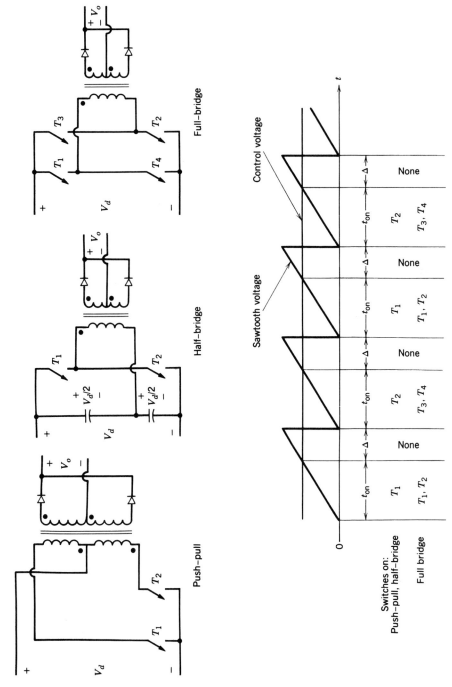

Figure 10-5 PWM Scheme used in dc–dc converters, where the converter output is rectified to produce a dc output.

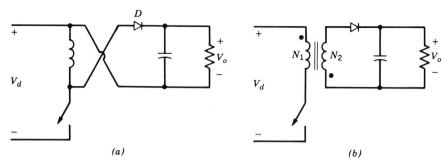

Figure 10-6   Flyback converter.

10-4-2   FLYBACK CONVERTERS (DERIVED FROM BUCK–BOOST CONVERTERS)

Flyback converters are derived from the buck–boost converter discussed in Chapter 7 and redrawn in Fig. 10-6a. By placing a second winding on the inductor, it is possible to achieve electrical isolation, as shown in Fig. 10-6b.

Figure 10-7a shows the converter circuit where the two-winding inductor is represented by its approximate equivalent circuit. When the switch is on, due to the winding polarities, the diode D in Fig. 10-7a becomes reverse biased. The continuous-current-conduction mode in a buck–boost converter corresponds to an incomplete demagnetization of the inductor core in the flyback converter. Therefore, as shown by the waveforms in Fig. 10-8, the inductor core flux increases linearly from its initial value $\phi(0)$, which is finite and positive:

$$\phi(t) = \phi(0) + \frac{V_d}{N_1}t \qquad 0 < t < t_{\text{on}} \tag{10-1}$$

and the peak flux $\hat{\phi}$ at the end of the on interval is given as

$$\hat{\phi} = \phi(t_{\text{on}}) = \phi(0) + \frac{V_d}{N_1}t_{\text{on}} \tag{10-2}$$

After t_{on}, the switch is turned off and the energy stored in the core causes the current to flow in the secondary winding through the diode D, as shown by Fig. 10-7b. The

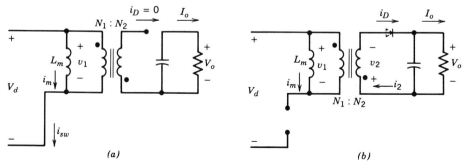

Figure 10-7   Flyback converter circuit states: (a) switch on; (b) switch off.

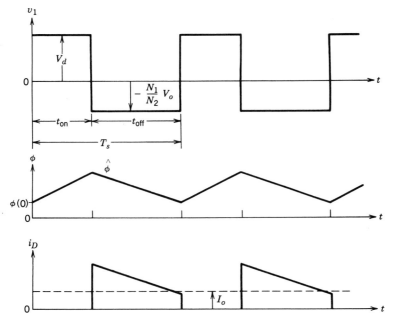

Figure 10-8   Flyback converter waveforms.

voltage across the secondary winding $v_2 = -V_0$, and therefore, the flux decreases linearly during t_{off}:

$$\phi(t) = \hat{\phi} - \frac{V_o}{N_2}(t - t_{\text{on}}) \qquad t_{\text{on}} < t < T_s \qquad (10\text{-}3)$$

and

$$\phi(T_s) = \hat{\phi} - \frac{V_o}{N_2}(T_s - t_{\text{on}}) \qquad (10\text{-}4)$$

$$= \phi(0) + \frac{V_d}{N_1}t_{\text{on}} - \frac{V_o}{N_2}(T_s - t_{\text{on}}) \quad \text{(using Eq. 10-2)} \qquad (10\text{-}5)$$

Since the net change of flux through the core over one time period must be zero in steady state,

$$\phi(T_s) = \phi(0) \qquad (10\text{-}6)$$

Therefore, from Eqs. 10-5 and 10-6

$$\frac{V_o}{V_d} = \frac{N_2}{N_1}\frac{D}{1 - D} \qquad (10\text{-}7)$$

where $D = t_{\text{on}}/T_s$ is the switch duty ratio. Equation 10-7 shows that the voltage transfer radio in a flyback converter depends on D in an identical manner as the buck–boost converter.

The voltage and current waveforms shown in Fig. 10-8 can be obtained from the equations below. During the on interval, the transformer primary voltage $v_1 = V_d$.

Therefore, the inductor current rises linearly from its initial value $I_m(0)$:

$$i_m(t) = i_{sw}(t) = I_m(0) + \frac{V_d}{L_m}t \qquad 0 < t < t_{on} \qquad (10\text{-}8)$$

and

$$\hat{I}_m = \hat{I}_{sw} = I_m(0) + \frac{V_d}{L_m}t_{on} \qquad (10\text{-}9)$$

During the off interval, the switch current goes to zero and $v_1 = -(N_1/N_2)V_o$. Therefore, i_m and the diode current i_D can be expressed during $t_{on} < t < T_s$ as

$$i_m(t) = \hat{I}_m - \frac{V_o(N_1/N_2)}{L_m}(t - t_{on}) \qquad (10\text{-}10)$$

and

$$i_D(t) = \frac{N_1}{N_2}i_m(t) = \frac{N_1}{N_2}\left[\hat{I}_m - \frac{V_o(N_1/N_2)}{L_m}(t - t_{on})\right] \qquad (10\text{-}11)$$

Since the average diode current equals I_o, from Eq. 10-11

$$\hat{I}_m = \hat{I}_{sw} = \frac{N_2}{N_1}\frac{1}{1-D}I_o + \frac{N_1}{N_2}\frac{(1-D)T_s}{2L_m}V_o \qquad (10\text{-}12)$$

The voltage across the switch during the off interval equals

$$v_{sw} = V_d + \frac{N_1}{N_2}V_o = \frac{V_d}{1-D} \qquad (10\text{-}13)$$

10-4-2-1 Other Flyback Converter Topologies

Two modifications of the flyback converter topology are shown in Fig. 10-9. Another flyback topology that is well suited for low-output-voltage applications is discussed in reference 5.

Two-Transistor Flyback Converter. Figure 10-9a shows a two-transistor version of a flyback converter where T_1 and T_2 are turned on and off simultaneously. The advantage of such a topology over a single-transistor flyback converter, discussed earlier, is that voltage rating of the switches is one-half of the single-transistor version. Moreover, since a current path exists through the diodes connected to the primary winding, a dissipative snubber across the primary winding is not needed to dissipate the energy associated with the transformer primary-winding leakage inductance (see reference 17).

Paralleling Flyback Converters. At high power levels, it may be beneficial to parallel two or more flyback converters rather than using a single higher power unit. Some of the advantages of paralleling, which are not limited just to the flyback converter are (a) that it provides higher system reliability due to redundancy, (b) that it increases the effective switching frequency and hence decreases current pulsations at the input and/or the output, and (c) that it allows low-power modules to be standardized where a number of these can be paralleled to provide a higher power capability.

The problem of current sharing among the parallel converters can be remedied by means of current-mode control, which is discussed later in this chapter.

Figure 10-9b shows two flyback converters in parallel; these operate at the same switching frequency, but the switches in the two converters are sequenced to turn on a

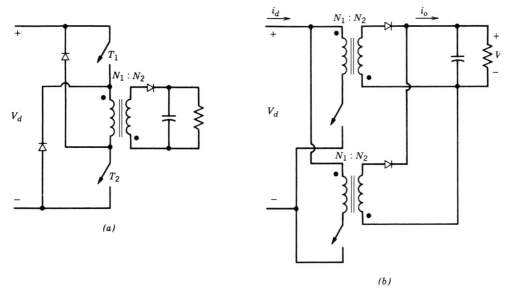

Figure 10-9 Other flyback topologies: (*a*) two-transistor flyback converter; (*b*) parallelled flyback converters.

half-time period apart from one another. This results in improved input and output current waveforms (see Problem 10-4).

10-4-3 FORWARD CONVERTER (DERIVED FROM STEP-DOWN CONVERTER)

Figure 10-10 shows an idealized forward converter. As will be discussed shortly, the transformer magnetizing current must be taken into account in these converters.

Initially, assuming a transformer to be ideal, when the switch is on, D_1 becomes forward biased and D_2 reverse biased. Therefore in Fig. 10-10,

$$v_L = \frac{N_2}{N_1}V_d - V_o \qquad 0 < t < t_{\text{on}} \qquad (10\text{-}14)$$

which is positive. Therefore, i_L increases. When the switch is turned off, the inductor current i_L circulates through the diode D_2, and

$$v_L = -V_o \qquad t_{\text{on}} < t < T_s \qquad (10\text{-}15)$$

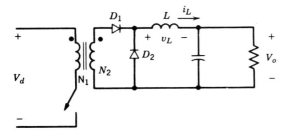

Figure 10-10 Idealized forward converter.

which is negative and therefore causes i_L to decrease linearly. Equating the integral of the inductor voltage over one time period to zero using Eqs. 10-14 and 10-15 yields

$$\frac{V_o}{V_d} = \frac{N_2}{N_1}D \tag{10-16}$$

Equation 10-16 shows that the voltage ratio in the forward converter is proportional to the switch duty ratio D, similar to the step-down converter. Another way to obtain the voltage ratio is to equate the average value of v_{oi} (defined in Fig. 10-11b) over one time period to V_o, recognizing that the average value of v_L is zero.

In a practical forward converter, the transformer magnetizing current must be taken into consideration for a proper converter operation. Otherwise, the stored energy in the transformer core would result in converter failure. An approach that allows the transformer magnetic energy to be recovered and fed back to the input supply is shown in Fig. 10-11a. It requires a third demagnetizing winding. Figure 10-11b shows the transformer in terms of its equivalent circuit, with the leakage inductances neglected. When the switch is on,

$$v_1 = V_d \qquad 0 < t < t_{on} \tag{10-17}$$

and i_m increases linearly from zero to $\hat{I}_m$ as shown in Fig. 10-11c. When the switch is turned off, $i_1 = -i_m$. With the current directions shown in Fig. 10-11b, $N_1 i_1 + N_3 i_3 = N_2 i_2$. Because of D_1, $i_2 = 0$ and therefore

$$i_3 = \frac{N_1}{N_3} i_m \tag{10-18}$$

which flows through D_3 into the input dc supply. During the time interval t_m in Fig. 10-11c when i_3 is flowing, the voltage across the transformer primary as well as L_m is

$$v_1 = -\frac{N_1}{N_3} V_d \qquad t_{on} < t < t_{on} + t_m \tag{10-19}$$

Once the transformer demagnetizes, $i_m = 0$ and $v_1 = 0$. The time interval t_m can be obtained by recognizing that the time integral of voltage v_1 across L_m must be zero over one time period. Using Eqs. 10-17 and 10-19,

$$\frac{t_m}{T_s} = \frac{N_3}{N_1}D \tag{10-20}$$

If the transformer is to be totally demagnetized before the next cycle begins, the maximum value t_m/T_s can attain is $1 - D$. Therefore, using Eq. 10-20, the maximum duty ratio D_{max} with a given turns ratio N_3/N_1 is

$$(1 - D_{max}) = \frac{N_3}{N_1} D_{max}$$

or

$$D_{max} = \frac{1}{1 + N_3/N_1} \tag{10-21}$$

The foregoing analysis shows that with an equal number of turns for the primary and the demagnetizing windings ($N_1 = N_3$, a common practice), the maximum duty ratio in such converters is limited to 0.5.

Note that since a large voltage isolation requirement does not exist between the primary and the demagnetizing windings, these two can be wound bifilar, in order to minimize the leakage inductance between the two windings. The demagnetizing winding

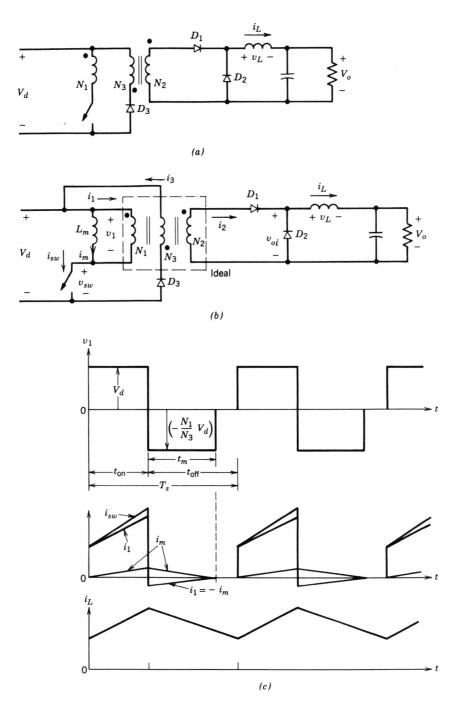

Figure 10-11 Practical forward converter.

requires a much smaller size of wire, since it has to carry only the demagnetizing current. It should be noted that when the transformer magnetizing inductance is included, the voltage transformation ratio V_o/V_d remains the same as given by Eq. 10-16, which was derived by assuming an ideal transformer. Instead of using a third demagnetizing winding, the energy in the core can be dissipated in the zener diode connected across the switch.

10-4-3-1 Other Forward Converter Topologies

Some of the common modifications of the forward converter topologies are shown in Fig. 10-12.

Two-Switch Forward Converter   As is shown in Fig. 10-12a, the two switches are turned on and off simultaneously. The voltage rating of each of the switches is one-half of that in a single-switch topology. More significantly, when the switches are off, the magnetizing current flows into the input supply through the diodes, thus eliminating the need for a separate demagnetizing winding or snubbers.

Paralleling Forward Converters   The same advantages can be gained by paralleling two or more forward converters as those discussed in the flyback converter section. Figure 10-12b shows two forward converters in parallel whose switches are sequenced to turn on a half-time period apart from one another. At the output, a common filter can be used, thus significantly reducing the size of the output filter capacitor and inductor (see Problem 10-7).

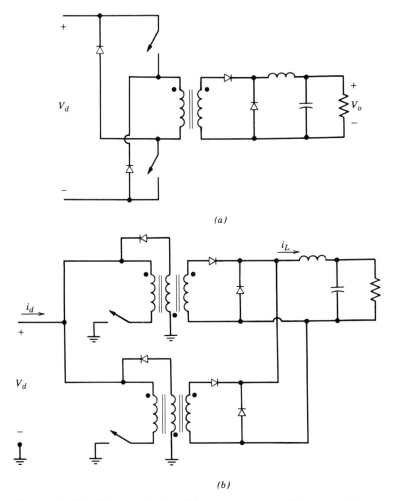

Figure 10-12   Other topologies of forward converter: (*a*) two-switch forward converter; (*b*) parallelled forward converters.

10-4-4 PUSH–PULL CONVERTER (DERIVED FROM STEP-DOWN CONVERTER)

Figure 10-13a shows the circuit arrangement for a push–pull dc–dc converter where the push–pull inverter of Chapter 8 is used to produce a square-wave ac at the input of the high-frequency transformer. The PWM switching scheme described by Fig. 10-5 is used to regulate the output voltage. A center-tapped secondary is used, which results in only one diode voltage drop on the secondary side.

In Fig. 10-13a, when T_1 is on, D_1 conducts and D_2 gets reverse biased. This results in $v_{oi} = (N_2/N_1)V_d$ in Fig. 10-13b. Therefore, the voltage across the filter inductor is given as

$$v_L = \frac{N_2}{N_1}V_d - V_o \qquad 0 < t < t_{on} \tag{10-22}$$

and i_L through D_1 increases linearly as shown by Fig. 10-13b.

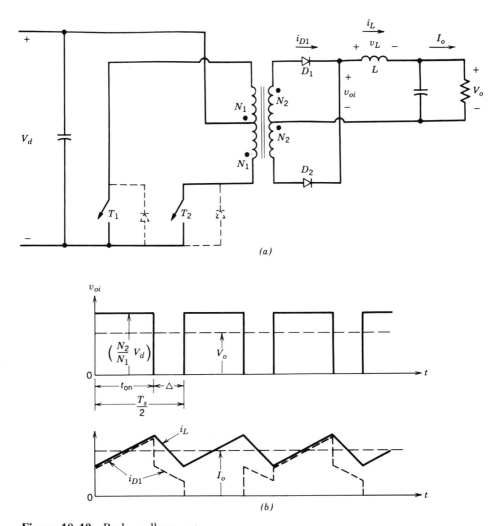

(a)

(b)

Figure 10-13 Push–pull converter.

During the interval Δ when both the switches are off, the inductor current splits equally between the two secondary half-windings and $v_{oi} = 0$. Therefore, during $t_{on} < t < t_{on} + \Delta$,

$$v_L = -V_o \tag{10-23}$$

and

$$i_{D1} = i_{D2} = \tfrac{1}{2}i_L \tag{10-24}$$

The next half-cycle consists of t_{on} (during which T_2 is on) and the interval Δ. The waveforms repeat with a period $\tfrac{1}{2}T_s$ and

$$t_{on} + \Delta = \tfrac{1}{2}T_s \tag{10-25}$$

Equating the time interval of the inductor voltage during one repetition period $\tfrac{1}{2}T_s$ to zero using Eqs. 10-22, 10-23, and 10-25 yields

$$\frac{V_o}{V_d} = 2\frac{N_2}{N_1}D \qquad 0 < D < 0.5 \tag{10-26}$$

where $D = t_{on}/T_s$ is the duty ratio of switches 1 and 2 and the maximum value it can attain is 0.5 (in practice, to maintain a small blanking time to avoid turning both the switches on simultaneously, D should be kept smaller than 0.5). The average value of the v_{oi} waveform in Fig. 10-13b equals V_o.

It should be noted that in the push–pull inverter of Chapter 8, the feedback diodes connected in antiparallel to the switches were required to carry the reactive current and their conduction interval depended inversely on the power factor of the output load. In the push–pull dc–dc converter, these antiparallel diodes shown dotted in Fig. 10-13a are needed to provide a path for the current required due to leakage flux of the transformer.

In push–pull circuits, due to a slight and unavoidable difference in the switching times of two switches T_1 and T_2, there is always an imbalance between the peak values of the two switch currents. This imbalance can be eliminated by means of current-mode control of the converter, which is discussed later in this chapter.

10-4-5 HALF-BRIDGE CONVERTER (DERIVED FROM STEP-DOWN CONVERTER)

Figure 10-14a shows a half-bridge dc–dc converter. As discussed in Chapter 8 in connection with the half-bridge inverters, the capacitors C_1 and C_2 establish a voltage midpoint between zero and the input dc voltage. The switches T_1 and T_2 are turned on alternatively, each for an interval t_{on}. With T_1 on, $v_{oi} = (N_2/N_1)(V_d/2)$ as shown in Fig. 10-14b and, therefore,

$$v_L = \frac{N_2}{N_1}\frac{V_d}{2} - V_o \qquad 0 < t < t_{on} \tag{10-27}$$

During the interval Δ, when both switches are off, the inductor current splits equally between the two secondary halves. Assuming ideal diodes, $v_{oi} = 0$, and therefore,

$$v_L = -V_o \qquad t_{on} < t < t_{on} + \Delta \tag{10-28}$$

In steady state, the waveforms repeat with a period $\tfrac{1}{2}T_s$ and

$$t_{on} + \Delta = \tfrac{1}{2}T_s \tag{10-29}$$

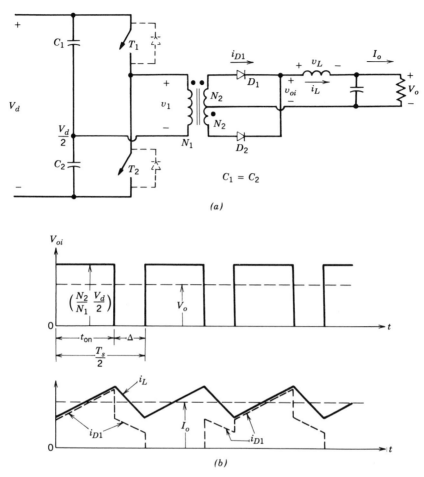

Figure 10-14 Half-bridge dc–dc converter.

Equating the time integral of the inductor voltage during one repetition period to zero using Eqs. 10-27 through 10-29 yields

$$\frac{V_o}{V_d} = \frac{N_2}{N_1}D \tag{10-30}$$

where $D = t_{on}/T_s$ and $0 < D < 0.5$. The average value of v_{oi} in Fig. 10-14b equals V_o.

The diodes in antiparallel with the switches T_1 and T_2 are used for switch protection, as in a push–pull converter.

10-4-6 FULL-BRIDGE CONVERTER (DERIVED FROM STEP-DOWN CONVERTER)

Figure 10-15a shows a full-bridge converter where (T_1, T_2) and (T_3, T_4) are switched as pairs alternatively at the selected switching frequency. When (T_1, T_2) or (T_3, T_4) are on, $v_{oi} = (N_2/N_1)V_d$, as shown in Fig. 10-15b, and therefore

$$v_L = \frac{N_2}{N_1}V_d - V_o \qquad 0 < t < t_{on} \tag{10-31}$$

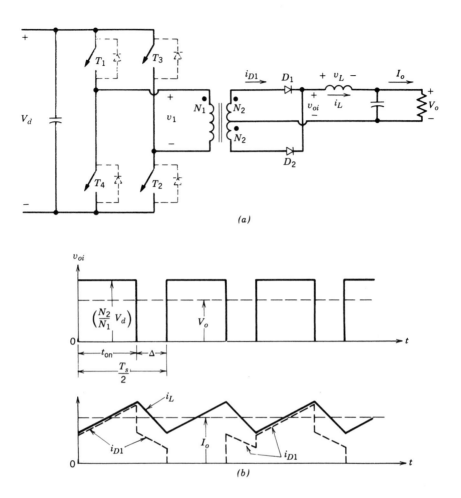

Figure 10-15   Full-bridge converter.

When both the switch pairs are off, the inductor current splits equally between the two secondary halves. Assuming ideal diodes, $v_{oi} = 0$ and therefore

$$v_L = -V_o \qquad t_{on} < t < t_{on} + \Delta \qquad (10\text{-}32)$$

Equating the time integral of the inductor voltage over one time period to zero in steady state and recognizing that $t_{on} + \Delta = \frac{1}{2}T_s$ give

$$\frac{V_o}{V_d} = 2\frac{N_2}{N_1}D \qquad (10\text{-}33)$$

where $D = t_{on}/T_s$ and $0 < D < 0.5$. In Fig. 10-15b, the average value of v_{oi} equals V_o.

The diodes connected in antiparallel to the switches (shown as dashed) provide a path to the current due to the energy associated with the primary-winding leakage inductance.

Comparison of the full-bridge (FB) converter with the half-bridge (HB) converter for identical input and output voltages and power ratings requires the following turns ratio:

$$\left(\frac{N_2}{N_1}\right)_{HB} = 2\left(\frac{N_2}{N_1}\right)_{FB} \qquad (10\text{-}34)$$

Neglecting the ripple in the current through the filter inductor at the output and assuming the transformer magnetizing current to be negligible in both circuits, the switch currents I_{sw} are given by

$$(I_{sw})_{HB} = 2(I_{sw})_{FB} \tag{10-35}$$

In both converters, the input V_d appears across the switches; however, they are required to carry twice as much current in the half-bridge compared with the full-bridge converter. Therefore, in large power ratings, it may be advantageous to use a full-bridge over a half-bridge converter to reduce the number of paralleled devices in the switch.

10-4-7 CURRENT-SOURCE dc–dc CONVERTERS

The dc–dc converters (derived from the step-down converter topology) in the previous sections are supplied with a voltage at their input and, therefore, are voltage source converters. By inserting an inductor at the input of a push–pull circuit, as shown in Fig. 10-16, and operating the switches at a duty ratio D of greater than 0.5, the converter is fed through a current source. Here D greater than 0.5 implies simultaneous conduction of the top switches, which was to be strictly avoided in the normal voltage source push–pull converter.

When both switches are on, the voltage across each primary half-winding becomes zero. The input current i_d builds up linearly and the energy is stored in the input inductor. When only one of the two switches is conducting, the input voltage and the stored energy in the input inductor supply the output stage. Therefore, this circuit operates in a manner similar to the step-up converter of Chapter 7.

In the continuous-current-conduction mode, its voltage transfer ratio can be derived to be (see Problem 10-9)

$$\frac{V_o}{V_d} = \frac{N_2}{N_1} \frac{1}{2(1-D)} \qquad D > 0.5 \tag{10-36}$$

which is similar to the voltage transfer ratio of a step-up converter.

Current-source converters have the disadvantage of having a low power-to-weight ratio compared to voltage-source converters.

10-4-8 TRANSFORMER CORE SELECTION IN dc–dc CONVERTERS WITH ELECTRICAL ISOLATION

It is desirable to have power transformers that are small in weight and size and have low power losses. The motivation for using high switching frequencies is to reduce the size of

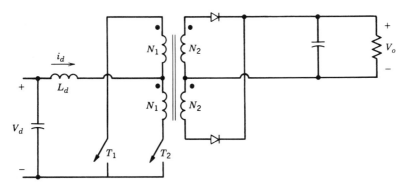

Figure 10-16 Current source converter $(D > 0.5)$.

the power transformer and the filter components. If this benefit is to be realized, the power loss in the transformer core should remain low even at high frequencies.

Ferrite materials such as 3C8 are commonly used to build transformer cores. Similar to Fig. 10-4a, Fig. 10-17a shows a typical B–H loop for such a material, where the maximum flux density B_m beyond which the saturation occurs is in a range of 0.2–0.4 Wb/m^2 and the remnant flux density B_r is in a range of 0.1–0.2 Wb/m^2. In Fig. 10-17b, the core loss per unit weight for several switching frequencies is plotted as a function of $(\Delta B)_{max}$, where $(\Delta B)_{max}$ is the peak swing in the flux density around its average value during each cycle of the switching frequency f_s. In general, the expression for the core loss per unit weight or per unit volume is given as

$$\text{Core loss density} = k f_s^a [(\Delta B)_{max}]^b \tag{10-37}$$

where the exponents a and b, and the coefficient k depend on the type of material.

A forward converter (with $N_1 = N_3$ in Fig. 10-11a) is chosen as an example of the unidirectional core excitation, and a full-bridge converter is chosen to represent a bidirectional core excitation. With a switch duty ratio of 0.5, the peak flux density excursions are calculated by using the waveforms shown in Figs. 10-18a and 10-18b, where v_1 is voltage across the primary winding. In both the converters

$$(\Delta B)_{max} = \frac{V_d}{4N_1 A_c f_s} \quad \text{(at } D = 0.5) \tag{10-38}$$

where A_c is the cross-sectional area of the core and N_1 is the number of turns in the primary winding. In the forward converter with a unidirectional core excitation, waveforms of Fig. 10-18a and the B–H loop in Fig. 10-4a dictate that

$$(\Delta B)_{max} < \tfrac{1}{2}(B_m - B_r) \tag{10-39a}$$

In the full-bridge converter with a bidirectional core excitation

$$(\Delta B)_{max} < B_m \tag{10-39b}$$

Based on the foregoing discussion, the following conclusions can be reached regarding the desired core properties:

1. A large value of maximum flux density B_m allows $(\Delta B)_{max}$ to be large and results in a small A_c in Eq. 10-38 and hence in a smaller core size.
2. At switching frequencies below 100 kHz, for example, $(\Delta B)_{max}$ is limited by B_m. Therefore, a higher switching frequency in Eq. 10-38 results in a smaller core area. However, at switching frequencies above 100 kHz, a smaller value of $(\Delta B)_{max}$ is chosen to limit the core losses given in Fig. 10-17b.
3. In a forward converter topology where the core is excited in only one direction, $(\Delta B)_{max}$ is limited by $B_m - B_r$. Therefore, it is important to use a core with a low remnant flux density B_r in such a topology unless a complex-core resetting mechanism is used. In practice, a small air gap is introduced in the core that linearizes the core characteristic and significantly lowers B_r (see Problem 10-11).

In the converters with bidirectional core excitation topologies, the presence of an air gap prevents core saturation under start-up and transient conditions but does not prevent core saturation if there is a volt-second imbalance during the two half-cycles of operation (a volt-second imbalance implies that a dc voltage component is applied to the transformer core). In a practical implementation, there are several causes of such a volt-second imbalance, such as unequal conduction voltage drops and unequal switching times of the switches. The preferable way to avoid core saturation due to these practical limitations is

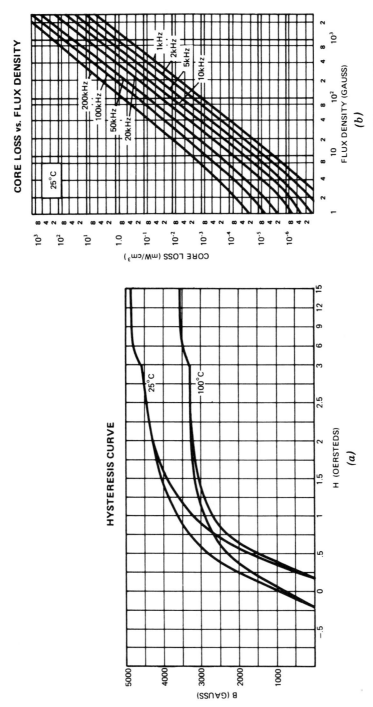

Figure 10-17 3C8 ferrite characteristic curves: (*a*) B–H loop; (*b*) core loss curves. (*Courtesy of Ferroxcube Division of Amperex Electronic Corporation.*)

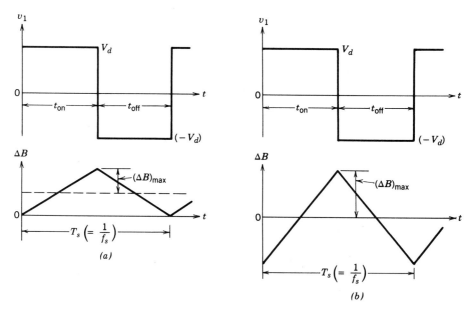

Figure 10-18 Core excitation: (a) forward converter, $D = 0.5$; (b) full-bridge converter, $D = 0.5$.

to monitor switch currents, as is done in the current-mode control discussed in a later section in this chapter. Use of an appropriate control integrated circuit (IC) also eliminates saturation under start-up and transient conditions. The other way to prevent core saturation due to voltage imbalance is to use a blocking capacitor in series with the primary winding of the half-bridge and the full-bridge inverters. The blocking capacitor should be chosen appropriately so that it is not too large as to be ineffective under transient conditions and not too small to cause a large ac voltage drop across it under steady-state operating condition. In the push–pull converters, the current-mode control is used to prevent the switch currents from becoming unequal.

In the core of a two-winding inductor of a flyback converter, an air gap must be present to provide energy storage capability. In the presence of this air gap, which is larger than that in the previous topologies, the remnant flux density B_r is essentially zero and the B–H characteristic becomes essentially linear.

The amount of inductance needed to operate only in the discontinuous mode (complete demagnetization mode) can be calculated from the given converter voltages and the switching frequency (see the problems at the end of the chapter).

10-5 CONTROL OF SWITCH-MODE dc POWER SUPPLIES

The output voltages of dc power supplies are regulated to be within a specified tolerance band (e.g., $\pm 1\%$ around its nominal value) in response to changes in the output load and the input line voltages. This is accomplished by using a negative-feedback control system, shown in Fig. 10-19a, where the converter output v_o is compared with its reference value $V_{o,\text{ref}}$. The error amplifier produces the control voltage v_c, which is used to adjust the duty ratio d of the switch(es) in the converter.

If the power stage of the switch-mode converter in Fig. 10-19a can be linearized, then the Nyquist stability criterion and the Bode plots can be used to determine the appropriate compensation in the feedback loop for the desired steady-state and transient responses.

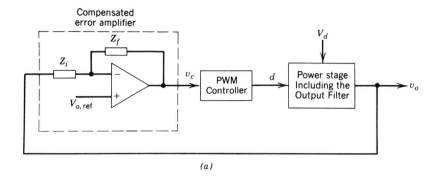

(a)

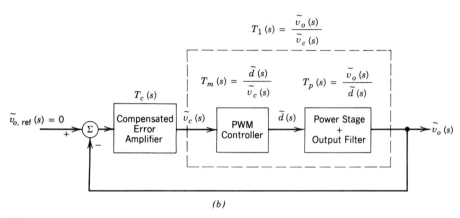

(b)

Figure 10-19 Voltage regulation: (a) feedback control system; (b) linearized feedback control system.

Middlebrook, Cúk, and their colleagues at the California Institute of Technology [10] have developed a state-space averaging technique that results in a linear model of the power stage including the output filter in Fig. 10-19a for small ac signals, linearized around a steady-state dc operating point. Similarly, the PWM controller in Fig. 10-19a can be linearized around a steady-state operating point. Therefore, each block in Fig. 10-19a can be represented by a transfer function as shown in Fig. 10-19b, where the small ac signals are represented by "~."

Another method to linearize the circuit is to use an average model of the PWM switch [see references 11 and 12].

10-5-1 LINEARIZATION OF THE POWER STAGE INCLUDING THE OUTPUT FILTER USING STATE-SPACE AVERAGING TO OBTAIN $\tilde{v}_o(s)/\tilde{d}(s)$

The goal of the following analysis is to obtain a small signal transfer function $\tilde{v}_o(s)/\tilde{d}(s)$, where $\tilde{v}_o$ and $\tilde{d}$ are small perturbations in the output voltage v_o and the switch duty ratio d, respectively, around their steady-state dc operating values V_o and D. Only a converter operating in a continuous-conduction mode is discussed. The procedure is as follows:

  Step 1 State-Variable Description for Each Circuit State. In a converter operating in a continuous-conduction mode, there are two circuit states: one state corresponds to when the switch is on and the other to when the switch is off. A third circuit state exists during the discontinuous interval, which is not considered in the following analysis because of the assumption of a continuous-conduction mode of operation.

During each circuit state, the linear circuit is described by means of the state-variable vector **x** consisting of the inductor current and the capacitor voltage. In the circuit description, the parasitic elements such as the resistance of the filter inductor and the equivalent series resistance (ESR) of the filter capacitor should also be included. Here V_d is the input voltage. A lowercase letter is used to represent a variable, which includes its steady-state dc value plus a small ac perturbation, for example, $v_o = V_o + \tilde{v}_o$. Therefore, during each circuit state, we can write the following state equations:

$$\dot{\mathbf{x}} = \mathbf{A}_1 \mathbf{x} + \mathbf{B}_1 v_d \quad \text{during } d \cdot T_s \tag{10-40}$$

and

$$\dot{\mathbf{x}} = \mathbf{A}_2 \mathbf{x} + \mathbf{B}_2 v_d \quad \text{during } (1 - d) \cdot T_s \tag{10-41}$$

where $\mathbf{A}_1$ and $\mathbf{A}_2$ are state matrices and $\mathbf{B}_1$ and $\mathbf{B}_2$ are vectors.

The output v_o in all converters can be described in terms of their state variables alone as

$$v_o = \mathbf{C}_1 \mathbf{x} \quad \text{during } d \cdot T_s \tag{10-42}$$

and

$$v_o = \mathbf{C}_2 \mathbf{x} \quad \text{during } (1 - d) \cdot T_s \tag{10-43}$$

where $\mathbf{C}_1$ and $\mathbf{C}_2$ are transposed vectors.

Step 2   Averaging the State-Variable Description Using the Duty Ratio d.   To produce an average description of the circuit over a switching period, the equations corresponding to the two foregoing states are time weighted and averaged, resulting in the following equations:

$$\dot{\mathbf{x}} = [\mathbf{A}_1 d + \mathbf{A}_2(1 - d)]\mathbf{x} + [\mathbf{B}_1 d + \mathbf{B}_2(1 - d)]v_d \tag{10-44}$$

and

$$v_o = [\mathbf{C}_1 d + \mathbf{C}_2(1 - d)]\mathbf{x} \tag{10-45}$$

Step 3   Introducing Small ac Perturbations and Separation into ac and dc Components.   Small ac perturbations, represented by "~", are introduced in the dc steady-state quantities (which are represented by the uppercase letters). Therefore,

$$\mathbf{x} = \mathbf{X} + \tilde{\mathbf{x}} \tag{10-46}$$

$$v_o = V_o + \tilde{v}_o \tag{10-47}$$

and

$$d = D + \tilde{d} \tag{10-48}$$

In general, $v_d = V_d + \tilde{v}_d$. However, in view of our goal to obtain the transfer function between voltage $\tilde{v}_o$ and the duty ratio $\tilde{d}$, the perturbation $\tilde{v}_d$ is assumed to be zero in the input voltage to simplify our analysis. Therefore

$$v_d = V_d \tag{10-49}$$

Using Eqs. 10-46 through 10-49 in Eqs. 10-44 and recognizing that in steady state, $\dot{\mathbf{X}} = 0$,

$$\dot{\mathbf{x}} = \mathbf{A}\mathbf{X} + \mathbf{B}V_d + \mathbf{A}\tilde{\mathbf{x}} + [(\mathbf{A}_1 - \mathbf{A}_2)\mathbf{X} + (\mathbf{B}_1 - \mathbf{B}_2)V_d]\tilde{d}$$
$$+ \text{ terms containing products of } \tilde{\mathbf{x}} \text{ and } \tilde{d} \text{ (to be neglected)}$$

$$(10\text{-}50)$$

where

$$\mathbf{A} = \mathbf{A}_1 D + \mathbf{A}_2(1 - D) \tag{10-51}$$

and

$$\mathbf{B} = \mathbf{B}_1 D + \mathbf{B}_2(1 - D) \tag{10-52}$$

The steady-state equation can be obtained from Eq. 10-50 by setting all the perturbation terms and their time derivatives to zero. Therefore, the steady-state equation is

$$\mathbf{A}\mathbf{X} + \mathbf{B}V_d = 0 \tag{10-53}$$

and therefore in Eq. 10-50

$$\dot{\tilde{\mathbf{x}}} = \mathbf{A}\tilde{\mathbf{x}} + [(\mathbf{A}_1 - \mathbf{A}_2)\mathbf{X} + (\mathbf{B}_1 - \mathbf{B}_2)V_d]\tilde{d} \tag{10-54}$$

Similarly, using Eqs. 10-46 through 10-49 in Eq. 10-45 results in

$$V_o + \tilde{v}_o = \mathbf{C}\mathbf{X} + \mathbf{C}\tilde{\mathbf{x}} + [(\mathbf{C}_1 - \mathbf{C}_2)\mathbf{X}]\tilde{d} \tag{10-55}$$

where

$$\mathbf{C} = \mathbf{C}_1 D + \mathbf{C}_2(1 - D) \tag{10-56}$$

In Eq. 10-55, the steady-state output voltage is given as

$$V_o = \mathbf{C}\mathbf{X} \tag{10-57}$$

and therefore,

$$\tilde{v}_o = \mathbf{C}\tilde{\mathbf{x}} + [(\mathbf{C}_1 - \mathbf{C}_2)\mathbf{X}]\tilde{d} \tag{10-58}$$

Using Eqs. 10-53 and 10-57, the steady-state dc voltage transfer function is

$$\frac{V_o}{V_d} = -\mathbf{C}\mathbf{A}^{-1}\mathbf{B} \tag{10-59}$$

Step 4   Transformation of the ac Equations into s-Domain to Solve for the Transfer Function.   Equations 10-54 and 10-58 consist of the ac perturbations. Using Laplace transformation in Eq. 10-54,

$$s\tilde{\mathbf{x}}(s) = \mathbf{A}\tilde{\mathbf{x}}(s) + [(\mathbf{A}_1 - \mathbf{A}_2)\mathbf{X} + (\mathbf{B}_1 - \mathbf{B}_2)V_d]\tilde{d}(s) \tag{10-60}$$

or

$$\tilde{\mathbf{x}}(s) = [s\mathbf{I} - \mathbf{A}]^{-1}[(\mathbf{A}_1 - \mathbf{A}_2)\mathbf{X} + (\mathbf{B}_1 - \mathbf{B}_2)V_d]\tilde{d}(s) \tag{10-61}$$

where $\mathbf{I}$ is a unity matrix. Using a Laplace transformation in Eq. 10-58 and expressing $\tilde{\mathbf{x}}(s)$ in terms of $\tilde{d}(s)$ from Eq. 10-61 results in the desired transfer function $T_p(s)$ of the power stages:

$$T_p(s) = \frac{\tilde{v}_o(s)}{\tilde{d}(s)} = \mathbf{C}[s\mathbf{I} - \mathbf{A}]^{-1}[(\mathbf{A}_1 - \mathbf{A}_2)\mathbf{X} + (\mathbf{B}_1 - \mathbf{B}_2)V_d] + (\mathbf{C}_1 - \mathbf{C}_2)\mathbf{X}$$

$$(10\text{-}62)$$

■ *Example 10-1*   Obtain the transfer function $\tilde{v}_o(s)/\tilde{d}(s)$ in a forward converter operating in a continuous-conduction mode. Assume $N_1/N_2 = 1$ for simplicity.

Solution   A forward converter is redrawn in Fig. 10-20a and the circuit states with the switch on and the switch off are shown in Figs. 10-20b and 10-20c, respectively. Here r_L is inductor resistance, r_c is the equivalent series resistance of the capacitor, and R is the load resistance.

Let x_1 and x_2 be as shown in Fig. 10-20. Then, in the circuit of Fig. 10-20b with the switch on,

$$-V_d + L\dot{x}_1 + r_L x_1 + R(x_1 - C\dot{x}_2) = 0 \qquad (10\text{-}63)$$

and

$$-x_2 - Cr_c\dot{x}_2 + R(x_1 - C\dot{x}_2) = 0 \qquad (10\text{-}64)$$

In matrix form, these two equations can be written as

$$\begin{bmatrix} \dot{x}_1 \\ \dot{x}_2 \end{bmatrix} = \begin{bmatrix} -\dfrac{Rr_c + Rr_L + r_c r_L}{L(R + r_c)} & -\dfrac{R}{L(R + r_c)} \\ \dfrac{R}{C(R + r_c)} & -\dfrac{1}{C(R + r_c)} \end{bmatrix} \begin{bmatrix} x_1 \\ x_2 \end{bmatrix} + \begin{bmatrix} \dfrac{1}{L} \\ 0 \end{bmatrix} V_d \qquad (10\text{-}65)$$

Comparing this equation with Eq. 10-40 yields

$$\mathbf{A}_1 = \begin{bmatrix} -\dfrac{Rr_c + Rr_L + r_c r_L}{L(R + r_c)} & -\dfrac{R}{L(R + r_c)} \\ \dfrac{R}{C(R + r_c)} & -\dfrac{1}{C(R + r_c)} \end{bmatrix} \qquad (10\text{-}66)$$

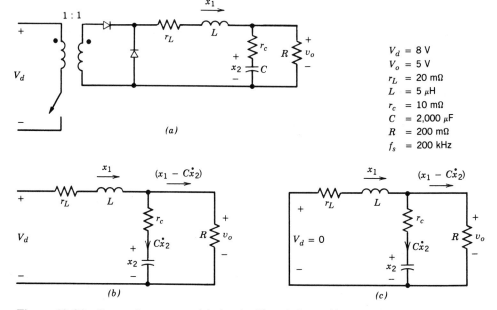

$V_d = 8$ V
$V_o = 5$ V
$r_L = 20$ mΩ
$L = 5$ μH
$r_c = 10$ mΩ
$C = 2{,}000$ μF
$R = 200$ mΩ
$f_s = 200$ kHz

Figure 10-20   Forward converter: (a) circuit; (b) switch on; (c) switch off.

and

$$\mathbf{B}_1 = \begin{bmatrix} \frac{1}{L} \\ 0 \end{bmatrix} \qquad (10\text{-}67)$$

The state equation for the circuit of Fig. 10-20c with the switch off can be written by observation, noting that the circuit of Fig. 10-20c is exactly the same as the circuit of Fig. 10-20b with V_d set to zero. Therefore, in Eq. 10-41,

$$\mathbf{A}_2 = \mathbf{A}_1 \qquad (10\text{-}68)$$

and

$$\mathbf{B}_2 = 0 \qquad (10\text{-}69)$$

The output voltage in both the circuit states is given as

$$v_o = R(x_1 - C\dot{x}_2) \qquad (10\text{-}70)$$

$$= \frac{Rr_c}{R + r_c}x_1 + \frac{R}{R + r_c}x_2$$

$$= \begin{bmatrix} \dfrac{Rr_c}{R + r_c} & \dfrac{R}{R + r_c} \end{bmatrix} \begin{bmatrix} x_1 \\ x_2 \end{bmatrix} \quad \text{(using } \dot{x}_2 \text{ from Eq. 10-64)}$$

Therefore, in Eqs. 10-42 and 10-43

$$\mathbf{C}_1 = \mathbf{C}_2 = \begin{bmatrix} \dfrac{Rr_c}{R + r_c} & \dfrac{R}{R + r_c} \end{bmatrix} \qquad (10\text{-}71)$$

Now, the following averaged matrices and vector can be obtained:

$$\mathbf{A} = \mathbf{A}_1 \quad \text{(from Eqs. 10-51 and 10-68)} \qquad (10\text{-}72)$$
$$\mathbf{B} = \mathbf{B}_1 D \quad \text{(from Eqs. 10-52 and 10-69)} \qquad (10\text{-}73)$$

and

$$\mathbf{C} = \mathbf{C}_1 \quad \text{(from Eqs. 10-56 and 10-71)} \qquad (10\text{-}74)$$

Model Simplification

In all practical circuits,

$$R \gg (r_c + r_L) \qquad (10\text{-}75)$$

Therefore, **A** and **C** are simplified as

$$\mathbf{A} = \mathbf{A}_1 = \mathbf{A}_2 = \begin{bmatrix} -\dfrac{r_c + r_L}{L} & -\dfrac{1}{L} \\ \dfrac{1}{C} & -\dfrac{1}{CR} \end{bmatrix} \qquad (10\text{-}76)$$

$$\mathbf{C} = \mathbf{C}_1 = \mathbf{C}_2 \simeq \begin{bmatrix} r_c & 1 \end{bmatrix} \qquad (10\text{-}77)$$

and **B** remains unaffected as

$$\mathbf{B} = \mathbf{B}_1 D = \begin{bmatrix} 1/L \\ 0 \end{bmatrix} D \qquad (10\text{-}78)$$

where $\mathbf{B}_2 = 0$. From Eq. 10-76,

$$\mathbf{A}^{-1} = \frac{LC}{1 + (r_c + r_L)/R} \begin{bmatrix} -\dfrac{1}{CR} & \dfrac{1}{L} \\ -\dfrac{1}{C} & -\dfrac{r_c + r_L}{L} \end{bmatrix} \tag{10-79}$$

Using Eqs. 10-76 through 10-79 in Eq. 10-59, the steady-state dc voltage transfer function is

$$\frac{V_o}{V_d} = D\frac{R + r_c}{R + (r_c + r_L)} \simeq D \tag{10-80}$$

Similarly, using Eqs. 10-76 through 10-79 in Eq. 10-62 yields

$$T_p(s) = \frac{\bar{v}_o(s)}{\bar{d}(s)} \simeq V_d\frac{1 + sr_cC}{LC\{s^2 + s[1/CR + (r_c + r_L)/L] + 1/LC\}} \tag{10-81}$$

The terms in the curly brackets in the denominator of Eq. 10-81 are of the form $s^2 + 2\xi\omega_0 s + \omega_0^2$, where

$$\omega_0 = \frac{1}{\sqrt{LC}} \tag{10-82}$$

and

$$\xi = \frac{1/CR + (r_c + r_L)/L}{2\omega_0} \tag{10-83}$$

Therefore, from Eq. 10-81 the transfer function $T_p(s)$ of the power stage and the output filter can be written as

$$T_p(s) = \frac{\bar{v}_0(s)}{\bar{d}(s)} = V_d\frac{\omega_0^2}{\omega_z}\frac{s + \omega_z}{s^2 + 2\xi\omega_0 s + \omega_0^2} \tag{10-84}$$

where a zero is introduced due to the equivalent series resistance of the output capacitor at the frequency

$$\omega_z = \frac{1}{r_cC} \tag{10-85}$$

Figure 10-21 shows the Bode plot for the transfer function in Eq. 10-84 using the numerical values given in Fig. 10-20a. It shows that the transfer function has a fixed gain and a minimal phase shift at low frequencies. Beyond the resonant frequency $\omega_0 = \sqrt{1/LC}$ of the LC output filter, the gain begins to fall with a slope of -40 dB/decade and the phase tends toward $-180°$. At frequencies beyond ω_z, the gain falls with a slope of -20 dB/decade and the phase angle tends toward $-90°$. The gain plot shifts vertically with V_d but the phase plot is not affected. ∎

In the flyback converter operating in a continuous mode, the transfer function is a nonlinear function $f(D)$ of the duty ratio D and is given as

$$\frac{\bar{v}_0(s)}{\bar{d}(s)} = V_df(D)\frac{(1 + s/\omega_{z1})(1 - s/\omega_{z2})}{as^2 + bs + c} \tag{10-86}$$

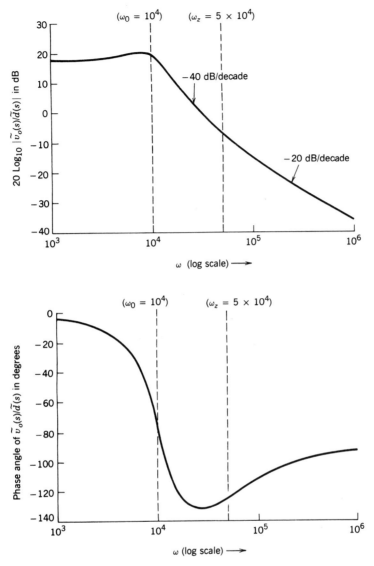

Figure 10-21   (a) Gain plot of the forward converter in Fig. 10-20a.
(b) Phase plot of the forward converter in Fig. 10-20a.

where the zero ω_{z2} in the transfer function appears in right half of the s plane. The
frequency of the right-half-plane zero depends on the load resistance and the effective
value of the filter inductance, where the effective value of the filter inductance is the filter
inductance times a nonlinear function of the steady-state dc duty ratio D. A Bode plot of
such a transfer function is drawn in Figs. 10-22a and 10-22b. Unlike the transfer function
of the converters derived from step-down converters and discussed earlier, the gain at low
frequencies is a nonlinear function of the dc operating point (i.e., of V_d). Also the
frequency at which the gain falls with a slope of -40 dB/decade depends on the dc
operating point. The phase associated with this gain slope tends toward $-180°$. Assuming
that $\omega_{z2} > \omega_{z1}$, at frequencies beyond the frequency ω_{z1} of the left-half-plane zero caused
by the equivalent series resistance of the capacitor, the gain falls with a slope of -20

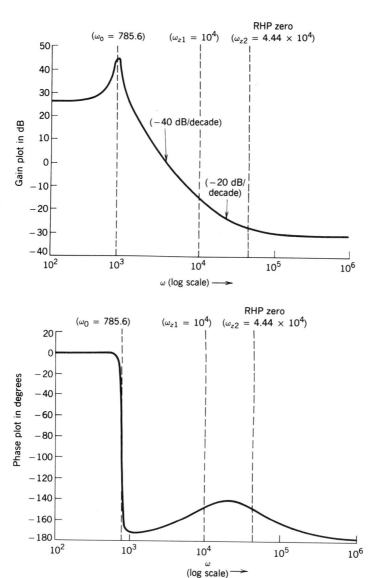

Figure 10-22   (a) Gain plot for a flyback converter. (b) Phase plot for a flyback converter.

$$\frac{\tilde{v}_o(s)}{\tilde{d}(s)} = 13.33 \times 10^6 \frac{(1 + s/\omega_{z1})(1 - s/\omega_{z2})}{s^2 + 2\xi\omega_0 s + \omega_0^2}$$

where

$$\omega_{z1} = 10^4 \qquad \omega_{z2} = 4.44 \times 10^4 \qquad \omega_0 = 785.6 \qquad \xi = 0.012$$

dB/decade and the phase angle tends toward $-90°$. At frequencies beyond the frequency ω_{z2} of the right-half-plane zero, the grain curve flattens out but the phase angle begins to decrease again. The additional phase lag introduced by the right-half-plane zero must be considered in designing the compensation of such a system to provide enough gain and phase margins.

The presence of the right-half zero can be explained by noting that in a flyback converter operating in a continuous mode, if the duty ratio d is increased instantaneously, the output voltage decreases momentarily because the inductor current has not had the time to increase, but the time interval $(1 - d)T_s$ during which the inductor transfers energy to the output stage has been suddenly decreased. This initial decline in the output voltage with the increase in d is opposite of what eventually takes place. This effect results in a zero in the right-half plane, which introduces phase lag in the transfer function $\tilde{v}_o(s)/\tilde{d}(s)$.

In a flyback converter operating in a discontinuous mode, the foregoing effect does not occur and the output voltage always increases with the increased duty ratio. Therefore, in the discontinuous mode of operation, the right-half-plane zero in the transfer function of Eq. 10-86 does not exist; thus, compensating the feedback loop to provide enough gain and phase margins is simpler.

10-5-2 TRANSFER FUNCTION $\tilde{d}(s)/\tilde{v}_c(s)$ OF THE DIRECT DUTY RATIO PULSE-WIDTH MODULATOR

In the direct duty ratio pulse-width modulator, the control voltage $v_c(t)$, which is the output of the error amplifier, is compared with a repetitive waveform $v_r(t)$, which establishes the switching frequency f_s, as shown in Fig. 10-23a. The control voltage $v_c(t)$ consists of a dc component and a small ac perturbation component

$$v_c(t) = V_c + \tilde{v}_c(t) \qquad (10\text{-}87)$$

where $v_c(t)$ is in a range between zero and $\hat{V}_r$, as shown in Fig. 10-23a. Here $\tilde{v}_c(t)$ is a sinusoidal ac perturbation in the control voltage at a frequency ω, where ω is much smaller than the switching frequency ω_s ($=2\pi f_s$). The ac perturbation in the control voltage can be expressed as

$$\tilde{v}_c(t) = a\,\sin(\omega t - \phi) \qquad (10\text{-}88)$$

by means of an amplitude a and an arbitrary phase angle ϕ.

In Fig. 10-23b, the instantaneous switch duty ratio $d(t)$ is as follows:

$$d(t) = \begin{cases} 1.0 & \text{if } v_c(t) \geq v_r(t) \\ 0 & \text{if } v_c(t) < v_r(t) \end{cases} \qquad \begin{matrix}(10\text{-}89)\\(10\text{-}90)\end{matrix}$$

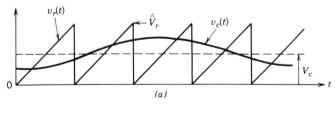

(a)

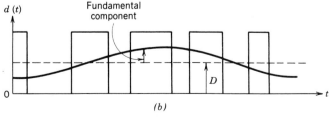

(b)

Figure 10-23 Pulse-width modulator.

Similar to the analysis of sinusoidal PWM carried out in Chapter 8, $d(t)$ in Fig. 10-23b can be expressed in terms of the Fourier series as

$$d(t) = \frac{V_c}{\hat{V}_r} + \frac{a}{\hat{V}_r}\sin(\omega t - \phi) + \text{other high-frequency components} \quad (10\text{-}91)$$

The higher frequency components in the output voltage v_o due to the high-frequency components in $d(t)$ are eliminated because of the low-pass filter at the output of the converter. Therefore, the high-frequency components in Eq. 10-91 can be ignored. In terms of its dc value and its ac perturbation

$$d(t) = D + \tilde{d}(t) \quad (10\text{-}92)$$

Comparing Eqs. 10-91 and 10-92 yields

$$D = \frac{V_c}{\hat{V}_r} \quad (10\text{-}93)$$

and

$$\tilde{d}(t) = \frac{a}{\hat{V}_r}\sin(\omega t - \phi) \quad (10\text{-}94)$$

From Eqs. 10-88 and 10-94, the transfer function $T_m(s)$ of the modulator is given by

$$T_m(s) = \frac{\tilde{d}(s)}{\tilde{v}_c(s)} = \frac{1}{\hat{V}_r} \quad (10\text{-}95)$$

Therefore, the theoretical transfer function of the pulse-width modulator is surprisingly simple, without any time delays. However, the time delay associated with the comparator can lead to a delay in the modulator response.

■ *Example 10-2* In practice, the transfer function of the modulator may not have to be calculated from Eq. 10-95. Figure 10-24 shows the approximate transfer function of a commonly used PWM integrated circuit, supplied as a part of the data sheets, in terms of the duty ratio d as a function of the control voltage v_c, where v_c is the output of the error amplifier.

Calculate the transfer function $\tilde{d}(s)/\tilde{v}_c(s)$ for this PWM integrated circuit.

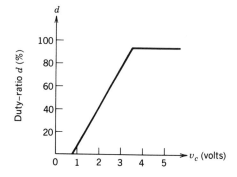

Figure 10-24 Pulse-width modulator transfer function.

Solution For this particular modulator, the duty ratio d increases from 0 (at $v_c = 0.8$ V) to 0.95 (at $v_c = 3.6$ V). Therefore, the slope of the transfer function in Fig. 10-24 is equal to the transfer function of this modulator:

$$\frac{\tilde{d}(s)}{\tilde{v}_c(s)} = \frac{\Delta d}{\Delta v_c}$$

$$= \frac{0.95 - 0}{3.6 - 0.8} \simeq 0.34 \qquad (10\text{-}96)$$

With this modulator, the transfer function between v_o and the control voltage v_c can be obtained as

$$T_1(s) = \frac{\tilde{v}_o(s)}{\tilde{v}_c(s)} = \frac{\tilde{v}_o(s)}{\tilde{d}(s)} \frac{\tilde{d}(s)}{\tilde{v}_c(s)} = T_p(s)T_m(s) \qquad (10\text{-}97)$$

The gain plot of the transfer function $\tilde{v}_o(s)/\tilde{v}_c(s)$ can be obtained by adjusting the gain curve in the Bode plot of Fig. 10-21a or Fig. 10-22a to account for a constant gain of 0.34 ($= -9.37$ dB) of the modulator. Assuming zero delay in the modulator, the phase plot of $\tilde{v}_o(s)/\tilde{v}_c(s)$ is the same as that of $\tilde{v}_o(s)/\tilde{d}(s)$. ∎

10-5-3 COMPENSATION OF THE FEEDBACK SYSTEM USING A DIRECT DUTY RATIO PULSE-WIDTH MODULATOR

In the switch-mode power supply shown in Fig. 10-19b, the overall open-loop transfer function is

$$T_{OL}(s) = T_1(s)T_c(s) \qquad (10\text{-}98)$$

where $T_1(s)$ is as given by Eq. 10-97 and

$$T_c(s) = \text{transfer function of compensated error amplifier} \qquad (10\text{-}99)$$

For a given $T_1(s)$, the transfer function of the compensated error amplifier $T_c(s)$ must be properly tailored so that $T_{OL}(s)$ meets the performance requirements expected of the power supply. Some of the desired characteristics of the open-loop transfer function $T_{OL}(s)$ are as follows:

1. The gain at low frequencies should be high to minimize the steady-state error in the power supply output.
2. The crossover frequency is the frequency at which the gain of $T_{OL}(s)$ falls to 1.0 (0 dB), as shown in Fig. 10-25. This crossover frequency ω_{cross} should be as high as possible but approximately an order of magnitude below the switching frequency to allow the power supply to respond quickly to the transients, such as a sudden change of load.
3. The phase margin (PM) is defined by means of Fig. 10-25 as

$$PM = \phi_{OL} + 180° \qquad (10\text{-}100)$$

where ϕ_{OL} is the phase angle of $T_{OL}(s)$ at the crossover frequency and is negative. The phase margin, which should be a positive quantity in Eq. 10-100, determines the transient response of the output voltage in response to sudden changes in the load and the input voltage. A phase margin in a range of 45°–60° is desirable.

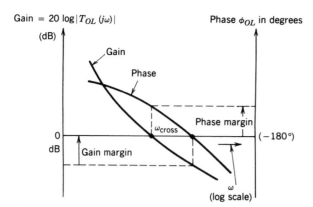

Figure 10-25 Gain and phase margins.

To meet these requirements simultaneously, a general error amplifier is shown in Fig. 10-26, where the amplifier can be assumed to be ideal. One of the inputs to the amplifier is the output voltage v_o of the converter; the other input is the desired (reference) value V_{ref} of v_o. The output of the error amplifier is the control voltage v_c. In terms of Z_i and Z_f in Fig. 10-26, the transfer function between the input and the output perturbations can be obtained as

$$\frac{\tilde{v}_c(s)}{\tilde{v}_o(s)} = -\frac{Z_f(s)}{Z_i(s)} = -T_c(s) \tag{10-101}$$

where $T_c(s)$ is as defined in Fig. 10-19b.

One of the options in the selection of $T_c(s)$ is to introduce a pole-zero pair in addition to a pole at the origin so that $T_c(s)$ is of the form

$$T_c(s) = \frac{A\ (s + \omega_z)}{s\ (s + \omega_p)} \tag{10-102}$$

where A is positive and $\omega_z < \omega_p$. In Eq. 10-102, due to the pole at the origin, the phase of $T_c(s)$ starts with $-90°$, as shown in Fig. 10-27a. The presence of the zero causes the phase angle to increase (or in other words provides a "boost") to be something greater than $-90°$. Eventually, because of the pole at ω_p, the phase angle of $T_c(s)$ comes back down to $-90°$. The gain plot is also shown in Fig. 10-27a. The parameters in Eq. 10-102 can be chosen such that the minimum phase lag in $T_c(s)$ occurs at the specified (desired) crossover frequency of the overall open-loop transfer function $T_{OL}(s)$.

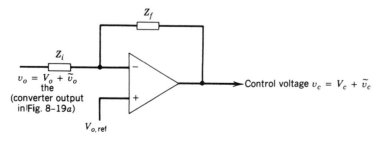

Figure 10-26 A general compensated error amplifier.

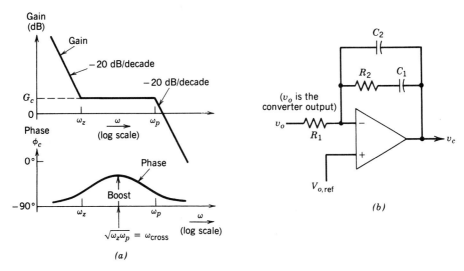

Figure 10-27   Error amplifier.

$$T_c(s) = \frac{1}{R_1 C_2} \frac{s + \omega_z}{s(s + \omega_p)}$$

The transfer function in Eq. 10-102 can be realized by means of the amplifier network shown in Fig. 10-27b, where

$$T_c(s) = \frac{1}{R_1 C_2} \frac{s + \omega_z}{s(s + \omega_p)} \qquad (10\text{-}103)$$

$$\omega_z = \frac{1}{R_2 C_1} \qquad (10\text{-}104)$$

$$\omega_p = \frac{C_1 + C_2}{R_2 C_1 C_2} \qquad (10\text{-}105)$$

A step-by-step explanation that is easy to follow in the selection of the foregoing parameters has been provided in reference 16 using a K-factor approach. This procedure suggests that as a first-step, the crossover frequency ω_{cross}, where $|T_{\text{OL}}(s)|$ would equal 0 dB, should be selected. This crossover frequency also defines the frequency in Fig. 10-27a, where the minimum phase lag occurs in the transfer function of $T_c(s)$. The K factor is used such that in the transfer function $T_c(s)$ of Eq. 10-103

$$\omega_z = \frac{\omega_{\text{cross}}}{K} \qquad (10\text{-}106)$$

$$\omega_p = K\omega_{\text{cross}} \qquad (10\text{-}107)$$

It is shown in reference 16 that K in Eqs. 10-106 and 10-107 is related to the boost (defined in Fig. 10-27a) in the following manner:

$$K = \tan\left(45° + \frac{\text{boost}}{2}\right) \qquad (10\text{-}108)$$

Therefore, the next step is to define the phase margin (PM) and, hence, the boost needed from the error amplifier at the crossover frequency to calculate K in Eqs. 10-106 and 10-107. From the definition of phase margin in Eq. 10-100

$$PM = 180° + \phi_1 + \phi_c \qquad (10\text{-}109)$$

where ϕ_c is the phase angle of $T_c(s)$ at the crossover frequency. From Eq. 10-97

$$\phi_1 = \phi_p(s) + \phi_m(s) \qquad (10\text{-}110)$$

where ϕ_1 is the phase angle of $T_1(s)$, $\phi_p(s)$ is the phase angle of the power stage $T_p(s)$, and $\phi_m(s)$ is the phase angle (if any) of the modulator $T_m(s)$. From the phase plot of the transfer function $T_c(s)$ shown in Fig. 10-27a

$$\phi_c = -90° + \text{boost} \qquad (10\text{-}111)$$

From Eqs. 10-109 and 10-111,

$$\text{boost} = PM - \phi_1 - 90° \qquad (10\text{-}112)$$

Therefore, once the phase margin (usually in a range of 45°–60°) is chosen, the boost is defined from Eq. 10-112 where ϕ_1 (assuming ϕ_m to be zero) can be obtained from Fig. 10-21b or Fig. 10-22b at the frequency chosen to be the crossover frequency. Knowing the boost, K can be calculated from Eq. 10-108.

The next step in the design procedure is to ensure that the gain G_{OL} of the overall open-loop is equal to 1 (i.e., $G_{OL} = |T_{OL}(s)| = 1$) at the chosen crossover frequency. This requires that from Eq. 10-98, the gain G_c of the compensated error amplifier at ω_{cross} be as follows:

$$G_c(\text{at } \omega_{cross}) = \frac{1}{G_1(\text{at } \omega_{cross})} \qquad (10\text{-}113)$$

where G_1 is the magnitude $|T_1(j\omega_{cross})|$ of the transfer function $T_1(s) = T_p(s)T_m(s)$ at ω_{cross}. Therefore, at $\omega = \omega_{cross}$, from Eq. 10-113 and by the substitution of Eqs. 10-104 through 10-107 into Eq. 10-103,

$$G_c = \frac{1}{KC_2R_1\omega_{cross}} = \frac{1}{G_1} \qquad (10\text{-}114)$$

In the circuit of Fig. 10-27b, R_1 is chosen arbitrarily and the rest of the circuit parameters can be calculated as follows from the K-factor procedure outlined before using Eqs. 10-104 through 10-107 and Eq. 10-114,

$$C_2 = \frac{G_1}{KR_1\omega_{cross}} \qquad (10\text{-}115)$$

$$C_1 = C_2(K^2 - 1) \qquad (10\text{-}116)$$

$$R_2 = \frac{K}{(C_1\omega_{cross})} \qquad (10\text{-}117)$$

For the converters, such as a flyback converter operating in a continuous mode, it may be necessary to use an error amplifier that has two pairs of poles and zeros in addition to the pole at the origin for a proper compensation.

10-5-4 VOLTAGE FEED-FORWARD PWM CONTROL

In the direct duty ratio PWM control discussed in the previous two sections, if the input voltage changes, an error is produced in the output voltage, which eventually gets corrected by the feedback control. This results in a slow dynamic performance in regulating the output in response to the changes in input voltage.

If the duty ratio could be adjusted directly to accommodate the change in the input voltage, then the converter output would remain unchanged. This can be accomplished by feeding the input voltage level to the PWM IC. The PWM switching strategy here is very similar to the one discussed in connection with the direct duty ratio PWM control except for one difference: the ramp (and, hence, the peak) of the sawtooth waveform does not stay constant but varies in direct proportion to the input voltage, as shown in Fig. 10-28. This shows how an increased input voltage (hence, increased $\hat{V}_r$) results in a decreased duty ratio, shown by dashed curves in Fig. 10-28. This type of control in step-down derived converters (e.g., forward converters) results in $\bar{v}_o(s)/\bar{v}_d(s)$ equal to zero and hence in an excellent inherent regulation for the changes in the input voltage. The same is true for a flyback converter operating in a complete demagnetization mode.

If this voltage feed-forward is implemented in a double-ended power supply (like push–pull, half-bridge, full bridge), then care must be taken to provide a dynamic volt–time balance so that the on times of the two switches are kept equal on a dynamic basis to prevent the saturation of the high-frequency isolation transformer.

10-5-5 CURRENT-MODE CONTROL

The PWM direct duty ratio control discussed so far is shown in Fig. 10-29a, where the control voltage v_c (amplified error signal between the actual output and the reference) controls the duty ratio of the switch by comparing the control voltage with a fixed-frequency sawtooth waveform. This control of the switch duty ratio adjusts the voltage across the inductor and hence the inductor current (which feeds the output stage) and eventually brings the output voltage to its reference value.

In a current-mode control, an additional inner control loop is used as shown in Fig. 10-29b, where the control voltage v_c directly controls the output inductor current that feeds the output stage and thus the output voltage. Ideally, the control voltage should act to directly control the *average* value of the inductor current for the fastest response, though, as we will see later, various types of current-mode controls tend to accomplish this differently. The fact that the current feeding the output stage is controlled directly in a current-mode control has a profound effect on the dynamic behavior of the negative-feedback control loop.

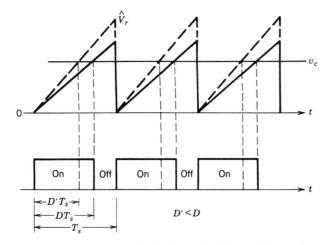

Figure 10-28 Voltage feed-forward: effect on duty ratio.

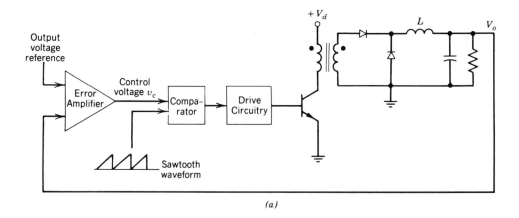

(a)

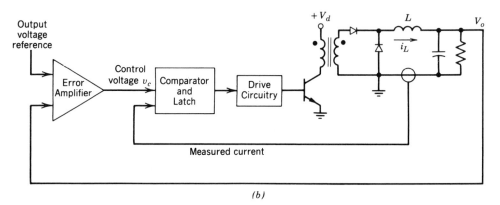

(b)

Figure 10-29 PWM duty ratio versus current-mode control: (a) PWM duty ratio control; (b) current-mode control.

There are three basic types of current-mode controls:

1. Tolerance band control
2. Constant-"off"-time control
3. Constant-frequency control with turn-on at clock time

In all these types of controls, either the inductor current or the switch current, which is proportional to the output inductor current, is measured and compared with the control voltage.

In the *tolerance band control,* the control voltage v_c dictates the average value of the inductor current as shown in Fig. 10-30a. The term ΔI_L is a design parameter. The switching frequency depends on ΔI_L, the converter parameters, and the operating conditions. This direct control over the average value of i_L is a very desirable feature of this type of control. However, this scheme works well only in the continuous-current-conduction mode. Otherwise, in the discontinuous-current-conduction mode, the inductor current becomes zero (though $\frac{1}{2}\Delta I_L$ would actually be demanding a negative i_L, which is not possible). If the controller is not designed to handle this discontinuous current when i_L is zero and i_L being demanded by the controller is negative, the switch will never turn on and the inductor current will decay to zero.

In the *constant-off-time control,* the control voltage dictates $\hat{I}_L$, as is shown in Fig. 10-30b. Once $\hat{I}_L$ is reached, the switch turns off for a fixed (constant) off time, which is

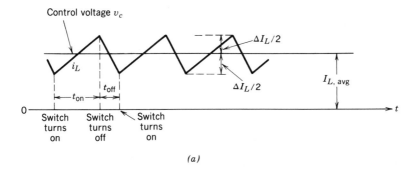

(a)

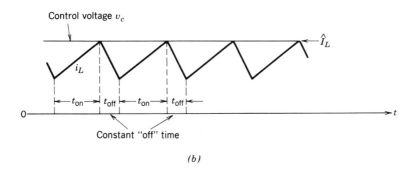

(b)

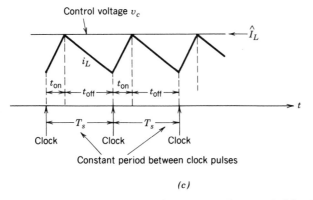

(c)

Figure 10-30 Three types of current-mode control: (*a*) tolerance band control; (*b*) constant-off-time control; (*c*) constant frequency with turn-on at clock time.

a design parameter. Here also, the switching frequency is not fixed and depends on the converter parameters and its operating condition.

The *constant-frequency control with a turn-on at clock time* is thus far the most common type of current-mode control. Here, the switch is turned on at the beginning of each constant-frequency switching time period. The control voltage dictates $\hat{I}_L$ and the instant at which the switch is turned off, as shown in Fig. 10-30*c*. The switch remains off until the beginning of the next switching cycle. A constant switching frequency makes it easier to design the output filter.

In the current-mode control in practice, a slope compensation is added to the control voltage, as shown in Fig. 10-31, to provide stability, to prevent subharmonic oscillations, and to provide a feed-forward property. Figure 10-31 shows the waveforms for a forward

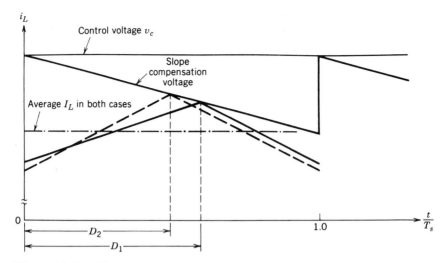

Figure 10-31   Slope compensation in current-mode control (D_2 is smaller for a higher input voltage with a constant V_o).

converter of the type shown in Fig. 10-29b, where the slope of the slope compensation waveform is one-half of the slope of the inductor current when the switch is off. With given input and output voltages, the duty ratio is D_1 and the waveform of the inductor current i_L is shown by the solid lines. If the input voltage is increased but the output voltage is to remain unaffected, the duty ratio decreases to D_2 and the inductor current waveform is shown by dashed lines. The average value of the inductor current, which equals the load current, remains the same in both cases in spite of a change in the input voltage. This shows the voltage feed-forward property of the current-mode control with a proper slope compensation.

The current-mode control has several advantages over the conventional direct duty ratio PWM control:

1. *It limits peak switch current.* Since either the switch current is directly measured or the current is measured somewhere in the circuit (like through the output inductor) where it represents the switch current without delay, the peak value of the switch current can be limited by simply putting an upper limit on the control voltage. This can be easily accomplished in the controllers that control $\hat{I}_L$.

2. *It removes one pole* (corresponding to the output filter inductor) from the control-to-output transfer function $\tilde{v}_o(s)/\tilde{v}_c(s)$, thus simplifying the compensation in the negative-feedback system, especially in the presence of the right-half-plane zero.

3. It allows a *modular design* of power supplies by equal current sharing where several power supplies can be operated in parallel and provide equal currents, if the same control voltage is fed to all the modules.

4. It results in a *symmetrical flux excursion* in a push–pull converter, thus eliminating the problem of transformer core saturation.

5. *It provides input voltage feed-forward.* As shown by Fig. 10-31, an input voltage feed-forward is automatically accomplished, resulting in an excellent rejection of input line transients.

10-5-6 DIGITAL PULSE-WIDTH MODULATION CONTROL

In recent years, there has been an ongoing attempt to implement the foregoing types of controls by means of digital controllers. The main advantages of a digital approach over its analog counterpart are (1) its lower sensitivity to the changes in the environment such as temperature, supply voltage fluctuations, aging of components, and so on, and (2) possibility of a lower parts count, thus improving the supply reliability. A proportional-integral-differential (PID) control can be implemented in hardware, software, or both to minimize the steady-state error and to yield a satisfactory transient response.

10-6 POWER SUPPLY PROTECTION

In addition to a stable control that provides appropriate steady-state and transient performance, it is important that the control of the power supply also provide its protection against abnormal operating conditions. These protective control features are explained by means of a direct duty ratio PWM IC belonging to the 1524 family, which has been used in a large number of power supplies.

The modulator UC1524A is an enhanced version of the original modulator. It can be used for switching frequencies of up to 500 kHz. The block diagram of the UC1524A is shown in Fig. 10-32a. The internal reference circuit provides a regulated 5 V output (pin 16) for the input supply voltage variations in a range of 8–40 V (pin 15).

An error amplifier (a transconductance type) allows the measured output voltage of the power supply (connected to pin 1) to be compared to the reference or the desired output voltage (connected to pin 2) of the amplifier. An appropriate feedback network to provide compensation and the loop gain can be connected from the error amplifier output (pin 9) to the inverting input (pin 1). If a separate error amplifier with compensation, as discussed in Section 10-5-3, is used, then this error amplifier can be wired to be a noninverting amplifier with a unity gain. The parameters R_T and C_T (between pins 6 and 7 to ground) determine the frequency of the oscillator, which produces a sawtooth waveform at pin 7. The oscillator frequency is determined as

$$\text{Oscillator frequency (kHz)} = \frac{1.15}{R_T \text{ (k}\Omega) \times C_T \text{ (}\mu\text{F)}} \qquad (10\text{-}118)$$

The sawtooth waveform is compared with the error amplifier output in a comparator to determine the duty ratio of the switches. The output of the oscillator (pin 3) is a narrow 3.5-V clock pulse of a 0.5-μs pulse width at a frequency determined by Eq. 10-118.

This integrated circuit allows PWM control of push–pull and bridge (half and full) converters where two switches (or switch pairs) need to be controlled alternatively. Comparison of the sawtooth waveform with the error amplifier output and the flip-flop (triggered by the oscillator clock pulse) and logic gates provide positive base drives alternatively to output switches A and B, which can be used to drive the power switches. One complete switching period requires two oscillator frequency cycles, and hence the switch operating frequency is one-half the oscillator frequency. The PWM latch ensures that only a single pulse is allowed to reach the appropriate output stage within each period.

There is another significant function of the oscillator output clock pulse. As the switch duty ratio begins to approach 0.5, this narrow pulse ensures a blanking time between the turning off and the turning on of the converter switches (or switch pairs). The selection of C_T determines the blanking time from 0.5 μs to as large as 4 μs. The

BLOCK DIAGRAM

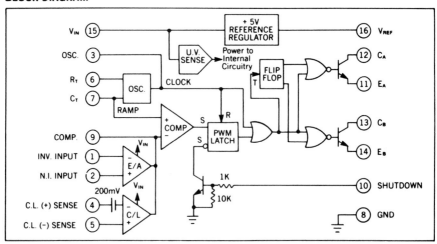

Figure 10-32   Pulse-width modulator UC1524A: (*a*) block diagram; (*b*) transfer function. (*Courtesy of Unitrode Integrated Circuits Corp.*)

oscillator frequency is determined by Eq. 10-118, by selecting R_T, whereas C_T is selected to yield a desired blanking time.

The transfer function between the duty ratio of one of the two outputs as a function of the input voltage at pin 9 (which is the output of the error amplifier internal to this IC) is shown in Fig. 10-32*b*. To control the single-switch converters (such as forward and flyback), both the IC output switches A and B can be connected in parallel and the switch duty ratio can be as high as 0.95.

Some of the protective features are explained in the following sections.

10-6-1 SOFT START

A soft start in switch-mode dc power supplies is provided by increasing the duty ratio and hence the output voltage slowly, subsequent to the input voltage switch-on. This can be provided by connecting a simple circuit to pin 9.

10-6-2 VOLTAGE PROTECTION

Overvoltage and undervoltage protection can be incorporated by adding a few external components to the shut-down pin 10.

10-6-3 CURRENT LIMITING

For protection against overcurrent at the output, the circuit output current can be sensed by measuring the voltage across a sensing resistor. This voltage is applied across pins 4 and 5. When this sensed voltage exceeds a temperature-compensated threshold of 200 mV, the output of the error amplifier is pulled toward ground and linearly decreases the output pulse width.

10-6-3-1 Foldback Current Limiting

In a constant current-limited power supply, if the gain of the current-limiting stage is high, the supply V_o–I_o characteristic can be as shown in Fig. 10-33a, where once a critical value of current I_{limit} is reached, I_o is not allowed to increase any more and the output voltage V_o depends on the load line. Therefore, a load resistance R_1 yields an output voltage V_{o1} and a load resistance R_2 yields V_{o2}, as shown in Fig. 10-33a. Even with a complete short circuit across the output supply, the output current does not exceed the current limit (a design parameter) by any appreciable amount. This may be a requirement in a dc power supply, which may be used to supply a constant current and hold the output current to a specified value once the output load resistance decreases below a certain value.

However, in many applications, the output current exceeding the critical value represents an abnormal load condition, and a foldback current limit is introduced where, as the load resistance decreases, the output current also decreases (along with the decreasing output voltage V_o), as shown in Fig. 10-33b. Here, in case of a short circuit across the output, the current will have a much smaller value I_{FB} in comparison to I_{limit}. The motivation for this foldback current limit is to reduce the current flowing through the supply unnecessarily and to bring it to a much smaller value under an abnormal load condition. Once the load recovers to its normal value, the supply once again begins to regulate V_o to its reference value. This foldback current limit can be implemented using a PWM controller such as that shown in Fig. 10-32a.

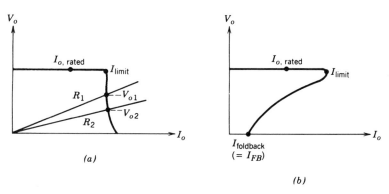

Figure 10-33 Current limiting: (a) constant current limiting; (b) foldback current limiting.

10-7 ELECTRICAL ISOLATION IN THE FEEDBACK LOOP

In an electrically isolated switching power supply, it is necessary to provide electrical isolation in the feedback path where the output voltage on the secondary side of the high-frequency power transformer is measured to control the power switches that are on the primary side of the high-frequency power transformer. Two options are presented in Figs. 10-34a and 10-34b.

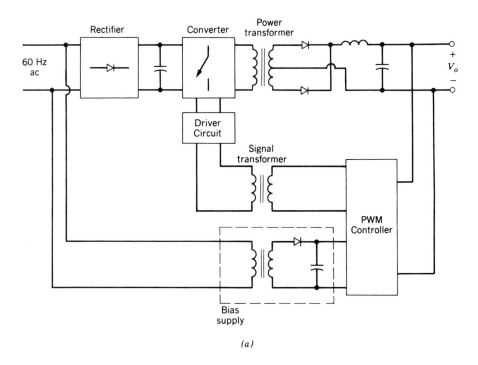

(a)

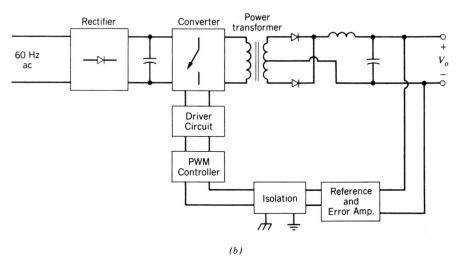

(b)

Figure 10-34 Electrical isolation in the feedback loop: (a) secondary-side control; (b) primary-side control.

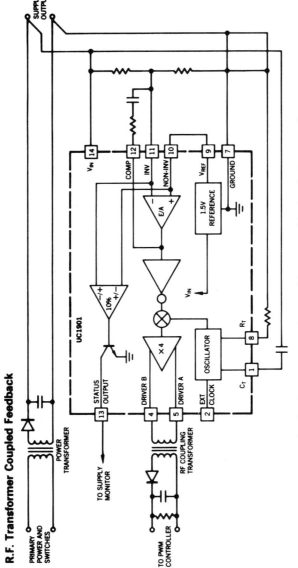

Figure 10-35 Isolated feedback generator UC1901. (*Courtesy of Unitrode Integrated Circuits Corp.*)

In the secondary-side control shown in Fig. 10-34a, the PWM controller, such as the UC1524A discussed earlier, is on the secondary side of the power transformer. Its supply (or bias) voltage is provided by a bias supply through an isolation transformer from the primary side. The signals to the switch driver circuit are provided through small signal transformers, thus maintaining isolation in the feedback loop.

As an alternative, the primary-side control is shown in Fig. 10-34b, where the PWM controller is on the primary side along with the power switches. This requires electrical isolation between the output voltage error amplifier and the PWM controller. The advantages of having the PWM controller on the same side of the switches are that it simplifies the interface with the switch driver circuit and it is possible to implement the input voltage feed forward control.

One way to implement isolation in the control of Fig. 10-34b is to use an optocoupler between the dc output of the error amplifier and the PWM controller. However, the optocoupler approach suffers from several drawbacks, namely, the stability of the optocoupler gain with temperature and time, thus making it difficult to guarantee the stability and the performance of the power supply.

The other alternative in the primary-side control is to use an amplitude-modulated oscillator such as the UC1901 shown in Fig. 10-35. The high-frequency oscillator output is coupled through a small high-frequency signal transformer to a demodulator that supplies the dc error voltage to the PWM controller. Reference 24 provides a detailed description of using such a feedback isolation technique.

10-8 DESIGNING TO MEET THE POWER SUPPLY SPECIFICATIONS

Power supplies have to meet several specifications. The considerations for meeting some of these specifications are discussed in the following sections.

10-8-1 INPUT FILTER

A simple low-pass filter such as a single-stage filter consisting of L and C as shown in Fig. 10-36 may be used at the input to the switch-mode supply to improve its power factor of operation and to reduce the conducted EMI. From the energy efficiency standpoint, this filter should have as little power loss as possible. However, one must consider the possibility of oscillations in the presence of such a low-loss filter.

A regulated switch-mode power supply appears as a negative resistance across the input filter capacitor. The reason for a negative resistance is the fact that with increasing input voltage, the input current decreases, since the output voltage is regulated, and hence, the output power and the input power do not change. Decreasing input current with an increasing input voltage implies a negative input resistance.

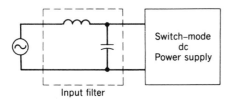

Input filter

Figure 10-36 Input filter.

If an adequate damping is not provided, a possibility of sustained oscillation exists. A useful design criterion requires that the resonant frequency of the input filter be a decade lower than the resonant frequency of the output filter to avoid interaction between the two. The input filter capacitor should be chosen to be as large as possible, and additional damping elements should be included.

It is also possible to provide active waveshaping of the input current as described in Chapter 18, which results in an essentially harmonic-free current at a unity power factor.

10-8-2 INPUT RECTIFIER BRIDGE

To be able to operate from a nominal rms ac voltage of either 115 or 230 V, it is possible to use a voltage-doubler circuit such as that of Fig. 5-27, which was discussed in Chapter 5.

10-8-3 BULK CAPACITOR AND THE HOLD-UP TIME

The dc link capacitor C_d, usually referred to as the bulk capacitor, reduces the voltage ripple in the input to the dc–dc converter. In addition, it also provides a hold-up time during which the regulated supply keeps on providing the regulated voltage output in the absence of the ac input voltage caused by a momentary power outage. The bulk capacitor C_d can be calculated as a function of the desired hold-up time:

$$C_d \simeq 2 \times \frac{\text{rated power output} \times \text{hold-up time}}{(V_{d,\text{nominal}}^2 - V_{d,\text{min}}^2) \times \eta} \tag{10-119}$$

where $V_{d,\text{min}}$ is chosen to be in the range of 60–75% of the nominal input voltage $V_{d,\text{nominal}}$ and η is the energy efficiency of the power supply.

It should be noted that for a given capacitance, the capacitor volume is roughly proportional to the voltage rating, and the maximum energy storage capability is proportional to the square of the voltage rating. This points out the significant advantage of switch-mode power supplies over linear power supplies, since energy storage in switch-mode supplies is at a much higher voltage compared with linear power supplies.

10-8-4 LIMITING INRUSH (SURGE) CURRENT AT INITIAL TURN-ON

When the power switch to the supply is initially turned on, the bulk capacitor C_d initially appears effectively as a short circuit across the ac source, which may result in an unacceptably large inrush current. To limit this inrush current, a series element between the dc side of the rectifier bridge and C_d can be used. This series element can be a thermister, which initially has a large resistance when it is cold, thus limiting the inrush current when the switch is on. As it heats up, its resistance goes down to a reasonably low value to yield a reasonable efficiency. However, it has a long thermal time constant, and therefore if a short-term power outage occurs that is long enough to discharge the bulk capacitor but not long enough to allow the thermistor to cool down, a large inrush current can result when the line power comes back on.

Another option is to use a current-limiting resistor, with a thyristor in parallel to make up the series element. Initially the thyristor is off and the current-limiting resistor limits the inrush current at turn-on. When the bulk capacitor voltage charges up, the thyristor is turned on, thus bypassing the current-limiting resistor. It is also possible to design the series element by a device such as a MOSFET or an IGBT. The device is slowly turned on, thus limiting the peak inrush current.

10-8-5 EQUIVALENT SERIES RESISTANCE OF OUTPUT FILTER CAPACITOR

The ESR of the output filter capacitor in Fig. 10-37 is required to be as low as possible. In high-switching-frequency applications, the ESR significantly contributes to the peak-to-peak and the rms values of the ripple in the output voltage (see Problem 10-16). The peak deviation in the output voltage from its steady-state value, following a step change in load, also depends on the capacitor ESR. For a step change of load, the output filter inductor in Fig. 10-37 acts as a source of constant current during this load transient, and the change in load current as a transient is supplied by the filter capacitor. Hence, following a load transient,

$$\Delta V_o = -\text{ESR} \times \Delta I_o \qquad (10\text{-}120)$$

10-8-6 SYNCHRONOUS RECTIFIER TO IMPROVE ENERGY EFFICIENCY

There is an increasing requirement in equipment such as computers for power supplies with voltages of even lower than 5 V, for example, in a 2–3-V range, as a consequence of increasing integration of logic gates on a single monolithic substrate. In switching power supplies with a low output voltage, the diodes in the output rectifier stage can be the biggest source of power loss. Even the commonly used Schottky diodes have a relatively large voltage drop and, hence, a large power loss in such low-output-voltage applications. As a remedy, low-voltage MOSFETs with a very low on-state resistance $r_{\text{DS(on)}}$ and low-voltage BJTs with a very low on-state voltage $V_{\text{CE(sat)}}$ can be used to replace the diodes in the output stage. These devices in this application are commonly referred to as synchronous rectifiers.

10-8-7 MULTIPLE OUTPUTS

For a multiple-output power supply, dynamic cross regulation refers to how well the power supply can regulate the voltage of its regulated output if a load change occurs on one of its unregulated outputs. For separate filter inductors used for each of the multiple outputs, the dynamic cross regulation is very poor. This is because a load change on one of the unregulated outputs takes a relatively long time for its effect to show up at the regulated output and for the controller to take the corrective action. If the output filter inductors are coupled (i.e., wound on a common core), the dynamic cross regulation is much better, since the change in the unregulated output voltage is immediately propagated to the regulated output, thus forcing the feedback controller to act.

10-8-8 EMI CONSIDERATIONS

Switching power supplies must meet the conducted and the radiated EMI specifications. These specifications and the EMI filters are discussed in Chapter 18.

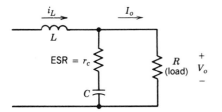

Figure 10-37 ESR in the output capacitor.

SUMMARY

In this chapter, principles behind a successful switching power supply design are discussed. Various topics unique to the design of switching power supply are covered. These topics include converter topologies, transformer core excitation, various types of controls and the compensation of the feedback loop in regulated supplies, power supply protection, providing isolation in the feedback loop, and design considerations to meet the power supply specifications.

PROBLEMS

LINEAR POWER SUPPLIES

10-1 A 12-V regulated linear power supply of the type shown in Fig. 10-1a is designed to operate with a 60-Hz ac voltage in a range of 120 V (+10%, −25%). At the maximum load, the peak-to-peak ripple in the capacitor voltage is 1.0 V. The power supply is designed such that $V_{d,\min} - V_o = 0.5$ V in Fig. 10-1b.

Calculate the loss of efficiency due to power losses in the transistor at full load, with the input voltage at its maximum. (*Hint:* Approximate the capacitor voltage waveform with straight-line segments.)

FLYBACK CONVERTER

10-2 A flyback converter is operating in a complete demagnetization mode. Derive the voltage transfer ratio V_o/V_d in terms of the load resistance R, switching frequency f_s, transformer inductance L_m, and duty ratio D.

10-3 In a regulated flyback converter with a 1 : 1 turns ratio, $V_o = 12$ V, V_d is 12–24 V, P_{load} is 6–60 W, and the switching frequency $f_s = 200$ kHz. Calculate the maximum value of the magnetizing inductance L_m that can be used if the converter is always required to operate in a complete demagnetization (equivalent to a discontinuous-conduction) mode. Assume ideal components.

10-4 A flyback converter is operating in an incomplete demagnetization mode with a duty ratio of 0.4. In the same application, another option may be to parallel two half-size flyback converters as shown in Fig. 10-9b. Compare the ripple in the input current i_d and the output stage current i_o waveforms in these two options, assuming a very large output capacitor such that $v_o(t) \simeq V_o$.

FORWARD CONVERTER

10-5 A switch-mode power supply is to be designed with the following specifications:

$$V_d = 48\ V \pm 10\ \%$$
$$V_o = 5\ V \quad \text{(regulated)}$$
$$f_s = 100\ \text{kHz}$$
$$P_{\text{load}} \text{ is 15–50 W}$$

A forward converter operating in a *continuous-conduction mode* with a demagnetizing winding ($N_3 = N_1$) is chosen. Assume all components to be ideal except for the presence of transformer magnetization inductance.

(a) Calculate N_2/N_1 if this turns ratio is desired to be as small as possible.

(b) Calculate the minimum value of the filter inductance.

10-6 A forward converter with a demagnetizing winding is designed to operate with a maximum duty ratio $D_{\max}$ of 0.7. Calculate the voltage rating of the switch in terms of the input voltage V_d.

10-7 In the circuit of Fig. 10-12b with two parallel forward converters, draw the input current i_d and i_L waveforms if each converter is operating at a duty ratio of 0.3 in a continuous-conduction mode.

Compare these two waveforms with those if a single forward converter (with twice the power rating but with the same value of the output filter inductance as in Fig. 10-12b) is used. Assume $v_o(t) \simeq V_o$.

PUSH–PULL CONVERTERS

10-8 In the push–pull converter of Fig. 10-13a, assume the losses to be zero and each switch duty ratio to be 0.25. The transformer has a finite magnetizing inductance and i_m is the magnetizing current.

(a) At a large load where $i_L(N_2/N_1) \gg i_m$, draw the i_m, i_{D1}, and i_{D2} waveforms.

(b) At essentially no load, draw the i_m waveform and show that the peak value of i_m is higher than in part (a).

CURRENT-SOURCE CONVERTERS

10-9 Derive the voltage transfer ratio given by Eq. 10-36 in the current-source converter of Fig. 10-16.

TRANSFORMER CORE

10-10 A transformer for a full-bridge converter is built with a ferrite material with properties similar to those shown in Figs. 10-17a and 10-17b. Assume $V_d = 170$ V, $f_s = 50$ kHz, and $(\Delta B)_{max} = 0.2$ Wb/m² with a switch duty ratio of 0.5. The peak magnetizing current is measured to be 1.0 A. Estimate the core losses in watts in the transformer at 25°C under the operating conditions described.

10-11 A toroidal transformer core is built with a material whose B–H loop is shown in Fig. 10-17a. An air gap is included whose length is one-hundredth of the length of the flux path in the core. Plot the B–H loop and calculate the remnant flux in the gapped core.

DIRECT DUTY RATIO CONTROL

10-12 The forward converter of Fig. 10-20a is to have a gain crossover frequency $\omega_{cross} = 10^5$ rad/s with a phase margin of 30°. Use the Bode plot of Figs. 10-21a and 10-21b for the transfer function $\bar{v}_o(s)/\bar{d}(s)$. The PWM transfer function is given by Fig. 10-24.

Calculate the values for R_2, C_1, and C_2 in the compensated error amplifier of Fig. 10-27, assuming $R_1 = 1$ kΩ.

10-13 Repeat Problem 10-12 for a flyback converter, assuming that the Bode plots in Figs. 10-22a and 10-22b are for its transfer function $\bar{v}_o(s)/\bar{d}(s)$. The crossover frequency ω_{cross} and the phase margin are required to be 5×10^3 rad/s and 30°, respectively.

CURRENT-MODE CONTROL

10-14 In a forward converter with $N_1/N_2 = 1$, the output voltage is regulated to be 6.0 V by means of a current-mode control, where the slope of the slope compensation ramp is one-half of the slope of the inductor current with the switch off.

Draw the waveforms as in Fig. 10-13 to show that the average inductor current remains the same if V_d changes from 10 to 12 V.

CAPACITOR HOLD-UP TIME

10-15 A 100-W power supply with a full-load efficiency of 85% has a hold-up time of 40 ms at full load when it is supplied with a nominal input voltage of 120 V at 60 Hz. A full-bridge rectifier is used at the input. If the power supply can operate only if the average dc voltage V_d is above 100 V, calculate the required value of the input capacitor C_d. (*Hint:* Assume that the capacitor voltage is charged approximately to the peak of the ac input voltage.)

ESR OF THE OUTPUT FILTER CAPACITOR

10-16 In the forward converter shown in Fig. 10-20a, use the numerical values given, except assume r_L to be zero. Under a steady-state operating condition, plot i_L, voltage across r_c, voltage across C, and the ripple in v_o. Compare the peak-to-peak ripple in the following three voltages: v_o, voltage across C, and voltage across r_c.

PSPICE SIMULATION

10-17 Using PSpice, simulate the forward converter shown in Fig. P10-17.

(a) Apply a step increase of 0.1 V in the nominal value of the output voltage V_o reference equal to 4 V at 150 μs. Observe the system response.

(b) Replace the power stage including the output filter by a transfer function given by the following equation:

$$T_p(s) = \frac{\tilde{v}_o(s)}{\tilde{d}(s)} = 1.6 \times 10^4 \frac{s + 5 \times 10^4}{s^2 + (0.85 \times 10^4)s + 10^8}$$

Apply the step change in V_o given in part a and compare the results.

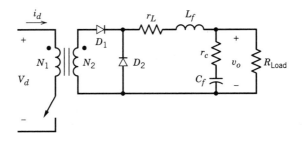

Nominal Values:

$r_c = 10$ mΩ, $C_f = 2,000$ μF, $R_{Load} = 200$ mΩ,
$V_d = 24$ V, $V_o = 4$ V, $r_L = 10$ mΩ, $L_f = 5$ μH,
$f_s = 200$ kHz, $N_1/N_2 = 3$.

$$T_c(s) = \frac{\tilde{v}_c(s)}{\tilde{v}_{o,\,ref}(s) - \tilde{v}_o(s)} = \frac{(27.5)s + 10^6}{[(6.05 \times 10^{-6})\, s^2 + 1.66\, s]}$$

$$T_m(s) = 0.34\ (-9.37\ \text{dB})$$

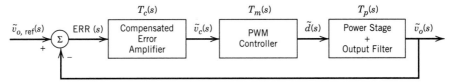

Figure P10-17 From reference 5 of Chapter 4, "Power Electronics: Computer Simulation, Analysis, and Education Using PSpice (evaluation, classroom version)," on a diskette with a manual, Minnesota Power Electronics, P.O. Box 14503, Minneapolis, MN 55414.

REFERENCES

SWITCH-MODE dc POWER SUPPLIES

1. R. P. Severns and G. E. Bloom, *Modern DC-to-DC Switch Mode Power Converter Circuits*, Van Nostrand Reinhold, New York, 1985.
2. K. Kit Sum, *Switch Mode Power Conversion—Basic Theory and Design*, Marcel Dekker, New York and Basel, 1984.
3. R. E. Tarter, *Principles of Solid State Power Conversion*, H. W. Sams Co., Indianapolis, IN, 1985.
4. G. Chryssis, *High Frequency Switching Power Supplies: Theory and Design*, McGraw-Hill, New York, 1984.

FLYBACK CONVERTERS

5. H. C. Martin, "Miniature Power Supply Topology for Low Voltage Low Ripple Requirements," U.S. Patent 4,618,919, 1986.

FORWARD CONVERTERS

6. B. Brakus, "100 Amp Switched Mode Charging Rectifier for Three-Phase Mains," Proceedings of the *IEEE / INTELEC 1984*, pp. 72–78.

PUSH–PULL CONVERTERS

7. R. Redl, M. Domb, and N. Sokal, "How to Predict and Limit Volt-Second Unbalance in Voltage-Fed Push–Pull Power Converters," *PCI Proceedings,* April 1983, pp. 314–330.

CURRENT SOURCE CONVERTERS

8. References 1 and 3.

TRANSFORMER CORE

9. Ferroxcube, "Ferroxcube Linear Ferrite Materials and Components," Ferroxcube Corporation, Saugerties, NY, 1988.

CONTROL LINEARIZATION

10. R. D. Middlebrook and S. Cúk, "A General Unified Approach to Modelling Switching—Converter Power Stages," *1976 IEEE Power Electronics Specialists Conference Record,* 1976, pp. 18–34.

11. V. Vorpérian, "Simplified Analysis of PWM Converters Using Model of PWM Switch. Part 1: Continuous Conduction Mode," *IEEE Transactions on Aerospace and Electronic Systems,* May 1990, pp. 490–496.

12. V Vorpérian, "Simplified Analysis of PWM Converters Using Model of PWM Switch. Part 2: Discontinuous Conduction Mode," *IEEE Transactions on Aerospace and Electronic Systems,* May 1990, pp. 497–505.

13. Reference 1.

14. R. D. Middlebrook, "Predicting Modulator Phase Lag in PWM Converter Feedback Loops," 8th International Solid-State Power Electronics Conference, Dallas, TX, April 27–30, 1981.

CONTROL, FEEDBACK COMPENSATION

15. K. Ogata, *Modern Control Engineering,* Prentice-Hall, Englewood Cliffs, NJ, 1970.

16. H. Dean Venable, "The *k*-Factor: A New Mathematical Tool for Stability Analysis and Synthesis," *Proceedings of Powercon 10,* San Diego, CA, March 22–24, 1983.

FEED-FORWARD CONTROL

17. Unitrode, "Switching Regulated Power Supply Design Seminar Manual," Unitrode Corporation, 1986.

CURRENT-MODE CONTROL

18. B. Holland, "Modeling, Analysis and Compensation of the Current-Mode Converter," *Proceedings of the Powercon 11,* 1984, pp. I-2-1–I-2-6.

19. R. Redl and N. Sokal, "Current-Mode Control, Five Different Types, Used with the Three Basic Classes of Power Converters," *1985 IEEE Power Electronics Specialists Conference Record,* 1985, pp. 771–785.

20. Reference 14.

DIGITAL CONTROL

21. C. P. Henze and N. Mohan, "Modeling and Implementation of a Digitally Controlled Power Converter Using Duty-Ratio Quantization," *Proceedings of ESA (European Space Agency) Sessions at the 1985 IEEE Power Electronics Specialists Conference,* 1985, pp. 245–255.

HOLD-UP TIME AND CAPACITOR ESR

22. B. Landon, "Myth—Holdup Is Free with SMPS," *Powerconversion International Magazine,* October 1981, pp. 72–80.

23. W. Chase, "Capacitors for Switching Regulator Filters," *Powerconversion International Magazine,* May 1981, pp. 57–60.

ELECTRICAL ISOLATION IN THE FEEDBACK LOOP

24. Unitrode, *Unitrode Applications Handbook 1987–1988,* Unitrode Corporation, Merrimack, NH 1987.

LIMITING INRUSH CURRENTS

25. R. Adair, "Limiting Inrush Current to a Switching Power Supply Improves Reliability, Efficiency," *Electronic Design News* (EDN), May 20, 1980.

EMI

26. D. L. Ingram, "Designing Switch-Mode Converter Systems for Compliance with FCC Proposed EMI Requirements," *Power Concepts,* 1977, pp. G1-1–G1-11.

CHAPTER 11

POWER CONDITIONERS AND UNINTERRUPTIBLE POWER SUPPLIES

11-1 INTRODUCTION

In the previous chapters, it was mentioned that power electronics converters produce EMI and inject current harmonics into the utility system. An interface between a power electronic system and the utility source that can minimize these potential problems is discussed in Chapter 18. In this chapter, the focus is on powerline disturbances and how power electronic converters can be utilized to prevent the power line disturbances from disrupting the operation of critical loads such as computers used for controlling important processes, medical equipment, and the like.

11-2 POWER LINE DISTURBANCES

Ideally, the voltage supplied by the utility system should be a perfect sine wave without any harmonics at its nominal frequency of 60 Hz and at its nominal magnitude. For a three-phase system, the voltages should form a balanced set, with each phase displaced by 120° with respect to the others.

11-2-1 TYPES OF DISTURBANCES

In practice, however, voltages can significantly depart from the ideal condition due to the power line disturbances listed below:

- **Overvoltage.** The voltage magnitude is substantially higher than its nominal value for a sustained period of a few cycles.
- **Undervoltage (brownout).** The voltage is substantially lower than its nominal value for a few cycles.
- **Outage (blackout).** The utility system voltage collapses for a few cycles or more.
- **Voltage spikes.** These are superimposed on the normal 60-Hz waveforms and occur occasionally (not on a repetitive basis). These can be either of a line-mode (differential-mode) or a common-mode type.
- **Chopped voltage waveform.** This refers to a repetitive chopping of the voltage waveform and the associated ringing, as shown in Fig. 11-1a.

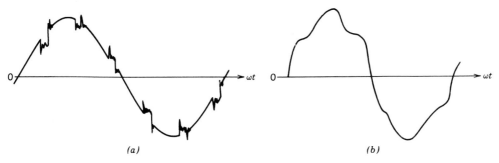

Figure 11-1   Possible distortions in input voltage: *(a)* chopped voltage waveform; *(b)* distorted voltage waveform due to harmonics.

- **Harmonics.** A distorted voltage waveform, as shown in Fig. 11-1*b*, contains harmonic voltage components at harmonic frequencies (usually low-order multiples of the line frequency). These harmonics exist on a sustained basis.
- **Electromagnetic interference.** This refers to high-frequency noise, which may be conducted on the power line or radiated from its source.

11-2-2   SOURCES OF DISTURBANCES

Sources that produce these disturbances are very diverse. Overvoltages may be caused by sudden decreases in the system load, thus causing the utility voltage to go up. Undervoltages may be caused by overload conditions, by start of induction motors, or for many other reasons. Occasional large voltage spikes may be a result of switching in or out of power factor correction capacitors, power lines, or even such things as pump/compressor motors in the vicinity. Chopping of the voltage waveform may be caused by ac-to-dc line-frequency thyristor converters of the type discussed in Chapter 6, if such converters are used to interface the power electronic equipment with the utility system. These converters produce a short circuit on the ac voltage source through the ac system impedance on a repetitive basis. The voltage harmonics may be caused by a variety of sources. These include magnetic saturation of power system transformers as well as the harmonic currents injected by power electronic loads. These harmonic currents flowing through the ac system impedances result in harmonic voltages. Electromagnetic interference is produced by most power electronics equipment due to rapid switching of voltages and currents, as will be discussed in Chapter 18.

11-2-3   EFFECT ON SENSITIVE EQUIPMENT

The effect of such power line disturbances on the sensitive equipment depends on the following factors: (1) type and magnitude of the power line disturbance, (2) type of equipment and how well it is designed, and (3) if any power conditioning equipment is used. Sustained overvoltages and undervoltages may cause the equipment to trip out, which is highly undesirable in certain applications. Large voltage spikes may cause a hardware failure in the equipment. Manufacturers of critical equipment often provide a certain degree of protection by incorporating surge arrestors such as metal–oxide varistors (MOVs) at the input to guard against such failures. However, spikes of very large magnitude in combination with a higher frequency of occurrence can still result in a hardware failure. Chopped voltage waveforms and voltage harmonics have the potential of interfering with the equipment if it is not designed to be immune from such effects. Power conditioners consisting of filters and an isolation transformer can correct such problems.

The effect of power system outage depends on the duration of the outage and the equipment design. As an example, a personal computer power supply may be designed such that for an outage of less than 100 ms, the power supply outputs to the digital circuitry ride through and hold their nominal values, and no effect is felt on the computer operation. For an outage of a longer duration, after a 100-ms interval a logic signal within the computer allows the computer central processing unit (CPU) an additional 50 ms to back up the existing information, beyond which time all power supply output voltages decrease rapidly. Figure 11-2 shows the tolerance of large mainframe computers to power line disturbances, beyond which a backup procedure may be initiated and a shutdown occurs for some time. Typical power quality specified by major computer manufacturers is listed in Table 11-1. In case of critical applications where such a shutdown is unacceptable, the backup to the utility grid is provided by means of uninterruptible power supplies (UPSs). Both the power conditioners and the UPSs are discussed in the following sections.

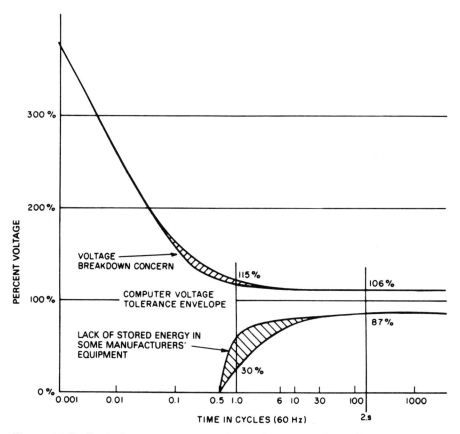

Figure 11-2 Typical computer system voltage tolerance envelope. (*Source:* IEEE Std. 446, "Recommended Practice for Emergency and Standby Power Systems for Industrial and Commercial Applications.")

Table 11-1 Typical Range of Input Power Quality and Load Parameters of Major Computer Manufacturers

Parameters[a]	Range or Maximum
1. Voltage regulation, steady state	+5, −10 to +10%, −15% (ANSI C84.1—1970 is +6, −13%)
2. Voltage disturbances	
a. Momentary undervoltage	−25 to −30% for less than 0.5 s, with −100% acceptable for 4−20 ms
b. Transient overvoltage	+150 to 200% for less than 0.2 ms
3. Voltage harmonic distortion[b]	3−5% (with linear load)
4. Noise	No standard
5. Frequency variation	60 Hz ± 0.5 Hz to ±1 Hz
6. Frequency rate of change	1 Hz/s (slew rate)
7. 3ɸ, Phase voltage unbalance[c]	2.5−5%
8. 3ɸ, Load unbalance[d]	5−20% maximum for any one phase
9. Power factor	0.8−0.9
10. Load demand	0.75−0.85 (of connected load)

[a]Parameters 1, 2, 5, and 6 depend on the power source, while parameters 3, 4, and 7 are the product of an interaction of source and load, and parameters 8, 9, and 10 depend on the computer load alone.

[b]Computed as the sum of all harmonic voltages added vectorially.

[c]Computed as follows:

$$Percent \text{ phase voltage unbalance} = \frac{3(V_{max} - V_{min})}{V_a + V_b + V_c} \times 100$$

[d]Computed as difference from average single-phase load.

Source: IEEE Std. 446, ''Recommended Practice for Emergency and Standby Power Systems for Industrial and Commercial Applications.''

11-3 POWER CONDITIONERS

Power conditioners provide an effective way of suppressing some or all of the electrical disturbances other than the power outages and frequency deviations from 60 Hz (frequency deviation is not a problem in an interconnected ac power system). Some of these power conditioners are listed:

- Metal−oxide varistors provide protection against line-mode voltage spikes.
- Electromagnetic interference filters help to prevent the effect of the chopped waveform on the equipment as well as to prevent the equipment from conducting high-frequency noise into the utility grid.
- Isolation transformers with electrostatic shields not only provide galvanic isolation but also filter the line-mode and the common-mode voltage spikes.
- Ferroresonant transformers provide voltage regulation as well as filtering of the line-mode spikes. They are also partially effective in filtering the common-mode noise.
- Linear conditioners are used in many sensitive applications to supply clean power.

Since none of these power conditioners employs switch-mode or resonant-mode power electronics, they are not discussed here any further.

For voltage regulation, an electronic tap changing scheme using triacs as shown in Fig. 11-3 can be used, where triacs or back-to-back connected thyristors replace a mechanical contact and allow a bidirectional current flow.

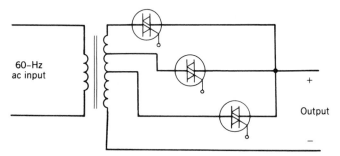

Figure 11-3 Electronic tap changer.

11-4 UNINTERRUPTIBLE POWER SUPPLIES (UPSs)

For supplying very critical loads such as computers used for controlling important processes, some medical equipment, and the like, it may be necessary to use UPSs. These provide protection against power outages as well as voltage regulation during power line overvoltage and undervoltage conditions. They are also excellent in terms of suppressing incoming line transient and harmonic disturbances.

Uninterruptible power supplies in their block diagram form are shown in Fig. 11-4. A rectifier is used for converting single-phase or three-phase ac input into dc, which supplies power to the inverter as well as to the battery bank to keep it charged.

In the normal mode of operation, the power to the inverter is provided by the rectifier. In case of a line outage, power comes from the battery bank. The inverter produces either a single-phase or a three-phase sinusoidal waveform depending on the UPS. The output voltage of the inverter is filtered, prior to being applied to the load.

11-4-1 RECTIFIER

For supplying power to the inverter and for keeping the battery bank charged, two rectifier arrangements are shown in Fig. 11-5. In a conventional arrangement shown in Fig. 11-5a, a phase-controlled rectifier as in Chapter 6 is used. It is also possible to use a diode rectifier bridge in cascade with a step-down dc–dc converter as in Chapter 7, as shown in Fig. 11-5b.

When an electrical isolation from the mains is required, it is possible to use a dc–dc converter with a high-frequency isolation transformer, as shown in Fig. 11-6. The dc–dc converter with electrical isolation may be similar to the ones used in the switch-mode dc power supplies of Chapter 10 or may utilize resonant converter concepts discussed in Chapter 9.

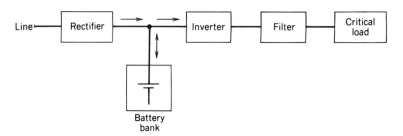

Figure 11-4 A UPS block diagram.

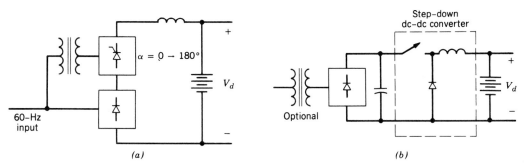

Figure 11-5 Possible rectifier arrangements.

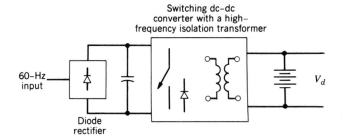

Figure 11-6 Rectifier consisting of a high-frequency isolation transformer.

Another rectifier arrangement is shown in Fig. 11-7, where the bulk of the power (supplied to the inverter) flows through the diode bridge and only the power required for charging of the battery bank flows through a single-phase phase-controlled thyristor rectifier. The voltage V_{charge} can be controlled in magnitude and polarity for proper charging of the battery bank. Thyristor T_1 normally remains off; it is turned on in the event of a power outage.

11-4-2 BATTERIES

There are many different types of battery systems. Of these, the conventional lead–acid batteries are commonly used for the UPS applications.

In the normal mode when the line voltage is present, the battery is trickle charged to offset the slight self-discharge by the battery. This requires that a constant trickle charge

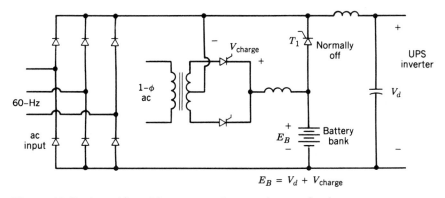

Figure 11-7 A rectifier with a separate battery charger circuit.

voltage be applied across the battery, and the battery continuously draws a small amount of current, thus maintaining itself in a fully charged state.

In the event of a line outage, the battery supplies the load. The capacity of a battery is expressed in ampere-hours, which is the product of a constant discharge current and the duration beyond which the battery voltage falls below a voltage level called the final discharge voltage. The battery voltage should not be allowed to fall below the final discharge voltage level; otherwise the battery life is shortened. Typically, a 10-h current is defined as the current in amperes that causes the fully charged battery to discharge in 10 hours to its final voltage level. Discharge currents in excess of the 10-h current cause the final discharge voltage to be reached sooner than their magnitudes would suggest. Therefore, the higher discharge currents reduce the effective battery capacity.

Once the line voltage is restored, the battery bank in a UPS is brought back to its fully charged state. This is accomplished by initially charging the battery at a constant charging current rate, as shown in Fig. 11-8. This causes the battery terminal voltage to increase to its trickle charge voltage level. Once the trickle charge voltage level is reached, the voltage applied is kept constant, as shown in Fig. 11-8, and the charging current finally decreases to the trickle charge current and stays at that level. It is possible to program the battery-charging characteristic to bring it to a full-charge state more quickly.

11-4-3 INVERTERS

The filtered output of the inverter is normally specified to contain very little harmonic distortion, even though most loads are highly nonlinear and, hence, inject larger harmonic currents into the UPS. Therefore, the inverter must allow almost instantaneous control over its output ac waveform. The output voltage harmonic content is specified by means of a term called total harmonic distortion (THD), which was defined in Chapter 3 as

$$\%\text{THD} = 10 \times \frac{\left(\sum_{h=2}^{\infty} V_h^2\right)^{1/2}}{V_1} \tag{11-1}$$

where V_1 is the fundamental-frequency rms value of the output voltage and V_h is the rms magnitude at the harmonic of order h. Typically, THD is specified to be less than 5%; each harmonic voltage as a ratio of V_1 is specified to be less than 3%.

Modern UPSs normally use the PWM dc-to-ac inverters of Chapter 8, with either a single-phase or three-phase ac output. A schematic is shown in Fig. 11-9a. An isolation

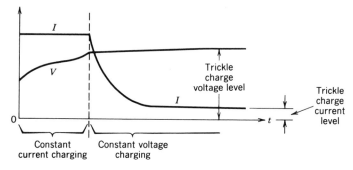

Figure 11-8 Charging of a battery after a line outage causes battery discharge.

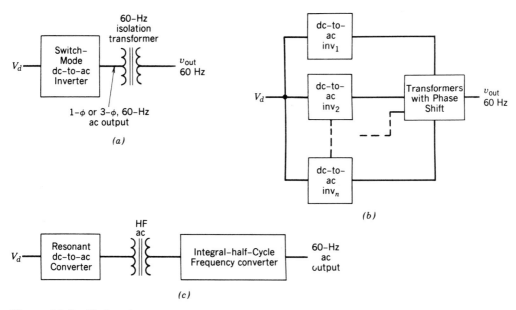

(a)

(b)

(c)

Figure 11-9 Various inverter arrangements.

transformer is generally used at the output. Large UPSs may employ a scheme where the outputs of two or more such inverters are paralleled through transformers with phase shift, as shown schematically in Fig. 11-9b. This allows the inverters to operate at a relatively lower switching frequency, utilizing either a low-frequency PWM, selective harmonic cancellation, or a square-wave switching scheme. As shown schematically in Fig. 11-9c, it is also possible to use resonant converters, high-frequency isolation transformers, and the integral-half-cycle frequency converter concepts discussed in Chapter 9.

It is important to minimize the harmonics content of the inverter output. This decreases the filter size, which not only results in cost savings but also results in an improved dynamic response of the UPS as the load changes. A feedback control is shown in Fig. 11-10, where the actual output waveform is compared with the sinusoidal reference. The error is used to modify the inverter switching. A control loop with a fast response is needed for a good dynamic performance.

Above a few kilowatts, most UPSs provide power to several loads connected in parallel. As shown in Fig. 11-11, each load is supplied through a fuse. In the event of a

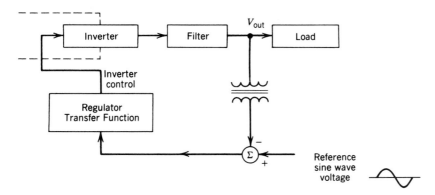

Figure 11-10 Uninterruptible power supply control.

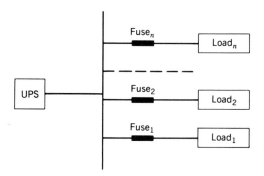

Figure 11-11 A UPS supplying several loads.

short circuit in one of the loads, it is important for the UPS to blow that particular fuse and to keep on supplying the rest of the loads. Therefore, the current rating of the UPS under a sustained short-circuit condition should be sufficient to blow the fuse of the faulted load. In this respect, a rotating-type UPS with a large short-circuit current capacity is far superior to the power electronics type UPS.

An alternative scheme, where the functions of battery charging and the inverter are combined, is shown in Fig. 11-12. In the normal mode, the switching converter operates as a rectifier, charging the battery bank. In addition, it can draw inductive or capacitive currents from the mains, thus providing a fine regulation of the voltage supplied to the load. In case of a utility outage, the utility is isolated and the switching converter operates as an inverter, supplying power to the load from the battery bank. This arrangement is usually referred to as the "standby power supply," as we discuss in the next section.

11-4-4 STATIC TRANSFER SWITCH

For additional reliability, the power line itself is used as a backup to the UPS, and a static transfer switch transfers the load from the UPS to the power line, as shown in Fig. 11-13 by means of a block diagram.

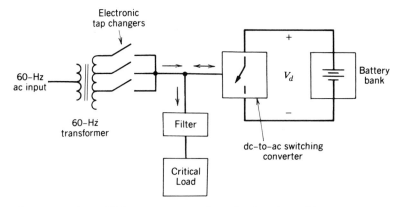

Figure 11-12 A UPS arrangement where the functions of battery charging and inverter are combined.

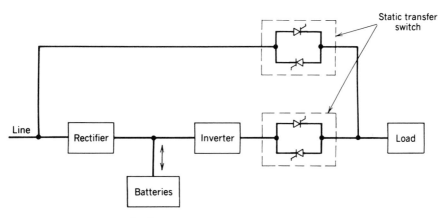

Figure 11-13 Line as backup.

As an alternative, in the normal mode the load is supplied by the power line in Fig. 11-13. In the event of a line outage, the static transfer switch transfers the load to the UPS. This arrangement is usually referred to as the *standby power supply*. When a static transfer switch is used, the inverter output should be synchronized to the line voltage. Therefore, transferring the load from one source to the other results in the least amount of disturbance seen by the load.

SUMMARY

There are many types of disturbances associated with the power line input. Power conditioners provide an effective way to protect sensitive electronic loads from these disturbances except for the power outages and frequency deviations.

For very critical loads, UPSs are used so that the power flow to the load is uninterrupted even in the event of a power outage. The storage capacity of the battery bank is sized based on the likelihood of an outage of a specified duration.

PROBLEMS

11-1 A UPS with transformer-coupled inverters is shown in Fig. 11-1. A programmed harmonic elimination switching scheme is used to eliminate the eleventh and the thirteenth harmonics, which also provides a control over the magnitude of the fundamental.

Show that third, fifth, and seventh harmonics are neutralized by the transformers, assuming that the voltage waveforms of inverter 2 lag that of inverter 1 by 30°.

11-2 What is the minimum switching frequency in Problem 11-1?

11-3 The UPS arrangement shown in Fig. 11-12 consists of taps to yield 95, 100, and 105% of the input voltage at no load. The transformer is rated at 120 V, 60 Hz, and 1 kVA. It has a leakage reactance (resistance can be neglected) of 6%.

Calculate the reactive power that the switch-mode converter must draw to bring the load voltage to 100% of the nominal value of 120 V if the utility voltage is 128 V. Assume that the critical load draws a sinusoidal current at a unity power factor.

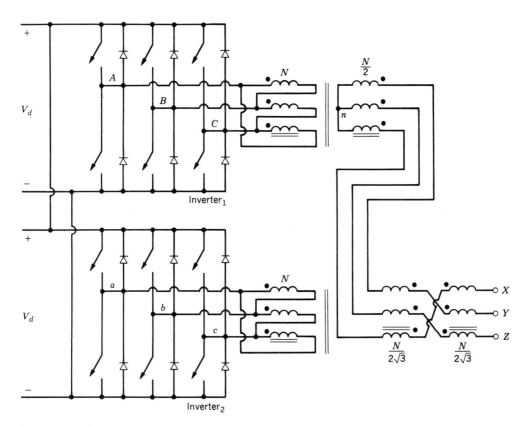

Figure P11-1

REFERENCES

1. T. S. Key, "Diagnosing Power Quality-related Computer Problems," *IEEE Transactions on Industry Applications,* Vol. IA-15, No. 4, July/August 1979, pp. 381–393.
2. "IEEE Recommended Practice for Emergency and Standby Power Systems for Industrial and Commercial Applications," ANSI/IEEE Std. 446, 1987.
3. K. Thorborg, *Power Electronics,* Prentice-Hall International (UK) Ltd., London, UK, 1988.
4. H. Gumhalter, *Power Supply Systems in Communications Engineering—Part I Principles,* Wiley, New York, 1984.
5. T. Kawabata, S. Doi, T. Morikawa, T. Nakamura, and M. Shigenobu, "Large Capacity Parallel Redundant Transistor UPS," *1983 IPEC-Tokyo Conference Record,* Vol. 1, 1983, pp. 660–671.
6. A. Skjellnes, "A UPS with Inverter Specially Designed for Nonlinear Loads," *IEEE/ INTELEC Conference Records, 1987.*
7. S. Manias, P. D. Ziogas, and G. Olivier, "Bilateral DC to AC Converter Employing a High Frequency Link," *1985 IEEE/IAS Conference Records,* 1985, pp. 1156–1162.

PART 4

MOTOR DRIVE APPLICATIONS

CHAPTER 12

INTRODUCTION TO MOTOR DRIVES

12-1 INTRODUCTION

Motor drives are used in a very wide power range, from a few watts to many thousands of kilowatts, in applications ranging from very precise, high-performance position-controlled drives in robotics to variable-speed drives for adjusting flow rates in pumps. In all drives where the speed and position are controlled, a power electronic converter is needed as an interface between the input power and the motor.

Above a few hundred watt power level, there are basically the following three types of motor drives, which are discussed in Chapters 13 through 15: (1) dc motor drives, (2) induction motor drives, and (3) synchronous motor drives.

A general block diagram for the control of motor drives is shown in Fig. 12-1. The process determines the requirements on the motor drive; for example, a servo-quality drive (called the servo drive) is needed in robotics, whereas only an adjustable-speed drive may be required in an air conditioning system, as explained further.

In servo applications of motor drives, the response time and the accuracy with which the motor follows the speed and position commands are extremely important. These servo systems, using one of these motor drives, require speed or position feedback for a precise control, as shown in Fig. 12-2. In addition, if an ac motor drive is used, the controller must incorporate sophistication, such as field-oriented control, to make the ac motor (through the power electronic converter) meet the servo drive requirements.

However, in a large number of applications, the accuracy and the response time of the motor to follow the speed command is not critical. As shown in Fig. 12-1, there is a

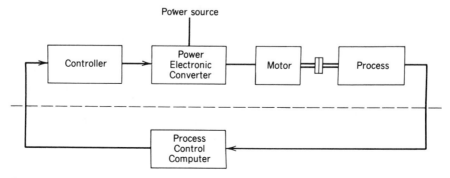

Figure 12-1 Control of motor drives.

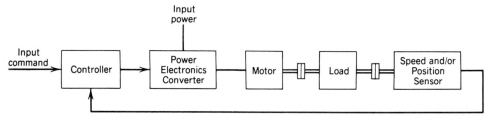

Figure 12-2   Servo drives.

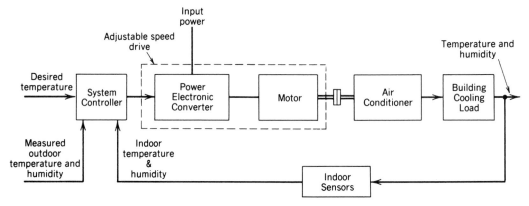

Figure 12-3   Adjustable-speed drive in an air conditioning system.

feedback loop to control the process, outside of the motor drive. Because of the large time constants associated with the process-control feedback loop, the motor drive's accuracy and the time of response to speed commands are not critical. An example of such an adjustable-speed drive is shown in Fig. 12-3 for an air conditioning system.

12-2   CRITERIA FOR SELECTING DRIVE COMPONENTS

As shown in Figs. 12-1 through 12-3, a motor drive consists of an electric motor, a power electronic converter, and possibly a speed and/or position sensor. In this section, criteria for optimum match between the mechanical load and the drive components are discussed in general terms.

12-2-1   MATCH BETWEEN THE MOTOR AND THE LOAD

Prior to selecting the drive components, the load parameters and requirements such as the load inertia, maximum speed, speed range, and direction of motion must be available. The motion profile as a function of time, for example as shown in Fig. 12-4a, must also be specified. By means of modeling the mechanical system, it is possible to obtain a load–torque profile. Assuming a primarily inertial load with a negligible damping, the torque profile, corresponding to the speed profile in Fig. 12-4a, is shown in Fig. 12-4b. The torque required by the load peaks during the acceleration and deceleration.

One way to drive a rotating load is to couple it directly to the motor. In such a direct coupling, the problems and the losses associated with a gearing mechanism are avoided. But the motor must be able to provide peak torques at specified speeds. The other option for a rotating load is to use a gearing mechanism. A coupling mechanism such as rack-

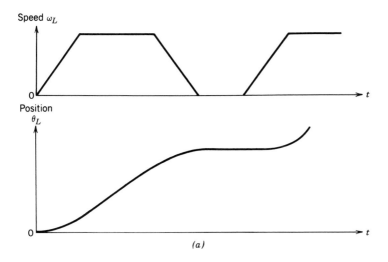

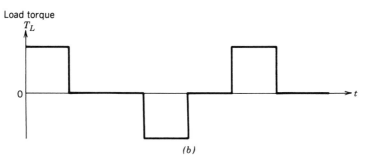

(a)

(b)

Figure 12-4   Load profile: (a) load–motion profile; (b) load–torque profile (assuming a purely inertial load).

and-pinion, belt-and-pulley, or feed-screw must be used to couple a load with a linear motion to a rotating motor. A gear and a feed-screw drive are shown in Figs. 12-5a and 12-5b, respectively. Assuming the energy efficiency of the gear in Fig. 12-5a to be 100%, the torques on the two sides of the gear are related as

$$\frac{T_m}{T_L} = \frac{\omega_L}{\omega_m} = \frac{\theta_L}{\theta_m} = \frac{n_m}{n_L} = a \qquad (12\text{-}1)$$

where the angular speed $\omega = \dot{\theta}$, n_m and n_L are the number of teeth, and a is the coupling ratio.

In the feed-screw drive of Fig. 12-5b, the torque and the force are related as

$$\frac{T_m}{F_L} = \frac{v_L}{\omega_m} = \frac{x_L}{\theta_m} = \frac{s}{2\pi} = a \qquad (12\text{-}2)$$

where the linear velocity $v_L = \dot{x}_L$, s is the pitch of the feed screw in m/turn, and a is the coupling ratio.

The electromagnetic torque T_{em} required from the motor can be calculated on the basis of energy considerations in terms of the inertias, required load acceleration, coupling ratio a, and the working torque or force. In Fig. 12-5a, T_{WL} is the working torque of the load and $\dot{\omega}_L$ is the load acceleration. Therefore,

$$T_{em} = \frac{\dot{\omega}_L}{a}\left[J_m + a^2 J_L\right] + aT_{WL} + \frac{\omega_L}{a}(B_m + a^2 B_L) \qquad (12\text{-}3a)$$

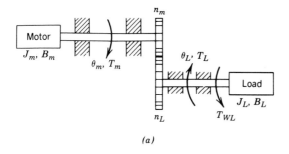

(a)

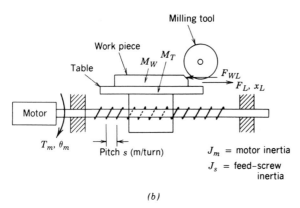

(b)

Figure 12-5 Coupling mechanisms: (a) gear; (b) feed screw.

This equation can be written in terms of the motor speed (recognizing that $\omega_m = \omega_L/a$), the equivalent total inertia $J_{eq} = J_m + a^2 J_L$, the equivalent total damping $B_{eq} = B_m + a^2 B_L$, and the equivalent working torque of the load $T_{Weq} = a T_{WL}$:

$$T_{em} = J_{eq}\dot{\omega}_m + B_{eq}\omega_m + T_{Weq} \qquad (12\text{-}3b)$$

Similarly, for the feed-screw system in Fig. 12-5b with F_{WL} as the working or the machining force and a as the coupling ratio calculated in Eq. 12-2 in terms of pitch s, T_{em} can be calculated as (see Problem 12-3)

$$T_{em} = \frac{\dot{v}_L}{a}\left[J_m + J_s + a^2(M_T + M_W)\right] + aF_{WL} \qquad (12\text{-}4)$$

where $\dot{v}_L$ is the linear acceleration of the load.

As indicated by Eqs. 12-1 and 12-2, the choice of the coupling ratio a affects the motor speed. At the same time, the value of a affects the peak electromagnetic torque T_{em} required from the motor, as is indicated by Eqs. 12-3a and 12-4. In selecting the optimum value of the coupling ratio a, the cost and losses associated with the coupling mechanism must also be included.

12-2-2 THERMAL CONSIDERATIONS IN SELECTING THE MOTOR

In the previous section, the match between the load and the motor is discussed that establishes the peak torque and the maximum speed required from the motor. This matching also establishes the motor–torque profile, which, for example, has the same form (but different magnitudes) as the load–torque profile of Fig. 12-4b.

As another example, the electromagnetic torque required from the motor as a function of time is obtained as shown in Fig. 12-6a. In electric machines, the electromagnetic torque produced by the motor is proportional to the motor current i, provided the flux in the air gap of the motor is kept constant. Therefore, the motor–current profile is identical to the motor–torque profile, as shown in Fig. 12-6b. The motor current in Fig. 12-6b during various time intervals is a dc current for a dc motor. For an ac motor, the motor current shown is approximately the rms value of the ac current drawn during various time intervals. The power loss P_R in the winding resistance R_M due to the motor current is a large part of the total motor losses, which get converted into heat. This resistive loss is proportional to the square of the motor current and, hence, proportional to T_{em}^2 during various time intervals in Figs. 12-6a and 12-6b. If the time period t_{period} in Fig. 12-6, with which the waveforms repeat, is short compared with the motor thermal time constant, then the motor heating and the maximum temperature rise can be calculated based on the resistive power loss P_R averaged over the time period t_{period}. Therefore, in Fig. 12-6, the rms value of the current over the period of repetition can be obtained as

$$P_R = R_M I_{rms}^2 \tag{12-5}$$

where

$$I_{rms}^2 = \frac{\sum_{k=1}^{m} I_k^2 t_k}{t_{period}} \tag{12-6}$$

and $m = 6$ in this example.

Because of the motor current being linearly proportional to the motor torque, the rms value of the motor torque over t_{period} from Fig. 12-6 and Eq. 12-6 is

$$T_{em,\ rms}^2 = k_1 \frac{\sum_{k=1}^{m} I_k^2 t_k}{t_{period}} \tag{12-7}$$

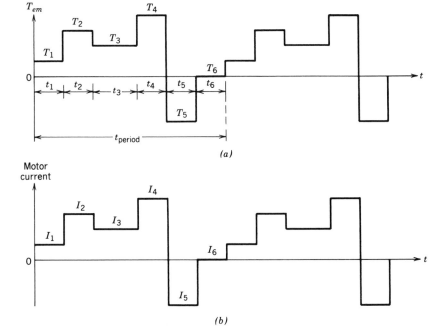

Figure 12-6 Motor torque and current.

and therefore,

$$T^2_{\text{em, rms}} = k_1 I^2_{\text{rms}} \tag{12-8}$$

where k_1 is a constant of proportionality.

From Eqs. 12-5 and 12-8, the average resistive power loss P_R is given as

$$P_R = k_2 T^2_{\text{em, rms}} \tag{12-9}$$

where k_2 is a constant of proportionality.

In addition to P_R, there are other losses within the motor that contribute to its heating. These are P_{FW} due to friction and windage, P_{EH} due to eddy currents and hysteresis within the motor laminations, and P_s due to switching frequency ripple in the motor current, since it is supplied by a switching power electronic converter rather than an ideal source. There are always some power losses called stray power losses P_{stray} that are not included with the foregoing losses. Therefore, the total power loss within the motor is

$$P_{\text{loss}} = P_R + P_{\text{FW}} + P_{\text{EH}} + P_s + P_{\text{stray}} \tag{12-10}$$

Under a steady-state condition, the motor temperature rise $\Delta\Theta$ in degrees centigrade is given as

$$\Delta\Theta = P_{\text{loss}} R_{\text{TH}} \tag{12-11}$$

where P_{loss} is in watts and the thermal resistance R_{TH} of the motor is in degrees centigrade per watt.

For a maximum allowable temperature rise $\Delta\Theta$, the maximum permissible value of P_{loss} in steady state depends on the thermal resistance R_{TH} in Eq. 12-11. In general, the loss components other than P_R in the right side of Eq. 12-10 increase with the motor speed. Therefore, the maximum allowable P_R and, hence, the maximum continuous motor torque output from Eq. 12-9 would decrease at higher speed, if R_{TH} remains constant. However, in self-cooled motors with the fan connected to the motor shaft, for example, R_{TH} decreases at higher speeds due to increased air circulation at higher motor speeds. Therefore, the maximum safe operating area in terms of the maximum rms torque available from a motor at various speeds depends on the motor design and is specified in the motor data sheets (specially in case of servo motors). For a motor torque profile like that shown in Fig. 12-6a, the motor should be chosen such that the rms value of the torque required from the motor remains within the motor's safe operating area in the speed range of operation.

12-2-3   MATCH BETWEEN THE MOTOR AND THE POWER ELECTRONIC CONVERTER

A match between the load and the various characteristics of the motor, such as its inertia and the peak and the rms torque capability, have been discussed in the previous two sections. Depending on the power rating, speed of operation, operating environment, reliability, various other performance requirements by the load, and the cost of the overall drive, one of the following three types of motor drive is selected: dc motor drive, induction motor drive, and synchronous motor drive. The advantages and the disadvantages associated with each of these motor drives are discussed in Chapters 13 through 15.

The power electronic converter topology and its control depend on the type of motor drive selected. In general, the power electronic converter provides a controlled voltage to the motor in order to control the motor current and, hence, the electromagnetic torque produced by the motor. Some of the considerations in matching the power electronic converter to the motor are discussed in the following sections.

12-2-3-1 Current Rating

As we discussed previously, the rms value of the torque that a motor can supply depends on its thermal characteristics. However, a motor can supply substantially larger peak torques (as much as four times the continuous maximum torque) provided that the duration of the peak torque is small compared with the thermal time constant of the motor. Since T_{em} is proportional to i, a peak torque requires a corresponding peak current from the power electronic converter. The current capability of the power semiconductor devices used in the converter is limited by the maximum junction temperature within the devices and other considerations. A higher current results in a higher junction temperature due to power losses within the power semiconductor device. The thermal time constants associated with the power semiconductor devices are in general much smaller than the thermal time constants of various motors. Therefore, the current rating of the power electronic converter must be selected based on both the rms and the peak values of the torque that the motor is required to supply.

12-2-3-2 Voltage Rating

In both dc and ac motors, the motor produces a counter-emf e that opposes the voltage v applied to it, as shown by a simplified generic circuit of Fig. 12-7. The rate at which the motor current and, hence, the torque can be controlled is given by

$$\frac{di}{dt} = \frac{v-e}{L} \qquad (12\text{-}12)$$

where L is the inductance presented by the motor to the converter.

To be able to quickly control the motor current and, hence, its torque, the output voltage v of the power electronic converter must be reasonably greater than the counter-emf e. The magnitude of e in a motor increases linearly with the motor speed, with a constant flux in the air gap of the motor. Therefore, the voltage rating of the power electronic converter depends on the maximum motor speed with a constant air-gap flux.

12-2-3-3 Switching Frequency and the Motor Inductance

In a servo drive, the motor current should be able to respond quickly to the load demand, thus requiring L to be small in Eq. 12-12. Also, the steady-state ripple in the motor current should be as small as possible to minimize the motor loss P_s in Eq. 12-10 and the ripple in the motor torque. A small current ripple requires the motor inductance L in Eq. 12-12 to be large. Because of the conflicting requirements on the value of L, the ripple in the motor current can be reduced by increasing the converter switching frequency. However, the switching losses in the power electronic converter increase linearly with the switching frequency. Therefore, a reasonable compromise must be made in selecting the motor inductance L and the switching frequency.

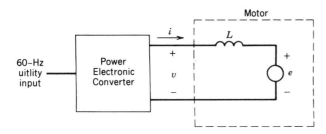

Figure 12-7 Simplified circuit of a motor drive.

12-2-4 SELECTION OF SPEED AND POSITION SENSORS

In selecting the speed and position sensors, the following items must be considered: direct or indirect coupling, sensor inertia, possibility and avoidance of torsional resonance, and the maximum sensor speed.

To control the instantaneous speed within a specified range, the ripple in the speed sensor should be small. This can be understood in terms of incremental position encoders, which are often used for measuring speed as well as position. If such a sensor is used at very low speeds, the number of pulse outputs per revolution must be large to provide instantaneous speed measurement with sufficient accuracy. Similarly, an accurate position information will require an incremental position encoder with a large number of pulse outputs per revolution.

12-2-5 SERVO DRIVE CONTROL AND CURRENT LIMITING

A block diagram of a servo drive was shown in Fig. 12-2. In most practical applications, a very fast response to a sudden change in position or speed command would require a large peak torque, which would result in a large peak current. This may be prohibitive in terms of the cost of the converter. Therefore, the converter current (same as the motor current) is limited by the controller. Figures 12-8a and 12-8b show two ways of implementing the current limit.

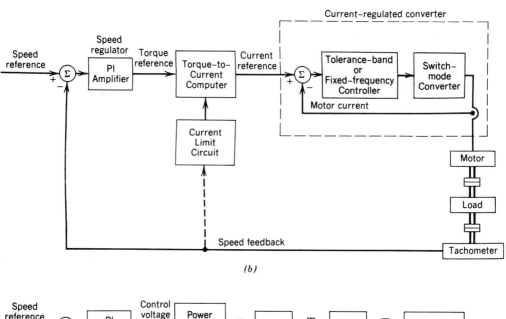

(b)

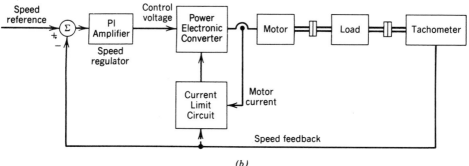

(b)

Figure 12-8   Control of servo drives: (a) inner current loop; (b) no inner current loop.

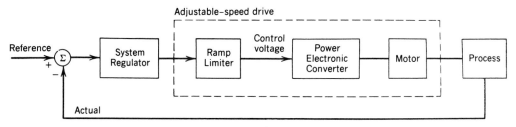

Figure 12-9 Ramp limiter to limit motor current.

In Fig. 12-8a, an inner current loop is used where the actual current is measured, and the error between the reference and the actual current controls the converter output current by means of a current-regulated modulation similar to that discussed in Section 8-6-3. Here, the power electronic converter operates as a current-regulated voltage source converter. An inner control loop improves the response time of the drive. As shown, the limit on the reference current may be dependent on the speed.

In the other control scheme shown in Fig. 12-8b, the error between the speed reference and the actual speed controls the converter through a proportional-integral (PI) amplifier. The output of the PI amplifier, which controls the converter, is suppressed only if the converter current exceeds the current limit. The current limit can be made to be speed dependent.

In a position control system, the speed reference signal in Figs. 12-8a and 12-8b is obtained from the position regulator. The input to such a position regulator will be the error between the reference position and the actual position.

12-2-6 CURRENT LIMITING IN ADJUSTABLE-SPEED DRIVES

In adjustable-speed drives such as that shown in Fig. 12-3, the current is kept from exceeding its limit by means of limiting the rate of change of control voltage with time in the block diagram of Fig. 12-9.

SUMMARY

1. Primary types of motor drives are dc motor drives, induction motor drives and synchronous motor drives.

2. Most of the applications of motor drives belong to one of the two categories: servo drives or adjustable-speed drives. In servo drive applications, the response time and the accuracy with which the motor follows the speed and/or position commands are extremely important. In adjustable-speed drive applications, response time to changes in speed command is not as critical; in fact, in many applications, it is not necessary to control the speed accurately where the process feedback loop has large time constants, relative to the response time of the drive.

3. A modeling of the mechanical system is necessary to determine the dynamics of the overall system and to select the motor and the power electronic converter of the appropriate ratings.

4. Servo drives require speed and/or position sensors to close the feedback loop. It is possible to operate with or without an inner current feedback loop. In an adjustable-speed drive, the current is kept within its limit by limiting the rate of change of control voltage to the power electronic converter with time.

PROBLEMS

12-1   In the system shown in Fig. 12-5a, the gear ratio $n_L/n_m = 2$, $J_L = 10$ kg-m^2, and $J_m = 2.5$ kg-m^2. Damping can be neglected. For the load–speed profile in Fig. P12-1, draw the torque profile and the rms value of the electromagnetic torque required from the motor.

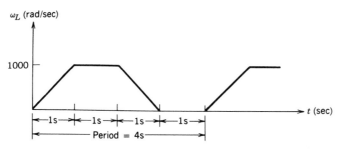

Figure P12-1

12-2   Consider the belt-and-pulley system shown in Fig. P12-2:

$$J_m = \text{motor inertia}$$
$$M = \text{mass of load}$$
$$r = \text{pulley radius}$$

Other inertias are negligible.

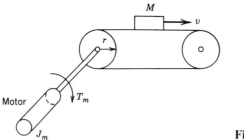

Figure P12-2

Calculate the torque T_{em} required from the motor to accelerate a load of 0.5 kg from rest to a velocity 1 m/s in a time of 3 s. Assume the motor torque to be constant during this interval, the pulley radius $r = 0.1$ m and the motor inertia $J_m = 0.006$ kg-m^2.

12-3   Derive Eq. 12-4.

12-4   In the system of Fig. 12-5a, assume a triangular velocity profile with equal acceleration and deceleration rates. The system is purely inertial and B_m, B_L, and T_{WL} can be neglected.

Assuming a gear efficiency of 100% and an optimum gear ratio (such that the reflected load inertia equals the motor inertia), calculate the time needed to rotate the load by an angle θ_L in terms of J_m, J_L, and the peak torque T_{em} that the motor must be capable of developing.

REFERENCES

1.   H. Gross (Ed.), *Electrical Feed Drives for Machine Tools,* Siemens and Wiley, New York, 1983.

2.   *DC Motors Speed Controls ServoSystem—An Engineering Handbook,* 5th ed., Electro-Craft Corporation, Hopkins, MN, 1980.

3.   A. E. Fitzgerald, C. Kingsley, Jr., and S. D. Umans, *Electric Machinery,* 4th ed. McGraw-Hill, New York, 1983.

4.   G. R. Slemon and A. Straughen, *Electric Machines,* Addison-Wesley, Reading, MA, 1980.

CHAPTER 13

dc MOTOR DRIVES

13-1 INTRODUCTION

Traditionally, dc motor drives have been used for speed and position control applications. In the past few years, the use of ac motor servo drives in these applications is increasing. In spite of that, in applications where an extremely low maintenance is not required, dc drives continue to be used because of their low initial cost and excellent drive performance.

13-2 EQUIVALENT CIRCUIT OF dc MOTORS

In a dc motor, the field flux ϕ_f is established by the stator, either by means of permanent magnets as shown in Fig. 13-1a, where ϕ_f stays constant, or by means of a field winding as shown in Fig. 13-1b, where the field current I_f controls ϕ. If the magnetic saturation in the flux path can be neglected, then

$$\phi_f = k_f I_f \tag{13-1}$$

where k_f is a field constant of proportionality.

The rotor carries in its slots the so-called armature winding, which handles the electrical power. This is in contrast to most ac motors, where the power-handling winding is on the stator for ease of handling the larger amount of power. However, the armature winding in a dc machine has to be on the rotor to provide a "mechanical" rectification of voltages and currents (which alternate direction as the conductors rotate from the influence

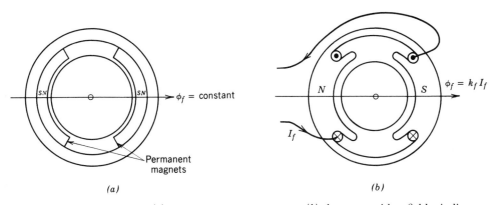

(a)   *(b)*

Figure 13-1 A dc motor: (*a*) permanent-magnet motor; (*b*) dc motor with a field winding.

of one stator pole to the next) in the armature-winding conductors, thus producing a dc voltage and a dc current at the terminals of the armature winding. The armature winding, in fact, is a continuous winding, without any beginning or end, and it is connected to the commutator segments. These commutator segments, usually made up of copper, are insulated from each other and rotate with the shaft. At least one pair of stationary carbon brushes is used to make contact between the commutator segments (and, hence, the armature conductors), and the stationary terminals of the armature winding that supply the dc voltage and current.

In a dc motor, the electromagnetic torque is produced by the interaction of the field flux ϕ_f and the armature current i_a:

$$T_{em} = k_t \phi_f i_a \tag{13-2}$$

where k_t is the torque constant of the motor. In the armature circuit, a back-emf is produced by the rotation of armature conductors at a speed ω_m in the presence of a field flux ϕ_f:

$$e_a = k_e \phi_f \omega_m \tag{13-3}$$

where k_e is the voltage constant of the motor.

In SI units, k_t and k_e are numerically equal, which can be shown by equating the electrical power $e_a i_a$ and the mechanical power $\omega_m T_{em}$. The electrical power is calculated as

$$P_e = e_a i_a = k_e \phi_f \omega_m i_a \quad \text{(using Eq. 13-3)} \tag{13-4}$$

and the mechanical power as

$$P_m = \omega_m T_{em} = k_t \phi_f \omega_m i_a \quad \text{(using Eq. 13-2)} \tag{13-5}$$

In steady state,

$$P_e = P_m \tag{13-6}$$

Therefore, from the foregoing equations

$$k_t \left[\frac{\text{Nm}}{\text{A} \cdot \text{Wb}} \right] = k_e \left[\frac{\text{V}}{\text{Wb} \cdot \text{rad/s}} \right] \tag{13-7}$$

In practice, a controllable voltage source v_t is applied to the armature terminals to establish i_a. Therefore, the current i_a in the armature circuit is determined by v_t, the induced back-emf e_a, the armature-winding resistance R_a, and the armature-winding inductance L_a:

$$v_t = e_a + R_a i_a + L_a \frac{di_a}{dt} \tag{13-8}$$

Equation 13-8 is illustrated by an equivalent circuit in Fig. 13-2.

The interaction of T_{em} with the load torque, as given by Eq. 12-3b of Chapter 12, determines how the motor speed builds up:

$$T_{em} = J \frac{d\omega_m}{dt} + B\omega_m + T_{WL}(t) \tag{13-9}$$

where J and B are the total equivalent inertia and damping, respectively, of the motor–load combination and T_{WL} is the equivalent working torque of the load.

Seldom are dc machines used as generators. However, they act as generators while braking, where their speed is being reduced. Therefore, it is important to consider dc

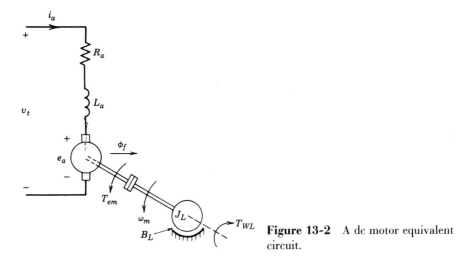

Figure 13-2 A dc motor equivalent circuit.

machines in their generator mode of operation. In order to consider braking, we will assume that the flux ϕ_f is kept constant and the motor is initially driving a load at a speed of ω_m. To reduce the motor speed, if v_t is reduced below e_a in Fig. 13-2, then the current i_a will reverse in direction. The electromagnetic torque T_{em} given by Eq. 13-2 now reverses in direction and the kinetic energy associated with the motor load inertia is converted into electrical energy by the dc machine, which now acts as a generator. This energy must be somehow absorbed by the source of v_t or dissipated in a resistor.

During the braking operation, the polarity of e_a does not change, since the direction of rotation has not changed. Equation 13-3 still determines the magnitude of the induced emf. As the rotor slows down, e_a decreases in magnitude (assuming that ϕ_f is constant). Ultimately, the generation stops when the rotor comes to a standstill and all the inertial energy is extracted. If the terminal-voltage polarity is also reversed, the direction of rotation of the motor will reverse. Therefore, a dc motor can be operated in either direction and its electromagnetic torque can be reversed for braking, as shown by the four quadrants of the torque–speed plane in Fig. 13-3.

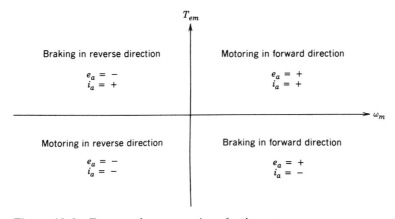

Figure 13-3 Four-quadrant operation of a dc motor.

13-3   PERMANENT-MAGNET dc MOTORS

Often in small dc motors, permanent magnets on the stator as shown in Fig. 13-1a produce a constant field flux ϕ_f. In steady state, assuming a constant field flux ϕ_f, Eqs. 13-2, 13-3, and 13-8 result in

$$T_{em} = k_T I_a \tag{13-10}$$

$$E_a = k_E \omega_m \tag{13-11}$$

$$V_t = E_a + R_a I_a \tag{13-12}$$

where $k_T = k_t \phi_f$ and $K_E = K_e \phi_f$. Equations 13-10 through 13-12 correspond to the equivalent circuit of Fig. 13-4a. From the above equations, it is possible to obtain the steady-state speed ω_m as a function of T_{em} for a given V_t:

$$\omega_m = \frac{1}{k_E}\left(V_t - \frac{R_a}{k_T}T_{em}\right) \tag{13-13}$$

The plot of this equation in Fig. 13-4b shows that as the torque is increased, the torque–speed characteristic at a given V_t is essentially vertical, except for the droop due to the voltage drop $I_a R_a$ across the armature-winding resistance. This droop in speed is quite small in integral horsepower dc motors but may be substantial in small servo motors. More importantly, however, the torque–speed characteristics can be shifted horizontally

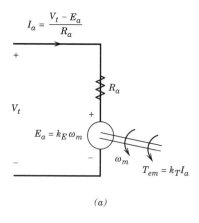

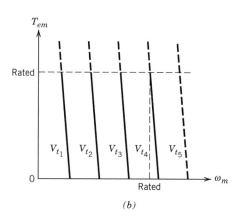

(a)   (b)

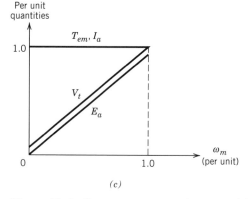

(c)

Figure 13-4   Permanent-magnet dc motor: (a) equivalent circuit; (b) torque-speed characteristics: $V_{t5} > V_{t4} > V_{t3} > V_{t2} > V_{t1}$, where V_{t4} is the rated voltage; (c) continuous torque–speed capability.

in Fig. 13-4*b* by controlling the applied terminal voltage V_t. Therefore, the speed of a load with an arbitrary torque–speed characteristic can be controlled by controlling V_t in a permanent-magnet dc motor with a constant ϕ_f.

In a continuous steady state, the armature current I_a should not exceed its rated value, and therefore, the torque should not exceed the rated torque. Therefore, the characteristics beyond the rated torque are shown as dashed in Fig. 13-4*b*. Similarly, the characteristic beyond the rated speed is shown as dashed, because increasing the speed beyond the rated speed would require the terminal voltage V_t to exceed its rated value, which is not desirable. This is a limitation of permanent-magnet dc motors, where the maximum speed is limited to the rated speed of the motor. The torque capability as a function of speed is plotted in Fig. 13-4*c*. It shows the steady-state operating limits of the torque and current; it is possible to significantly exceed current and torque limits on a short-term basis. Figure 13-4*c* also shows the terminal voltage required as a function of speed and the corresponding E_a.

13-4 dc MOTORS WITH A SEPARATELY EXCITED FIELD WINDING

Permanent-magnet dc motors are limited to ratings of a few horsepower and also have a maximum speed limitation. These limitations can be overcome if ϕ_f is produced by means of a field winding on the stator, which is supplied by a dc current I_f, as shown in Fig. 13-1*b*. To offer the most flexibility in controlling the dc motor, the field winding is excited by a separately controlled dc source v_f, as shown in Fig. 13-5*a*. As indicated by Eq. 13-1, the steady-state value of ϕ_f is controlled by $I_f (= V_f/R_f)$, where R_f is the resistance of the field winding.

Since ϕ_f is controllable, Eq. 13-13 can be written as follows:

$$\omega_m = \frac{1}{k_e\phi_f}\left(V_t - \frac{R_a}{k_t\phi_f}T_{em}\right) \qquad (13\text{-}14)$$

recognizing that $k_E = k_e\phi_f$ and $k_T = k_t\phi_f$. Equation 13-14 shows that in a dc motor with a separately excited field winding, both V_t and ϕ_f can be controlled to yield the desired torque and speed. As a general practice, to maximize the motor torque capability, ϕ_f (hence I_f) is kept at its rated value for speeds less than the rated speed. With ϕ_f at its rated value, the relationships are the same as given by Eqs. 13-10 through 13-13 of a permanent-magnet dc motor. Therefore, the torque–speed characteristics are also the same as those for a permanent-magnet dc motor that were shown in Fig. 13-4*b*. With ϕ_f constant and equal to its rated value, the motor torque–speed capability is as shown in Fig. 13-5*b*, where this region of constant ϕ_f is often called the constant-torque region. The required terminal voltage V_t in this region increases linearly from approximately zero to its rated value as the speed increases from zero to its rated value. The voltage V_t and the corresponding E_a are shown in Fig. 13-5*b*.

To obtain speeds beyond its rated value, V_t is kept constant at its rated value and ϕ_f is decreased by decreasing I_f. Since I_a is not allowed to exceed its rated value on a continuous basis, the torque capability declines, since ϕ_f is reduced in Eq. 13-2. In this so-called field-weakening region, the maximum power E_aI_a (equal to ω_mT_{em}) into the motor is not allowed to exceed its rated value on a continuous basis. This region, also called the constant-power region, is shown in Fig. 13-5*b*, where T_{em} declines with ω_m and V_t, E_a, and I_a stay constant at their rated values. It should be emphasized that Fig. 13-5*b* is the plot of the maximum continuous capability of the motor in steady state. Any

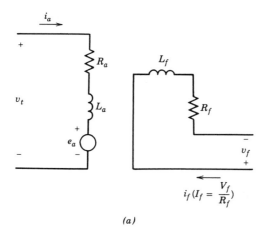

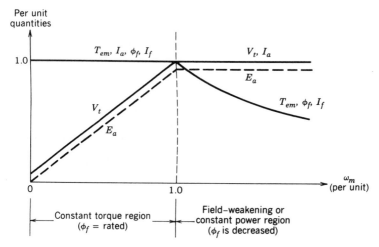

Figure 13-5   Separately excited dc motor: (*a*) equivalent circuit; (*b*) continuous torque–speed capability.

operating point within the regions shown is, of course, permissible. In the field-weakening region, the speed may be exceeded by 50–100% of its rated value, depending on the motor specifications.

13-5   EFFECT OF ARMATURE CURRENT WAVEFORM

In dc motor drives, the output voltage of the power electronic converter contains an ac ripple voltage superimposed on the desired dc voltage. Ripple in the terminal voltage can lead to a ripple in the armature current with the following consequences that must be recognized: the form factor and torque pulsations.

13-5-1   FORM FACTOR

The form factor for the dc motor armature current is defined as

$$\text{Form factor} = \frac{I_a(\text{rms})}{I_a(\text{average})} \qquad (13\text{-}15)$$

The form factor will be unity only if i_a is a pure dc. The more i_a deviates from a pure dc, the higher will be the value of the form factor. The power input to the motor (and hence the power output) varies proportionally with the average value of i_a, whereas the losses in the resistance of the armature winding depend on $I_a^2(\text{rms})$. Therefore, the higher the form factor of the armature current, the higher the losses in the motor (i.e., higher heating) and, hence, the lower the motor efficiency.

Moreover, a form factor much higher than unity implies a much larger value of the peak armature current compared to its average value, which may result in excessive arcing in the commutator and brushes. To avoid serious damage to the motor that is caused by large peak currents, the motor may have to be derated (i.e., the maximum power or torque would have to be kept well below its rating) to keep the motor temperature from exceeding its specified limit and to protect the commutator and brushes. Therefore, it is desirable to improve the form factor of the armature current as much as possible.

13-5-2 TORQUE PULSATIONS

Since the instantaneous electromagnetic torque $T_{em}(t)$ developed by the motor is proportional to the instantaneous armature current $i_a(t)$, a ripple in i_a results in a ripple in the torque and hence in speed if the inertia is not large. This is another reason to minimize the ripple in the armature current. It should be noted that a high-frequency torque ripple will result in smaller speed fluctuations, as compared with a low-frequency torque ripple of the same magnitude.

13-6 dc SERVO DRIVES

In servo applications, the speed and accuracy of response is important. In spite of the increasing popularity of ac servo drives, dc servo drives are still widely used. If it were not for the disadvantages of having a commutator and brushes, the dc motors would be ideally suited for servo drives. The reason is that the instantaneous torque T_{em} in Eq. 13-2 can be controlled linearly by controlling the armature current i_a of the motor.

13-6-1 TRANSFER FUNCTION MODEL FOR SMALL-SIGNAL DYNAMIC PERFORMANCE

Figure 13-6 shows a dc motor operating in a closed loop to deliver controlled speed or controlled position. To design the proper controller that will result in high performance (high speed of response, low steady-state error, and high degree of stability), it is im-

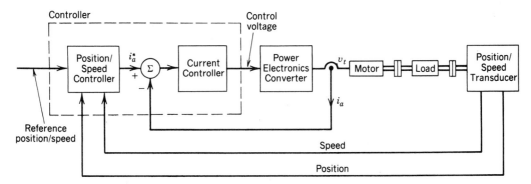

Figure 13-6 Closed-loop position/speed dc servo drive.

portant to know the transfer function of the motor. It is then combined with the transfer function of the rest of the system in order to determine the dynamic response of the drive for changes in the desired speed and position or for a change in load. As we will explain later on, the linear model is valid only for small changes where the motor current is not limited by the converter supplying the motor.

For analyzing small-signal dynamic performance of the motor–load combination around a steady-state operating point, the following equations can be written in terms of small deviations around their steady-state values:

$$\Delta v_t = \Delta e_a + R_a \, \Delta i_a + L_a \frac{d}{dt}(\Delta i_a) \tag{13-16}$$

$$\Delta e_a = k_E \, \Delta \omega_m \tag{13-17}$$

$$\Delta T_{em} = k_T \, \Delta i_a \tag{13-18}$$

$$\Delta T_{em} = \Delta T_{WL} + B \, \Delta \omega_m + J \frac{d(\Delta \omega_m)}{dt} \quad \text{(from Eq. 13-9)} \tag{13-19}$$

If we take the Laplace transform of these equations, where the Laplace variables represent only the small-signal Δ values in Eqs. 13-16 through 13-19,

$$\begin{aligned}
V_t(s) &= E_a(s) + (R_a + sL_a)I_a(s) \\
E_a(s) &= k_E \omega_m(s) \\
T_{em}(s) &= k_T I_a(s) \\
T_{em}(s) &= T_{WL}(s) + (B + sJ)\omega_m(s) \\
\omega_m(s) &= s\theta_m(s)
\end{aligned} \tag{13-20}$$

These equations for the motor–load combination can be represented by transfer function blocks, as shown in Fig. 13-7. The inputs to the motor–load combination in Fig. 13-7 are the armature terminal voltage $V_t(s)$ and the load torque $T_{WL}(s)$. Applying one input at a time by setting the other input to zero, the superposition principle yields (note that this is a linearized system)

$$\omega_m(s) = \frac{k_T}{(R_a + sL_a)(sJ + B) + k_T k_E} V_t(s) - \frac{R_a + sL_a}{(R_a + sL_a)(sJ + B) + k_T k_E} T_{WL}(s) \tag{13-21}$$

This equation results in two closed-loop transfer functions:

$$G_1(s) = \left. \frac{\omega_m(s)}{V_t(s)} \right|_{T_{WL}(s)=0} = \frac{k_T}{(R_a + sL_a)(sJ + B) + k_T k_E} \tag{13-22}$$

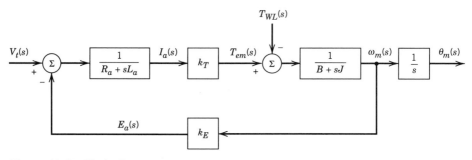

Figure 13-7   Block diagram representation of the motor and load (without any feedback).

$$G_2(s) = \frac{\omega_m(s)}{T_{WL}(s)}\bigg|_{V_t(s)=0} = -\frac{R_a+sL_a}{(R_a + sL_a)(sJ + B) + k_Tk_E} \tag{13-23}$$

As a simplification to gain better insight into the dc motor behavior, the friction term, which is usually small, will be neglected by setting $B = 0$ in Eq. 13-22. Moreover, considering just the motor without the load, J in Eq. 13-22 is then the motor inertia J_m. Therefore

$$G_1(s) = \frac{k_T}{sJ_m(R_a + sL_a) + k_Tk_E} = \frac{1}{k_E\left(s^2\dfrac{L_aJ_m}{k_Tk_E} + s\dfrac{R_aJ_m}{k_Tk_E} + 1\right)} \tag{13-24}$$

We will define the following constants:

$$\tau_m = \frac{R_aJ_m}{k_Tk_E} = \text{mechanical time constant} \tag{13-25}$$

$$\tau_e = \frac{L_a}{R_a} = \text{electrical time constant} \tag{13-26}$$

Using τ_m and τ_e in the expression for $G_1(s)$ yields

$$G_1(s) = \frac{1}{k_E(s^2\tau_m\tau_e + s\tau_m + 1)} \tag{13-27}$$

Since in general $\tau_m \gg \tau_e$, it is a reasonable approximation to replace $s\tau_m$ by $s(\tau_m + \tau_e)$ in the foregoing expression. Therefore

$$G_1(s) = \frac{\omega_m(s)}{V_t(s)} \approx \frac{1}{k_E(s\tau_m + 1)(s\tau_e + 1)} \tag{13-28}$$

The physical significance of the electrical and the mechanical time constants of the motor should also be understood. The electrical time constant τ_e determines how quickly the armature current builds up, as shown in Fig. 13-8, in response to a step change Δv_t in the terminal voltage, where the rotor speed is assumed to be constant.

The mechanical time constant τ_m determines how quickly the speed builds up in response to a step change Δv_t in the terminal voltage, provided that the electrical time constant τ_e is assumed to be negligible and, hence, the armature current can change

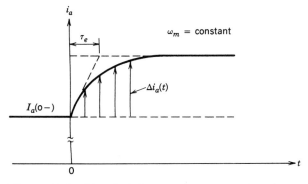

Figure 13-8 Electrical time constant τ_e; speed ω_m is assumed to be constant.

instantaneously. Neglecting τ_e in Eq. 13-28, the change in speed from the steady-state condition can be obtained as

$$\omega_m(s) = \frac{V_t(s)}{k_E(s\tau_m + 1)} = \frac{\Delta v_t}{k_E s(s\tau_m + 1)} = \frac{\Delta v_t}{k_E} \frac{1/\tau_m}{s(s + 1/\tau_m)} \tag{13-29}$$

recognizing that $V_t(s) = \Delta v_t/s$. From Eq. 13-29

$$\Delta\omega_m(t) = \frac{\Delta v_t}{k_E} (1 - e^{-t/\tau_m}) \tag{13-30}$$

where τ_m is the mechanical time constant with which the speed changes in response to a step change in the terminal voltage, as shown in Fig. 13-9a. The corresponding change in the armature current is plotted in Fig. 13-9b. Note that if the motor current is limited by the converter during large transients, the torque produced by the motor is simply $k_T I_{a,max}$.

13-6-2   POWER ELECTRONIC CONVERTER

Based on the previous discussion, a power electronic converter supplying a dc motor should have the following capabilities:

1. The converter should allow both its output voltage and current to reverse in order to yield a four-quadrant operation as shown in Fig. 13-3.

2. The converter should be able to operate in a current-controlled mode by holding the current at its maximum acceptable value during fast acceleration and deceleration. The dynamic current limit is generally several times higher than the continuous steady-state current rating of the motor.

3. For accurate control of position, the average voltage output of the converter should vary linearly with its control input, independent of the load on the motor. This item is further discussed in Section 13-6-5.

4. The converter should produce an armature current with a good form factor and should minimize the fluctuations in torque and speed of the motor.

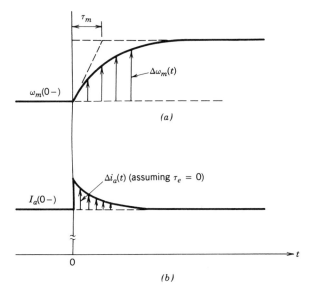

(a)

(b)

Figure 13-9
Mechanical time constant τ_m; load torque is assumed to be constant.

5. The converter output should respond as quickly as possible to its control input, thus allowing the converter to be represented essentially by a constant gain without a dead time in the overall servo drive transfer function model.

A linear power amplifier satisfies all the requirements listed above. However, because of its low energy efficiency, this choice is limited to a very low power range. Therefore, the choice must be made between switch-mode dc–dc converters of the type discussed in Chapter 7 or the line-frequency-controlled converters discussed in Chapter 6. Here, only the switch-mode dc–dc converters are described. Drives with line-frequency converters can be analyzed in the same manner.

A full-bridge switch-mode dc–dc converter produces a four-quadrant controllable dc output. This full-bridge dc–dc converter (also called an H-bridge) was discussed in Chapter 7. The overall system is shown in Fig. 13-10, where the line-frequency ac input is rectified into dc by means of a diode rectifier of the type discussed in Chapter 5 and filtered by means of a filter capacitor. An energy dissipation circuit is included to prevent the filter capacitor voltage from becoming large in case of braking of the dc motor.

As discussed in Chapter 7, all four switches in the converter of Fig. 13-10 are switched during each cycle of the switching frequency. This results in a true four-quadrant operation with a continuous-current conduction, where both V_t and I_a can smoothly reverse, independent of each other. Ignoring the effect of blanking time, the average voltage output of the converter varies linearly with the input control voltage $v_{control}$, independent of the load:

$$V_t = k_c v_{control} \tag{13-31}$$

where k_c is the gain of the converter.

As discussed in Sections 7-7-1 and 7-7-2 of Chapter 7, either a PWM bipolar voltage-switching scheme or a PWM unipolar voltage-switching scheme can be used. Thus, the converter in Fig. 13-6 can be replaced by an amplifier gain k_c given by Eq. 13-31.

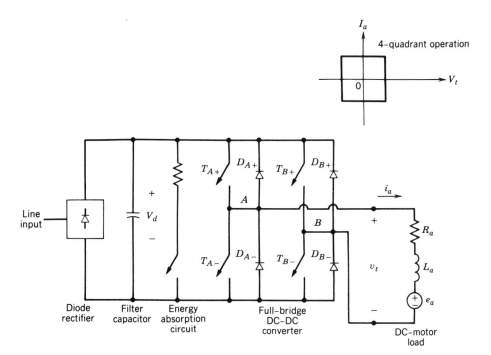

Figure 13-10 A dc motor servo drive; four-quadrant operation.

13-6-3   RIPPLE IN THE ARMATURE CURRENT i_a

In Chapter 7, it was mentioned that the current through a PWM full-bridge dc–dc converter supplying a dc motor load flows continuously even at small values of I_a. However, it is important to consider the peak-to-peak ripple in the armature current because of its impact on the torque pulsations and heating of the motor. Moreover, a larger current ripple requires a larger peak current rating of the converter switches.

In the system of Fig. 13-10 under a steady-state operating condition, the instantaneous speed ω_m can be assumed to be constant if there is sufficient inertia, and therefore $e_a(t) = E_a$. The terminal voltage and the armature current can be expressed in terms of their dc and the ripple components as

$$v_t(t) = V_t + v_r(t) \tag{13-32}$$

$$i_a(t) = I_a + i_r(t) \tag{13-33}$$

where $v_r(t)$ and $i_r(t)$ are the ripple components in v_t and i_a, respectively. Therefore, in the armature circuit, from Eq. 13-8,

$$V_t + v_r(t) = E_a + R_a[I_a + i_r(t)] + L_a \frac{di_r(t)}{dt} \tag{13-34}$$

where

$$V_t = E_a + R_a I_a \tag{13-35}$$

and

$$v_r(t) = R_a i_r(t) + L_a \frac{di_r(t)}{dt} \tag{13-36}$$

Assuming that the ripple current is primarily determined by the armature inductance L_a and R_a has a negligible effect, from Eq. 13-36

$$v_r(t) \simeq L_a \frac{di_r(t)}{dt} \tag{13-37}$$

The additional heating in the motor is approximately $R_a I_r^2$, where I_r is the rms value of the ripple current i_r.

By means of an example and Fig. 7-30, it was shown in Chapter 7 that for a PWM bipolar voltage switching, the ripple voltage is maximum when the average output voltage is zero and all switches operate at equal duty ratios. Applying these results to the dc motor drive, Fig. 13-11a shows the voltage ripple $v_r(t)$ and the resulting ripple current $i_r(t)$ using Eq. 13-37. From these waveforms, the maximum peak-to-peak ripple can be calculated as

$$(\Delta I_{\text{p-p}})_{\text{max}} = \frac{V_d}{2L_a f_s} \tag{13-38}$$

where V_d is the input dc voltage to the full-bridge converter.

The ripple voltage for a PWM unipolar voltage switching is shown to be maximum when the average output voltage is $\frac{1}{2} V_d$. Applying this result to a dc motor drive, Fig. 13-11b shows $i_r(t)$ waveform, where

$$(\Delta I_{\text{p-p}})_{\text{max}} = \frac{V_d}{8L_a f_s} \tag{13-39}$$

Equations 13-38 and 13-39 show that the maximum peak-to-peak ripple current is inversely proportional to L_a and f_s. Therefore, careful consideration must be given to the

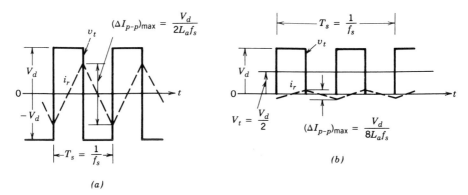

Figure 13-11   Ripple i_r in the armature current: (a) PWM bipolar voltage switching, $V_t = 0$; (b) PWM unipolar voltage switching, $V_t = \frac{1}{2}V_d$.

selection of f_s and L_a, where L_a can be increased by adding an external inductor in the series with the motor armature.

13-6-4   CONTROL OF SERVO DRIVES

A servo system where the speed error directly controls the power electronic converter is shown in Fig. 13-12a. The current-limiting circuit comes into operation only when the drive current tries to exceed an acceptable limit $I_{a,\max}$ during fast accelerations and decelerations. During these intervals, the output of the speed regulator is suppressed and the current is held at its limit until the speed and position approach their desired values.

To improve the dynamic response in high-performance servo drives, an internal current loop is used as shown in Fig. 13-12b, where the armature current and, hence, the torque are controlled. The current control is accomplished by comparing the actual measured armature current i_a with its reference value i_a^* produced by the speed regulator. The current i_a is inherently controlled from exceeding the current rating of the drive by limiting the reference current i_a^* to $I_{a,\max}$.

The armature current provided by the dc–dc converter in Fig. 13-12b can be controlled in a similar manner as the current-regulated modulation in a dc-to-ac inverter, discussed in Section 8-6-3. The only difference is that the reference current in steady state in a dc–dc converter is a dc rather than a sinusoidal waveform as in Section 8-6-3. Either a variable-frequency tolerance band control, discussed in Section 8-6-3-1, or a fixed-frequency control, discussed in Section 8-6-3-2, can be used for current control.

13-6-5   NONLINEARITY DUE TO BLANKING TIME

In a practical full-bridge dc–dc converter, where the possibility of a short circuit across the input dc bus exists, a blanking time is introduced between the instant at which a switch turns off and the instant at which the other switch in the same leg turns on. The effect of the blanking time on the output of dc-to-ac full bridge PWM inverters was discussed in detail in Section 8-5. That analysis is also valid for PWM full-bridge dc–dc converters for dc servo drives. Equation 8-77 and Fig. 8-32b show the effect of blanking time on the output voltage magnitude. Recognizing that the output voltage of the converter is proportional to the motor speed ω_m and the output current i_a is proportional to the torque T_{em}, Fig. 8-32b is redrawn as in Fig. 13-13. If at an arbitrary speed ω_m, the torque and, hence, i_a are to be reversed, there is a dead zone in $v_{control}$, as shown in Fig. 13-13, during which

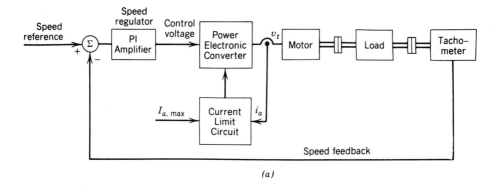

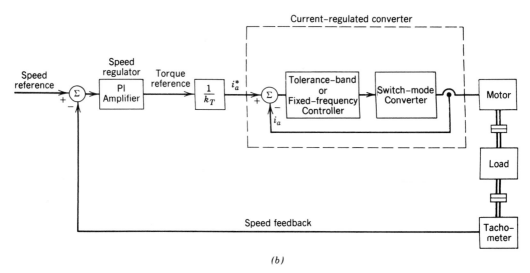

(b)

Figure 13-12 Control of servo drives: (a) no internal current-control loop; (b) internal current-control loop.

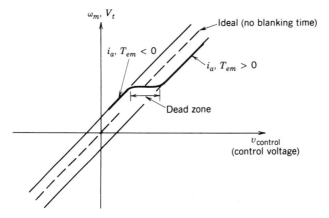

Figure 13-13 Effect of blanking time.

i_a and T_{em} remain small. The effect of this nonlinearity due to blanking time on the performance of the servo system is minimized by means of the current-controlled mode of operation discussed in the block diagram of Fig. 13-12b, where an internal current loop directly controls i_a.

13-6-6 SELECTION OF SERVO DRIVE PARAMETERS

Based on the foregoing discussion, the effects of armature inductance L_a, switching frequency f_s, blanking time t_Δ, and switching times t_c of the solid state devices in the dc–dc converter can be summarized as follows:

1. The ripple in the armature current, which causes torque ripple and additional armature heating, is proportional to L_a/f_s.
2. The dead zone in the transfer function of the converter, which degrades the servo performance, is proportional to $f_s t_\Delta$.
3. Switching losses in the converter are proportional to $f_s t_c$.

All these factors need to be considered simultaneously in the selection of the appropriate motor and the power electronic converter.

13-7 ADJUSTABLE-SPEED dc DRIVES

Unlike servo drives, the response time to speed and torque commands is not as critical in adjustable-speed drives. Therefore, either switch-mode dc–dc converters as discussed for servo drives or the line-frequency controlled converters discussed in Chapter 6 can be used for speed control.

13-7-1 SWITCH-MODE dc–dc CONVERTER

If a four-quadrant operation is needed and a switch-mode converter is utilized, then the full-bridge converter shown in Fig. 13-10 is used.

If the speed does not have to reverse but braking is needed, then the two-quadrant converter shown in Fig. 13-14a can be used. It consists of two switches, where one of the switches is on at any time, to keep the output voltage independent of the direction of i_a. The armature current can reverse, and a negative value of I_a corresponds to the braking mode of operation, where the power flows from the dc motor to V_d. The output voltage V_t can be controlled in magnitude, but it always remains unipolar. Since i_a can flow in both directions, unlike in the single-switch step-down and step-up dc–dc converters discussed in Chapter 7, i_a in the circuit of Fig. 13-14a will not become discontinuous.

For a single-quadrant operation where the speed remains unidirectional and braking is not required, the step-down converter shown in Fig. 13-14b can be used.

13-7-2 LINE-FREQUENCY CONTROLLED CONVERTERS

In many adjustable-speed dc drives, especially in large power ratings, it may be economical to utilize a line-frequency controlled converter of the type discussed in Chapter 6. Two of these converters are repeated in Fig. 13-15 for single-phase and three-phase ac inputs. The output of these line-frequency converters, also called the phase-controlled converters, contains an ac ripple that is a multiple of the 60-Hz line frequency. Because of this low frequency ripple, an inductance in series with the motor armature may be

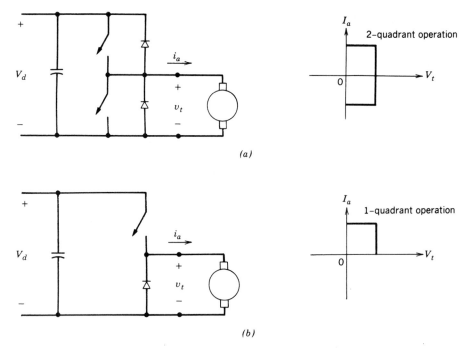

(a)

(b)

Figure 13-14 (*a*) Two-quadrant operation; (*b*) single-quadrant operation.

required to keep the ripple in i_a low, to minimize its effect on armature heating and the ripple in torque and speed.

A disadvantage of the line-frequency converters is the longer dead time in responding to the changes in the speed control signal, compared to high-frequency switch-mode dc–dc converters. Once a thyristor or a pair of thyristors is triggered on in the circuits of Fig. 13-15, the delay angle α that controls the converter output voltage applied to the motor terminals cannot be increased for a portion of the 60-Hz cycle. This may not be a problem in adjustable-speed drives where the response time to speed and torque commands is not too critical. But it clearly shows the limitation of line-frequency converters in servo drive applications.

The current through these line-frequency controlled converters is unidirectional, but the output voltage can reverse polarity. The two-quadrant operation with the reversible voltage is not suited for dc motor braking, which requires the voltage to be unidirectional

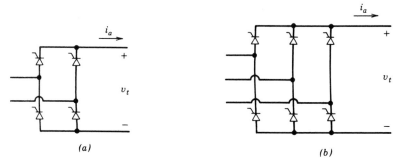

(a)   *(b)*

Figure 13-15 Line-frequency-controlled converters for dc motor drives: (*a*) single-phase input; (*b*) three-phase input.

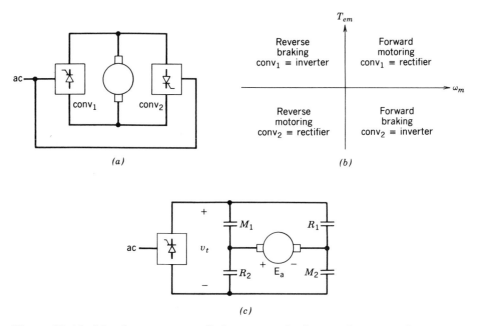

Figure 13-16 Line-frequency-controlled converters for four-quadrant operation:
(*a*) back-to-back converters for four-quadrant operation (without circulating current);
(*b*) converter operation modes; (*c*) contactors for four-quadrant operation.

but the current to be reversible. Therefore, if regenerative braking is required, two back-to-back connected thyristor converters can be used, as shown in Fig. 13-16*a*. This, in fact, gives a capability to operate in all four quadrants, as depicted in Fig. 13-16*b*.

An alternative to using two converters is to use one phase-controlled converter together with two pairs of contactors, as shown in Fig. 13-16*c*. When the machine is to be operated as a motor, the contactors M_1 and M_2 are closed. During braking when the motor speed is to be reduced rapidly, since the direction of rotation remains the same, E_a is of the same polarity as in the motoring mode. Therefore, to let the converter go into an inverter mode, contactors M_1 and M_2 are opened and R_1 and R_2 are closed. It should be noted that the contactors switch at zero current when the current through them is brought to zero by the converter.

13-7-3 EFFECT OF DISCONTINUOUS ARMATURE CURRENT

In line-frequency phase-controlled converters and single-quadrant step-down switch-mode dc–dc converters, the output current can become discontinuous at light loads on the motor. For a fixed control voltage $v_{control}$ or the delay angle α, the discontinuous current causes the output voltage to go up. This voltage rise causes the motor speed to increase at low values of I_a (which correspond to low torque load), as shown generically by Fig. 13-17. With a continuously flowing i_a, the drop in speed at higher torques is due to the voltage drop $R_a I_a$ across the armature resistance; additional drop in speed occurs in the phase-controlled converter-driven motors due to commutation voltage drops across the ac-side inductance L_s, which approximately equal $(2\omega L_s/\pi)I_a$ in single-phase converters and $(3\omega L_s/\pi)I_a$ in three-phase converters, as discussed in Chapter 6. These effects results in poor speed regulation under an open-loop operation.

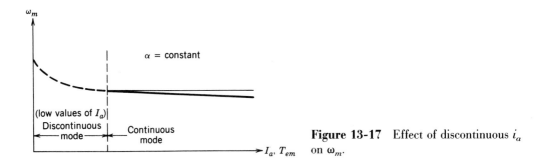

Figure 13-17 Effect of discontinuous i_a on ω_m.

13-7-4 CONTROL OF ADJUSTABLE-SPEED DRIVES

The type of control used depends on the drive requirements. An open-loop control is shown in Fig. 13-18 where the speed command ω^* is generated by comparing the drive output with its desired value (which, e.g., may be temperature in case of a capacity-modulated heat pump). A d/dt limiter allows the speed command to change slowly, thus preventing the rotor current from exceeding its rating. The slope of the d/dt limiter can be adjusted to match the motor–load inertia. The current limiter in such drives may be just a protective measure, whereby if the measured current exceeds its rated value, the controller shuts the drive off. A manual restart may be required. As discussed in Section 13-6, a closed-loop control can also be implemented.

13-7-5 FIELD WEAKENING IN ADJUSTABLE-SPEED dc MOTOR DRIVES

In a dc motor with a separately excited field winding, the drive can be operated at higher than the rated speed of the motor by reducing the field flux ϕ_f. Since many adjustable-speed drives, especially at higher power ratings, employ a motor with a wound field, this capability can be exploited by controlling the field current and ϕ_f. The simple line-frequency phase-controlled converter shown in Fig. 13-15 is normally used to control I_f through the field winding, where the current is controlled in magnitude but always flows in only one direction. If a converter topology consisting of only thyristors (such as in Fig. 13-15) is chosen, where the converter output voltage is reversible, the field current can be decreased rapidly.

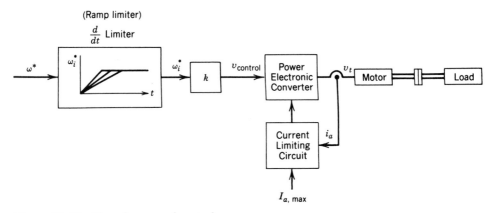

Figure 13-18 Open-loop speed control.

13-7-6 POWER FACTOR OF THE LINE CURRENT IN ADJUSTABLE-SPEED DRIVES

The motor operation at its torque limit is shown in Fig. 13-19a in the constant-torque region below the rated speed and in the field-weakening region above the rated speed. In a switch-mode drive, which consists of a diode rectifier bridge and a PWM dc–dc converter, the fundamental-frequency component I_{s1} of the line current as a function of speed is shown in Fig. 13-19b. Figure 13-19c shows I_{s1} for a line-frequency phase-controlled thyristor drive. Assuming the load torque to be constant, I_{s1} decreases with

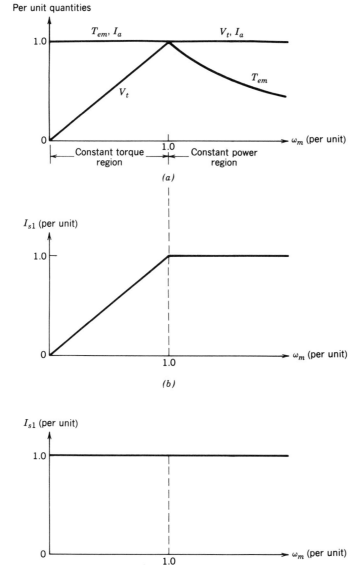

Figure 13-19 Line current in adjustable-speed dc drives: (a) drive capability; (b) switch-mode converter drive; (c) line-frequency thyristor converter drive.

decreasing speed in a switch-mode drive. Therefore, the switch-mode drive results in a good displacement power factor. On the other hand, in a phase-controlled thyristor drive, I_{s1} remains essentially constant as speed decreases, thus resulting in a very poor displacement power factor at low speeds.

As discussed in Chapters 5 and 6, both the diode rectifiers and the phase-controlled rectifiers draw line currents that consist of large harmonics in addition to the fundamental. These harmonics cause the power factor of operation to be poor in both types of drives. The circuits described in Chapter 18 can be used to remedy the harmonics problem in the switch-mode drives, thus resulting in a high power factor of operation.

SUMMARY

1. Because of mechanical contact between the commutator segments and brushes, dc motors require periodic maintenance. Because of arcing between these two surfaces, dc motors are not suitable for certain environments.

2. In a dc motor, the field flux is established by either a field winding supplied through a dc current or permanent magnets located on the stator. The magnitude of the electromagnetic torque is directly proportional to the field flux and the armature current magnitude. This makes a dc motor ideal for servo drive applications.

3. The induced back-emf across the armature-winding terminals is directly proportional to the field flux magnitude and the rotational speed of the rotor.

4. A simple transfer function model can be obtained for a dc motor to obtain its dynamic performance.

5. The form factor of the armature current is defined as the ratio of its rms value to its average value. A poor armature current waveform with a high form factor results in excessive armature heating, arcing across commutator segments and brushes, and large torque pulsations. Therefore, an appropriate remedial action should be taken to avoid damage to the dc motor.

6. The dc motor drives utilize either the line-frequency controlled converters or the dc–dc switch-mode converters. By field weakening in a wound-field dc motor, the speed can be controlled beyond its rated value, without exceeding the rated armature voltage.

7. The power factor at which a dc motor drive operates from the utility grid and the current harmonics injected into the utility grid depend on the type of converter used: line-frequency controlled converter or switch-mode dc–dc converter.

PROBLEMS

13-1 Consider a permanent-magnet dc servo motor with the following parameters:

$$T_{rated} = 10 \text{ N-m}$$
$$n_{rated} = 3700 \text{ rpm}$$
$$k_T = 0.5 \text{ N-m/A}$$
$$k_E = 53\text{V}/1000 \text{ rpm}$$
$$R_a = 0.37 \text{ }\Omega$$
$$\tau_e = 4.05 \text{ ms}$$
$$\tau_m = 11.7 \text{ ms}$$

Calculate the terminal voltage V_t in steady state if the motor is required to deliver a torque of 5 N-m at a speed of 1500 rpm.

13-2 $G_1(s) = [\omega_m(s)/V_t(s)]$ is the transfer function of an unloaded and uncontrolled dc motor. Express $G_1(s)$ given by Eq. 13-27 in the following form:

$$G_1(s) = \frac{1/k_E}{1 + 2sD/\omega_n + s^2/\omega_n^2}$$

Calculate D and ω_n for the servomotor parameters given in Problem 13-1. Plot the magnitude and the phase of $G_1(s)$ by means of a Bode plot.

13-3 Using the servomotor parameters given in Problem 13-1, calculate and plot the change in ω_m as a function of time for a step increase of 10 V in the terminal voltage of that uncontrolled, unloaded servomotor.

13-4 The servomotor of Problem 13-1 is driven by a full-bridge dc–dc converter operating from a 200-V dc bus. Calculate the peak-to-peak ripple in the motor current if a PWM bipolar voltage-switching scheme is used. The motor is delivering a torque of 5 N-m at a speed of 1500 rpm. The switching frequency is 20 kHz.

13-5 Repeat Problem 13-4 if a unipolar voltage-switching scheme is used.

13-6 In the servo drive of Problem 13-1, a PI regulator is used in the speed loop to obtain a transfer function of the following form in Fig. P13-6:

$$F_\omega(s) = \frac{\omega(s)}{\omega^*(s)} = \frac{1}{1 + s\,(2D/\omega_n) + s^2/\omega_n^2}$$

where $D = 0.5$ and $\omega_n = 300$ rad/s.

(a) Draw the Bode plot of the closed-loop transfer function $F_\theta(s) = [\theta(s)/\theta^*(s)]$ if a gain $k_p = 60$ is used for the proportional position regulator in Fig. P13-6.

(b) What is the bandwidth of the above closed-loop system?

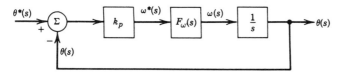

Figure P13-6

13-7 Consider the servomotor of Problem 13-1 in a speed-control loop. If an internal current loop is not used, the block diagram is as shown in Fig. P13-7a, where only a proportional control is used. If an internal current loop is used, the block diagram without the current limits is as shown in Fig. P13-7b, where ω_n is 10 times that in part a.

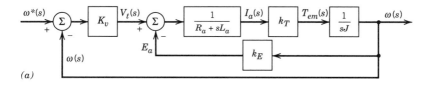

(a)

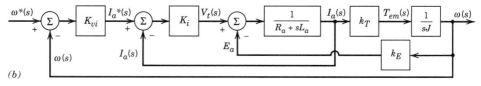

(b)

Figure P13-7

Design the controllers (K_v, K_{vi}, K_i) to yield a control loop with slightly underdamped response ($D = 0.7$). Compare the two control schemes in terms of bandwidth and transient performance, assuming that the current limit is not reached in either of them.

REFERENCES

1. A. E. Fitzgerald, C. Kingsley, Jr., and S. D. Umans, *Electric Machinery,* 4th ed., McGraw-Hill, New York, 1983.
2. P. C. Sen, *Thyristor DC Drives,* Wiley, New York, 1981.
3. G. R. Slemon and A. Straughen, *Electric Machines,* Addison-Wesley, Reading, MA, 1980.
4. T. Kenjo and S. Nagamori, *Permanent-Magnet and Brushless DC Motors,* Clarendon, Oxford, 1985.
5. *DC Motors · Speed Controls · Servo System—An Engineering Handbook,* 5th ed., ElectroCraft Corporation, Hopkins, MN, 1980.

CHAPTER 14

INDUCTION MOTOR DRIVES

14-1 INTRODUCTION

Induction motors with squirrel-cage rotors are the workhorse of industry because of their low cost and rugged construction. When operated directly from the line voltages (60 Hz utility input at essentially a constant voltage), an induction motor operates at a nearly constant speed. However, by means of power electronic converters, it is possible to vary the speed of an induction motor. The induction motor drives can be classified into two broad categories based on their applications:

1. *Adjustable-speed drives.* One important application of these drives is in process control by controlling the speed of fans, compressors, pumps, blowers, and the like.
2. *Servo drives.* By means of sophisticated control, induction motors can be used as servo drives in computer peripherals, machine tools, and robotics.

The emphasis in this chapter is on understanding the behavior of induction motors and how it is possible to control their speed where the dynamics of speed control need not be very fast and precise. This is the case in most process control applications where induction motor drives are used. As a side benefit, use of induction motor drives results in energy conservation, as discussed below.

Consider a simple example of an induction motor driving a centrifugal pump as shown in Fig. 14-1a, where the motor and the pump operate at a nearly constant speed. To reduce the flow rate, the throttling valve is partially closed. This causes loss of energy across the throttling valve. This energy loss can be avoided by eliminating the throttling valve and driving the pump at a speed that results in the desired flow rate, as in Fig. 14-1b.

In the system of Fig. 14-1b, the input power decreases significantly as the speed is decreased to reduce flow rate. This decrease in power requirement can be calculated by recognizing that in a centrifugal pump,

$$\text{Torque} \simeq k_1 (\text{speed})^2 \qquad (14\text{-}1)$$

and therefore, the power required by the pump from the motor is

$$\text{Power} \simeq k_2 (\text{speed})^3 \qquad (14\text{-}2)$$

where k_1 and k_2 are the constants of proportionality.

If the motor and the pump energy efficiencies can be assumed to be constant as their speed and loadings change, then the input power required by the induction motor would

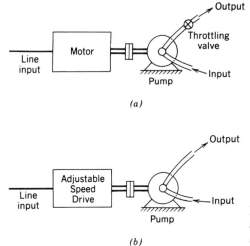

(a)

(b)

Figure 14-1   Centrifugal pump: (a) constant-speed drive: (b) adjustable-speed drive.

also vary as the speed cubed. Therefore, in comparison with a throttling valve to control the flow rate, the variable-speed-driven pump can result in significant energy conservation, where reduced flow rates are required for long periods of time. Moreover, pump systems are usually designed to provide a flow margin of 20–30% over the maximum values of their actual flow. Therefore, an adjustable-speed pump can result in substantial energy conservation. This conclusion is valid only if it is possible to adjust the motor speed in an energy-efficient manner. As we will see in this chapter, energy efficiency associated with power electronic inverters (described in Chapter 8) used for controlling induction motor speeds is high over wide speed and load ranges.

14-2   BASIC PRINCIPLES OF INDUCTION MOTOR OPERATION

In a large majority of applications, induction motor drives incorporate a three-phase, squirrel-cage motor. Therefore, the discussion here also assumes a three-phase, squirrel-cage induction motor. The stator of an induction motor consists of three phase windings distributed in the stator slots. These three windings are displaced by 120° in space, with respect to each other. The squirrel-cage rotor consists of a stack of insulated laminations. It has electrically conducting bars inserted through it, close to the periphery in the axial direction, which are electrically shorted at each end of the rotor by end rings, thus producing a cagelike structure. This also illustrates the simple, low-cost, and rugged nature of the rotor.

The objective of the following analysis is to explain as simply as possible the interaction between the induction motor and the power electronic converter. With this objective in mind, the details of proportionalities between various motor variables are simply expressed as k_j (where the subscript j is assigned arbitrary numeric values). Moreover, the motor is assumed to operate without any magnetic saturation.

If a balanced set of three-phase sinusoidal voltages at a frequency $f = \omega/2\pi$ are applied to the stator, it results in a balanced set of currents, which establishes a flux density distribution B_{ag} in the air gap with the following properties: (1) it has a constant amplitude and (2) it rotates with a constant speed, also called the synchronous speed, of

ω_s radians per second. The synchronous speed in a p-pole motor, supplied by frequency f, can be obtained as

$$\omega_s = \frac{2\pi/(p/2)}{1/f} = \frac{2}{p}(2\pi f) = \frac{2}{p}\omega \quad (\text{rad}/s) \qquad (14\text{-}3)$$

which is synchronized to the frequency f of the applied voltages and currents to the stator windings. In terms of the revolutions per minute (rpm), the synchronous speed is

$$n_s = 60 \times \frac{\omega_s}{2\pi} = \frac{120}{p}f \qquad (14\text{-}4)$$

The air gap flux ϕ_{ag} (due to the flux density distribution B_{ag}) rotates at a synchronous speed relative to the stationary stator windings. As a consequence, a counter-emf, often called the air gap voltage E_{ag} is induced in each of the stator phases at frequency f. This can be illustrated by means of the per-phase equivalent circuit shown in Fig. 14-2a, where $\mathbf{V}_s$ is the per-phase voltage (equal to the line–line rms voltage V_{LL} divided by $\sqrt{3}$) and $\mathbf{E}_{ag}$ is the air gap voltage. Here R_s is the resistance of the stator winding and L_{ls} is the leakage inductance of the stator winding. The magnetizing component $\mathbf{I}_m$ of the stator current $\mathbf{I}_s$ establishes the air gap flux. From the magnetic circuit analysis, it can be seen that

$$N_s\phi_{ag} = L_m i_m \qquad (14\text{-}5)$$

where N_s is an equivalent number of turns per phase of the stator winding and L_m is the magnetizing inductance shown in Fig. 14-2a.

From Faraday's law

$$e_{ag} = N_s\frac{d\phi_{ag}}{dt} \qquad (14\text{-}6)$$

With the air gap flux linking the stator phase winding to be $\phi_{ag}(t) = \phi_{ag}\sin\omega t$, Eq. 14-6 results in

$$e_{ag} = N_s\omega\phi_{ag}\cos\omega t \qquad (14\text{-}7)$$

which has an rms value of

$$E_{ag} = k_3 f\phi_{ag} \qquad (14\text{-}8)$$

where k_3 is a constant.

The torque in an induction motor is produced by interaction of the air gap flux and the rotor currents. If the rotor is rotating at the synchronous speed, there will be no relative motion between ϕ_{ag} and the rotor, and hence there will be no induced rotor voltages, rotor currents, and torque. At any other speed ω_r of the rotor in the same direction of the air

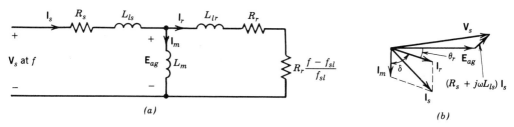

Figure 14-2 Per-phase representation: (*a*) equivalent circuit; (*b*) phasor diagram.

gap flux rotation, the motor is "slipping" with respect to the air gap flux at a relative speed called the slip speed ω_{sl}, where

$$\omega_{sl} = \omega_s - \omega_r \qquad (14\text{-}9)$$

This slip speed, normalized by the synchronous speed, is simply called the "slip" s:

$$\text{Slip } s \atop \text{(in per unit)} = \frac{\text{slip speed}}{\text{synchronous speed}} = \frac{\omega_s - \omega_r}{\omega_s} \qquad (14\text{-}10)$$

Therefore, the speed of the air gap flux with respect to the motor is calculated as

$$\text{Slip speed } \omega_{sl} = \omega_s - \omega_r = s\omega_s \qquad (14\text{-}11)$$

From Faraday's law, the induced voltages in the rotor circuit are at a slip frequency f_{sl}, which is proportional to the slip speed:

$$f_{sl} = \frac{\omega_{sl}}{\omega_s} f = sf \qquad (14\text{-}12)$$

The magnitude E, of this slip-frequency voltage that is induced in any of the rotor conductors can be obtained in a similar manner as the induced voltages in the stator phases. The same air gap flux ϕ_{ag} links the rotor conductors as the one that links the stator windings. However, the flux density distribution in the air gap rotates at a slip speed ω_{sl} with respect to the rotor conductors. Therefore, the induced emf E_r in the rotor conductors can be obtained by replacing f in Eq. 14-8 by the slip frequency f_{sl}. By assuming the squirrel-cage rotor to be represented by a three-phase short-circuited winding with the same equivalent number of turns N_s per phase as on the stator, we get

$$E_r = k_3 f_{sl}\phi_{ag} \qquad (14\text{-}13)$$

where k_3 is the same as in Eq. 14-8.

Since the rotor squirrel-cage winding is short circuited by the end rings, these induced voltages at the slip frequency result in rotor currents I_r at the slip frequency f_{sl}:

$$\mathbf{E}_r = R_r\mathbf{I}_r + j2\pi f_{sl}L_{lr}\mathbf{I}_r \qquad (14\text{-}14)$$

where R_r and L_{lr} are the resistance and the leakage inductance of the per-phase equivalent rotor winding. The slip-frequency rotor currents produce a field that rotates at the slip speed with respect to the rotor and, hence, at the synchronous speed with respect to the stator (since $\omega_{sl} + \omega_r = \omega_s$). The interaction of ϕ_{ag} and the field produced by the rotor currents results in an electromagnetic torque. Losses in the rotor winding resistance are

$$P_r = 3R_rI_r^2 \qquad (14\text{-}15)$$

Multiplying both sides of Eq. 14-14 by f/f_{sl} and using Eqs. 14-8 and 14-13 give

$$\mathbf{E}_{ag} = \frac{f}{f_{sl}}\mathbf{E}_r = f\frac{R_r}{f_{sl}}\mathbf{I}_r + j2\pi fL_{lr}\mathbf{I}_r \qquad (14\text{-}16)$$

as shown in Fig. 14-2a, where fR_r/f_{sl} is represented as a sum of R_r and $R_r(f - f_{sl})/f_{sl}$. In Eq. 14-16, all rotor quantities are referred to N_s (the stator number of turns). By multiplying both sides of Eq. 14-16 by $\mathbf{I}_r^*$ and taking the real part $\text{Re}[\mathbf{E}_r\mathbf{I}_r^*]$, the power crossing the air gap, often called the air gap power P_{ag}, is

$$P_{ag} = 3\frac{f}{f_{sl}}R_rI_r^2 \qquad (14\text{-}17)$$

From Eqs. 14-17 and 14-15, the electromechanical power P_{em} is calculated as

$$P_{em} = P_{ag} - P_r = 3R_r\frac{f - f_{sl}}{f_{sl}}I_r^2 \qquad (14\text{-}18a)$$

and

$$T_{em} = \frac{P_{em}}{\omega_r} \qquad (14\text{-}18b)$$

From Eqs. 14-9, 14-17, 14-18a, and 14-18b

$$T_{em} = \frac{P_{ag}}{\omega_s} \qquad (14\text{-}18c)$$

In the equivalent circuit of Fig. 14-2a, the loss in the rotor resistance and the per-phase electromechanical power are shown by splitting the resistance $f(R_r/f_{sl})$ in Eq. 14-16 into R_r and $R_r(f - f_{sl})/f_{sl}$.

The total current $\mathbf{I}_s$ drawn by the stator is the sum of the magnetizing current $\mathbf{I}_m$ and the equivalent rotor current $\mathbf{I}_r$ ($\mathbf{I}_r$ here is the component of the stator current that cancels out the ampere-turns produced by the actual rotor current):

$$\mathbf{I}_s = \mathbf{I}_m + \mathbf{I}_r \qquad (14\text{-}19)$$

The phasor diagram for the stator voltages and currents is shown in Fig. 14-2b. The magnetizing current $\mathbf{I}_m$, which produces ϕ_{ag}, lags the air gap voltage by 90°. The current $\mathbf{I}_r$, which is responsible for producing the electromagnetic torque, lags $\mathbf{E}_{ag}$ by the power factor angle θ_r of the rotor circuit:

$$\theta_r = \tan^{-1}\frac{2\pi f_{sl}L_{lr}}{R_r} = \tan^{-1}\frac{2\pi f L_{lr}}{R_r f/f_{sl}} \qquad (14\text{-}20)$$

From electromagnetic theory, the torque produced is

$$T_{em} = k_4\phi_{ag}I_r\sin \delta \qquad (14\text{-}21)$$

where

$$\delta = 90° + \theta_r \qquad (14\text{-}22)$$

is the torque angle between the magnetizing current $\mathbf{I}_m$, which produces ϕ_{ag}, and $\mathbf{I}_r$, which represents the rotor field. The applied per-phase stator voltage V_s is given as

$$V_s = \mathbf{E}_{ag} + (R_s + j2\pi f L_{ls})\mathbf{I}_s \qquad (14\text{-}23)$$

In induction motors of normal design, the following condition is true in the rotor circuit at low values of f_{sl} corresponding to normal operation:

$$2\pi f_{sl}L_{lr} \ll R_r \qquad (14\text{-}24)$$

Therefore, θ_r in Eq. 14-20 approximately equals zero and the torque angle δ in Eq. 14-22 equals 90°. Therefore, in Eq. 14-21

$$T_{em} \simeq k_4\phi_{ag}I_r \qquad (14\text{-}25)$$

From Eqs. 14-13 and 14-14, using the approximation in Eq. 14-24,

$$I_r \simeq k_5\phi_{ag}f_{sl} \qquad (14\text{-}26)$$

Combining Eqs. 14-25 and 14-26 yields

$$T_{em} \simeq k_6\phi_{ag}^2 f_{sl} \qquad (14\text{-}27)$$

The approximation in Eq. 14-24 also allows the relationship in Eq. 14-19 to result in

$$I_s \simeq \sqrt{I_m^2 + I_r^2} \tag{14-28}$$

For normal motor parameters, except at low values of operating frequency f, as will be discussed later, in Eq. 14-23

$$V_s \approx E_{ag} \tag{14-29}$$

Using Eq. 14-8 in Eq. 14-29 yields

$$V_s \simeq k_3 \phi_{ag} f \tag{14-30}$$

From Eqs. 14-15 and 14-18a, the ratio of the power loss in the rotor to the electromechanical output power P_{em} is

$$\%P_r = \frac{P_r}{P_{em}} = \frac{f_{sl}}{f - f_{sl}} \tag{14-31}$$

The important equations for a frequency-controlled induction motor are summarized in Table 14-1, some of which assume that the condition in Eq. 14-24 is valid.

The following important observations can be drawn from these relationships:

1. The synchronous speed can be varied by varying the frequency f of the applied voltages.
2. Except at low values of f, the percentage of power loss in the motor resistance is small, provided f_{sl} is small. Therefore, in steady state, the slip frequency f_{sl} should not exceed its rated value (corresponding to the motor operation at the rated conditions listed on its nameplate).
3. With small f_{sl}, except at low values of f, the slip s is small and the motor speed varies approximately linearly with the frequency f of the applied voltages.
4. For the torque capability to equal the rated torque at any frequency, ϕ_{ag} should be kept constant and equal to its rated value. This requires that V_s must vary proportionately to f (the voltage boost needed at low values of f is discussed later on).
5. Since I_r is proportional to f_{sl}, to restrict the motor current I_s from exceeding its rated value, the steady-state slip frequency f_{sl} should not exceed its rated value.

Based on the preceding observations, it can be concluded that the motor speed can be varied by controlling the applied frequency f, and the air gap flux should be kept constant

Table 14-1 Important Relationships

$$\omega_s = k_7 f$$
$$s = \frac{\omega_s - \omega_r}{\omega_s}$$
$$f_{sl} = sf$$
$$\%P_r = \frac{f_{sl}}{f - f_{sl}}$$
$$V_s \simeq k_3 \phi_{ag} f$$
$$I_r \simeq k_5 \phi_{ag} f_{sl}$$
$$T_{em} \simeq k_6 \phi_{ag}^2 f_{sl}$$
$$I_m = k_8 \phi_{ag} \quad \text{(from Eq. 14-5)}$$
$$I_s \simeq \sqrt{I_m^2 + I_r^2}$$

at its rated value by controlling the magnitude of the applied voltages in proportion to f. If an induction motor is controlled in such a way, then the motor is capable of supplying its rated torque while f_{sl}, I_r, I_s, and the percentage losses in the rotor circuit all remain within their respective rated values.

14-3   INDUCTION MOTOR CHARACTERISTICS AT RATED (LINE) FREQUENCY AND RATED VOLTAGE

Typical characteristics of an induction motor under nameplate values of frequency and voltage are shown in Figs. 14-3 and 14-4 where T_{em} and I_r, respectively, are plotted as functions of rotor speed and f_{sl}. At low values of f_{sl}, T_{em} and I_r vary linearly with f_{sl}. As f_{sl} becomes larger, T_{em} and I_r no longer increase linearly with f_{sl} for the following reasons: (1) the rotor circuit inductive reactance term is no longer negligible compared to R_r in Eq. 14-14; (2) θ_r in Eq. 14-20 becomes significant, thus causing δ to depart from its optimum value of 90°; and (3) large values of $\mathbf{I}_r$, and hence $\mathbf{I}_s$, cause significant voltage drop across the stator winding impedance in Eq. 14-23 and hence cause ϕ_{ag} ($=E_{ag}/f$) to decline for a fixed supply input V_s at frequency f. All of these effects take place simultaneously, and the resulting torque and current characteristics for large f_{sl} are shown as dashed in Figs. 14-3 and 14-4. The maximum torque that the motor can produce is called the pull-out torque.

   It should be emphasized that in the commonly used induction motor drives, which are discussed in detail in this chapter, f_{sl} is kept small, and hence, the dashed portions of the torque and the current characteristics of Figs. 14-3 and 14-4 are not used. However, if an induction motor is started from the line-voltage supply without a power electronic con-

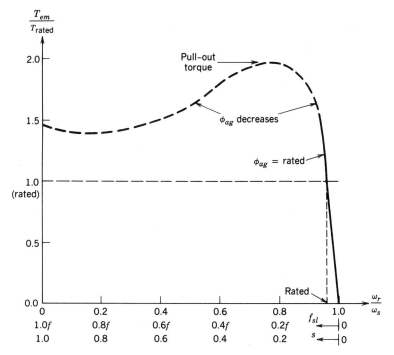

Figure 14-3   A typical torque–speed characteristic; V_s and f are constant at their rated values.

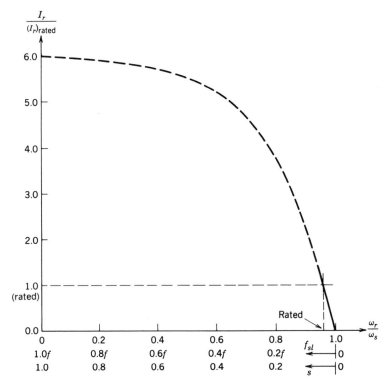

Figure 14-4   Plot of I_r versus f_{sl}; V_s and f are constant at their rated values.

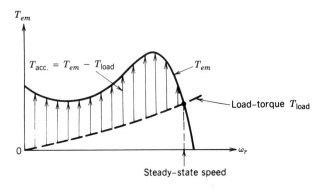

Figure 14-5   Motor start-up; V_s and f are constant at their rated values.

troller, it would draw 6 to 8 times its rated current at start-up as shown in Fig. 14-4. Figure 14-5 shows the available acceleration torque ($T_{\text{em}} - T_{\text{load}}$) for the motor to accelerate from standstill. Here, an arbitrary torque–speed characteristic of the load is assumed, and the intersection of the motor and the load characteristics determines the steady-state point of operation.

14-4   SPEED CONTROL BY VARYING STATOR FREQUENCY AND VOLTAGE

The discussion in Section 14-2 suggested that the speed can be controlled by varying f, which controls the synchronous speed (and, hence, the motor speed, if the slip is kept small), keeping ϕ_{ag} constant by varying V_s in a linear proportion to f. We will examine other speed control techniques later on, but varying the stator frequency and voltage is the preferred technique in most variable-speed induction motor drive applications, and hence, we will discuss it in detail.

14-4-1 TORQUE–SPEED CHARACTERISTICS

From the relationships in Table 14-1 for small values of f_{sl}, keeping ϕ_{ag} constant results in a linear relationship between T_{em} and f_{sl} at any value of f:

$$T_{em} \simeq k_9 f_{sl} \qquad (14\text{-}32)$$

which represents the solid portion of the torque–speed characteristic in Fig. 14-3. Since f is varied, it is preferable to express T_{em} as a function of the slip speed ω_{sl}. From Eqs. 14-3 and 14-12

$$\omega_{sl} = \frac{f_{sl}}{f}\,\omega_s = \frac{4\pi}{p}f_{sl} \qquad (14\text{-}33)$$

From Eqs. 14-32 and 14-33

$$T \simeq k_{10}\omega_{sl} \qquad (14\text{-}34)$$

Such a characteristic is shown in Fig. 14-6 for frequency f equal to f_1 with a corresponding synchronous speed ω_{s1}.

The torque–speed characteristics shift horizontally in parallel, as shown in Fig. 14-6 for four different values of f. To explain this, consider two frequencies f_1 and f_2. The synchronous speeds ω_{s1} and ω_{s2} are in proportion to f_1 and f_2. If an equal load torque is to be delivered at both these frequencies, from Eq. 14-34, $\omega_{sl1} = \omega_{sl2}$. Therefore, in the torque–speed plane of Fig. 14-6, equal torques and equal slip speeds (at f_1 and f_2) result in parallel but horizontally shifted characteristics.

Note that at a constant load torque, the slip frequency (which is the frequency of the induced voltages and currents in the rotor circuit in hertz) is constant, but from Eq. 14-12 the slip s goes up as frequency f goes down. From Eq. 14-31, the percentage power loss in the rotor increases as f is decreased to reduce the motor speed. However, in many loads such as the centrifugal pumps, compressors, and fans, the load torque varies by the square of the speed, as given by Eq. 14-1. In such cases, f_{sl} as well as s declines with decreasing frequency, as shown in Fig. 14-7. Hence, the rotor losses remain small.

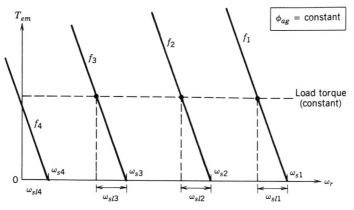

Figure 14-6 Torque–speed characteristics at small slip with a constant ϕ_{ag}; constant load torque.

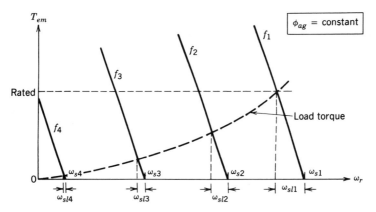

Figure 14-7 Centrifugal load torque; torque varies as the speed squared.

■ *Example 14-1* A four-pole, 10-hp, 460-V motor is supplying its rated power to a centrifugal load at a 60-Hz frequency. Its rated speed is 1746 rpm.

Calculate its speed, slip frequency, and slip when it is supplied by a 230-V, 30-Hz source.

Solution

At 60 Hz,

$$n_s = 1800 \text{ rpm (four-pole)}$$

$$s_{rated} = \frac{1800 - 1746}{1800} = 3\%$$

$$(f_{sl})_{rated} = s_{rated}f = 0.03 \times 60 = 1.8 \text{ Hz}$$

$$(n_{sl})_{rated} = 1800 - 1746 = 54 \text{ rpm}$$

At 30 Hz, keeping V_s/f constant,

$$T_{em} \simeq \frac{1}{4}T_{rated} \quad \text{(centrifugal load; using Eq. 14-1)}$$

$$f_{sl} = \frac{1}{4}(f_{sl})_{rated} = \frac{1.8}{4} = 0.45 \text{ Hz} \quad \text{(using Eq. 14-32)}$$

$$n_{sl} = \frac{120}{\text{poles}}f_{sl} = \frac{120}{4} \times 0.45 = 13.5 \text{ rpm}$$

$$n_s = 900 \text{ rpm}$$

$$\therefore n_r = n_s - n_{sl} = 900 - 13.5 = 886.5 \text{ rpm}$$

$$s = \frac{f_{sl}}{f} = \frac{0.45}{30} = 1.5\%$$ ■

14-4-2 START-UP CONSIDERATIONS

For a solid-state inverter-driven induction motor, it is an important consideration to keep the current draw from becoming large during start-up. This can be achieved by considering the following relationship: for a constant ϕ_{ag} from Eq. 14-26

$$I_r \simeq k_{11}f_{sl} \qquad (14\text{-}35)$$

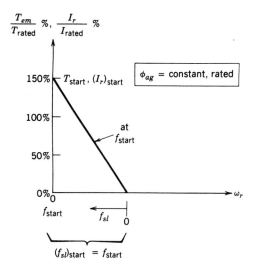

Figure 14-8   Frequency at start-up.

Using Eqs. 14-32 and 14-35, T_{em} and I_r are plotted in Fig. 14-8 to show how the motor can be started at a small applied frequency $f\,(=f_{start})$. Since at start-up f_{sl} equals f_{start}, I_r can be limited by selecting an appropriate f_{start}. With a constant I_m due to a constant ϕ_{ag}, the stator current I_s is therefore kept from becoming large.

For example, if the starting torque is required to be 150% of the rated torque and the solid-state drive can withstand a current overload of 150% on a short-term basis, the starting frequency f_{start} can be determined from the motor nameplate ratings. For the motor in Example 14-1, the starting frequency for 150% torque (and, hence, current) based on the rated speed of 1746 rpm at 60 Hz is calculated by using Fig. 14-8 as

$$f_{start} = \frac{T_{start}}{T_{rated}} (f_{sl})_{rated} \qquad (14\text{-}36)$$
$$= 1.5 \times 1.8 = 2.7 \text{ Hz}$$

In practice, the stator frequency f is increased continuously at a preset rate, as shown in Fig. 14-9, which does not let the current I_s exceed a specified limit (like 150% of the rated) until the final desired speed has been achieved. This rate is decreased for higher inertia loads to allow the rotor speed to catch up.

14-4-3   VOLTAGE BOOST REQUIRED AT LOW FREQUENCIES

The effect of R_s at low values of operating frequency f cannot be neglected, even if f_{sl} is small. This can be easily seen if the following observation is made: in induction motors of normal design, $2\pi f L_{lr}$ is negligible in comparison to $R_r(f/f_{sl})$ in the equivalent circuit

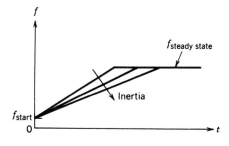

Figure 14-9   Ramping of frequency f at start-up.

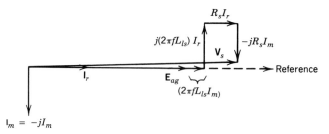

$$I_m = -jI_m$$

Figure 14-10   Phasor diagram at a small value of f_{sl}.

of Fig. 14-2a. Therefore, $\mathbf{I}_r$ will be in phase with $\mathbf{E}_{ag}$. Using $\mathbf{E}_{ag}$ as the reference phasor, $\mathbf{I}_s = I_r - jI_m$. Therefore, Eq. 14-23 can be written as

$$\mathbf{V}_s \simeq [E_{ag} + (2\pi f L_{ls})I_m + R_s I_r] + j[(2\pi f L_{ls})I_r - R_s I_m] \qquad (14\text{-}37)$$

and represented by the phasor diagram of Fig. 14-10. As shown by Fig. 14-10, the second term in the right-hand side of Eq. 14-37 corresponds to a phasor that is almost perpendicular to $\mathbf{V}_s$, and therefore, its influence on the magnitude of V_s can be neglected:

$$V_s \simeq E_{ag} + (2\pi f L_{ls})I_m + R_s I_r \qquad (14\text{-}38a)$$

If ϕ_{ag} is kept constant, E_{ag} varies linearly with f. If ϕ_{ag} is kept constant, I_m is also constant. Therefore, the additional voltage required due to L_{ls} in Eq. 14-38a is also proportional to the operating frequency f. Therefore, for a constant ϕ_{ag}, Eq. 14-38a can be written as

$$V_s \approx k_{12}f + R_s I_r \qquad (14\text{-}38b)$$

Equation 14-38b shows that the additional voltage required to compensate for the voltage drop across R_s to keep ϕ_{ag} constant does not depend on f but depends on I_r. Recognizing that I_r is proportional to T_{em}, the terminal voltage V_s required to keep ϕ_{ag} constant at the rated torque is shown by a solid line in Fig. 14-11. A voltage proportional to f, with the rated voltage at the rated frequency, is indicated by a dashed line in Fig. 14-11. The voltage boost required to maintain a constant air gap flux for a given T_{em} can be obtained from Eq. 14-38b and Fig. 14-11. Figure 14-11 shows that to keep ϕ_{ag} constant, a much

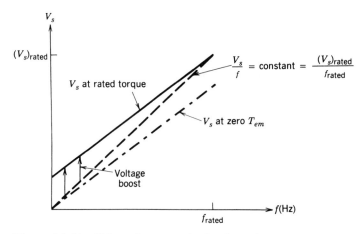

Figure 14-11   Voltage boost required to keep ϕ_{ag} constant.

higher percentage voltage boost is required at low operating frequencies due to the voltage drop across R_s, whereas at large values of f, the voltage drop across R_s can be neglected in comparison to E_{ag}. The voltage required at no load is shown by a short–long dashed line.

14-4-4 INDUCTION MOTOR CAPABILITY: BELOW AND ABOVE THE RATED SPEED

Speed control by means of frequency (and voltage) variation also allows the capability to operate the motor not only at speeds below the rated speed but also at above the rated speed. This capability is very attractive in many applications, since most induction motors, because of their rugged construction, can be operated up to twice the rated speed without mechanical problems. However, the torque and power capabilities as a function of rotor speed need to be clearly established.

The motor torque–speed characteristics are shown in Fig. 14-12a. In Fig. 14-12b, V_s, I_r, I_m, and T_{em} are plotted as functions of the normalized rotor speed; f_{sl} and s are plotted in Fig. 14-12c. It should be noted that in large motors at the limit of motor capability, $I_s \approx I_r$, since the contribution of I_m to I_s in Eq. 14-28 is small.

14-4-4-1 Below the Rated Speed: Constant-Torque Region

In the region of speed below its rated value, the solid curves in Fig. 14-12a show the motor torque–speed characteristics at low values of f_{sl} where ϕ_{ag} is kept constant by controlling V_s/f. The stator voltage magnitude is decreased approximately in proportion to the frequency from its rated value down to very low values, as shown in Fig. 14-12b. If ϕ_{ag} is maintained constant, the motor can deliver its rated torque (on a continuous basis) by drawing its rated current at a constant f_{sl} as shown in Fig. 14-12. Therefore, this region (below the rated speed) is called the constant-torque region.

In this region, f_{sl} remains constant at its full-load (rated) value while delivering the rated torque. Figure 14-12c shows f_{sl} and s.

At the constant rated torque, the power loss $P_r = 3R_r I_r^2$ in the rotor resistances is also constant, where I_r stays constant. However, in practice, getting rid of this rotor heat due to P_r becomes a problem at low speeds due to reduced cooling. Therefore, unless the motor has a constant-speed fan or is designed to be totally enclosed and nonventilated, the torque capability drops off at very low speeds. It should be noted that this is of no concern in centrifugal loads where the torque requirement is very low at low speeds.

14-4-4-2 Beyond the Rated Speed: Constant-Power Region

By increasing the stator frequency above its nominal (rated) value, it is possible to increase the motor speed beyond the rated speed. In most adjustable-speed drive applications, the motor voltage is not exceeded beyond its rated value, unlike that explained in Section 14-4-4-4. Therefore, by keeping V_s at its rated value, increasing the frequency f results in a reduced V_s/f and, hence, a reduced ϕ_{ag}. From Eqs. 14-27, 14-30, and 14-33 in this region

$$T_{em} \approx \frac{k_{13}}{f^2} \omega_{sl} \qquad (14\text{-}39)$$

which results in torque–speed curves whose slopes are proportional to $(1/f)^2$, as shown in Fig. 14-12a for higher than rated frequencies.

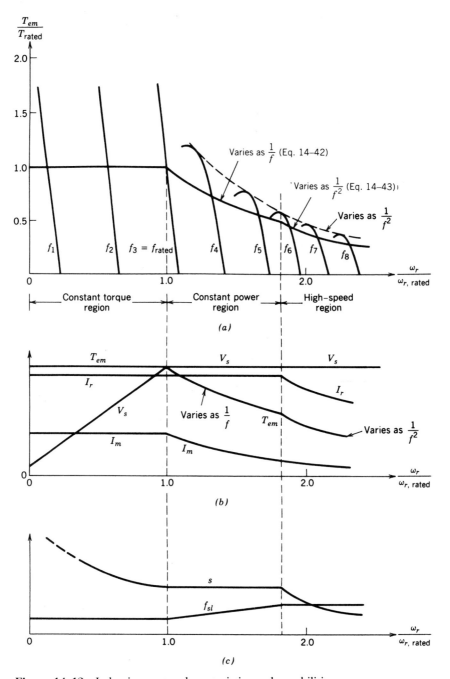

Figure 14-12   Induction motor characteristics and capabilities.

At the limit of the motor capability in this region, I_r equals its rated value, similar to the previous region. This corresponds to a constant $s = f_{sl}/f$ in this region, which can be shown by using Eqs. 14-12, 14-26, and 14-30:

$$I_r \simeq k_{14}\frac{f_{sl}}{f} \simeq k_{14}s = \text{constant} \qquad (14\text{-}40)$$

The slip frequency f_{sl} now increases with f, as shown in Fig. 14-12c. At a constant slip

$$\omega_r = (1 - s)\omega_s = k_{15}f \qquad (14\text{-}41)$$

Using both V_s and f_{sl}/f as constants, the maximum torque in this region can be calculated from Eqs. 14-27 and 14-30 in terms of the rated torque and the rated frequency:

$$T_{em,max} = \frac{f_{rated}}{f} T_{rated} \qquad (14\text{-}42)$$

Therefore $P_{em,max} = \omega_r T_{em,max}$ can be held constant at its rated value, recognizing from Eq. 14-41 that ω_r is proportional to f. Hence, this region of operation is called the constant-power region. In Fig. 14-12b, V_s, I_r, I_m, and the maximum steady-state T_{em} are plotted.

In practice, the motor can deliver higher than its rated power by noting that (1) I_m goes down as a result of decreased ϕ_{ag}, and therefore, I_s equal to its rated value allows a higher value of I_r and hence higher torque and power, and (2) since I_m is decreased, the core losses are reduced and, at the same time, there is better cooling at higher speeds.

14-4-4-3 High-Speed Operation: Constant-f_{sl} Region

With V_s equal to its rated value, depending on the motor design, beyond a speed somewhere in a range of 1.5–2 times the rated speed, ϕ_{ag} is reduced so much that the motor approaches its pull-out torque, as is shown in Fig. 14-12a. At still higher speeds, the motor can deliver only a fixed percentage of the pull-out torques, as shown graphically by Fig. 14-12a, and $\omega_{sl}(f_{sl})$ becomes constant. Therefore, the torque capability declines as

$$T_{em,max} \simeq k_{16}\frac{1}{f^2} \qquad (14\text{-}43)$$

Both the torque and the motor current decline with speed, as is shown in Fig. 14-12b. Keeping V_s constant, the motor torque in this region is not limited by the current-handling capability of the motor, since the current at the limit is less than its rated value and declines with speed, as is shown in Fig. 14-12b; rather, it is limited by the maximum torque produced by the motor.

14-4-4-4 Higher Voltage Operation

In most motors, the voltage insulation level is much higher than the specified rated voltage of the motor. Therefore, by means of a proper solid-state power source, it is possible to apply a higher than rated voltage at speeds above the rated speed of the motor. This is particularly easy to see in the case of a dual-voltage motor, for example, a 230/460-V motor. With the motor connected for a 230-V operation, if 460 V is applied at twice the rated frequency, the motor operates at the rated air gap flux and, hence, can deliver its rated torque without exceeding its current rating. Since the motor under these conditions will run at twice its rated speed, it will also deliver twice its rated power. Before using a motor in this manner, the motor manufacturer ought to be consulted.

14-4-5 BRAKING IN INDUCTION MOTORS

In many applications, it is repeatedly required to quickly reduce the motor speed or to bring it to a halt. One of the advantages of using a variable-frequency controller for speed control is that it can accomplish this in a controlled manner. This provides a controlled alternative to the following ways of speed reduction, which are not as attractive: mechanical brakes, which waste energy associated with load–motor inertia and whose brake pads wear out with repeated use; letting the motor coast to a halt, which could take a long time; and "plugging" where the phase sequence of the utility supply to the motor is suddenly reversed, causing large currents to flow into the utility source and bringing the motor to a halt in an uncontrolled manner.

To understand braking in induction motors, it should be realized that it is possible to operate an induction machine as a generator by mechanically driving it above the synchronous speed (which is related to the supply voltage frequency). As shown in Fig. 14-13a, a rotor speed higher than ω_s results in a negative slip speed ω_{sl} and a negative slip s. The torque–speed characteristic corresponding to $\omega_r > \omega_s$ is shown in Fig. 14-13b, where the electromagnetic torque developed in this mode is negative and acts in an opposite direction to the direction of rotating magnetic field.

Note that for the induction machine to operate in a generator mode, the ac voltages must be present at the stator terminals, that is, the machine will not generate, for example, if only a resistor bank is connected to the stator terminals and the shaft is turned; there is no source to establish the rotating magnetic field in the air gap.

The generation mechanism discussed before is used to provide braking in variable-frequency induction motor drives. Figure 14-14 shows the motor torque–speed characteristics at two frequencies, assuming a constant ϕ_{ag}. These curves are extended beyond the corresponding synchronous speeds. Consider that the motor is initially operating with a stator frequency f_0 at a rotor speed of ω_{r0} below ω_{s0}. If the stator frequency is decreased to f_1, the new synchronous speed is ω_{s1}. The slip speed becomes negative and thus T_{em} becomes negative, as is shown in Fig. 14-14. This negative T_{em} causes the motor speed to decrease quickly and some of the energy associated with the motor–load inertia is fed into the source connected to the stator.

In practice, the stator frequency (keeping ϕ_{ag} constant) is reduced slowly to avoid large currents through the variable-frequency controller. This procedure, if used to bring the motor to a halt, can be viewed as opposite of the start-up procedure. It should be noted that the variable-frequency controller must be capable of handling the energy supplied by the motor in the braking mode.

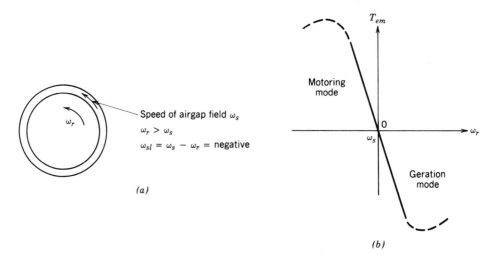

(a)

Speed of airgap field ω_s

$\omega_r > \omega_s$

$\omega_{sl} = \omega_s - \omega_r$ = negative

(b)

Figure 14-13 Generation mode.

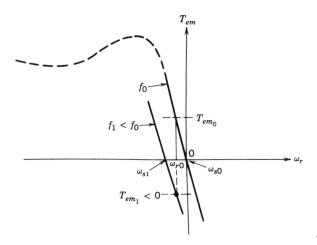

Figure 14-14 Braking (initial motor speed is ω_{r0} and the applied frequency is instantaneously decreased from f_0 to f_1).

14-5 IMPACT OF NONSINUSOIDAL EXCITATION ON INDUCTION MOTORS

In the preceding section, it was assumed that the induction motors are supplied from a three-phase, balanced, and sinusoidal set of voltages. In practice, the inverters used in variable-frequency controllers produce three phase voltages or currents that are identical in each phase, except for the 120° phase displacement. Unfortunately, these are not purely sinusoidal and contain higher frequency components that are harmonics of the fundamental frequency, as discussed in Chapter 8. In the following analysis, we will assume that the motor is supplied by three-phase voltage sources, as in the case of a voltage source inverter. This analysis can be easily modified to three-phase current sources, as in the case of a current source inverter.

14-5-1 HARMONIC MOTOR CURRENTS

As a first-order approximation, the motor currents in the presence of harmonic voltage components can be found by calculating each harmonic current component i_h (at harmonic h) from the per-phase equivalent circuit of Fig. 14-2a. Then the motor currents can be obtained by using the principle of superposition and adding the fundamental and all other harmonic current components.

At a harmonic h (which, in practice, will be odd and not a multiple of 3), the flux produced by the voltage components (v_{ah}, v_{bh}, v_{ch}) rotates in the air gap at a speed of

$$\omega_{sh} = h\omega_s \qquad (14\text{-}44)$$

where the direction may be the same or in opposition to the rotor's direction of rotation. It can be easily verified that the flux produced by the harmonics $h = 6n - 1$ (where, $n = 1, 2, 3, \ldots$) has an opposite phase rotation compared with the fundamental. Therefore, these harmonics result in a flux rotation opposite to that of the rotor. Harmonics $h = 6n + 1$ (where $n = 1, 2, 3, \ldots$) produce a flux rotation in the same direction as the rotor.

Under a variable-frequency operation for speed control, the motor rotates at a reasonably small value of slip; thus, to a first-order approximation, the rotor speed can be assumed to be the same as the fundamental-frequency synchronous speed:

$$\omega_r \simeq \omega_s \qquad (14\text{-}45)$$

Therefore, at a harmonic h in the equivalent circuit of Fig. 14-2a, using Eqs. 14-44 and 14-45, the rotor slip relative to the synchronous speed at the harmonic frequency is

$$\text{Slip } s_h = \frac{\omega_{sh} \pm \omega_r}{\omega_{sh}} \simeq \frac{h \pm 1}{h} \approx 1 \qquad (14\text{-}46)$$

where the plus-or-minus sign corresponds to the direction of air gap flux rotation in opposition to or the same as the rotor's direction of rotation, respectively. Recognizing that $\omega_r \approx \omega_s$ and $s_h \simeq 1$, an approximate equivalent circuit at the frequency of harmonic h is shown in Fig. 14-15, which is obtained by the equivalent circuit of Fig. 14-2a by neglecting L_m. If the motor is excited from a voltage source and the harmonic components of the stator voltage v_s are known, the corresponding harmonic components in the motor current i_s can be obtained by using the principle of superposition and the harmonic equivalent circuit of Fig. 14-15 for each harmonic, one at a time.

For calculating the harmonic current components, the magnetizing components can generally be neglected and the harmonic current magnitude is primarily determined by the leakage reactances at the harmonic frequency, which dominate over R_s and R_r:

$$I_h \approx \frac{V_h}{h\omega(L_{ls} + L_{lr})} \qquad (14\text{-}47)$$

Equation 14-47 for I_h shows that by increasing the frequencies at which the harmonic voltages occur in the converter output (which is accomplished by increasing the switching frequency, as discussed in Chapter 8), the magnitudes of harmonic currents can be reduced. Note that the foregoing procedure for calculating harmonic currents is at best a first-order approximation, since the motor leakage reactances and resistances vary with frequency.

14-5-2 HARMONIC LOSSES

The per-phase additional power loss in the copper of stator and rotor windings due to these harmonic currents can be approximated as

$$\Delta P_{\text{cu}} = \sum_{h=2}^{\infty} (R_s + R_r)I_h^2 \qquad (14\text{-}48)$$

where R_s and R_r increase in a nonlinear manner with harmonic frequencies. It is tedious to estimate the additional core losses due to harmonic frequency eddy currents and hysteresis. These and the additional stray losses depend on the motor geometry, magnetic material used, lamination thickness, which may have been optimized for the 60-Hz frequency, and so on. These additional losses, which may be significant, can be measured, and the estimation procedures have been discussed in the literature. In general, these additional losses are in a range of 10–20% of the total power losses at the rated load.

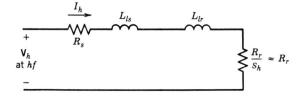

Figure 14-15   Per-phase harmonic equivalent circuit.

14-5-3 TORQUE PULSATIONS

Presence of harmonics in the stator excitation results in a pulsating-torque component. If the pulsating torques are at low frequencies, they can cause troublesome speed fluctuations and shaft fatigue.

Considering the lowest harmonic frequencies, which are fifth and seventh in a three-phase square-wave inverter, the generation of a pulsating-torque component can best be explained by considering these harmonic excitations one at a time.

In Fig. 14-16a, the seventh harmonic excitation results in an air gap flux component rotating at a speed $7\omega_s$ in the same direction as the fundamental air gap flux and the rotor. Assuming the rotor speed to be approximately equal to ω_s, it is easy to see that the rotor field produced consists of the fundamental component B_{r1} at a speed of ω_s and the seventh harmonic component B_{r7} at a speed of $7\omega_s$, as is shown in Fig. 14-16a. Fields ϕ_{ag1} and B_{r1} rotate at the same speed and, hence, result in a nonpulsating torque. The same is true for the interaction of ϕ_{ag7} and B_{r7}, which rotate at the same speed. However, the relative speed between ϕ_{ag7} and B_{r1} is $6\omega_s$. Similarly, the relative speed between ϕ_{ag1} and B_{r7} is $6\omega_s$. Therefore, both these interactions produce torque components that pulsate at a sixth harmonic frequency.

In Fig. 14-16b, the fifth harmonic excitation results in an air gap flux that rotates at a speed of $5\omega_s$ in a direction opposite to the rotor. The induced rotor fields are shown in Fig. 14-16b. Here ϕ_{ag5} interacts with B_{r1} and ϕ_{ag1} interacts with B_{r5} to produce torque components, both of which pulsate at a sixth harmonic frequency.

The above discussion shows that both fifth and seventh harmonic excitations combine to produce a torque that pulsates at the sixth harmonic frequency. Similar calculations can be made for other harmonic frequency excitations.

The effect of torque ripple on the ripple in the rotor speed can be written as follows, assuming no resonance occurs:

$$\text{Amplitude of speed ripple} = k_{17} \frac{\text{amplitude of torque ripple}}{\text{ripple frequency} \times \text{inertia}} \qquad (14\text{-}49)$$

which shows that a given amplitude of torque ripple may result in negligible speed ripple at high ripple frequencies.

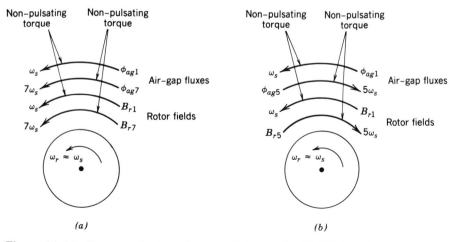

Figure 14-16 Torque pulsations: (a) seventh harmonic; (b) fifth harmonic.

14-6 VARIABLE-FREQUENCY CONVERTER CLASSIFICATIONS

Based on the discussion in the previous section, the variable-frequency converters, which act as an interface between the utility power system and the induction motor, must satisfy the following basic requirements:

1. Ability to adjust the frequency according to the desired output speed
2. Ability to adjust the output voltage so as to maintain a constant air gap flux in the constant-torque region
3. Ability to supply a rated current on a continuous basis at any frequency

Except for a few special cases of very high power applications where cycloconverters are used (these are briefly discussed in Chapter 15), variable-frequency drives employ inverters with a dc input, as discussed in Chapter 8. Figure 14-17 illustrates the basic concept where the utility input is converted into dc by means of either a controlled or an uncontrolled rectifier and then inverted to provide three phase voltages and currents to the motor, adjustable in magnitude and frequency. These converters can be classified based on the type of rectifier and inverter used in Fig. 14-17:

1. Pulse-width-modulated voltage source inverter (PWM-VSI) with a diode rectifier
2. Square-wave voltage source inverter (square-wave VSI) with a thyristor rectifier
3. Current source inverter (CSI) with a thyristor rectifier

As the names imply, the basic difference between the VSI and the CSI is the following: In the VSI, the dc input appears as a dc voltage source (ideally with no internal impedance) to the inverter. On the other hand, in the CSI, the dc input appears as a dc current source (ideally with the internal impedance approaching infinity) to the inverter.

Figure 14-18a shows the schematic of a PWM-VSI with a diode rectifier. In the square-wave VSI of Fig. 14-18b, a controlled rectifier is used at the front end and the inverter operates in a square-wave mode (also called the six-step). The line voltage may be single phase or three phase. In both VSI controllers, a large dc bus capacitor is used to make the input to the inverter appear as a voltage source with a very small internal impedance at the inverter switching frequency.

From the schematic of VSI converters shown in Figs. 14-18a and 14-18b, it is recognized that the switch-mode, dc-to-ac VSIs have been discussed previously in Chapter 8 in both square-wave and PWM modes of operation. It should be noted that, in practice, only three-phase motors are controlled by means of variable frequency. Therefore, only the dc-to-three-phase-ac inverters are applicable here. Also, the controlled and uncontrolled (diode) rectification of single-phase and three-phase ac inputs to dc has been discussed in detail in Chapters 5 and 6. Therefore, the main emphasis in this chapter will be on the interaction of VSIs with induction motor type of loads.

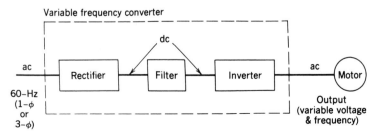

Figure 14-17 Variable-frequency converter.

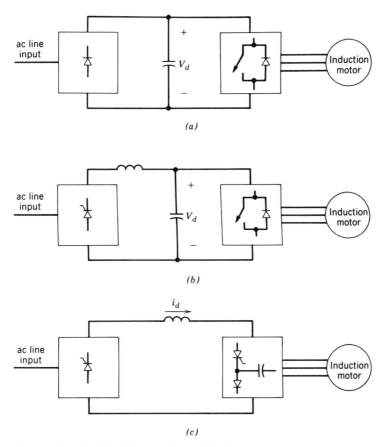

Figure 14-18 Classification of variable-frequency converters:
(a) PWM-VSI with a diode rectifier; (b) square-wave VSI with a
controlled rectifier; (c) CSI with a controlled rectifier.

Figure 14-18c shows the schematic of a CSI drive where a line-voltage-commutated
controlled converter (discussed in Chapter 6) is used at the front end. Because of a large
inductor in the dc link, the input to the inverter appears as a dc current source. The
inverter utilizes thyristors, diodes, and capacitors for forced commutation.

14-7 VARIABLE-FREQUENCY PWM-VSI DRIVES

Figure 14-19a shows the schematic of a PWM-VSI drive, assuming a three-phase utility
input. As a brief review of what has already been covered in Chapter 8, a PWM inverter
controls both the frequency and the magnitude of the voltage output. Therefore, at the
input, an uncontrolled diode bridge rectifier is generally used. One possible method of
generating the inverter switch control signals is by comparing three sinusoidal control
voltages (at the desired output frequency and proportional to the output voltage magnitude) with a triangular waveform at a selected switching frequency, as shown in Fig.
14-19b.

As discussed in Chapter 8, in a PWM inverter, the harmonics in the output voltage
appear as sidebands of the switching frequency and its multiples. Therefore, a high

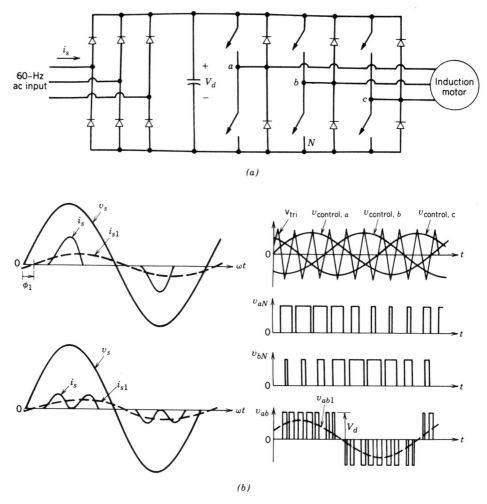

Figure 14-19   PWM-VSI: (*a*) schematic; (*b*) waveforms.

switching frequency results in an essentially sinusoidal current (plus a superimposed small ripple at a high frequency) in the motor.

Since the ripple current through the dc bus capacitor is at the switching frequency, the *input dc source* impedance seen by the inverter would be smaller at higher switching frequencies. Therefore, a small value of capacitance suffices in PWM inverters, but this capacitor must be able to carry the ripple current. A small capacitance across the diode rectifier also results in a better input current waveform drawn from the utility source. However, care should be taken in not letting the voltage ripple in the dc bus voltage become too large, which would cause additional harmonics in the voltage applied to the motor.

14-7-1   IMPACT OF PWM-VSI HARMONICS

In a PWM inverter output voltage, since the harmonics are at a high frequency, the ripple in the motor current is usually small due to high leakage reactances at these frequencies. Since these high-frequency voltage harmonics can have as high or even higher amplitude compared to the fundamental-frequency component, the iron losses (eddy current and hysteresis in the stator and the rotor iron) dominate. In fact, the total losses due to harmonics may even be higher with a PWM inverter than with a square-wave inverter. This comparison would of course depend on the motor design class, magnetic material property, and switching frequency. Because of these additional harmonic losses, it is generally recommended that a standard motor with a 5–10% higher power rating be used.

In a PWM drive, the pulsating torques developed are small in amplitude and are at high frequencies (compared to the fundamental). Therefore, as shown in Eq. 14-49, they produce little speed pulsations because of the motor inertia.

14-7-2 INPUT POWER FACTOR AND CURRENT WAVEFORM

The input ac current drawn by the rectifier of a PWM-VSI drive contains a large amount of harmonics similar to that discussed in Chapter 5. Its waveform is shown in Fig. 14-19b for a single-phase and a three-phase input. As discussed in Chapter 5, the input inductance L_s improves the input ac current waveform somewhat. Also, a small dc-link capacitance will result in a better waveform.

The power factor at which the drive operates from the utility system is essentially independent of the motor power factor and the drive speed. It is only a slight function of the load power, improving slightly at a higher power. The displacement power factor (DPF) is approximately 100%, as can be observed from the input current waveforms of Fig. 14-19b.

14-7-3 ELECTROMAGNETIC BRAKING

As we discussed in Section 14-4-5, the power flow during electromagnetic braking is from the motor to the variable-frequency controller. During braking, the voltage polarity across the dc-bus capacitor remains the same as in the motoring mode. Therefore, the direction of the dc bus current to the inverter gets reversed. Since the current direction through the diode rectifier bridge normally used in PWM-VSI drives cannot reverse, some mechanism must be implemented to handle this energy during braking; otherwise the dc-bus voltage can reach destructive levels.

One way to accomplish this goal is to switch on a resistor in parallel with the dc-bus capacitor, as is shown in Fig. 14-20a, if the capacitor voltage exceeds a preset level, in order to dissipate the braking energy.

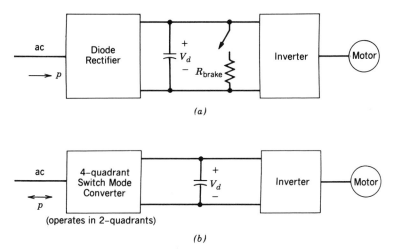

Figure 14-20 Electromagnetic braking in PWM-VSI: (a) dissipative braking; (b) regenerative braking.

An energy-efficient technique is to use a four-quadrant converter (switch-mode or a back-to-back connected thyristor converter) at the front end in place of the diode bridge rectifier. This would allow the energy recovered from the motor–load inertia to be fed back to the utility supply, as shown in Fig. 14-20b, since the current through the four-quadrant converter used for interfacing with the utility source can reverse in direction. This is called regenerative braking since the recovered energy is not wasted. The decision to employ regenerative braking over dissipative braking depends on the additional equipment cost versus the savings on energy recovered and the desirability of sinusoidal currents and unity power factor operation from the utility source.

14-7-4   ADJUSTABLE-SPEED CONTROL OF PWM-VSI DRIVES

In VSI drives (both PWM and square-wave type), the speed can be controlled without a speed feedback loop, where there may be a slower acting feedback loop through the processor controller, as explained in Chapter 12. Figure 14-21 shows such a control. The frequency f of the inverter output voltages is controlled by the input speed reference signal ω_{ref}. The input command ω_{ref} is modified for protection and improved performance, as will be discussed shortly, and the required control inputs (ω_s or f and V_s signals) to the PWM controller in Fig. 14-21 are calculated. The PWM controller can be realized by analog components, as discussed in Chapter 8 and indicated by Fig. 14-19b. The control signals (e.g., $v_{a,\text{control}}$) can be calculated from the f and V_s signals and by knowing V_d and $\hat{V}_{\text{tri}}$.

As discussed in Chapter 8, a synchronous PWM must be used. This requires that the switching frequency vary in proportion to f. To keep the switching frequency close to its maximum value, there are jumps in m_f and, hence, in f_s as f decreases, as shown in Fig. 14-22. To prevent jittering at frequencies where jumps occur, a hysteresis must be provided. Digital ICs such as HEF5752V are commercially available that incorporate many of the functions of the PWM controller described earlier.

For protection and better speed accuracy, current and voltage feedback may be employed. These signals are required anyway for starting/stopping of the drive, to limit

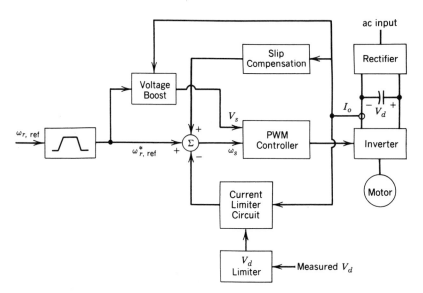

Figure 14-21   Speed control circuit. Motor speed is not measured.

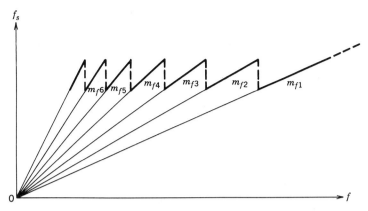

Figure 14-22 Switching frequency versus the fundamental frequency.

the maximum current through the drive during acceleration/deceleration or under heavy load conditions, and to limit the maximum dc link voltage during braking of the induction motor. Because of slip, the induction motor operates at a speed lower than the synchronous speed. It is possible to approximately compensate for this slip speed, which increases with torque, without measuring the actual speed. Moreover, a voltage boost is required at lower speeds. To meet these objectives, the motor currents and the dc link voltage V_d across the capacitor are measured. To represent the instantaneous three-phase ac motor currents, a current i_o at the inverter input, as shown in Fig. 14-21, is measured. The following control options are described:

1. *Speed control circuit.* As shown in Fig. 14-21, a speed control circuit accepts the speed reference signal $\omega_{r,\text{ref}}$ as the input that controls the frequency of the inverter output voltages. By the ramp limiter, the maximum acceleration/deceleration rates can be specified by the user through potentiometers that adjust the rate-of-change allowed to the speed reference signal. During the acceleration/deceleration condition, it is necessary to keep the motor current i_o and the dc-bus voltage V_d within limits.

    If the speed regulation is to be improved, to be more independent of the load torque, it also accepts an input from the slip compensation subcircuit, as shown in Fig. 14-21 and explained in item 3 below.

2. *Current-limiting circuit.* A current-limiting circuit is necessary if a speed ramp limiter as in Fig. 14-21 is not used. In the motoring mode, if ω_s is increased too fast compared to the motor speed, then ω_{sl} and, hence, i_o would increase. To limit the maximum rate of acceleration so that the motor current stays below the current limit, the actual motor current is compared with the current limit, and the error, through a controller, acts on the speed control circuit by reducing the acceleration rate (i.e., by reducing ω_s).

    In the braking mode, if ω_s is reduced too fast, the negative slip would become large in magnitude and would result in a large braking current through the motor and the inverter. To restrict this current to the current limit during the braking, the actual current is compared with the current limit, and the error, fed through a controller, acts on the speed control circuit by decreasing the deceleration rate (i.e., by increasing ω_s). During braking, the dc-bus capacitor voltage must be kept within a maximum limit. If there is no regenerative braking, a dissipation resistor is switched on in parallel with the dc-bus capacitor to provide a dynamic braking capability. If the energy recovered is larger than that lost through various losses, the capacitor voltage could become excessive. Therefore, if the voltage limit is exceeded, the control circuit decreases the deceleration rate (by increasing ω_s).

3. *Compensation for slip.* To keep the rotor speed constant, a term must be added to the applied stator frequency, which is proportional to the motor torque T_{em}, as can be seen from Fig. 14-6:

$$\omega_s = \omega_{r,ref} + k_{18}T_{em} \qquad (14\text{-}50)$$

The second term in Eq. 14-50 is calculated by the slip compensation block of Fig. 14-21. One option is to estimate T_{em}. This can be done by measuring the dc power to the motor and subtracting the losses in the inverter and in the stator of the motor to get the air-gap power P_{ag}. From Eqs. 14-3 and 14-18c, T_{em} can be calculated.

4. *Voltage boost.* To keep the air gap flux ϕ_{ag} constant, the motor voltage must be (as found by combining Eqs. 14-38b and 14-25)

$$V_s = k_{19}\omega_s + k_{20}T_{em} \qquad (14\text{-}51)$$

Using T_{em} as calculated in item 3 above and knowing ω_s, the required voltage can be calculated from Eq. 14-51. This provides the necessary voltage boost in Fig. 14-21.

It should be noted that, if needed, the speed can be precisely controlled by measuring the actual speed and thereby using the actual slip in the block diagram of Fig. 14-21. By knowing the slip, the actual torque can be calculated from Eq. 14-27, thereby allowing the voltage boost to be calculated more accurately.

14-7-5 INDUCTION MOTOR SERVO DRIVES

In the previous sections, the emphasis has been on controlling the speed of induction motors. Recently, because of the ready availability of the digital signal processors (DSPs), induction motors are beginning to be used for servo drives. In servo drives, the torque developed by the motor should respond quickly and precisely to the torque command without oscillation, at all speeds including at rest, since these drives are used for position control.

The control of induction motor servo drives is normally done by field-oriented space-vector-based calculations of what the stator currents of the induction motor should be to provide an electromagnetic torque T_{em} equal to the torque command specified by the speed regulator. In these calculations, a model of the induction motor is needed; therefore, the motor parameters must be plugged into the model in the I_s calculator block shown in Fig. 14-23. Many of these models rely on an accurate knowledge of the rotor resistance R_r. As resistance of copper varies 40% when temperature varies by 100°C, such knowledge is not easy to come by during motor operation. Most of these models also need the actual speed information; this is, however, not a serious restriction, since speed is measured in servo drives anyway. Adaptive control with parameter estimation is often utilized.

As shown in Fig. 14-23, the field-oriented controller calculates the three-phase reference currents that must be delivered by the power converter to the motor. A current-regulated VSI (CR-VSI) inverter, as discussed in Chapter 8, can be utilized where the inverter switch control signals are obtained by comparing the reference currents with the actual phase currents measured.

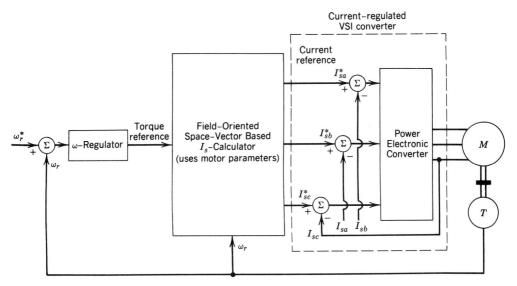

Figure 14-23 Field-oriented control for induction motor servo drive.

14-8 VARIABLE-FREQUENCY SQUARE-WAVE VSI DRIVES

The schematic of such a drive was shown in Fig. 14-18*b*. The inverter operates in a square-wave mode, which results in phase-to-motor-neutral voltage, as shown in Fig. 14-24*a*. With the square-wave inverter operation, described in Chapter 8, each inverter switch is on for 180° and a total of three switches are on at any instant of time. The resulting motor current waveform is also shown in Fig. 14-24*b*. Because of the inverter operating in a square-wave mode, the magnitude of the motor voltages is controlled by controlling V_d in Fig. 14-18*b* by means of a line-frequency phase-controlled converter.

Voltage harmonics in the inverter output decrease as V_1/h with $h = 5, 6, 11, 13,$ . . ., where V_1 is the fundamental-frequency phase-to-neutral voltage. Because of substantial magnitudes of low-order harmonics, harmonic currents calculated from Eq. 14-47 are significant. These harmonic currents result in large torque ripple, which can produce troublesome speed ripple at low operating speeds.

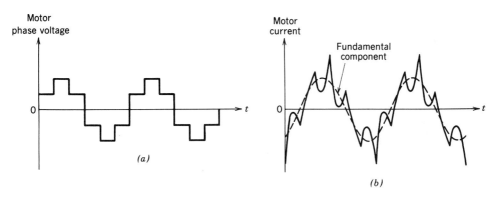

Figure 14-24 Square-wave VSI waveforms.

The line rectifier in Fig. 14-18b is similar to the line-frequency phase-controlled converters described in Chapter 6. Assuming a continuously flowing current through the rectifier, and for simplicity, ignoring the line-side inductances,

$$V_d = 1.35 V_{LL} \cos \alpha \qquad (14\text{-}52)$$

where V_{LL} is the line–line rms line voltage. From Eq. 8-58, the motor line–line voltage for a given V_d is

$$V_{LL1}^{motor} = 0.78 V_d \qquad (14\text{-}53)$$

From Eqs. 14-52 and 14-53,

$$V_{LL1}^{motor} = 1.05 V_{LL} \cos \alpha \simeq V_{LL} \cos \alpha \qquad (14\text{-}54)$$

which shows that the maximum line–line fundamental-frequency motor voltage (at $\alpha = 0$) is approximately equal to V_{LL}. Note that the same maximum motor voltage (equal to the line voltage) can be approached in PWM-VSI drives only by overmodulation, as described in Chapter 8. Therefore, in both PWM and square-wave VSI drives, the maximum available motor voltage in Fig. 14-12b is approximately equal to the line voltage. This allows the use of standard 60-Hz motors, since the inverter is able to supply the rated voltage of the motor at its rated frequency of 60 Hz.

In a square-wave drive, from Eq. 14-54 and assuming V_s/f to be constant.

$$\frac{\omega_r}{\omega_{r,\text{rated}}} \approx \frac{V_{LL1}^{motor}}{V_{LL}} \approx \cos \alpha \qquad (14\text{-}55)$$

From Eqs. 6-47a and 14-55, the drive operates at the following power factor from the line (assuming that a sufficiently large filter inductor is present in Fig. 14-18b at the rectifier output):

$$\text{Line power factor} \approx 0.955 \cos \alpha \approx 0.955 \, \frac{\omega_r}{\omega_{r,\text{rated}}} \qquad (14\text{-}56)$$

which shows that the line power factor at the rated speed is better than that of an induction motor supplied directly by the line. At low speed, however, the line power factor of a square-wave drive can become quite low, as seen from Eq. 14-56. This can be remedied by replacing the thyristor rectifier by a diode rectifier bridge in combination with a step-down dc–dc converter.

14-9   VARIABLE-FREQUENCY CSI DRIVES

Figure 14-18c shows the schematic of a CSI drive. Basically it consists of a phase-controlled rectifier, a large inductor, and a dc-to-ac inverter. A large inductor is used in the dc link, which makes the input appear as a current source to the inverter.

Since the induction motor operates at a lagging power factor, circuits for forced commutation of the inverter thyristors are needed, as shown in Fig. 14-25a. These forced-commutation circuits consist of diodes, capacitors, and the motor leakage inductances. This requires that the inverter be used with the specific motor for which it is designed. At any time, only two thyristors conduct: one of the thyristors connected to the positive dc bus and the other connected to the negative dc bus. The motor current and the resulting phase voltage waveform are shown in Fig. 14-25b. In a CSI drive, the regenerative braking can be easily provided without any additional circuits.

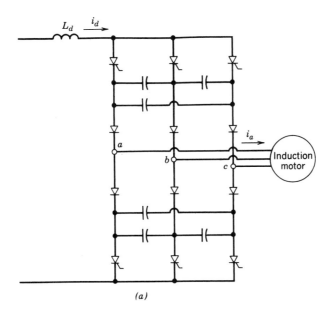

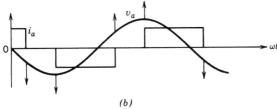

Figure 14-25   CSI drive: (a) inverter; (b) idealized phase waveforms.

In the past, the fact that line-frequency thyristors with simple commutation circuits act as the inverter switches was a very important asset of CSI drives. With the availability of controllable switches in ever-increasing power ratings, nowadays CSI drives are used mostly in very large horsepower applications.

14-10 COMPARISON OF VARIABLE-FREQUENCY DRIVES

It is possible to use all three types of drives (PWM-VSI, square-wave VSI, and CSI) with general-purpose induction motors. All three can provide a constant-torque capability, from the rated speed down to some small speed where the reduced cooling in the motor dictates that the torque capability will decline. Motor derating as a percentage of the nameplate horsepower rating is essentially independent of the drive type.

In spite of the somewhat similar nature of these three drive types, there are certain basic differences, which are compared in Table 14-2, where "+" is a positive and "−" is a negative attribute. It should be kept in mind that this comparison addresses the inherent capability of each drive. By means of additional circuits, most of the limitations can be overcome.

Table 14-2 Comparison of Adjustable Frequency Drives

Parameter	PWM	Square Wave	CSI
Input power factor	+	−	− −
Torque pulsations	+ +	−	−
Multimotor capability	+	+	−
Regeneration	−	−	+ +
Short-circuit protection	−	−	+ +
Open-circuit protection	+	+	−
Ability to handle undersized motor	+	+	−
Ability to handle oversized motor	−	−	−
Efficiency at low speeds	−	+	+
Size and weight	+	+	− −
Ride-through capability	+	−	−

Some general comments can be made about the trend in applying these drives. For retrofit applications, PWM-VSI is preferred over CSI, which requires a better match between the inverter and the motor. In sizes below a few hundred horsepower, there is an increasing trend to use PWM-VSI.

To make these solid-state controllers more reliable, a host of other protective features are incorporated. These include instantaneous overcurrent trip, input circuit breakers, current-limiting fuses, line reactors or isolation transformers at the input, output disconnect switch between the VSI and the motor, motor thermal protection incorporated with the controller, trips in case of overvoltage, undervoltage, or loss of a phase, and so on.

14-11 LINE-FREQUENCY VARIABLE-VOLTAGE DRIVES

The variable-frequency variable-voltage drives we described earlier are the most energy efficient and versatile way to control the speed of squirrel-cage induction motors. However, in some applications, it may be cheaper to use line-frequency variable-voltage drives as discussed in this chapter.

In the equivalent circuit of Fig. 14-2a, with f equal to the line frequency and a fixed value of f_{sl}, the power in any resistive element is proportional to V_s^2. Therefore, using Eqs. 14-17 and 14-18, the torque T_{em} will be proportional to V_s^2 for a value of rotor speed ω_r determined by f and f_{sl}:

$$T_{em} = k_{21}V_s^2 \qquad (14-57)$$

Based on Eq. 14-57, Fig. 14-26a shows the motor torque–speed curves at various values of V_s for a normal induction motor with a small value of the rated slip. The load torque of a fan- or a pump-type load varies approximately as the square of speed. Therefore, only a small torque is required at low speeds, and as Fig. 14-26a shows, the speed can be controlled over a wide range. Because of the heavy dependence of the load torque on speed, the motor operating point, for example A (intersection of the load and the motor torque–speed curves) in Fig. 14-26a, is stable. The operating point would not be stable if the load torque remained constant with speed.

For a load requiring a constant torque with speed, it is necessary to use a motor with a higher motor resistance whose torque–speed characteristics are shown in Fig. 14-26b. A motor with a high rotor resistance has a large value of slip at which the pull-out torque is developed. Such a motor allows the speed to be controlled over a wide range even when supplying a constant-torque load.

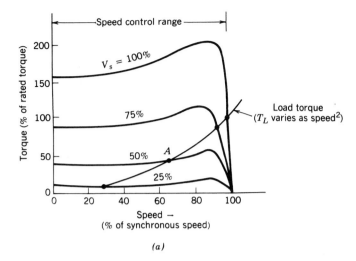

(a)

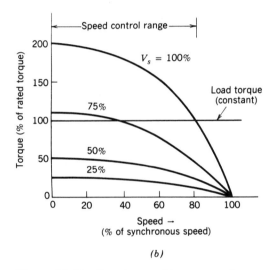

(b)

Figure 14-26 Speed control by stator voltage control: (a) motor with a low value of s_{rated}, fan-type load; (b) motor with a large s_{rated}, constant-torque load.

Speed control by controlling the stator voltage results in a very poor energy efficiency at low speeds because of high rotor losses caused by large slips. In a motor, the actual rotor loss must be below its rated rotor loss (which occurs while the motor is supplying the rated torque, supplied from the rated voltage). As seen from Figs. 14-26a and 14-26b, speed control by adjusting the stator voltage results in a large slip and, hence, in a large rotor power loss at reduced speeds. Therefore, the motor selected for this application must have a high enough rating so its rated rotor loss is larger than the maximum rotor loss encountered by using this technique. This technique is widely used in fractional-horse-power fan or pump drives. These fractional-horsepower motors are generally single-phase motors, but the analysis presented earlier for the three-phase motor is applicable. This speed control technique is also used for cranes and hoists (which have speed-independent load torque) where the high-slip, high-power-loss operation is required for only a small portion of the load duty cycle.

A practical circuit for controlling the stator voltage of a three-phase induction motor is shown in Fig. 14-27a. It consists of three pairs of back-to-back connected thyristors.

Because of the interaction between the phases, it is not possible to analyze this circuit on a per-phase basis. It has been shown in the literature that, for analysis purposes, each phase of the motor can be represented by a sinusoidal back-emf in series with an inductance. Figure 14-27b shows one of the motor phase voltages, v_{an}, and the phase current i_a. As shown in Fig. 14-27b, the motor currents are no longer sinusoidal and their harmonic components result in a pulsating torque and in a higher power loss compared with a sinusoidal supply. These losses are in addition to the rotor circuit losses due to a high-slip operation at low speeds. These nonsinusoidal currents also flow into the utility system. Because of high rotor losses, this technique for controlling the speed is limited to low-horsepower or intermittent-load applications.

14-12 REDUCED VOLTAGE STARTING ("SOFT START") OF INDUCTION MOTORS

The circuit of Fig. 14-27a can also be used in constant-speed drives to reduce the motor voltages at start-up, thereby reducing the starting currents. In normal (low-slip) induction motors, the starting currents can be as large as six to eight times the full-load current. To reduce these large starting currents, the motor can be started at reduced voltages obtained from the circuit of Fig. 14-27a. Provided the torque developed at reduced voltage is sufficient to overcome the load, the motor accelerates (slip s will decrease) and the motor current decreases. During the steady-state operation, each thyristor conducts for an entire half-cycle. Then, these thyristors can be shorted out by mechanical contactors connected in parallel with the back-to-back connected thyristor pairs, to eliminate the power losses in the thyristors due to a finite $(1-2$ V) conduction voltage drop across the thyristors.

The circuit of Fig. 14-27a can also be used in constant-speed drives to minimize motor losses. In an induction motor (single phase and three phase) at a given torque output, the motor losses vary with the stator voltage V_s. The stator voltage at which the minimum power loss occurs decreases with decreasing load. Therefore, it is possible to use the circuit of Fig. 14-27a to reduce V_s at reduced loads and, hence, save energy.

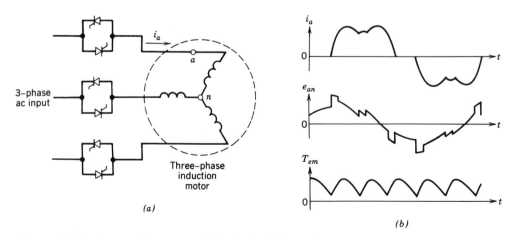

Figure 14-27 Stator voltage control: (a) circuit; (b) waveforms.

The amount of energy saved is significant (compared with the extra losses in the motor due to current harmonics and in the thyristors due to a finite conduction voltage drop) only if the motor operates at very light loads for substantial periods of time. In applications where reduced voltage starting (soft start) is required, the power switches are already implemented and only the control for the minimum power loss needs to be added. In such cases, this concept may be economical.

14-13 SPEED CONTROL BY STATIC SLIP POWER RECOVERY

From the induction motor equivalent circuit, it is possible to obtain torque–speed curves for various values of rotor resistances. In the equivalent circuit of Fig. 14-2a, if R_r/s is kept constant (i.e., R_r and s are increased by the same factor), I_r and, hence, T_{em} remain constant. This results in the characteristics shown in Fig. 14-28 for various values of rotor resistance R_r. In a wound-rotor induction motor, the total resistance R_r in the rotor phases can be varied by adding an external resistance through the slip rings.

For the load-torque versus speed curve shown in Fig. 14-28, it is quite apparent that the speed of operation can be continuously varied by controlling the external resistance in the rotor circuit (the steady-state speed is given by the intersection of the load and the motor torque–speed curves as in all motor drives). However, high rotor losses (due to high slips) may be unacceptable.

The static slip power recovery scheme provides an alternative to the historical Scherbius and Kramer drives, both of which require a second rotating machine to recover the rotor circuit electrical power. In the static slip-power recovery system, rather than dissipating the slip power in the rotor external resistances, these resistances are simulated by means of a diode rectifier and the energy recovered is fed back to the ac source by a means of a line-voltage-commutated inverter, as is shown in Fig. 14-29.

This scheme requires a wound-rotor motor with slip rings. Such a motor is not as inexpensive and as maintenance-free as its squirrel-cage counterpart. However, in very large power ratings, this scheme may compete with the adjustable-frequency drive if the speed needs to be controlled only in a small range around its nominal value. A small speed range results in a smaller rating of the solid-state converter required, thus making this scheme competitive.

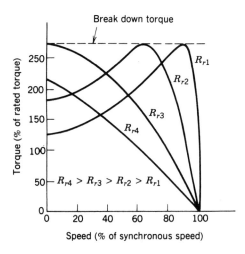

Figure 14-28   Torque–speed curves for a wound-rotor induction motor.

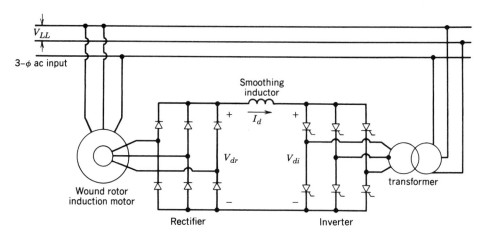

Figure 14-29   Static slip recovery.

SUMMARY

1. Induction motors are the workhorse of industry because of their low cost and rugged construction. When operated directly from the line voltages, an induction motor operates at nearly a constant speed. By means of power electronic converters, induction motors can be used for adjustable-speed and servo drive applications. A major application of adjustable-speed induction motor drives is for improving the energy efficiency in various residential, industrial, and electrical utility systems.

2. In a three-phase induction motor, the resultant field distribution in the air gap is sinusoidal and rotates at a synchronous speed $\omega_s = (2\pi f)2/p$ radians per second for a p-pole winding when it is excited by three-phase voltages and currents at a frequency f.

3. The speed of an induction motor can be controlled by varying the stator frequency f, which controls the synchronous speed and, hence, the motor speed, since the slip s is kept small. The air gap flux ϕ_{ag} is kept constant by V_s in linear proportion to f. This technique allows the induction motor to deliver its rated torque at speeds up to its rated speed. Beyond the rated speed, the motor torque capability declines, though the motor can deliver its rated output power up to a certain speed.

4. For braking in an induction motor to reduce its speed, the stator frequency f is decreased so that the synchronous speed at which the air gap magnetic field rotates is less than the rotor speed.

5. Switch-mode dc-to-ac inverters, as discussed in Chapter 8, are used to supply adjustable-frequency, adjustable-magnitude three-phase ac voltages for induction motor speed control. The harmonics in the inverter output voltages result in harmonics in the motor current, harmonic losses in the motor, and possibly the motor torque pulsations. Therefore, care must be taken in selecting the inverter and the inverter switching frequency.

6. The inverters used for the induction motor speed control can be classified as pulse-width-modulated voltage source inverters, square-wave voltage source inverters, and currents source inverters. The comparative advantages and disadvantages of these inverters are given in Table 14-2.

7. By means of field-oriented vector control, induction motor drives can be used for servo applications.

8. There are other means of controlling the speed of induction motors. Some of these are (a) stator pole changing, (b) pole amplitude modulation, (c) stator voltage control at the line frequency, and (d) static slip power recovery. In certain applications, one of these techniques rather than the stator frequency control may be preferable.

PROBLEMS

14-1 A three-phase, 60-Hz, four-pole, 10-hp, 460-V (line–line, rms) induction motor has a full-load speed of 1746 rpm. Assume the torque–speed characteristic in a range of 0–150% rated torque to be linear. It is driven by an adjustable-frequency sinusoidal supply such that the air gap flux is held constant. Plot its torque–speed characteristics at the following values of frequency f: 60, 45, 30, and 15 Hz.

14-2 The drive in Problem 14-1 is supplying a centrifugal pump load, which at the full-load speed of the motor requires the rated torque of the motor. Calculate and plot speed, frequency f, slip frequency f_{sl}, and slip s at the following percentage values of pump rated torque: 100, 75, 50, and 25%.

14-3 A 460-V, 60-Hz, four-pole induction motor develops its rated torque by drawing 10 A at a power factor of 0.866. The other parameters are as follows:

$$R_s = 1.53 \ \Omega \qquad X_{ls} = 2.2 \ \Omega \qquad X_m = 69.0 \ \Omega$$

If such a motor is to produce a rated torque at frequencies below 60 Hz while maintaining a constant air gap flux, calculate and plot the required line-to-line voltage as a function of frequency.

14-4 The motor in Problem 14-3 has a full-load speed of 1750 rpm. Calculate f_{start}, I_{start}, and $(V_{LL})_{start}$ if the motor is to develop a starting torque equal to 1.5 times its rated torque. Assume the effect of L_{lr} to be negligibly small and the air gap flux to be at its rated value.

14-5 The idealized motor of Problem 14-1 is initially operating at its rated conditions at 60 Hz. If the supply frequency is suddenly decreased by 5% while maintaining a constant air gap flux, calculate the braking torque developed as a percentage of its rated torque.

14-6 In a three-phase 60-Hz, 460-V induction motor, $R_s + R_r = 3.0 \ \Omega$ and $X_{ls} + X_{lr} = 5.0 \ \Omega$. The motor is driven by a square-wave voltage source inverter that supplies a 460-V line–line voltage at the frequency of 60 Hz. Estimate the harmonic currents and the additional copper losses due to these harmonic currents by including fifth, seventh, eleventh, and thirteenth harmonics.

14-7 For harmonic frequency analysis, an induction motor can be represented by a per-phase equivalent circuit as shown in Fig. P14-7, which includes a fundamental-frequency counter-emf or Thèvenin voltage E_{TH}. Also, $R_{TH} = 3.0 \ \Omega$ and $X_{TH} = 5.0 \ \Omega$. It is supplied by a voltage source inverter, which produces a 60-Hz line–line voltage component of 460 V. The load on the motor is such that the fundamental-frequency current drawn by the motor is 10 A, which lags the fundamental-frequency voltage by an angle of 30°.

   Obtain and plot the current drawn by the motor as a function of time, if it is driven by a square-wave VSI. What is the peak current that the inverter switches must carry?

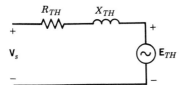

Figure P14-7

14-8 Repeat Problem 14-7 if the induction motor is driven by a PWM-VSI with an amplitude modulation ratio $m_a = 1.0$ and frequency modulation ratio $m_f = 15$. Compare the peak switch currents with those in Problem 14-7.

14-9 A square-wave VSI drive supplies 460 V line–line at a frequency of 60 Hz to an induction motor that develops a rated torque of 50 N-m at 1750 rpm. The motor and the inverter efficiencies can be assumed to be constant at 90 and 95%, respectively, while operating at the rated torque of the motor.

   If the motor is operated at its rated torque and the rated air gap flux, determine the equivalent resistance R_{eq} that can represent the inverter–motor combination in Fig. P14-9 at the motor frequencies of 60, 45, 30, and 15 Hz.

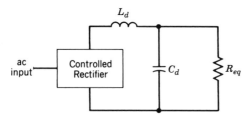

Figure P14-9

14-10 Repeat Problem 14-9 if a PWM-VSI drive with an uncontrolled rectifier is used, where $m_a = 1.0$ at 60 Hz output.

14-11 A CSI-driven induction motor is supplying a constant-torque load equal to the rated torque of the motor. The CSI drive is supplied from a 460-V, three-phase, 60-Hz input. It supplies the motor a 460-V (line–line) voltage at a 60 Hz frequency with a fundamental frequency current of 100 A that lags behind the fundamental-frequency voltage by an angle of 30°.

   If the motor displacement power factor angle remains constant at 30°, estimate and plot the input power factor and the displacement power factor at the motor frequencies of 60, 45, 30, and 15 Hz. Idealize the motor current waveforms to be as shown in Fig. 14-25b and assume a constant air gap flux in the motor. Neglect losses in the motor and the inverter.

14-12 Show that in a voltage-controlled induction motor supplying a constant load torque, the power loss in the rotor circuit at a voltage V_s as a ratio of the power loss at the rated voltage condition can be approximated as

$$\frac{P_r}{(P_r)_{\text{rated voltage}}} \approx \left(\frac{V_{s,\text{rated}}}{V_s}\right)^2$$

for reasonably small values of slip.

REFERENCES

1. N. Mohan and R. J. Ferraro, "Techniques for Energy Conservation in AC Motor Driven Systems," Electric Power Research Institute Final Report EM-2037, Project 1201-13, September 1981, Palo Alto, CA.
2. A. E. Fitzgerald, C. Kingsley, Jr., and S. D. Umans, *Electric Machinery*, 4th ed., McGraw-Hill, New York, 1983.
3. G. R. Slemon and A. Straughen, *Electrical Machines*, Addison-Wesley, Reading, MA, 1980.
4. B. K. Bose, *Power Electronics and AC Drives*, Prentice-Hall, Englewood Cliffs, NJ, 1986.
5. W. Leonard, *Control of Electrical Drives*, Springer-Verlag, New York, 1985.

CHAPTER 15

SYNCHRONOUS MOTOR DRIVES

15-1 INTRODUCTION

Synchronous motors are used as servo drives in applications such as computer peripheral equipment, robotics, and adjustable-speed drives in a variety of applications such as load-proportional capacity-modulated heat pumps, large fans, and compressors. In low-power applications up to a few kilowatts, permanent-magnet synchronous motors are used (see Fig. 15-1a). Such motors are often referred to as "brushless dc" motors or electronically commutated motors. Synchronous motors with wound rotor field are used in large power ratings (see Fig. 15-1b).

15-2 BASIC PRINCIPLES OF SYNCHRONOUS MOTOR OPERATION

The field winding on the rotor produces a flux ϕ_f in the air gap. This flux rotates at a synchronous speed ω_s rad/s, which is the same as the rotor speed. The flux ϕ_{fa} linking one of the stator phase windings, for example phase a, varies sinusoidally with time:

$$\phi_{fa}(t) = \phi_f \sin \omega t \qquad (15\text{-}1)$$

where

$$\omega = 2\pi f = \frac{p}{2}\,\omega_s \qquad (15\text{-}2)$$

and p is the number of poles in the motor. If we assume N_s as an equivalent number of turns in each stator phase winding, the emf induced in phase a from Eq. 15-1 is

$$e_{fa}(t) = N_s \frac{d\phi_{fa}}{dt} = \omega N_s \phi_f \cos \omega t \qquad (15\text{-}3)$$

This induced voltage in the stator winding is called the excitation voltage, whose rms value is

$$E_{fa} = \frac{\omega N_s}{\sqrt{2}}\,\phi_f \qquad (15\text{-}4)$$

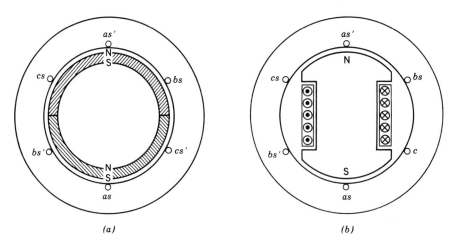

Figure 15-1 Structure of synchronous motors: (*a*) permanent-magnet rotor (two-pole); (*b*) salient-pole wound rotor (two-pole).

In accordance with the normal convention, the amplitudes of voltage and current phasors are represented by their rms values; the amplitudes of flux phasors are represented by their peak values. Being sinusoidal with time, e_{fa} and ϕ_{fa} can be represented as phasors at $\omega t = 0$, where $\mathbf{E}_{fa} = E_{fa}$ is the reference phasor in Fig. 15-2*a*, and from Eq. 15-1

$$\phi_{fa} = -j\phi_f \tag{15-5}$$

From Eqs. 15-3 through 15-5 and Fig. 15-2*a*

$$\mathbf{E}_{fa} = j\frac{\omega N_s}{\sqrt{2}}\,\phi_{fa} = E_{fa} \tag{15-6}$$

In synchronous motor drives, the stator is supplied with a set of balanced three-phase currents, whose frequency is controlled to be *f*, which from Eq. 15-2 is

$$f = \frac{p}{4\pi}\,\omega_s \tag{15-7}$$

The fundamental-frequency components of these stator currents produce a constant amplitude flux ϕ_s in the air gap, which rotates at the synchronous speed ω_s. The amplitude of ϕ_s is proportional to the amplitudes of the fundamental-frequency components in the stator currents.

In this three-phase motor, the flux linking with phase *a* due to ϕ_s produced by all three stator currents is $\phi_{sa}(t)$. As shown in reference 1, $\phi_{sa}(t)$ is proportional to the phase *a* current $i_a(t)$:

$$N_s\phi_{sa}(t) = L_a i_a(t) \tag{15-8}$$

where the armature inductance L_a is 3/2 times the self-inductance of phase *a*. Therefore, the voltage induced in phase *a* due to $\phi_{sa}(t)$, from Eq. 15-8, is

$$e_{sa}(t) = N_s\frac{d\phi_{sa}}{dt} = L_a\frac{di_a}{dt} \tag{15-9}$$

Assuming the fundamental component of the supplied current to the stator phase *a* to be

$$i_a(t) = \sqrt{2}I_a\sin(\omega t + \delta) \tag{15-10}$$

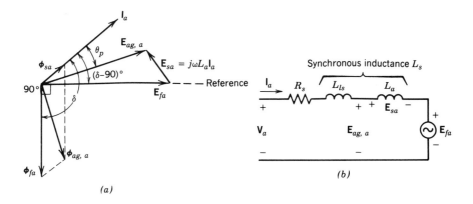

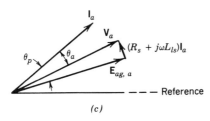

Figure 15-2 Per-phase representation: (a) phasor diagram; (b) equivalent circuit; (c) terminal voltage.

yields

$$e_{sa}(t) = \sqrt{2}\omega L_a I_a \cos(\omega t + \delta) \qquad (15\text{-}11)$$

from Eq. 15-9, where δ is defined later on to be the torque angle. Here i_a and e_{sa} can be represented as phasors, which at $\omega t = 0$

$$\mathbf{I}_a = I_a e^{j(\delta - \pi/2)} \qquad (15\text{-}12)$$

and as shown in Fig. 15-2a,

$$\mathbf{E}_{sa} = j\omega L_a \mathbf{I}_a = \omega L_a I_a e^{+j\delta} \qquad (15\text{-}13)$$

The resultant air gap flux $\phi_{ag,a}(t)$ linking the stator phase a is the sum of $\phi_{fa}(t)$ and $\phi_{sa}(t)$:

$$\phi_{ag,a}(t) = \phi_{fa}(t) + \phi_{sa}(t) \qquad (15\text{-}14)$$

which can be represented as a phasor,

$$\boldsymbol{\phi}_{ag,a} = \boldsymbol{\phi}_{fa} + \boldsymbol{\phi}_{sa} \qquad (15\text{-}15)$$

The air gap voltage $e_{ag,a}(t)$ due to the resultant air gap flux linking phase a is

$$e_{ag,a}(t) = N_s \frac{d\phi_{ag,a}}{dt} = e_{fa}(t) + e_{sa}(t) \qquad (15\text{-}16)$$

from Eqs. 15-14, 15-3, and 15-9. Equations 15-6 and 15-13 combined with Eq. 15-16 result in

$$\mathbf{E}_{ag,a} = \mathbf{E}_{fa} + \mathbf{E}_{sa} = \mathbf{E}_{fa} + j\omega L_a \mathbf{I}_a \qquad (15\text{-}17)$$

All of these phasors are drawn in Fig. 15-2a. Based on Eq. 15-17 and the phasor diagram, a per-phase equivalent circuit of a synchronous motor is shown in Fig. 15-2b, where R_s

and L_{ls} are the stator winding resistance and leakage inductance, respectively. Including the voltage drop across R_s and L_{ls}, the per-phase terminal voltage in phase a is

$$\mathbf{V}_a = \mathbf{E}_{ag,a} + (R_a + j\omega L_s)\mathbf{I}_a \tag{15-18}$$

The phasor diagram corresponding to Eq. 15-18 is shown in Fig. 15-2c, where θ_a is the angle between the current and the terminal voltage phasors.

From the per-phase equivalent circuit of Fig. 15-2b and the phasor diagram of Fig. 15-2a, the electromagnetic torque T_{em} can be obtained as follows: the electrical power that gets converted into the mechanical power P_{em} is

$$P_{em} = 3E_{fa}I_a\cos\left(\delta - \frac{1}{2}\pi\right) \tag{15-19}$$

and

$$T_{em} = \frac{P_{em}}{\omega_s} \tag{15-20}$$

Using Eqs. 15-19, 15-20, and 15-4,

$$T_{em} = k_t\phi_f I_a\sin\delta \tag{15-21}$$

where the angle δ between ϕ_{fa} and $\mathbf{I}_a$ is called the torque angle and k_t is the constant of proportionality.

In the phasor diagram of Fig. 15-2c, $\mathbf{I}_a$ leads $\mathbf{V}_a$. This leading power factor operation is required if the synchronous motor is supplied by a drive where the current through the inverter thyristors is commutated by the synchronous motor voltages.

A torque angle δ equal to 90° results in a decoupling of the field flux ϕ_f and the field due to the stator currents, which is important in high-performance servo drives. With $\delta = 90°$, a constant field flux ϕ_f, and the amplitudes of the stator phase currents equal to I_s, Eq. 15-21 can be written as

$$T_{em} = k_T I_s \tag{15-22}$$

where k_T is the motor torque constant. A phasor diagram corresponding to $\delta = 90°$ is shown in Fig. 15-3, where $\mathbf{I}_a$ must lead ϕ_{fa} by 90°. This condition for servo drives implies that the current i_a must become positive maximum, $\omega t = 90°$ or $t = (\pi/2)(p/2)\omega_s$ seconds, before ϕ_{fa} reaches its positive maximum value. Another observation from the phasor diagram of a servo drive is that $\mathbf{I}_a$ is at a lagging power factor. Therefore, the inverter of the drive must consist of self-controlled switches.

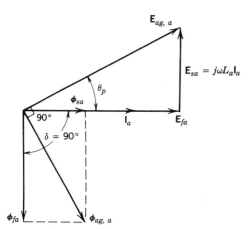

Figure 15-3 Phasor diagram with $\delta = 90°$.

In the previous analysis, the rotor saliency was ignored. The effect of rotor saliency cannot be represented by a per-phase equivalent circuit because of a different magnetic permeance along the rotor pole axis (called the d axis) and along the axis midway between two rotor poles (called the q axis). This d- and q-axis analysis is beyond the scope of this book, but in qualitative terms, an additional reluctance torque component is present because of the difference in reactances in the d and q axes. This component is usually small, but not negligible, compared with the electromagnetic torque component discussed above, assuming a nonsalient round rotor.

15-3 SYNCHRONOUS SERVOMOTOR DRIVES WITH SINUSOIDAL WAVEFORMS

The air gap flux density distribution and the induced excitation voltages in the stator phase windings in such a motor are nearly sinusoidal. In this regard, the description of this motor is identical to that presented in the previous section. Moreover, the torque angle δ is maintained at 90°. For controlling such a synchronous servo drive, the rotor field position is measured by means of an absolute position sensor with respect to a stationary axis, for example, as shown in Fig. 15-4 for a two-pole motor. Recognizing that at $\theta = 0$ in the two-pole motor of Fig. 15-4, i_a should be at its positive peak yields

$$i_a(t) = I_s \cos[\theta(t)] \qquad (15\text{-}23)$$

where the amplitude I_s is obtained from Eq. 15-22. For a p-pole motor, in general, if θ is the mechanical angle measured, then the electrical angle θ_e is calculated as

$$\theta_e(t) = \frac{p}{2}\,\theta(t) \qquad (15\text{-}24)$$

If we use Eqs. 15-23 and 15-24 and recognize that $i_b(t)$ and $i_c(t)$ are delayed by 120° and 240°, respectively,

$$i_a(t) = I_s \cos[\theta_e(t)] \qquad (15\text{-}25)$$

$$i_b(t) = I_s \cos[\theta_e(t) - 120°] \qquad (15\text{-}26)$$

$$i_c(t) = I_s \cos[\theta_e(t) - 240°] \qquad (15\text{-}27)$$

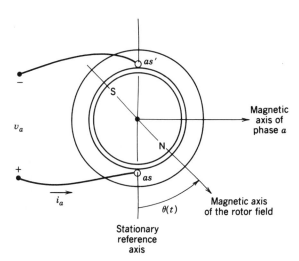

Figure 15-4 Measured rotor position θ at time t.

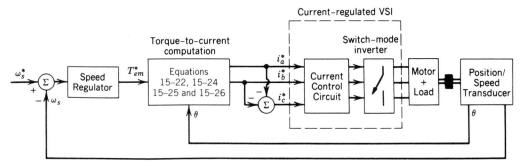

Figure 15-5   Synchronous motor servo drive.

This control strategy can also be used for induction motor drives, as described in reference 10.

With the frequency of the stator currents "locked" or synchronized to the rotor position, which is continuously measured, there is no possibility of losing synchronism, and the torque angle δ remains at its optimal value of 90°. If a holding torque is required at zero speed to overcome the load torque and, hence, to keep the load from moving from a position where θ is constant, as is often the case in servo drives, a synchronous servomotor drive provides this torque by applying dc currents to the stator as given by Eqs. 15-24 to 15-27.

Figure 15-5 shows the overall block diagram of a synchronous servomotor drive with sinusoidal waveforms. The absolute rotor field position is sensed by means of an absolute position sensor such as a high-accuracy resolver, which is mechanically prealigned to measure the rotor field position θ with respect to a known axis, for example as indicated in Fig. 15-4. By using prestored cosine tables in read-only memory (ROM), cosine functions required in Eqs. 15-25 to 15-27 are generated for two of the three phase currents, for example, a and b. The stator current amplitude I_s is determined by the torque–speed loop using Eq. 15-22. Once the reference currents i_a^* and i_b^* are defined for phases a and b, $i_c^* = -i_a^* - i_b^*$ in a three-wire motor. As discussed in Chapter 8, a current-regulated voltage source inverter is used to force the motor currents to equal the reference currents.

15-4   SYNCHRONOUS SERVOMOTOR DRIVES WITH TRAPEZOIDAL WAVEFORMS

The motors described in the previous section are designed such that the induced excitation emf's in the stator due to the field flux are sinusoidal and the stator currents produce a sinusoidal field. The motors described in this section are designed with concentrated coils, and the magnetic structure is shaped such that the flux density of the field because of the permanent magnets and the induced excitation voltages have trapezoidal waveforms.

Figure 15-6a shows the induced emf $e_{fa}(t)$ in phase a, where the rotor is rotating in a counterclockwise direction at a speed of ω_s radians per second and θ is measured with respect to the stator as shown in Fig. 15-4. The electrical angle θ_e is defined by Eq. 15-24 for a p-pole motor. This emf waveform has a flat portion, which occurs for at least 120° (electrical) during each half-cycle. The amplitude $\hat{E}_f$ is proportional to the rotor speed:

$$\hat{E}_f = k_E \omega_s \tag{15-28}$$

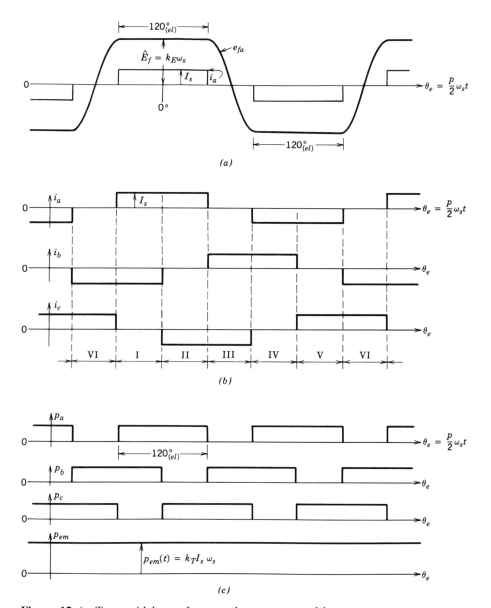

Figure 15-6 Trapezoidal-waveform synchronous motor drive.

where k_E is the motor voltage constant. Similar voltage waveforms are induced in phases b and c, displaced by 120° and 240°, respectively.

To produce as ripple-free torque as possible in such a motor, the phase currents supplied should have rectangular waveforms, as is shown in Fig. 15-6b, which results in instantaneous electrical power $p_a(t) = e_{fa}(t) \cdot i_a(t)$, and so on, as shown in Fig. 15-6c. Since $p_{\text{total}}(t) = p_a(t) + p_b(t) + p_c(t) = P_{\text{em}}$ is independent of time, the instantaneous electromagnetic torque is also independent of time and depends only on the current amplitude I_s:

$$T_{\text{em}}(t) = \frac{P_{\text{em}}}{\omega_s} = k_T I_s \tag{15-29}$$

where k_T and k_E of Eqs. 15-28 and 15-29 are related (see Problem 15-5). In practice, because of the finite time required for the phase currents to change, T_{em} contains ripple.

A current-regulated VSI similar to that shown in Fig. 15-5 is used where the sinusoidal reference currents are replaced by rectangular current references, which are shown in Fig. 15-6b. One complete cycle is divided into six intervals of 60 electrical degrees each. In each interval, the current through two of the phases is constant and proportional to the torque command. To obtain these current references, the rotor position is usually measured by Hall effect sensors that indicate the six current commutation instants per electrical cycle of waveforms. In non-servo applications, it is possible to use the three-phase induced emf's to determine the current commutation instants, thereby eliminating the need for any rotor position sensor.

15-5   LOAD-COMMUTATED INVERTER DRIVES

In very large power ratings in excess of 1000 hp, load-commutated inverter (LCI) synchronous motor drives become competitive with the induction motor drives in adjustable-speed applications. The circuit diagram of a LCI drive is shown in Fig. 15-7a. Each phase of the synchronous motor is represented by an internal voltage in series with the motor inductance, as discussed in the previous sections, assuming a nonsalient pole motor.

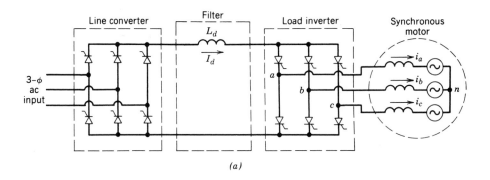

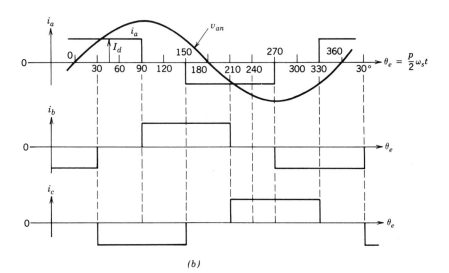

Figure 15-7   An LCI drive: (a) circuit; (b) idealized waveforms.

As an overview, the LCI drive is the source of three-phase currents to the motor. The frequency and the phase of these currents are synchronized to the rotor position. The current commutation in the load inverter to supply currents to the motor phases in an appropriate sequence is provided by the induced emf's in the motor. The amplitude of the currents supplied to the motor is controlled by the phase-controlled line converter through a filter inductance L_d. The filter inductance reduces the current harmonics and ensures that the input to the load inverter and, hence, to the motor appears as a current source.

Note that the line converter in Fig. 15-7a is identical to the phase-controlled, line-frequency converters discussed in Chapter 6. By controlling the firing angle of the converter, its dc output voltage and, hence, the current magnitude (in the dc link as well as to the motor) can be controlled. Normally, it operates in a rectifier mode.

The load inverter is identical to the line converter, that is, it also consists of only thyristors, but it normally operates in an inverter mode. The current commutation is provided by the internally induced emf's in the synchronous motor. The presence of these three-phase emfs facilitates the current commutation in the load converter in an identical manner as in a line-frequency thyristor converter operating in an inverter mode. The idealized motor current waveforms are shown in Fig. 15-7b.

At start-up and at low speeds (less than 10% of the full speed) the induced emf in the synchronous motor is not sufficient to provide current commutation in the load converter. Under this condition, the current commutation is provided by the line converter by going into an inverter mode and forcing I_d to become zero, thus providing turn-off of thyristors in the load inverter.

There are many control possibilities. Considering first the range below the rated speed, one possibility is to keep the field excitation current I_f constant. Also, the turn-off time t_{off} available to the thyristors in the load converter is kept constant. The dc-link current, which is proportional to the motor current, is varied with torque at a given speed. The voltage waveforms at the motor terminals are measured to calculate the rotor field position as a function of time. The measured three-phase voltages are rectified to provide a dc signal proportional to the instantaneous rotor speed. Keeping I_f and t_{off} constant, the actual speed is compared with the reference speed, as is shown in Fig. 15-8. The amplified error signal determines the I_d reference. If the actual I_d is less than its reference value, the line converter increases the dc voltage applied to the link, thus increasing I_d and, hence, the torque produced by the motor. In response to increased T_{em}, the motor speed goes up. Based on I_d and the information obtained from the measured terminal voltage waveforms, the firing pulses to the thyristor gates of the load inverter are provided such as to keep t_{off}

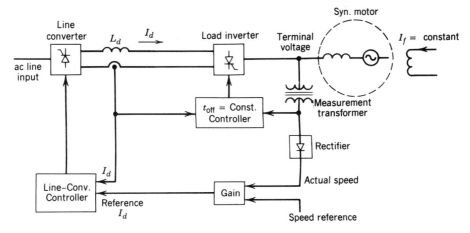

Figure 15-8 An LCI drive controller.

constant. In practice, I_f is not kept constant. Rather, it is controlled as a function of torque and speed to result in the rated air gap flux in the motor.

For speeds above the rated speed, the motor torque capability declines but the drive can supply the rated power. For operation above the rated speed, the field flux needs to be reduced by reducing I_f. Therefore, this region is also called the flux-weakening region.

Some of the other important properties of LCI drives are described as follows:

1. Use of synchronous motors in very large horsepower ratings ($>$1000 hp) results in overall drive efficiencies exceeding 95% at the rated power, a few percentage points higher than what can be accomplished in the induction motor drives.

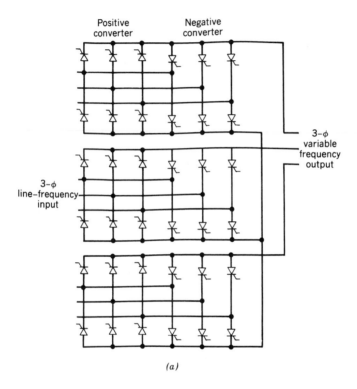

(a)

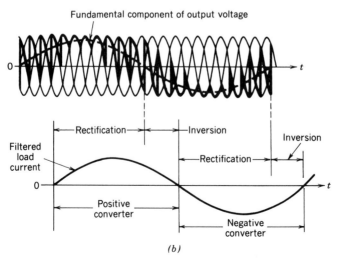

(b)

Figure 15-9   Three-phase cycloconverter.

2. The load-commutated inverter is much simpler and has lower losses compared to the inverter used in CSI induction motor drives. Eliminating the requirement for self-controlled switches is a distinct advantage at high voltage and current ratings.

3. As in any power-electronic motor drive, there is no inrush current at start-up, unlike line-started motors. By designing the pole face (or the damper cage) winding to provide a sufficient torque, a synchronous motor can be line started on an induction motor principle. Once it reaches a speed close to the synchronous speed, the rotor field is excited, thus making it operate as a synchronous motor. Being able to line start and operate at the line frequency provides additional reliability in case of the inverter failure, where the line voltages act as backup, as in the induction motor drive (though the drive would operate only at one speed).

4. The LCI drives have an inherent capability to provide regenerative braking by making the synchronous motor operate as a generator, rectifying the motor voltages by means of the load converter and feeding the power into the utility grid by operating the line converter in an inverter mode.

15-6 CYCLOCONVERTERS

In low-speed and very large horsepower applications, it is possible to use cycloconverters to control the speed of synchronous and induction motors. A basic cycloconverter circuit that utilizes line-frequency-commutated converters is shown in Fig. 15-9a. The three-phase 60-Hz input is through isolation transformers. Each phase consists of two back-to-back connected line-frequency thyristor converters as discussed in Chapter 6. The firing (or delay) angles of the two converters in each phase are cyclically controlled to yield a low-frequency sinusoidal output. One of the phase outputs is shown in Fig. 15-9b where the operating mode (rectification or inversion) of the positive and the negative converters depends on the direction of the output load current.

The cycloconverter output is derived directly from the line-frequency input without an intermediate dc link. The maximum output frequency is limited to about one-third of the input ac frequency to maintain an acceptable waveform with a low harmonic content.

SUMMARY

1. In synchronous motors, the flux is produced by the rotor either by means of permanent magnets or by a field winding excited by a dc current. This feature of synchronous motors allows them to offer higher efficiencies compared to induction motors of similar ratings. In large horsepower ratings, a wound-field construction is used, whereas permanent-magnet rotors are used at smaller power ratings.

2. The synchronous motor drives can be categorized as (a) precision servo drives for computer peripheral equipment and robotics and (b) adjustable-speed drives for controlling the speed such as in load-proportional capacity-modulated air conditioners and heat pumps. In very large (>1000 hp) power ratings, the LCI synchronous motor drives may be used due to their higher efficiency and a simpler inverter compared with the induction motor drives.

3. A per-phase equivalent circuit can be drawn in terms of the induced emf $\mathbf{E}_{fa}$, $\mathbf{I}_a$, and the per-phase inductances, as shown in Fig. 15-2 for phase a.

4. For an optimal torque condition in servo drives, the torque angle δ is kept at 90°. This results in a decoupling between the stator field and the rotor field. Thereby, the torque T_{em} required by the load is supplied by controlling the stator current amplitude I_s, without affecting the field flux ϕ_f. This makes T_{em} linearly proportional to I_s.

5. Synchronous motors used for servo drives can be broadly classified as (a) sinusoidal-waveform motors and (b) trapezoidal-waveform motors.

6. In the sinusoidal-waveform synchronous motor drives, the rotor position θ with respect to a stationary axis is accurately measured by means of an absolute position encoder, for example a resolver. The three stator phase currents i_a, i_b, and i_c are calculated based on θ as given by Eqs. 15-25 to 15-27 where the amplitude I_s is determined by the torque requirement. Figure 15-5 shows the overall block diagram of such a servo drive.

7. The trapezoidal-waveform synchronous motor drives are used for both the servo and the adjustable-speed applications. Here, the induced emfs, for example e_{fa}, in phase a have a trapezoidal waveform with a 120° long flat portion during each half-cycle. The stator currents have a rectangular waveform as shown in Fig. 15-6a. Since the stator currents are rectangular, their on and off instants are determined by the rotor position, which is determined by three Hall-effect sensors in servo drives or by measuring the three-phase terminal voltage waveforms in non-servo drives.

8. In the LCI drives in very large horsepower ratings, a current source thyristor inverter is used. This inverter is similar to the line-frequency inverter discussed in Chapter 6 since the current commutation from one phase to the next is provided by the induced emfs in the motor.

9. In low-speed and very large horsepower applications, cycloconverters can be used.

PROBLEMS

15-1 A brushless, permanent-magnet, four-pole, three-phase motor has the following parameters:

$$\text{Torque constant} = 0.229 \text{ N-m/A}$$
$$\text{Voltage constant} = 24.0 \text{ V/1000 rpm}$$
$$\text{Phase-to-phase resistance} = 8.4 \text{ } \Omega$$
$$\text{Phase-to-phase winding inductance} = 16.8 \text{ mH}$$

The above motor produces a trapezoidal back-emf. The torque constant is obtained as a ratio of the maximum torque produced to the current flowing through two of the phases. The voltage constant is the ratio of the peak phase-to-phase voltage to the rotational speed. If the motor is operating at a speed of 3000 rpm and delivering a torque of 0.25 N-m, plot the idealized phase current waveforms.

15-2 In a sinusoidal-waveform, three-phase, two-pole brushless dc motor with a permanent-magnet rotor, $k_T = 0.5$ N-m/A, where k_T is defined by Eq. 15-22. Calculate i_a, i_b, and i_c if the motor is required to supply a holding torque of 0.75 N-m (to keep the load from moving) at the rotor position of $\theta = 30°$, where θ is defined in Fig. 15-4.

15-3 Estimate the minimum dc input voltage to the switch-mode converter required to supply the motor in Problem 15-1 if the maximum speed is 5000 rpm and the torque is 0.25 N-m.

15-4 In a sinusoidal-waveform, permanent-magnet brushless servo motor, phase-to-phase resistance is 8.0 Ω and the phase-to-phase inductance is 16.0 mH. The voltage constant, which is the ratio of the peak phase voltage induced to the rotational speed, is 25 V/1000 rpm; $p = 2$ and $n = 10,000$ rpm. Calculate the terminal voltage if the load is such that the motor draws 10 A rms per phase. Calculate the power factor of operation.

15-5 Show the relationship between k_E and k_T in Eqs. 15-28 and 15-29 for a trapezoidal-waveform brushless motor. Compare the result with the ratio of the torque constant to the voltage constant of the motor specified in Problem 15-1.

REFERENCES

1. A. E. Fitzgerald, C. Kingsley, and S. D. Umans, *Electrical Machinery,* 4th ed., McGraw-Hill, New York, 1983.
2. B. K. Bose, *Power Electronics and AC Drives,* Prentice-Hall, Englewood Cliffs, NJ, 1986.
3. T. Kenjo and S. Nagamori, *Permanent-Magnet and Brush-less DC Motors,* Claredon, Oxford, 1985.
4. *DC Motors · Speed Controls · Servo Systems—An Engineering Handbook,* 5th ed., Electro-Craft Corporation, Hopkins, MN, 1980.
5. D. M. Erdman, H. B. Harms, and J. L. Oldenkamp, "Electrically Commutated DC Motors for the Appliance Industry," *IEEE/IAS 1984 Annual Meeting Record,* 1984, pp. 1339–1345.
6. S. Meshkat and E. K. Persson, "Optimum Current Vector Control of a Brushless Servo Amplifier Using Microprocessor," *IEEE/IAS 1984 Annual Meeting Record,* pp. 451–457.
7. R. H. Comstock, "Trends in Brushless Permanent Magnet and Induction Motor Servo Drives," *Motion Magazine,* Second Quarter, 1985, pp. 4–12.
8. P. Zimmerman, "Electronically Commutated DC Feed Drives for Machine Tools," *Drives and Controls International,* Oct./Nov. 1982, pp. 13–19.
9. L. Gyugyi and B. R. Pelly, *State Power Frequency Changers,* Wiley, New York, 1975.
10. T. Undeland, S. Midttveit, and R. Nilssen, "Phasor-applied Control (PAC) of Induction Motors: A New Concept for Servo-Quality Dynamic Performance," paper presented at 1986 Conference on Applied Motion Control, Minneapolis, MN, pp. 1–8.

PART 5

OTHER APPLICATIONS

CHAPTER 16

RESIDENTIAL AND INDUSTRIAL APPLICATIONS

16-1 INTRODUCTION

Power electronic converters are described in a generic manner in Chapters 1–9. Their applications in dc and ac power supplies are described in Chapters 10 and 11, respectively and in motor drives in Chapters 12 to 15. The objectives of this chapter are twofold: (1) to give a brief overview of various residential power electronic applications and (2) to describe some additional industrial applications of power electronics such as welding and induction heating.

16-2 RESIDENTIAL APPLICATIONS

Residential homes and buildings are the endpoint of approximately 35% of the total electricity generated in the United States, which corresponds to approximately 8.5% of the total primary energy usage. The residential applications include space heating and air conditioning, refrigeration and freezing, water heating, lighting, cooking, television, clothes washer and dryer, and many other miscellaneous appliances.

The role of power electronics in residential applications is to provide energy conservation, reduced operating cost, increased safety, and greater comfort. Benefits of incorporating power electronics into some of the dominant residential applications are discussed here.

16-2-1 SPACE HEATING AND AIR CONDITIONING

Approximately 25–30% of the electric energy in an all-electric home is used for space heating and air conditioning. Heat pumps are now being used in one out of every three new homes. Incorporating load-proportional capacity modulation can increase the heat pump efficiency by as much as 30% over the conventional single-speed heat pumps. In a conventional heat pump, the compressor operates essentially at a constant speed when the motor is running. The compressor output in this system is matched to the building heating or cooling load by cycling the compressor on or off. In a load-proportional capacity-modulated heat pump (shown in Fig. 16-1), the speed of the compressor motor and hence the compressor output are adjusted to match the building heating or cooling load, thus

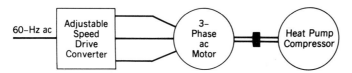

Figure 16-1 Load-proportional capacity-modulated heat pump.

eliminating the on or off cycling of the compressor. Either an induction motor drive, discussed in Chapter 14 or a self-synchronous motor drive, discussed in Chapter 15, is used to adjust the compressor speed in proportion to the building load.

The benefits of eliminating on or off cycling of the compressor are discussed by means of Fig. 16-2. In a conventional heat pump system in the cooling mode, if the sensed temperature of the building exceeds the upper limit of the thermostat setting, the compressor motor is turned on. The electric motor almost immediately begins to draw its maximum electrical power, but the compressor output increases slowly, as shown in Fig. 16-2. Therefore, the shaded area in the plot of the compressor output represents a loss in the compressor output and hence a loss in the energy efficiency of the system. When the building temperature reaches the lower limit of the thermostat temperature setting, the motor and the compressor are turned off. By this on–off cycling, the average compressor output, shown by a dashed line in the plot of the compressor output in Fig. 16-2, is matched to the building load, and the building temperature is maintained within a tolerance band around the thermostat temperature setting.

The loss in the compressor output due to on–off cycling is eliminated in the load-proportional capacity-modulated heat pump, where the speed of the compressor and, hence, the compressor output are adjusted to equal the building load. In spite of some losses in the power electronic converter used in this system, the overall electric energy consumed can be reduced by as much as 30% compared with the conventional single-speed heat pumps. Moreover, the building temperature can be maintained in a narrower band, thus resulting in increased comfort.

16-2-2 HIGH-FREQUENCY FLUORESCENT LIGHTING

Lighting consumes approximately 15% of the energy in residential buildings and 30% in commercial buildings. Fluorescent lamps are three to four times more energy efficient compared with the incandescent lamps. The energy efficiency of fluorescent lamps can be further increased by 20–30% by operating them at a high frequency (>25 kHz), compared to the conventional 60-Hz fluorescent lamps.

Fluorescent lamps exhibit a negative resistance characteristic. This requires that an inductive ballast (also called a choke) be used in series for stable operation, as shown in the simplified schematic of Fig. 16-3a. Since the lamp impedance is essentially resistive, the three voltages in the circuit of Fig. 16-3a are related as

$$V_{\text{ballast}}^2 + V_{\text{lamp}}^2 = V_s^2 \qquad (16-1)$$

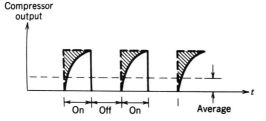

Figure 16-2 Conventional heat pump waveforms.

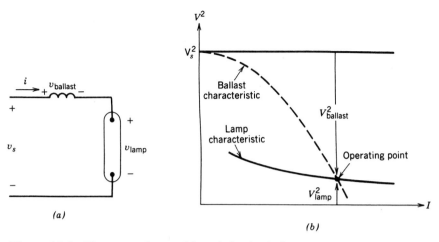

Figure 16-3   Fluorescent lamp with an inductive ballast.

The lamp and the ballast characteristics are plotted in Fig. 16-3b in terms of V^2 and I. The intersection of the two characteristics provides a stable operating point.

Figure 16-4a shows a circuit schematic for the conventional 60-Hz rapid-start system consisting of two lamps in series. In this system, the lamp cathodes are continuously heated by the cathode heater windings A, B, and C. The circuit is redrawn in Fig. 16-4b without the heating windings to explain the basic operation. The input voltage is boosted by the autotransformer (primary in series with the secondary). The leakage inductances of the primary and the secondary transformer windings provide the ballast inductance needed for a stable operation. The starting capacitor has a low impedance compared to an

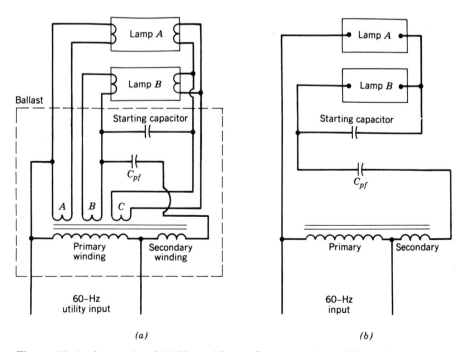

Figure 16-4   Conventional 60-Hz rapid-start fluorescent lamp: (a) circuit schematic; (b) simplified schematic.

unignited lamp and a high impedance compared to an ignited lamp. Therefore at start-up, the starting capacitor provides a shunt across lamp B, and nearly all of the input voltage appears across lamp A, thus striking an arc. Once the arc discharge is established in lamp A, a high voltage appears across lamp B, which ignites an arc in lamp B. Then, the series combination of lamps A and B is in series with a power factor correction capacitor C_{pf}, which is used to correct an otherwise poor power factor of operation.

The high-frequency fluorescent lighting system is shown in a block diagram form in Fig. 16-5a. The high-frequency electronics ballast converts the 60-Hz input to a high-frequency output, usually in a range of 25–40 kHz. The block diagram of the high-frequency electronics ballast shown in Fig. 16-5b consists of a diode rectifier bridge discussed in Chapter 5 and a dc-to-high-frequency-ac inverter. The inversion of dc to high-frequency ac can be obtained in one of several ways: for example, a class E resonant converter discussed in Chapter 9 can be used to produce sinusoidal lamp voltage and current; another possibility is to use a switch-mode converter, for example, a half-bridge topology, as discussed in Chapter 10, but without the isolation transformer and the output rectifying stage. An EMI filter is used before the rectifier bridge to suppress the conducted EMI. As in most power electronics equipment, the current drawn by the ballast from the utility system will contain significant harmonics, and hence the electronic ballast will operate at a poor power factor. The problem of harmonics can be remedied efficiently by the input current waveshaping circuit described in Chapter 18.

Because a large electromagnetic ballast associated with a standard 60-Hz fluorescent system is not required, the electronic ballasts in general are more energy efficient compared to the standard ballasts. A dimming control can be incorporated in the 60-Hz as well as the high-frequency lighting systems to compensate for the daylight coming in through the windows. In addition, a dimming control can lead to significant energy savings in the following manner: the lumen capacity of a lamp diminishes with time. Therefore, the new lamps are selected to have a lumen capacity that is approximately 30% higher than the nominal requirement. With a dimming control, new lamps can be operated at a reduced power to deliver the nominal requirement, thus resulting in energy savings during the period while the lamps have a high lumen capacity.

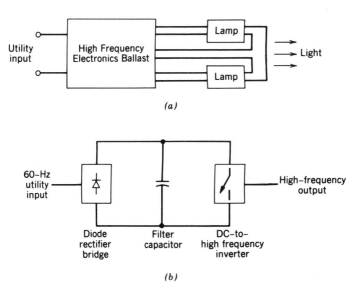

(a)

(b)

Figure 16-5 High-frequency fluorescent lighting system: (*a*) system block diagram; (*b*) ballast block diagram.

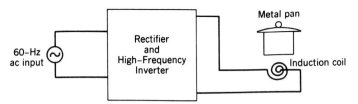

Figure 16-6 Induction cooking

16-2-3 INDUCTION COOKING

In a standard electric or gas cooking range, a significant amount of heat escapes to the surroundings, thus resulting in poor thermal efficiency. This can be avoided by means of induction cooking, which is shown in Fig. 16-6 in a block diagram form. The 60-Hz ac input is converted to a high-frequency ac in a range of 25–40 kHz, which is supplied to an induction coil. This induces circulating currents in the metal pan on top of the induction coil, thus directly heating the pan. Similar circuits as discussed in connection with the high-frequency electronics ballasts for fluorescent lights can be used to convert 60-Hz ac input to high-frequency ac.

16-3 INDUSTRIAL APPLICATIONS

Industrial applications such as induction heating and welding are discussed here in terms of the converter circuits discussed in the previous chapters.

16-3-1 INDUCTION HEATING

In induction heating, the heat in the electrically conducting workpiece is produced by circulating currents caused by electromagnetic induction. Induction heating is clean, quick, and efficient. It allows a defined section of the workpiece to be heated accurately. The magnitude of the induced currents in workpiece decreases exponentially with the distance x from the surface, as given by the equation

$$I(x) = I_0 e^{-x/\delta} \tag{16-2}$$

where I_0 is the current at the surface and δ is the penetration depth at which the current is reduced to I_0 times a factor $1/e$ ($\simeq 0.368$). The penetration depth is inversely proportional to the square root of frequency f and proportional to the square root of the workpiece resistivity ρ

$$\delta = k \sqrt{\frac{\rho}{f}} \tag{16-3}$$

where k is a constant. Therefore, the induction frequency is selected based on the application. A low frequency such as the utility frequency may be used for induction melting of large workpieces. High frequencies of up to a few hundred kilohertz are used for forging, soldering, hardening, and annealing.

The circulating currents are caused in the workpiece by the currents in the induction coil. The induction coil is inductive, and the induction load can be represented by an equivalent resistance in series with the coil inductance or by an equivalent parallel resistance as shown in Figs. 16-7a and 16-7b, respectively. A resonant capacitor is used to supply a sinusoidal current to the induction coil and to compensate for the poor power factor due to the coil inductance. This leads to the following two basic circuit configurations:

1. Voltage-source, series-resonant inverters, as shown in Fig. 16-7a
2. Current-source, parallel-resonant inverters, as shown in Fig. 16-7b

The voltage-source series-resonant inverter configuration of Fig. 16-7a is similar to the series-loaded resonant (SLR) converters discussed in Chapter 9. The inverter input is a dc voltage and the output is a square-wave voltage at the desired frequency. If the operating frequency is chosen to be near the resonant frequency, then the current i will be essentially sinusoidal due to the impedance characteristic shown in Fig. 9-7 of Chapter 9 for a series-resonant circuit. Up to a few tens of kilohertz, it is possible to use thyristors as switches in the inverter. This will require that the operating frequency be below the resonant frequency so that the circuit impedance is capacitive and the current through the thyristors is naturally commutated. The power to the load can be controlled by controlling the inverter frequency.

The current-source, parallel-resonant inverters of Fig. 16-7b for induction heating were discussed in Chapter 9 in connection with resonant converters.

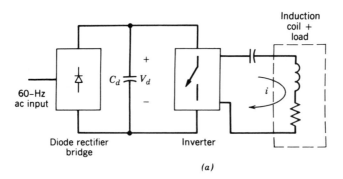

(a)

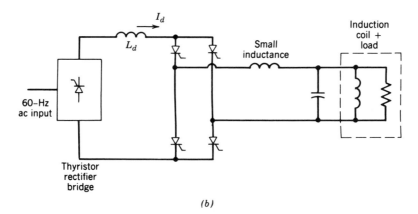

(b)

Figure 16-7   Induction heating: (a) voltage-source series-resonant induction heating; (b) current-source parallel-resonant induction heating.

16-3-2 ELECTRIC WELDING

In electric arc welding, the melting energy is provided by establishing an arc between two electrodes, one of which is the metallic workpiece being welded.

The voltage–current characteristic of the welder depends on the type of welding process employed. Typical rated voltage and current are 50 V and 500 A dc, respectively. It is desirable to have a very low ripple in the current once an arc is established. In all welding applications, the output needs to be electrically isolated from the utility input. This electrical isolation is provided by either a 60-Hz power transformer or a high-frequency transformer.

In welders with a 60-Hz power transformer, the input ac voltage is first stepped down to a suitably low voltage. Then, it is converted to a controlled dc by means of one of the three schemes shown in Fig. 16-8 in block diagram form. In Fig. 16-8a, a full-bridge thyristor rectifier is used. A large inductor is needed at the input to limit the current ripple. The other alternative, as shown in Fig. 16-8b, is to use a diode rectifier bridge that produces an uncontrolled dc. This uncontrolled dc voltage is controlled by means of a transistor series regulator. The transistor operates in its active region and acts as an adjustable resistor in order to regulate the welder's output. In the scheme of Fig. 16-8c, a switch-mode, step-down dc–dc converter is used to control the output voltage and current. All of these schemes suffer from the weight, size, and losses in the 60-Hz power transformer. The energy efficiency is particularly low in the scheme of Fig. 16-8b with a series regulator, where a substantial power loss takes place in the transistor operating in its active region.

The block diagram of a switch-mode welder is shown in Fig. 16-9, where the electrical isolation is provided by a high-frequeney transformer. The various blocks shown in Fig. 16-9 are very similar to those used for the switching dc power supplies discussed in Chapter 10. One of the resonant concepts discussed in Chapter 9 may be used

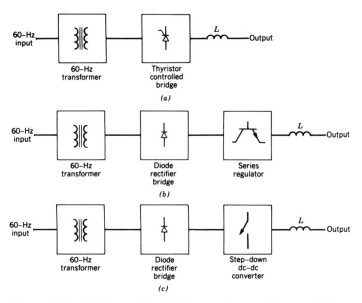

Figure 16-8 Welders with a 60-Hz transformer: (*a*) controlled thyristor bridge; (*b*) series regulator; (*c*) step-down dc–dc converter.

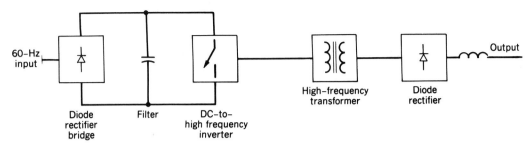

Figure 16-9   Switch-mode welder.

to invert dc into a high-frequency ac. A small inductance is needed at the output to limit the output current ripple at high frequencies. The efficiency of such a welder is in the 85–90% range, in addition to a much smaller weight and size compared with the welders employing a 60-Hz power transformer.

16-3-3 INTEGRAL HALF-CYCLE CONTROLLERS

In industrial applications requiring resistive heating or melting where the thermal time constants of the process are much longer than the 60-Hz time period, it is possible to employ an integral half-cycle control. This is shown in Fig. 16-10a for a resistive Y-connected load supplied through three triacs or back-to-back connected thyristors. If the neutral wire is accessible, this circuit can be analyzed on a per-phase basis, as shown in

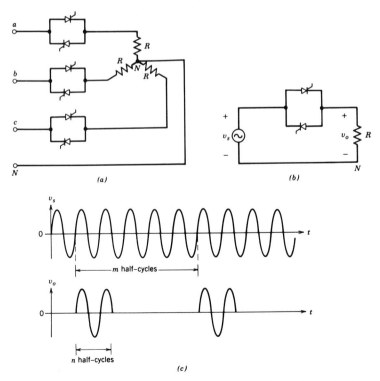

Figure 16-10   Integral half-cycle controllers: (a) three-phase circuit; (b) per-phase circuit; (c) waveforms.

Fig. 16-10*b*. The waveforms are drawn in Fig. 16-10*c*. By controlling the ratio *n/m*, keeping *m* constant, the average power supplied to the load is controlled.

SUMMARY

Some of the residential and industrial applications of power electronics are discussed. The associated power quality issues of the injected harmonic currents and the power factor of operation are discussed in Chapter 18.

PROBLEMS

16-1 In Fig. 16-2 for a single-speed heat pump, assume that each on and off period is 10 min long, that is, there are three cycles per hour. When the compressor is turned on, its output increases exponentially, reaching 99% of its maximum capacity at the end of the 10-min on interval. Once the compressor is turned off, the heating (or cooling) decays with a much smaller time constant and can be assumed to be instantaneous.

(a) If the rated electrical power is drawn throughout the on interval, calculate the loss in efficiency due to the exponential rise in the compressor output.

(b) A load-proportional capacity-modulated heat pump is used to eliminate the above on–off cycling. The efficiency of the solid-state controller is 96% and the motor efficiency is lower by 1 percentage point because of reduced speed, reduced load operation, and inverter harmonics. Assume that the compressor efficiency remains unchanged.

Compare the system efficiency with the single-speed compressor system of part a if the motor efficiency in part a can be assumed to be 85%.

REFERENCES

SPACE HEATING AND AIR CONDITIONING
1. N. Mohan and J. W. Ramsey, *Comparative Study of Adjustable-Speed Drives for Heat Pumps,* EPRI Final Report, EPRI EM-4704, Project 2033-4, Aug. 1986, Palo Alto, CA.

HIGH-FREQUENCY FLUORESCENT LIGHTING
2. E. E. Hammer and T. K. McGowan, "Characteristics of Various F40 Fluorescent Systems at 60 Hz and High Frequency," *IEEE/IAS Transactions,* Vol. IA-21, No. 1, Jan./Feb. 1985, pp. 11–16.
3. Illuminating Engineering Society (IES), *Lighting Handbook,* IES, 1981 Reference Volume.

INDUCTION HEATING AND ELECTRIC WELDING
4. Siemens and John Wiley & Sons, *Electrical Engineering Handbook,* John Wiley & Sons, New York, 1985.

CHAPTER 17

ELECTRIC UTILITY APPLICATIONS

17-1 INTRODUCTION

Power electronic systems that have unique electric utility applications such as high-voltage dc transmission, static var compensators, and the interconnection of renewable energy sources and energy storage systems to the utility grid are discussed in this chapter.

In recent years, as the semiconductor devices have improved in their voltage- and current-handling capabilities, new applications of power electronics in power systems are being investigated. Two such examples are the flexible ac transmission systems (FACTSs), discussed in reference 5 of Chapter 1, and active filters to improve power quality.

Several universities have added new courses to specifically discuss these high-power applications of power electronics, where EMTP (discussed in Section 4-6-2 of Chapter 4) is used as the simulation tool in reference 1. In keeping with the objective of this book, we have only provided an overview of these applications.

17-2 HIGH-VOLTAGE dc TRANSMISSION

Electrical plants generate power in the form of ac voltages and currents. This power is transmitted to the load centers on three-phase, ac transmission lines. However, under certain circumstances, it becomes desirable to transmit this power over dc transmission lines. This alternative becomes economically attractive where a large amount of power is to be transmitted over a long distance from a remote generating plant to the load center. This breakeven distance for HVDC overhead transmission lines usually lies somewhere in a range of 300–400 miles and is much smaller for underwater cables. In addition, many other factors, such as the improved transient stability and the dynamic damping of the electrical system oscillations, may influence the selection of dc transmission in preference to the ac transmission. It is possible to interconnect two ac systems, which are at two different frequencies or which are not synchronized, by means of an HVDC transmission line.

Figure 17-1 shows a typical one-line diagram of an HVDC transmission system for interconnecting two ac systems (where each ac system may include its own generation and load) by an HVDC transmission line. Power flow over the transmission line can be

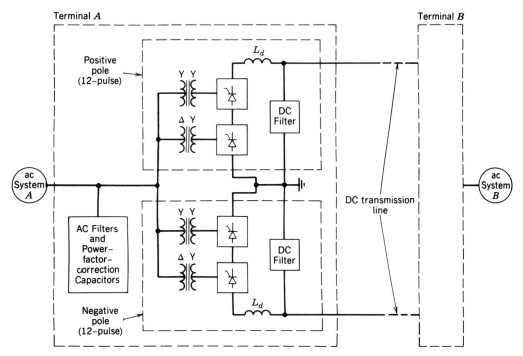

Figure 17-1   A typical HVDC transmission system.

reversed. If we assume the power flow to be from system A to B, the system A voltage, in the range of 69–230 kV, is transformed up to the transmission level and then rectified by means of the converter terminal A and applied to the HVDC transmission line. At the receiving end, the dc power is inverted by means of the converter terminal B, and the voltage is transformed down to match the ac voltage of system B. The power received over the HVDC transmission line is then transmitted over ac transmission and distribution lines to wherever it is needed in system B.

Each converter terminal in Fig. 17-1 consists of a positive pole and a negative pole. Each pole consists of two 6-pulse, line-frequency bridge converters connected through a Y–Y and a Δ–Y transformer to yield a 12-pulse converter arrangement. On the ac side of the converter, the filters are required to reduce the current harmonics generated by the converters from entering the ac system. Moreover, the power factor correction capacitors are included along with the ac filter banks to supply the lagging reactive power (or the inductive vars) required by the converter in the rectifier as well as in the inverter mode of operation. On the dc side of the converter, the ripple in the dc voltage is prevented from causing excessive ripple in the dc transmission line current by means of smoothing inductors L_d and the dc-side filter banks, as shown in Fig. 17-1.

17-2-1   TWELVE-PULSE LINE-FREQUENCY CONVERTERS

The 6-pulse line-frequency controlled converters were discussed in detail in Chapter 6. Because of high power levels associated with the HVDC transmission application, it is important to reduce the current harmonics generated on the ac side and the voltage ripple produced on the dc side of the converter. This is accomplished by means of a 12-pulse converter operation, which requires two 6-pulse converters connected through a Y–Y and a Δ–Y transformer, as is shown in Fig. 17-2. The two 6-pulse converters are connected

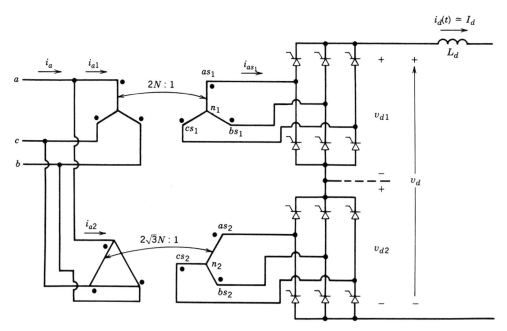

Figure 17-2 Twelve-pulse converter arrangement.

in series on the dc side and in parallel on the ac side. The series connection of two 6-pulse converters on the dc side is important to meet the high voltage requirement of an HVDC system.

In Fig. 17-2, $V_{as_1n_1}$ leads $V_{as_2n_2}$ by 30°. The voltage and current waveforms can be drawn by assuming the current I_d on the dc side of the converter to be a pure dc in the presence of the large smoothing inductor L_d shown in Fig. 17-2. Initially, for simplicity, we will assume that the per-phase ac-side commutating inductance L_s is negligible, thus resulting in rectangular current pulses. In practice, however, substantial commutating inductances are present as a result of the transformer leakage inductances. The effects of these commutating inductances on the 12-pulse waveform are discussed later.

With the foregoing assumptions of $L_s = 0$ and $i_d(t) \simeq I_d$ and recognizing that $V_{as_1n_1}$ leads $V_{as_2n_2}$ by 30°, we can draw the current waveforms as in Fig. 17-3a. Each 6-pulse converter operates at the same delay angle α. The waveform of the total per-phase current $i_a = i_{a1} + i_{a2}$ clearly shows that it contains fewer harmonics than either i_{a1} or i_{a2} drawn by the 6-pulse converters. In terms of their Fourier components

$$i_{a1} = \frac{2\sqrt{3}}{2N\pi}I_d(\cos\theta - \tfrac{1}{5}\cos 5\theta + \tfrac{1}{7}\cos 7\theta - \tfrac{1}{11}\cos 11\theta + \tfrac{1}{13}\cos 13\theta \ \ldots)$$

(17-1)

$$i_{a2} = \frac{2\sqrt{3}}{2N\pi}I_d(\cos\theta + \tfrac{1}{5}\cos 5\theta - \tfrac{1}{7}\cos 7\theta - \tfrac{1}{11}\cos 11\theta + \tfrac{1}{13}\cos 13\theta \ \ldots)$$

(17-2)

where $\theta = \omega t$ and the transformer turns ratio N is indicated in Fig. 17-2. Therefore, the combined current drawn is

$$i_a = i_{a1} + i_{a2} = \frac{2\sqrt{3}}{N\pi}I_d(\cos\theta - \tfrac{1}{11}\cos 11\theta + \tfrac{1}{13}\cos 13\theta \ \ldots)$$ (17-3)

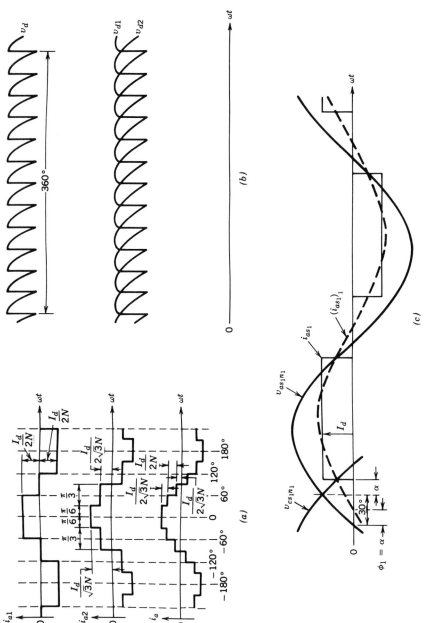

Figure 17-3 Idealized waveforms assuming $L_s = 0$.

This Fourier analysis shows that the combined line current has harmonics of the order

$$h = 12k \pm 1 \quad \text{(where } k = \text{an integer)} \tag{17-4}$$

resulting in a 12-pulse operation, as compared with a 6-pulse operation where the ac current harmonics are of the order $6k \pm 1$ (where $k =$ an integer). The harmonic current amplitudes in Eq. 17-3 for a 12-pulse converter are inversely proportional to their harmonic order and the lowest order harmonics are the eleventh and the thirteenth. The currents on the ac side of the two 6-pulse converters add, confirming that the two converters are effectively in parallel on the ac side.

On the dc side, the voltage waveforms v_{d1} and v_{d2} for the two 6-pulse converters are shown in Fig. 17-3b. These two voltage waveforms are shifted by 30° with respect to each other. Since the two 6-pulse converters are connected in series on the dc side, the total dc voltage $v_d = v_{d1} + v_{d2}$ has 12 ripple pulses per fundamental-frequency ac cycle. This results in the voltage harmonics of the order h in v_d, where

$$h = 12k \quad \text{(} k = \text{an integer)} \tag{17-5}$$

and the twelfth harmonic is the lowest order harmonic. Magnitudes of the dc-side voltage harmonics vary significantly with the delay angle α.

In practice, L_s is substantial because of the leakage inductance of the transformers. The presence of L_s does not change the order of characteristic harmonics produced either on the ac side or on the dc side, provided that the two 6-pulse converters operate under identical conditions. However, the harmonic magnitudes depend significantly on L_s, delay angle α, and dc current I_d. The effect of L_s on the ac current waveform and harmonics was discussed in Chapter 6.

Based on the derivation in Chapter 6, the average dc voltage can be written as

$$V_{d1} = V_{d2} = \frac{V_d}{2} = \frac{3\sqrt{2}}{\pi} V_{LL} \cos \alpha - \frac{3\omega L_s}{\pi} I_d \tag{17-6}$$

where V_{LL} is the line-to-line rms voltage applied to each of the 6-pulse converters and L_s is the per-phase leakage inductance of each of the transformers, referred to their converter side.

As we explained in Chapter 6, $\alpha > 90°$ corresponds to an inverter mode of operation with a transfer of power from the dc to the ac side of the converter.

17-2-2 REACTIVE POWER DRAWN BY CONVERTERS

As was alluded to earlier, the line-frequency, line-voltage-commutated converters operate at a lagging power factor and, hence, draw reactive power from the ac system. Even though the ac-side currents associated with the converter contain harmonics in addition to their fundamental-frequency components, the harmonic currents are "absorbed" by the ac-side filters, whose design must be based on the magnitude of the generated harmonic current magnitudes, as we will discuss later. Therefore, only the fundamental-frequency components of the ac currents are considered for the real power transfer and the reactive power drawn. It is necessary to consider only one of the two 6-pulse converters, since the real and the reactive power for the 12-pulse converter arrangement making up a pole are twice the per-converter values.

17-2-2-1 Rectifier Mode of Operation

With the initial assumption that $L_s = 0$ in Fig. 17-2, Fig. 17-3c shows the phase-to-neutral voltage $v_{as_1 n_1}$ and the current i_{as_1} (corresponding to converter 1 in Fig. 17-3) with $i_d(t) \simeq$

I_d at a delay angle α. The fundamental-frequency current component $(i_{as_1})_1$ shown by the dashed curve lags behind the phase voltage $v_{as_1n_1}$ by the displacement power factor angle ϕ_1, where

$$\phi_1 = \alpha \tag{17-7}$$

Therefore, the three-phase reactive power (lagging) required by the 6-pulse converter because of the fundamental-frequency reactive current components, which lag their respective phase voltages by 90°, equals

$$Q_1 = \sqrt{3}\, V_{LL}(I_{as_1})_1 \sin\alpha \tag{17-8}$$

where V_{LL} is the line-to-line voltage on the ac side of the converter.

From the Fourier analysis of i_{as_1} in Fig. 17-3c, the rms value of its fundamental-frequency component is

$$(I_{as_1})_1 = \frac{\sqrt{6}}{\pi}I_d \simeq 0.78 I_d \tag{17-9}$$

Therefore, from Eqs. 17-8 and 17-9

$$Q_1 = \sqrt{3}\, V_{LL}\left(\frac{\sqrt{6}}{\pi}I_d\right)\sin\alpha = 1.35 V_{LL}I_d\sin\alpha \tag{17-10}$$

The real power transfer through each of the 6-pulse converters can be calculated from Eq. 17-6 with $L_s = 0$ as

$$P_{d1} = V_{d1}I_d = 1.35 V_{LL}I_d\cos\alpha \tag{17-11}$$

For a desired power transfer P_{d1}, the reactive power demand Q_1 should be minimized as much as possible. Similarly, I_d should be kept as small as possible to minimize I^2R losses on the dc transmission line. To minimize I_d and Q_1, noting that V_{LL} is essentially constant in Eqs. 17-10 and 17-11, we should choose a small value for the delay α in the rectifier mode of operation. For practical reasons, the minimum value of α is chosen in a range of 10°–20°.

17-2-2-2 Inverter Mode of Operation

In the inverter mode, the dc voltage of the converter acts like a counter-emf in a dc motor. Therefore, it is convenient to define the dc voltage polarity as shown in Fig. 17-4a, so that the dc voltage is positive when written specifically for the inverter mode of operation. In Chapter 6, the extinction angle γ for the inverter was defined in terms of α and u as

$$\gamma = 180° - (\alpha + u) \tag{17-12}$$

where α is the delay angle and u is the commutation or the overlap angle. The inverter voltage in Fig. 17-4 can be obtained as (see Problems 17-7)

$$V_{d1} = V_{d2} = \frac{V_d}{2} = 1.35 V_{LL}\cos\gamma - \frac{3\omega L_s}{\pi}I_d \tag{17-13}$$

Again with the assumption that $L_s = 0$ for simplicity, Fig. 17-4b shows the idealized waveforms for $v_{as_1n_1}$ and i_{as_1} at an $\alpha > 90°$, corresponding to the inverter mode of operation. The fundamental-frequency component $(i_{as_1})_1$ of the phase current is shown by the dashed curve. In the phasor diagram of Fig. 17-4c, the fundamental-frequency reactive current component lags behind the phase-to-neutral voltage, indicating that even in the inverter mode, where the direction of power flow through the converter has reversed, the converter requires reactive power (lagging) from the ac system.

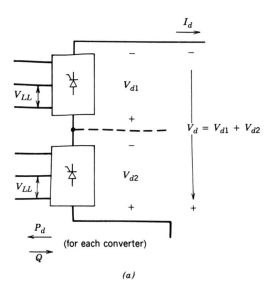

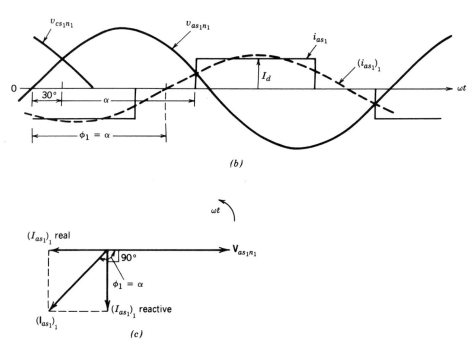

Figure 17-4 Inverter mode of operating (assuming $L_s = 0$).

With $L_s = 0$, $u = 0$ in Eq. 17-12 and $\gamma = 180° - \alpha$. Therefore, the expressions for per-converter Q_1 and P_{d1} in Eqs. 17-10 and 17-11 can be obtained specifically for the inverter mode in terms of γ as

$$Q_1 = 1.35 V_{LL} I_d \sin \gamma \qquad (17\text{-}14)$$

and

$$P_{d1} = 1.35 V_{LL} I_d \cos \gamma \qquad (17\text{-}15)$$

where the directions of the reactive power (lagging) and the real power are as shown in Fig. 17-4a.

In Eqs. 17-14 and 17-15, γ should be as small as possible for a given power transfer level to minimize I^2R losses in the transmission line due to I_d and to minimize the reactive power demand by the converter. As we discussed in Chapter 6, the minimum value that γ is allowed to attain is called the minimum extinction angle γ_{min} that is based on allowing sufficient turn-off time to the thyristors.

In a 12-pulse converter arrangement, the reactive power requirement is the sum of the reactive powers required by each of the two 6-pulse converters. The ac-side filter banks and the power factor correction capacitors partially provide the reactive power demand of the converters, as discussed in Section 17-2-4.

17-2-3 CONTROL OF HVDC CONVERTERS

It is possible to discuss the control of converters in an HVDC system on a per-pole basis, since both the positive and the negative poles are operated under identical conditions. Figure 17-5a shows the positive pole, for example, consisting of the 12-pulse converters A and B. Terminal A is assumed to be operating as a rectifier, and its dc voltage is defined as V_{dA}. Terminal B is assumed to be operating as an inverter, and its dc voltage V_{dB} is shown with a polarity that is specific to the inverter mode of operation, so that V_{dB} has a positive value.

In steady state in Fig. 17-5a

$$I_d = \frac{V_{dA} - V_{dB}}{R_{dc}} \qquad (17\text{-}16)$$

where R_{dc} is the dc resistance of the positive transmission line conductor. In practice, R_{dc} is small and I_d results as a consequence of a small difference between two very large voltages in Eq. 17-16. Therefore, one converter is assigned to control the voltage on the transmission line and the other to control I_d. Since the inverter should operate at a constant $\gamma = \gamma_{min}$, it is natural to choose the inverter (converter B in Fig. 17-5a) to control V_d. Then, I_d and, hence, the power level are controlled by the rectifier (converter A in Fig. 17-5a).

Figure 17-5b shows the rectifier and the inverter control characteristics in the V_d–I_d plane, where V_d is chosen to be the voltage at the rectifier, that is, $V_d = V_{dA}$. At the constant extinction angle $\gamma = \gamma_{min}$, the inverter produces a voltage V_d in Fig. 17-5a, which is given as

$$V_d = 2 \times \left[1.35 V_{LL} \cos \gamma_{min} - \frac{3\omega L_s}{\pi} I_d \right] + R_{dc} I_d$$

$$= 2 \times 1.35 V_{LL} \cos \gamma_{min} - \left(\frac{6\omega L_s}{\pi} - R_{dc} \right) I_d \qquad (17\text{-}17)$$

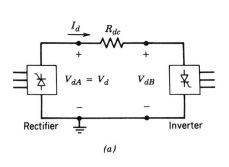

(a)

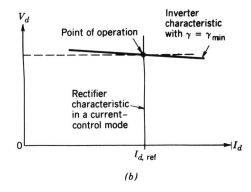

(b)

Figure 17-5 Control of HVDC system.

Assuming the quantity within the bracket in Eq. 17-17 to be positive, the constant extinction angle operation of the inverter results in a V_d–I_d characteristic as shown in Fig. 17-5b.

The rectifier can be controlled to maintain I_d equal to its commanded or reference value $I_{d,\text{ref}}$. The actual current I_d is measured, and the error $(I_d - I_{d,\text{ref}})$, if positive, increases the rectifier delay angle α; if the error is negative, α is decreased. A high-gain current controller results in a nearly vertical rectifier characteristic in Fig. 17-5b at $I_{d,\text{ref}}$. The intersection of the two characteristics in Fig. 17-5b establishes the transmission line voltage V_d and the current I_d.

The foregoing discussion shows how the power flow $P_d = V_d I_d$ from terminal A to terminal B can be controlled in Fig. 17-5a by controlling I_d, while maintaining the transmission line voltage as high as possible to minimize $I_d^2 R_{\text{dc}}$ power loss in the transmission line. This type of control also results in a small value of α in the rectifier and a small $\gamma = \gamma_{\text{min}}$ in the inverter, thus minimizing the reactive power demand by both the rectifier and the inverter. In practice, the transformers at both the terminals consist of tap changers, which can control the ac voltage V_{LL} supplied to the converters in a small range, thus providing an additional degree of control.

The control characteristics shown in Fig. 17-5b can be extended for negative values of V_d so that the power flow can be controlled smoothly in magnitude as well as in direction. A detailed discussion can be found in reference 2. This capability to be able to reverse the power flow is useful if the two ac systems interconnected by the dc transmission line have loads that vary differently with seasons or the time of day. The same may be the case if one of the ac systems contains hydro generation whose output depends on seasons. Another application of this control capability is to modulate the power flow on the dc line to damp out the ac system oscillations.

17-2-4 HARMONIC FILTERS AND POWER FACTOR CORRECTION CAPACITORS

17-2-4-1 dc-Side Harmonic Filters

To minimize the inductively coupled harmonic interference produced in the telephone system and other types of control/communication channels in parallel with the HVDC transmission lines, it is important to minimize the magnitudes of the current harmonics on the dc transmission line. The voltage harmonics are of the order $12k$, where k is an integer. The magnitudes of the harmonic voltages depend on α, L_s, and I_d for a given ac system voltage. Under a balanced 12-pulse operating condition, the 12-pulse converter can be represented by an equivalent circuit as shown in Fig. 17-6a, where the harmonic voltages are connected in series with the dc voltage V_d.

A large smoothing inductor L_d of the order of several hundred millihenries is used in combination with a high-pass filter, as shown in Fig. 17-6a, in order to limit the flow of harmonic currents on the transmission line. The impedance of the high-pass filter in Fig. 17-6a is plotted in Fig. 17-6b, where the filter is designed specifically to provide a low impedance at the dominant twelfth harmonic frequency.

17-2-4-2 ac-Side Harmonic Filters and Power Factor Correction Capacitors

In a 12-pulse converter, the ac currents consist of the characteristic harmonics of the order $12k \pm 1$ (k = an integer), as given by Eq. 17-4. The harmonic currents can be represented

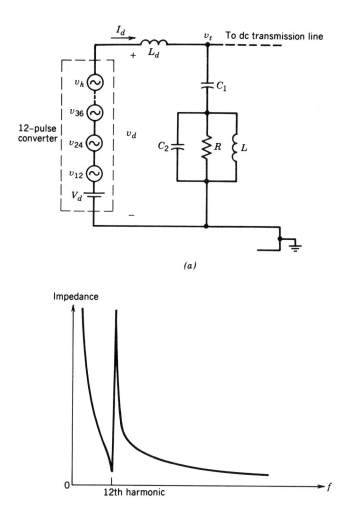

Figure 17-6   Filter for dc-side voltage harmonics: (*a*) dc-side equivalent circuit; (*b*) high-pass filter impedance vs. frequency.

by means of an equivalent circuit, as shown in Fig. 17-7*a*. It is desirable to prevent these harmonic currents from entering the ac network, where they cause power losses and may also cause interference with the other electronic communication equipment. For this purpose, the per-phase filters shown in Fig. 17-7*a* are commonly used. The series-tuned filters are used for the two lower order harmonics: the eleventh and the thirteenth. A high-pass filter as shown in Fig. 17-7*a* is used to eliminate the rest of the higher order harmonics. The combined impedance of all the harmonic filters is plotted in Fig. 17-7*b*.

Note that the filter design very much depends on the ac system impedance at the harmonic frequencies in order to provide adequate filtering and to avoid certain resonance conditions. The system impedance depends on the system configuration based on loads, generation pattern, and the transmission lines in service. Therefore, the filter design must anticipate the changes to occur in the foreseeable future (e.g., an additional interconnection), which may alter the ac system impedance.

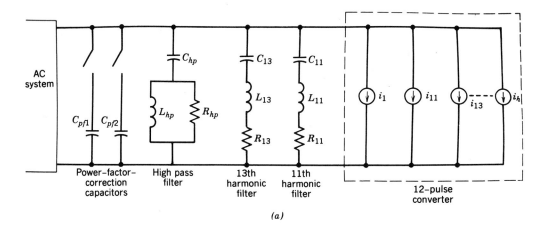

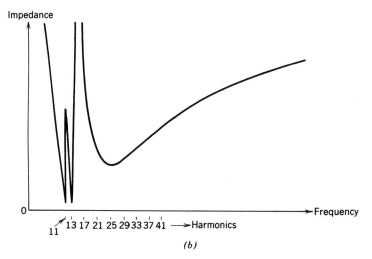

(b)

Figure 17-7 The ac side filters and power factor correction capacitors: (a) per-phase equivalent circuit; (b) combined per-phase filter impedance vs. frequency.

The harmonic filters also provide a large percentage of the reactive power required by the converters in the rectifier and the inverter mode. In the ac-side filters discussed earlier (both series tuned and high pass), the capacitive impedance dominates at the 60-Hz frequency over the inductive elements connected in series with the capacitor. Therefore, the effective shunt capacitance offered per phase by the ac filter bank at the fundamental or line frequency can be approximated as

$$C_f \simeq C_{11} + C_{13} + C_{hp} \qquad (17\text{-}18)$$

At the 60-Hz fundamental system frequency, the per-phase reactive power (vars) supplied by the filter bank equals

$$Q_f \simeq 377 C_f V_s^2 \qquad (17\text{-}19)$$

where V_s is the rms phase voltage applied across the filters. Thus, the ac filters play an important role in meeting the reactive power demand of the converters, in addition to filtering the current harmonics.

As was discussed in Section 17-2-3, which deals with the control of HVDC systems, the power flow on the dc line is controlled by adjusting I_d in Fig. 17-5. Therefore, the reactive power demand by the converters increases with the increasing power transfer level, as is suggested by Eq. 17-10. The filter capacitors are chosen such that the reactive power supplied by them does not exceed the reactive power demand of the converters at the minimum power level of the HVDC system operation. The reason is that if the reactive vars supplied by the filters exceed the converter demand, the problem of system over-voltage, which tends to occur at the light-load conditions in the first place, will increase. Therefore, to compensate for the higher reactive power demand by the converters at higher power transfer levels, additional power factor correction capacitors C_{pf} are switched in, as shown in Fig. 17-7a.

17-3   STATIC var COMPENSATORS

In an electric utility network, it is desirable to regulate the voltage within a narrow range of its nominal value. Most utilities attempt to maintain the voltage deviations in a +5%, −10% range around their nominal values. Moreover, it is desirable to have a balanced load on all three phases to eliminate negative- and zero-sequence currents that can have undesirable consequences such as additional heating in electrical equipment, torque pul-sations in generators and turbines, and so on. The load on the power system fluctuates and can result in voltages outside of their acceptable limits. In view of the fact that the internal impedance of the ac system seen by the load is mainly inductive (since the transmission and distribution lines, transformers, generators, etc., have mainly inductive impedance at the line frequency of 60 Hz), it is the reactive power change in the load that has the most adverse effect on the voltage regulation.

Consider the simple per-phase system equivalent circuit shown in Fig. 17-8a by means of the ac system Thévenin equivalent, where the internal impedance of the ac system is assumed to be purely inductive. Figure 17-8b shows the phasor diagram for a lagging power factor load $P + jQ$ with a current $\mathbf{I} = I_p + jI_q$, which lags the terminal voltage $\mathbf{V}_t$. The terminal voltage magnitude is assumed to be at its nominal value. An increase ΔQ in the lagging vars drawn by the load causes the reactive current component to increase to $I_q + \Delta I_q$, while I_p is assumed to be unchanged. The phasor diagram for the increased Q is indicated by "primed" quantities in Fig. 17-8b, where the terminal voltage is again chosen to be the reference phasor for simplicity; the magnitude of the internal system voltage V_s remains the same as before. The phasor diagram of Fig. 17-8b shows a drop in the terminal voltage by ΔV_t caused by an increase in the lagging reactive power drawn by the load. In this case, even if I_p remains constant, the real power P will decrease because of the reduction in V_t. For comparison purposes, Fig. 17-8c shows the phasor diagram where the percentage change in I_p is the same as the percentage change in I_q in Fig. 17-8b, while I_q is assumed to be unchanged. Figure 17-8c shows that the voltage change ΔV_t is small due to a change in I_p.

Most utility systems utilize power factor correction capacitor banks, which are switched in and out by means of mechanical contactors to compensate for the slow changes in the reactive power of the load in order to keep the overall load power factor as close to unity as possible. The reasons for the power factor correction are twofold: (1) it facilitates regulation of the system voltage within a range of +5%, −10% around its nominal value and (2) a close to unity power factor of the load–capacitor combination results in the lowest magnitude of the current drawn for a given real power demand. This in turn reduces the I^2R losses in the various equipment within the ac system. Moreover, the capacity of the equipment, which is rated in terms of current-handling capability, is more effectively utilized.

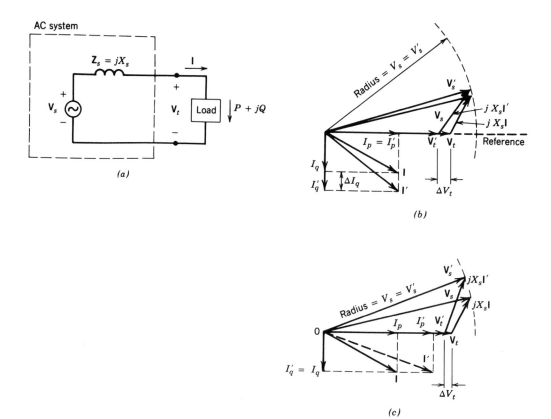

Figure 17-8 Effect of I_p and I_q on V_t: (a) equivalent circuit; (b) change in I_q; (c) change in I_p.

In this section, however, the objective is to discuss static var controllers, which by means of a power electronic interface can provide a quick control over the reactive power. These static var controllers are used to prevent annoying voltage flickers caused by industrial loads such as arc furnaces, which cause very rapid changes in the reactive power and also introduce a fluctuating load unbalance between the three phases. Another use for the static var controllers is to provide a dynamic voltage regulation to enhance the stability of the interconnection between two ac systems.

There are primarily three types of static var controllers:

1. Thyristor-controlled inductors (TCIs)
2. Thyristor-switched capacitors (TSCs)
3. Switching converters with minimum energy storage elements

A hybrid arrangement of a TCI with a TSC minimizes the no-load losses.

17-3-1 THYRISTOR-CONTROLLED INDUCTORS

Thyristor-controlled inductors act as variable inductors where the inductive vars supplied can be varied very quickly. The system may require either inductive or capacitive vars, depending on the system conditions. This requirement can be met by paralleling TCIs with a capacitor bank.

The basic principle of a TCI can be understood by considering the per-phase circuit of Fig. 17-9a, where an inductor L is connected to the ac source through a bidirectional

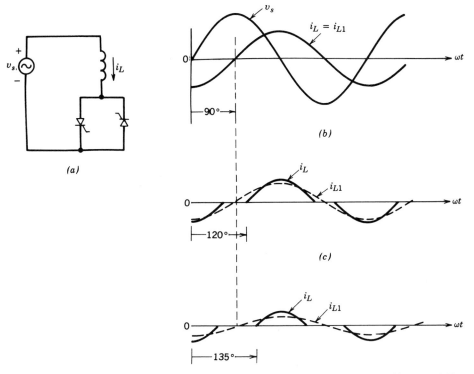

Figure 17-9 A TCI, basic principle: (a) per-phase TCI; (b) $0 < \alpha < 90°$; (c) $\alpha = 120°$; (d) $\alpha = 135°$.

switch, consisting of two back-to-back connected thyristors. If the resistive component of the inductor is assumed to be negligibly small, the current through the inductor in steady state can be obtained as a function of the thyristor delay angle α. An equal value of α is used for both thyristors.

As a base case, Fig. 17-9b shows the current waveform where the thyristor gate pulses are always present, corresponding to $\alpha = 0$, as if thyristors were replaced by diodes. With $\omega = 2\pi f$, this results in a sinusoidal current i_L whose rms value equals

$$I_L = I_{L1} = \frac{V_s}{\omega L}(\omega = 2\pi f) \qquad (17\text{-}20)$$

where the inductor current consists solely of the fundamental-frequency component without any harmonics. Since i_L lags behind V_s by 90° as shown in Fig. 17-9b, the delay angle α in a range of 0–90° has no control over i_L and its rms value remains the same as given by Eq. 17-20.

If α is increased beyond 90°, i_L can be controlled as is shown in Figs. 17-9c and 17-9d corresponding to α values of 120° and 135°, respectively. Clearly as α is increased, I_{L1} is reduced, thus allowing a control over the effective value of inductance connected to the utility voltage, since

$$L_{\text{eff}} = \frac{V_s}{\omega I_{L1}} \qquad (17\text{-}21)$$

where by Fourier analysis (see Problem 17-8)

$$I_{L_1} = \frac{V_s}{\pi \omega L}(2\pi - 2\alpha + \sin 2\alpha) \qquad \tfrac{1}{2}\pi \le \alpha \le \pi \qquad (17\text{-}22)$$

Therefore, the lagging reactive power drawn by the per-phase TCI at the fundamental frequency is

$$Q_I = V_s I_{L1} = \frac{V_s^2}{\omega L_{\text{eff}}} \tag{17-23}$$

The inductor current is not a pure sine wave at $\alpha > 90°$, as can be seen by the waveforms in Fig. 17-9c and 17-9d. A Fourier analysis of the inductor current waveform shows that i_L consists of odd harmonics h of the order 3, 5, 7, 9, 11, 13, . . . whose amplitudes as a ratio of I_{L1} depend on α. To prevent the third-order and the multiples of third-order harmonics, it is a common practice to connect three-phase TCI in a Δ so that these harmonics circulate through the inductors and do not enter the ac system. The capacitor, in parallel with the TCI to meet the system var requirements, filters out high-frequency harmonics. Similar to the discussion in Section 17-2-4-2, the fifth and the seventh harmonics are filtered out by series-tuned filters. These series-tuned filters also provide the capacitive vars as described by Eqs. 17-18 and 17-19.

17-3-2   THYRISTOR-SWITCHED CAPACITORS

Figure 17-10 shows the basic arrangement where several (three or four) capacitors can be connected to the supply voltage through a bidirectional switch consisting of back-to-back connected thyristors. Unlike the phase control used in TCIs to vary the effective value of the inductor, TSCs employ integral half-cycle control where the capacitor is either fully in or out of the circuit.

The capacitor bank can be switched out by blocking the gate pulses to both the thyristors. The current flow stops at the instant of its zero crossing, which also corresponds to the capacitor voltage equal to the maximum ac system voltage. The polarity of the capacitor voltage depends on the instant when the thyristor gate pulses are blocked. At switch-on, the thyristor must be gated at the proper instant of maximum ac voltage to avoid large overcurrents. Moreover, inductors, shown dashed in Fig. 17-10, are used to limit overcurrents at switch-on. By using a large number of thyristor-switched small capacitor banks, it is possible to vary the reactive power Q_c in small but still discrete steps.

17-3-3   INSTANTANEOUS var CONTROL USING SWITCHING CONVERTERS WITH MINIMUM ENERGY STORAGE

The static var control schemes discussed in the previous sections consist of large energy storage inductors and capacitors to meet the var demand. Moreover, they cannot provide an instantaneous var control because of their inherent time delays.

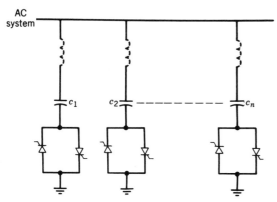

Figure 17-10 A TSC arrangement.

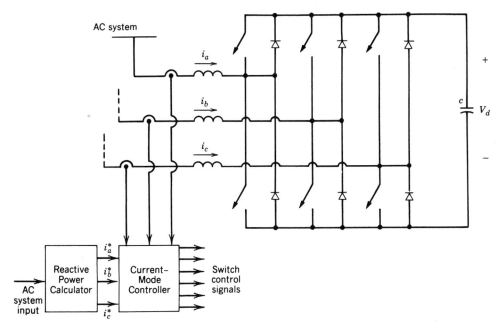

Figure 17-11 Instantaneous var controller.

In Chapter 8, dealing with the switch-mode converters (inverters and rectifiers), it was shown that their ac current can be controlled by operating them in a current-control mode. Such a converter is shown in Fig. 17-11. The ac current of such switch-mode converters can be quickly controlled in magnitude as well as in their phase relationship (leading or lagging) to the ac voltages. Since the average power drawn or supplied by these converters is desired to be zero, a dc source at the dc input of the converter is not necessary. Only a small capacitor with a minimum energy storage is sufficient, and its dc voltage is maintained by the switch-mode converter, which transfers sufficient real power from the ac system to compensate for its own losses and to maintain a constant dc voltage across the capacitor, in addition to controlling the vars.

The discussion of instantaneous reactive power can be found in reference 8, where the reactive power calculator calculates the instantaneous reference currents i_a^*, i_b^*, and i_c^*, which the switch-mode converter of Fig. 17-11 supplies under a current-mode control. Note that the resonant converter concepts discussed in Chapter 9 can be used to reduce switching losses in the converter.

17-4 INTERCONNECTION OF RENEWABLE ENERGY SOURCES AND ENERGY STORAGE SYSTEMS TO THE UTILITY GRID

A power electronic interface is needed to connect renewable energy sources such as photovoltaic, wind, and small hydro to the utility system. The same is true for the interconnection of energy storage systems for utility load-leveling (also called load-peak shaving), such as batteries, fuel cells, and superconductive energy storage inductors. Some of these systems are briefly discussed here.

17-4-1 PHOTOVOLTAIC ARRAY INTERCONNECTION

A large number of solar cells connected in series and parallel make up the photovoltaic or solar arrays. These cells produce a dc voltage when they are exposed to sunlight. Figure

17-12 shows the $i-v$ characteristics of such a cell for various insolation (sunlight intensity) levels and temperatures. Figure 17-12 shows that the cell characteristic at a given insolation and temperature basically consists of two segments: (1) the constant-voltage segment and (2) the constant-current segment. The current is limited as the cell is short-circuited. The maximum power condition occurs at the knee of the characteristic where the two segments meet. It is desirable to operate at the maximum power point. Ideally, a pure dc current should be drawn from the solar array, though the reduction in delivered power is not very large even in the presence of a fair amount of ripple current. As an example, a 5% peak ripple current results in a power reduction of less than 1%.

To ensure that the array keeps on operating at the maximum power point, a perturb-and-adjust method (also called the "dithering" technique) is used where at regular intervals (once every few seconds) the amount of current drawn is perturbed and the resulting power output is observed. If an increased current results in a higher power, it is further increased until power output begins to decline. On the other hand, if an increase in current results in less power than before, then the current is decreased until the power output stops increasing and begins to go down.

The solar array is interconnected to the utility grid through an electrical isolation. The current supplied to the utility grid should be sinusoidal at nearly a unity power factor. The guidelines for the harmonic currents injected into the utility system and the total harmonic distortion (THD) are discussed in Chapter 18.

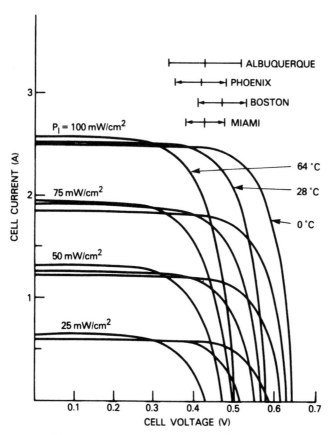

Figure 17-12 The $I-V$ characteristics of solar cells. (*Source:* reference 10.)

17-4-1-1   Single-Phase Interconnection

It is possible to use a line-frequency, phase-controlled converter of the type discussed in Chapter 6, where the converter always operates in an inverter mode and the electrical isolation is provided by a 60-Hz transformer. However, ac-side filters and the reactive power compensation would be needed, since the output current will contain harmonics and will be at a lagging power factor. Alternatively, a switch-mode pulse-width-modulated converter of the type discussed in Chapter 8 can be used where the electrical isolation is again provided by a 60-Hz transformer. However, since the current output is controlled to be in phase with the utility voltage, it may be economically feasible to utilize a high-frequency transformer to provide electrical isolation.

A circuit diagram of an interface that utilizes a high-frequency transformer is shown in Fig. 17-13. The dc voltage input is inverted to produce a high-frequency ac across the primary of the high-frequency transformer. Its secondary voltage is rectified and the resulting dc is interfaced with the line voltage through a line-frequency line-voltage-commutated thyristor inverter of the type discussed in Chapter 6. Since the line current is required to be sinusoidal and in phase with the line voltage, the line-voltage waveform is measured to establish the reference waveform for the sinusoidal line current i_s^*, whose amplitude is determined by the maximum power controller using a "dithering" scheme discussed earlier. The current i_s^* multiplied by the transformer turns ratio acts as the reference current at the switch-mode inverter output. The inverter can be controlled to deliver the reference current by means of current-regulated control, as described in Chapter 8. One way to implement this control is described in references 9 and 10. The line-frequency thyristor converter in Fig. 17-13 can be operated at a very small value of extinction angle γ since the current through it is controlled to be very small near the zero crossing of the ac system voltage.

17-4-1-2   Three-Phase Interconnection

At a power level above a few kilowatts, it is preferable to use a three-phase interconnection. Sinusoidal ac currents at a unity power factor can be delivered by using a switching-mode dc-to-ac inverter of the type discussed in Chapter 8 under a current-mode control. A 60-Hz three-phase transformer would be required to provide electrical isolation.

17-4-2   WIND AND SMALL HYDRO INTERCONNECTION

In the case of wind, the power available varies with the cube of the wind velocity. For small hydro, the power available depends on the pressure head and the flow. For both

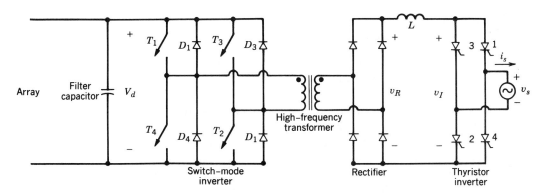

Figure 17-13   High-frequency photovoltaic interface.

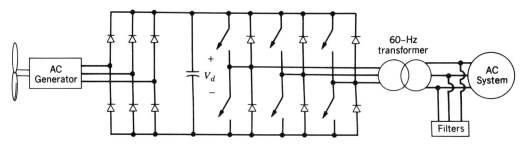

Figure 17-14   Interconnection of wind/hydro generator.

wind and small hydro, to extract the maximum amount of power, it is desirable to let the turbine speed vary over a wide range to an optimum value dependent on the operating conditions. This would not be possible if a synchronous generator were directly connected to the utility (60-Hz) system that dictated a constant speed (synchronous speed). The induction generators connected to the utility system would allow the speed to vary only in a very narrow speed range. Therefore, to allow the generator-turbine speed to vary to optimize efficiency of power generation, the three-phase generator output is rectified into dc and then interfaced with the three-phase utility source by means of a switch-mode converter of the type discussed in Chapter 8. A block diagram is shown in Fig. 17-14, where a 60-Hz isolation transformer is included.

Because of the medium power levels (a few tens of kilowatts and higher) usually associated with the wind and small hydro generators, a three-phase utility interconnection is preferable.

17-4-3   MINNESOTA INTERFACE: A NEW TOPOLOGY UTILITY INTERFACE FOR PHOTOVOLTAIC, WIND, AND FUEL CELL SYSTEMS

A unique topology interface, using a thyristor inverter and only two controlled switches, has been developed to supply power to the three-phase utility from photovoltaic (PV), wind, and fuel cell systems. The utility currents are sinusoidal at nearly a unity power factor. The circuit diagram is shown in Fig. 17-15a, and the resulting voltage and current waveforms are shown in Fig. 17-15b, where the current has a THD of only 3.4%. The operating principle of this interface is described in reference 12. The recent advances in its control allow the displacement power factor of the current to be adjustable (lagging, unity, or leading) and completely eliminate the possibility of commutation failures.

17-4-4   INTERCONNECTION OF ENERGY STORAGE SYSTEMS FOR UTILITY LOAD LEVELING

Fuel cells, batteries, and superconductive energy storage inductors are some of the means under consideration to reduce utility load peaks. As a brief explanation, it is desirable to operate the most efficient utility generators (such as nuclear and the newer high-efficiency coal-fired plants) at their rated capacity at all times. However, the load on the utility system does not remain constant with the time of day and also fluctuates based on the weather conditions. To meet the peak load, either oil- or gas-fired generators, also called the peaking plants, have to be used, which are expensive to operate because of the high cost of fuel. An alternative to this is to store the electrical energy generated by the efficient generating plants during low-load conditions and to supply it back to the utility grid under peak-load conditions, thus reducing or eliminating the need for peaking plants powered by

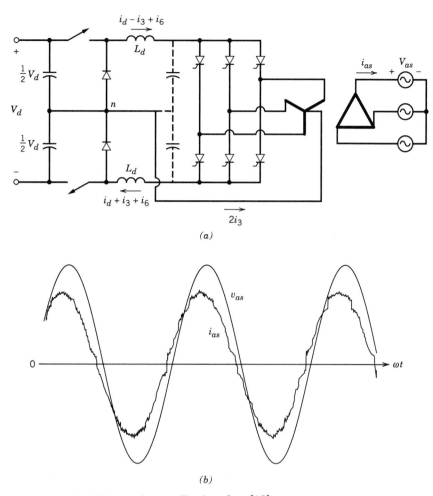

(a)

(b)

Figure 17-15 New topology, utility interface [12].

gas or oil. The electric energy can be stored in batteries or in the form of a magnetic field in a superconductive inductor. Another option is to use the electric energy during low-load conditions to produce oxygen by electrolysis, which can then be used in fuel cells to provide electrical output.

Both batteries and fuel cells produce a dc voltage. To interconnect them to the utility system, a scheme similar to that used for single-phase or three-phase interconnection of photovoltaic arrays can be used.

The most economical way to interconnect large superconductive inductors would be to use 12-pulse line-commutated converters, as shown in Fig. 17-16. Under the control of

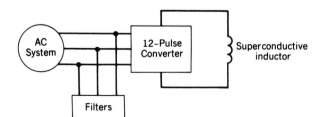

Figure 17-16 Superconductive energy storage inductor interconnection.

the delay angle, the converter operation can be continuously varied from a full-rectifier (charge) mode to the full-inverter (discharge) mode, while the current flows continuously in the same direction.

17-5   ACTIVE FILTERS

The next chapter (Chapter 18) deals with the concerns due to the harmonic components in the utility current drawn by the power electronics equipment. It also describes the corrective measures within the power electronics equipment that would result in sinusoidal currents at nearly a unity power factor. In this section, we would discuss active filters, which can prevent harmonic currents from entering the utility system, if harmonic-current-producing nonlinear loads are being supplied by the utility.

Figure 17-17 shows a one-line diagram of how an active filter functions. The current drawn by the nonlinear load(s) consists of a fundamental-frequency component i_{L1} and a distortion component $i_{L,\text{distortion}}$. The load current is sensed and filtered to provide a signal proportional to the distortion component $i_{L,\text{distortion}}$. Under a current-mode control, as discussed in Section 8-6-3 of Chapter 8, a switch-mode dc-to-ac converter is operated to deliver the current $i_{L,\text{distortion}}$ to the utility. Therefore, in an ideal case, the harmonics in the utility current are eliminated. On the dc side of the converter, only a capacitor with a minimum energy storage is needed, as discussed in Section 17-3-3, because the dc voltage across it is maintained by the switch-mode converter that transfers real power from the utility to compensate for its own losses by drawing a small current $i_{1,\text{loss}}$.

Active filters have been studied for a long time. Now due to improved power-handling capabilities of power semiconductor devices, they are being considered seriously in novel active-hybrid filter topologies, as discussed in references [14-16].

SUMMARY

In this chapter, high-power electric utility applications of power electronics are discussed. These include HVDC transmission, static var control, interconnection of renewable energy sources, and the energy storage systems for the utility load leveling.

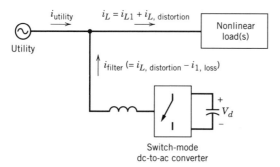

Figure 17-17   One-line diagram of an active filter.

PROBLEMS

17-1 Verify the current waveforms in Fig. 17-3a and the expressions given by Eqs. 17-1 through 17-3.

17-2 With a constant-input ac voltage V_{LL} and a constant dc current I_d, plot the locus in the $P-Q$ plane as the delay angle α of the converters in Fig. 17-2 is varied. Repeat this for a family of I_d values.

17-3 A dc transmission link interconnects two 230-kV ac systems. It has four bridges at each terminal (two per pole) with each pole rated at ± 250 kV, 1000 A. The parameters for each pole of the dc link are given in the following table:

	Rectifier	Inverter
Actual open-circuit voltage ratio for line–line voltages on primary and secondary sides of converter transformers; secondary voltage divided by primary voltage	0.468	0.435
Number of three-phase converter bridges in series on the dc side	2	2
Converter transformer leakage reactance per bridges in ohms, referred to secondary side	16.28	14.27
dc line resistance per pole = 15.35 Ω		
Minimum extinction angle of inverter = 18°		

In this system, $\gamma_I = \gamma_{\min} = 18°$. At the rectifier terminal, the voltage is as close to 250 kV as possible. $I_d = 1000$ A.

Calculate all the currents, voltages, real and reactive powers, and angles at each end of the dc link.

17-4 Repeat Problem 17-3 if each converter transformer is equipped with a tap changer. Now, it is possible to operate the rectifier at a firing angle as close to 18° as possible, while the inverter operates as close to the minimum extinction angle of 18° (but $\gamma_I \geq 18°$) as possible. The tap information is given below where the nominal line–line, primary voltage for each converter transformer is 230 kV (rms):

(a) Maximum value of the converter transformer tap ratio in per unit: 1.15 at the rectifier terminal and 1.10 at the inverter terminal

(b) Minimum value of the converter transformer tap ratio in per unit: 0.95 at the rectifier and 0.90 at the inverter terminal

(c) Converter transformer tap step in per unit: 0.0125 at both terminals

17-5 The nominal line-to-line voltage at a bus in a three-phase ac system is 230 kV (rms) when it is supplying a three-phase inductive load of $P + jQ = 1500$ MW $+ j750$ Mvar.

The per-phase ac system impedance Z_s seen by the bus can be approximated to be purely inductive with $Z_s = j5.0$ Ω.

(a) Calculate the percentage change in the bus voltage magnitude for a 10% increase in P.

(b) Calculate the percentage change in the bus voltage magnitude for a 10% increase in Q.

17-6 A hybrid arrangement of a TCI and a TSC is connected at the ac bus in Problem 17-5. The TCI can draw a maximum of 50 Mvars per phase, whereas the TSC consists of four-capacitor banks, each with a per-phase rating of 50 Mvars. Holding the ac bus voltage to its nominal value for a 10% increase in Q in Problem 17-5b, calculate the number of capacitor banks that should be switched in, the delay angle α at which the TCI should operate, and the per-phase effective inductance of the TCI.

17-7 Derive Eq. 17-13.

17-8 Derive Eq. 17-22.

REFERENCES

NEW APPLICATIONS

1. "Power Electronics in Power Systems: Analysis and Simulation Using EMTP," Course notes, University of Minnesota.

HVDC TRANSMISSION

2. E. W. Kimbark, *Direct Current Transmission,* Vol. I, Wiley-Interscience, New York, 1971.
3. C. Adamson and N. G. Hingorani, *High Voltage Direct Current Transmission,* Garraway, London, 1960, available from University Microfilms, Ann Arbor, MI.

STATIC var CONTROL: TCI AND TSC

4. L. Gyugyi and W. P. Matty, "Static VAR Generator with Minimum No Load Losses for Transmission Line Compensation," *Proceedings of the 1979 American Power Conference.*
5. T. J. E. Miller (Ed.), *Reactive Power Control in Electric Systems*, Wiley-Interscience, New York, 1982.
6. L. Gyugyi and E. R. Taylor, "Characteristic of Static, Thyristor-controlled Shunt Compensators for Power Transmission System Applications," *IEEE Transactions on Power Apparatus and Systems,* Vol. PAS-99, No. 5, Sept./Oct. 1980, pp. 1795–1804.

STATIC var CONTROL: MINIMUM ENERGY STORAGE

7. Y. Sumi et al., "New Static Var Control using Force-commutated Inverters," *IEEE Transactions on Power Apparatus and Systems,* Vol. PAS-100, No. 9, Sept. 1981, pp. 4216–4224.
8. H. Akagi, Y. Kanazawa, and A. Nabae, "Instantaneous Reactive Power Compensators Comprising Switching Devices without Energy Storage Components," *IEEE Transactions on Industry Applications,* Vol. IA-20, No. 3, May/June 1984, pp. 625–630.

PHOTOVOLTAIC ARRAY INTERCONNECTION

9. R. L. Steigerwald, A. Ferraro, and F. G. Turnbull, "Application of Power Transistors to Residential and Intermediate Rating Photovoltaic Array Power Conditioners," *IEEE Transactions on Industry Applications,* Vol. IA-19, No. 2, March/April 1983, pp. 254–267.
10. R. L. Steigerwald, A. Ferraro, and R. E. Tompkins, "Final Report—Investigation of a Family of Power Conditioners Integrated into Utility Grid—Residential Power Level," DOE Contract DE-AC02-80ET29310, Sandia National Lab., Report No. SAND81-7031, 1981.
11. K. Tsukamoto and K. Tanaka, "Photovoltaic Power System Interconnected with Utility," *1986 Proceedings of the American Power Conference,* 1986, pp. 276–281.

NEW TOPOLOGY, UTILITY INTERFACE

12. R. Naik and N. Mohan, "A Novel Grid Interface for Photovoltaic, Wind-Electric, and Fuel-Cell Systems with a Controllable Power Factor," IEEE-APEC'95, pp. 995–998.

SUPERCONDUCTIVE ENERGY STORAGE

13. H. A. Peterson, N. Mohan, and R. W. Bloom, "Superconductive Energy Storage Inductor–Converter Units for Power Systems," *IEEE Transactions on Power Apparatus and Systems,* Vol. 94, No. 4, July/Aug. 1975, pp. 1337–1348.

ACTIVE FILTERS

14. M. Rastogi, N. Mohan, and A-A Edris, "Filtering of Harmonic Currents and Damping of Resonances in Power Systems with a Hybrid-Active Filter," IEEE-APEC'95, pp. 607–612.
15. G. Kamath, N. Mohan, and V. Albertson, "Hardware Implementation of a Novel Reduced VA Rating Filter for Nonlinear Loads in 3-Phase, 4-wire Systems," IEEE-APEC'95, pp. 984–989.
16. C. Quinn, N. Mohan, and H. Mehta, "A Four-Wire, Current-Controlled Converter Provides Harmonic Neutralization in Three-Phase, Four-Wire Systems," IEEE-APEC'93.

CHAPTER 18

OPTIMIZING THE UTILITY INTERFACE WITH POWER ELECTRONIC SYSTEMS

18-1 INTRODUCTION

In Chapter 11, we discussed various power line disturbances and how power electronic converters can perform as power conditioners and uninterruptible power supplies to prevent these power line disturbances from disrupting the operation of critical loads such as computers used for controlling important processes, medical equipment, and the like. However, as discussed in the previous chapters, all power electronic converters (including those used to protect critical loads) can add to the inherent power line disturbances by distorting the utility waveform due to harmonic currents injected into the utility grid and by producing EMI. To illustrate the problems due to current harmonics i_h in the input current i_s of a power electronic load, consider the simple block diagram of Fig. 18-1. Due to the finite (nonzero) internal impedance of the utility source, which is simply represented by L_s in Fig. 18-1, the voltage waveform at the point of common coupling to the other loads will become distorted, which may cause them to malfunction. In addition to the voltage waveform distortion, some other problems due to the harmonic currents are as follows: additional heating and possibly overvoltages (due to resonance conditions) in the utility's distribution and transmission equipment, errors in metering and malfunction of utility relays, interference with communication and control signals, and so on. In addition to these problems, phase-controlled converters discussed in Chapter 6 cause notches in the utility voltage waveform and many draw power at a very low displacement power factor, which results in a very poor power factor of operation.

The foregoing discussion shows that the proliferation of power electronic systems and loads has the potential for significant negative impact on the utilities themselves as well

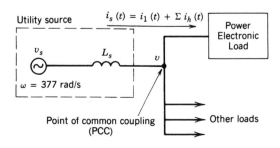

Figure 18-1 Utility interface.

as on their customers. One approach to minimize this impact is to filter the harmonic currents and the EMI produced by the power electronic loads, as discussed in Section 17-5 of the previous chapter. An alternative, in spite of a small increase in the initial cost, is to design the power electronic equipment such that the harmonic currents and the EMI are prevented or minimized from being generated in the first place. Both the concerns about the utility interface and the design of power electronic equipment to minimize these concerns are discussed in this chapter.

18-2   GENERATION OF CURRENT HARMONICS

In most power electronic equipment, such as switch-mode dc power supplies, uninterruptible power supplies (UPSs), and ac and dc motor drives, ac-to-dc converters are used as the interface with the utility voltage source. Commonly, a line-frequency diode rectifier bridge as shown in Fig. 18-2 is used to convert line frequency ac into dc. The rectifier output is a dc voltage whose average magnitude V_d is uncontrolled. A large filter capacitor is used at the rectifier output to reduce the ripple in the dc voltage v_d. The dc voltage v_d and the dc current i_d are unipolar and unidirectional, respectively. Therefore, the power flow is always from the utility ac input to the dc side. These line-frequency rectifiers with a filter capacitor at the dc side were discussed in detail in Chapter 5.

A class of power electronic systems utilizes line-frequency thyristor-controlled ac-to-dc converters as the utility interface. In these converters, which were discussed in detail in Chapter 6, the average dc output voltage V_d is controllable in magnitude and polarity, but the dc current i_d remains unidirectional. Because of the reversible polarity of the dc voltage, the power flow through these converters is reversible. As was pointed out in Chapter 6, the trend is to use these converters only at very high power levels such as in high-voltage dc transmission systems. Because of the very high power levels, the techniques to filter the current harmonics and to improve the power factor of operation are quite different in these converters, as discussed in Chapter 17, than those for the line-frequency diode rectifiers. In the general discussion presented in this chapter, therefore, only the diode rectifiers, where the dc voltage V_d remains essentially constant, are considered.

The diode rectifiers are used to interface with both the single-phase and the three-phase utility voltages. Typical ac current waveforms with minimal filtering were shown in Chapter 5. Typical harmonics in a single-phase input current waveform are listed in Table 18-1, where the harmonic currents I_h are expressed as a ratio of the fundamental current I_1. As is shown by Table 18-1, such current waveforms consist of large harmonic magnitudes. Therefore, for a finite internal per-phase source impedance L_s, the voltage distortion at the point of common coupling in Fig. 18-1 can be substantial. The higher the internal source inductance L_s, the greater would be the voltage distortion.

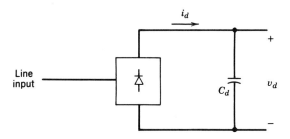

Figure 18-2   Diode rectifier bridge.

Table 18-1 Typical Harmonics in a Single-Phase Input Current
Waveform with No Line Filtering

h	3	5	7	9	11	13	15	17
$\left(\dfrac{I_h}{I_1}\right)\%$	73.2	36.6	8.1	5.7	4.1	2.9	0.8	0.4

18-3 CURRENT HARMONICS AND POWER FACTOR

As we discussed in Chapter 3, the power factor PF at which an equipment operates is the
product of the current ratio I_1/I_s and the displacement power factor DPF:

$$\text{PF} = \frac{\text{power}}{\text{volt-ampere}} = \frac{I_1}{I_s} \cdot \text{DPF} \qquad (18\text{-}1)$$

In Eq. 18-1, the displacement power factor equals the cosine of the angle ϕ_1 by which the
fundamental-frequency component in the current waveform is displaced with respect to
the input voltage waveform. The current ratio I_1/I_s in Eq. 18-1 is the ratio of the rms value
of the fundamental-frequency current component to the rms value of the total current. The
power factor indicates how effectively the equipment draws power from the utility; at a
low power factor of operation for a given voltage and power level, the current drawn by
the equipment will be large, thus requiring increased volt-ampere ratings of the utility
equipment such as transformers, transmission lines, and generators. The importance of
the high power factor has been recognized by residential and office equipment manufacturers for their own benefit to maximize the power available from a wall outlet. For
example, from a 120-V, 15-A electrical circuit in a building, the maximum power available is 1.8 kW, provided the power factor is unity. The maximum power that can be
drawn without exceeding the 15-A limit decreases with decreasing power factor. The
foregoing arguments indicate the responsibility and desirability on the part of the equipment manufacturers and users to design power electronic equipment with a high power
factor of operation. This requires that the displacement power factor should be high in Eq.
18-1. Moreover, the current harmonics should be low to yield a high current ratio I_1/I_s in
Eq. 18-1.

18-4 HARMONIC STANDARDS AND RECOMMENDED PRACTICES

In view of the proliferation of the power electronic equipment connected to the utility
system, various national and international agencies have been considering limits on harmonic current injection to maintain good power quality. As a consequence, various
standards and guidelines have been established that specify limits on the magnitudes of
harmonic currents and harmonic voltage distortion at various harmonic frequencies. Some
of these are as follows:

1. EN 50 006, "The Limitation of Disturbances in Electricity Supply Networks
   caused by Domestic and Similar Appliances Equipped with Electronic Devices,"
   European Standard prepared by Comité Européen de Normalisation Electrotechnique, CENELEC.
2. IEC Norm 555-3, prepared by the International Electrical Commission.

Table 18-2 Harmonic Current Distortion (I_h/I_1)

I_{SC}/I_1	Odd Harmonic Order h (%)					Total Harmonic Distortion (%)
	h < 11	11 ≤ h < 17	17 ≤ h < 23	23 ≤ h < 35	35 ≤ h	
<20	4.0	2.0	1.5	0.6	0.3	5.0
20–50	7.0	3.5	2.5	1.0	0.5	8.0
50–100	10.0	4.5	4.0	1.5	0.7	12.0
100–1000	12.0	5.5	5.0	2.0	1.0	15.0
>1000	15.0	7.0	6.0	2.5	1.4	20.0

Note: Harmonic current limits for nonlinear load connected to a public utility at the point of common coupling (PCC) with other loads at voltages of 2.4–69 kV. I_{sc} is the maximum short-circuit current at PCC. I_1 is the maximum fundamental-frequency load current at PCC. Even harmonics are limited to 25% of the odd harmonic limits above.

Source: Reference 1.

3. West German Standards VDE 0838 for household appliances, VDE 0160 for converters, and VDE 0712 for fluorescent lamp ballasts.

4. *IEEE Guide for Harmonic Control and Reactive Compensation of Static Power Converters*, ANSI/IEEE Std. 519-1981, which was revised in 1992 to 519-1992.

CENELEC, IEC, and VDE standards specify the limits on the voltages (as a percentage of the nominal voltage) at various harmonic frequencies of the utility frequency, when the equipment-generated harmonic currents are injected into a network whose impedances are specified.

The revised IEEE-519, which contains recommended practices and requirements for harmonic control in electric power systems, specifies requirements on the user as well as on the utility. Table 18-2 lists the limits on the harmonic currents that user of power electronic equipment and other nonlinear loads is allowed to inject into the utility system. Table 18-3 lists the quality of voltage that the utility must furnish the user. A utility will be able to furnish the voltage as listed in Table 18-3, provided that the harmonic currents injected by the users on a distribution feeder are limited in accordance with Table 18-2. Tables 18-2 and 18-3 are very broad in their scope and apply to wide voltage and power ranges. They are primarily intended for three-phase systems but can also be used as a guide to limit distortion in single-phase systems.

The principal justification for the harmonic limits specified in Table 18-2 is explained below. The voltage distortion at the point of common coupling (PCC) in Fig. 18-1 depends on the internal impedance of the ac source and the magnitudes of the injected current harmonics. In practice, the internal impedance of the source is highly inductive

Table 18-3 Harmonic Voltage Limits (V_h/V_1) (%) for Power Producers (Public Utilities or Cogenerators)

	2.3–69 kV	69–138 kV	> 138 KV
Maximum for individual harmonic	3.0	1.5	1.0
Total harmonic distortion	5.0	2.5	1.5

Note: This table lists the quality of the voltage that the power producer is required to furnish a user. It is based on the voltage level at which the user is supplied.

Source: Reference 1.

and therefore is represented by L_s in Fig. 18-1. At a harmonic h of the line frequency ω, the rms harmonic voltage at PCC is

$$V_h = (h\omega L_s)I_h \qquad (18\text{-}2)$$

where I_h is the h harmonic current injected into the ac source.

The internal inductance L_s in Eq. 18-2 is often specified in terms of the short-circuit current I_{sc} at the PCC. On a per-phase basis, I_{sc} will be the per-phase rms current supplied by the ac source to the fault if all three phases are shorted to ground at PCC:

$$I_{sc} = \frac{V_s}{\omega L_s} \qquad (18\text{-}3)$$

where V_s is the rms value of the per-phase internal voltage of the ac source, which is assumed to be sinusoidal. A large I_{sc} represents a large capacity of the ac system at PCC. From Eqs. 18-2 and 18-3, the harmonic voltage can be expressed as a ratio of the nominal system voltage V_s in percentage

$$\%V_h = \frac{V_h}{V_s} \times 100 = h\frac{I_h}{I_{sc}} \times 100 \qquad (18\text{-}4)$$

If I_1 is the line-frequency component of the current drawn by the power electronic load, then dividing both I_h and I_{sc} in Eq. 18-4 by I_1 results in

$$\%V_h = h\frac{I_h/I_1}{I_{sc}/I_1} \times 100 \qquad (18\text{-}5)$$

In Eq. 18-5, I_{sc}/I_1 represents the capacity of the utility system with respect to the fundamental-frequency volt-amperes of the load. Equation 18-5 shows that for an acceptable harmonic voltage distortion in percentage, the harmonic current ratio I_h/I_1 can be higher (although not in a linear proportion) for a higher I_{sc}/I_1 ratio in Table 18-2. Moreover, because the internal impedance of the ac system is mostly inductive, the harmonic voltage distortion in Eq. 18-5 is proportional to the harmonic order h. Therefore, the maximum allowable harmonic current ratio I_h/I_1 decreases (although not linearly) with increasing value of h, as shown in Table 18-2. As in Chapter 5, the total harmonic distortion (THD) in the input current is defined as

$$\text{THD} = \frac{\sqrt{\displaystyle\sum_{h=2}^{\infty} I_h^2}}{I_1} \qquad (18\text{-}6)$$

The total current harmonic distortion allowed in Table 18-2 increases with I_{sc}/I_1. It should be noted that there are other factors such as higher losses at high frequencies that also contribute to the allowable limits in Table 18-2.

The THD in the voltage can be calculated in a manner similar to that given by Eq. 18-6. Table 18-3 specifies the individual harmonics and the THD limits on the voltage that the utility supplies to the user at the PCC.

18-5 NEED FOR IMPROVED UTILITY INTERFACE

Because of the large harmonic content as indicated by Table 18-1, typical diode rectifiers used for interfacing power electronic equipment with the utility system may exceed the limits on individual current harmonics and THD specified in Table 18-2. In addition to the effect on the power line quality, the poor waveform of the input current also affects the power electronic equipment itself in the following ways:

- The power available from the wall outlet is reduced to approximately two-thirds.
- The dc-side filter capacitor in Fig. 18-2 is severely stressed due to large peak pulse currents.
- The losses in the diodes of the rectifier bridge are higher due to a current-dependent forward voltage drop across the diodes.
- The components in the EMI filter used at the input to the rectifier bridge must be designed for higher peak pulse currents.
- If a line-frequency transformer is used at the input, it must be highly overrated.

In view of these drawbacks, some of the alternatives for improving the input current waveforms are discussed, along with their relative advantages and disadvantages.

18-6 IMPROVED SINGLE-PHASE UTILITY INTERFACE

Various options for improving the single-phase utility interface of power electronic equipment are discussed.

18-6-1 PASSIVE CIRCUITS

Inductors and capacitors can be used in conjunction with the diode rectifier bridge to improve the waveform of the current drawn from the utility grid. The simplest approach is to add an inductor on the ac side of the rectifier bridge in Fig. 18-2. This added inductor results in a higher effective value of the ac-side inductance L_s, which improves the power factor and reduces harmonics as shown by means of Fig. 5-18 in Chapter 5. The impact of adding an inductor can be summarized as follows:

- Because of an improved current waveform, the power factor is improved from very poor to somewhat acceptable.
- The output voltage V_d is dependent on the output load and is substantially ($\sim 10\%$) lower compared with the no-inductance case.
- Inductance and C_d together in Fig. 18-2 form a low-pass filter, and therefore, the peak-to-peak ripple in the rectified output voltage v_d is less.
- The overall energy efficiency remains essentially the same; there are additional losses in the inductor, but the conduction losses in the diodes are lower.

It is possible to further improve the input current waveform (Fig. 18-3b) by using a circuit arrangement, as is shown in Fig. 18-3a. In Fig. 18-3a, C_{d1} directly across the rectifier bridge is small relative to C_d. This allows a larger ripple in v_{d1} but results in an improved waveform of i_s. The ripple in v_{d1} is filtered out by the low-pass filter consisting of L_d and C_d. Obvious disadvantages of such an arrangement are cost, size, losses, and the significant dependence of the average dc voltage V_d on the power drawn by the load.

18-6-2 ACTIVE SHAPING OF THE INPUT LINE CURRENT

By using a power electronic converter for current shaping, as is shown in the diagram of Fig. 18-4a, it is possible to shape the input current drawn by the rectifier bridge to be

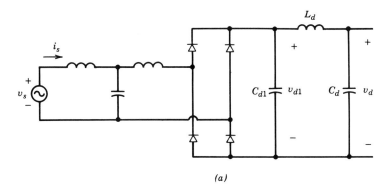

(a)

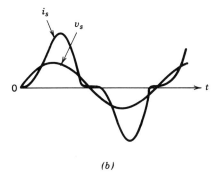

(b)

Figure 18-3 Passive filters to improve i_s waveform: (a) passive filter arrangement; (b) current waveform.

sinusoidal and in phase with the input voltage. The choice of the power electronic converter is based on the following considerations:

- In general, electrical isolation between the utility input and the output of the power electronic system either is not needed (e.g., in ac and dc motor drives) or it can be provided in the second converter stage, as in the switch-mode dc power supplies.
- In most applications it is acceptable, and in many cases desirable, to stabilize the dc voltage V_d slightly in excess of the peak of the maximum of the ac input voltage.
- The input current drawn should ideally be at a unity power factor so that the power electronic interface emulates a resistor supplied by the utility source. This also implies that the power flow is always unidirectional, from the utility source to the power electronic equipment.
- The cost, power losses, and size of the current shaping circuit should be as small as possible.

Based on these considerations, a line-frequency transformer isolation is ruled out. Also, it is acceptable to have $V_d > \hat{V}_s$, where $\hat{V}_s$ is the peak of the ac input voltage. Therefore, the obvious choice for the current shaping circuit is a step-up dc–dc converter, similar to that discussed in Chapter 7. This converter is shown in Fig. 18-4a, where C_d is used to minimize the ripple in v_d and to meet the energy storage requirement of the power electronic system. As in Chapter 5, a dc current I_{load} represents the power supplied to the rest of the system (the high-frequency component in the output current is effectively filtered out by C_d). For simplicity, the internal inductance L_s of the utility source is not included in Fig. 18-4a.

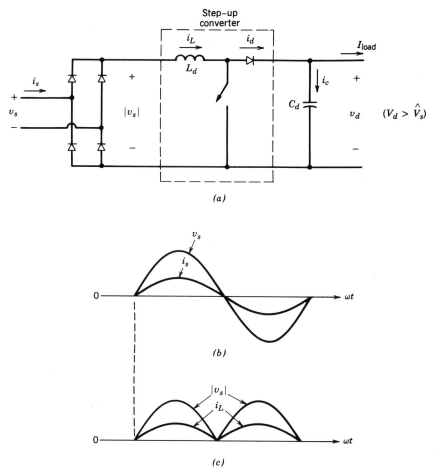

Figure 18-4   Active harmonic filtering: (*a*) step-up converter for current shaping; (*b*) line waveforms; (*c*) v_s and i_L.

The basic principle of operation is straightforward. At the utility input, the current i_s is desired to be sinusoidal and in phase with v_s, as is shown in Fig. 18-4*b*. Therefore, at the full-bridge rectifier output in Fig. 18-4*a*, i_L and $|v_s|$ have the same waveform as shown in Fig. 18-4*c*. In practice, the power losses in the rectifier bridge and the step-up dc–dc converter are fairly small. These are neglected in the following theoretical analysis. From the waveforms of Fig. 18-4*b*, where $\hat{V}_s = \sqrt{2}\, V_s$ and $\hat{I}_s = \sqrt{2}\, I_s$, the input power $p_{\text{in}}(t)$ from the ac source is

$$p_{\text{in}}(t) = \hat{V}_s|\sin \omega t|\ \hat{I}_s|\sin \omega t| = V_s I_s - V_s I_s \cos 2\omega t \qquad (18\text{-}7)$$

Because of a fairly large capacitance C_d, the voltage v_d can be initially assumed to be dc, that is, $v_d(t) = V_d$. Therefore, the output power is

$$p_d(t) = V_d i_d(t) \qquad (18\text{-}8)$$

where in Fig. 18-4*a*

$$i_d(t) = I_{\text{load}} + i_c(t) \qquad (18\text{-}9)$$

If the step-up converter in Fig. 18-4*a* is idealized and can be assumed to be operating at a switching frequency approaching infinity, then required L_d would be negligibly small.

This allows the assumption in Fig. 18-4a that $p_{in}(t) = p_d(t)$ on an instantaneous basis. Therefore, from Eqs. 18-7 through 18-9,

$$i_d(t) = I_{load} + i_c(t) = \frac{V_s I_s}{V_d} - \frac{V_s I_s}{V_d}\cos 2\omega t \qquad (18\text{-}10)$$

where the average value of i_d is

$$I_d = I_{load} = \frac{V_s I_s}{V_d} \qquad (18\text{-}11)$$

and the current through the capacitor is

$$i_c(t) = -\frac{V_s I_s}{V_d}\cos 2\omega t = -I_d\cos 2\omega t \qquad (18\text{-}12)$$

Even though this analysis is carried out by assuming the voltage across the capacitor to be ripple-free dc, the ripple in v_d can be estimated from Eq. 18-12 as

$$v_{d,\text{ripple}}(t) \approx \frac{1}{C_d}\int i_c\, dt = -\frac{I_d}{2\omega C_d}\sin 2\omega t \qquad (18\text{-}13)$$

which can be kept low by selecting a suitably large value of C_d. A series-tuned LC filter tuned for twice the ac frequency according to Eq. 18-12 may be put in parallel with C_d to minimize the ripple in the dc voltage. It should be noted that the switching-frequency components of currents in i_d and the high-frequency components in the load current will also flow through C_d.

Because the input current to the step-up converter is to be shaped, the step-up converter is operated in a current-regulated mode, as discussed in Chapter 8 in connection with dc-to-ac inverters. The feedback control is shown in a block diagram form in Fig. 18-5, where i_L^* is the reference or the desired value of the current i_L in Fig. 18-4a. Here i_L^* has the same waveform as $|v_s|$. The amplitude of i_L^* should be such as to maintain the output voltage at a desired or reference level V_d^*, in spite of the variation in load and the fluctuation of the line voltage from its nominal value. The waveform of i_L^* in Fig. 18-5 is obtained by measuring $|v_s|$ in Fig. 18-4a by means of a resistive potential divider and multiplying it with the amplified error between the reference value V_d^* and the actual measured value of V_d. The actual current i_L is sensed, usually by measuring the voltage across a small resistor inserted in the return path of i_L. The status of the switch in the step-up converter is controlled by comparing the actual current i_L with i_L^*.

Once i_L^* and i_L in Fig. 18-5 are available, there are various ways to implement the current-mode control of the step-up converter. Some of these were discussed in connection with the current-mode control of switch-mode dc power supplies in Chapter 10. Four such control modes are discussed as follows where f_s is the switching frequency and I_{rip}

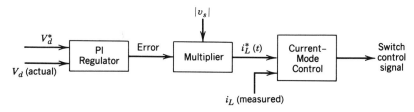

Figure 18-5 Control block diagram.

is the peak-to-peak ripple in i_L during one time period of the switching frequency. Only the constant-frequency control is described in some detail.

1. *Constant-frequency control.* Here, the switching frequency f_s is kept constant. When i_L reaches i_L^*, the switch in the step-up converter is turned off. The switch is turned on by a clock at a fixed frequency f_s, which results in i_L as shown in Fig. 18-6a. As discussed in Chapter 10, a slope compensation ramp must be used; otherwise i_L will be irregular at switch duty ratios in excess of 0.5. Normalized I_{rip} is plotted in Fig. 17-6b.

2. *Constant-tolerance-band control.* Here, the current i_L is controlled such that the peak-to-peak ripple I_{rip} in i_L remains constant. With a preselected value of I_{rip}, i_L is forced to be within the tolerance band $i_L^* + \frac{1}{2}I_{rip}$ and $i_L^* - \frac{1}{2}I_{rip}$ by controlling the switch status.

3. *Variable-tolerance-band control.* Here, the peak-to-peak ripple current I_{rip} is increased in proportion to the instantaneous value of $|v_s|$. Otherwise, this approach is similar to the constant-tolerance-band control.

4. *Discontinuous-current control.* In this scheme, the switch is turned off when i_L reaches $2i_L^*$. The switch is kept off until i_L reaches zero, at which instant the switch is turned back on. This can be considered as a special case of a variable-tolerance-band control.

During a switching-frequency time period, the output voltage is assumed to be constant as V_d and the input voltage to the step-up converter is assumed to be constant at that instant of time; I_{rip} is the peak-to-peak ripple current during one time period of the

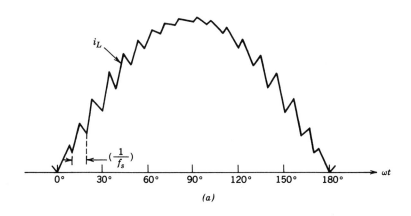

(a)

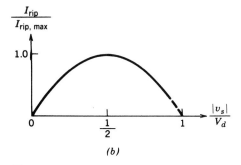

(b)

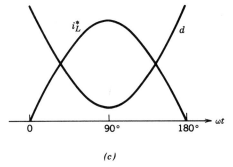

(c)

Figure 18-6 Constant-frequency control.

switching frequency. The following equations can be written from Fig. 18-4a during the on interval t_{on} and the off interval t_{off} of the switch

$$t_{on} = \frac{L_d I_{rip}}{|v_s|} \tag{18-14}$$

$$t_{off} = \frac{L_d I_{rip}}{V_d - |v_s|} \tag{18-15}$$

where the switching frequency f_s is given as

$$f_s = \frac{1}{t_{on} + t_{off}} = \frac{(V_d - |v_s|)|v_s|}{L_d I_{rip} V_d} \tag{18-16}$$

In a *constant-frequency control scheme*, f_s in Eq. 18-16 is constant and hence

$$I_{rip} = \frac{(V_d - |v_s|)|v_s|}{f_s L_d V_d} \tag{18-17}$$

Figure 18-6b shows the plot of the normalized I_{rip} as a function of $|v_s|/V_d$, noting that in a step-up converter $|v_s|/V_d$ must be less than or equal to 1. The maximum ripple current is given as

$$I_{rip,max} = \frac{V_d}{4 f_s L_d} \quad \text{when } |v_s| = \tfrac{1}{2} V_d \tag{18-18}$$

In the active current-shaping circuit using a step-up dc–dc converter, the following additional observations are made:

- The output voltage v_d across the capacitor C_d contains a 120-Hz ripple at twice the line frequency. The feedback control circuit used to control V_d at a desired value cannot compensate this voltage ripple without distorting the input line current.
- If the switching-frequency ripple in i_L is kept to a small amplitude, then a laminated iron core inductor can possibly be used that will be smaller in size due to its higher saturation flux density compared to the high-frequency ferrite materials.
- A higher switching frequency allows a lower value of L_d and an increased ease of filtering high-frequency ripple. However, the switching frequency is chosen as a compromise between the foregoing advantages and the increased switching losses.
- The voltage V_d much larger than 10% beyond the peak input ac voltage $\hat{V}_s$ will cause efficiency to decline.
- To limit the in-rush current at start-up, a current-limiting resistor in series with L_d can be used. Subsequent to the initial transient, the resistor is bypassed by a contactor or a thyristor in parallel with the current-limiting resistor.
- The step-up converter topology is well suited for the input current shaping because when the switch is off, the input current directly (through the diode) feeds the output stage. In a constant-frequency current control, as an example, the switch duty ratio d is as shown in Fig. 18-6c as a function of ωt, recognizing from Chapter 7 that in a step-up converter with an input voltage $|v_s|$ and an output voltage V_d, $(|v_s|/V_d) = 1 - d$. Therefore

$$d = 1 - \frac{|v_s|}{V_d} \tag{18-19}$$

Figure 18-6c shows that d is smallest at the peak of i_L^*. Thus, the large values of i_L flow through the switch only during a small fraction of the switching time period.

- A small filter capacitor must be used across the output of the diode rectifier bridge to prevent the ripple in i_L from entering the utility system. An EMI filter at the input is still required as in a conventional circuit without the active current shaping.

In addition to an almost sinusoidal input waveform at nearly a unity power factor, the other advantages of an active input current shaping can be summarized as follows:

- The dc voltage V_d can be stabilized to a nearly constant value for large variations in the line voltage. With V_d equal to 1.1 times the peak of the nominal input voltage, for example, this circuit will continue to draw sinusoidal current for line overvoltages of up to 10%. By selecting proper current ratings of the components, this current-shaping circuit can easily handle large undervoltages at the utility input.
- Since V_d is stabilized to a nearly constant value, the volt-ampere ratings of the semiconductor devices in the converter fed from V_d are significantly reduced.
- Because of the absence of large peaks in the input current, the size of the EMI filter components is smaller.
- For the equal ripple in v_d, only one-third to one-half the capacitance C_d is needed compared with the conventional circuit, thus resulting in a reduced size.
- The energy efficiency from v_s to V_d of such a circuit is typically 96% compared with the efficiency of 99% in the conventional arrangement without active current shaping.

At present, the cost, slightly higher power losses, and complexity of active current shaping have prevented their widespread usage. This may change in the future because of increased device integration leading to lower semiconductor cost, a strict enforcement of harmonic standards, and some of the advantages mentioned above. Another factor in favor of the active line-current shaping is as follows: in power supplies to computers, a sinusoidal line current is important to avoid the added kilovolt-ampere (kVA) rating and, hence, the increased cost of UPSs, and standby diesel generators, which often supply computer systems. In such applications, the current-shaping techniques described above are being applied. Therefore, ICs and other components suitable for these applications have become available, which will lower the cost of development and the components of the active current-shaping circuit.

18-6-3 INTERFACE FOR A BIDIRECTIONAL POWER FLOW

In certain applications, for example, in motor drives with regenerative braking, the power flow through the utility interface converter reverses during the regenerative braking while the kinetic energy associated with the inertias of the motor and load is recovered and fed back to the utility system. One approach used in the past is to employ two back-to-back connected line-frequency thyristor converters, as is shown in Fig. 18-7. During the normal mode, converter 1 acts as a rectifier and the power flows from the ac input to the dc side. During regenerative braking, the gate pulses to the thyristors of converter 1 are blocked and converter 2 operates in an inverter mode where the polarity of v_d remains the same but the direction of i_d is reversed. Each of these converters is similar to those discussed in detail in Chapter 6. There are several drawbacks associated with this approach: (1) the input current i_s has a distorted waveform and the power factor is low, (2) the dc voltage V_d is limited in the inverter mode because of the minimum extinction angle requirement of converter 2 while it operates in an inverter mode, and (3) there is a possibility of commutation failure in the inverter mode due to ac line disturbances.

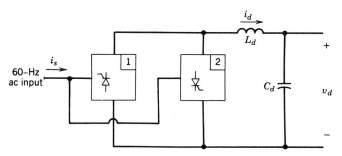

Figure 18-7   Back-to-back connected converters for bidirectional power flow.

It is possible to overcome these limitations by using a switch-mode converter, as shown in Fig. 18-8. This converter is identical to the four-quadrant switch-mode inverters discussed in Chapter 8, where the inverter mode in which the power flows from the dc to the ac side was discussed in detail. The rectifier mode was only briefly discussed in Section 8-7.

The rectifier being the dominant mode of operation, i_s is defined with a direction, as shown in Fig. 18-8. An inductance L_s (which augments the internal inductance of the utility source) is included to reduce the ripple in i_s at a finite switching frequency. In the circuit of Fig. 18-8,

$$v_s = v_{\text{conv}} + v_L \qquad (18\text{-}20)$$

where

$$v_L = L_s \frac{di_s}{dt} \qquad (18\text{-}21)$$

Assuming v_s to be sinusoidal, the fundamental-frequency components of v_{conv} and i_s in Fig. 18-8 can be expressed as phasors $\mathbf{V}_{\text{conv1}}$ and $\mathbf{I}_{s1}$, respectively. Choosing $\mathbf{V}_s$ arbitrarily as the reference phasor $\mathbf{V}_s = V_s e^{j0^\circ}$, at the line frequency $\omega = 2\pi f$

$$\mathbf{V}_s = \mathbf{V}_{\text{conv1}} + \mathbf{V}_{L1} \qquad (18\text{-}22)$$

where

$$\mathbf{V}_{L1} = j\omega L_s \mathbf{I}_{s1} \qquad (18\text{-}23)$$

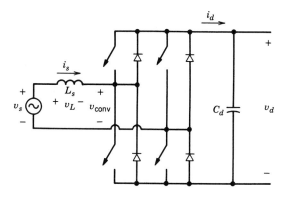

Figure 18-8   Switch-mode converter for the utility interface.

A phasor diagram corresponding to Eqs. 18-22 and 18-23 is shown in Fig. 18-9a, where $\mathbf{I}_{s1}$ lags $\mathbf{V}_s$ by an arbitrary phase angle θ. The real power P *supplied* by the ac source to the converter is

$$P = V_s I_{s1} \cos\theta = \frac{V_s^2}{\omega L_s}\left(\frac{V_{\text{conv1}}}{V_s}\sin\delta\right) \tag{18-24}$$

since in Fig. 18-9a, $V_{L1}\cos\theta = \omega L_s I_{s1}\cos\theta = V_{\text{conv1}}\sin\delta$.

In the phasor diagram of Fig. 18-9a, the reactive power Q *supplied* by the ac source is positive. It can be expressed as

$$Q = V_s I_{s1} \sin\theta = \frac{V_s^2}{\omega L_s}\left(1 - \frac{V_{\text{conv1}}}{V_s}\cos\delta\right) \tag{18-25}$$

since in Fig. 18-9a, $V_s - \omega L_s I_{s1}\sin\theta = V_{\text{conv1}}\cos\delta$. Note that Q is the sum of the reactive power absorbed by the converter and the reactive power consumed by the inductance L_s. However, at very high switching frequencies, L_s can be made to be quite small; thus, Q can be approximated as the reactive power absorbed by the converter.

The important equations are summarized below:

$$P = \frac{V_s^2}{\omega L_s}\left(\frac{V_{\text{conv1}}}{V_s}\sin\delta\right) \quad \text{(Eq. 18-24, repeated)}$$

$$Q = \frac{V_s^2}{\omega L_s}\left(1 - \frac{V_{\text{conv1}}}{V_s}\cos\delta\right) \quad \text{(Eq. 18-25, repeated)}$$

and

$$\mathbf{I}_{s1} = \frac{\mathbf{V}_s - \mathbf{V}_{\text{conv1}}}{j\omega L_s} \tag{18-26}$$

From these equations it is clear that for a given line voltage v_s and the chosen inductance L_s, desired values of P and Q can be obtained by controlling the magnitude and the phase of v_{conv1}. Figure 18-9a shows how $\mathbf{V}_{\text{conv1}}$ can be varied, keeping the magnitude

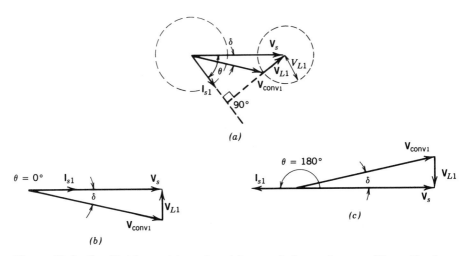

(a)

(b)

(c)

Figure 18-9   Rectification and inversion: (a) general phasor diagram; (b) rectification at unity power factor; (c) inversion at unity power factor.

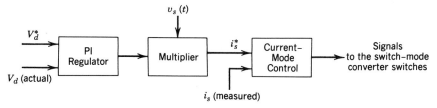

Figure 18-10 Control of the switch-mode interface.

of $\mathbf{I}_{s1}$ constant. The two circles in Fig. 18-9a are traced by the loci of $\mathbf{I}_{s1}$ and $\mathbf{V}_{\mathrm{conv1}}$ phasors.

In the general analysis discussed earlier, two cases are of special interest: rectification and inversion at a unity power factor. These two are shown by phasor diagrams in Figs. 18-9b and 18-9c. In both cases

$$V_{\mathrm{conv1}} = [V_s^2 + (\omega L_s I_{s1})^2]^{1/2} \qquad (18\text{-}27)$$

If a high switching frequency is used, only a small inductance L_s is needed. Therefore, from Eq. 18-27

$$V_{\mathrm{conv1}} \approx V_s \qquad (18\text{-}28)$$

For the desirable magnitude and the direction of power flow as well as Q, the magnitude V_{conv1} and the phase angle δ with respect to the line voltage must be controlled. In the circuit of Fig. 18-8, V_d is established by charging the capacitor C_d through the switch-mode converter. The value of V_d should be of a sufficiently large magnitude so that v_{conv1} at the ac side of the converter is produced by a PWM that corresponds to a PWM in a linear region (i.e., $m_a \leq 1.0$ as discussed in Chapter 8). This is necessary to limit ripple in the input current i_s. Therefore, from Eq. 8-19 of Chapter 8 and Eq. 18-28, V_d must be greater than the peak of the input ac voltage, that is,

$$V_d > \sqrt{2}V_s \qquad (18\text{-}29)$$

The control circuit to regulate V_d in Fig. 18-8 at its reference value V_d^* and to achieve a unity power factor of operation is shown in Fig. 18-10. The amplified error between V_d and V_d^* is multiplied with the signal proportional to the input voltage v_s waveform to produce the reference current signal i_s^*. A current-mode control such as a tolerance band control or a fixed-frequency control as discussed in Chapter 8 can be used to deliver i_{s1} equal to i_s^* and in phase or 180° out of phase with the line voltage v_s. The magnitude and direction of power flow are automatically controlled by regulating V_d at its desired value. It is possible to obtain a phase shift between v_s and i_{s1} and hence a finite reactive power flow by introducing the corresponding phase shift in the signal proportional to v_s in the control circuit of Fig. 18-10. The steady-state waveforms in the circuit of Fig. 18-8 for a unity power factor rectifier operation are shown in Fig. 18-11.

As we discussed in the previous section, in a single-phase circuit with i_s nearly sinusoidal and V_d essentially dc, the current i_d in Fig. 18-8 consists of a ripple at twice the line frequency. This ripple in i_d results in a ripple in the dc voltage across C_d, which can be minimized by means of an LC filter (series tuned at twice the line frequency) in parallel with C_d. This is done in high-power applications such as electric locomotives.

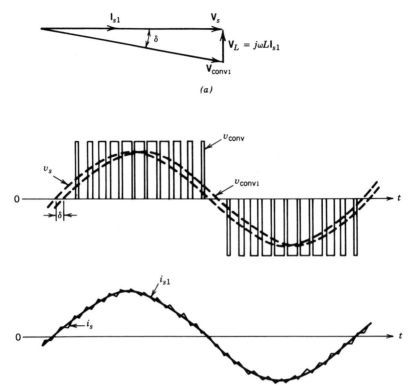

Figure 18-11   Waveforms in the circuit of Fig. 18-8 at unity power factor of operation: (*a*) phasor diagram; (*b*) circuit waveforms.

18-7   IMPROVED THREE-PHASE UTILITY INTERFACE

Three-phase diode rectifier bridges with a filter capacitor on the dc side were discussed in Chapter 5. The input current in these rectifiers is also highly distorted. One of the ways to improve the input current waveform is to increase the ac-side inductor L_s between the rectifier and the utility. Improvement in the input power factor due to increased L_s was shown in Fig. 5-37 in Chapter 5.

In a three-phase case, active input line current shaping can be achieved by three separate transformer-isolated current-shaping circuits. At least two of the current-shaping circuits must have their outputs electrically isolated from their inputs, since the outputs of the three current-shaping circuits feed the same output capacitor. The requirement of electrical isolation can be met by using the high-frequency transformer-isolated dc–dc converters discussed in Chapter 10 in the current-shaping circuit. Because the three-phase input is generally used for higher power level equipment, an alternative would be to use 60-Hz isolation transformers at the input.

Since in many applications the electrical isolation between the utility input and the output is not needed, the requirement of transformer isolation will cause unnecessary losses in the transformers and will be more expensive. In such cases, therefore, a better alternative is to use a four-quadrant switch-mode inverter as shown in Fig. 18-12. This converter is capable of supplying nearly sinusoidal input current at a unity power factor; in addition, the power flow through such a converter is reversible. This three-phase switch-mode dc-to-ac inverter was considered in detail in Chapter 8. Its operation in the rectifier mode was briefly discussed in Section 8-7 of Chapter 8 on a per-phase basis.

The block diagram for controlling such a converter is the same as Fig. 18-10 for the single-phase case in Section 18-6-4 where the dc voltage V_d is regulated to its reference value V_d^*. For the converter to be capable of controlling the input current waveforms to be sinusoidal, V_d^* should be appropriately chosen. If a high switching frequency is used, the ac-side inductances L_s in Fig. 18-12 can be minimized. Therefore, the voltage drops across L_s are small and the rms voltages

$$(V_{\text{conv}})_{LL} \approx V_{LL} \tag{18-30}$$

If the converter in Fig. 18-12 is to be pulse-width-modulated in a linear range with $m_a \leq 1.0$ to control the input currents to be sinusoidal, then from Eq. 8-57

$$V_d > 1.634 V_{LL} \tag{18-31}$$

An important difference between the dc current i_d in the three-phase converters as compared with single-phase converters is that it consists of a dc component I_d and the high-switching-frequency component (the ripple at twice the line frequency does not exist as in the single-phase case). As discussed in Chapter 8,

$$I_d = \frac{3V_s I_s}{V_d} \cos \phi_1 \tag{18-32}$$

where V_s and I_s are the sinusoidal per-phase line quantities and ϕ_1 is the angle by which the phase current lags the phase voltage. For rectification at a unity power factor, $\phi_1 = 0$ and

$$I_d = \frac{3V_s I_s}{V_d} \tag{18-33}$$

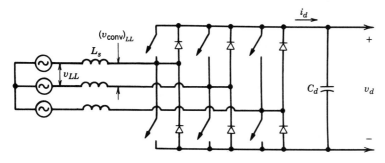

Figure 18-12 Three-phase, switch-mode converter.

Since only the high-switching-frequency current flows through C_d, only a small capacitance is needed.

18-7-1 MINNESOTA RECTIFIER

A new topology rectifier that draws nearly sinusoidal currents at a unity power factor from the utility is discussed in references 10 and 11. It requires a 6-diode rectifier and only two controlled switches which can be switched at zero current.

18-8 ELECTROMAGNETIC INTERFERENCE

Because of rapid changes in voltages and currents within a switching converter, power electronic equipment is a source of EMI with other equipment as well as with its own proper operation. The EMI is transmitted in two forms: radiated and conducted. The switching converters supplied by the power lines generate conducted noise into the power lines that is usually several orders of magnitude higher than the radiated noise into free space. Metal cabinets used for housing power converters reduce the radiated component of the EMI.

Conducted noise as shown in Fig. 18-13 consists of two categories commonly known as the differential mode and the common mode. The differential-mode noise is a current or a voltage measured between the lines of the source, that is, a line-to-line voltage or the line current i_{dm} in Fig. 18-13. The common-mode noise is a voltage or current measured between the power lines and ground, such as i_{cm} in Fig. 18-13. Both differential-mode and common-mode noises are present in general on both the input lines and the output lines. Any filter design has to take into account both of these modes of noise.

18-8-1 GENERATION OF EMI

Switching waveforms such as that shown in Fig. 18-14, for example, are inherent in all switching converters. Because of short rise and fall times, these waveforms contain significant energy levels at harmonic frequencies in the radio frequency (RF) region, several orders above the fundamental frequency.

The transmission of the differential-mode noise is through the input line to the utility system and through the dc-side network to the load on the power converter. Moreover, conduction paths through stray capacitances between components and due to magnetic coupling between circuits must also be considered.

The transmission of the common-mode noise is entirely through "parasitic" or stray capacitors and stray electric and magnetic fields. These stray capacitances exist between

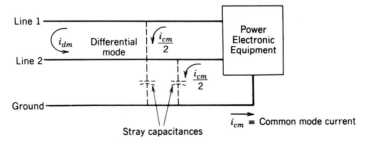

Figure 18-13 Conducted interference.

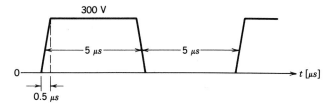

Figure 18-14 Switching waveform.

various system components and between components and ground. For safety reasons, most power electronic equipment has a grounded cabinet. The noise appearing on the ground line contributes significantly to the EMI.

18-8-2 EMI STANDARDS

There are various CISPR, IEC, VDE, FCC, and military standards that specify the maximum limit on the conducted EMI. Figure 18-15 shows the FCC and VDE standards for the RF equipment used in industrial, commercial, and residential equipment. To compare against these limits, the conducted noise is measured by means of a specified impedance network called LISN (Line Impedance Stabilization Network). Standards for the radiated EMI are also specified by the various agencies.

18-8-3 REDUCTION OF EMI

As is discussed in reference 14, the most cost-effective way of dealing with EMI is to prevent the EMI from being generated at its source, which can significantly reduce radiated and conducted interference before the application of filters, shielding, and the

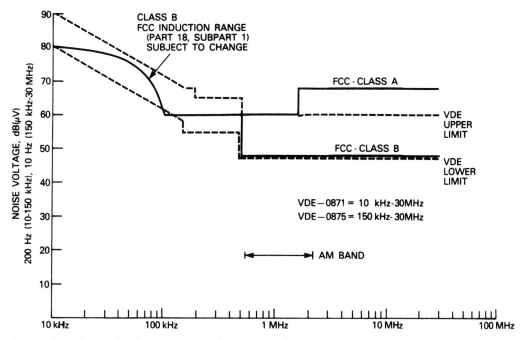

Figure 18-15 The FCC and VDE standards for conducted EMI.

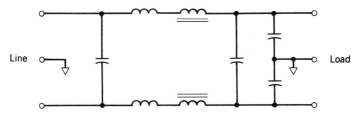

Figure 18-16 Filter for conducted EMI.

like. Another advantage to this approach is that a system that is not generating a high level of EMI will not be susceptible to its own noise and therefore will be more reliable.

From the point of view of EMI reduction, a properly designed snubber is quite effective, since it reduces both the *dv/dt* and the *di/dt* of the circuit. The snubber must be connected directly on the component being snubbed with as short leads as possible. Another approach to reducing EMI generation is to use the resonant converter concepts discussed in Chapter 9. In addition to these, the magnitudes of the coupling fields should be reduced by proper mechanical layout, wiring, and shielding.

To reduce magnetic fields, it is important to minimize the net area enclosed by a current loop. All current loops with switching transients should be made to have as small an area as possible. All current-carrying conductors should be run in close proximity to the return wire, such as by copper strips, discussed in Chapter 28. A twisted pair of wires will reduce the generated external field to a minimum.

To reduce stray capacitances, the area of exposed metal at the switching potential should be minimized and kept as far from ground as possible by proper mechanical design.

In addition to these steps to minimize the generation of EMI, EMI filters such as that shown in Fig. 18-16 are used to meet the conducted EMI limits.

Generally, the radiated noise is effectively shielded by the metal cabinets used for housing the power electronic equipment. Additional steps may be necessary if the power electronic equipment is operating near sensitive communication or medical equipment.

SUMMARY

1. Power electronic equipment is a source of current harmonics and EMI. Appropriate steps must be taken to prevent these from degrading power quality.

2. Harmonic standards and recommended practices have been established by various agencies to limit the harmonics injected by power electronic equipment.

3. In power electronic equipment with diode bridge rectifiers, input inductance reduces the input current harmonics.

4. In circuits with a single-phase utility input, the input line current can be actively shaped to be sinusoidal.

5. In single-phase and three-phase circuits, if bidirectional power flow is needed, the switch-mode converters such as those discussed in Chapter 8 can be used to interface with the utility system. These produce sinusoidal utility currents at a unity power factor.

6. EMI-generating mechanisms, EMI standards, and EMI reduction techniques are discussed.

PROBLEMS

18-1 A residential load of a 240-V, 5-kW load-proportional, capacity-modulated heat pump is supplied through a single-phase, 25-kVA secondary distribution transformer with a leakage reactance of 4%. The input current harmonic components are given in Table P18-1 as a percentage of the fundamental-frequency current component. The displacement power factor is approximately 1.

(a) Calculate the short-circuit capacity of the system at the point of coupling, assuming the impedance of the rest of the system other than the secondary distribution transformer to be negligible.

(b) Are the harmonics and THD within the limits specified in Table 18-2?

(c) Calculate the power factor of the load.

Table P18-1

h	3	5	7	9	11	13	15	17	19	21	23
$\left(\dfrac{I_h}{I_1}\right)\%$	34.0	5.3	1.8	1.8	1.6	1.2	0.9	0.8	0.8	0.4	0.4

18-2 In a single-phase, 240-V, 60-Hz, 2-kW diode rectifier interface, the displacement power factor is essentially 1.0 and the input current harmonics are listed in Table 18-1 as a percentage of the fundamental-frequency component. Calculate the rms value of the ripple current through the filter capacitor C_d due to the input interface. Neglect all losses.

18-3 In the single-phase interface for a bidirectional power flow shown in Fig. 18-8, $V_s = 240$ V (rms) at 60 Hz and $L_s = 2.5$ mH. Neglect all losses and assume that the switch-mode converter is pulse-width modulated in its linear range with $m_a \leq 1.0$. The converter is controlled such that i_s is either in phase or out of phase with v_s. Calculate the minimum value of V_d if the power flow through the converter is 2 kW (a) from the ac to the dc side and (b) from the dc to the ac side.

REFERENCES

1. C. K. Duffey and R. P. Stratford, ''Update of Harmonic Standard IEEE-519 IEEE Recommended Practices and Requirements for Harmonic Control in Electric Power System,'' *IEEE/ IAS Transactions,* Nov./Dec. 1989, pp. 1025–1034.

2. EN 50006, ''The Limitation of Disturbances in Electric Supply Networks Caused by Domestic and Similar Appliances Equipped with Electronic Devices,'' European Standards prepared by Comité Européen de Normalisation Electrotechnique, CENELEC.

3. IEC Norm 555-3 prepared by the International Electrical Commission.

4. VDE Standards 0838 for Household Appliances and 0712 for Fluorescent Lamp Ballasts, West Germany.

5. *IEEE Guide for Harmonic Control and Reactive Compensation of Static Power Converters,* IEEE Project No. 519/05, July 1979.

6. C. P. Henze and N. Mohan, ''A Digitally Controlled AC to DC Power Conditioner that Draws Sinusoidal Input Current,'' presented 1986 IEEE Power Electronics Specialist Conference, pp. 531–540.

7. M. Herfurth, ''TDA 4814—Integrated Circuit for Sinusoidal Line Current Consumption,'' Siemens Components, 1987.

8. M. Herfurth, ''Active Harmonic Filtering for Line Rectifiers of Higher Power Output,'' Siemens Components, 1986.

9. N. Mohan, T. Undeland, and R. J. Ferraro, ''Sinusoidal Line Current Rectification with a 100 kHz B-SIT Step-up Converter,'' paper presented at 1984 IEEE Power Electronics Specialists Conference, pp. 92–98.

10. N. Mohan, "System and Method for Reducing Harmonic Currents by Current Injection," U.S. Patent No. 5,345,375, Sept. 6, 1994.

11. M. Rastogi, N. Mohan and C. Henze, "Three-Phase Sinusoidal Current Rectifier with Zero-Current Switching," IEEE/APEC Records, 1994, Orlando, FL, pp. 718–724.

12. VDE Standards 0875/6.77 for radio interference suppression of electrical appliances and systems.

13. VDE Standards 0871/6.78 for radio interference suppression of radio frequency equipment for industrial, scientific, and medical (ISM) and similar purposes.

14. N. Mohan, "Techniques for Energy Conservation in AC Motor Drive Systems," Electric Power Research Institute Final Report EM-2037, September 1981.

15. L. M. Schneider, "Take the Guesswork out of Emission Filter Design," *EMC Technology*, April–June 1984, pp. 23–32.

PART 6

SEMICONDUCTOR DEVICES

CHAPTER 19

BASIC SEMICONDUCTOR PHYSICS

19-1 INTRODUCTION

In the previous chapters, it has been assumed that the semiconductor power devices had nearly ideal characteristics. These properties included:

1. Large breakdown voltages
2. Low on-state voltages and resistances
3. Fast turn-on and turn-off
4. Large power dissipation capability

In spite of significant progress in the development of power devices, there are none available that simultaneously have all of these properties. In all device types, there is a trade-off between breakdown voltages and on-state losses. In bipolar (minority carrier) devices, there is also a trade-off between on-state losses and switching speeds.

Such trade-offs mean that there is not one device type that can be used for all applications. The requirement of the specific application must be matched to the capabilities of the available devices. This often requires clever and innovative design approaches. For example, several devices may have to be combined in parallel or series connections in order to control larger amounts of power.

In this environment, it is important for the user to have a firm qualitative understanding of the physics of power devices and of how they are fabricated and packaged. A superficial knowledge of low-power solid-state devices is insufficient for the effective use of high-power devices. Indeed, a naive extrapolation of low-power device knowledge to the high-power regime could lead to the damage or even destruction of the high-power device.

In the next several chapters, the operation, fabrication, and packaging of high-power devices will be explored. In this chapter, the fundamental properties of semiconductors will be reviewed.

19-2 CONDUCTION PROCESSES IN SEMICONDUCTORS

19-2-1 METALS, INSULATORS, AND SEMICONDUCTORS

Electrical current will flow in a material if there are charge carriers (usually electrons) in the material that are free to move in response to an applied electric field. The number of free carriers in various materials varies over an extraordinarily wide range. In metals such as copper or silver, the free-electron density is on the order of 10^{23} cm^{-3}, whereas in insulators such as quartz or aluminum oxide the free-electron density is less than 10^3 cm^{-3}. This difference in free-electron densities is the reason why the electrical conductivity in metals can be on the order of 10^6 mhos-cm^{-1}, whereas it is on the order of 10^{-15} mhos-cm^{-1} or smaller in a good insulator. A material such as silicon or gallium arsenide, which has a free-carrier density intermediate between that of an insulator and a metal $(10^8-10^{19}$ cm$^{-3})$, is termed a semiconductor.

In a metal or an insulator, the free-carrier density in a constant of the material and cannot be changed to any significant degree. However, in a semiconductor the free-carrier density can be changed by orders of magnitude either by introduction of impurities into the material or by the application of electric fields to appropriate semiconductor structures. This ability to manipulate the free-carrier density by large amounts is what makes the semiconductor such a unique and useful material for electrical applications.

19-2-2   ELECTRONS AND HOLES

A single crystal of a semiconductor such as silicon, which has four valance electrons, is composed of a regular array or lattice of silicon atoms. Each silicon atom is bonded to four nearest neighbors, as illustrated in Fig. 19-1, by covalent bonds composed of electrons shared between the two adjacent atoms. At temperatures above absolute zero, some of these bonds are broken by energy carried by the silicon atom due to its random thermal motion about its equilibrium position. This process, known as thermal ionization, creates a free electron and leaves behind a fixed positive charge on the nucleus of the silicon atom where the bond was broken, as is shown in Fig. 19-1.

At some later time such as t_2 indicated in Fig. 19-2b another free electron may be attracted to the positive charge and become trapped in the bond that was broken at an earlier time t_1, as indicated in Fig. 19-2a, thus becoming bound. However, the silicon atom where this free electron originated now has a positive charge, and the end result of the transaction is the movement of the positive charge, as shown in Fig. 19-2c at time t_3.

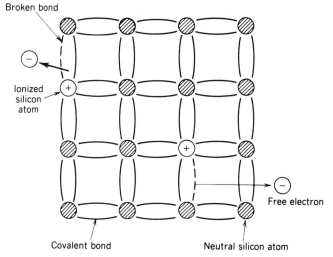

Figure 19-1   A silicon lattice showing thermal ionization and the creation of free electrons.

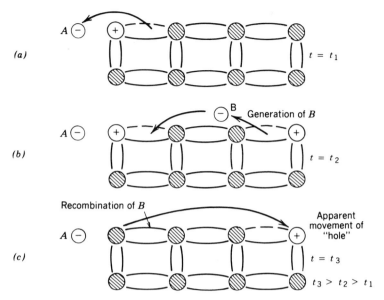

Figure 19-2 Hole movement in semiconductors.

This moving positive charge is termed a hole because it originates from an empty bond normally occupied by an electron.

The thermal ionization mechanism generates an equal number of electrons and holes. The thermal equilibrium density of electrons and holes, n_i, in a pure (intrinsic) semiconductor is given by

$$n_i^2 \approx C \exp \frac{-qE_g}{kT} \qquad (19\text{-}1)$$

where E_g is the energy gap of the semiconductor (1.1 eV for silicon), q is the magnitude of the electron charge, k is Boltzmann's constant, T is the temperature in degrees Kelvin, and C is a constant of proportionality. At room temperature (300 K), $n_i \approx 10^{10}$ cm^{-3} in silicon.

19-2-3 DOPED SEMICONDUCTORS

The thermal equilibrium density of electrons and holes can be changed by adding appropriate impurity atoms to the semiconductor. In the case of silicon, the appropriate impurities are elements from column III of the periodic table, such as boron, or from column V, such as phosphorus.

Elements such as boron have only three electrons (valance electrons) available for bonding to other atoms in a crystal, and thus when boron is introduced into a silicon crystal, it needs an additional electron to bond to the four neighboring silicon atoms, as shown in Fig. 19-3a. The boron will very quickly acquire or accept the needed electron from the silicon lattice by capturing a free electron. This immobilizes a free electron and leaves a hole free to move through the crystal. The result is that the silicon now has more free holes, now termed majority carriers, than free electrons, now termed minority carriers. The silicon is said to be doped *p*-type with an acceptor impurity.

Column V elements, such as phosphorus, have five valance electrons but only four are needed for bonding in a silicon lattice. Such atoms are easily thermally ionized when

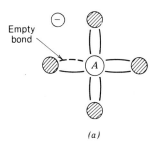

Empty bond

(a)

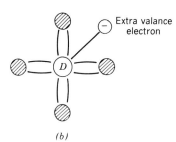

Extra valance electron

(b)

Figure 19-3 Doping by acceptors (a) and donors (b) to create p-type and n-type material: (a) p-type silicon; (b) n-type silicon.

placed in a silicon crystal and the fifth electron becomes free, as is illustrated in Fig. 19-3b. The resulting positive charge on the donor impurity (so named because it donates the fifth electron to the silicon lattice) represents a trapped or bound hole. Electrons are now the majority carriers and holes are the minority carriers. The silicon is said to be doped n-type.

The impurity levels commonly used in semiconductor devices (10^{19} cm^{-3} or less) are orders of magnitude smaller than the density (about 10^{23} cm^{-3}) of semiconductor atoms. Thus, the presence of impurities in a semiconductor will not affect the rate at which covalent bonds are broken by thermal ionization and subsequently refilled by free electrons (electron–hole recombination). This means that the product of the thermal equilibrium electron density, now termed n_0, and the thermal equilibrium hole density, now termed p_0, must still equal n_i^2, as is shown below:

$$p_0 n_0 = n_i^2 \qquad (19\text{-}2)$$

even though p_0 and n_0 are no longer equal. This relationship (Eq. 19-2) is sometimes called the law of mass action or the principle of detailed balance.

A doped (extrinsic) semiconductor is electrically neutral even though n_0 is no longer equal to p_0. The positive charge per unit volume in the extrinsic material is the sum of the hole density p_0 and the ionized donor density N_d, whereas the negative-charge density is the sum of the electron density n_0 and the ionized acceptor density N_a. The space charge neutrality condition in the general case where both donors and acceptors are assumed to be present in the material thus becomes

$$p_0 + N_d = n_0 + N_a \qquad (19\text{-}3)$$

Equations 19-2 and 19-3 can be solved simultaneously to find p_0 and n_0 separately. In a p-type material, the simultaneous solution of Eqs. 19-2 and 19-3, assuming $N_a \gg n_i$, yields the approximate result

$$n_0 \approx \frac{n_i^2}{N_a} \quad \text{and} \quad p_0 \approx N_a \qquad (19\text{-}4)$$

Similar expressions can be developed for strongly n-type material $N_d \gg n_i$. In either type of material, the minority carrier density is proportional to the square of the intrinsic carrier density (see Eq. 19-4) and thus is strongly temperature dependent.

19-2-4 RECOMBINATION

Fixed numbers of free electrons and holes require that mechanisms exist for the disappearance or recombination of them at the same rate as they are generated in thermal equilibrium. These mechanisms include direct recombination of electrons and holes (capture of a free electron in an empty covalent bond) and the trapping of carriers by impurities or imperfections in the crystal. In our largely qualitative examination of device physics, a simple rate equation describing the approximate time behavior of the excess carrier density (δn, the free-carrier density in excess of p_0 and n_0) is sufficient for our purposes. Space charge neutrality forces the excess hole density δp to equal the excess electron density δn.

This rate equation is given by

$$\frac{d(\delta n)}{dt} = -\frac{\delta n}{\tau} \tag{19-5}$$

assuming that there is no generation of excess carriers during the time interval when this equation is to be applied. If $\delta n > 0$ at $t = 0$, Eq. 19-5 predicts that $\delta n(t)$ decays exponentially with time for $t > 0$. The characteristic decay time or time constant τ is termed the excess-carrier lifetime, an important characteristic of minority-carrier devices.

In most situations it is convenient to consider the lifetime as a constant of the material. However, in two situations encountered in power semiconductor devices, the lifetime varies with device operating conditions. First, the excess-carrier lifetime will increase somewhat as the internal temperature of the devices increases. This will lead to a lengthening of the switching times of some devices (the minority-carrier or bipolar devices such as BJTs, thyristors, and GTOs). In simplistic terms, the minority carriers are more energetic at higher temperatures and thus are somewhat less likely to be captured by a recombination center. The details of why the lifetime increases with temperature are beyond the scope of this discussion.

Second, at large excess-carrier densities δn, the carrier lifetime becomes dependent on the value of δn. As excess-carrier densities approach a value n_b approximately equal to 10^{17} cm^{-3} and larger, another recombination process, Auger recombination, becomes important and causes the lifetime to decrease as δn increases. At these large carrier densities, the lifetime is given by

$$\tau = \frac{\tau_0}{1 + (\delta n)^2/n_b^2} \tag{19-6}$$

where τ_0 is the lifetime for $\delta n \ll n_b$. This decrease in excess-carrier lifetime will increase the on-state losses of some power devices at high current levels and thus will be a limiting factor in the operation of these devices.

The value of the excess-carrier lifetime has important effects on the characteristics of minority-carrier (also called bipolar) power devices. Larger values of the lifetime minimize the on-state losses but also tend to slow down the switching transition from on to off and vice versa. Hence, the device manufacturer strives for fairly precise and reproducible control of the lifetime during the fabrication process. Two commonly used methods of lifetime control are the use of gold doping and the use of electron irradiation. Gold is an impurity in silicon devices that acts as a recombination center. The higher the gold-doping density, the shorter the lifetimes will be. When electron irradiation is used, high-energy

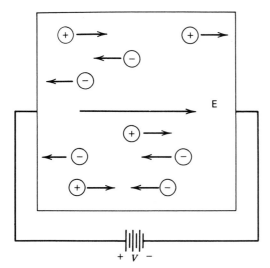

Figure 19-4 Drift of electrons and holes under the influence of an applied electric field.

(a few million electron volts of kinetic energy) electrons penetrate deeply (the depth of penetration is a function of the energy) into a semiconductor before they collide with the crystalline lattice. When a collision occurs, imperfections in the crystalline lattice are created that act as recombination centers. The impinging dose of high-energy electrons is easily controlled, so the final density of recombination centers and thus the lifetime is under good control. In recent years this method has become the preferred method of lifetime control because it can be applied during the final stages of fabrication as a final "tuning" or "tweaking" of the device characteristics that depend on the value of the lifetime.

19-2-5 DRIFT AND DIFFUSION

The flow of current in a semiconductor is the sum of the net flow of holes in the direction of the current and the net flow of electrons in the opposite direction. The free carriers can move via two mechanisms, drift and diffusion.

When an electric field is impressed across a semiconductor, the free holes are accelerated by the field and acquire a velocity component parallel to the field while electrons acquire a velocity component antiparallel to the field, as shown in Fig. 19-4. This velocity is termed the drift velocity and is proportional to the strength of the electric field. The drift component of current is given by

$$J_{drift} = q\mu_n nE + q\mu_p pE \tag{19-7}$$

where E is the applied electric field, μ_n is the electron mobility, μ_p is the hole mobility, and q is the charge on an electron. At room temperature in moderately doped silicon (less than 10^{15} cm^{-3}), $\mu_n \approx 1500$ cm^2/V-s and $\mu_p \approx 500$ cm^2/V-s. The carrier mobilities decrease with increasing temperature T (approximately T^{-2}).

If there is a variation in the spatial density of the free carriers such as is illustrated in Fig. 19-5, then there will be a movement of carriers from regions of higher concentration to regions of lower concentration. This movement is termed diffusion and is due to the random thermal velocity that each free carrier has. Such a spatial variation in carrier density could be obtained by a variety of methods including a variation in doping density.

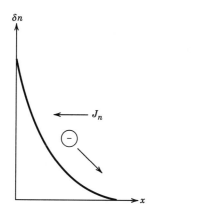
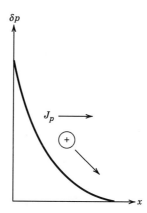

Figure 19-5 Carrier movement and current flow by diffusion. The electron current J_n is the direction of positive current flow (opposite to the electron flow).

The movement of carriers by diffusion will produce a component of current density that, in one dimension, is given by

$$J_{\text{diff}} = J_n + J_p = qD_n\frac{dn}{dx} - qD_p\frac{dp}{dx} \tag{19-8}$$

where D_n is the electron diffusion constant and D_p is the hole diffusion constant. The diffusion constants and mobilities are related by the Einstein relation, which is given by

$$\frac{D_p}{\mu_p} = \frac{D_n}{\mu_n} = \frac{kT}{q} \tag{19-9}$$

At room temperature, $kT/q = 0.026$ eV. In a particular situation, current flow will usually be either predominantly by drift or by diffusion. In the general case, current flow by both mechanisms may have to be considered simultaneously.

19-3 *pn* JUNCTIONS

A *pn* junction is formed when an *n*-type region in a silicon crystal is adjacent to or abuts a *p*-type region in the same crystal, as illustrated in Fig. 19-6. Such a junction can be formed by diffusing acceptor impurities into an *n*-type silicon crystal, for example. The opposite sequence (diffusing donors into *p*-type silicon) can also be used.

The junction is often characterized by how the doping changes from *n*-type to *p*-type as the junction is crossed. A so-called step or abrupt junction is shown in Fig. 19-7*a*. A

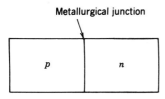

Figure 19-6 A *pn* junction.

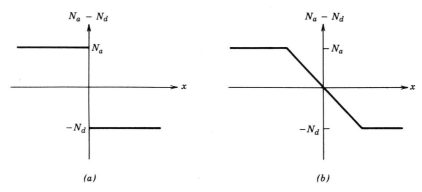

Figure 19-7   Impurity density versus position for (*a*) an abrupt (step) junction
and (*b*) a linearly graded junction.

more gradual change in doping density is the linearly graded junction shown in Fig.
19-7*b*. The junction is also characterized by the relative doping densities on each side of
the junction. If the acceptor density on the *p*-type side is very large compared to the donor
density on the *n*-type side, the junction is sometimes termed a p^+n junction. If the donor
density is not much larger than n_i in the previous example, the junction might be termed
a p^+n^- junction. Other variations on this theme are possible and are found in power
semiconductor devices.

19-3-1   POTENTIAL BARRIER AT THERMAL EQUILIBRIUM

Some of the majority carriers on either side of the junction will diffuse across it to the
opposite side, where they are in the minority. This will create a space charge layer on
either side of the junction, as is illustrated in Fig. 19-8, because the diffusing carriers will
leave behind ionized impurities that are immobile and that are now not screened by
enough free carriers for electrical neutrality. The resulting space charge density ρ, whose
spatial variation for a step junction is shown in Fig. 19-9*a*, gives rise to an electric field.

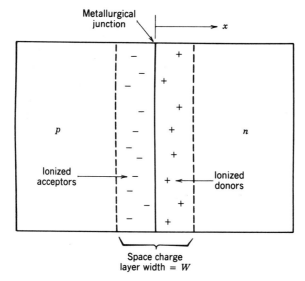

Figure 19-8   A *pn* junction with
a space charge or depletion layer
shown.

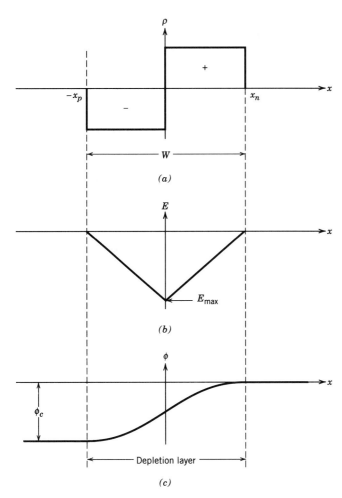

(a)

(b)

E_{max}

Figure 19-9 Spatial variation of (*a*) the space charge density ρ, (*b*) the electric field *E*, and (*c*) the potential φ across the space charge layer of a step *pn* junction. The electric field and potential are shown as negative because the applied voltages are positive when the *p* region is biased positive with respect to the *n* region (see Fig. 19-10).

(c)

The electric field can be estimated using Poisson's equation, which for a step junction is given by

$$\frac{dE}{dx} = \frac{-qN_a}{\epsilon} \qquad -x_p < x < 0: \qquad (19\text{-}10)$$

$$= \frac{qN_d}{\epsilon} \qquad 0 < x < x_n$$

Integrating from $x = -x_p$ to $x = x_n$, the respective distances the depletion region penetrates into the *p*-type and *n*-type sides of the junction, yields

$$E(x) = \frac{-qN_a(x + x_p)}{\epsilon} \qquad -x_p < x < 0 \qquad (19\text{-}11a)$$

$$= \frac{qN_d(x - x_n)}{\epsilon} \qquad 0 < x < x_n$$

$$(19\text{-}11b)$$

This electric field, which is plotted in Fig. 19-9*b*, gives rise to a potential barrier that is plotted in Fig. 19-9*c*. Integrating the electric field across the depletion layer yields the magnitude of the barrier (often termed the contact potential), which for the step junction of Figs. 19-8 and 19-9 is given by

$$-\int_{-x_p}^{x_n} E(x)\, dx = \phi_c = \frac{qN_ax_p^2 + qN_dx_n^2}{2\epsilon} \tag{19-12}$$

The electric field increases in strength as more ionized impurities are exposed because of the diffusing carriers. The field, however, tends to retard the diffusion process because it acts to push the electrons back to the n-type side and holes back to the p-type side. An equilibrium is reached when the carrier flux caused by diffusion is counterbalanced by the carrier flux due to the electric field (drift). In equilibrium the hole flux and electron flux separately sum to zero, rather than merely the total current being zero. Otherwise, there would be a buildup of electrons and holes on one side of the junction. Setting the hole current density J_p equal to zero yields

$$J_p = q\mu_p p \left[-\frac{d\phi}{dx}\right] - qD_p\frac{dp}{dx} = 0 \tag{19-13}$$

Separating variables in Eq. 19-13 and integrating across the depletion regions yields another expression for the contact potential:

$$\phi_c = \int_{\phi(-x_p)}^{\phi(x_n)} d\phi = -\frac{D_p}{\mu_p}\int_{p(-x_p)}^{p(x_n)} p(x)\, dx = \frac{kT}{q}\ln\left[\frac{N_aN_d}{n_i^2}\right] \tag{19-14}$$

In Eq. 19-14, $p(-x_p) = N_a$, $p(x_n) = n_i^2/N_d$, and use is made of the Einstein relation (Eq. 19-9). This contact potential cannot be measured directly with a voltmeter because equal and opposite contact potentials are developed when any measuring probes are attached to the n and p sides of the junction. At room temperature, $\phi_c < E_g$. For example, in a silicon pn junction at room temperature with $N_a = N_d = 10^{16}$ cm^{-3}, $\phi_c = 0.72$ eV.

19-3-2 FORWARD AND REVERSE BIAS

When an external voltage is applied between the p and n regions, as shown in Fig. 19-10a, it appears entirely across the space charge region because of the large resistance of the depletion layer compared to the rest of the material. If the applied potential is positive on the p side, it opposes the contact potential and reduces the height of the potential barrier, and the junction is said to be forward biased.

When the applied voltage makes the n side more positive, as is indicated in Fig. 19-11a, the junction is said to be reverse biased and the barrier height is increased to $V + \phi_c$. The width $W(V) = x_n(V) + x_p(V)$ of the space charge layer (or depletion region as it is also termed because of the absence of free carriers) must grow or shrink as the height of the potential barrier either grows or shrinks. This occurs because a change in the potential requires a change in the magnitude of the total amount of charge on each side of the junction. Since the charge density equals the impurity density (a constant), a change in the total charge can occcur only if there is a change in the dimensions of the depletion region. This variation in the dimensions of the depletion region with applied voltage has important consequences for the design of power semiconductor devices.

Under reverse-bias conditions, the step junction relationship (Eq. 19-12) between the potential barrier height and the depletion layer partial widths x_n and x_p is changed by replacing ϕ_c with $V + \phi_c$. The total negative charge $qN_ax_p(V)$ in the p-type region of the depletion layer must equal the total positive charge $qN_dx_n(V)$ in the n-type region of the depletion layer, that is,

$$qN_ax_p(V) = qN_dx_n(V) \tag{19-15}$$

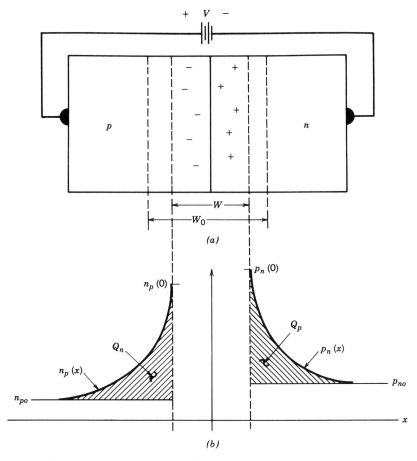

Figure 19-10 A *pn* junction with voltage applied in the forward direction: (a) forward-biased *pn* junction; (b) minority-carrier densities vs. position in the forward-biased *pn* junction.

Solving Eq. 19-15 simultaneously with the modified form of Eq. 19-12 yields separate expressions for $x_p(V)$ and $x_n(V)$. Combining these separate expressions into an equation for the total step junction depletion layer width $W(V)$ yields

$$W(V) = W_0\sqrt{1 - (V/\phi_c)} \tag{19-16}$$

where W_0 is the depletion layer width at zero bias and V is the applied diode voltage (negative for reverse bias). Here W_0 is given by

$$W_0 = \sqrt{\frac{2\epsilon\phi_c(N_a + N_d)}{qN_aN_d}} \tag{19-17}$$

where ϵ is the dielectric constant of the semiconductor ($\epsilon = 11.7\epsilon_0$ for silicon where $\epsilon_0 = 8.85 \times 10^{-14}$ F/cm). The electric field is a maximum at the metallurgical junction and is given by

$$E_{\max} = \frac{2\phi}{W_0}\sqrt{1 - V/\phi_c} \tag{19-18}$$

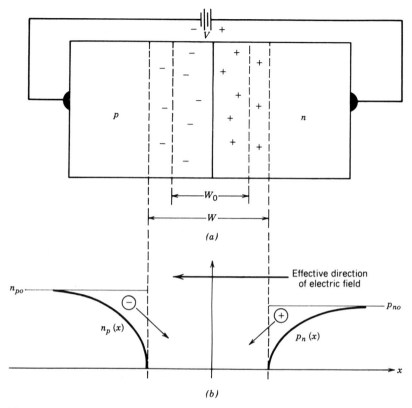

Figure 19-11 A *pn* junction with voltage applied in the reverse direction:
(*a*) reversed-biased *pn* junction; (*b*) minority-carrier densities vs. position in
reverse-biased *pn* junction.

Other doping profiles will produce slightly different functional dependencies of W and
E_{max} with applied voltage and doping levels, but qualitatively their behavior is the same
as the step junction. Thus, we will use the step junction as our model for discussing how
the properties of a *pn* junction affect the operation and performance of power semiconductor devices.

19-4   CHARGE CONTROL DESCRIPTION OF *pn*-JUNCTION OPERATION

A reverse-bias voltage increases the potential barrier, which in turn makes the probability
of any carrier diffusing across the junction vanishingly small. The minority carrier densities become nearly zero at the edge of the depletion region, as is shown in Fig. 19-11*b*.
The small gradients in the minority-carrier densities will cause a small flux of diffusing
minority carriers toward the depletion region, as is indicated in the figure. When these
diffusing carriers reach the depletion layer, the large electric fields in the space charge
layer will immediately sweep them across the layer into the electrical neutral region on the
other side of the junction. Electrons will be swept to the *n*-type side and holes to the *p*-type
side. This will constitute a small leakage current termed the reverse saturation current I_s,
which is diagrammed in the *i–v* curve shown in Fig. 19-12. This current is independent

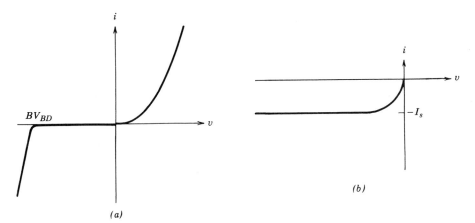

Figure 19-12 The *I–V* characteristic of a *pn* junction in both forward and reverse bias (*a*). The reverse bias portion is redrawn in (*b*) because the reverse saturation current is too small to be seen on the same linear scales as the forward current.

of the reverse voltage. There will also be a contribution to the leakage current by the electrons and holes created in the space charge layer by thermal ionization processes.

A forward-bias voltage lowers the potential barrier and upsets the equilibrium between drift and diffusion in favor of diffusion. As is shown in Fig. 19-10*b*, this results in an enormous increase in the minority-carrier densities (electrons and holes) in the electrically neutral regions on both sides of the junction immediately adjacent to the depletion regions. This increase, sometimes termed carrier injection, is an excess carrier density, that is, above the thermal equilibrium values, and it has a significant impact on the diode's characteristics, especially the switching behavior.

The injected minority carriers eventually recombine with the majority carriers as they diffuse farther into the electrically neutral drift regions shown schematically in the one-dimensional diode model illustrated in Fig. 19-10*b*. This leads to an exponential decrease in the excess-minority-carrier density with distance, which is illustrated in the figure. The characteristic decay length is termed the minority-carrier diffusion length, which for electrons in *p*-type material is given by

$$L_n = \sqrt{D_n \tau_n} \tag{19-19}$$

and for holes in *n*-type material is given by

$$L_p = \sqrt{D_p \tau_p} \tag{19-20}$$

In these equations, D_n and D_p are the electron and hole diffusion constants and τ_n and τ_p are the corresponding minority-carrier lifetimes, which can range from nanoseconds to tens of microseconds depending on how the diode is fabricated.

The minority-carrier density at the edge of the depletion region, $p_n(0)$ in Fig. 19-10*b*, is exponentially dependent on the forward-bias voltage and is given by

$$p_n(0) = \frac{n_i^2}{N_d} e^{qV/kT} \tag{19-21}$$

A similar expression exists for $n_p(0)$. At zero bias, Eq. 19-21 reduces to the correct thermal equilibrium value. As a result of this voltage dependence, the minority-carrier densities will vary by orders of magnitude as the voltage is changed by relatively small amounts. This will result in large gradients in the minority-carrier densities in the regions adjacent to the depletion layer (within a few diffusion lengths) and consequently large diffusion currents (see Eq. 19-8). In fact, in these regions the diffusion component of the current is the dominant component of the total current (the drift component is negligible in this region because the large injected carrier densities short out the electric field needed for any substantial drift current).

In the steady-state (dc) forward-bias situation, the excess-carrier distributions shown in Fig. 19-10b neither grow nor decay in time. For this to happen, the carriers that are lost per unit time from the distribution via recombination must be replaced. They are replaced by new carriers that are injected across the junction by the forward-bias current density $J(J = I/A$, where I is the terminal current and A is the cross-sectional area of the diode). This leads to the simple relation

$$J = \frac{Q_n}{\tau_n} + \frac{Q_p}{\tau_p} \tag{19-22}$$

where Q_n and Q_p are the areas under the excess-minority-carrier distributions shown in Fig. 19-10b. Since the excess-carrier distributions depend exponentially on position (decaying with distance away from the junction), they can easily be integrated to find Q_n and Q_p. The resulting expressions are

$$Q_p = q\left[p_n(0) - \frac{n_i^2}{N_d}\right]L_p \tag{19-23}$$

$$Q_n = q\left[n_p(0) - \frac{n_i^2}{N_a}\right]L_n \tag{19-24}$$

It should be noted in Eqs. 19-23 and 19-24 that the excess-minority-carrier density at the edge of the depletion region [$p_n(0) - n_i^2/N_d$ in Eq. 19-23 and $n_p(0) - n_i^2/N_a$ in Eq. 19-24] is used rather than the total minority-carrier density [$p_n(0)$ in Eq. 19-23 or $n_p(0)$ in Eq. 19-24]. This of course is because only the excess minority carriers have any spatial density gradient that can lead to diffusion currents. Putting the expressions for Q_p and Q_n into Eq. 19-22 using Eq. 19-21 yields

$$J = qn_i^2\left[\frac{L_n}{N_a\tau_n} + \frac{L_p}{N_d\tau_p}\right]\left[e^{qV/(kT)} - 1\right] \tag{19-25}$$

Equation 19-25 is plotted in Fig. 19-12 and is the $i-v$ characteristic of the pn-junction diode for both forward and reverse bias. The term $qn_i^2[\{L_n/(N_a\tau_n)\} + \{L_p/(N_d\tau_p)\}]$ is readily identified as the reverse saturation current J_s shown in Fig. 19-12b. This current has an extreme temperature sensitivity because of its dependence on n_i^2 (see Eq. 19-1). This temperature sensitivity must be considered in all diode applications. In the forward direction the current depends exponentially on the forward voltage. The large increase in current in Fig. 19-12a at the reverse bias voltage BV_{BD} is caused by impact ionization and is not predicted by Eq. 19-25.

19-5 AVALANCHE BREAKDOWN

The rapid increase in current at the reverse-bias voltage BV_{BD} shown in Fig. 19-12a is termed reverse breakdown or avalanche breakdown. Operation of the diode in breakdown must be avoided because the product of large voltage and large current leads to excessive power dissipation that will quickly destroy the device if it is not reduced. The breakdown is caused by a physical mechanism termed impact ionization.

19-5-1 IMPACT IONIZATION

If a free electron with sufficient kinetic energy strikes a silicon atom, it can break a covalent bond and liberate an electron from the bond. If the kinetic energy is gained from an applied electric field, such as reverse voltages applied across a space charge layer, the liberation of the electron from the bond is termed impact ionization. This process is important because the newly liberated electron can gain enough energy from the applied field to break a covalent bond when it strikes a silicon atom, thus liberating an additional electron. This process can cascade (avalanche) very quickly in a chain reaction–like manner, producing a large number of free electrons and thus a large current, and the large power dissipation will quickly destroy the device, as previously mentioned.

An approximately constant value of electric field, E_{BD}, is required to cause appreciable impact ionization according to experimental observations. This value can be estimated from a very simple model, which highlights the important mechanism of impact ionization. The amount of kinetic energy required to break a bond is the energy gap E_g, assuming that all of the kinetic energy of the incident free electron is transferred into breaking the bond and that both the incident and liberated electrons have little kinetic energy after the collision. If it is now assumed that these electrons start essentially at rest and are accelerated by the electric field until their next collisions and that the time between collisions is t_c, then the value of E_{BD} for impact ionization is calculated as

$$E_{BD} = \sqrt{\frac{2E_g m}{q t_c^2}} \qquad (19\text{-}26)$$

In this equation m is the mass of the electron (about 10^{-27} g) and q is the electron charge. The average time between collisions of electrons with the lattice is on the order of 10^{-12}–10^{-14} s. Taking an intermediate value of 10^{-13} s as typical for silicon and using it in Eq. 19-26 predicts a value of about 300,000 V/cm for E_{BD}. This estimate is surprisingly close to the experimental value of 200,000 V/cm determined from avalanche breakdown measurements in power devices.

A more correct picture of the impact ionization process would have to take into account that not all electrons will lose their entire kinetic energy at each collision. Also, electrons will have a wide range of kinetic energies obtained from the thermal energy in the lattice. This considerably complicates the calculation of the electric field needed for appreciable ionization, although it has been done. However, the more correct theory yields a value for E_{BD} that is close to the value estimated in the preceding paragraph. Thus it is concluded that the simple picture of impact ionization given is adequate for the qualitative study of voltage breakdown in power semiconductor devices.

19-5-2 BREAKDOWN VOLTAGE ESTIMATE

The reverse-bias voltage is dropped entirely across the depletion region, and the larger the voltage, the larger the electric field in the region and the closer it approaches the value E_{BD} where substantial impact ionization begins. The reverse-bias voltage magnitude that generates substantial impact ionization is termed the avalanche breakdown voltage BV_{BD}, and it depends on the doping profile (step, linearly graded, diffused, etc.) of the junction and on the magnitudes of the doping densities. The breakdown voltage of the step junction is sufficiently representative of all pn junctions that a detailed examination of it will yield the basic features of breakdown that are needed for a qualitative study of power semiconductor devices without the quantitative details of more complicated doping profiles obscuring the essential characteristics. Setting E_{max} in Eq. 19-18 to E_{BD} and the voltage

$V = -\mathrm{BV_{BD}}$ (reverse-bias voltages are negative) and solving for the breakdown voltage yields

$$\mathrm{BV_{BD}} = \phi_c\left[\frac{W_0 E_{BD}}{2\phi_c}\right]^2 - \phi_c \approx \frac{\epsilon(N_a + N_d)E_{BD}^2}{2\, qN_aN_d} \tag{19-27}$$

SUMMARY

In this chapter the basic properties of semiconductors were reviewed. The important concepts are as follows:

1. Current in a semiconductor is carried by both electrons and holes.
2. Electrons and holes move by both drift and diffusion.
3. Intentional doping of the semiconductor with impurities will cause the density of holes and electrons to be vastly different.
4. The density of minority carriers increases exponentially with temperature.
5. A pn junction can be formed by doping one region n-type and the adjacent region p-type.
6. A potential barrier is set up across a pn junction in thermal equilibrium that balances out the drift and diffusion of carriers across the junction so that no net current flows.
7. In reverse bias a depletion region forms on both sides of the pn junction and only a small current can flow by drift.
8. In forward bias large numbers of electrons and holes are injected across the pn junction and large currents flow by diffusion with small applied voltages.
9. Large numbers of excess electron–hole pairs are created by impact ionization if the electric field in the semicondutor exceeds a critical value.
10. Avalanche breakdown occurs when the reverse-bias voltage is large enough to generate the critical electric field E_{BD}.

PROBLEMS

19-1 The intrinsic temperature T_i of a semiconductor device is that temperature at which n_i equals the doping density. What is T_i of a silicon pn junction that has 10^{18} cm^{-3} acceptors on the p-type side and 10^{14} cm^{-3} donors on the n-type side?

19-2 What are the resistivities of the p region and n region of the pn junction described in Problem 19-1?

19-3 Estimate p_0 and n_0 in a silicon sample where both donor and acceptor impurities are simultaneously present and $N_d - N_a = 10^{13}$ cm^{-3}.

19-4 What change in temperature ΔT doubles the minority-carrier density in the n-type side of the pn junction described in Problem 19-1 compared with the room temperature value?

19-5 Show that a decade increase in the forward current of a pn junction is accompanied by an increase in the forward voltage of about 60 mV.

19-6 Consider a step silicon pn junction with 10^{14} cm^{-3} donors on the n-type side and 10^{15} cm^{-3} acceptors on the p-type side:
(a) Find the width of the depletion layer on each side of the junction.
(b) Sketch and dimension the electric field distribution versus position through the depletion layer.
(c) Estimate the contact potential ϕ_c.
(d) Using a parallel-plate capacitor formalism, estimate the capacitance per unit area of the junction at 0 V and at -50 V.

(e) Estimate the current flowing through this junction if the forward voltage is 0.7 V, the excess-barrier lifetime is 1 μs, and the cross-sectional area is 100 μm square.

19-7 A bar of *n*-type silicon with 10^{14} cm^{-3} donors is 200 μm long and 1×1 mm square. What is its resistance at room temperature and at 250°C? Assume temperature independent mobilities.

19-8 Estimate the breakdown voltage of the *pn* junction diode described in Problem 19-6.

19-9 Show that for a step junction, the width of the space charge layer when the reverse-bias voltage is equal to the avalanche breakdown value BV$_{BD}$ can be written as

$$W(\text{BV}_{BD}) = \frac{2\text{BV}_{BD}}{E_{BD}}$$

[*Hint:* Examine the plot of electric field vs. position in Fig. 19-9*b* with $E_{max} = E_{BD}$ and $W = W(\text{BV}_{BD})$]. Use this result to find the depletion layer width of the *pn* junction of Problem 19-6 when it is biased at avalanche breakdown.

19-10 What are the carrier diffusion lengths, L_n and L_p, for the *pn* junction of Problem 19-6, part e?

19-11 A one-sided step junction with a *p*-side doping level N_a much greater than the *n*-side doping level N_d, conducts a current I when forward-biased by a voltage V_F. The current is to be increased to twice this value (to 2I) at the same voltage V_F by adjusting the carrier lifetime. What adjustment is required in the lifetime τ to realize this change in current?

19-12 Find the minimum conductivity achievable in silicon by choice of doping level and conditions under which it occurs. Assume room temperature and mobilities independent of the doping levels.

19-13 What is the resistivity of intrinsic silicon at room temperature (300 °K)?

REFERENCES

1. B. G. Streetman, *Solid State Electronic Devices,* 2nd ed., Prentice-Hall, Englewood Cliffs, NJ, 1980, Chapters 1–6.
2. S. K. Ghandhi, *Semiconductor Power Devices,* John Wiley, New York, 1977, Chapters 1–3.
3. A. S. Grove, *Physics and Technology of Semiconductor Devices,* John Wiley, New York, 1967, Chapters 4–6.
4. A. S. Sedra and K. C. Smith, *Microelectronics Circuits,* 2nd ed., Holt, Rinehart, and Winston, New York, 1987, Chapter 4.

CHAPTER 20

POWER DIODES

20-1 INTRODUCTION

Power semiconductor devices, even diodes, are more complicated in structure and operational characteristics than their low-power counterparts with which most of us have some degree of familiarity. The added complexity arises from the modifications made to the simple low-power devices to make them suitable for high-power applications. These modifications are essentially generic in nature, that is, the same basic modifications are made to all low-power semiconductor devices in order to scale up their respective power capabilities. Thus, if the modifications can be understood in the context of one specific type of device, then it will be much easier to see the effects of these modifications in the other types of power devices.

The study of power semiconductor devices begins with the diode, both pn-junction and Schottky barrier devices, because they are the simplest of all semiconductor devices. The modifications for high-power operation will thus be most easily considered first in these devices. Additionally, the diode or pn junction is the basic building block of all other power semiconductor devices. A comprehension of the other power devices will be more easily obtained if first the characteristics of the diode are clearly understood.

20-2 BASIC STRUCTURE AND $I-V$ CHARACTERISTICS

The ideal pn-junction diode geometry was discussed in Chapter 19. The practical realization of the diode for power applications is shown in Fig. 20-1. It consists of a heavily doped n-type substrate on top of which is grown a lightly doped n^- epitaxial layer of specified thickness. Finally the pn junction is formed by diffusing in a heavily doped p-type region that forms the anode of the diode. Typical layer thicknesses and doping levels are shown in Fig. 20-1. The cross-sectional area A of the diode will vary according to the amount of total current the device is designed to carry. For diodes that can carry several thousand amperes, the area can be several square centimeters (wafers with diameters as large as 4 in. are used in the production of power devices, and for the largest diodes, only one is made from the wafer). The circuit symbol for the diode is shown in Fig. 20-1 and is the same as that used for low-power signal level diodes.

The n^- layer in Fig. 20-1, which is often termed the drift region, is the prime structural feature not found in low-power diodes. Its function is to absorb the depletion layer of the reverse-biased p^+n^- junction. This layer can be quite wide at large reverse voltages. The drift region establishes what the reverse breakdown voltage will be. This relatively long lightly doped region would appear to add significant ohmic resistance to the diode when it is forward biased, a situation that would apparently lead to unacceptably

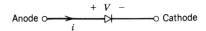

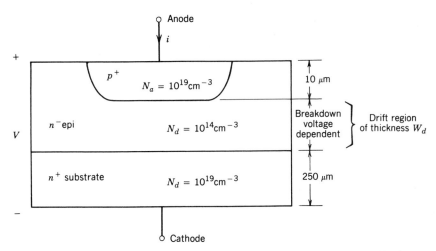

Figure 20-1 Cross-sectional view of a *pn*-junction diode intended for power applications. The circuit symbol is also shown with anode and cathode designations. The cross-sectional view of a high-power diode with additional provisions for high breakdown voltages is shown in Figs. 20-6 and 20-7.

large power dissipation in the diode when it is conducting current. However, other mechanisms to be described shortly greatly reduce this apparent problem.

The *i–v* characteristic of the diode is shown in Fig. 20-2. In the forward direction it appears the same as that discussed in Chapter 19 except that the current grows linearly with the forward-bias voltage rather than exponentially. The large currents in a power diode create ohmic drops that mask the exponential *i–v* characteristic. The voltage drop in the lightly doped drift region accounts for part of this ohmic resistance. The nature of the on-stage voltage drop will be discussed in a later section of this chapter.

In reverse bias only a small leakage current, which is independent of the reverse voltage, flows until the reverse breakdown voltage BV_{BD} is reached. When breakdown is reached, the voltage appears to remain essentially constant while the current increases dramatically, being limited only by the external circuit. The combination of a large

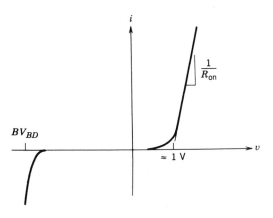

Figure 20-2 The *I–V* characteristic of a *pn*-junction diode. The reverse-bias portion of the characteristic shows avalanche breakdown at BV_{BD}. The exponential *i–v* relationship in forward bias expected from signal-level diode characteristics is masked by the ohmic resistance R_{on} in power diodes.

voltage at breakdown and a large current leads to excessive power dissipation that can quickly destroy the device. Operation of the diode in breakdown must therefore be avoided.

20-3   BREAKDOWN VOLTAGE CONSIDERATIONS

20-3-1   BREAKDOWN VOLTAGE OF NON-PUNCH-THROUGH DIODES

In designing a diode for a particular breakdown voltage rating, one of two variants of the structure shown in Fig. 20-1 are used. If the length W_d of the lightly doped drift region is longer than the depletion layer width at breakdown, then the structure is termed a non-punch-through diode, that is, the depletion layer has not reached through (or punched through) the lightly doped drift region and reached the highly doped n^+ substrate. For the non-punch-through diode using a step junction doping profile, the breakdown voltage is given by Eq. 19-27. Since the drift region is much more lightly doped than the p-type side, Eq. 19-27 simplifies to

$$\text{BV}_{\text{BD}} \approx \frac{\epsilon E_{\text{BD}}^2}{2qN_d} \tag{20-1}$$

and the depletion layer is contained almost entirely on the much more lightly doped drift region side. Putting in the numerical values of E_{BD} (2×10^5 V/cm) and the dielectric constant of silicon (1.05×10^{-12} F/cm) yields

$$\text{BV}_{\text{BD}} \approx \frac{1.3 \times 10^{17}}{N_d} \tag{20-2}$$

where the doping density is in number per cubic centimeter. The corresponding depletion layer width (in centimeters) of the step junction at breakdown is (using Eqs. 19-11 and 19-12 and N_d as given by Eq. 20-2)

$$W_d \geq W(\text{BV}_{\text{BD}}) \approx \frac{2\text{BV}_{\text{BD}}}{E_{\text{BD}}} = 1 \times 10^{-5}\text{BV}_{\text{BD}} \tag{20-3}$$

Recall from Fig. 20-1 that W_d is the drift region thickness.

Two basic facts are evident from Eqs. 20-2 and 20-3. First, large breakdown voltages require lightly doped junctions, at least on one side. Second, the drift layer in the diode must be fairly long in high-voltage devices to accommodate the long depletion layers. For example, a breakdown voltage of 1000 V requires a doping level of approximately 10^{14} cm^{-3} or less and a minimum drift region thickness W_d of about 100 μm to accommodate the depletion region. These requirements are satisfied by the lightly doped drift region shown in the cross-sectional diagram of the power diode illustrated in Fig. 20-1.

20-3-2   BREAKDOWN VOLTAGE OF PUNCH-THROUGH DIODE

In some situations, it is possible to have shorter drift region lengths and still have the diode block large reverse voltages. Consider the situation shown in Fig. 20-3, where the depletion region has extended all the way across the drift region and is in contact with the n^+ layer. When this occurs (which is commonly termed punch-through and hence punch-through diode), further increases in reverse voltage will not cause the depletion region to widen any further because the large doping density in the n^+ layer effectively blocks

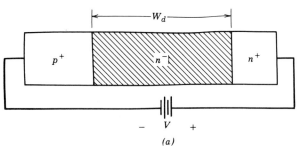

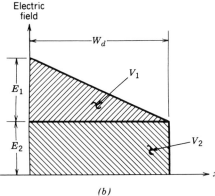

Figure 20-3 Punch-through in a reverse-biased power diode: (a) reverse-biased diode with the depletion layer extending completely across the drift region, sometimes called the punch-through condition; (b) electric field profile of the punch-through condition in a reverse-biased diode.

further growth of the depletion layer. Instead the electric field profile begins to flatten out, as is shown in Fig. 20-3b, becoming less triangular and more rectangular.

As shown in the figure, the electric field profile can be considered to be composed of a triangular-shaped component with a peak electric field value of E_1 at the junction and a rectangular-shaped component of constant electric field value E_2. The triangular-shaped component is due to the ionized donors in the drift region, and hence E_1 is given by

$$E_1 = \frac{qN_dW_d}{\epsilon} \tag{20-4}$$

and the area under the triangular component represents a voltage V_1, which is given by

$$V_1 = \frac{qN_dW_d^2}{2\epsilon} \tag{20-5}$$

The area under the rectangular component is a voltage given by

$$V_2 = E_2W_d \tag{20-6}$$

When the junction having this punch-through profile breaks down, the following conditions exist:

$$E_1 + E_2 = E_{BD} \tag{20-7}$$

and

$$BV_{BD} = V_1 + V_2 \qquad (20\text{-}8)$$

Substituting Eq. 20-5 for V_1 into Eq. 20-8 and simultaneously substituting $V_2 = [E_{BD} - E_1]W_d$ with E_1 being given by Eq. 20-4 yield

$$BV_{BD} = E_{BD}W_d - \frac{qN_dW_d^2}{2\epsilon} \qquad (20\text{-}9)$$

If the doping density in the drift region is made much smaller than the values permitted by Eq. 20-2, then V_1 will be much less than V_2 (the electrical field profile will be essentially constant and independent of position) so that

$$BV_{BD} \approx E_{BD}W_d \quad \text{or} \quad W_d \approx \frac{BV_{BD}}{E_{BD}} \qquad (20\text{-}10)$$

which is about one-half the value of W_d given by Eq. 20-3 for the same value of breakdown voltage. The low doping density means that the ohmic resistivity of the punch-through drift region is much larger than the non-punch-through drift region resistivity.

The higher resistivity of the drift region of the punch-through structure has no significant effect on the operation of the diode because the conductivity modulation that occurs during on-state operation (a subject to be discussed shortly) shorts out the drift region. As we will discuss in detail in later sections of this chapter, the shorter drift region length of the punch-through diode permits this diode to have lower on-state voltages compared with the conventional non-punch-through diode, assuming the same lifetime in each diode and the same breakdown voltage. However, the punch-through structure cannot be used in majority-carrier devices because there is no conductivity modulation in the on-state operation of these devices. Hence, the large resistance of the punch-through drift region will not be shorted out and there will be large amounts of on-state power dissipation.

20-3-3   DEPLETION LAYER BOUNDARY CONTROL

The plane parallel junctions that have been implicitly assumed thus far in these discussions are an unrealistic idealization in practice. Junctions in actual devices are formed via masked diffusions of impurities (such as acceptors shown in Fig. 20-4) that will inevitably cause the resulting *pn* junction to have some degree of curvature, as shown in Fig. 20-4. The amount of curvature, which is specified by the radius of curvature (indicated in Fig. 20-4), will depend on the size of the diffusion mask, the length of the diffusion time, and the magnitude of the diffusion temperature. The curvature is caused by the fact that the impurities diffuse as fast laterally as they do vertically into the substrate. In such junctions, the plane parallel description becomes inaccurate when the radius of curvature becomes comparable to the depletion layer width. In these circumstances, the electric field in the depletion layer becomes spatially nonuniform and has its largest magnitude where the radius of curvature is the smallest. This will lead to a smaller breakdown voltage compared to a plane parallel junction of similar doping.

The obvious step in combating this potential reduction in breakdown voltage is to keep the radius of curvature as large as possible. Modeling studies with a cylindrical *pn* junction of radius R indicate that the radius must be six or more times larger than the depletion layer thickness, at breakdown, of a comparable plane parallel junction to keep

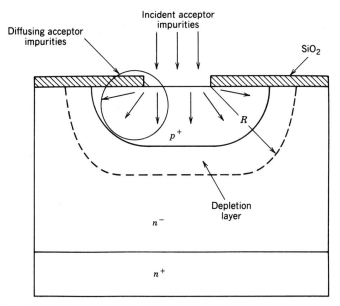

Figure 20-4 A *pn* junction formed by a masked diffusion of impurities into the substrate. The lateral diffusion of impurities gives rise to a curvature of the *pn*-junction boundary and thus to the depletion layer. The smaller the radius of curvature *R*, the lower the breakdown voltage becomes.

the breakdown voltage of the cylindrical junction within 90% of the plane parallel junction breakdown voltage. In high-voltage diodes where BV_{BD} is 1000 V and larger, the required radius of curvature *R*, using the estimate of 100 μm for depletion layer thickness, at breakdown, of the plane parallel junction given earlier, would be 600 μm. The realization of such large values would require deep diffusions (depths comparable to *R*) into the substrate, which would be impractical because of impossibly long diffusion times.

Thus, the radius of curvature of the depletion layer boundary must be controlled by other means. One method is via the use of electrically floating field plates that are illustrated in Fig. 20-5. The field plates act as an equipotential surface, and by their proper

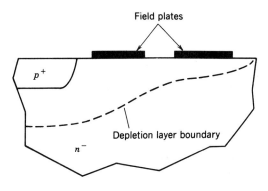

Figure 20-5 Use of field plates in a *pn*-junction diode to control the depletion layer boundary curvature in order to keep breakdown voltages from being reduced.

placement, as is indicated in the figure, they can redirect the electric field lines and prevent the depletion layer from having too small a radius of curvature. The price to be paid for this approach is that the field plates require a considerable amount of silicon real estate.

Another method of controlling the depletion layer is the use of guard rings, illustrated for a simple *pn* junction in Fig. 20-6. The *p*-type guard rings are allowed to float electrically. The depletion layers of the guard rings merge with the growing depletion layer of the reverse-biased *pn* junction, which prevents the radius of curvature from getting too small. Since the guard rings are electrically floating, they do not acquire the full reverse-bias voltage, and thus, breakdown will not occur across their depletion layers even though their radii of curvature may be somewhat small, as Fig. 20-6 implies.

In some devices the metallurgical junction extends to the surface of the silicon and the high field depletion layer intersects the semiconductor–air boundary, as illustrated in Fig. 20-7a. This situation will also cause curvature of the depletion layer boundary even if the metallurgical junction itself is a plane parallel structure. The fringing electric fields at or near the surface may cause premature breakdown or interact with surface impurities that will ultimately degrade the performance of the device. Experience indicates that this causes a 20–30% reduction in E_{BD} at the surface compared to the bulk. Thus, it is often necessary to shape the topological contours of the device to minimize the surface electric fields. An example of beveling is shown in Fig. 20-7b. Additionally coating the sample surfaces with appropriate materials such as silicon dioxide or some other insulator will aid in controlling the electric fields at the surface.

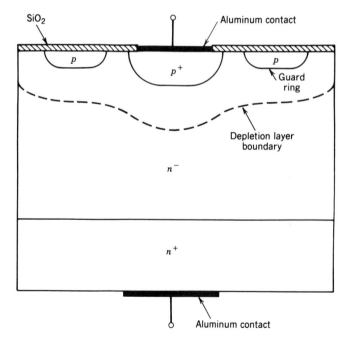

Figure 20-6   A *pn*-junction diode with both an *n*-type drift region and guard rings to improve breakdown voltage capabilities. The guard rings help to prevent the depletion layer from having too small a radius of curvature.

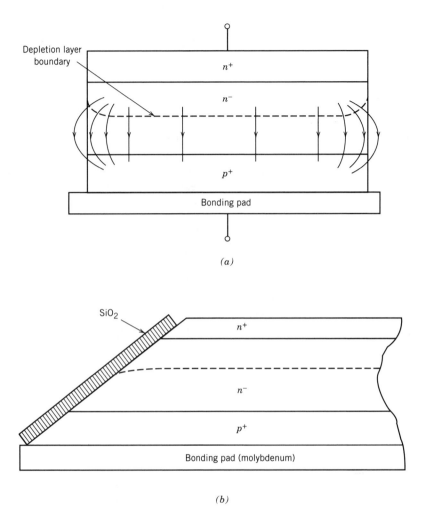

(a)

(b)

Figure 20-7 (a) Depletion layer intersection with a semiconductor surface and attendant field crowding. (b) The use of topological contouring (beveling in this example) to minimize field crowding due to depletion layer curvature.

20-4 ON-STATE LOSSES

20-4-1 CONDUCTIVITY MODULATION

Nearly all of the power dissipated in a diode occurs when it is in the on state (forward biased). At high switching frequencies a significant amount of dissipation can occur during the switching transient from one state to the other, a subject that will be considered later. In low-power applications, the forward-bias voltage is the junction voltage and is often considered approximately constant, since the voltage depends logarithmically on the current. In a silicon diode this constant voltage is $0.7-1.0$ V and the on-state dissipation would then be $P = 0.7\,I$, where I is the current through the diode.

In power diodes, this estimate would be satisfactory only at low current levels. At large current levels, it would severely underestimate the total dissipation because the dissipation in the drift region of the power diode is ignored. It is these losses that limit the diode's ultimate power capability. However, in estimating the power dissipated in the drift region, care must be exercised because the effective value of resistance of this region in the on state is much less than the apparent ohmic value calculated on the basis of the geometric size and the thermal equilibrium carrier densities. In the on state there is a substantial reduction in the resistance of the drift region because of the large amount of excess-carrier injection into the drift region. This conductivity modulation, as it is sometimes termed, substantially reduces the power dissipation over what would be estimated on the basis of the thermal equilibrium conductivity of the drift region.

Consider the one-dimensional power diode model shown in Fig. 20-8. In the on state the forward-biased pn junction injects holes into the n-type drift region. At low injection levels, $\delta p \ll n_{no}$, the thermal equilibrium electrons n_{no} easily neutralize the space charge of the holes. But at high injection levels, $\delta p > n_{no}$, the hole space charge is large enough to attract electrons from the n^+ region into the drift region. This leads to the injection of electrons across the $n^+ n^-$ interface into the drift region with densities $\delta n = \delta p$. These injected electrons and holes diffuse into the drift region toward each other, recombining as they diffuse. This is the origin of so-called double injection.

If the diffusion length L, where $L = \sqrt{D\tau}$ (see Eqs. 19-19 and 19-20), is greater than the drift region length W_d, then the spatial distribution of the excess carriers will be fairly flat, as shown in Fig. 20-8, and equal to an average value n_a. Since the doping density n^- of the drift region is quite small, typically 10^{14} cm^{-3}, n_a will be typically much greater than n^-, a condition termed high-level injection. The conductivity of the drift region will thus be greatly enhanced over its ohmic or low injection level.

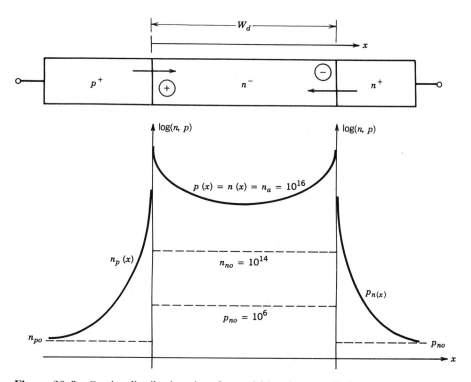

Figure 20-8 Carrier distributions in a forward-biased power diode structure with a lightly doped drift region. Note how the excess carriers are injected into the drift region from both ends. The average value of the excess carriers, n_a is large compared with n_{no}, the majority-carrier density in the drift region.

Under these conditions the current in the drift region can be written approximately as

$$I_F \approx \frac{q(\mu_n + \mu_p)n_a A V_d}{W_d} \tag{20-11}$$

where V_d/W_d is the average electric field in the conductivity-modulated region, A is the cross-sectional area of the diode, and V_d is the approximate voltage drop across the drift region. Now the excess carrier stored in the drift region during the on state can be related to the current I_F using a stored charge formulation (Q_F being the stored charge in the drift region) so that

$$I_F \approx \frac{qAW_d n_a}{\tau} \approx \frac{Q_F}{\tau} \tag{20-12}$$

Setting Eq. 20-11 equal to Eq. 20-12 yields

$$V_d \approx \frac{W_d^2}{(\mu_n + \mu_p)\tau} \tag{20-13}$$

Note that the voltage given by this equation is not the total forward-bias diode voltage. The total voltage $V = V_j + V_d$, where V_j is the voltage across the pn junction and can be estimated using an equation similar to Eq. 19-25.

The voltage V_d cannot be reduced to arbitrarily small values. Two mechanisms become active, that cause V_d to increase with increasing current density. First, as the excess-carrier density n_a gets large enough, on the order of 10^{17} cm^{-3}, the lifetime τ begins to decrease because of Auger recombination, as discussed in Chapter 19 (see Eq. 19-6). The lifetime reduction caused by this mechanism will cause the voltage drop across the drift region to increase substantially at large current densities where the carrier density is correspondingly large. Second, at about the same excess-carrier density, the carrier mobilities begin to decrease with increasing excess-carrier density, becoming inversely proportional to n_a, as is shown in Eq. 20-14a:

$$\mu_n + \mu_p = \frac{\mu_o}{1 + n_a/n_b} \tag{20-14a}$$

where μ_o is the low injection level value of $\mu_n + \mu_p$. This decrease occurs because the carrier densities are large enough that the free carriers collide with each other almost as often as they do with the crystalline lattice. These additional collisions reduce the mobilities just as do collisions with the lattice. This carrier–carrier scattering represents another mechanism that causes V_d to increase at large current densities.

Insertion of the density-dependent lifetime given by Eq. 19-6 into Eq. 20-12 yields

$$\frac{I_F}{A} = J_F = \frac{qW_d n_a}{\tau_o}\left(1 + \frac{n_a^2}{n_b^2}\right) \tag{20-14b}$$

Insertion of Eq. 20-14a for the density-dependent mobilities into Eq. 20-11 and solving for n_a as a function of V_d, J, and other parameters and using this resulting expression for n_a in Eq. 20-14b yield, after a few manipulations

$$\left(V_d - \frac{J_F W_d}{q\mu_o n_b}\right)^2\left(V_d - \frac{W_d^2}{\mu_o \tau_o}\right) = \frac{J_F^2 W_d^4}{q^2 \mu_o^3 n_b^2 \tau_o} \tag{20-15}$$

As n_a approaches n_b, V_d will become larger than $W_d^2/\mu_o \tau_o$ so that Eq. 20-15 becomes approximately

$$V_d \approx \frac{J_F W_d}{q\mu_o n_b} + \sqrt[3]{\frac{J_F^2 W_d^4}{q^2 \mu_o^3 n_b^2 \tau_o}} \tag{20-16}$$

Thus at large currents where J_F is large, the drift region voltage has an ohmic-like dependence upon the current density J_F. Thus it is not surprising that the total on-state diode voltage drop is approximately given by

$$V \approx V_j + R_{on}I \tag{20-17}$$

as is indicated in Fig. 20-2.

20-4-2 IMPACT ON ON-STATE LOSSES

This simplified analysis summarizes several important features about the on-state losses of not only diodes but also other minority-carrier-based devices such as bipolar junction transistors and thyristors. First, if the lifetime can be made large enough so that the diffusion length L is compatible with the drift region length W_d, then the voltage drop across the drift region can be made quite small and approximately independent of the current. This means that the power dissipation in bipolar devices will be much less than for majority carrier devices such as MOSFETs or JFETs carrying about the same current density. Second, the price of the reduced on-state losses is a large amount of stored charge, which will compromise the switching times, a subject to be examined in the next section. Third, the larger the breakdown voltage of the device, the larger the voltage drop V_d across the drift region will be since this voltage is proportional to the square of the length W_d, and this length must be larger for larger breakdown voltages.

A natural question to ask at this point is how much lower are the on-state losses in a bipolar device compared with a majority-carrier device. One way to address the question is to estimate the current density that can flow in the drift region of both types of devices as a function of the breakdown voltage rating of the device and the maximum desired voltage drop V_d across the drift region. The drift region is chosen for comparison because in both types of devices, this is where most of the on-state losses will occur. The current density for a minority-carrier device can be developed by using Eq. 20-11 and expressing the drift region length W_d in terms of the breakdown voltage BV_{BD} using Eq. 20-3. Using this with $\mu_n + \mu_p \approx 900$ cm^2/V-s (the approximate value at excess-carrier densities of 10^{17} cm^{-3}) yields

$$J(\text{minority}) \approx 1.4 \times 10^6 \frac{V_d}{BV_{BD}} \tag{20-18}$$

In the drift region of a majority-carrier device, the current density J(majority) can be written as

$$J(\text{majority}) \approx \frac{q\mu_n N_d V_d}{W_d} \tag{20-19}$$

where the drift region is assumed to be n-type in order to take advantage of the higher mobility of electrons. Using Eq. 20-2 to express N_d in terms of BV_{BD} and Eq. 20-3 to express W_d in terms of BV_{BD} and setting $\mu_n \approx 1500$ cm^2/V-s, the value of μ_n in silicon at doping densities of 10^{14} cm^{-3}, yield

$$J(\text{majority}) \approx 3.1 \times 10^6 \frac{V_d}{BV_{BD}^2} \tag{20-20}$$

These equations clearly show that as breakdown voltages increase, bipolar devices and majority-carrier devices suffer reductions in their current-carrying capabilities. However, the reduction in bipolar devices is less severe compared to majority-carrier devices, and hence bipolar devices are generally the device of choice at larger blocking voltages (several hundred volts and larger).

20-5 SWITCHING CHARACTERISTICS

20-5-1 OBSERVED SWITCHING WAVEFORMS

A power diode requires a finite time to switch from the blocking state (reverse bias) to the on state (forward bias) and vice versa. The user must be concerned not only with the time required for the transitions but also with how the diode current and voltage vary during the transitions. Both the transition times and the shapes of the waveforms are affected by the intrinsic properties of the diode and by the circuit in which the diode is embedded.

The switching properties of a diode are often given on specification sheets for diode currents with a specified time rate of change, di/dt, as is shown in Fig. 20-9. The reason for this selection is that power diodes are very often used in circuits containing inductances that control the rate of change of the current, or the diodes are used as free-wheeling diodes where the turn-off of a solid-state device controls di/dt. The resulting diode voltage and current versus time are shown in Fig. 20-9. The features of particular interest in these waveforms are the voltage overshoot during turn-on and the sharpness of the fall of the reverse current during the turn-off phase. The overshoot of the voltage during turn-on is not observed with signal-level diodes.

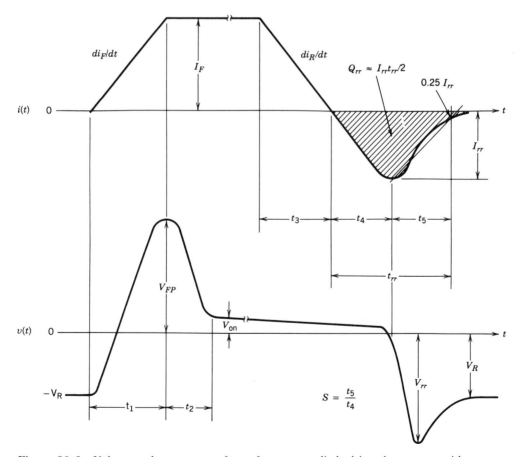

Figure 20-9 Voltage and current waveforms for a power diode driven by currents with a specified rate of rise during turn-on and a specified rate of fall during turn-off.

20-5-2 TURN-ON TRANSIENT

The turn-on portion of the diode waveforms in Fig. 20-9 is encompassed by the times labeled t_1 and t_2. During these intervals two physical processes occur in sequence. First the space charge stored in the depletion region (located mainly in the drift region) because of the large reverse-bias voltage is removed (discharged) by the growth of the forward current. When the depletion layer is discharged to its thermal equilibrium level, the metallurgical junction becomes forward biased and the injection of excess carriers across the junction into the drift region commences at time t_1, thus marking the start of the second phase and the end of the first. During the second phase, the excess-carrier distribution in the drift region grows toward the steady-state value that can be supported by the forward diode current I_F. The approximate growth of the excess-carrier distribution in time is diagrammed in Fig. 20-10. Note that excess carriers are injected into the drift region from both ends with holes being injected from the p^+n^- junction and electrons from the n^+n^- junction.

A simple interpretation of this sequence of events would lead one to expect that the diode voltage would rise smoothly and monotonically from its initial large negative value to a steady-state forward-bias value of about 1.0 V. Only one distinct time interval would be observed, which would be the discharge of the space charge layer, which is analogous to the discharge of a capacitance. Indeed, the depletion layer under reverse-bias conditions is often modeled as a capacitor (space charge capacitance) whose value (per square centimeter using a parallel plate capacitor formalism) is given by

$$C_{sc}(V) \approx \frac{\epsilon}{W(V)} \tag{20-21}$$

where the depletion layer with width $W(V)$ is given by Eq. 19-6. The interval duration should scale with I_F and inversely with di/dt.

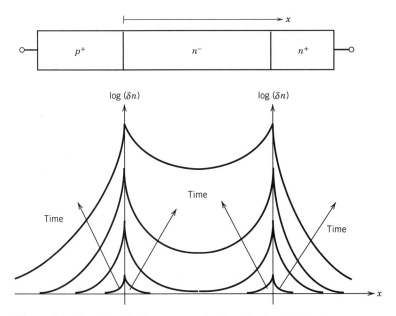

Figure 20-10 Growth of excess-carrier distribution during the turn-on of a power pn-junction diode. Note that the carriers are injected into the drift region from both ends.

However, this interpretation does not account for the voltage overshoot shown in the waveform in Fig. 20-9. The shortcoming of the interpretation is that it fails to consider the effect of the ohmic resistance of the drift region and the inductance of the silicon wafer and of the bonding wires attached to it. As the forward current grows in time, there is an increasingly large voltage drop across the drift region, since there is no conductivity modulation of the region until the space charge layer is discharged to its thermal equilibrium value. The inductance also adds a significant voltage drop if large values of di/dt are applied. The combined effect of these two factors is an overshoot that can be as large as several tens of volts, which is large enough to seriously affect the operation of some power electronic circuits.

The growth of the diode voltage slows and eventually turns over as the drift region becomes shorted out by the large amount of carrier injection into it. In addition, the inductive contribution ends when the diode current stabilizes at I_F. The interval during which the voltage falls from the peak overshoot value to the steady-state forward value marks the completion of the transient growth of the excess charge distribution in the drift region.

The duration of the space charge layer discharge and the growth of the excess-carrier distribution in the drift region is governed by both the intrinsic properties of the diode and the external circuit in which the diode is embedded. A large value of di/dt will minimize the time needed to discharge the space charge layer. However, a large value of I_F and of carrier lifetime in the drift region will lengthen the time needed for the excess-carrier portion of the transient to be completed. Typical values for these switching times in high-voltage diodes are in the hundreds of nanoseconds for t_1 and in the microsecond range for t_2. Devices with faster turn-on times are available, but their improved performance in this respect is achieved only by reducing the lifetime, as was explained previously. Thus, there is an inherent trade-off between shorter turn-on transients and higher on-state losses.

20-5-3 TURN-OFF TRANSIENT

The turn-off portion of the switching waveform is encompassed by the times labeled t_3, t_4, and t_5 and is essentially the inverse of the turn-on process. First the excess carriers stored in the drift region must be removed before the metallurgical junctions can become reverse biased. The decay of the excess-carrier distribution is illustrated in Fig. 20-11. Once the carriers are removed by the combined action of recombination and sweepout by negative diode currents, the depletion layer acquires a substantial amount of space charge from the reverse-bias voltage and expands into the drift region from both ends (junctions).

As long as there are excess carriers at the ends of the drift region, the p^+n^- and n^+n^- junctions must be forward biased. Thus, the diode voltage will be little changed from its on-state value except for a small decrease due to ohmic drops caused by the reverse current. But after the current goes negative and carrier sweepout has proceeded for a sufficient time (t_4) to reduce the excess-carrier density at one or both of the junctions to zero, the junction or junctions become reverse biased. At this point the diode voltage goes negative and rapidly acquires substantial negative values as the depletion regions from the two junctions expand into the drift region toward each other. At this time the negative diode current demanded by the stray inductance of the external circuit cannot be supported by the excess-carrier distribution because too few carriers remain. The diode current ceases its growth in the negative direction and quickly falls, becoming zero after a time t_5. The reverse current has its maximum reverse value, I_{rr}, at the end of the t_4 interval.

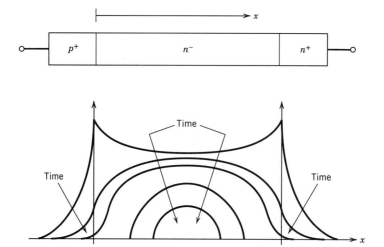

Figure 20-11   Decay of the excess-carrier distributions during turn-off of a power *pn*-junction diode. Note that the excess carriers decay to zero at the *pn* junction while substantial number of excess carriers remain in the central part of the drift region.

20-5-4   REVERSE RECOVERY

The time interval $t_{rr} = t_4 + t_5$ shown in Fig. 20-9 is often termed the reverse-recovery time. Its characteristics are important in almost all power electronic circuits where diodes are used. Diode specification sheets often give detailed plots of t_{rr}, reverse-recovery charge Q_{rr}, and "snappiness" factor S (all defined in Fig. 20-9) as functions of the time rate of change of the reverse current, di_R/dt. These quantities are all interrelated to each other and to other diode parameters such as breakdown voltage and on-state voltage drop across the drift region.

A useful quantitative description of these relationships can be obtained from the following considerations. From Fig. 20-9, we note that I_{rr} can be written as

$$I_{rr} = \frac{di_R}{dt}t_4 = \frac{di_R}{dt}\frac{t_{rr}}{S+1} \tag{20-22}$$

since $t_4 = t_{rr} - t_5 = t_{rr}/(S+1)$. From Fig. 20-9, $Q_{rr} \approx \frac{1}{2}I_{rr}t_{rr}$ so that

$$Q_{rr} = \frac{di_R}{dt}\frac{t_{rr}^2}{2(S+1)} \tag{20-23}$$

Solving Eq. 20-23 for the reverse-recovery time yields

$$t_{rr} = \sqrt{\frac{2Q_{rr}(1+S)}{di_R/dt}} \tag{20-24}$$

Using Eq. 20-24 in Eq. 20-22 yields

$$I_{rr} = \sqrt{\frac{2Q_{rr}(di_R/dt)}{S+1}} \tag{20-25}$$

The charge Q_{rr} represents the portion of the total charge Q_F (the charge stored in the diode during forward bias), which is swept out by the reverse current and not lost to internal recombination. Most of Q_F is stored in the drift region (especially in higher voltage diodes) and is given by $Q_F = \tau I_F$ (see Eq. 19-22 and 20-12). Since Q_{rr} must be less than Q_F (because of internal recombination in the diode), Eqs. 20-24 and 20-25 can be rewritten as

$$t_{rr} < \sqrt{\frac{2\tau I_F}{di_R/dt}} \tag{20-26}$$

$$I_{rr} < \sqrt{2\tau I_F \frac{di_R}{dt}} \tag{20-27}$$

In these last two equations we have made use of the observation that $S < 1$ in most diodes.

It was explained earlier that the diffusion length $L = \sqrt{D\tau}$ must be at least as large as the drift region length W_d to have small voltage drops across the drift region. This yields an expression for the lifetime using Eq. 19-9, which is

$$\tau = \frac{W_d^2}{(kT/q)\{\mu_n + \mu_p\}} \tag{20-28}$$

In this equation, the width of the drift region W_d must be at least as large as the depletion layer width at the diode breakdown voltage (because the drift region must contain the depletion layer). Setting W_d equal to $W(BV_{BD})$ using Eq. 20-3 and the result in Eq. 20-28 along with $\mu_n + \mu_p \approx 900$ cm^2/V-s yields

$$\tau \approx 4 \times 10^{-12} BV_{BD}^2 \tag{20-29}$$

Insertion of this expression for the carrier lifetime into Eqs. 20-26 and 20-27 gives

$$t_{rr} \approx 2.8 \times 10^{-6} BV_{BD} \sqrt{\frac{I_F}{di_R/dt}} \tag{20-30}$$

$$I_{rr} \approx 2.8 \times 10^{-6} BV_{BD} \sqrt{I_F di_R/dt} \tag{20-31}$$

In Eqs. 20-29 to 20-31 the times are in seconds, the currents in amperes, the voltages in volts, and the time derivative of current in amperes per second.

These last three equations are approximate estimates in several respects. First, they are based only on an approximate analysis of one type of diode, the abrupt junction. Second, it is assumed that the drift region width is the minimum allowable width given by Eq. 20-3. A larger value of W_d will make the estimates larger by the same amount. Third, they are based on the approximation that $Q_{rr} = Q_F$, a result that is most accurate for large values of di_R/dt (short values of t_{rr}), which minimizes the excess carriers lost to recombination in the diode. Thus, numerical estimates made with these equations may not be precise, but they do indicate the general trends. Most important, they summarize the trade-offs that must be made in the design of high-voltage pn-junction diodes between low on-state losses (small V_d), faster switching times (small carrier lifetime τ and short values of t_{rr}), and larger breakdown voltages BV_{BD}.

20-6 SCHOTTKY DIODES

20-6-1 STRUCTURE AND I–V CHARACTERISTICS

A Schottky diode is formed by placing a thin film of metal in direct contact with a semiconductor. The metal film is usually deposited on an n-type semiconductor as is shown in Fig. 20-12, although appropriate metal films on p-type material could also be

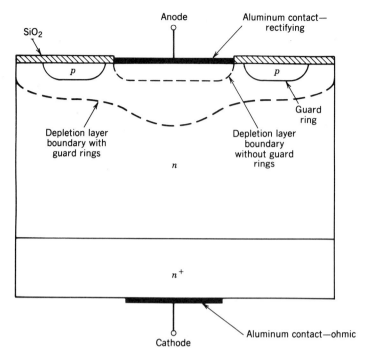

Figure 20-12 Cross-sectional view of a Schottky diode. A guard ring structure is also shown for improving the breakdown voltage capability of the diode.

used. In Fig. 20-12 the metal film is the positive electrode and the semiconductor is the cathode.

Such a structure has a rectifying $i-v$ characteristic very similar to that of a pn-junction diode. Although the fundamental physics of the Schottky diode is different than that of the pn junction, the $i-v$ characteristic of the Schottky diode can be expressed by the same equation as for the pn junction. Thus, the basic Schottky diode $i-v$ curve is

$$I = I_s[e^{qV/kT} - 1] \tag{20-32}$$

However, the on-state voltage is significantly lower, typically $0.3-0.4$ V, than that of a silicon diode. Thus, the Schottky diode may be preferable for use in some power applications such as those discussed in Chapter 10. In the reverse direction, the Schottky diode has a reverse leakage current that is larger than that of a comparable silicon pn-junction diode. The breakdown voltage of a Schottky diode at present cannot be made reliably larger than $100-200$ V.

20-6-2 PRINCIPLE OF OPERATION

The key to the operation of the Schottky diode is the fact that electrons in different materials have different absolute potential energies compared with electrons at rest in free space (the potential energies are lower in materials, indicating that the electrons are bound in the solid). Consider Fig. 20-12, where an n-type semiconductor is in contact with a metal whose electrons have a lower absolute potential energy than the electrons in the semiconductor. There is a flow of electrons in both directions across the metal–semiconductor interface when contact is first made. However, the flux of electrons from the semiconductor into the metal will be much larger because of the higher absolute potential energy of the electrons in the semiconductor. As a consequence, the metal will become negatively charged and the semiconductor will acquire a positive charge by forming a depletion region adjacent to the interface. The overall picture is quite similar to that shown in Figs. 19-8 and 19-9 except that the metal replaces the *p*-type side and the negative space charge comes from free electrons. The electrostatic potential barrier that accompanies the space charge region will grow in magnitude and oppose the continued flow of electrons from the semiconductor into the metal.

Eventually the potential barrier gets large enough so that the flux of electrons from the semiconductor into the metal is just equal to the flux from the metal to the semiconductor. At this point thermal equilibrium is established and there is no net current flow across the interface. Note that in establishing this equilibrium, no minority carriers, holes in this situation, were involved. Only majority carriers, electrons in this example, took part. This is the key difference between a Schottky diode and a *pn*-junction diode. For this reason, Schottky diodes are termed majority-carrier devices and *pn* junctions are labeled minority-carrier devices or bipolar devices, since they use both electrons and holes in their basic operation.

When a voltage is applied to the structure of Fig. 20-12 that biases the metal positive with respect to the semiconductor, it opposes the built-in potential and makes it easier for current to flow. Biasing the metal negative with respect to the semiconductor increases the potential barrier to majority-carrier current flow. Thus, the metal–semiconductor interface has rectifying characteristics similar to those of a *pn* junction. The major difference is that at any given forward current, the voltage across the Schottky diode is smaller than that across a *pn* junction. The difference amounts to roughly 0.3 V. The reason for the smaller voltage drop across the Schottky diode is that the reverse saturation current of a Schottky diode is significantly larger than that of a *pn*-junction diode of the same cross-sectional area. The details of why the Schottky diode reverse current is larger is beyond the scope of this discussion. Suffice it to say that the lower forward voltage of the Schottky means that it is less lossy than a conducting *pn*-junction diode.

20-6-3 OHMIC CONTACTS

The metal–semiconductor structure can form ohmic contacts (e.g., the cathode contact in Fig. 20-12) to semiconductor materials of all types. Such ohmic contacts are used in all types of semiconductor power devices. By an ohmic contact, we mean a contact that has no rectifying characteristics. The slope of the i–v characteristic of the contact is extremely steep (low resistance) and the same regardless of the voltage and current polarity.

The possibility of ohmic contacts utilizing a Schottky geometry arises because not all metals have electrons with lower absolute potential energies than electrons in a semiconductor. If such a metal is brought into contact with a semiconductor, an electric field and hence a potential barrier is set up across the interface that opposes the movement of electrons from the metal to the semiconductor. The source of the barrier is the accumulation of electrons in the semiconductor in the vicinity of the interface. This accumulation is extremely large and greatly increases the conductivity of the interface. The enhancement of the conductivity is so great that it obscures the nonlinearities (rectification) of the junction and makes the voltage drop across the interface proportional to the current through it. Naturally such a junction is termed an ohmic contact.

Even the cathode structure of Fig. 20-12 can be made into an ohmic contact if the doping in the *n* region is made very heavy. In this circumstance the depletion region that is set up is extremely narrow (examine Eqs. 20-2 and 20-3) and the electric field that is set up is very large, approaching impact ionization values. Under these circumstances, electrons move very easily across the interface under the influence of small applied voltages. The mechanism that leads to this is tunneling, a quantum-mechanical effect that is beyond the scope of this discussion. This is why the cathode end of the Schottky diode in Fig. 20-12 is an ohmic contact and not a rectifying contact.

20-6-4   BREAKDOWN VOLTAGE

The same factors that determine the breakdown voltage of a *pn*-junction diode also govern the breakdown voltage of a Schottky diode. In modeling the reverse-bias behavior of a Schottky diode including space charge capacitance and breakdown voltage, the Schottky is often modeled as a one-sided step junction, that is, a p^+–n–n^+ with the middle *n*-type region being a drift region whose width is determined by breakdown voltage considerations. However, the breakdown voltage of the Schottky diode cannot be reliably made larger than about 200 V. The reasons for this are threefold. First, the basic geometry of a Schottky diode leads to depletion layers, which have an extremely small radius of curvature at the edges of the contact metal, as illustrated in Fig. 20-12. This leads to electric field crowding and low breakdown voltages, as explained earlier in this chapter. The use of field plates can alleviate this problem to some degree.

Another contributing factor is the lower breakdown field strength of the silicon at the surface. Several factors, some of which are process related and others more fundamental in nature contribute to the lower breakdown electric field strength. Since the geometry of the Schottky diode places the depletion layer (where large electric fields occur) right at the silicon surface, the reduced breakdown voltages compared to *pn*-junction diodes are not surprising. Improvements in device processing and fabrication have reduced the surface imperfections and contaminants, which in turn have led to improvements in the breakdown voltage to the present-day levels.

A third contribution to the smaller breakdown voltages in Schottky diodes is the lack of any stored minority carriers that can short out the ohmic resistance of the drift region that must support the depletion region in reverse bias. This basic trade-off was discussed in Section 20-4-2 in conjunction with on-state losses, but it bears repeating here because keeping on-state losses in a Schottky diode within set limits means using doping densities much larger than say 10^{14} cm^{-3} and so the achievable breakdown voltage is circumscribed. In comparison, the on-state losses and breakdown voltage of a *pn*-junction diode can be designed much more independently of each other.

20-6-5   SWITCHING CHARACTERISTICS

A Schottky diode turns on and off faster than a comparable *pn*-junction diode. The basic reason is that Schottky diodes are majority-carrier devices and have no stored minority carriers that must be injected into the device during turn-on and pulled out during turn-off.

The lack of any significant stored charge changes the shape of observed switching waveforms in important ways. During turn-off, there will be no reverse current associated with removal of stored charge. However, reverse current, associated with the growth of the depletion layer charge in reverse bias, will flow. This current may be comparable to the reverse current observed during switching of a *pn* junction because the space charge capacitance of a Schottky diode is larger (by as much as a factor of 5) than in a comparable *pn* junction. The reason is that the depletion layer in a Schottky diode is thinner than that of a *pn* junction because of the heavier doping used in the *n* region of the Schottky to keep the ohmic losses under control.

Schottky diodes have much less voltage overshoot during device turn-on than comparable *pn*-junction diodes. The basic reason is that the ohmic resistance of the drift regions in a Schottky diode must be made much less than that of a *pn*-junction diode in order to carry the same forward current because there is no excess-carrier injection to short out high-resistivity drift regions. Some voltage overshoot associated with parasitic inductance will be observed if *di/dt* is large.

SUMMARY

This chapter has explored the characteristics of *pn*-junction diodes intended for power applications. The characteristics of power Schottky diodes were also briefly discussed. The important points are as follows:

1. Power diodes are constructed with a vertically oriented structure that includes a n^- drift region to support large blocking voltages.
2. The breakdown voltage is approximately inversely proportional to the doping density of the drift region, and the required minimum length of the drift region scales with the desired breakdown voltage.
3. Achievement of large breakdown voltages requires special depletion layer boundary-shaping techniques.
4. Conductivity modulation of the drift region in the on state keeps the losses in the diode to manageable levels even for large on-state currents.
5. Low on-state losses require long carrier lifetimes in the diode drift region.
6. Minority-carrier devices have lower on-state losses than majority-carrier devices such as MOSFETs at high blocking voltage ratings.
7. During the turn-on transient the forward voltage in a diode may have a substantial overshoot, on the order of tens of volts.
8. Short turn-off times require short carrier lifetimes, so a trade-off between switching times and on-state losses must be made by the device designer.
9. During turn-off, fast reverse recovery may lead to large voltage spikes because of stray inductance.
10. The problems with the reverse-recovery transient are most severe in diodes with large blocking voltage ratings.
11. Schottky diodes turn on and off faster than *pn*-junction diodes and have no substantial reverse-recovery transient.
12. Schottky diodes have lower on-state losses than *pn*-junction diodes but also have low breakdown voltage ratings, rarely exceeding 100 V.

PROBLEMS

20-1 The silicon diode shown in Fig. 20-1 is to have a breakdown voltage of 2500 V. Estimate what the doping density of the drift region should be and what the minimum width of the drift region should be. The diode is a non-punch-through device.

20-2 A silicon diode similar to that shown in Fig. 20-1 has a drift region doping density of 5×10^{13} cm^{-3} donors and a drift region width of 50 μm. What is the breakdown voltage?

20-3 The diode in Problem 20-2 has a cross-sectional area of 2 cm^2 and carrier lifetime τ_0 of 2 μs. Approximately sketch and dimension the on-state voltage including the junction drop versus forward current. Consider currents as large as 3000 A.

20-4 The diode of Problem 20-3 is forward biased by 1000 A current that rises at the rate of 250 A/μs.

   (a) Assuming that no carrier injection takes place until the current reaches its steady-state value, sketch and dimension the forward voltage for $0 < t < 4$ μs.

   (b) Now assume that carrier injection commences at $t = 0$ and that this causes the drift region to drop linearly in time from its ohmic value at $t = 0$ to its on-state value at $t = 4$ μs. Sketch and dimension the forward voltage for $0 < t < 4$ μs.

20-5 A silicon diode with a breakdown voltage of 2000 V that is conducting a forward current of 2000 A is turned off with a constant $di_R/dt = 250$ A/μs. Roughly estimate the time required for the diode to turn off.

20-6 A Schottky diode having the p^+–n–n^+ structure shown in Fig. 20-12 is to be designed to have a breakdown voltage of 150 V. What should be the donor doping density in the drift region and what should be the length of the drift region? (*Hint:* Recall that a Schottky diode can be modeled as a one-sided step junction.)

20-7 Consider a Schottky diode that has an n-type drift region with a donor doping density of 10^{15} cm^{-3} and a drift region length of 20 μm. The diode is to carry 100 A of current in the on state with a maximum drift region voltage drop of 2 V. What should the cross-sectional area of the diode be?

20-8 A punch-through geometry is to be used for a power pn-junction diode. The breakdown voltage is to 300 V and the drift region length is to be 20 μm. What should the doping level be in the n-type drift region? You may assume that the pn junction is a p^+n^- junction (i.e., a one-sided step junction).

20-9 Does a punch-through diode of a specified breakdown voltage BV_{BD} have a larger or smaller value of drift region ohmic resistance than a non-punch-through diode having the same breakdown voltage and cross-sectional area? Answer the question quantitatively by assuming one-sided step junction doping profiles for the diodes. This question is of particular importance in minimizing the on-state power losses in majority-carrier devices such as Schottky diodes. [*Hint:* The n-type drift region resistance per unit area $R/A = W_d/q\mu_n N_d$, where W_d is the drift region length. For a non-punch-through diode use Eqs. 20-1 and 20-3 for $W(BV_{BD})$, the space charge width at breakdown, and N_d and assume $W_d = W(BV_{BD})$. For a punch-through diode use Eq. 20-9 to find W_d, which then gives $W_d(N_d)$ for a fixed BV_{BD}. Use $W_d(N_d)$ in the expression for resistance and find the N_d value that minimizes the resistance.]

20-10 A pn-junction diode (step junction) and a Schottky diode are to both have a breakdown voltage of 150 V and a drift region voltage drop of 2 V when carrying a rated current of 300 A. What is the zero-bias space charge capacitance C_{jo} of each diode? Assume a contact potential of 0.7 V for each diode and an excess-carrier lifetime of 100 ns in the pn-junction diode drift region.

20-11 The breakdown voltage of a cylindrical abrupt junction (with $N_a \gg N_d$) is given by

$$BV_{cyl} = BV_{pp}\{2\,\rho^2(1 + \rho^{-1})\ln[1 + \rho^{-1}] - 2\,\rho\}$$

where BV_{cyl} = cylindrical junction breakdown voltage

BV_{pp} = plane parallel junction breakdown voltage = $\dfrac{\epsilon E_{BD}{}^2}{2qN_d}$

$\rho = \dfrac{R}{2W_n}$; R = radius of cylindrical p-type region

$W_n = \dfrac{\epsilon E_{BD}}{qN_d}$ = depletion layer thickness (at breakdown) of plane parallel abrupt junction having same doping levels as cylindrical junction.

Plot $\dfrac{BV_{cyl}}{BV_{pp}}$ versus ρ for $0.2 < \rho < 10$.

20-12 How large must R be if $BV_{pp} = 1000$ V and BV_{cyl} is to be 950 V?

REFERENCES

1. W. McMurray, "Optimum Snubbers for Power Semiconductors," IEEE Trans. on Indus. Appl. Vol. IA8, No. 5, pp. 503-510, 1972

2. *SCR Manual*, 6th ed., General Electric Company, Syracuse, NY, 1979.

3. M. H. Rashid, *Power Electronics: Circuits, Devices, and Applications,* Prentice-Hall, Englewood Cliffs, NJ, 1988, Chapter 15.

4. B. G. Streetman, *Solid State Electronic Devices,* 2nd ed., Prentice-Hall, Englewood Cliffs, NJ, 1980, Chapter 5.

5. S. K. Ghandhi, *Semiconductor Power Devices,* Wiley, New York, 1977, Chapters 2–3.

6. B. M. Bird and K. G. King, *An Introduction to Power Electronics,* Wiley, New York, 1983, Chapter 6.

7. M. S. Adler and V. A. K. Temple, "Analysis and Design of High-Power Rectifiers," *Semiconductor Devices for Power Conditioning,* R. Sittig and P. Roggwiller Ed., Plenum, New York, 1982.

8. R. J. Grover, "Epi and Schottky Diodes," *Semiconductor Devices for Power Conditioning,* R. Sittig and P. R. (Eds.), Plenum, New York, 1982.

9. B. W. Williams, *Power Electronics, Devices, Drivers, and Applications,* Wiley, New York, 1987, Chapters 1–4.

CHAPTER 21

BIPOLAR JUNCTION TRANSISTORS

21-1 INTRODUCTION

The need for a large blocking voltage in the off state and a high current-carrying capability in the on state means that a power bipolar junction transistor (BJT) must have a substantially different structure than its logic-level counterpart. The modified structure leads to significant differences in the $i-v$ characteristics and switching behavior between the two types of devices. In this chapter these and other topics will be explored for both power BJTs and monolithic Darlington-connected devices.

21-2 VERTICAL POWER TRANSISTOR STRUCTURES

A power transistor has a vertically oriented four-layer structure of alternating p-type and n-type doping such as the npn transistor shown in Fig. 21-1. The transistor has three terminals, as is indicated in the figure, and they are respectively labeled collector, base, and emitter. In most power applications, the base is the input terminal, the collector is the output terminal, and the emitter is common between input and output (the so-called common emitter configuration). The circuit symbol for the BJT is shown in the same figure. A pnp transistor, whose circuit symbol is also shown in Fig. 21-1, would have the opposite type of doping in each of the layers shown in the figure. npn transistors are much more widely used than pnp transistors as power switches.

The vertical structure is preferred for power transistors because it maximizes the cross-sectional area through which the current in the device is flowing. This minimizes the on-state resistance and thus the power dissipation in the transistor. In addition, having a large cross-sectional area minimizes the thermal resistance of the transistor, thus also helping to keep power dissipation problems under control.

The doping levels in each of the layers and the thickness of the layers have a significant effect on the characteristics of the device. The doping in the emitter layer is quite large (typically 10^{19} cm^{-3}), whereas the base doping is moderate (10^{16} cm^{-3}). The n^- region that forms the collector half of the C–B (collector–base) junction is usually termed the collector drift region and has a light (10^{14} cm^{-3}) doping level. The n^+ region that terminates the drift region has a doping level similar to that found in the emitter. This region serves as the collector contact to the outside world. The thickness of the drift region

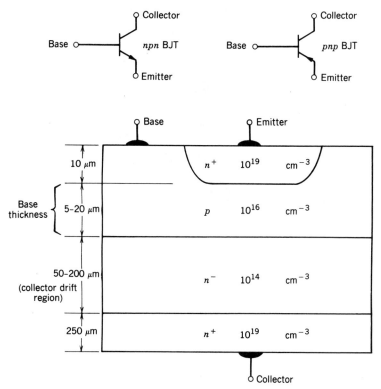

Figure 21-1   Vertical cross section of a typical *npn* power BJT. The circuit symbol for the transistor is also shown.

determines the breakdown voltage of the transistor and thus can range from tens to hundreds of micrometers in extent. The base thickness is made as small as possible in order to have good amplification capabilities, as will be explained in later sections. However, if the base thickness is too small, the breakdown voltage capability of the transistor is compromised, as will be explained. Thus, base thicknesses in power devices are a compromise between these two competing considerations and are typically several micrometers to a few tens of micrometers in thickness, compared with the small fraction of a micrometer in thickness for logic-level transistors.

Practical power transistors have their emitters and bases interleaved as narrow fingers, as is shown in Fig. 21-2. The purpose of this arrangement is principally to reduce the effects of current crowding, a phenomenon that can lead to second breakdown and possible device failure. These topics will be considered later in this chapter. This multiple-emitter layout also reduces the parasitic ohmic resistance in the base current path, which helps to reduce power dissipation in the transistor.

The relatively thick base found in power transistor structures causes the current gain, $\beta = I_C/I_B$, to be rather small, typically 5–10. This is undesirably small for some applications and, hence, monolithic designs for Darlington-connected BJT pairs shown in Fig. 21-3 have been developed. The current gain of a Darlington pair is given by

$$\beta = \beta_M\beta_D + \beta_M + \beta_D \qquad (21\text{-}1)$$

so that even though each individual transistor has a small beta, the effective beta of the pair can still be quite large. The vertical cross section of a monolithic Darlington is shown

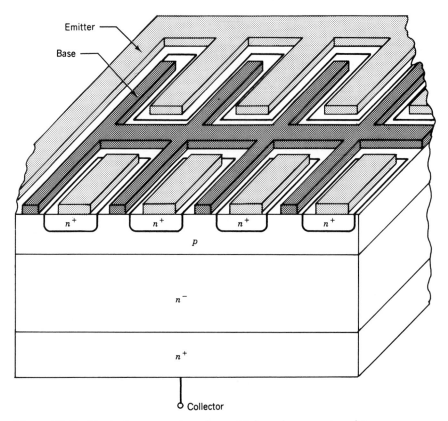

Figure 21-2   Vertical cross section of a multiple-emitter *npn* transistor.

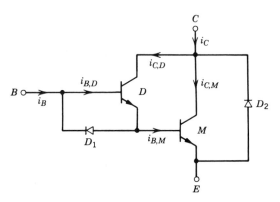

Figure 21-3   Power transistors in a Darlington configuration in order to obtain a larger effective current gain beta. The discrete diodes are added to aid turn-off (D_1) and for full-bridge applications (D_2).

in Fig. 21-4. A discrete diode D_1 is added, as shown in Fig. 21-3, to speed up the turn-off time of the main transistor, as will be explained shortly. The discrete diode D_2, also shown in Fig. 21-3, is added for half- and full-bridge circuit applications.

21-3   *I–V* CHARACTERISTICS

The output characteristics (i_C versus v_{CE}) of a typical *npn* power transistor are shown in Fig. 21-5. The various curves are distinguished from each other by the value of the base

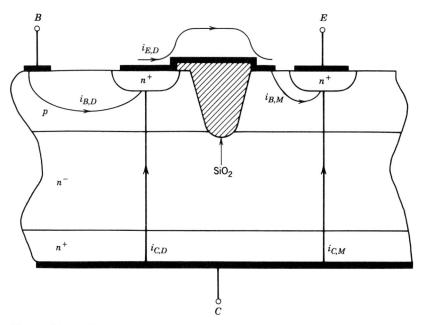

Figure 21-4 Vertical cross section of a pair of monolithic Darlington-connected bipolar transistors. The silicon dioxide protrusion through the upper *p*-layer (the base region of both transistors) electrically isolates the two bases from each other.

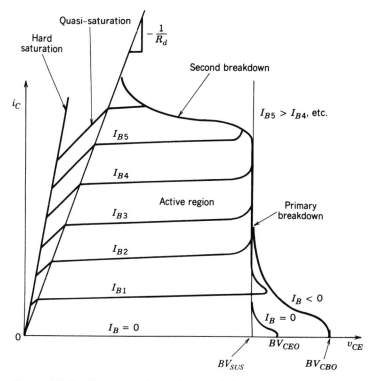

Figure 21-5 Current–voltage characteristics of an *npn* power BJT showing second breakdown and quasi-saturation.

current. The characteristics of monolithic Darlingtons are quite similar to those shown in the figure. Several features of the characteristics should be noted. First, there is a maximum collector–emitter voltage that can be sustained across the transistor when it is carrying substantial collector current. This voltage is usually labeled BV_{SUS}. In the limit of zero base current, the maximum voltage between collector and emitter that can be sustained increases somewhat to a value labeled BV_{CEO}, the collector–emitter breakdown voltage when the base is open circuited. This latter voltage is often used as the measure of the transistor's voltage standoff capability because usually the only time the transistor will see large voltages is when the base current is zero and the BJT is in cutoff. The voltage BV_{CBO} is the collector–base breakdown voltage when the emitter is open circuited. The fact that this voltage is larger than BV_{CEO} is used to advantage in so-called open-emitter transistor turn-off circuits.

The region labeled primary breakdown is due to conventional avalanche breakdown of the C–B junction and the attendant large flow of current. This region of the characteristics is to be avoided because of the large power dissipation that clearly accompanies such breakdown. The region labeled second breakdown must also be avoided because large power dissipation also accompanies it, particularly at localized sites within the semiconductor. The origin of second breakdown is different from that of avalanche breakdown and will be considered in detail later in this chapter. Bipolar junction transistor failure is often associated with second breakdown.

The major observable difference between the $i–v$ characteristics of a power transistor and those of a logic-level transistor is the region labeled quasi-saturation on the power transistor characteristics of Fig. 21-5. As we will explain in detail in later sections of this chapter, quasi-saturation is a consequence of the lightly doped collector drift region found in the power transistor. Logic-level transistors do not have this drift region and so do not exhibit quasi-saturation. Otherwise all of the major features of the power transistor characteristic are also found on those of logic-level devices.

21-4 PHYSICS OF BJT OPERATION

21-4-1 BASIC GAIN MECHANISM AND BETA

An understanding of how the BJT provides current (power) amplification is most easily obtained by considering the simplified one-dimensional transistor structure shown in Fig. 21-6a. In this model, which is essentially the structure of a logic-level transistor, there is no lightly doped collector drift region. It is further assumed that the transistor is in the active region. In the active mode of BJT operation, the drift region does not play a major role, and retaining it would needlessly complicate the discussion. The effect of the drift region will be considered in detail when breakdown voltage, on-state losses, and switching times are considered since the presence of the drift region has a significant effect on these items but not on the active mode of operation.

In the active region, the B–E (base–emitter) junction is forward biased and the C–B junction is reverse biased. Electrons are injected into the base from the emitter and holes are injected from the base into the emitter. This produces the minority-carrier distributions shown in Fig. 21-6b. These distributions have large density gradients, especially in the base region, which support significant diffusion currents. In fact, the total current flowing across the B–E junction will be almost entirely diffusion current, the same as for the *pn*-junction diode described in Chapter 19. Unlike the forward-biased *pn*-junction diode of Chapter 19, the base current entering the *p*-type side of the B–E junction from the B–E junction bias source will not equal the current (emitter current) that leaves the *n*-type side.

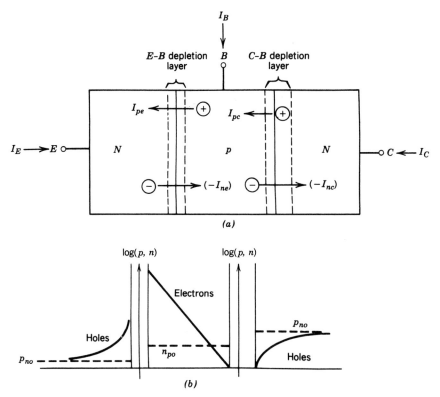

Figure 21-6 (a) Simplified model of a BJT. (b) Stored charge distribution that exists in the BJT biased in the normal active region. The internal current components that flow in the active region are also shown on the model.

The currents will be unequal because the transistor structure provides an alternative besides the base terminal for electrons injected into the base from the emitter to exit the region.

Electrons injected into the base from the emitter are most likely to exit the base via the collector rather than the base terminal for three reasons. First, the thickness of the base region is made quite small compared with the electron diffusion length $L_{nb} = (D_{nb}\tau_{nb})^{1/2}$ in the base so they are unlikely to recombine there. Second, the area of the collector is made much larger than that of the emitter or the base contact, as is shown in Fig. 21-1, so that the electrons diffusing away from the emitter are much more likely to encounter the collector than anything else because of the short distance between emitter and collector. Third, the density of electrons at the C–B junction is essentially zero, as is shown in Fig. 21-6b, because the high electric fields in the reverse-biased C–B junction sweep all the diffusing electrons at the edge of the space charge region across the junction and into the collector region. The large density of injected electrons at the B–E junction and essentially zero excess electrons at the collector side of the base means that a very large gradient of electrons exist in the base, as shown in Fig. 21-6b. This density gradient carries most of the injected electrons and very few of them exit the base region via the base lead.

This means that the base current will be much less than the emitter current, and the collector current will almost equal the emitter current. A small base current causes the flow of a much larger current between the collector and emitter, and thus a substantial amount of gain is obtained between the input base current and the output collector current. This is the basic gain mechanism of the BJT. The gain is quantitatively characterized by the ratio of the collector current i_C to the base current i_B (beta or β of the transistor).

The features of the transistor structure that lead to large values of beta can be more clearly understood by considering the currents that flow internally in the transistor. These currents can be conveniently divided into the four components shown in Fig. 21-6a. The current I_{pe} is the diffusion current that originates from holes injected from the base into the emitter to sustain the hole distribution in the emitter. Similarly, I_{ne} is the diffusion current that originates from electron injection from the emitter into the base in order to support the electron distribution in the base. As is explained in the preceding paragraphs, these electrons then diffuse across the base, and those that survive recombination arrive at the C–B depletion layer. These excess electrons, along with a much smaller number that are thermally generated in the depletion layer, are swept across the depletion layer by the large electric fields into the collector layer. This flow of electrons is the I_{nc} current. The current I_{pc} arises from holes that are thermally generated in the C–B depletion layer and then are swept into the base layer by the large electric fields in the depletion layer. It is much smaller than the other current components because the hole density in the C–B space charge layer is much less than the carrier densities in the other regions. Hence, I_{pc} will be neglected in all further discussions.

The terminal currents of the transistor, I_C and I_B, can be expressed in terms of these internal currents. The collector current is given by

$$I_C = I_{nc} \tag{21-2}$$

and the base current is given by

$$I_B = -I_C - I_E = -I_{nc} + I_{ne} + I_{pe} \tag{21-3}$$

Beta can be expressed in terms of the internal currents as

$$\frac{I_B}{I_C} = \frac{1}{\beta} = \frac{I_{ne} - I_{nc}}{I_{nc}} + \frac{I_{pe}}{I_{nc}} \tag{21-4}$$

In order for beta to be large, the numerators of the two terms in Eq. 21-4 must be small compared with I_{nc}. The I_{pe} term can be minimized by doping the emitter very heavily so that the stored hole distribution there is made small (see Eq. 19-23). The term $I_{ne} - I_{nc}$ represents the difference between the electrons injected into the base at the B–E junction and those swept across the C–B junction into the collector. This difference is caused by the recombination of some of the injected electrons in the base region, and it is minimized by having a large electron lifetime in the base region and by making the base thickness small (fractions of a micrometer in logic-level BJTs) compared with the electron diffusion length (see Eq. 19-19).

In summary, there are three prime requirements for large values of beta in a BJT. These are (1) heavy doping of the emitter, (2) long minority-carrier lifetimes in the base, and (3) short base thicknesses. These factors will conflict with other characteristics desired for the transistor, and hence, a trade-off will be required between large gain and other parameters such as fast switching times. The consequence of these trade-offs is that the base thickness in a power transistor is larger than in a logic-level transistor and the beta of power transistors is typically 5–20.

A feature of BJTs including power transistors not predicted by the foregoing discussion is the fall-off of the current gain at collector current values larger than some value that is characteristic for a specific type of transistor. This fall-off, which is illustrated in Fig. 21-7, commences at currents of less than 1 A in logic-level transistors and at current levels of 100 A in some power devices. Several different mechanisms operative in the transistor

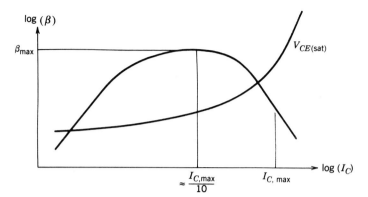

Figure 21-7   Variation in the BJT current gain β and $V_{CE(\text{sat})}$ as a function of the dc collector current showing the fall-off of beta and increase in $V_{CE(\text{sat})}$ at large collector currents.

simultaneously contribute to this fall-off in beta, of which two of the most significant are conductivity modulation in the base and emitter current crowding. Conductivity modulation of the base is essentially the same as the conductivity modulation of the diode drift region discussed in the previous chapter. In the case of the BJT base layer, conductivity modulation occurs when the minority-carrier density in the base becomes comparable to the majority-carrier doping density. For example, in the BJT diagram of Fig. 21-1, high-level injection in the base would occur when the excess electron density reaches about 10^{16} cm^{-3}. When the electron density gets this large, excess holes of the same density must also be injected into the base, and the only way for this to occur is for the base current to supply them. This represents an increase in the base current without a similar increase in the collector current and, hence, a fall in the value of beta. At the larger values of collector current, beta is approximately inversely proportional to the collector current.

Emitter current crowding is another mechanism that causes a decrease in beta. Consider the simplified BJT cross section shown in Fig. 21-8a, where both the base current and collector current flow paths are shown, assuming the BJT is in the active region. Because of the device geometry, there is a lateral ohmic voltage drop in the base region, as indicated in the figure, which is caused by the lateral flow of base current. This lateral ohmic voltage drop subtracts from the externally applied B–E voltage, and this means that the voltage drop across the B–E junction is larger at the emitter periphery near the base contact than it is in the center of the emitter area. This in turn causes a larger current density to flow across the junction at the emitter edge near the base terminal compared with the current density in the center of the emitter area. The current crowding clearly will mean that the onset of high-level injection and the attendant reduction in beta will occur at lower total currents than if the current density were uniformly spread over the entire emitter area. As we mentioned earlier, modern power BJTs have their emitters separated into many narrow rectangular areas, as is shown in Fig. 21-2, to minimize current crowding.

21-4-2   QUASI-SATURATION

To understand the phenomenon of quasi-saturation, the one-dimensional model of the BJT is now generalized to include a collector drift region, as is shown in Fig. 21-9. As in the previous section, it is assumed that the transistor is initially in the active region and now

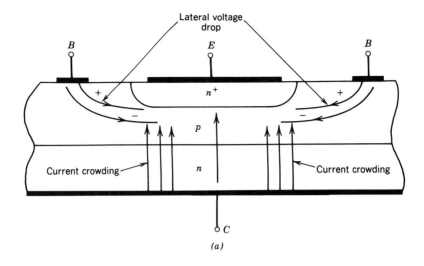

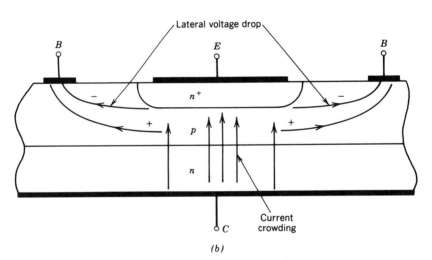

Figure 21-8 (*a*) Illustration of emitter current crowding in forward bias and (*b*) reverse bias (turn-off transient) caused by lateral voltage drops induced by large base currents.

the base current is allowed to increase. As the collector current rises in response to the base current, the C–E voltage drops because of the increased voltage drop across the collector load. However, there is a simultaneous increase in the voltage drop in the drift region as a result of its ohmic resistance because of the increase in i_C. This means that the reverse bias across the actual C–B junction, the n^-p junction, is getting smaller and at some point the junction will become forward biased.

When this occurs, injection of holes from the base into the collector drift region commences. At the same time, space charge neutrality requires that electrons also be injected into the drift region in about the same numbers as the holes. These electrons are conveniently obtained from the very large number of electrons being supplied to the C–B junction via injection from the emitter and subsequent diffusion across the base. As this excess carrier build-up in the drift region beings to occur, the quasi-saturation region of the i–v characteristic is entered. If the ohmic resistance of the drift region is R_d, then

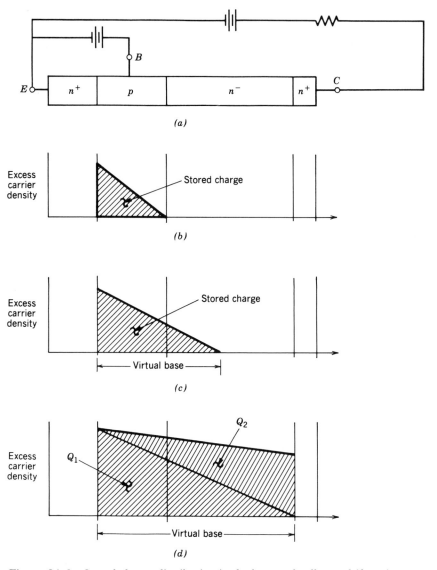

Figure 21-9 Stored charge distribution in the base and collector drift regions of a power BJT: (a) power transistor cross section; (b) active; (c) quasi-saturation; (d) hard saturation (Q_1 is the amount of stored charge that puts the BJT at the edge of hard saturation and Q_2 drives the transistor deeper into hard saturation).

the boundary between the quasi-saturation region and the active region in Fig. 21-5 is given by

$$i_C = \frac{v_{CE}}{R_d} \tag{21-5}$$

In quasi-saturation, double injection is occurring in the drift region in a manner similar to that in the drift region of the forward-biased power diode. However, the stored charge accumulates in the drift region from only one side of the drift region, the C–B junction side (or pn^- side), as is diagrammed in Fig. 21-9. In the transistor, electron injection across the n^-n^+ junction is much less noticeable because there is a much more plentiful supply of electrons at the pn^- junction (due to electrons injected from the emitter, as discussed in the previous paragraph) compared to the situation in the pn-junction diode where there is no such supply of electrons. As the injected carriers increase, the drift region is gradually shorted out and the voltage across the drift region drops even though the collector current is large. It is also apparent from Fig. 21-9c that as the hole injection from the base across the C–B junction commences, the thickness of the effective or virtual base is increasing. This means that the effective value of beta decreases and, hence, the collector current magnitude that a given base current can support must also decrease, as diagrammed in the $i-v$ characteristics of Fig. 21-5. In quasi-saturation, the drift region is not completely shorted out by high-level injection; hence, the power dissipation in the BJT is larger than when hard saturation is entered.

Hard saturation is obtained when the excess-carrier density reaches the other side (n^+ side) of the drift region, as is diagrammed in Fig. 21-9d. This requires a minimum amount of stored charge Q_1, which is indicated on the figure. In this situation, the effective base thickness is approximately the sum of the normal base thickness plus the length of the drift region. Any additional stored charge, such as Q_2, illustrated also on Fig. 21-9d, will drive the transistor deeper into hard saturation. The voltage drop across the drift region is small, roughly given by Eq. 20-16, and the on-state power dissipation is minimized compared to quasi-saturation.

21-5   SWITCHING CHARACTERISTICS

21-5-1   BJT TURN-ON

From the basic description of how the transistor works given in the previous section, we know that to switch the transistor from the off state to the on state, charge must be supplied to the transistor so that stored charge distributions similar to those shown in Fig. 21-9 are established and maintained in the transistor. The characteristics of the transistor and of the external circuit in which the device is embedded interact to determine just how fast the stored charge can be injected and, thus, how fast the device can turn on. To make this interaction as clear as possible, we shall assume that the BJT is embedded in the diode-clamped circuit shown in Fig. 21-10 that arises in the converters discussed in Chapters 7 and 8.

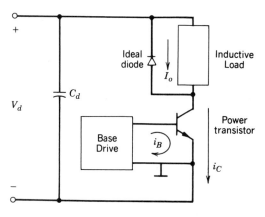

Figure 21-10 Inductively loaded BJT switching circuit with a free-wheeling diode clamp. The L/R time constant in the inductive load is large compared with the switching frequency so that it approximates a constant current source I_o. Note that the power BJT is a four-terminal device having two emitter leads. one for the large collector current and the other for the base current.

The external circuit determines the collector current that can flow in the on state. This value of collector current together with the carrier lifetimes in the transistor, particularly in the collector drift region, determines what minimum amount of stored charge must be maintained in the BJT in order that it be on. The current gain of the transistor then establishes what minimum base current must be provided to the device in order to establish and maintain this stored charge distribution. Forward-bias base currents in excess of this minimum amount will build up the stored charge distributions faster and thus shorten the switching times from the off state to the on state. However, such a base current overdrive will build up the stored charge to values larger than that needed to just maintain hard saturation.

The approximate manner in which the stored charge distribution grows during turn-on is shown in Fig. 21-11 for a forward-bias base current applied at $t = 0$. The input voltage that drives the base current, the resulting collector current, and other transistor voltages and currents of interest as functions of time are shown in Fig. 21-12. For an initial time period called the turn-on delay time $t_{d(on)}$, there is no build-up of stored charge because the negative charge on the B–E space charge capacitance must be discharged and the junction forward biased so that carrier injection can commence. During this interval only base current flows and only the B–E voltage changes.

After the $t_{d(on)}$ interval, the B–E junction is forward biased and the growth of the stored charge proceeds as diagrammed in Fig. 21-11 and the collector current rises quickly, reaching its on-state value in a time t_{ri}, the current rise time. The voltage v_{CE} is unchanged during this interval because of the diode clamp so that the transistor is still in the active region. After the t_{ri} interval, the collector–emitter voltage falls quickly since the diode no longer can act as a clamp (no forward-bias current through the diode). After a short interval labeled t_{fv1}, quasi-saturation is entered as carrier injection into the drift region begins from the C–B junction. During quasi-saturation, the rate of the collector voltage fall slows because of the reduction in beta that accompanies transistor operation in quasi-saturation. Hard saturation commences when the excess carriers have completely

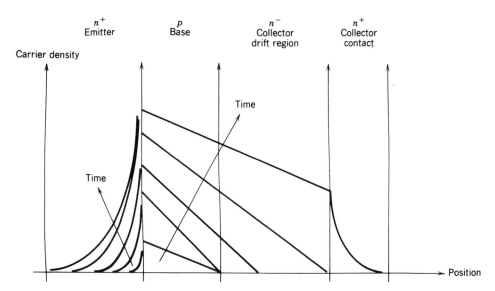

Figure 21-11 Growth of the stored charge distribution in a power BJT during the turn-on transient.

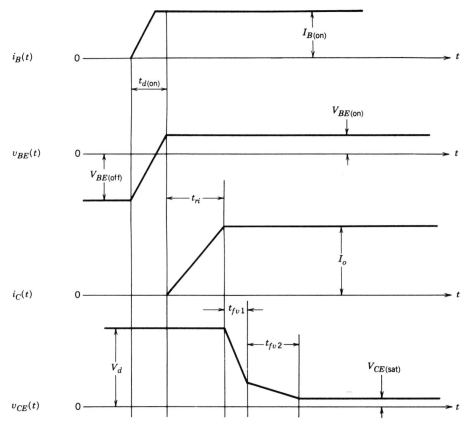

Figure 21-12   Power BJT current and voltage waveforms as the transistor turns on in the clamped inductive load circuit of Fig. 21-10.

swept across the drift region, which occurs after the time interval t_{fv2} indicated on the switching waveforms. The shaded area labeled Q_1 in Fig. 21-9 represents the stored charge that just puts the transistor at the edge of hard saturation. The area labeled Q_2 in Fig. 21-9 represents the excess stored charge that puts the transistor deeper into hard saturation and, in a sense, represents the overdrive of the transistor.

21-5-2   TRANSISTOR TURN-OFF

Turn-off of the transistor involves removing all of the stored charge in the transistor. This could be accomplished by merely reducing the base current to zero and relying on the internal recombination processes in the transistor to remove the charge. However, this would take far too long for practical applications, so the base current is driven negative to speed up the charge removal by carrier sweep-out processes. The process is initiated at $t = 0$, when the base current is either abruptly (step function) or more gradually (ramped with a controlled di_B/dt) changed to negative-bias value, as is indicated in Fig. 21-13. The other transistor voltages and currents of interest are shown in the same figure.

For a time interval labeled the storage time t_s in Fig. 21-13, the collector current remains at its on-state value while the excess stored charge Q_2 (refer to Fig. 21-9) is removed. After the t_s interval, quasi-saturation is entered and the voltage begins to rise

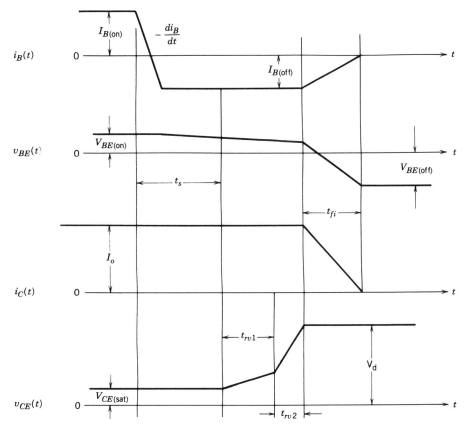

Figure 21-13 Power BJT current and voltage waveforms as the transistor turns off in the clamped inductively loaded circuit of Fig. 21-10.

with a rather shallow slope. When the stored charge distribution is reduced to zero at the C–B end of the drift region after a time interval t_{rv1}, the transistor enters the active region. The increased beta of the transistor causes v_{CE} to complete its growth to the power supply voltage with a much steeper slope as the still constant collector current charges up the space charge capacitance of the C–B junction. The growth of v_{CE} ends after the t_{rv2} interval shown in Fig. 21-13, and the collector current begins to fall as current is commutated into the diode clamp. After a time interval t_{fi}, the rest of the stored charge is removed from the transistor and the collector current becomes zero. The BJT now enters cutoff and the B–E space charge capacitance acquires a negative charge as v_{BE} goes negative.

The waveforms shown in Fig. 21-13 are predicated on the base current making a controlled transition from positive to negative values. If a large negative base current with a fast transition is made at $t = 0$, as is shown in Fig. 21-14, there will be significant changes in the collector current response. The t_s interval would be shortened, as would the two v_{CE} time intervals because of the larger negative base current at earlier times in the transient. Significantly more of the stored charge in the base region would be removed compared to the ramped i_B transient, as is shown in the stored charge distributions plotted in Fig. 21-15 for this situation. However, the amount of charge removed from the drift region would not be increased by the same proportion. Most of the drift region charge is

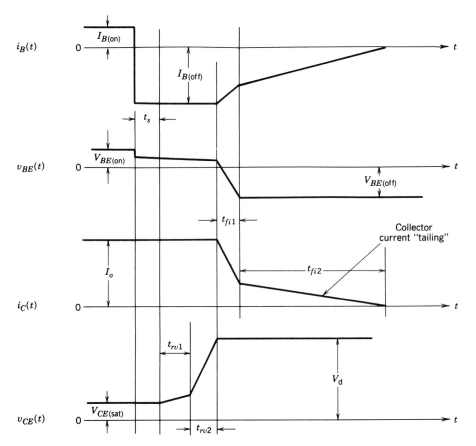

Figure 21-14 Power BJT current and voltage waveforms as the transistor turns off in the clamped inductively loaded circuit of Fig. 21-10 with a large step function reverse base current. Note the long "tail" current that leads to excessive power dissipation.

removed by the collector current and not by the increased base current. The smaller amount of stored charge left in the base means that a shorter time would be required to remove sufficient stored charge in the base so that the B–E junction could become reverse biased with the emitter current going to zero.

If this happens, there will still be stored charge left in the drift region (shown shaded in Fig. 21-15) that must be removed before the collector current can become zero and the BJT can enter cutoff. The only ways for this remaining charge to be removed is via internal recombination and by the negative base current that is flowing. The carrier sweep-out rate in this circumstance will be slow compared to the previous situation because the collector current must equal the negative base current and not beta times larger, as was the previous case. This loss of current gain produces the long tail in the collector current waveform shown in Fig. 21-14 during the time interval labeled t_{fi2}. This long "tailing" time is undesirable because it can lead to increased switching losses.

21-5-3 SWITCHING OF MONOLITHIC DARLINGTONS

The turn-on transient behavior of a monolithic Darlington (MD) embedded in the circuit of Fig. 21-10 will have the same qualitative features as those just described for the single BJT. However, there are two important quantitative differences. First, the main transistor

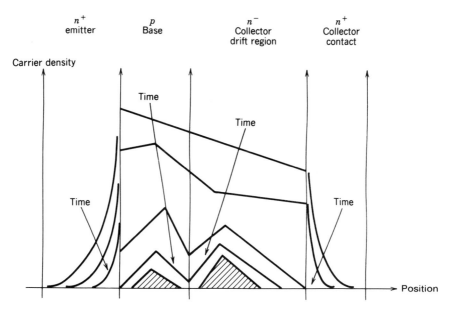

n^+ emitter p Base n^- Collector drift region n^+ Collector contact

Carrier density

Time

Time

Time

Time

Position

Figure 21-15 Decay of the stored charge distributions in a power BJT during a turn-off transient by an abrupt step function reverse base current $I_{B(off)}$. The shaded area represents stored charge remaining in the transistor after the B–E junction cuts off.

cannot go into hard saturation because the on-state voltage of the driver transistor keeps the voltage across the C–B terminals of the main transistor large enough so it stays in quasi-saturation. This means that the on-state power dissipation of MDs will be larger than those of an otherwise comparable single power BJT. Second, the overall switching time to the on state will be faster for the MD because the main transistor will be driven by a larger base current than a comparable single BJT. The base current to the main BJT is β_D (beta of the drive BJT) times larger than what base current would be provided to a single BJT in the same circuit shown in Fig. 21-10.

The most significant differences show up during the turn-off transient, as the waveforms in Fig. 21-16 illustrate. Once the driver transistor turns off, the base current of the main transistor goes negative and its collector current increases, since it must now carry that portion of the load current that the driver BJT had been carrying. The negative base current to the main transistor soon depletes enough stored charge out of the base and collector drift regions that the transistor goes active and completes the traverse of the switching locus to cutoff. As in the case of the conventional power transistor, a controlled rate of change of negative base current should be considered in preference to step function changes. Regardless of which type of turn-off base current drive is selected, the overall turn-off time of the MD will be somewhat longer than that of a conventional power BJT in the same circuit.

If the diode D_1 shown in the circuit of Fig. 21-3 was not present, the turn-off time would be much longer. This would occur because in the circuit of the MD, once the driver transistor cuts off, negative base current for the main transistor would not be able to flow. The only mechanism that would remain active for the removal of the stored change in the main transistor would be internal recombination, which would take far longer than the removal of the charge by carrier sweep-out via the negative base current and collector current.

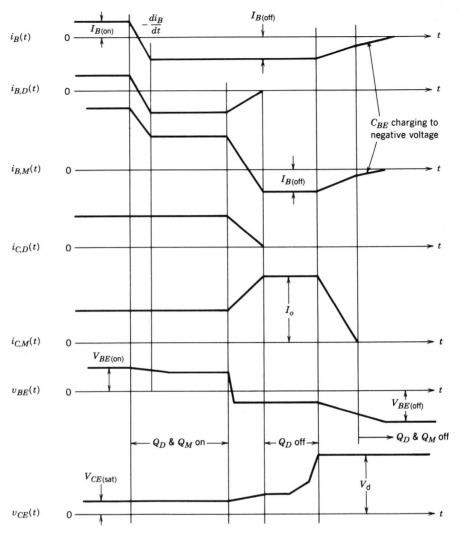

Figure 21-16   Current and voltage waveforms in a power Darlington during turn-off in the clamped inductive load circuit of Fig. 21-10.

21-6   BREAKDOWN VOLTAGES

When the BJT is in the blocking state, the C–B junction must withstand the applied voltage. A BJT cannot block the opposite polarity voltage because the B–E junction has a much lower breakdown voltage than the C–B junction because of the very heavy emitter doping used to increase the beta. Typical B–E breakdown voltages are 5–20 V.

In designing a transistor to withstand a specified voltage, the doping in the drift region on the collector side of the C–B junction is made much smaller than the base doping. This is done so that the depletion region will be predominantly on the collector side, where there is room for it. As in the case of the high-voltage diodes, a lightly doped drift region (the collector drift region) is carefully designed (Eqs. 20-2 and 20-3 apply approximately to this situation) to accommodate the width of the depletion layer at the maximum applied voltage without being overly long, which would lead to increased on-state losses. The base thickness must be kept small so that respectable values of beta can be realized. This means that no significant encroachment of the base by the C–B depletion layer can be tolerated. Some encroachment is unavoidable and decreases the effective base thickness and causes an apparent increase in beta. This effect is known as base thickness modulation and shows up as the finite slope in the active region portion of the $i_C - v_{CE}$ curves, such as is shown in Fig. 21-5. If the transistor base has significant encroachment by the C–B depletion layer, then it must be made thicker, which has the undesirable effect of lowering the beta.

However, the principal reason for strictly limiting the C–B depletion layer encroachment into the base is to avoid reach-through. This occurs when the depletion layer from the C–B junction stretches completely through the base layer to the E–B junction. If this happens, the enormous number of electrons in the emitter (or holes for a *pnp* BJT) will be drawn from the emitter into the base by the large electric fields in the depletion layer. This will lead to large current flows and to a breakdown-like behavior and attendant large power dissipation. To avoid reach-through, the base thickness must be large enough to accommodate the expected depletion layer encroachment and the doping level in the base must be large enough to keep the encroachment small. The thickness of the base in a power BJT is thus a compromise between being small for large beta and being large to minimize reach-through problems. The compromise leads to larger base thicknesses than are found in logic-level transistors and consequently smaller betas, with values of 5–10 being typical.

As we indicated in Section 21-3, in the common emitter configuration the breakdown voltage BV_{CEO} is smaller than BV_{CBO}. There is a semi-empirical relationship between these two parameters, which is given by

$$BV_{CEO} = \frac{BV_{CBO}}{\beta^{1/n}} \qquad (21-6)$$

where $n = 4$ for *npn* transistors and $n = 6$ for *pnp* transistors. The consequence of this relationship is that transistors with high breakdown voltages will have small values of beta. For high-voltage *npn* transistors where beta is between 10 and 20, the value of BV_{CEO} will be about one-half of BV_{CBO}.

The lowering of BV_{CEO} compared with BV_{CBO} is the result of excess-carrier injection into the base from the emitter (note that in the common emitter configuration, the B–E junction is forward biased even with $I_B = 0$ due to the reverse-bias current of the C–B junction). These excess carriers effectively increase the reverse-bias current (termed I_{CEO} in the common emitter configuration) of the C–B junction over the reverse-bias current I_{CBO} of the same junction when the emitter is open. Qualitatively the larger value of I_{CEO} compared with I_{CBO} means that more carriers are crossing the C–B depletion region at any given value of voltage. Consequently, the rate of impact ionization must be larger in the common emitter mode compared with the emitter open mode. A larger rate of impact ionization means that the breakdown voltage will be lower.

21-7 SECOND BREAKDOWN

Bipolar junction transistors and to some degree other types of minority-carrier devices have a potential failure mode, usually termed second breakdown. It appears on the output characteristics of the BJT as a precipitous drop in the collector–emitter voltage at large collector currents. As the collector voltage drops, there is often a significant increase in the collector current and a substantial increase in the power dissipation. What makes this situation particularly dangerous for the BJT is that the dissipation is not uniformly spread over the entire volume of the device but is concentrated in highly localized regions where the local temperature may grow very quickly to unacceptably high values. If this situation is not terminated in a very short time, device destruction results. When devices that have been so destroyed are analyzed, they often show dramatic evidence of the localized power dissipation and attendant heating in the form of melted and then recrystallized silicon.

Second breakdown does not originate from impact ionization and an attendant avalanche breakdown of a *pn* junction. This is clear from the fact that a drop in voltage accompanies second breakdown, whereas no such drop is observed in avalanche breakdown.

Several intrinsic aspects of the transistor combine to give the BJT its susceptibility to second breakdown. First there is the general propensity of minority-carrier devices to thermal runaway when the voltage across them is held approximately constant as the device temperature increases. Minority-carrier devices have a negative temperature coefficient of resistivity (the resistivity drops as the temperature increases because the minority-carrier densities are proportional to the intrinsic carrier density n_i, which increases exponentially with temperature; see Eqs. 19-1 and 19-25). This means the power dissipation will increase as the resistance drops as long as the voltage remains constant. If the rate of increase in power dissipation with temperature is greater than linear with temperature (the rate of heat removal is linear with temperature, i.e., characterized by a thermal resistance), then an unstable situation will result when the power dissipated exceeds the rate at which heat energy can be removed (a function of the thermal resistance). The situation becomes a classic case of positive feedback in which the power dissipation leads to an increase in temperature, which leads to further increases in power dissipation, and so on, until device destruction results. It is often and quite appropriately termed thermal runaway.

This potential for thermal runaway is made much more dangerous if the current density in the device is nonuniform across the device cross section. Current filaments where the current density is substantially larger than in surrounding areas may occur and localized thermal runaway becomes likely. Consider the situation illustrated in Fig. 21-17, where the current density J_A in region A is assumed to be greater than the current density J_B in region B. The power dissipation density will be greater in A than in B, which will lead to an increase in the temperature T_A compared to T_B. This in turn will lead to further increases in J_A compared with J_B and the temperature T_A will increase further. If the local temperature T_A exceeds the intrinsic temperature T_i (the temperature at which the intrinsic-carrier density n_i equals the majority-carrier doping density), then thermal runaway will be in progress in that local region and will lead to intense localized heating and device failure if not terminated very quickly.

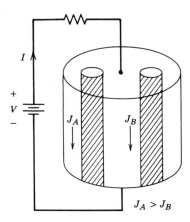

Figure 21-17 Semiconductor device with regions of current density nonuniformities that could lead to the formation of current filaments and possible second breakdown.

The formation of the current filaments and subsequent localized thermal runaway requires only a nonuniformity in the current density and enough localized power dissipation to cause a substantial rise in the temperature of the filament. Indeed, the increase in carrier density in the current filament may often cause a drop in the external voltage across the device if the external resistance in series with the device is appreciable, and yet device destruction may still occur. When the shorting effect of the filament is strong enough to cause this voltage drop, the device is said to be in second breakdown.

The key to avoiding second breakdown would then seem to be (1) keeping the total power dissipation under control and, more important, (2) avoiding any current density nonuniformities, especially during turn-on and turn-off, when the instantaneous power dissipation is largest. However, as we have already seen, the basic construction of the transistor leads to current constrictions via mechanisms such as emitter current crowding. While current crowding can be postponed until specific current levels are reached, once these levels are exceeded, the current constriction can be severe enough to lead to the formation of a current filament and to possible localized thermal runaway.

When device turn-off is initiated, the flow of negative base current induces a lateral voltage drop of the opposite polarity to that described in Fig. 21-8a. This causes a crowding of the emitter current toward the center of the emitter, as is shown in Fig. 21-8b, and once again the conditions are favorable for thermal runaway. If, however, the width of the emitter is made smaller, then the lateral voltage drop will be smaller (less resistance for the lateral flowing base current to develop a voltage drop across). This means that the severity of current crowding will be less and the attendant possibility for second breakdown will be smaller. For this reason, power BJTs are constructed with many narrow emitter fingers, as is shown in Fig. 21-2, in parallel rather than a few very large cross-sectional area emitters. Other measures to reduce the possibility of second breakdown include the use of a controlled rate of change of base current during turn-off, the use of protective circuitry such as snubbers and free-wheeling diodes, and the positioning of the switching trajectory within the safe operating area (SOA) boundaries, a topic to be discussed shortly.

21-8 ON-STATE LOSSES

Except at high switching frequencies, nearly all the power dissipated in the switch-mode operation of a BJT occurs when the transistor is in the on state, usually hard saturation. In this circumstance the power dissipation is given by (ignoring base current losses)

$$P_{on} = I_C V_{CE(sat)} \tag{21-7}$$

The collector–emitter saturation voltage $V_{CE(sat)}$ increases with increasing collector current.

Several internal voltage drops in a power transistor contribute to $V_{CE(sat)}$. These voltage drops and some of their origins are indicated schematically on the vertical cross section of a power transistor, as shown in Fig. 21-18. Adding these together yields

$$V_{CE(sat)} = V_{BE(on)} - V_{BC(sat)} + V_d + I_C(R_e + R_c) \tag{21-8}$$

The voltages $V_{BE(on)}$ and $V_{BC(sat)}$ are the voltages appearing across the forward-biased B–E and C–B junctions, respectively. These voltages differ from each other by 0.1–0.2 V because the two junctions are significantly different from each other. The C–B junction is much larger in area than the B–E junction and the doping levels are much lower across the C–B junction compared with the B–E junction. This voltage difference is relatively independent of collector current.

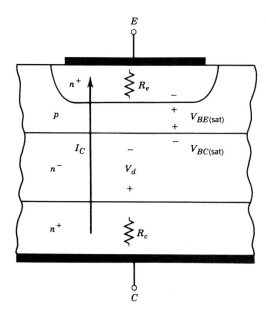

Figure 21-18   Vertical cross section of a power BJT showing the origins of the components of the on-state collector–emitter voltage $V_{CE(\text{sat})}$.

The resistances R_e and R_c represent the ohmic resistance of the heavily doped emitter and collector regions, respectively. At low to moderate collector currents, the voltage drops across these resistances are negligible. But at larger currents, these drops become important and add to the increase in $V_{CE(\text{sat})}$ with collector current.

The major contribution to the increase in $V_{CE(\text{sat})}$ with collector current is the voltage drop V_d across the collector drift region. This voltage can be made reasonably small because of conductivity modulation and relatively independent of collector current, as was described in Chapter 20 for the pn-junction power diode, which has an analogous drift region. However, at larger collector currents, the excess-carrier densities in the drift region approach large enough values that the carrier lifetime begins to decrease (Auger recombination) as well as the mobility (carrier–carrier scattering). When this occurs, V_d begins to increase significantly, thus increasing $V_{CE(\text{sat})}$. As was true with the diode, the magnitude of V_d is dependent on the excess-carrier lifetime (see Eq. 20-13), whose value is a compromise between large values that minimize V_d and shorter values that minimize switching times.

The increase in V_d with collector current will be most severe and commence at lower values of collector current in high-voltage BJTs because of the long drift region that these transistors must have to hold off large collector–emitter voltages in the off state. This aspect of the transistor behavior is analogous to the behavior of high-voltage pn-junction diodes. Based on these observations, we would expect the current capabilities of the BJT to be similar to those of diodes with the same voltage rating.

However, BJTs have a significantly lower current density capability versus break-down voltage than this optimistic estimate. Mechanisms such as emitter current crowding and conductivity modulation of the base, which lower the value of beta, commence at current densities lower than current densities that diodes can handle. The decrease in beta with increasing collector current means that the base current must be increased at a greater rate than the collector current to maintain the device in hard saturation or, at least, near it so that the voltage drop across the transistor is not too large. The transistor cannot be allowed to enter very far into quasi-saturation at larger collector currents because v_{CE} (and, hence, the power dissipation) increases very rapidly, as is shown in Fig. 21-7. Since there is a practical limit to how much base current the user is willing to put into the transistor, there is a practical upper limit to the collector current and, hence, to the current density. Approximately speaking, the transistor is usable to collector currents about 10 times larger than the value at which the current gain peaks and begins to fall with increasing collector current (see Fig. 21-7).

Emitter current crowding and other mechanisms that reduce beta at large collector currents are so significant that the transistor manufacturer designs the transistor structure with a specific value of collector current in mind at which the beta begins to decrease. Once this current level has been determined, the device designer then adjusts the carrier lifetime in the drift region so that the voltage drop across it is kept at the desired levels. In this design approach, the same basic trade-offs between breakdown voltage, on-state losses, and switching speeds described for the power diode are still valid; the only basic change is that they occur at lower current densities.

21-9 SAFE OPERATING AREAS

Safe operating areas, or SOAs, a concept described in earlier chapters of this book, are a very convenient and compact method of summarizing maximum values of current and voltage to which the BJT should be subjected. Two separate SOAs are used in conjunction with BJTs and both are commonly given on specification sheets. The so-called forward-bias safe operating area (FBSOA) is shown in Fig. 21-19 and the reverse-bias safe operating area (RBSOA) is shown in Fig. 21-20. The terms forward bias and reverse bias refer to whether the base current bias source forward biases the B−E junction or reverse biases it (which would be appropriate for turning off the BJT).

Several different physical mechanisms are active in determining the boundaries of the FBSOA shown in Fig. 21-19. The current I_{CM} is the maximum collector current even as a pulse that should be applied to the transistor. Exceeding this current may cause bonding wires or metalizations on the wafer to vaporize or otherwise fail. The thermal limit is a

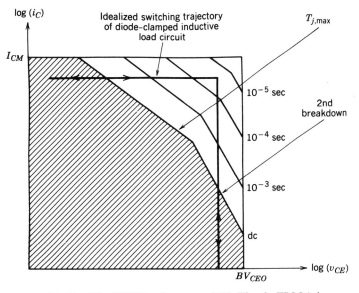

Figure 21-19 The FBSOA of a power BJT. The dc FBSOA is shown as shaded and the expansion of the area for pulsed operation of the BJT is shown with shorter switching times leading to a larger FBSOA.

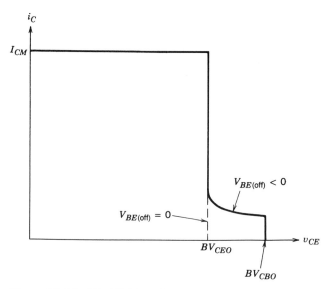

Figure 21-20   The RBSOA of a power BJT. Reverse bias refers to the base current being in the opposite direction to the normal on-state direction.

power dissipation limit set by the thermal resistance of the transistor and the maximum allowable junction temperature. The second breakdown boundary represents the maximum permissible combinations of voltage and current without getting into the region of the $i_C - v_{CE}$ plane where second breakdown may occur. The final portion of the boundary of the FBSOA is breakdown voltage limit BV_{CEO}.

If the transistor is operated as a switch, then the boundaries of the FBSOA expand, as is indicated in the figure. Crudely speaking, the expansion of the SOA occurs for switch-mode operation because the silicon wafer and its packaging have a thermal capacitance and, hence, an ability to absorb a finite amount of energy without the junction temperature rising to excessive levels. If the transistor turns on in a few microseconds or less, the amount of energy that is absorbed is too small to cause any appreciable rise in the junction temperature, and as a result, the FBSOA is essentially square, being limited only by I_{CM} and BV_{CEO}. The SOA is particular useful when the switching trajectory, such as the one for the circuit of Fig. 21-10, is plotted on it, as in Fig. 21-19, because such a construction makes it immediately clear if the circuit operation puts the transistor outside of its specification range.

In a similar fashion, the RBSOA shown in Fig. 21-20 is constructed. The area encompassed by the RBSOA, which is a pulsed SOA, is somewhat larger than the FBSOA because of the extension of the area of higher voltages than BV_{CEO}, up to BV_{CBO}, at low collector currents. The operation of the transistor up to the higher voltage is possible because the combination of low collector current and reverse base current has made the beta small so that the breakdown voltage rises toward BV_{CBO}, as Eq. 21-6 predicts.

SUMMARY

This chapter has explored the structure and operating characteristics of BJTs intended for power switch-mode applications. The important conclusions are listed below:

1. The power BJT has a vertically oriented structure with a highly interdigitated B–E structure and a lightly doped collector drift region.

2. The drift region determines the blocking voltage rating of the BJT and also causes the so-called quasi-saturation region of the I–V characteristics.

3. The BJT is a normally-off device that is turned on by the application of a sufficiently large base current to cause injection of large numbers of minority carriers into the base from the emitter region. The subsequent diffusion of these carriers across the base to the collector forms the collector current.

4. Power BJTs have low current gain, especially at larger breakdown voltage ratings. This has led to the development of monolithic Darlington transistors, which have larger current gains.

5. Lateral current flow in the base is the basic limiting factor in BJT performance. It causes lateral voltage drops, which lead to emitter current crowding, which in turn causes decreases in current gain. If the current crowding is excessive, second breakdown and device destruction will occur.

6. Heavy conductivity modulation of the drift region in order to minimize on-state losses requires large carrier lifetimes. But this leads to long turn-off times, so a trade-off must be made in the design of the BJT between lower on-state losses or shorter switching times.

7. Turn-off of some types of BJTs should be done with a controlled rate of change of negative base current in order to avoid isolating excessive stored charge in the BJT, which would result in excessively long turn-off times and large power dissipation.

8. The SOAs of the BJT are limited by second breakdown. The RBSOA is normally the limiting factor.

9. BJTs with limited SOAs may require that their switching trajectory be controlled with snubber circuits during both turn-on and turn-off.

PROBLEMS

21-1 Plot BV_{CEO} as a function of beta (β) for $5 < \beta < 100$ for identical *npn* and *pnp* silicon transistors. Assume that both BJTs have the same value of BV_{CBO}.

21-2 When emitter-open switching (see Ch. 28) is used to turn off a power BJT, the BJT is less susceptible to second breakdown compared with the normal turn-off situation, where negative base current flows while emitter current is still flowing. Qualitatively explain, with the aid of diagrams, why this is true.

21-3 Consider the step-down converter circuit of Fig. 21-10. The free-wheeling diode is ideal and the power transistor has the following parameters: $\beta = 10$, $V_{CE(on)} = 2$ V, $R_{\theta j-a} = 1°C/W$, $T_{j,max} = 150°C$, $t_{ri} = t_{fi} = 200$ ns, $t_{fv1} = t_{fv2} = t_{rv1} = t_{rv2} = 50$ ns, and $t_{d(on)} = t_{d(off)} = 100$ ns. The BJT is driven by a square wave (50% duty cycle) of variable frequency. Assume $I_o = 40$ A and $V_d = 100$ V.

(a) Sketch and dimension are the average power dissipated in the transistor versus the switching frequency.

(b) Estimate the maximum permissible switching frequency.

21-4 The transistor in the circuit of Problem 21-3 is driven by a 25-kHz square wave (50% duty cycle). The switching times increase by 40% as the junction temperature increases by 100°C (from 25 to 125°C). If the circuit is operating in an ambient temperature of 50°C, estimate the allowable range of values of the thermal resistance $R_{\theta j-a}$ that will keep the junction temperature T_j less than 110°C.

21-5 A transistor similar to that shown in Fig. 21-1 has an effective emitter area A of 1 cm² and a base width of 3 μm. At approximately what value of collector current does the beta of the device begin to drop as the current is increased? *Hint:* Recall from the discussion in Section 21-4-1 that beta begins to fall when high-level injection conditions are obtained in the base. Also recall that $I_C \approx -I_{ne} = qD_n A \, dn_b(x)/dx$.

21-6 Consider a BJT and a *pn*-junction diode each having the same cross-sectional area and same drift region length (and thus the same carrier lifetimes and blocking voltage capabilities). Which device can carry the larger forward current and why?

21-7 A BJT similar to that shown in Fig. 21-1 has been designed by a novice device designer. The voltage rating of the device is supposed to be 1000 V. The drift region doping is as indicated in the figure, and the drift region length is 100 μm. But the base doping density is 10^{15} cm^{-3} and the base width is 3 μm. What is the actual voltage rating of this device? Assume all junctions are step junctions.

21-8 A bipolar NPN transistor is to be designed for a breakdown voltage (BV$_{CEO}$) of 1000 V. The base-emitter breakdown voltage (BV$_{BEO}$) is to be 10 V. Find the required base doping density, collector doping density, base width, and collector drift region width. The emitter doping density is 10^{19} cm^{-3}. Assume that the base-emitter and base-collector junctions are step junctions and that beta = 5.

21-9 A Darlington pair has an effective beta of 150. The driver BJT has a beta of 20. What is beta of the main BJT?

21-10 A power BJT with a beta of 10 is characterized in the on-state by a $V_{BE,\text{sat}} = 0.8$ V, $V_{BC,\text{sat}} = 0.6$ V, and $R_{\text{on}} = 0.02$ ohms. Two such BJTs are used in a Darlington configuration which must conduct a current of 100 A in the on-state. What is the on-state power dissipation of the pair?

21-11 Find the zero-bias values of the collector-base space charge capacitance, C_{CBO}, and base-emitter space charge capacitance, C_{EBO}, of the transistor described in problem 21-8. Assume a base-emitter area A_E of 0.3 cm^2 and a base-collector area A_C of 3 cm^2.

21-12 The transistor described in Problems 21-8 and 21-11 is to be used in a step-down converter such as is shown in Fig. 21-10. The base drive circuit consists of an ideal voltage source in series with a resistance of 10 ohms. When the BJT is to be turned on the voltage changes from -8 V to $+8$ V. Estimate the turn-on delay time. Assume the dc voltage powering the step-down converter is 100 V.

REFERENCES

1. S. K. Ghandhi, *Semiconductor Power Devices,* John Wiley & Sons, New York, 1987, Chapter 4.
2. P. L. Hower, "Bipolar Transistors," *Semiconductor Devices for Power Conditioning,* Roland Sittig and P. Roggwiller (Eds.), Plenum, New York, 1982.
3. A. Blicher, *Field Effect and Bipolar Power Transistor Physics,* Academic, New York, 1981, Chapters 6–10.
4. A. S. Grove, *Physics and Technology of Semiconductor Devices,* John Wiley & Sons, New York, 1967, Chapter 7.
5. A. S. Sedra and K. C. Smith, *Microelectronics Circuits,* 2nd ed., Holt, Rinehart, and Winston, New York, 1987, Chapter 5.
6. M. H. Rashid, *Power Electronics: Circuits, Devices, and Applications,* Prentice-Hall, Englewood Cliffs, NJ, 1988, Chapter 15.
7. *Power Transistor in Its Environment,* Thompson-CSF, Semiconductor Division, 1978.
8. B. W. Williams, *Power Electronics, Devices, Drivers, and Applications,* John Wiley & Sons, New York, 1987, Chapters 3, 4, 7, 9.
9. B. Jayant Baliga and D. Y. Chen (Ed.), *Power Transistors, Device Design and Applications,* IEEE Press, Institute of Electrical and Electronic Engineers, New York, 1984, Part I, *Power Bipolar Transistors,* pp. 19–122.
10. M. S. Adler, K. W. Owyang, B. Jayant Baliga, and R. A. Kokosa, "The Evolution of Power Device Technology," *IEEE Transactions on Electron Devices,* Vol. ED-31, No. 11, Nov. 1984, pp. 1570–1591.

CHAPTER 22

POWER MOSFETs

22-1 INTRODUCTION

Metal–oxide–semiconductor field effect transistors (MOSFETs) with appreciable on-state current-carrying capability and off-state blocking voltage capability and, thus, potential for power electronic applications have been available since the early 1980s. They have become as widely used as power BJTs and in fact are replacing BJTs in many applications, especially those where high switching speeds are important. MOSFETs operate on different physical mechanisms than BJTs, and a clear understanding of these differences is essential for the effective utilization of both BJTs and MOSFETs. This chapter considers the basic physical mechanisms that govern the operation of MOSFETs, the factors that establish the current and voltage limits of the MOSFET, and possible failure modes if these limits are exceeded.

22-2 BASIC STRUCTURE

A power MOSFET has the vertically oriented four-layer structure of alternating p-type and n-type doping shown in Fig. 22-1a for a single cell of the many paralleled cells of a complete device. The $n^+pn^-n^+$ structure is termed an enhancement mode n-channel MOSFET (for reasons that will become apparent shortly). A structure with the opposite doping profile can also be fabricated and is termed a p-channel MOSFET. The doping in the two n^+ end layers, labeled source and drain in Fig. 22-1, is approximately the same in both layers and is quite large, typically 10^{19} cm^{-3}. The p-type middle layer is usually termed the body and is the region where the channel (to be discussed in the next section) is established between source and drain and is typically doped at 10^{16} cm^{-3}. The n^- layer is the drain drift region and is typically doped at 10^{14}–10^{15} cm^{-3}. This drift region determines the breakdown voltage of the device.

At first glance, it would appear that there is no way that current can flow between the drain and source terminals of the device because one of the pn junctions (either the body–source junction or the drain–body junction) will be reverse biased by either polarity of applied voltage between the drain and source. There can be no injection of minority carriers into the body region via the gate terminal because the gate is isolated from the body by a layer of silicon dioxide [usually termed the gate oxide and typically about 1000 Å (angstroms) thick], which is a very good insulator and, hence, there is no BJT operation. However, an application of a voltage that biases the gate positive with respect to the source will convert the silicon surface beneath the gate oxide into an n-type layer or channel, thus connecting the source to the drain and allowing the flow of appreciable currents. The thickness of the gate oxide, the width of the gate (as diagrammed in

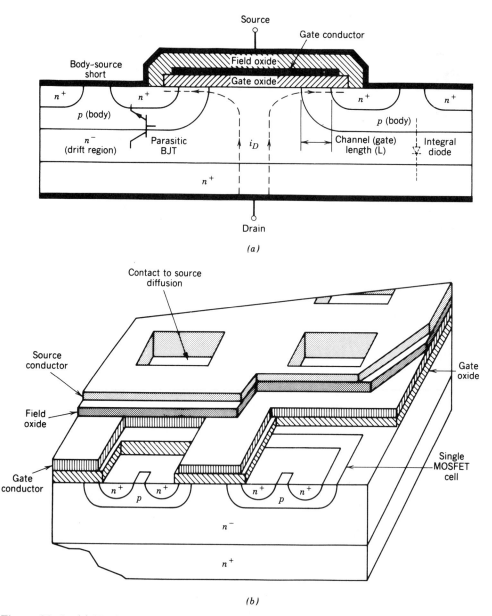

Figure 22-1 (*a*) Vertical cross-section and (*b*) perspective view of an *n*-channel power MOSFET. A complete MOSFET is composed of many thousands of cells connected in parallel to achieve large gain and low on-state resistance. Some of the layers in the perspective view have been cut away to enhance the clarity of the drawing.

Fig. 22-1), and the number of gate/source regions connected electrically in parallel are important in determining how much current will flow for a given gate-to-source voltage. The mechanisms that lead to the creation of the channel will be discussed later in this chapter.

The structure shown in Fig. 22-1 is usually termed VDMOS, meaning vertical diffused MOSFET. The name crudely describes the fabrication sequence of the device. The starting substrate is usually the n^+ drain onto which the n^- drift region of specified thickness is grown epitaxially. Then the *p*-type body region is diffused into the wafer from the source side of the wafer, followed by the n^+ source diffusion. These two diffusions are masked diffusions, meaning that portions of the wafer are protected by silicon dioxide so that the dopants cannot reach the wafer where the SiO_2 has been left. The remaining steps involve the deposition of the gate and source metallization and final packaging steps.

Several other aspects of the MOSFET structure shown in Fig. 22-1 should be noted. First, the source is constructed of many (thousands) small polygon-shaped areas that are connected in parallel and surrounded by the gate region. The geometric shape of the source regions to some degree influences the on-state resistance of the MOSFET, and some manufacturers even advertise their particular line of MOSFET devices by the shape of the source region (i.e., International Rectifier's HEXFET). The basic reason for the many small source regions is to maximize the width (the lateral dimension perpendicular to the direction of current flow in the channel) of the gate region compared to its length (the channel length). The gate width W of the MOSFET is the peripheral length of each cell times the number of cells that make up the device. A very large gate width-to-length ratio is desired because this maximizes the gain of the device.

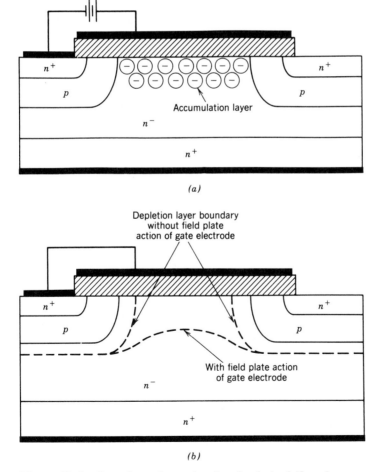

Figure 22-2 Gate electrode overlapping the drain drift region (*a*) to create an accumulation layer in the on state and (*b*) to act as a field plate in the off state.

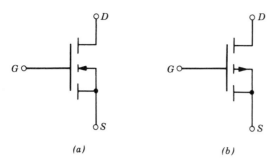

Figure 22-3 Circuit symbols for (*a*) an *n*-channel and (*b*) a *p*-channel MOSFET.

Second, there is a parasitic *npn* BJT between the source and drain contacts, as shown in Fig. 22-1, with the *p*-type body region serving as the base of the parasitic BJT. To minimize the possibility that this transistor is ever turned on, the *p*-type body region is shorted to the source region by overlapping the source metallization onto the *p*-type body region, as in Fig. 22-1. As a result of this body-to-source short, there is a parasitic diode connected between the drain and source of the MOSFET, as shown in Fig. 22-1. This integral diode can be used in half-bridge and full-bridge converters.

Third, there is the overlap of the gate metallization across the n^- drift region, where it protrudes to the surface of the wafer. This overlapping of the gate metallization serves two purposes. First, it tends to enhance the conductivity of the drift region at the $n^- - SiO_2$ interface by forming an accumulation layer (a region of enhanced conductivity to be discussed in later sections), as is shown in Fig. 22-2*a*, thus helping to minimize the on-state resistance. Second, the metallization tends to act as a field plate when the MOSFET is off that keeps the radius of curvature of the depletion region of the drain–body *pn* junction from getting too small and thus reducing the breakdown voltage of the device. This field plate function is diagrammed in Fig. 22-2*b*.

The circuit symbol for an *n*-channel MOSFET is shown in Fig. 22-3*a* and for a *p*-channel MOSFET in Fig. 22-3*b*. The direction of the arrow on the lead that goes to the body region indicates the direction of current flow if the body–source *pn* junction were forward biased by breaking the short between the two and a forward-bias voltage were applied. Thus, an *n*-channel MOSFET that has a *p*-type body region has the arrow pointing into the MOSFET symbol, as is shown in Fig. 22-3*a*, and the arrow points outwardly for a *p*-channel device.

22-3 *I–V* CHARACTERISTICS

The MOSFET, like the BJT, is a three-terminal device where the input, the gate in the case of the MOSFET, controls the flow of current between the output terminals, the source, and drain. The source terminal is usually common between the input and output of a MOSFET. The output characteristics, drain current i_D as a function of drain-to-source voltage v_{DS} with gate-to-source voltage V_{GS} as a parameter, are shown in Fig. 22-4*a* for an *n*-channel MOSFET. The output characteristics for a *p*-channel device are the same except that the current and voltage polarities are reversed so that the characteristics for the *p*-channel device would appear in the third quadrant of the $i_D - v_{DS}$ plane rather than the first quadrant, as do the characteristics of Fig. 22-4*a*.

In power electronic applications, the MOSFET is used as a switch to control the flow of power to the load in a manner analogous to the usage of the BJT. In these applications the MOSFET traverses the $i_D - v_{DS}$ characteristics from cutoff through the active region to

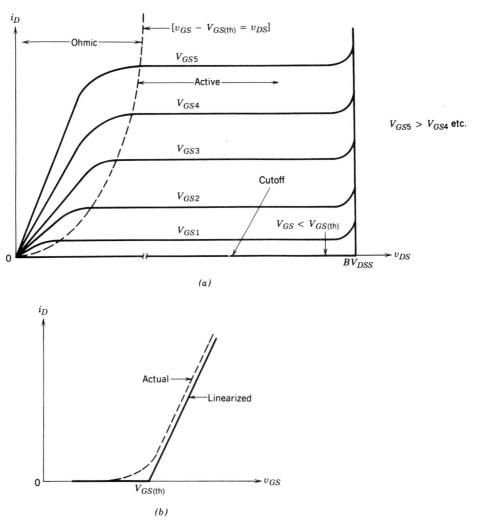

Figure 22-4 Current–voltage characteristic of an *n*-channel enhancement mode MOSFET: (*a*) output (i_D–v_{DS}) characteristics; (*b*) transfer curve.

the ohmic region as the device turns on and back again when it turns off. The cutoff, active, and ohmic regions of the characteristics are shown on Fig. 22-4*a*.

The MOSFET is in cutoff when the gate–source voltage is less than the threshold voltage $V_{GS(\text{th})}$, which is typically a few volts in most power MOSFETs. The device is an open circuit and must hold off the power supply voltage applied to circuit. This means that the drain–source breakdown voltage BV_{DSS} shown on Fig. 22-4 must be larger than the applied drain–source voltage to avoid breakdown and the attendant high power dissipation. When breakdown occurs, it is due to the avalanche breakdown of the drain–body junction.

When the device is driven by a large gate–source voltage, it is driven into the ohmic region (the reason for this designation is twofold, first having to do with the physical mechanisms operative in the MOSFET, which will be discussed in the next section, and second, to avoid confusion with the terminology of saturation, which means one thing when applied to BJTs but another when applied to MOSFETs) where the drain–source voltage $V_{DS(on)}$ is small. In this region the power dissipation can be kept within reasonable bounds by minimizing $V_{DS(on)}$ even if the drain current is fairly large. The MOSFET is in the ohmic region when

$$v_{GS} - V_{GS(th)} > v_{DS} > 0 \tag{22-1}$$

In the active region the drain current is independent of the drain–source voltage and depends only on the gate–source voltage. The current is sometimes said to have saturated, and consequently this region is sometimes called the saturation region or the pentode region. We will term this region the active region to avoid the use of the term saturation and the attendant possible confusion with saturation in BJTs. Simple first-order theory predicts that in the active region the drain current is given approximately by

$$i_D = K(v_{GS} - V_{GS(th)})^2 \tag{22-2}$$

where K is a constant that depends on the device geometry. At the boundary between the ohmic region and active region where $v_{GS} - V_{GS(th)} = v_{DS}$, Eq. 22-2 becomes

$$i_D = Kv_{DS}^2 \tag{22-3}$$

which is a convenient way of delineating the boundary between the two regions, as is in Fig. 22-4a.

The relationship expressed by Eq. 22-2 is followed reasonably well by logic-level MOSFETs. However, a plot of i_D versus v_{GS} (with the MOSFET in the active region) in Fig. 22-4b, usually termed the transfer curve, shows that this equation is followed only at lower values of drain current in power MOSFETs. Overall the transfer curve of a power MOSFET is quite linear, in contrast to the parabolic transfer curve of the logic-level device. The reasons for the different behavior of the transfer curve between logic-level and power MOSFETs will be considered in the next section.

22-4 PHYSICS OF DEVICE OPERATION

22-4-1 INVERSION LAYERS AND THE FIELD EFFECT

The gate portion of the MOSFET structure shown in Fig. 22-1 is the key to understanding how the MOSFET operates. The gate region is composed of the gate metallization, the silicon dioxide underneath the gate conductor, which is termed the gate oxide, and the silicon beneath the gate oxide. This region forms a high-quality capacitor, as is shown in Fig. 22-5, and it is sometimes termed a MOS capacitor. Although the capacitor shown is usually composed of aluminum metallization, SiO_2 insulator, and silicon bottom layer, the same basic structure can be fabricated on other semiconductors such as gallium arsenide and other insulators such as aluminum nitride or silicon nitride can be used for the insulator. The top metallization layer can also be other conducting materials. Poly-silicon, refractory metals such as tungsten, and other metals have been used on MOSFET devices.

When a small positive gate–source voltage is applied to the capacitor structure in the simplified diagram of the n-channel MOSFET shown in Fig. 22-5a, a depletion region forms at the interface between the SiO_2 and the silicon. The positive charge induced on the upper metallization (the gate side) by the applied voltage requires an equal negative charge on the lower plate, which is the silicon side of the gate oxide. The electric field from the positive charge repels the majority-carrier holes from the interface region and thus exposes the negatively charged acceptors, thus creating a depletion region.

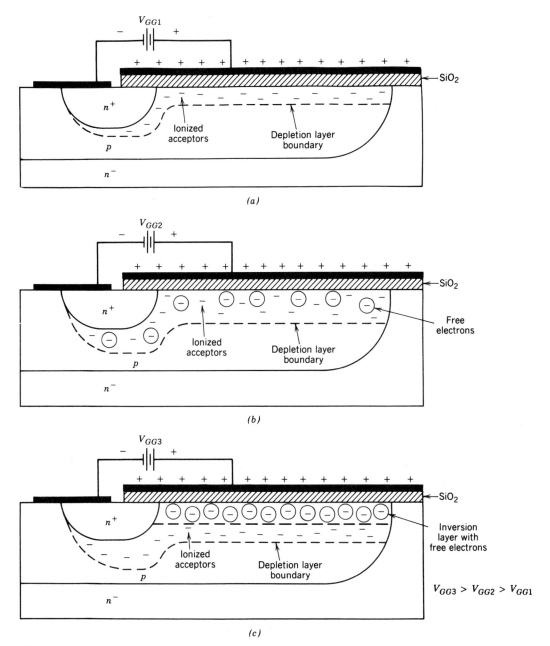

Figure 22-5 (*a*) Formation of the depletion layer. (*b*, *c*) Inversion layer at the Si–SiO$_2$ interface as the gate–source voltage is increased.

Further increases in v_{GS} cause the depletion layer to grow in thickness, as is shown in Fig. 22-5*b*, to provide the additional negative charge. The growth of this depletion layer can be modeled approximately as a one-sided step junction, such as was considered in Chapter 19. As the voltage is increased, the electric field at the oxide–silicon interface gets larger and begins to attract free electrons as well as repelling free holes. The immediate source of the electrons is electron–hole generation via thermal ionization with the free holes being pushed into the semiconductor bulk ahead of the depletion region. The extra holes are neutralized by electrons that are attracted from the n^+ source by the positive charge of the holes.

Eventually, as the bias voltage is increased, the density of free electrons at the interface will become equal to the free hole density in the bulk of the body region away from the depletion layer. The layer of free electrons at the interface will be highly conducting and will have all the properties of an n-type semiconductor. At this point the layer of free electrons is termed an inversion layer, as is illustrated in Fig. 22-5c. This n-type layer is a conductive path or channel between the n^+ drain and source regions (thus the terminology channel), which permits the flow of current between source and drain. This ability to modify the conductivity type of the semiconductor immediately beneath the gate insulator by means of an applied voltage or electric field is termed the field effect. The field effect enhances the conductivity of the interface and, hence, the name enhancement mode field effect transistor, which is based on this mechanism.

The value of v_{GS} where the inversion layer is considered to have formed is termed the threshold voltage $V_{GS(\text{th})}$. As v_{GS} is increased beyond $V_{GS(\text{th})}$, the inversion layer gets somewhat thicker and, more important, more conductive as the density of free electrons gets larger as the bias voltage increases. The inversion layer screens the depletion layer adjacent to it from the further bias voltage increases so that the depletion layer thickness now remains constant. The value of the threshold voltage is a function of several factors. A major factor is the oxide capacitance per unit area C_{ox}, which is given by

$$C_{\text{ox}} = \frac{\epsilon_{\text{ox}}}{t_{\text{ox}}} \tag{22-4}$$

where ϵ_{ox} is the dielectric constant of the silicon dioxide (1.05×10^{-12} F/cm) and t_{ox} is the thickness of the gate oxide (typically 1000 Å). The threshold voltage is inversely proportional to C_{ox}. Other factors that influence $V_{GS(\text{th})}$ are the work functions of the silicon and the gate metal, any charge bound or trapped in the silicon dioxide, impurities at the interface or in the silicon dioxide, as well as others. In spite of the complex array of factors that influence $V_{GS(\text{th})}$, device manufacturers can adjust its value to whatever value is desired (typically a few volts). Major adjustments are done by the choice of gate metallization, doping density of the body region, and thickness of the gate oxide. Minor adjustments of the threshold voltage during fabrication are done via ion implantation of impurities in the body region just beneath the gate oxide.

22-4-2 GATE CONTROL OF DRAIN CURRENT FLOW

We now embed the n-channel MOSFET of Fig. 22-5 into the circuit shown in Fig. 22-6, which has both a gate–source bias supply V_{GS} and a drain–source bias supply V_{DD}. Initially it is assumed that V_{GS} is greater than $V_{GS(\text{th})}$ and that V_{DD} is small. The MOSFET will be in the ohmic region with a relatively small value of I_D, and the inversion layer will have a spatially uniform thickness, as is shown in Fig. 22-6a. Now V_{DD} is slowly increased to ever larger values while V_{GS} is held constant. The drain current will initially increase in proportion to the increase in V_{DD}, since the inversion layer appears as an ohmic resistance connecting the drain to the source. This current increase will cause a voltage drop along the channel, which is shown in Fig. 22-6a as $V_{CS}(x)$ (channel-to-source voltage), with x being the distance from the source to the location x in the channel where the voltage is being specified.

In the discussion of the inversion layer formation in the previous section, the gate-to-body voltage, which is the voltage drop across the oxide, was greater than $V_{GS(\text{th})}$. It was also implicitly assumed that this voltage was spatially uniform along the length of the

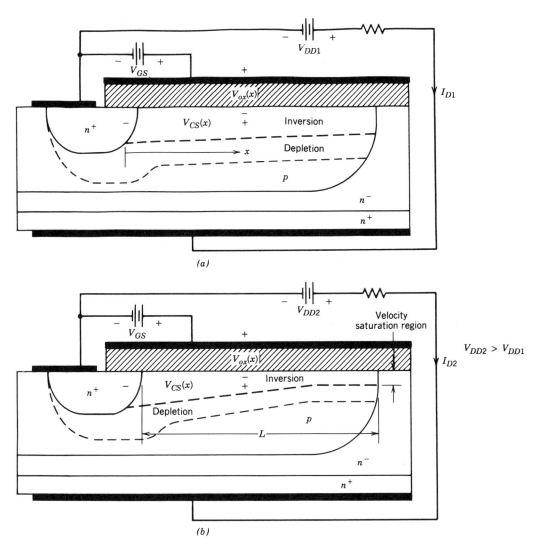

Figure 22-6 Change in the inversion layer thickness from being (a) spatially uniform at low drain current to being (b) spatially nonuniform at larger drain current values.

oxide from source to drain so that the inversion layer thickness would also be uniform. But in the structure of Fig. 22-6, the oxide voltage is actually $V_{GS} - V_{CS}(x)$, as indicated in the figure. As I_D increases, $V_{CS}(x)$ increases and the voltage across the oxide at the position x decreases. Since $V_{CS}(x)$ has its largest value, V_{DS}, at $x = L$ (the drain end of the channel), the voltage drop across the oxide, which determines the parameters of the inversion layer, will have its smallest value of $V_{GS} - V_{DS}$. The decrease in the oxide voltage from source to drain when I_D is flowing means that the thickness of the inversion layer must also decrease from source to drain, as indicated in Fig. 22-6b.

As the inversion layer thins out at the drain end of the channel, its resistance increases and the curve of I_D versus V_{DS} for a constant V_{GS} begins to flatten out, as is shown in Fig. 22-4a. This produces the concave curvature in the ohmic region curves shown in Fig. 22-4a. The larger the drain current becomes, the flatter the I_D versus V_{DS} characteristic.

But now an apparent dilemma develops. If I_D is made large enough then $V_{GS} - V_{DS}$ will get reduced to $V_{GS(th)}$. The inversion layer would essentially disappear at the drain end and no current would be able to flow. In reality the situation is more complicated. As I_D increases, the inversion layer does narrow down in thickness, as indicated. However, since the total current is the same everywhere in the channel, the current density at the drain end is higher, since the inversion layer thickness is less. Since the current is flowing by drift (there is no injection of minority carriers, since the junctions are shorted out by the inversion layer), the electric field parallel to the current flow is also larger at the drain end. (Recall $J = \sigma E$ and σ is constant.)

This larger electric field at the drain end is important in two respects. First, as it becomes larger, the electric field (due to V_{GS}) across the gate oxide at the drain end is becoming too small to maintain the inversion layer. The large electric field due to the constricted area of current flow takes over the maintenance of a minimum thickness of inversion layer at the drain end and thus circumvents the apparent dilemma mentioned in the preceding paragraph. Second, the velocity of the charge carriers is a function of the electric field with the velocity saturating at a constant value as the field is increased, as is indicated in Fig. 22-7. At the point where the field at the drain end is large enough to saturate the carrier velocity, the oxide voltage is approximately at the threshold value so that $V_{GS} - V_{DS} = V_{GS(th)}$ and the device is about to enter the active region. Further increases in V_{DD} will increase the electric field in the narrowest part of the channel and will lead to a growth in the length of the minimum thickness region of the channel toward the source as diagrammed in Fig. 22-6b. The voltage drop across the oxide will remain fixed and thus so will the inversion layer thickness. Hence, for $V_{DS} > V_{GS} - V_{GS(th)}$, the drain current remains relatively constant, as is indicated in the i–v characteristic of Fig. 22-4.

If V_{GS} is larger, then the inversion layer thickness is larger and a larger total current is required before the electric field at the drain end is large enough to cause carrier velocity saturation. Simple first-order theoretical calculations indicate that in the active region i_D is given by Eq. 22-2, where the constant K is given by

$$K = \mu_n C_{ox} \frac{W}{2L} \tag{22-5}$$

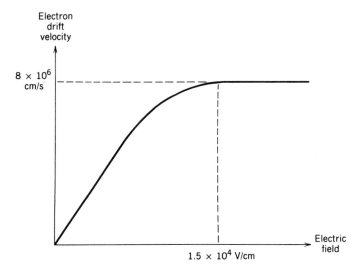

Figure 22-7 Electron drift velocity in silicon versus electric field intensity illustrating velocity saturation. The electron mobility is the incremental slope of the velocity versus electric field curve.

with μ_n being the inversion layer majority-carrier mobility and the other parameters are defined below. This equation points out one of the most important design considerations in the fabrication of MOSFETs, that in order to have appreciable gain, it is essential that the gate width W be much larger than the channel length L. In modern power MOSFETs the length L is kept to a minimum, consistent with breakdown voltage requirements, to minimize the on-state losses (typical channel lengths are a few micrometers). The W/L ratio is typically 10^5 or greater and is achieved by using many thousands of small source regions, as is indicated by Fig. 22-1. The effective gate width in this structure is the total peripheral distance around all the source regions.

The square law i_D–v_{GS} relationship is not maintained at larger values of drain current but instead becomes a linear relationship, as indicated in Fig. 22-4b. The reason for the changeover to a linear relationship is that the mobility in Eq. 22-5 does not remain constant as i_D increases, but instead decreases as the electric field in the inversion layer increases with increasing current. The mobility decreases both because of the velocity–electric field relationship diagrammed in Fig. 22-7 and because larger values of V_{GS} increase the free electron density in the channel. At larger carrier densities the mobility decreases because of what is termed carrier–carrier scattering (see Eq. 20-14a).

22-5   SWITCHING CHARACTERISTICS

22-5-1   MOSFET CIRCUIT MODELS

MOSFETs are intrinsically faster than bipolar devices because they have no excess minority carriers that must be moved into or out of the device as it turns on or off. The only charges that must be moved are those on the stray capacitances and depletion layer capacitances, which are shown in the cross-sectional view of the MOSFET given in Fig. 22-8. These capacitances can be modeled by the equivalent circuit shown in Fig. 22-9a, which is valid when the MOSFET is in cutoff or in the active region. Circuit models such as this are needed for a detailed study of the turn-on and turn-off characteristics of the MOSFET so that appropriate gate drive circuits can be designed.

The drain–source capacitance shown in Fig. 22-8 is not included in the equivalent circuit because it does not materially affect any of the switching characteristics or waveforms. However, it should be considered when designing snubbers, for example, in Chapter 9 where a lossless capacitor snubber is needed in zero-voltage switching converter topologies. There C_{ds} can be a part of the total snubber capacitance requirement.

The gate-voltage-controlled current source shown in the equivalent circuit is equal to zero when $v_{GS} < V_{GS(th)}$ and is equal to $g_m(v_{GS} - V_{GS(th)})$ when the device is in the active region. This method of accounting for the flow of drain current in the active region is suggested by the fact that the transfer characteristic shown in Fig. 22-4b is linear over most of its range. The slope of the transfer characteristic in the active region is the transconductance g_m.

The MOSFET enters the ohmic region when v_{DS} is equal to or less than $v_{GS} - V_{GS(th)}$. In switch-mode power applications $v_{GS} \gg V_{GS(th)}$ when the device is on, so that the criteria for entering the ohmic region can be simplified to $v_{DS} < v_{GS}$. In the ohmic region the dependent current source model is no longer valid because the inversion layer is no longer nearly pinched off at the drain end of the channel but instead has a nearly spatially uniform thickness since v_{DS} is quite small. The inversion layer essentially shorts the drain to the source and so the drain end of C_{gd} is shown in the ohmic region equivalent circuit of Fig. 22-9b as grounded. An on-state resistance $r_{DS(on)}$ is added to the equivalent circuit

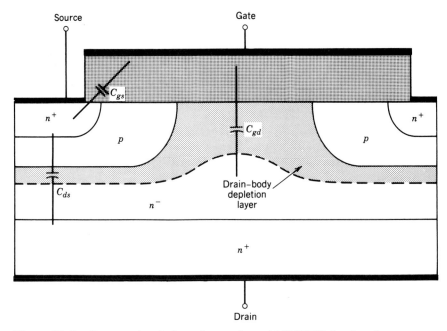

Figure 22-8 Cross-sectional view of an *n*-channel MOSFET showing the approximate origin of the parasitic capacitances that govern the switching speed of the device.

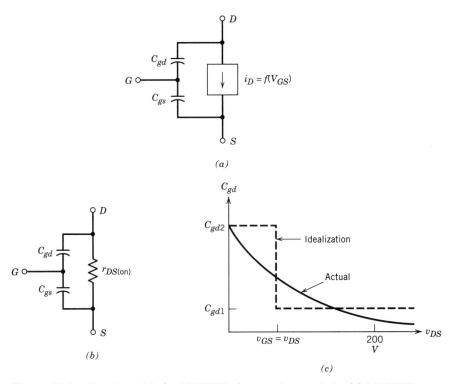

Figure 22-9 Circuit models for MOSFETs for transient analysis: (*a*) MOSFET equivalent circuit for transient analysis in cutoff and active regions; (*b*) MOSFET equivalent circuit in the ohmic region; (*c*) variation in gate–drain capacitance with drain–source voltage.

to account for the ohmic losses, which arise principally from the drain drift region. There are other contributions to the on-state resistance such as ohmic losses in the channel, but they are usually small compared with the drain drift region contribution except for low-breakdown-voltage devices. These other contributions will be discussed in later sections of this chapter.

Note that the capacitances C_{gs} and C_{gd} are not constant but vary with the voltage across them because part of the capacitance is contributed by depletion layers. For example, the gate–source capacitance is the combination of the electrostatic capacitance of the oxide layer in series with the capacitance of the depletion layer that forms at the Si–SiO$_2$ interface. The most significant change in capacitance occurs in C_{gd} because the voltage change across it, v_{DS}, is much larger than the voltage change across C_{gs}. The change in C_{gd} with v_{DG} ($\approx v_{DS}$), which is diagrammed in Fig. 22-9c, can be as large as a factor of 10 to 100. For approximate calculations of switching waveforms, C_{gd} is approximated by the two discrete values C_{gd1} and C_{gd2} shown in Fig. 22-9c with the change in value occurring at $v_{DS} = v_{GS}$ where the MOSFET is either entering or leaving the ohmic region. The gate–source capacitance will be assumed to be constant.

22-5-2 SWITCHING WAVEFORMS

The turn-on behavior of MOSFET embedded in a step-down dc–dc converter, a commonly encountered circuit in power electronics, will be examined. As was done for the analogous BJT circuit, the inductive load is modeled as a constant current source I_o in parallel with a diode D_f as shown in Fig. 22-10. The MOSFET is replaced in Fig. 22-10 by its active region equivalent circuit. The gate is driven by an ideal voltage source, which is assumed to be a step voltage between zero and V_{GG} in series with an external gate resistance R_G. To keep the explanation simple, we assume that the free-wheeling diode in Fig. 22-10 is ideal with a zero reverse-recovery current.

The turn-on waveforms are shown in Fig. 22-11, where the gate drive voltage changes in a step function manner at $t = 0$ from zero to V_{GG}, which is well above $V_{GS(th)}$. During the turn-on delay time $t_{d(on)}$ the gate–source voltage v_{GS} rises from zero to $V_{GS(th)}$

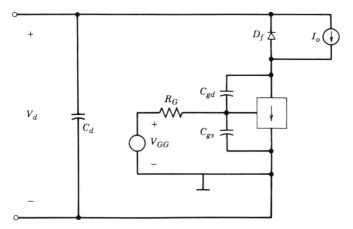

Figure 22-10 A MOSFET used to switch a diode-clamped inductive load. The circuit is essentially that of a step-down dc–dc converter; the equivalent circuit is valid for cutoff and active region transient analysis.

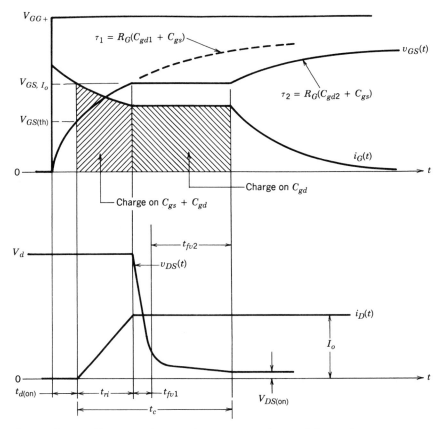

Figure 22-11 Turn-on voltage and current waveforms of the MOSFET embedded in a circuit having a diode-clamped inductive load with an ideal (zero reverse-recovery current) free-wheeling diode.

because of the currents flowing through C_{gs} and C_{gd}, as is shown in Fig. 22-12a. The rate of rise of v_{GS} in this region is almost linear, although it is a part of an exponential curve shown dashed in Fig. 22-11, which has a time constant $\tau_1 = R_G(C_{gs} + C_{gd1})$. Beyond $V_{GS(th)}$, v_{GS} continues to rise as before, and the drain current begins to increase according to the linearized transfer curve shown in Fig. 22-4b. Therefore, the equivalent circuit shown in Fig. 22-12b applies. The drain–source voltage remains at V_d as long as $i_D < I_o$ and the free-wheeling diode D_f is conducting. The time required for i_D to build up from zero to I_o is the current rise time t_{ri}.

Once the MOSFET is carrying the full load current I_o but is still in the active region, the gate–source voltage becomes temporarily clamped at V_{GS,I_o}, which is the gate–source voltage from the transfer curve of Fig. 22-4b needed to maintain $i_D = I_o$. The entire gate current i_G, which is given by

$$i_G = \frac{V_{GG} - V_{GS,I_o}}{R_G} \tag{22-6}$$

flows through C_{gd}, as is indicated in the equivalent circuit of Fig. 22-12c. This causes the drain–source voltage to drop at a rate

$$\frac{dv_{DG}}{dt} = \frac{dv_{DS}}{dt} = \frac{i_G}{C_{gd}} = \frac{V_{GG} - V_{GS,I_o}}{R_G C_{gd}} \tag{22-7}$$

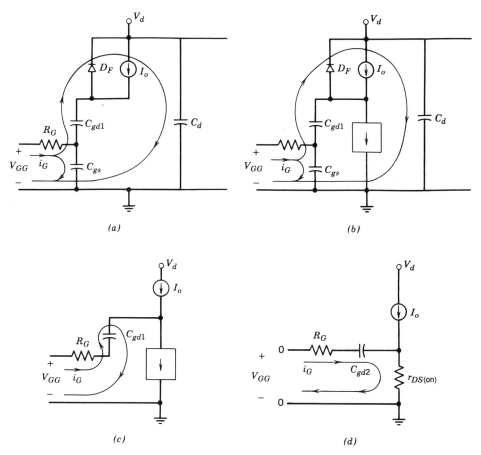

Figure 22-12 Equivalent circuits used to estimate the turn-on current and voltage waveforms of the MOSFET used in the diode-clamped inductive load circuit: (a) equivalent circuit during $t_{d(\text{on})}$; (b) equivalent circuit during t_{ri}; (c) equivalent circuit for the t_{fv1} interval; (d) equivalent circuit during t_{fv2}.

(Recall that $v_{GS} = V_{GS,I_o}$ during this interval so $dv_{GS}/dt = 0$.) The decrease in v_{DS} occurs in two distinct time intervals t_{fv1} and t_{fv2}. The first time interval corresponds to the traverse through the active region where $C_{gd} = C_{gd1}$. The second time interval corresponds to the completion of the transient in the ohmic region where the equivalent circuit shown in Fig. 22-12d applies and $C_{gd} = C_{gd2}$.

Once the drain–source voltage has completed its drop to the on-state value of $I_o r_{DS(\text{on})}$, the gate–source voltage becomes unclamped and continues its exponential growth to V_{GG}. This part of the growth occurs with a time constant $\tau_2 = R_G(C_{gs} + C_{gd2})$, and simultaneously the gate current decays toward zero with the same time constant, as is shown in the waveforms of Fig. 22-11.

If the free-wheeling diode D_f is not ideal but has a reverse-recovery current, then the switching waveforms are modified, as is shown in Fig. 22-13a. During the current rise time interval, the drain current increases beyond I_o to $I_o + I_{rr}$ because of the reverse-recovery current of D_f. This causes v_{GS} to increase beyond V_{GS,I_o}, as is shown in Fig. 22-13a. When the diode current snaps off and recovers to zero, there is a rapid decrease in v_{GS} to V_{GS,I_o}, and this rapid decrease provides an additional current to C_{gd} in addition to i_G, as is shown in the equivalent circuit of Fig. 22-13b. This additional current causes

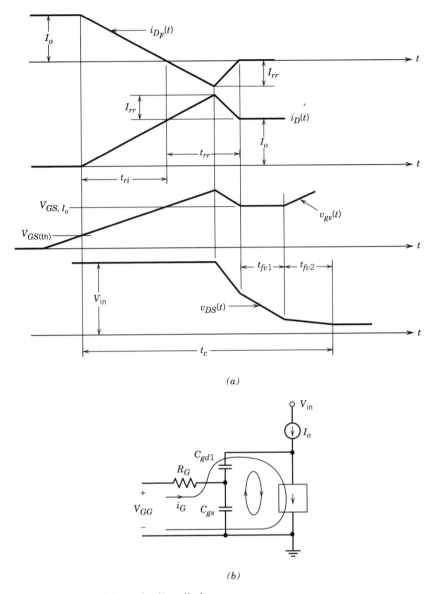

(a)

(b)

Figure 22-13 Effect of free-wheeling diode reverse-recovery current on MOSFET current waveforms at turn-on: (a) MOSFET turn-on waveforms modified by free-wheeling diode turn-off; (b) equivalent circuit for estimating effect of free-wheeling diode reverse recovery.

v_{DG} and v_{DS} to decrease very rapidly during this recovery interval, as is indicated in the waveforms of Fig. 22-13a. Once the reverse-recovery interval is over, the drain current is back to I_o, and the rest of the transient proceeds as in the ideal diode case shown in Fig. 22-11.

The turn-off of the MOSFET involves the inverse sequence of events that occurred during turn-on. The same basic analytical approach used to find the turn-on switching waveforms can be used to find the turn-off waveforms. The turn-off waveforms and associated time intervals are shown in Fig. 22-14 for an assumed step change in the gate drive voltage at $t = 0$ from V_{GG} to zero. The actual values of the switching times will vary

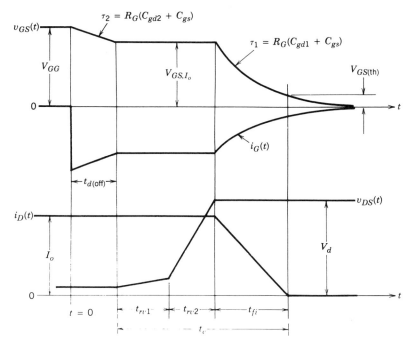

Figure 22-14 The MOSFET current and voltage waveforms at turn-off in the diode-clamped inductive load circuit. The free-wheeling diode is assumed to be ideal.

depending on whether the gate drive voltage is set to zero or made negative to speed up the transient. Moreover, the value of R_G used during turn-off may be different from that used during turn-on.

During turn-on and turn-off, the instantaneous power loss occurs primarily during the crossover time t_c indicated in Figs. 22-11, 22-13, and 22-14 where $p(t) = v_{DS}i_D$ is high. Since the MOSFET capacitances do not vary with the junction temperature, the switching power losses in the MOSFET are also independent of the junction temperature. However, the on-state resistance does vary with temperature, and thus the conduction loss will vary with junction temperature.

22-6 OPERATING LIMITATIONS AND SAFE OPERATING AREAS

22-6-1 VOLTAGE BREAKDOWN

MOSFETs have two voltage ratings that should not be exceeded: $V_{GS(\text{max})}$ and BV_{DSS}. The maximum allowable gate–source voltage $V_{GS(\text{max})}$ is determined by the requirement that the gate oxide not be broken down by large electric fields. Good-quality thermally grown SiO_2 breaks down at electric field values on the order of 5×10^6–10×10^6 V/cm. This means that a gate oxide 1000 Å thick can theoretically withstand a gate–source voltage of 50–100 V. Typical specifications for $V_{GS(\text{max})}$ are 20–30 V, which indicates that device manufacturers put a margin of safety into their ratings. This is done because the breakdown of the gate oxide means permanent failure of the device. Note that even static charge inadvertently put on the gate oxide by careless handling may be sufficient to rupture the oxide. The device user should carefully ground himself before handling any

MOSFET to avoid any static charge problems. If transient gate–source voltages in excess of $V_{GS(max)}$ are a possibility, the gate should be protected by a series connection of two zener diodes connected back to back between the gate and source terminals. The breakdown voltage of the zeners should be less than $V_{GS(max)}$.

The maximum allowable drain–source voltage BV_{DSS} is the largest voltage the MOSFET can hold off without avalanche breakdown of the drain–body *pn* junction. Large values of breakdown voltage are achieved by the use of the lightly doped drain drift region. The lightly doped drain drift region is used to contain the depletion layer of the reverse-biased drain–body junction. The length of the drift region is determined by the desired breakdown voltage rating and is given roughly by Eq. 20-2. The light doping of the drift region compared to the heavy doping of the body region ensures that the depletion layer of the junction does not extend far into the body toward the source region so that breakdown via reach-through (described in conjunction with the BJT) is avoided.

The relatively sharp curvature of the diffused *p*-type body region shown in Fig. 22-1 might lead to a reduction of BV_{DSS} if proper corrective steps are not taken. The extension of the gate metallization over the drain drift region that was pointed out in earlier sections of this chapter acts as a field plate that reduces the curvature of the depletion region. This in turn prevents a severe reduction of the breakdown voltage.

22-6-2   ON-STATE CONDUCTION LOSSES

Except at higher switching frequencies, nearly all of the power dissipated in a MOSFET in a switch-mode power application occurs when the device is in the on state. The instantaneous power dissipation in the on state of the MOSFET is given by

$$p_{on} = I_o^2 r_{DS(on)} \tag{22-8}$$

The on-state resistance has several components, as is illustrated in Fig. 22-15. At lower breakdown voltages (a few hundred volts or less), all of these resistance components contribute more or less equally to the total on-state resistance. The device manufacturer

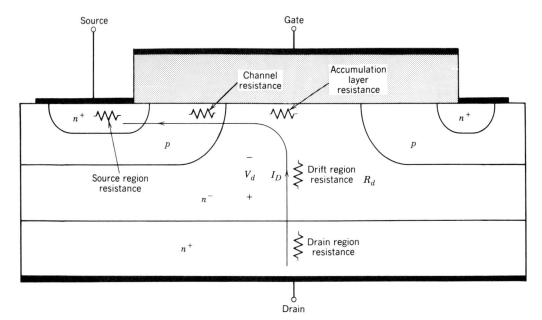

Figure 22-15   On-state resistance components in an *n*-channel enhancement mode MOSFET.

attempts to minimize all of the contributions by using the heaviest doping in each region consistent with other requirements such as breakdown voltage requirements. An example of the detailed attention paid to these resistance contributions is given by the extension of the gate metallization over the drain drift region where it protrudes to the silicon surface between the p-type body regions. This overlapping of the gate metal allows the gate-source bias to enhance the conductivity of the drift region at the interface region between the drift region and the gate oxide by attracting additional free electrons to the interface and creating an accumulation layer.

A premium is also placed on dimensional control of the MOSFET features so that the lengths of current paths in highly resistive regions are minimized. Generally speaking, the design of the source cell and its dimensional tolerances have the greatest impact on $r_{DS(on)}$ in low-breakdown-voltage MOSFETs. The significant progress that has been made in reducing the on-state losses in lower voltage MOSFETs is attested to by the fact that except at perhaps very high current levels, MOSFETs can have lower conduction losses than BJTs for breakdown voltage ratings below a few hundred volts.

Two of the resistance components, the channel resistance and the accumulation layer resistance, are affected by the gate–source bias as well as doping and dimensional considerations. In both of these components larger values of gate–source bias will lower these resistances. Hence, it is desirable to use as large a value of gate–source drive voltage as possible consistent with other considerations such as gate oxide breakdown.

For BV_{DSS} greater than a few hundred volts, the drain drift region resistance R_d dominates the on-state resistance. An optimistic estimate of the specific resistance (Ω-cm^2) of the drift region as a function of breakdown voltage rating can be obtained using Eq. 20-20. The result is

$$\frac{V_d}{J} = R_d A \approx 3 \times 10^{-7} BV_{DSS}^2 \qquad (22\text{-}9)$$

where A is the cross-sectional area through which the drain current flows. Experimental results and more exact theoretical treatments indicate that the dependence is actually $BV_{DSS}^{2.5-2.7}$. Because of the steep dependence of R_d on the breakdown voltage rating, MOSFETs will have higher on-state losses at the larger blocking voltages compared with BJTs.

The on-state resistance increases significantly with increasing junction temperature. This in turn means that the on-state power dissipation will increase with temperature in most power electronic applications like that illustrated in Fig. 22-10. The positive temperature coefficient of the on-state resistance arises from the decrease of the carrier mobility as the semiconductor temperature increases. This occurs because at higher temperatures the charge carriers undergo more collisions per unit time with the semiconductor lattice because each lattice atom has a larger amplitude of vibration at higher temperatures. The mobility is approximately inversely proportional to the number of collisions per unit time with the lattice, and $r_{DS(on)}$ is inversely proportional to the mobility.

22-6-3 PARALLELING OF MOSFETs

MOSFETs can be paralleled very easily, as are the two shown in Fig. 22-16, because of the positive temperature coefficient of their on-state resistance. For the same junction temperature, if $r_{DS(on)}$ of T_2 exceeds that of T_1, then during the on state, T_1 will have a higher current and thus higher power loss compared to T_2, since the same voltage appears across both transistors. Therefore, the junction temperature of T_1 will increase along with its on-state resistance. This will cause its share of current to decrease and, hence, there is a thermal stabilization effect.

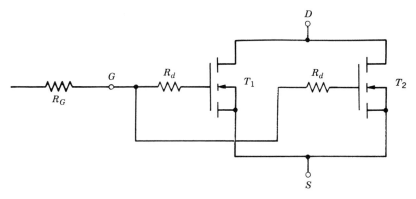

Figure 22-16   Parallel connection of MOSFETs. A small damping resistance should be included in series with the gate of each MOSFET to minimize any possible high-frequency oscillation.

During switching, the current in each MOSFET is determined by the transfer characteristic, as was explained in previous sections (see Fig. 22-4). The variation in the transfer characteristic from one device to another of the same part number is modest. Hence, it is best to keep the gate–source voltage of the paralleled transistors the same during switching so that they share approximately equal currents during switching. However, the gates cannot be directly connected together, but rather a small resistance or ferrite bead must be used in series with the individual gate connections, as is shown in Fig. 22-16. This is because the gate inputs are highly capacitive with almost no losses, although some stray inductance is always present. The stray inductances in combination with the gate capacitances can result in unwanted high-frequency oscillations in the MOSFETs, which are avoided by the damping resistance shown in Fig. 22-16. Another consideration to keep in mind while paralleling MOSFETs is that their layout should be symmetrical.

22-6-4   PARASITIC BJT

The MOSFET has a parasitic BJT as an integral part of its structure, as is shown in Fig. 22-1. The body region of the MOSFET serves as the base of the BJT, the source as the BJT emitter, and the drain as the BJT collector. The beta of this parasitic BJT may be significantly greater than 1 because the length of the body region where the channel of the MOSFET is formed is kept as short as possible to help minimize the on-state resistance.

It is imperative that this BJT be kept cut off at all times by keeping the potential of the parasitic base as close to the source potential as possible. This is the purpose of the body–source short mentioned previously and shown in Fig. 22-1. If the base of the parasitic BJT were allowed to float, two potential problems would arise. First, the breakdown voltage of the MOSFET would be reduced from $BV_{DSS} = BV_{CBO}$ to BV_{CEO}, a drop that might be as large as 50%. This drop in breakdown voltage could lead to excessive power dissipation because of large breakdown currents if the drain–source voltage exceeded the reduced breakdown voltage.

Second, the base–emitter potential may get large enough for the BJT to turn on and possibly go into saturation, a condition termed latchup. This situation is dangerous because there will be significant power dissipation, and worse yet, the BJT cannot be turned off via the base terminal because the base is not accessible. The only way to turn off the

BJT once latchup has occurred is to interrupt the flow of drain current external to the device.

Although the body–source short is quite effective in preventing BJT turn-on from a static viewpoint, it does not guarantee that turn-on cannot occur during high-speed turn-off of the MOSFET. The base of the parasitic BJT is connected to the drain terminal by a portion of the drain–gate capacitance, as is shown in Fig. 22-17a and 22-17b. If the rate of rise of v_{DS} is large enough during turn-off of the MOSFET, the displacement current through the coupling capacitor in Fig. 22-17 to the base of the BJT may be large enough to induce a voltage drop in the parasitic resistance connected between the base and emitter, as is shown in Fig. 22-17a, to turn the BJT on. This parasitic resistance arises from the distributed nature of the base (body) region and the remote location of the body–source short compared to the interior of the body region. This potential BJT turn-on mechanism imposes a maximum rate of rise dv_{DS}/dt on the MOSFET. Fortunately, this potential problem has been recognized by device designers and has been largely eliminated. Modern power MOSFETs have dv_{DS}/dt capabilities in excess of 10,000 V/μs. This potential problem can be easily circumvented by slowing down the switching speed of the MOSFET by using larger values of gate resistance R_G and smaller values of turn-off gate drive.

This unwanted turn-on is most likely to happen in bridge circuits such as shown in Fig. 22-17c where the integral or parasitic diode (the base–collector diode of the parasitic BJT) mentioned previously is used as the free-wheeling diode. When T_+ and T_- are both off, the integral diode of T_- is carrying the load current. At the end of the blanking time, T_+ is turned on and the integral diode of T_- undergoes reverse recovery and the reverse-recovery current flows through the C_{gd} of T_-. If the reverse recovery (snap-off) of this integral diode is very fast, a large positive dv_{DS}/dt will be imposed on C_{gd} and the combination of this displacement current plus the reverse-recovery current may be sufficient to turn on the parasitic BJT of T_-, as discussed in the preceding paragraph. This will result in the destruction of T_- since the parasitic BJT will be conducting a large current while the full voltage between the positive and negative power supply rails is applied across the drain–source terminals of T_-.

This problem can be solved by using the circuit shown in Fig. 22-17d, where an external free-wheeling diode is used and another diode D_L is put in series with each MOSFET, as is indicated in the figure. The D_L diodes prevent any current flowing through the parasitic diode when the MOSFET is controlled to be off and so all of the current is forced to flow through the external free-wheeling diode. In the off state the D_L diodes need only block the on-state voltage of the free-wheeling diodes, which will be small (a few volts) so Schottky diodes could be used for D_{L+} and D_{L-}. Fortunately MOSFETs with BV_{DSS} less than about 200 V are now available where the reverse-recovery current of the parasitic diodes is minimized and the parasitic body–source resistance is small enough so that the MOSFETs can be used, as shown in Fig. 22-17c.

22-6-5 SAFE OPERATING AREA

The SOA of a power MOSFET is shown in Fig. 22-18. Three factors determine the SOA of the MOSFET: the maximum drain current I_{DM}; the internal junction temperature T_j, which is governed by the power dissipation in the device; and the breakdown voltage BV_{DSS}. These limiting factors have been discussed to some extent already in this chapter and are analogous to those of the BJT and so need no further elaboration. The MOSFET does not have any second breakdown limitations as does the BJT, and so none show up on the SOA.

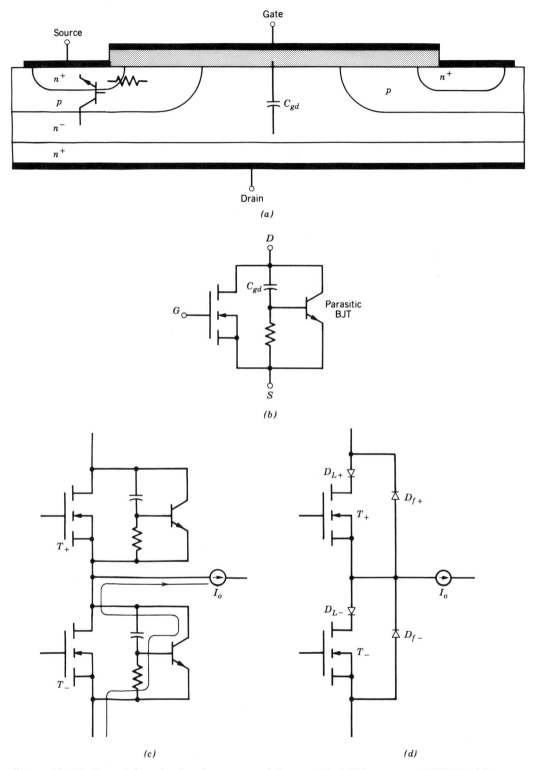

Figure 22-17 Potential mechanism for turn-on of the parasitic BJT in a power MOSFET: (*a*) cross section of the MOSFET showing how the parasitic BJT might be turned on by a fast-rising drain–source voltage; (*b*) equivalent circuit of the MOSFET showing the potential turn-on of the parasitic BJT; (*c*) mechanism of possible destruction of MOSFETs used in a bridge configuration. This latchup can be prevented using diodes D_{L+} and D_{L-} instead of directly connecting D_{f+} and D_{f-} to the MOSFET drains.

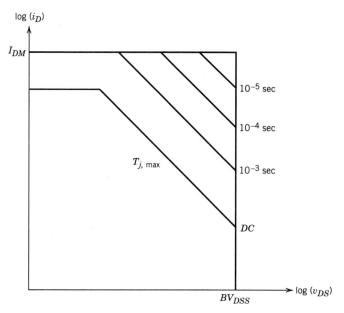

Figure 22-18 The SOA of an *n*-channel enhancement mode
MOSFET. Note the absence of second breakdown.

For switch-mode applications, the SOA of the MOSFET is square, as is indicated in
Fig. 22-18. There is no distinction between forward-bias and reverse-bias SOAs for the
MOSFET; they are identical.

SUMMARY

This chapter has studied the structure and characteristics of MOSFETs designed for power
applications:

1. The MOSFETs have a vertically oriented structure with a lightly doped drain drift
   region and a highly interdigitated gate–source structure.
2. The MOSFET is a normally-off device, and it is turned on by the application of a
   sufficiently large gate–source voltage to induce an inversion layer in the MOSFET
   channel region that shorts the drain to the source.
3. The MOSFET turns on and off very rapidly because it is a majority-carrier device and
   there is no stored charge that must be injected into or removed from it as there is with
   the BJT.
4. On-state losses in a MOSFET rise much more rapidly with blocking voltage rating than
   do those in a BJT.
5. Because the MOSFET is a majority-carrier device, its on-state resistance has a positive
   temperature coefficient, which makes it easy to parallel MOSFETs for increased
   current-handling capability.
6. The SOA of a MOSFET for switch-mode applications is large (rectangular) because it
   is not subject to second breakdown.

7. The MOSFET has a parasitic BJT in its structure that may latch in the on state in extreme circumstances, such as very large rates of increase in the drain–source voltage. Latchup of MOSFETs is most likely to occur in bridge circuits.

8. The large SOA of the MOSFET means that in most situations snubber circuits are not needed for the device.

PROBLEMS

22-1 The small-signal gate–source capacitance of a MOSFET decreases as V_{GS} is increased from zero volts. For $V_{GS} > V_{GS(th)}$, the small-signal capacitance is constant. Qualitatively explain the reasons for this behavior.

21-2 An n-channel MOSFET is to be used in a step-down converter circuit. The dc voltage $V_d = 300$ V, the load current $I_o = 10$ A, the free-wheeling diode is ideal, and the MOSFET is driven by a 15-V (base-to-peak) square wave (50% duty cycle and zero dc value) in series with 50 Ω. The MOSFET characteristics are $V_{GS(th)} = 4$ V, $I_D = 10$ A at $V_{GS} = 7$ V, $C_{gs} = 1000$ pF, $C_{gd} = 150$ pF, and $r_{DS(on)} = 0.5$ Ω.

   (a) Sketch and dimension $v_{DS}(t)$ and $i_D(t)$.

   (b) Estimate the power dissipation at a switching frequency of 20 kHz.

22-3 The switching times of a MOSFET with $V_{GS(th)} = 4$ V, $g_m = 1$ mho, and $C_{gs} = 1000$ pF are measured in a resistively loaded test circuit having a load resistance $R_D = 25$ Ω and a power supply voltage $V_d = 25$ V. The MOSFET is driven by a unipolar square wave of 15 V in series with 5 Ω. The measured switching times are $t_{fv} = t_{ri} = 30$ ns and $t_{fi} = t_{rv} = 70$ ns. The MOSFET is to be used in a resistively loaded circuit having $R_D = 150$ Ω, $V_d = 300$ V, and the drive circuit is a 15-V unipolar square wave in series with 100 Ω. What will the switching times be in this circuit?

22-4 The MOSFET used in Problem 22-3 has an on-state resistance $r_{DS(on)} = 2$ Ω at a junction temperature $T_j = 25°$C. This resistance increases linearly with increasing T_j, becoming equal to 3 Ω at $T_j = 100°$C. Plot the power dissipated in the MOSFET versus T_j when the MOSFET is used in the circuit of Problem 22-3 ($R_D = 150$ Ω and $V_d = 300$ V). Assume a switching frequency of 10 kHz.

22-5 Three MOSFETS, each rate at 2 A, are to be used in parallel to sink 5 A load current when they are on. The nominal on-state resistance of the MOSFETs is 2 Ω at $T_j = 25°$C, but measurements indicate that the actual values for each MOSFET are $r_{DS(on)1} = 1.8$ Ω, $r_{DS(on)2} = 2.0$ Ω, and $r_{DS(on)3} = 2.2$ Ω. The on-state resistance increases linearly with temperature and is 1.8 times larger at $T_j = 125°$C. How much power is dissipated in each MOSFET at a junction temperature of 105°C? Assume a 50% duty cycle.

22-6 A hybrid power switch composed of a MOSFET and a BJT connected in parallel is to be used in a switch-mode power application. Explain what the advantages of such a hybrid switch are and what the relative timing of the turn-on and turn-off of the two devices should be.

22-7 Design a MOSFET similar to that shown in Fig. 22-1a for a breakdown voltage $BV_{DSS} = 750$ volts. Specify the channel length (drain-source distance), drift region doping density and length. You may assume that the body region doping density $N_{body} = 5 \times 10^{16}$ cm^{-3}.

22-8 Consider the equivalent circuit for the power MOSFET shown in Fig. 22-17b which includes the parasitic BJT. Approximately estimate the rate of rise of the drain-source voltage, dv_{DS}/dt, that will turn on the BJT. Express your answer in terms of the gate-drain capacitance, C_{gd}, the body resistance, R_{body}, and other pertinent circuit parameters.

22-9 The gate oxide for a power MOSFET is 500 angstroms thick. What should be the maximum gate-source voltage, $V_{gs,max}$? Include a 50% factor of safety. Assume the breakdown field strength of the gate dielectric is 5×10^6 V/cm.

22-10 The transfer curve, i_D versus v_{GS}, of a power MOSFET operating in the active region is approximately given by $i_D = \mu_n C_{ox} \dfrac{W}{2L} [v_{GS} - V_{GS(th)}]$ where C_{ox} is given by Eq. (22-4) and $W = NW_{cell}$.

N is the number of identical cells connected in parallel and W_{cell} is the effective width of the gate in each cell. A power MOSFET is to be designed to conduct a drain current of 100 A when $V_{GS} = 15$ V and the threshold voltage $V_{GS(th)} = 4$ V. Assume $L = 1$ micron, $W_{\text{cell}} = 20$ microns, and $t_{\text{ox}} = 1000$ angstroms.

(a) How many cells are required for this MOSFET?

(b) How much current is carried by each cell?

22-11 A power MOSFET with a breakdown rating $BV_{DSS} = 800$ V must conduct 10 amps when on with a maximum on-state voltage $V_{DS,\text{on}} = 4$ V. If the current density in the MOSFET must be limited to 200 A/cm², estimate the conducting area which is required.

22-12 What is the approximate gate-source capacitance of the MOSFET described in Problem 22-10? The gate oxide is 1000 angstroms thick.

22-13 A MOSFET step-down converter such as shown in Fig. 22-10 operates at a switching frequency of 30 kHz with a 50% duty cycle at an ambient temperature of 50°C. The power supply $V_d = 100$ V and the load current $I_o = 100$ A. The free-wheeling diode is ideal but a stray inductance of 100 nanohenries is in series with the diode. The MOSFET characteristics are listed below:

$BV_{DSS} = 150$ V; $T_{j,\text{max}} = 150$°C; $R_{\theta,j\text{-}a} = 1$°C/W; $r_{DS(\text{on})} = 0.01$ ohm
$t_{ri} = t_{fi} = 50$ ns; $t_{rv} = t_{fv} = 200$ ns; $I_{D,\text{max}} = 125$ A

Is the MOSFET overstressed in this application and if so, how? Be specific and quantitative in your answer.

22-14 Consider the MOSFET step-down converter circuit shown in Fig. 22-10. The voltage rise and fall times, t_{rv} and t_{fv}, are significantly larger than current rise and fall times, t_{ri} and t_{fi} particularly at larger power supply voltages (>100 V). Briefly explain why this is so. Assume that the values of C_{gs} and C_{gd} are constant and independent of V_{DS}, that $C_{gd1} = C_{gd2} = C_{gd}$, and that the free-wheeling diode D_f is ideal.

REFERENCES

1. B. Jayant Baliga, *Modern Power Devices,* Wiley, New York, 1987, Chapter 6.
2. P. Aloisi, *Power Switch,* Motorola Inc., 1986, Section 2.
3. E. S. Oxner, *Power FETs and Their Applications*, Prentice-Hall, Englewood Cliffs, NJ, 1982, Chapters 1–4, 9.
4. B. W. Williams, *Power Electronics, Devices, Drivers, and Applications*, Wiley, New York, 1987, Chapters 3, 4, 7–10.
5. A. Blicher, *Field Effect and Bipolar Power Transistor Physics*, Academic, New York, 1981, Chapters 12, 13.
6. B. Jayant Baliga and D. Y. Chen (Eds.), *Power Transistors, Device Design and Applications*, IEEE Press, Institute of Electrical and Electronic Engineers, New York, 1984, Part III, *Power MOSFET Field Effect Transistors*, pp. 197–290.
7. M. S. Adler, K. W. Owyang, B. Jayant Baliga, and R. A. Kokosa, "The Evolution of Power Device Technology," *IEEE Transactions on Electron Devices*, Vol. ED-31, No. 11, Nov. 1984, pp. 1570–1591.

CHAPTER 23

THYRISTORS

23-1 INTRODUCTION

Thyristors (sometimes termed SCRs, meaning semiconductor-controlled rectifiers) are one of the oldest (1957 in General Electric Research Laboratories) types of solid-state power device and still have the highest power-handling capability. They have a unique four-layer construction and are a latching switch that can be turned on by the control terminal (gate) but cannot be turned off by the gate. The characteristics of the thyristor (particularly their large power-handling capability) ensures that they will always have important power electronic applications. Thus, the designer and user of power electronic devices and circuits must have a working knowledge of these devices.

23-2 BASIC STRUCTURE

The vertical cross section of a generic thyristor is shown in Fig. 23-1a. The approximate thicknesses of each of the four alternating layers of p-type and n-type doping that comprise the structure are indicated as well as the approximate doping densities. In terms of their lateral dimensions, thyristors are among the largest semiconductor devices made. A complete silicon wafer as large as 10 cm in diameter may be used to make a single high-power thyristor. Plan views of two different gate and cathode layouts are shown in Fig. 23-1b. The distributed gate structure is for a large-diameter (10-cm) thyristor while the localized gate electrode structure is for a smaller diameter thyristor. In general, the layout of the gates and cathodes of thyristors varies greatly depending on the diameter of the thyristor, the intended di/dt capability, and the intended range of switching speeds.

The circuit symbol for the thyristor is shown in Fig. 23-1c. It is essentially the symbol of a diode (or rectifier) with a third control terminal, the gate, added to it. The voltage and current conventions for the thyristor are shown in the figure.

The vertical cross section of the thyristor appears similar to that of the BJT, including some of the doping densities and layer thicknesses. The cathode is in the same location as the emitter of the BJT and the thyristor gate location is analogous to the base of the BJT. The n^- region of the thyristor absorbs the depletion layer of the junction that blocks the applied voltage when the thyristor is in the off state, thus performing the same function as the n^- collector drift region of the BJT.

The p layer that forms the anode of the thyristor is a feature of the thyristor structure not found in the BJT. This anode layer causes the thyristor to have characteristics quite different from those of the BJT.

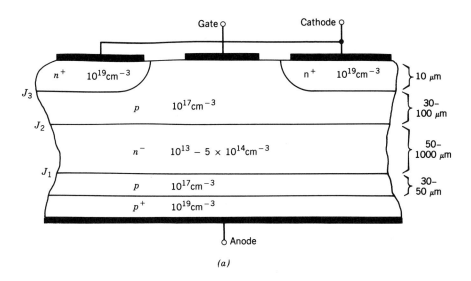

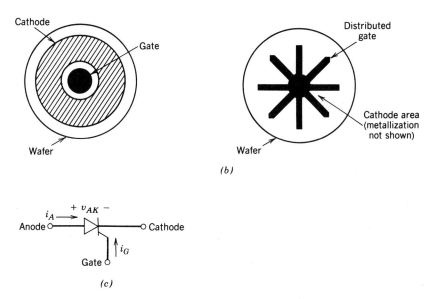

Figure 23-1 Structural details of a generic thyristor: (*a*) vertical cross section; (*b*) gate and cathode layouts; (*c*) circuit symbol.

23-3 *I–V* CHARACTERISTICS

The uniqueness of the thyristor lies principally in its *i–v* characteristic (anode current i_A as a function of the anode-to-cathode voltage v_{AK}), which is shown in Fig. 23-2. In the reverse direction the thyristor appears similar to a reverse-biased diode, which conducts very little current until avalanche breakdown occurs. For a thyristor the maximum reverse working voltage is termed V_{RWM} (devices with values of V_{RWM} as large as 7000 V are available). In the forward direction the thyristor has two stable states or modes of operation that are connected together by an unstable mode that appears as a negative resistance

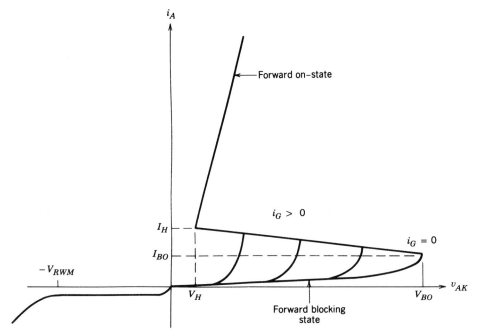

Figure 23-2   Current–voltage characteristic of a thyristor.

on the i–v characteristic. The low-current, high-voltage region is the forward-blocking state or the off state, and the low-voltage, high-current mode is the on state. Both of these states are indicated on the i–v characteristic. In the on state a high-power thyristor can conduct average currents as large as 2000–3000 A with on-state voltage drops of only a few volts.

Specific voltage and current values in the forward-bias quadrant of the i–v characteristic are of interest to the user and appear on specification sheets. The holding current I_H represents the minimum current that can flow through the thyristor and still maintain the device in the on state. This current value and the accompanying voltage across the device, termed the holding voltage V_H, represent the lowest possible extension of the on-state portion of the i–v characteristic. For the forward-blocking state, the quantities of interest are the forward-blocking voltage V_{BO} (sometimes termed the breakover voltage V_{BO} because the i–v curve breaks over and goes to the on-state portion of the characteristic) and the accompanying breakover current I_{BO}.

The breakover voltage and current are defined for zero gate current, that is, the gate is open-circuited. If a positive gate current is applied to the thyristor, then the transition or breakover to the on state will occur at smaller values of anode-to-cathode voltage, as indicated in Fig. 23-2. As indicated in this figure, the thyristor will switch to the on state at low values of v_{AK} if the gate current is reasonably large. Although not indicated on the i–v characteristic, the gate current does not have to be a dc current, but instead can be a pulse of current having some minimum time duration. This ability to switch the thyristor on by means of a current pulse has been the basis of the widespread applications of the device.

However, once the thyristor is in the on state, the gate cannot be used to turn the device off. The only way to turn off a thyristor is for the external circuit to force the current through the device to be less than the holding current for a minimum specified time period. Special designs of thyristors termed GTO (gate turn-off) devices have been developed where the gate can be used to turn off the device. These devices will be discussed in the next chapter.

23-4 PHYSICS OF DEVICE OPERATION

23-4-1 BLOCKING STATES

In describing how the thyristor operates from a physical point of view, it is convenient to consider the device as the idealized one-dimensional structure shown in Fig. 23-3a. An approximate low-frequency equivalent circuit composed of a *pnp* and an *npn* transistor is shown in Fig. 23-3b, which is easily derived from the one-dimensional model.

In the reverse-blocking state, the anode is biased negative with respect to the cathode. Junctions J_1 and J_3 indicated in Fig. 23-3a are reverse biased and J_2 is forward biased. Junction J_1 must support the reverse voltage because J_3 has a low breakdown voltage as a consequence of the heavy doping on both sides of the junction (examine Fig. 23-1). The reverse-blocking capability of junction J_1 is usually limited by the length of the n^- (n_1) region, which is approximately set by the avalanche breakdown limit given by Eq. 20-3.

In the forward-blocking state, the junctions J_1 and J_3 are forward biased and J_2 is reverse biased. The doping densities in each of the layers are such that the n^- layer (n_1 layer) is where the depletion region of the reverse-biased J_2 junction appears, and thus this region again determines the blocking voltage capability, this time for the forward-blocking state. Generally, the thyristor is designed so that the forward-blocking voltage V_{BO} will be about the same as the reverse-blocking voltage V_{RWM}.

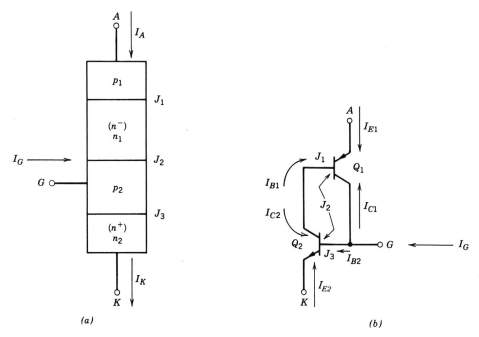

Figure 23-3 Simplified models of a thyristor: (*a*) one-dimensional model of a thyristor; (*b*) two-transistor equivalent circuit of a thyristor.

The values of V_{BO} and V_{RWM} begin to fall rapidly with increasing temperature as the junction temperature rises above roughly 150°C. The detailed explanation for this temperature dependence is beyond the scope of this discussion. Suffice it to say, most device manufacturers specify the maximum junction temperature for their thyristors at 125°C.

On the basis of the bias conditions of the three junctions in the forward-blocking state, it can be concluded that both transistors Q_1 and Q_2 in the thyristor equivalent circuit, Fig. 23-3b, are in the active region. Recognition of this fact makes possible a relatively simple, although admittedly qualitative, explanation of the forward-bias i–v characteristic of the thyristor. A BJT in the active region can be described, at low frequencies, by the Ebers–Moll equations. For transistors Q_1 and Q_2 in the active region, these equations are (the reader is referred to the references for a detailed derivation of these relations)

$$I_{C1} = -\alpha_1 I_{E1} + I_{CO1} \qquad (23\text{-}1)$$

$$I_{C2} = -\alpha_2 I_{E2} + I_{CO2} \qquad (23\text{-}2)$$

where the leakage current I_{CO} is given by

$$I_{CO} = I_{CS}[1 - \alpha_f \alpha_r] \qquad (23\text{-}3)$$

In Eq. 23-3, I_{CS} is the reverse saturation current of the collector–base diode with the emitter open-circuited and α_f and α_r are the forward active mode and reverse active mode base transport coefficients (or alphas where $\alpha \approx \beta/[1 + \beta]$), respectively, of the transistors. (In the description of how an npn BJT operates, which is given in Chapter 21, the base transport factor is the ratio of the electron current at the collector divided by the injected electron current at the emitter, or using the notation of Eqs. 21-2 through 21-4, $\alpha = I_{nc}/I_{ne}$.) If we note that $I_A = I_{E1}$ and that $I_K = -I_{E2} = I_A + I_G$ and setting the sum of all the currents into one of the transistors to zero, it can be shown that I_A is given by

$$I_A = \frac{\alpha_2 I_G + I_{CO1} + I_{CO2}}{1 - (\alpha_1 + \alpha_2)} \qquad (23\text{-}4)$$

In the blocking state the sum of $\alpha_1 + \alpha_2$ must be much less than unity so that the anode current I_A can be kept quite small, ranging from microamperes for low-current devices to a few hundred milliamperes for high-current thyristors.

23-4-2 TURN-ON PROCESS

Equation 23-4 indicates that if $\alpha_1 + \alpha_2$ approaches unity, the anode current will be arbitrarily large. If this occurs, the thyristor will be at the breakover point where it is about to go into the negative-resistance state, where the current gain (β) of both BJTs is greater than 1. The negative-resistance region is unstable because of the regenerative (positive-feedback) connection of the two transistors. Once in this region, the device will quickly carry itself to the stable on state. The key to the turn-on process is then understanding how the base transport coefficients (alphas) of the BJTs can be increased from the small values required for blocking state operation to the point where their sum is unity.

The mechanism that causes alphas to increase is the growth of the depletion region of junction J_2 into the n_1 layer as the anode–cathode voltage is increased. This causes the effective base thickness of the $p_1 n_1 p_2$ BJT to get smaller and, hence, for α_{pnp} to increase. The extension of the J_2 depletion layer into the p_2 region (the base region of the npn transistor) will likewise cause an increase in α_{npn}.

The combination of the positive-feedback connection of the npn and pnp BJTs and the current-dependent base transport factors is what makes it possible for the gate terminal to effect a turn-on of the thyristor. If a positive gate current of sufficient magnitude is applied to the thyristor, a significant amount of electron injection across the forward-biased J_3

junction into the p_2 base layer of the *npn* transistor will occur. The electrons will diffuse across the base and be swept across junction J_2 into the n_1 base layer of the *pnp* transistor.

These extra electrons in the n_1 layer will have two simultaneous effects. First, the depletion layer of junction J_2 will grow in thickness because additional positive space charge from ionized donors is needed to partially compensate for the negative space charge of the electrons. This growth of the depletion layer into the n_1 layer will reduce the effective base thickness of the *pnp* transistor and thus cause α_{pnp} to get larger. Second, the injection of majority carriers (electrons) into the base of the *pnp* transistor will cause the injection of holes into this base layer (because of space charge neutrality requirements) via injection from the base–emitter junction of the *pnp* transistor (the p_1n_1 junction). These injected holes will diffuse across the base and be swept across the depletion region of the J_2 junction and into the base region of the *npn* BJT. These holes will cause a further increase in electron injection (because the positive space charge of the holes attracts the electrons from the n_2 layer) into the base region of the *npn* BJT. These additional injected electrons will go through the same cycle just described for the initial injection of electrons. This positive reinforcement is thus a regenerative process.

This regenerative action carries the thyristor into the on state. The large current flow between the anode and cathode injects enough carriers into the base regions to keep the BJTs saturated without any continuous gate current flow. This is the origin of the latching action of the thyristor described earlier.

23-4-3 ON-STATE OPERATION

In the on state there is strong minority-carrier injection in all four regions of the thyristor structure. The stored charge distribution in the four regions is shown schematically in Fig. 23-4. Junction J_2 is forward biased, and the BJTs in the equivalent circuit are saturated.

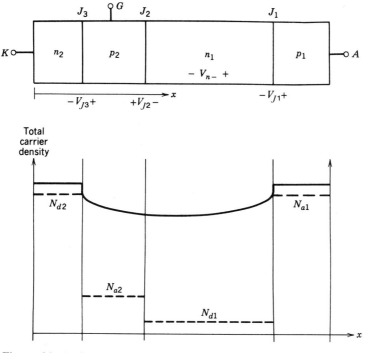

Figure 23-4 On-state carrier distributions in a thyristor.

In this situation large forward currents can flow (dictated by the external circuit), and only a small forward voltage drop occurs because of the large conductivity modulation represented by the stored charge distribution illustrated in Fig. 23-4. The on-state voltage is given approximately by

$$V_{AK(on)} \simeq V_{J1} - V_{J2} + V_{J3} + V_{n-} \qquad (23\text{-}5)$$

where the V_J's are the forward-bias junction voltages (0.7–0.9 V) and V_{n-}, which is on the order of a few tenths of a volt, is relatively independent of the current through the device and is given approximately by Eq. 20-13. The value of $V_{AK(on)}$ in Eq. 23-5 is quite similar to the expression for the on-state voltage of a diode (Eq. 20-17). At large current densities, the on-state voltage will increase with increasing current because of Auger recombination and reduction in carrier mobilities as described in Chapter 20 as well as parasitic ohmic resistances. The net effect of these factors is an on-state resistance.

23-4-4 TURN-OFF PROCESS

As we mentioned previously, once the thyristor is latched into the on state, the gate terminal no longer has any control over the state of the device. In particular, the gate cannot be used to turn the thyristor off. Turn-off can be accomplished only the external circuit, reducing the anode current below the holding current for a minimum specified period of time. During this time period, the simultaneous action of internal recombination and carrier sweep-out will remove enough stored charge so that the BJTs are pulled out of saturation and into the active region. When this occurs, the device will turn off via the regenerative connection of the transistors.

In the standard thyristor, a negative gate current cannot turn off the device because the cathode region is much greater in area than the gate area. When a negative gate current flows, it can only locally reverse bias the gate–cathode (base–emitter of the *npn* transistor) junction, as is shown in Fig. 23-5. Lateral voltage drops in the p_2 region caused by the negative i_G will result in current crowding toward the center of the cathode region similar to emitter current crowding during the turn-off of a BJT. The anode current that

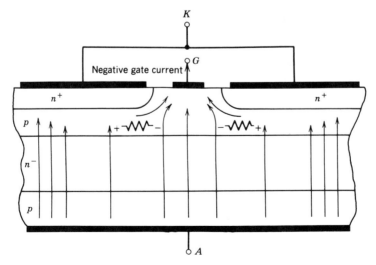

Figure 23-5 Current density distribution in a thyristor during attempted turn-off with a negative gate current illustrating current constriction at the center of the cathode.

now flows more in the center of each cathode region keeps the gate–cathode junction forward biased in this center portion and the thyristor stays on in spite of a negative gate current.

23-5 SWITCHING CHARACTERISTICS

23-5-1 TURN-ON TRANSIENT AND *di/dt* LIMITATIONS

In describing the turn-on transient of a thyristor, we will assume that the device is embedded in the circuit of Fig. 23-6, which is a simplified diagram of a multiple-phase controlled rectifier. Turn-on of a thyristor is accomplished by applying a pulse of current of specified magnitude and duration to the gate of the device. The gate current is applied at $t = 0$, as is shown in Fig. 23-7, to thyristor T_A in Fig. 23-6 with $t = 0$ corresponding to a time when the voltage in phase A is larger than in the other two phases. The resulting waveforms for the anode current and anode–cathode voltage are shown in Fig. 23-7. The anode current increases at a fixed rate di_F/dt, which is set by the external circuit because of the switching times of other devices or because of stray inductance in the circuit. Three distinct time intervals can be defined: the turn-on delay time $t_{d(\text{on})}$, the rise time t_r, and the spreading time t_{ps}.

During the turn-on delay time, the thyristor appears to remain in the blocking state. However, the gate current during this time is injecting excess carriers into the p_2 layer (the base of the *npn* transistor in the equivalent circuit) in the vicinity of the gate contact, as is illustrated in Fig. 23-8*a*. This increase in excess carriers causes the sum of the base transport factors $\alpha_1 + \alpha_2$ to increase until they equal unity. At this point, the thyristor is at breakover and heavy electron injection into the p_2 layer from the n_2 cathode layer and hole injection from the p_1 layer into the n_1 layer commences in the vicinity of the gate regions, as is illustrated in Fig. 23-8*b*. The anode current begins to increase, and this marks the end of the turn-on delay time and the start of the rise time interval.

During the rise time interval, a large excess-carrier density or plasma is built up in the vicinity of the gate regions, which then spreads laterally across the face of the cathode until the entire cross-sectional area of the thyristor is filled with a high excess-carrier

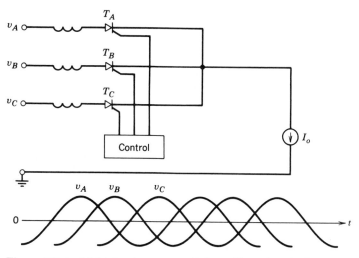

Figure 23-6 Multiple-phase-controlled rectifier using thyristors.

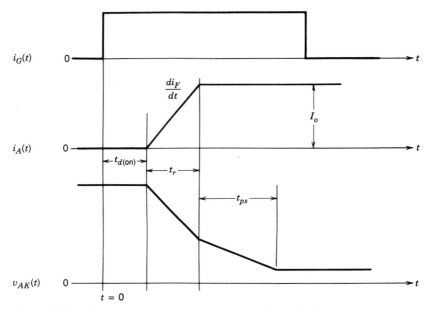

Figure 23-7 Thyristor voltage and current waveforms during turn-on.

density. Simultaneously, there is the commencement and growth of carrier injection from the p_1 anode region into the n layer that forms the base of the pnp transistor. The rate of current rise is usually large enough that the anode current reaches its constant on-state value in significantly shorter times than are required for the excess-carrier injection to spread laterally across the entire face of the cathode region. The attainment of the on-state value of the anode current marks the end of the rise time interval.

As the excess-carrier density becomes established and grows, as is shown in Fig. 23-8b, the anode–cathode voltage begins to drop. During the rise time interval, the voltage drop is fairly rapid because the localized regions of high excess-carrier densities in the vicinity of the gate regions provide a significant reduction in the blocking capabilities of the thyristor. Even after the rise time interval is over, the plasma still continues to spread over the lateral area of the thyristor, as is shown in Fig. 23-8b, until the thyristor is completely shorted out by the large excess-carrier densities. The time required for the plasma to spread from the initial regions around the gate terminals to the entire device cross section is the plasma-spreading time t_{ps}. Typical rates of plasma spreading are given in terms of the plasma-spreading velocity, which has values in the range of 20–200 $\mu m/\mu s$. For large-area devices having diameters of centimeters, it can take several hundred microseconds to completely turn on the device if the plasma must spread from a single gate electrode such as is shown in Fig. 23-1b. The rate of voltage drop during the plasma-spreading time is slower because t_{ps} is larger than t_r and because most of the drop occurs during the t_r interval.

It is important that the rate of rise of the anode current be kept less than a maximum value given on the thyristor specification sheet. If di_F/dt exceeds this maximum rate, the device may be damaged or even fail. Such damage may occur because large rates of current growth mean that rise time will be short and, consequently, the turned-on area around the gate region of the thyristor will be quite small at the end of the rise time interval compared with the cross-sectional area of the device. A small turn-on area further means that the voltage across the thyristor will not have fallen very far from the blocking state value during the rise time interval. Hence, the instantaneous power dissipation

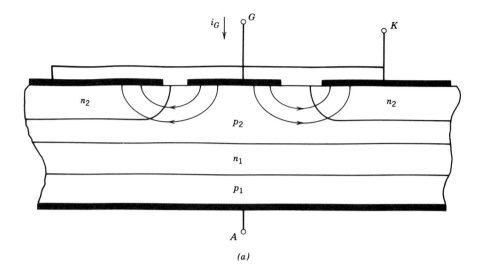

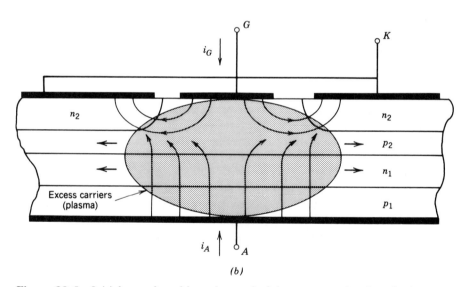

Figure 23-8 Initial growth and lateral spread of the excess carriers in a thyristor at turn-on illustrating the need to limit di_F/dt: (a) injection of minority carriers into the p_2 base region by the gate current during the turn-on delay time that initiates the regenerative switching action; (b) initial turned-on areas of the thyristor in the vicinity of the gate electrode shortly after the turn-on delay time. The further lateral expansion of this area is also shown.

during the t_r interval will both be large and confined to a relatively small volume. In such a situation, the ability to remove the heat generated by the dissipation will be less than the rate of dissipation, and thus the internal temperature of the region may grow so large that thermal runaway in the turned-on area will occur and, hence, device damage or failure.

Larger values of gate current during the $t_{d(\mathrm{on})}$ and t_r intervals will increase the size of the turned-on area by providing a larger amount of excess carriers. A larger turned-on area will reduce the peak instantaneous power dissipation. For this reason, gate current applied

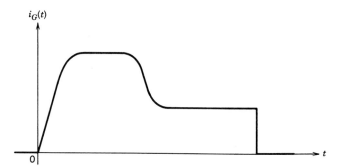

Figure 23-9 A gate current with an initial large value in order to maximize the initial turned-on areas of the thyristor. The current is then reduced to smaller values for a sufficient period of time to guarantee device turn-on.

is often large at the start of the turn-on interval and is gradually reduced as time proceeds, as is shown in Fig. 23-9. There are also structural modifications that can improve the di_F/dt rating, which will be discussed in a later section.

23-5-2 TURN-OFF TRANSIENT

Turn-off of the thyristor requires that it be reverse biased by the external circuit for a minimum time period. For thyristor T_A in Fig. 23-6, turn-off commences when thyristor T_B is turned on, which would be done when the voltage in phase B is larger than it is in phase A. The larger phase B voltage clearly will reverse bias T_A as soon as T_B is fully on. The turn-on time of a thyristor is appreciably shorter than the turn-off time; hence, as far as the discussion of the turn-off of T_A is concerned, the turn-on of T_B in nearly instantaneous.

However, the commutation of current from T_A to T_B will not be instantaneous but will occur over an extended time period, as is shown in the turn-off waveforms for T_A in Fig. 23-10. The current through T_A starts decreasing at $t = 0$ at a fixed rate di_R/dt, which is governed by the external circuit. The overall turn-off process is quite similar to the turn-off to the power diode described in Chapter 20. As the current decreases, the excess carrier in the four regions of the thyristor are decreasing from the steady-state values shown in the carrier distribution of Fig. 23-4 by a combination of internal recombination and carrier sweep-out.

As time proceeds, the current continues to decrease and soon passes through zero at a time t_1 and then grows toward negative values, as shown in Fig. 23-10. The voltage across the thyristor remains small and positive until either junction J_1 or J_3 starts to become reverse biased, an event that does not occur until the excess-carrier density at the junction has decayed to zero. Usually J_3 will become reversed biased first, occurring at time t_2 in Fig. 23-10, and then very quickly goes into avalanche breakdown as the anode–cathode voltage goes negative because the heavy doping in the n_2 and p_2 layers means that the reverse-blocking capability of this junction is not very large (20–30 V typically).

At or shortly after time t_2 the excess-carrier distribution in the thyristor is no longer large enough to sustain the ever-growing negative anode current, and so the current attains its peak negative value I_{rr} and begins to decay back toward zero. At about the same time the excess-carrier density at junction J_3 goes to zero, and it becomes reverse biased. The

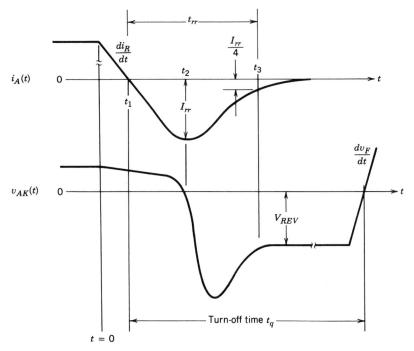

Figure 23-10 Thyristor voltage and current waveforms during turn-off. A reapplied forward-blocking voltage must not be impressed on the thyristor until a specified time period, the recovery time t_q, has elapsed. The rate of rise of the reapplied forward voltage dv_F/dt must be kept below a specified value.

growth of the negative anode–cathode voltage, which began at t_2, continues and overshoots the value $V_{REV} = V_B - V_A$, which will eventually be imposed on thyristor T_A by the circuit. The voltage overshoot arises from the inductance of the circuit and is governed by how rapidly the anode current decays to zero from its peak reverse value of I_{rr}. This will be described in more detail in Chapter 27.

23-5-3 TURN-OFF TIME AND REAPPLIED dv_F/dt LIMITATIONS

In the case of the power diode the reverse-recovery transient was defined to be over when the reverse current had decayed to some conveniently small value such as $\frac{1}{4}I_{rr}$ or to $\frac{1}{10}I_{rr}$, which is marked on the anode current waveform in Fig. 23-10 as time t_3. However, such a definition is not suitable for thyristors. Even at such a time as t_3, there are still substantial excess carriers remaining in the n_1 and p_2 regions of the thyristor. If a forward voltage is reapplied to the thyristor at a rate dv_F/dt, as is shown in Fig. 23-11, a pulse of decaying forward current, a forward recovery current, would flow as the remaining excess carriers just mentioned continued to simultaneously recombine internally and be swept out by the growing forward voltage.

The pulse of forward recovery current may have the same consequence as a deliberately applied pulse of gate current. If the forward-recovery current is large enough, it can turn the thyristor on even though turn-on was not intended. Since the larger the value of dv_F/dt the larger the peak forward-recovery current will be, two things must be done to prevent accidental turn-on. First, the time that the thyristor is maintained in the reverse-blocking mode must be lengthened beyond the time t_3. Device manufacturers specify a

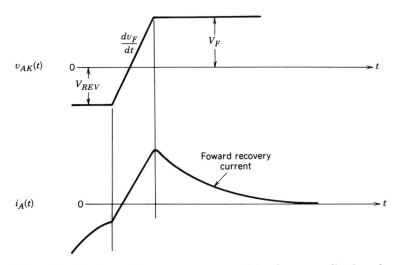

Figure 23-11   Forward-recovery current resulting from reapplication of a forward voltage across the thyristor before the end of the specified recovery time. If the forward-recovery current is too large, inadvertent turn-on of the thyristor may result.

turn-off time t_q for their thyristors, which represents the minimum time their thyristor should remain in the reverse-blocking mode before any forward voltage is reapplied. This turn-off time is usually several excess-carrier lifetimes in length.

Second, the rate of growth dv_F/dt of the reapplied forward voltage should be kept below a maximum value, which is also specified by the device manufacturer. This maximum value is usually arrived at on the basis of how large of a displacement current a given dv_F/dt can drive through the space charge capacitance of junction J_2. If this displacement current $C_{j2} \, dv_F/dt$ exceeds the breakover current I_{BO}, device turn-on may result. Hence, the reapplied dv_F/dt should be limited to

$$dv_F/dt|_{\max} < I_{BO}/C_{j2} \qquad (23\text{-}6)$$

Maximum values of dv_F/dt range from perhaps 100 V/μs for slow devices intended for low-frequency phase control applications to several thousand volts per microsecond or larger in devices intended for higher frequency inverter applications and high-voltage dc.

23-6   METHODS OF IMPROVING di/dt AND dv/dt RATINGS

23-6-1   IMPROVEMENTS IN di/dt

The key to improving the di/dt rating is to increase the amount of initial area of cathode conduction, since this is the factor that limits the rate of rise of the anode current. One way to increase this area is to increase the gate current as has already been mentioned. But it is desirable to do this without requiring the gate drive circuit to deliver substantially higher gate currents. One way of accomplishing this goal is to use a smaller auxiliary or pilot thyristor to provide large gate currents to the main thyristor, as diagrammed in Fig. 23-12. Furthermore, this pilot thyristor can be integrated onto the same silicon wafer as the main device.

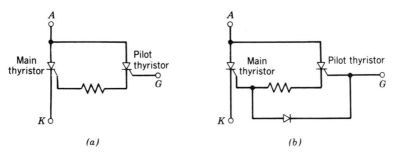

Figure 23-12 Thyristor modifications for more rapid turn-on and turn-off: (*a*) thyristor with an auxiliary thyristor to provide large turn-on gate currents, (*b*) gate-assisted turn-off thyristor (GATT).

Another improvement that can be made is to modify the gate–cathode geometry so that there are many small cathode and gate regions intermixed together, much the same as the base and emitter regions of a power BJT are constructed. By such intermixing or interdigitating using a variety of complex geometries (e.g., some devices use a complicated involute gate structure), the gate–cathode periphery is made large compared with the cathode area. The distributed gate structure shown in Fig. 23-1*b* is a step in this direction. The greater gate–cathode periphery leads to a significant increase in the initial conducting area of the thyristor and hence to a larger *di/dt* capability.

The use of an interdigitated gate–cathode structure can help to shorten the turn-off time of the thyristor. The large-area cathode and relatively small gate–cathode periphery in a conventional phase control thyristor makes negative gate currents ineffectual in turning off the device, as we explained earlier. However, a highly interdigitated gate–cathode structure where the center of the cathode region is not too far from the gate–cathode boundary makes current crowding toward the center of the cathode much weaker and, thus, allows a negative gate current to be more effective in sweeping out stored charge in the n_2 and p_2 regions, which will in turn shorten the turn-off time.

Since a highly interdigitated gate–cathode structure is usually used in conjunction with a pilot thyristor, a further modification, the addition of a diode as is shown in Fig. 23-12*b*, is necessary. If this diode is not added, then only the pilot thyristor would benefit from the negative gate current and no negative gate current would be drawn from the main thyristor. The circuit shown in Fig. 23-12*b*, which combines a pilot thyristor, a diode, and a highly interdigitated gate–cathode structure, is sometimes termed a gate-assisted turn-off thyristor (GATT). Even in this device reverse biasing of the anode–cathode terminals by the external circuit is required to turn off the thyristor. Turn-off times of 10 μs or less have been achieved in devices with forward-blocking voltages of 2000 V and on-state currents of 1000–2000 A. Such devices can be used at switching frequencies of a few tens of kilohertz.

23-6-2 CATHODE SHORTS

One useful way to reduce the effects of displacement currents that limit dv_F/dt is by means of cathode shorts illustrated in Fig. 23-13. These shorts, which are realized by overlapping the cathode metallization over portions of the gate region (p_2 region), can partially intercept the displacement current, as diagrammed in the figure. Any portion of the displacement current that is diverted to the cathode short does not pass through the gate–cathode junction and, hence, does not cause carrier injection into the *p*-type base

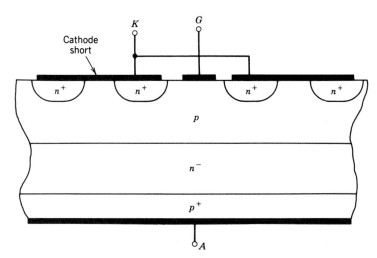

Figure 23-13   Thyristor with cathode shorts for enhancing the dv_F/dt rating of the device.

region. This in turn means that the total displacement current and thus dv_F/dt can be larger without turning on the device. It should also be evident that a highly interdigitated gate–cathode structure will make the use of cathode shorts more effective than in the conventional geometry of a phase control thyristor.

SUMMARY

This chapter has considered the structure and physical principles of operation of the thyristor. The unique switching characteristics of the thyristor were studied. The important conclusions are as follows:

1. The thyristor has a unique four-layer construction of alternating p-type and n-type regions.

2. The forward-bias portion of the thyristor's i–v characteristic has two stable operating regions, one being the on state and the other the off state. The reverse-bias portion of the characteristic is a blocking state.

3. A current pulse applied to the gate will latch the thyristor on, but then the gate cannot turn the device off. The external power circuit must reverse bias the thyristor in order to turn it off.

4. The thyristor is a minority-carrier device and has the highest blocking voltage capabilities and the largest current conduction capabilities of any of the solid-state switching devices.

5. The thyristor is inherently a slow switching device compared to BJTs or MOSFETs because of the long carrier lifetimes used for low on-state losses and because of the large amount of stored charge. It is therefore normally used at lower switching frequencies.

6. The rate of rise of the on-state current must be kept within bounds because the slow spread of the plasma during the turn-on transient leads to current crowding that could result in device failure if di/dt is too large.

7. The rate of rise of the reapplied forward-blocking voltage after turn-off must be limited or the device may be triggered back into the on state by induced displacement currents. Furthermore, the forward voltage must not be reapplied too soon after turn-off or the device will turn back on.

8. Special structure modifications, such as highly interdigitated gate–cathode layouts and the use of cathode shorts, can substantially improve the di/dt and dv/dt ratings.

9. Thyristors will have large reverse-recovery currents.

PROBLEMS

23-1 A thyristor is connected in series with a load resistor R_L and a 60-Hz sinusoidal voltage source with an rms voltage of V. A phase control circuit is used to set the trigger angle α so that a specified amount of power is delivered to the load. The on-state voltage of the thyristor is given by $V_{on} = 1.0 + R_{on}i(t)$ where R_{on} is the on-state resistance of the thyristor and $i(t)$ is the current flowing in the circuit through R_L and the thyristor. Develop an expression for the average power dissipated in the thyristor as a function of the trigger angle α.

23-2 In Problem 23-1, $R_L = 1\ \Omega$ and $V_s = 220$ V. The thyristor characteristics are listed below. The thyristor must operate in ambient temperatures as high as 120°F. How much power can be delivered to the load and what is the trigger angle?

$$R_{on} = 0.002\ \Omega \qquad V_{RWM} = V_{BO} = 800\ V \qquad I_{A(max)} = 1000\ A$$
$$T_{j(max)} = 125°C \qquad R_{\theta j\text{-}a} = 0.1°C/W$$

23-3 A crude estimate of the di/dt limitation in a thyristor can be obtained from the following model. Assume a gate–cathode structure such as is shown in Fig. 23-1 for a phase control thyristor and the radius of the central gate is r_o. When the current begins to rise at turn-on, it starts out in a small conducting area of radius r_o and spreads radially outward as is illustrated in Fig. 23-8, with a velocity of u_s. The current rises at a constant rate di/dt for a time t_f and the anode–cathode voltage $v_{AK} = V_{AK}(1 - t/t_f)$ during the current rise interval.

If it is assumed that all of the power dissipated in the thyristor during the transient goes into raising the temperature T_j of the turned-on area and none gets to the heat sink, then the temperature rise ΔT_j is given by

$$\Delta T_j = \frac{1}{C_v}\int_0^{t_f} P(t)\, dt$$

where C_v is the specific heat of silicon and $P(t)$ is the instantaneous power dissipation in the turned-on area. Assuming the following numerical values, find the maximum allowable di/dt:

$$V_{AK} = 1000\ V \qquad u_s = 100\ \mu m/\mu s \qquad r_o = 0.5\ cm \qquad t_f = 20\ \mu s \qquad T_{j(max)} = 125°C$$

23-4 A modification to the basic structure of a thyristor has been proposed that consists of putting an n^+ layer between the p_1 and n_1 layers shown in Fig. 23-3a. This means that a punch-through structure has been set up for junction J_2 when it is reverse biased. Explain the advantages and disadvantages this structure would have on the thyristor characteristics.

23-5 A thyristor is not completely latched in the on state until the plasma has spread across the entire cross section of the device. If the thyristor of Problem 23-3 has a diameter of 8 cm, how long does it take to reliably be latched in the on state?

23-6 Why can thyristors be made to have larger current-carrying capabilities in the on-state than BJTs? Answer qualitatively using simple diagrams to explain.

23-7 The doping levels in the simplified thyristor structure shown in Fig. 23-3a are $p_1 = 10^{19}$ cm^{-3}, $n_1 = ?$, $p_2 = 10^{17}$ cm^{-3}, and $n_2 = 10^{19}$ cm^{-3}.
   (a) What should be the approximate thickness and doping density of the n_1 layer if $BV_{BO} = 2000$ V?

(b)   Approximately estimate the required excess carrier lifetime in the n_1 region.

(c)   The thyristor is to carry a maximum current of 2000 A with a maximum voltage drop across the drift region (n_1 region) of 2 volts. Estimate the required cross-sectional conducting area of the thyristor.

23-8   A thyristor carrying a maximum on-state current of 3000 A is designed so that the maximum current density is 200 A/cm^2. The cathode and gate structures both occupy the same side of the silicon wafer such as is shown in Fig. 23-1. The cathode occupies 65% of the area and the gate occupies 35%. Estimate the total silicon wafer area required for this thyristor and express the area as the diameter of an equivalent circular area.

23-9   A thyristor has an effective conduction area of 10 cm^2 and a breakover current of 50 mA. The doping levels are $p_1 = 10^{19}$ cm^{-3}, $n_1 = 10^{14}$ cm^{-3}, $p_2 = 10^{17}$ cm^{-3}, and $n_2 = 10^{19}$ cm^{-3}. Approximately estimate the $\dfrac{dv}{dt}$ rating of this thyristor.

REFERENCES

1.   S. K. Ghandhi, *Semiconductor Power Devices*, Wiley, New York, 1977, Chapter 5.

2.   *SCR Manual*, 6th ed., General Electric Company, Syracruse, NY, 1979.

3.   B. M. Bird and K. G. King, *Power Electronics*, Wiley, New York, 1983, Chapters 1, 6.

4.   M. S. Adler, K. W. Owyang, B. Jayant Baliga, and R. A. Kokosa, "The Evolution of Power Device Technology," *IEEE Transactions on Electron Devices*, Vol. ED-31, No. 11, Nov. 1984, pp. 1570–1591.

5.   M. H. Rashid, *Power Electronics: Circuits, Devices, and Applications*, Prentice-Hall, Englewood Cliffs, NJ, 1988, Chapters 14–15.

6.   B. W. Williams, *Power Electronics, Devices, Drivers, and Applications*, Wiley, New York, 1987, Chapters 3, 4, 8–10.

7.   R. L. Avant and F. C. Lee, "A Unified SCR Model for Continuous Topology CADA," *IEEE Transactions or Industrial Electronics*, Vol. IE-31, No. 4, Nov. 1984, pp. 352–361.

CHAPTER 24

GATE TURN-OFF THYRISTORS

24-1 INTRODUCTION

In several respects, thyristors are nearly the ideal switch for use in power electronic applications. They can block high voltages (several thousand volts) in the off state and conduct large currents (several thousand amperes) in the on state with only a small on-state voltage drop (a few volts). Most useful of all is their capability of being switched on when desired by means of a control signal applied at the gate of the thyristor.

However, the thyristor has a serious deficiency that prevents its use in switch-mode applications: the inability to turn off the device by application of a control signal at the thyristor gate. The inclusion of a turn-off capability in a thyristor requires device modifications and some compromises in the operational capabilities of the device. This chapter describes the structure and operation of thyristors that have a gate turn-off capability, the so-called GTO thyristor usually abreviated as GTO, and the performance compromises required to achieve the turn-off capability. Drive circuits and snubber circuits commonly used with GTOs are also described.

24-2 BASIC STRUCTURE AND *I–V* CHARACTERISTICS

The vertical cross section of a GTO with its highly interdigitated gate–cathode structure is shown in Fig. 24-1. The GTO retains the basic four-layer structure of the thyristor described in the previous chapter and its doping profile. The thickness of the p_2 base layer is generally somewhat smaller in a GTO than in a conventional thyristor.

There are three significant differences between a GTO and a conventional thyristor. First, the gate and cathode structures are highly interdigitated, with various types of geometric forms being used to lay out the gates and cathodes, including complicated involute structures. The basic goal is to maximize the periphery of the cathode and minimize the distance from the gate to the center of a cathode region.

Second, the cathode areas are usually formed by etching away the silicon surrounding the cathodes so that they appear as islands or mesas, as indicated in Fig. 24-1. When the GTO is packaged, the cathode islands are directly contacted to a metal heat sink, which also forms the cathode connection to the outside world.

A third major difference is noted in the anode region of the GTO. At regular intervals, n^+ regions penetrate the p-type anode (p_1 layer) to make contact with the n^- region that forms the n_1 base layer. The n^+ regions are overlaid with the same metallization that

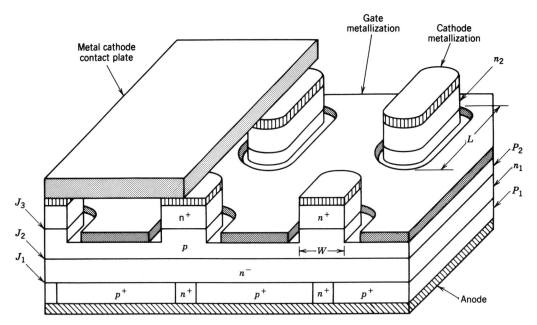

Figure 24-1 Vertical cross section and perspective view of a GTO.

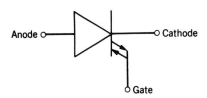

Figure 24-2 Circuit symbol for a GTO.

contacts the p-type anode resulting in a so-called anode short, as is shown in Fig. 24-1. The anode-short structure is used to speed up the turn-off of the GTO as will be explained in a later section of this chapter. Some GTOs are made without this anode short so that the device can block reverse voltages.

The $i-v$ characteristic of a GTO in the forward direction is identical to that of a conventional thyristor. However, in the reverse direction, the GTO has virtually no blocking capability because of the anode-short structure. The only junction that blocks in the reverse direction is junction J_3, and it has a rather low breakdown voltage (20–30 V typically) because of the large doping densities on both sides of the junction. The circuit symbol for the GTO is shown in Fig. 24-2. The two-way arrow convention on the gate lead distinguishes the GTO from the conventional thyristor.

24-3 PHYSICS OF TURN-OFF OPERATION

24-3-1 TURN-OFF GAIN

The basic operation of the GTO is the same as that of the conventional thyristor. The principal differences between the two devices lie in the modifications made in the basic thyristor structure to achieve a gate turn-off capability. Hence, in this chapter we discuss only this aspect of the GTO operation in any detail.

A convenient starting point for appreciating why the GTO structure differs from the conventional thyristor and what performance compromises must be made is to analyze the turn-off conditions in the two-transistor model of the thyristor given in Fig. 23-3b. In the equivalent circuit both Q_1 and Q_2 are saturated in the thyristor on state. However, if the base current to Q_2 could briefly be made less than the value needed to maintain saturation ($I_{B2} < I_{C2}/\beta_2$), then Q_2 would go active and the thyristor would begin to turn off because of the regenerative action present in the circuit when one or both of the transistors are active.

Using the equivalent circuit of Fig. 23-3b we can write I_{B2} in terms of the thyristor terminal currents as

$$I_{B2} = \alpha_1 I_A - I_G' \qquad (24\text{-}1)$$

where I_G' is the negative of the normal gate current. From the equivalent circuit it is clear that a negative gate current I_G' is the only way that Q_2 can be brought out of saturation. The collector current I_{C2} can be expressed as

$$I_{C2} = (1 - \alpha_1)I_A \qquad (24\text{-}2)$$

Setting up the inequality $I_{B2} < I_{C2}/\beta_2$ with $\beta_2 = \alpha_2/(1 - \alpha_2)$ and using Eqs. 24-1 and 24-2 yield

$$I_G' > \frac{I_A}{\beta_{off}} \qquad (24\text{-}3)$$

The parameter β_{off} is the turn-off gain and is given by

$$\beta_{\text{off}} = \frac{\alpha_2}{\alpha_1 + \alpha_2 - 1} \qquad (24\text{-}4)$$

24-3-2 REQUIRED STRUCTURAL MODIFICATIONS AND PERFORMANCE COMPROMISES

The first step in converting a conventional thyristor into a GTO is to make the turn-off gain as large as feasible so that overlarge values of negative gate current can be avoided. This means that α_2 should be near unity and α_1 should be small. Making α_2 near unity involves the use of a narrow p_2 layer for the npn transistor Q_2 and the use of heavy doping in the n_2 cathode layer (emitter of Q_2). Note that these are the same steps needed to achieve a large value of beta in a conventional BJT and that they are the normal steps used in the fabrication of a conventional thyristor.

To make α_1 small, the n_1 thyristor layer (the base of transistor Q_1) should be as thick as possible and the carrier lifetime in this layer should be short. A thick n_1 layer is standard in thyristor fabrication because this layer must accommodate the depletion layer of junction J_2 during device operation in the forward-blocking state. However, the need for a short carrier lifetime conflicts with the need for a long lifetime to minimize on-state power dissipation in this region. To achieve gate turn-off action, some reduction in carrier lifetimes must be accepted and, consequently, a GTO will have a higher on-state voltage drop at a given current level than a conventional thyristor.

Fortunately, the conflicting requirements on carrier lifetimes are substantially resolved by the anode-shorting structure shown in Fig. 24-1. For the GTO to turn off, the excess carriers, especially holes, must be removed from the n_1 layer. The initial distribution of excess carriers at the start of turn-off is shown in Fig. 23-4. Because of the shorted anode structure, there can be no reverse anode–cathode voltages and thus no reverse anode currents to remove the excess carriers by carrier sweep-out. The only way to remove excess carriers is via internal recombination and by diffusion. Unfortunately, the conventional thyristor structure shown in Fig. 23-4 along with the carrier distributions suppresses the diffusion of holes from the n_1 layer because either side of this layer is a heavily doped p-type region where the equilibrium hole densities are larger than the excess hole density in the n_1 region. This suppresses the diffusion of holes out of the n_1 layer into either of two p-type layers (p_1 and p_2). Thus, the only way to remove holes is via internal recombination.

However, the n^+ regions in the GTO anode structure remove the barrier to hole diffusion and provide a sink for excess holes. This permits hole diffusion to occur at a substantial rate so that excess holes in the n_1 layer are removed at least as much by diffusion as by internal recombination. The net result is that the total stored charge is removed much faster during device turn-off, and the GTO thus has both a desirably shorter turn-off and forward-recovery times compared with conventional thyristors without severely compromising its on-state losses. This shorted anode structure is so effective in reducing turn-off and recovery times that it is sometimes used in special thyristor structures called reverse-conducting thyristors (RCTs), which have short turn-off and recovery times like GTO but cannot be turned off by a negative gate current because some of the other needed structural modifications are not included.

The essential modification for gate turn-off capability is the use of a highly interdigitated gate and cathode structure that minimizes the lateral voltage drops in the p_2 layer during turn-on and turn-off. Such lateral voltage drops, which were described in the previous chapter (see Fig. 23-5), are especially noticeable in conventional thyristor structures. Such lateral voltage drops lead to current crowding problems and di/dt limitations. However, as has already been mentioned, the use of highly interdigitated gate–cathode structures, which have relatively short distances between the gate contacts and the center of the cathode regions, minimizes these problems. The need for such interdigitation is a two-dimensional consideration and is not predicted by the one-dimensional two-transistor model. To avoid significant voltage drops in the gate metallization with large gate turn-off currents, contacts to the gate metal are uniformly spaced over the wafer surface.

24-4 GTO SWITCHING CHARACTERISTICS

24-4-1 INCLUSION OF SNUBBER AND DRIVE CIRCUITS

In describing the switching characteristics of other semiconductor devices, we have embedded the device in a typical switching circuit application without any snubber circuits and with only the most general of driving circuits. The motivation was to describe the switching characteristics of the device without the complications that are added by the presence of the snubber circuits. However, GTOs must normally be used with snubbers, as will be detailed later, and so any realistic description of the GTO switching behavior must include the effects of the snubber circuits. Similarly, we will include the gating circuit in describing the switching waveforms of the GTO.

The step-down converter circuit shown in Fig. 24-3, which uses a GTO as the switching element, will be used to describe the switching waveforms. It is important to realize that GTO are only used in medium- to high-power applications where not only are the voltage and current levels large, but also the other solid-state components that may be used in conjunction with the GTO are likely to be rather slow. Thus, the free-wheeling

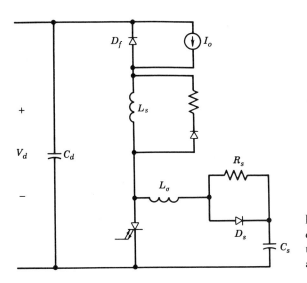

Figure 24-3 Step-down converter circuit using a GTO as the switching device with turn-on and turn-off snubbers.

diode shown in the circuit of Fig. 24-3 will not be a very fast recovery diode. On the other hand, the GTO will have a fast current rise time at turn-on compared with the diode's reverse-recovery time because of the GTO's highly interdigitated gate–cathode structure. The consequence of this is that without protective circuits, very large overcurrents would flow in both the GTO and the diode because of the relatively slow reverse recovery of the diode. Hence, the snubber inductor shown in Fig. 24-3 is included in the circuit to act as a turn-on snubber, as discussed in Chapter 27.

When the GTO is being turned off, the rate of anode–cathode voltage growth, dv/dt, must be limited to specified levels. Otherwise retriggering of the GTO back into the on state may occur, as was described in Chapter 23 with conventional thyristors. For this reason, a turn-off snubber is included as a part of the switching circuit, as is illustrated in Fig. 24-3. Detailed description of the operation of turn-on and turn-off snubbers is given in Chapter 27. A gate drive circuit capable of meeting the recommended gating conditions suggested by GTO thyristor manufacturers is shown in Fig. 24-4.

24-4-2 GTO TURN-ON TRANSIENT

When the GTO in Fig. 24-3 is off, the current free-wheels through the diode D_f. Turn-on is initiated by a pulse of gate current shown in Fig. 24-5. The sequence of events occurring inside the GTO during the turn-on process are essentially the same as those described for the conventional thyristor in the previous chapter and so will not be repeated here. During turn-on, both the rate of gate current increase, di_G/dt, and the peak gate current I_{GM} should be large to ensure that all cathode islands begin to conduct and that there is good dynamic sharing of the anode current. Otherwise only a small number of islands might be carrying the total current and localized thermal runaway could occur, resulting in the destruction of the GTO.

A large I_{GM} value is supplied for a long enough time period, for example 10 μs, to ensure that the turn-on process is complete. After completion of turn-on, there must be a minimum continuous gate current I_{GT} flowing during the entire on-state period to prevent unwanted turn-off. The current I_{GT} is sometimes termed the "backporch" current. If the gate current is zero and the anode current gets too low, some of the cathode islands may

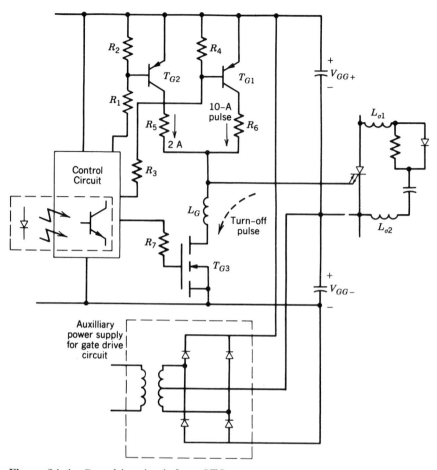

Figure 24-4 Gate drive circuit for a GTO.

stop conducting, and if the anode current were to subsequently increase, the remaining conducting islands may not be able to handle the current and the GTO may be destroyed by the ensuing thermal runaway.

The initial large pulse of gate current is provided by the gate drive circuit of Fig. 24-4 by turning on both transistors T_{G1} and T_{G2}. The stray inductance in the positive gate drive circuit should be kept to a minimum to achieve a large di_G/dt value at turn-on. After a time duration t_{w1}, the gate current is reduced from I_{GM} to I_{GT} by turning T_{G1} off.

During the growth of the anode current, the input voltage is shared between the turn-on snubber inductance and the GTO. If the di/dt of the anode current is limited by this inductance because of its large value, then the voltage across the GTO will drop quickly to a fairly low value, as is shown in Fig. 24-5. The overshoot in the anode current comes from the reverse recovery of the free-wheeling diode D_f.

24-4-3 GTO TURN-OFF TRANSIENT

The GTO is turned off by applying a large negative gate current, as is shown in Fig. 24-6. The resulting current and voltage waveforms for the GTO in the circuit of Fig. 24-3 are

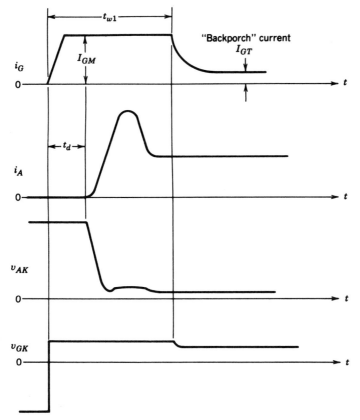

Figure 24-5 Turn-on waveforms for a GTO embedded in a step-down converter with turn-on and turn-off snubbers.

shown in Fig. 24-6. There are several distinct intervals during the turn-off, which are described in the following paragraphs.

The gate drive circuit of Fig. 24-4 supplies the negative gate current by turning on transistor T_{G3}. The gate current must be very large, on the order of $\frac{1}{5}-\frac{1}{3}$ (corresponding to turn-off gains of 3–5) of the anode current being turned off, but fortunately this large negative current is required for only a relatively short time. Low-voltage MOSFETs are a nearly ideal choice for T_{G3}. The negative di_G/dt must be large in order to have a short storage time and a short anode current fall time and to reduce the gate power dissipation. However, too large a value of negative di_G/dt will result in the anode tail current to be described shortly. Hence, di_G/dt should be kept in the range specified by the device manufacturer.

The negative di_G/dt is controlled by V_{GG-} and L_G of the negative gate drive portion of the circuit of Fig. 24-4. Here V_{GG-} must be chosen to be less than the gate–cathode junction breakdown voltage. Knowing V_{GG-}, L_G is selected to give the specified di_G/dt. For large GTO, the stray inductance in the negative gate drive circuit may equal the required L_G.

During the first time interval, the storage time t_s, the growing negative gate current is removing charge stored in the p_2 and n_2 layers at the periphery of the cathode islands,

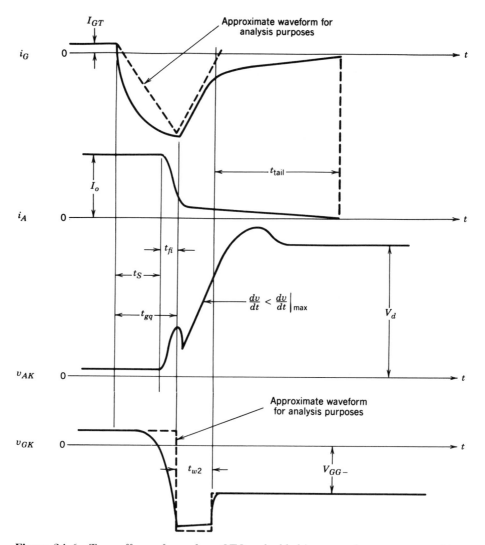

Figure 24-6   Turn-off waveforms for a GTO embedded in a step-down converter with turn-on and turn-off snubbers.

as shown in Fig. 24-7. As the stored charge continues to be removed from the periphery, the size of the plasma-free region grows as it expands in the lateral direction toward the centers of the cathode islands with a so-called squeezing velocity. In essence this removal of the plasma is the inverse of how it was established during turn-on. When a sufficient amount of stored charge has been removed, the regenerative action in the GTO is stopped and the anode current begins to fall. This marks the end of the storage time interval.

Once the regenerative action of the GTO is stopped, the anode current begins to fall rapidly. The current $I_o - i_A$ commutates to the turn-off snubber capacitor C_s, which is fairly large in GTO applications. There is simultaneously a rapid rise in voltage across the GTO because of stray inductance in the turn-off snubber circuit loop. This stray inductance (L_σ in Fig. 24-3) should be kept to a minimum to keep the peak of the voltage spike during the anode current fall time interval to a specified value. The anode current fall time t_{fi} interval ends when the excess carriers at the gate–cathode junction have been swept out and the junction recovers its reverse-blocking capability.

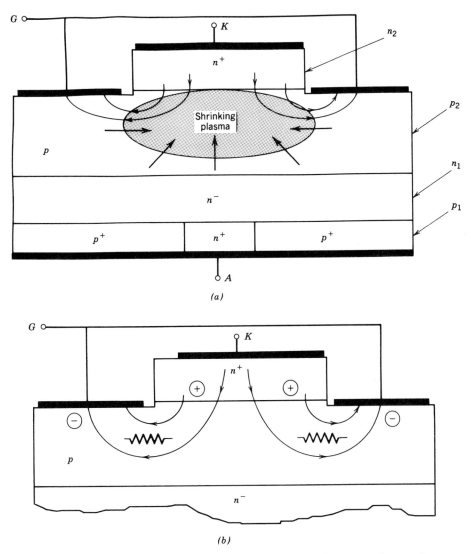

Figure 24-7 Mechanisms that determine the maximum anode current that can be turned off by a negative gate current: (*a*) negative gate current squeezing the excess-carrier plasma down to a small volume at the center of the cathode island the farthest possible distance away from the gate electrode; (*b*) lateral ohmic resistance in the p_2 base layer limiting the maximum gate current.

As the gate–cathode junction recovers its reverse-blocking capability, the gate–cathode voltage begins to grow toward negative values and the negative gate current thus begins to decrease rapidly in magnitude, as is shown in Fig. 24-6. The voltage induced in the inductance L_G forces the gate current to keep flowing and the gate–cathode junction goes into avalanche breakdown. The gate–cathode junction now operates as a zener diode, and the di_G/dt during this interval is given by

$$\frac{di_G}{dt} = \frac{V_{GK,\text{breakdown}} - V_{GG-}}{L_G} \tag{24-5}$$

This avalanche breakdown is desirable for a short duration t_{w2} (the gate–cathode junction avalanche breakdown time) to sweep out as much stored charge from the gate and p_2 layer as possible. This interval depends on di_G/dt, which is controlled by the selection of L_G and V_{GG-}. The interval t_{w2} should be kept less than a maximum specified value in order to avoid destruction of the gate–cathode junction.

At the end of the t_{w2} interval there will still be some excess stored charge in the two base regions (n_1 and p_2 layers) of the GTO. A small anode current, usually termed the anode tail current, will continue to flow between the anode and the negatively biased gate, which is due to the sweep-out of this remaining stored charge. This current is driven by the growing voltage difference between the anode and gate. The time interval during which this tail current flows is termed the anode tail-current time t_{tail}. During most of the tail time interval, the gate voltage is at V_{GG-}, the value it will have during the entire off-state interval.

During the tail time interval, the voltage across the GTO grows at a constant rate given by

$$\frac{dv_{AK}}{dt} \approx \frac{I_o}{C_s} \tag{24-6}$$

This interval will contribute a major part of the turn-off losses because this interval is relatively long and the voltage across the GTO during the interval is fairly large.

The overvoltage at turn-off shown in the anode–cathode voltage waveform in Fig. 24-6 is due to stray inductance in the power circuit. The anode–cathode overvoltage can be reduced by means of an overvoltage snubber of the type described in Chapter 27.

24-4-4 MINIMUM ON- AND OFF-STATE TIMES

It is strongly recommended that the GTO not be turned on until is has been off for a specified time because of the possibility of poor current sharing between the various cathode islands. Some excess minority carriers will remain in the GTO for fairly long times because of the long lifetimes, and these remaining carriers will cause the few cathode islands in the vicinity of the carriers to have a better conduction characteristic than the rest of the cathode islands. Hence, if turn-on is attempted before all of the carriers have recombined or been swept out, then most of the current will be carried by these few islands (poor current sharing) and device destruction may occur.

Similarly, the GTO should be maintained in the on state for a specified time period before turn-off is initiated. Again, the reason is because otherwise there may be poor current sharing between the various cathode islands.

The circuit designer should also recognize that the turn-on and turn-off snubbers also require a minimum off-state and minimum on-state time, respectively, to operate properly. This is described in detail in Chapter 27, where the design of these snubbers is discussed in detail.

24-4-5 MAXIMUM CONTROLLABLE ANODE CURRENT

The excess carriers, principally those in the p_2 layer, are the source of carriers for the negative gate current. As the negative gate current grows and the plasma-free region grows, as is diagrammed in Fig. 24-7, there is a build-up of a substantial voltage across the gate–cathode junction, as is indicated because of the lateral flow of gate current in the p_2 layer. The voltage across the junction is largest at the cathode periphery nearest the gate contact. If this voltage exceeds the breakdown voltage of the junction, then the negative gate current will flow only at the cathode periphery where breakdown has occurred and none of the remaining stored charge will be removed, and hence, the GTO will not be turned off. For this reason, the voltage V_{GG-} must be kept less than the breakdown voltage of the gate–cathode junction.

The limitation on the negative gate–cathode voltage means that there is a maximum gate current that can be pulled out of the GTO. As the removal of stored charge enters its final phase, the region of excess carriers has shrunk to a small area near the center of the cathode island and is the greatest distance from the gate contact. Under these circumstances the reverse voltage across the junction is at its greatest value. The lateral ohmic resistance shown in Fig. 24-7b, which is a function of the device geometry, and the doping level of the p_2 layer along with the junction breakdown voltage determine how large the maximum negative gate current can be. This also means that there is a maximum anode current that can be turned off since, by Eq. 24-3, $I_A < \beta_{\text{off}}I_{G,\text{max}}$. The maximum controllable anode current is given on GTO specification sheets by the device manufacturer.

24-5 OVERCURRENT PROTECTION OF GTOs

In a MOSFET and a BJT, an accidental overcurrent causes the device to go out of saturation and enter the active region. The device itself limits the maximum current, but the voltage across the device becomes large. Thus, the overcurrent condition can be easily detected by measuring the on-state voltage. The overcurrents can also be detected by a current sensor or, in the case of a MOSFET, by using a SENSEFET. Once the overcurrent is detected, the BJTs and the MOSFETs can be protected by turning them off within a few microseconds.

The overcurrent protection of a GTO is more complicated. As is shown in Fig. 24-8a, the GTO is designed for an allowable peak operating current that is chosen to be less than the maximum controllable current by a safety factor. The overcurrent in a GTO must be

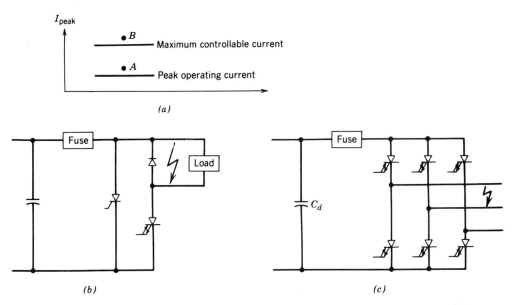

Figure 24-8 Overcurrent protection methods for GTO: (a) definition of overcurrents; (b) overcurrent protection by "crowbarring"; (c) overcurrent protection by turning on all the GTO in the bridge to share the current until the fuse opens up.

detected by current sensing. If the detected overcurrent is less than the maximum controllable current, for example, at point A in Fig. 24-8a, then the GTO can be turned off by a negative gate current.

However, if the detected overcurrent is greater than the maximum controllable current, for example, at point B in Fig. 24-8a, then an attempt to turn off the GTO by a negative gate current will result in the failure of the GTO. Hence, the GTO is protected by the so-called crowbarring technique, where a thyristor in parallel with the GTO as is shown in Fig. 24-8b is turned on quickly, which then blows the fuse. Without crowbarring, the only way to protect the GTO in the circuit of Fig. 24-8b would be to use a GTO of much larger current rating, which is quite expensive.

In the case of a three-phase configuration, which is shown in Fig. 24-8c, the crowbarring can be obtained by turning on all six GTO simultaneously. As was discussed in Chapter 8, under normal operation the current through the fuse is in the same range as the current through one GTO. By turning on all three legs simultaneously, the current through the fuse is shared by three legs, which the GTO will be capable of carrying until the fuse blows.

SUMMARY

This chapter has examined the structure and characteristics of the GTO. The important conclusions are listed:

1. The GTO has the same four-layer structure as the conventional thyristor, but special modifications are made to the structure to enable the gate to turn off the device.

2. The major modifications include a highly interdigitated gate–cathode structure with small cathode and gate widths, the use of anode shorts, and a shorter carrier lifetime in the drift region than is used in a conventional thyristor.

3. The forward-bias portion of the GTO i–v characteristic is the same as for the conventional thyristor, but GTO with anode shorting has very limited reverse-blocking capability.

4. The turn-off gain of the GTO is not large (typically 5 or less), so large negative gate current pulses are required to turn off the device.

5. The magnitude of negative gate current that can be applied is limited by current crowding phenomena, and hence, there is a maximum anode current that can be safely turned off.

6. Special gating requirements for the GTO include not only large positive and negative gate current pulses but a continuous on-state gate current to ensure complete turn-on of all the cathode islands.

7. Gate turn-off thyristors are used almost exclusively for medium- and high-power applications. Turn-off snubber circuits must be used. The GTO must particularly be protected against overcurrents because the gate cannot turn off currents that exceed a specified maximum value.

PROBLEMS

24-1 Each cathode island in the GTO diagram of Fig. 24-1 has a width W and a length L. The p_2 layer has a thickness t and a resistivity ρ_{p_2}. The GTO has a turn-off gain of β_{off}, N cathode islands connected in parallel, and junction J_3 has a breakdown voltage of BV_{J_3}. Develop an approximate expression for the maximum controllable anode current I_{AM}.

24-2 The stray inductance L_σ in the turn-off snubber circuit of Fig. 24-3 will cause an overvoltage across the GTO. Estimate the maximum stray inductance that can be tolerated in the circuit if the overvoltage is not to exceed $1.5\,V_d$. Express the estimate in terms of the circuit parameters. Assume that the turn-on snubber circuit acts like a constant current source of value I_o during the GTO current fall time $t_{fi} = 1$ μs.

24-3 The drive circuit of Fig. 24-4 is used to turn off the GTO in the circuit of Figure 24-3. The dc input voltage $V_d = 500$ V, the load current $I_o = 500$ A, and the switching waveform is a 1-kHz square wave. The GTO has a gate–cathode breakdown voltage $\mathrm{BV}_{J_3} = 25$ V and a turn-off gain $\beta_{off} = 5$. Assume that the negative voltage V_{GG-} in the drive circuit is 15 V and that the time interval t_{gq} defined in Fig. 24-6 is equal to 5 μs. Estimate the required value of the inductor L_G and the magnitude of the time interval t_{w2} defined in Fig. 24-6. Use the approximate waveforms for $i_G(t)$ and $v_{AK}(t)$ shown in Fig. 24-6 to simplify the analysis.

REFERENCES

1. S. K. Ghandhi, *Semiconductor Power Devices*, Wiley, New York, 1977, Chapter 5.
2. Thyristor Application Notes, "Applying International Rectifier's Gate Turn-off Thyristors," AN-315A, International Rectifier, El Segundo, CA, 1984.
3. R. Sittig and P. Roggwiller (Eds.), *Semiconductor Devices for Power Conditioning*, Plenum, New York, 1982, pp. 91–120.
4. *SCR Manual*, 6th ed., General Electric Company, Syracuse, NY, 1979.
5. B. Jayant Baliga and D. Y. Chen (Ed.), *Power Transistors: Device Design and Applications*, IEEE Press, Institute of Electrical and Electronics Engineers, New York, 1984, Part II, *Gate Turn-off Thyristors/Latching Transistors*, pp. 123–189.
6. M. S. Adler, K. W. Owyank, B. Jayant Baliga, and R. A. Kokosa, "The Evolution of Power Device Technology," *IEEE Transactions on Electron Devices*, Vol. ED-31, No. 11, Nov. 1984, pp. 1570–1591.
7. M. H. Rashid, *Power Electronics: Circuits, Devices, and Applications*, Prentice-Hall, Englewood Cliffs, NJ, 1988, Chapters 14–15.
8. B. W. Williams, *Power Electronics, Devices, Drivers, and Applications*, Wiley, New York, 1987, Chapters 3, 4, 8, 9.

CHAPTER 25

INSULATED GATE
BIPOLAR TRANSISTORS

25-1 INTRODUCTION

Bipolar junction transistors and MOSFETs have characteristics that complement each other in some respects. BJTs have lower conduction losses in the on state, especially in devices with larger blocking voltages, but have longer switching times, especially at turn-off. MOSFETs can be turned on and off much faster, but their on-state conduction losses are larger, especially in devices rated for higher blocking voltages (a few hundred volts and greater). These observations have led to attempts to combine BJTs and MOS-FETs monolithically on the same silicon wafer to achieve a circuit or even perhaps a new device that combines the best qualities of both types of devices.

These attempts have led to the development of the insulated gate bipolar transistor (IGBT), which is becoming the device of choice in most new applications. Other names for this device include GEMFET, COMFET (conductivity-modulated field effect transistor), IGT (insulated gate transistor), and bipolar-mode MOSFET or bipolar-MOS transistor. This chapter describes the basic structure and physical operation of the IGBT and the operating limitations that should be observed in using this new device.

25-2 BASIC STRUCTURE

The vertical cross section of a generic n-channel IGBT is shown in Fig. 25-1a. This structure is quite similar to that of the vertical diffused MOSFET shown in Fig. 22-1. The principal difference is the presence of the p^+ layer that forms the drain of the IGBT. This layer forms a pn junction (labeled J_1 in the figure), which injects minority carriers into what would appear to be the drain region of the vertical MOSFET. The gate and source of the IGBT are laid out in an interdigitated geometry similar to that used for the vertical MOSFET.

The doping levels used in each of the IGBT layers are similar to those used in the comparable layers of the vertical MOSFET structures except for the body region, as is explained later. It is also feasible to make p-channel IGBTs, and this would be done by changing the doping type in each of the layers of the device.

It is shown in Fig. 25-1a that the IGBT structure has a parasitic thyristor. Turn-on of this thyristor is undesirable, and several structural details of a practical IGBT geometry, principally in the p-type body region that forms junctions J_2 and J_3, are different from the simple geometry shown in Fig. 25-1a to minimize the possible activation of this thyristor.

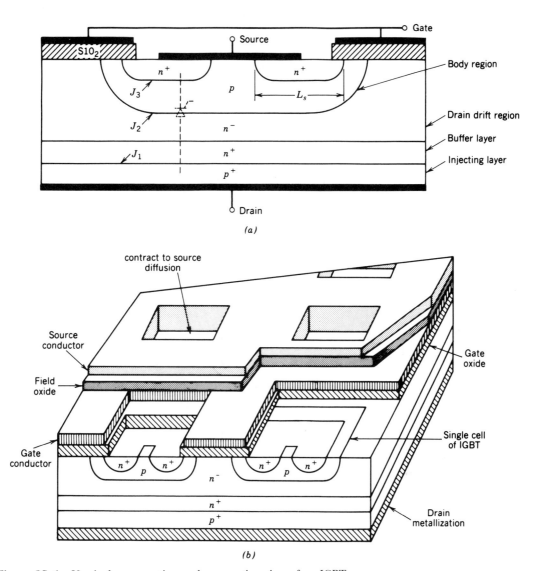

Figure 25-1 Vertical cross section and perspective view of an IGBT.

These structural changes will be discussed in later sections of this chapter. The IGBT does retain the extension of the source metallization over the body region that is also used in power MOSFETs, such as is illustrated in Fig. 22-1. The body–source short in the IGBT helps to minimize the possible turn-on of the parasitic thyristor, as we explain later.

The n^+ buffer layer between the p^+ drain contact and the n^- drift layer is not essential for the operation of the IGBT, and some IGBTs are made without it (sometimes termed non-punch-through, NPT-IGBTs, whereas those with this buffer layer are termed punch-through, PT-IGBTs). If the doping density and thickness of this layer are chosen appropriately, the presence of this layer can significantly improve the operation of the IGBT. The influence of the buffer layer on the characteristics of the IGBT will be discussed in a later section of this chapter.

A circuit symbol for an n-channel IGBT is shown in Fig. 25-2c. The directions of the arrowheads would be reversed in a p-channel IGBT. This symbol is essentially the same

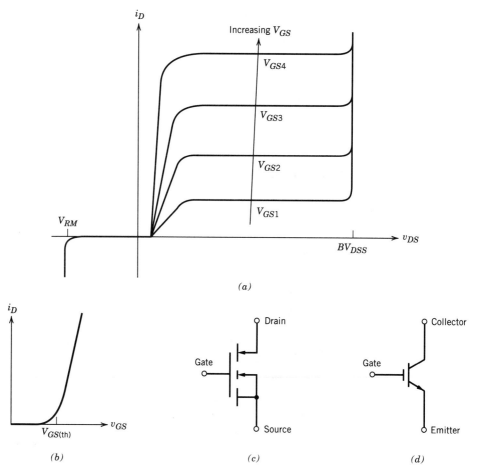

Figure 25-2 The IGBT current–voltage characteristics and circuit symbol: (*a*) output characteristics; (*b*) transfer characteristics; (*c*) and (*d*) *n*-channel IGBT circuit symbols.

as that used for an *n*-channel MOSFET, but with the addition of an arrowhead in the drain lead pointing into the body of the device, indicating the injecting contact. There is some disagreement in the engineering community over the proper symbol and nomenclature to use with the IGBT. Some prefer to consider the IGBT as basically a BJT with a MOSFET gate input and, thus, to use the modified BJT symbol for the IGBT shown in Fig. 25-2*d*. This symbol device has a collector and emitter rather than a drain and source. The symbol and nomenclature shown in Fig. 25-2*c* is the one we have adopted.

25-3 *I–V* CHARACTERISTICS

The *i–v* characteristics of an *n*-channel IGBT is shown in Fig. 25-2*a*. In the forward direction they appear qualitatively similar to those of a logic-level BJT except that the controlling parameter is an input voltage, the gate–source voltage, rather than an input current. The characteristics of a *p*-channel IGBT would be the same except that the polarities of the voltages and currents would be reversed.

The junction labeled J_2 in Fig. 25-1*a* blocks any forward voltages when the IGBT is off. The reverse-blocking voltage indicated on the *i–v* characteristic can be made as large as the forward-blocking voltage if the device is fabricated without the n^+ buffer layer. Such a reverse-blocking capability is useful in some types of ac circuit applications. The junction labeled J_1 in Fig. 25-1a is the reverse-blocking junction. However, if the n^+ buffer layer is used in the device construction, the breakdown voltage of this junction is lowered significantly, to a few tens of volts, because of the heavy doping now present on both sides of this junction, and the IGBT no longer has any reverse-blocking capability.

The transfer curve i_D–v_{GS} shown in Fig. 25-2b is identical to that of the power MOSFET. The curve is reasonably linear over most of the drain current range, becoming nonlinear only a low drain currents where the gate–source voltage is approaching the threshold. If v_{GS} is less than the threshold voltage $V_{GS(th)}$, then the IGBT is in the off state. The maximum voltage that should be applied to the gate–source terminals is usually limited by the maximum drain current that should be permitted to flow in the IGBT, as will be discussed in Section 25-7.

25-4 PHYSICS OF DEVICE OPERATION

25-4-1 BLOCKING STATE OPERATION

Since the IGBT is basically a MOSFET, the gate–source voltage controls the state of the device. When v_{GS} is less than $V_{GS(th)}$, there is no inversion layer created to connect the drain to the source and, hence, the device is in the off state. The applied drain–source voltage is dropped across the junction labeled J_2 and only a very small leakage current flows. This blocking state operation is essentially identical to that of the MOSFET.

The depletion region of the J_2 junction extends principally into the n^- drift region, since the p-type body region is purposely doped much more heavily than the drift region. The thickness of drift region is large enough to accommodate the depletion layer so that the depletion layer boundary does not touch the p^+ injecting layer. This type of IGBT is sometimes termed a symmetrical IGBT or non-punch-through IGBT, and it can block reverse voltages as large in magnitude as the forward voltages it is designed to block. This reverse-blocking capability is useful in some ac circuit applications.

However, it is possible to reduce the required thickness of the drift region by approximately a factor of 2 if a so-called punch-through structure similar to that described in Chapter 20 for the power diode and illustrated in Fig. 20-3 is used. In this geometry, the depletion layer is allowed to extend all the way across the drift region at voltages significantly below the desired breakdown voltage limit. The reach-through of the depletion layer to the p^+ layer is prevented by inserting an n^+ buffer layer between the drift region and the p^+ region, as is shown in Fig. 25-1a. This type of IGBT structure is sometimes termed an antisymmetric or punch-through IGBT. The shorter drift region length means lower on-state losses, but the presence of the buffer layer means that the reverse-blocking capability of this punch-through geometry will be quite low (a few tens of volts) and therefore nonexistent as far as circuit applications are concerned.

25-4-2 ON-STATE OPERATION

When the gate–source voltage exceeds the threshold, an inversion layer forms beneath the gate of the IGBT. This inversion layer shorts the n^- drift region to the n^+ source region exactly as in the MOSFET. An electron current flows through this inversion layer, as is diagrammed in Fig. 25-3, which in turn causes substantial hole injection from the p^+ drain contact layer into the n^- drift region, as also indicated in the figure. The injected holes move across the drift region by both drift and diffusion, taking a variety of paths,

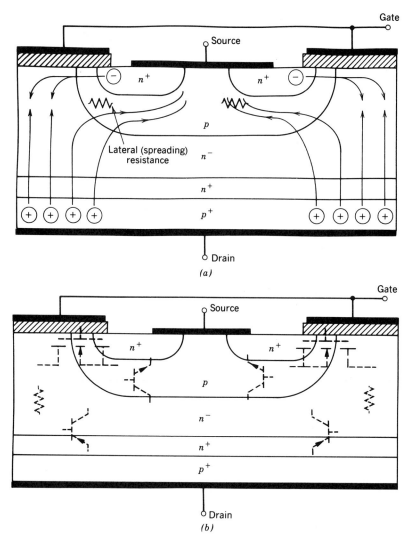

Figure 25-3   Vertical cross section of an IGBT showing (*a*) the on-state current flow paths and (*b*) the effective MOSFET and BJT operating portions of the structure.

as is indicated in Fig. 25-3, and reach the *p*-type body region that surrounds the n^+ source region. As soon as the holes are in the *p*-type body region, their space charge attracts electrons from the source metallization that contacts the body region, and the excess holes are quickly recombined.

The junction formed by the *p*-type body region and the n^- drift region is "collecting" the diffusing holes and thus functions as the collector of a thick base *pnp* transistor. This transistor, diagrammed in Fig. 25-3*b*, has the p^+ drain contact layer as an emitter, a base composed of the n^- drift region, and a collector formed from the *p*-type body region. From this description an equivalent circuit for modeling the operation of the IGBT can be developed, which is shown in Fig. 25-4*a*. This circuit models the IGBT as a Darlington circuit with the *pnp* transistor as the main transistor and the MOSFET as the driver device. The MOSFET portion of the equivalent circuit is also diagrammed in Fig. 25-4*a* along

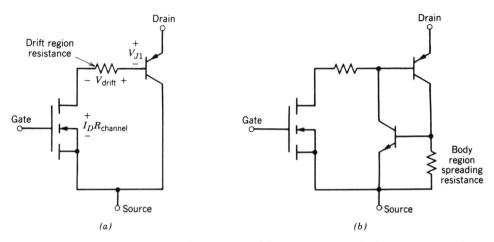

Figure 25-4 Equivalent circuits for the IGBT: (*a*) approximate equivalent circuit valid for normal operating conditions; (*b*) more complete IGBT equivalent circuit showing the transistors comprising the parasitic thyristor.

with the BJT portion. The resistance between the *pnp* base and the MOSFET drain represents the resistance of the n^- drift region.

Unlike the conventional Darlington circuit, the driver MOSFET in the equivalent circuit of the IGBT carries most of the total terminal current. This unequal division of the total current flow is desirable for reasons having to do with potential turn-on of the parasitic thyristor, a subject that we will discuss shortly. In this situation the on-state voltage $V_{DS(\text{on})}$, using the equivalent circuit of Fig. 25-4*a*, can be expressed as

$$V_{DS(\text{on})} = V_{J1} + V_{\text{drift}} + I_D R_{\text{channel}} \qquad (25\text{-}1)$$

The voltage drop across the injection junction J_1 is a typical forward-bias voltage drop across a *pn* junction, which depends exponentially on the current and to first order has an approximately constant value of $0.7-1.0$ V. The drop across the drift region is similar to that developed across the drift region in a high-power *pn* junction and is approximately constant and given approximately by an equation similar to Eq. 20-13 in Chapter 20. The V_{drift} voltage is much less in the IGBT than in the MOSFET because of the conductivity modulation of the drift region and is what makes the overall on-state voltage of the IGBT much less than that of a comparable power MOSFET. The use of the punch-through structure also aids in keeping V_{drift} small. The voltage drop across the channel is due to the ohmic resistance of the channel and is similar to the comparable drop in the power MOSFET discussed in Chapter 22.

25-5 LATCHUP IN IGBTs

25-5-1 CAUSES OF LATCHUP

The paths traveled by the holes injected into the drift region (or *pnp* transistor base) are crucial to the operation of the IGBT. A component of the hole current travels in fairly straight-line paths directly to the source metallization. However, most of the holes are attracted to the vicinity of the inversion layer by the negative charge of the electrons in the layer. This results in a hole current component that travels laterally through the *p*-type body layer, as is diagrammed in Fig. 25-3*a*. This lateral current flow will develop a lateral voltage drop in the ohmic resistance of the body layer (modeled as the spreading resistance in Fig. 25-3a), as indicated in the figure. This will tend to forward bias the n^+p junction (labeled J_3 in Fig. 25-1a) with the largest voltage across the junction occurring where the inversion layer meets the n^+ source.

If the voltage is large enough, substantial injection of electrons from the source into the body region will occur and the parasitic *npn* transistor diagrammed in Fig. 25-3b will be turned on. If this occurs, then both the parasitic *npn* and *pnp* transistors will be on, and hence, the parasitic thyristor composed of these transistors will latch on and latchup of the IGBT will have occurred. For a given IGBT with a specified geometry, there is a critical value of drain current that will cause a large enough lateral voltage drop to activate the thyristor. Hence, the device manufacturer specifies the peak allowable drain current I_{DM} that can flow without latchup occurring. There is also a corresponding gate–source voltage that permits this current to flow that should not be exceeded.

Once the IGBT is in latchup, the gate no longer has any control of the drain current. The only way to turn off the IGBT in this situation is by forced commutation of the current, exactly the same as for a conventional thyristor. If latchup is not terminated quickly, the IGBT will be destroyed by the excessive power dissipation. A more complete equivalent circuit for the IGBT, which includes the parasitic *npn* transistor and spreading resistance of the body layer, is shown in Fig. 25-4b.

The description of latchup just presented is the so-called static latchup mode because it occurs when the continuous on-state current exceeds a critical value. Unfortunately, under dynamic conditions when the IGBT is switching from on to off, it may latch up at drain current values less than the static current value. Consider the IGBT embedded in a step-down converter circuit. When IGBT is turned off, the MOSFET portion of the device turns off quite rapidly, and the portion of the total device current that it carries goes to zero. There is a corresponding rapid buildup of drain–source voltage, as will be described in detail in the next section, which must be supported across the J_2 drift–body junction. This results in a rapid expansion of the depletion layer because of its low doping. This increases the base transport factor α_{pnp} of the *pnp* transistor, which means that a greater fraction of the holes injected into the drift region will survive the traverse of the drift region and will be collected by the J_2 junction. The magnitude of the lateral hole current flow will then increase and, hence, the lateral voltage will increase. As a consequence, the conditions for latchup will be satisfied even though the on-state current prior to the start of turn-off was below the static value needed for latchup. The value of I_{DM} specified by the device manufacturer usually is given for the dynamic latchup mode.

25-5-2   AVOIDANCE OF LATCHUP

There are several steps that can be taken by the device user to avoid latchup and that the device manufacturer can take to increase the critical current required for the initiation of latchup. The user has the responsibility to design circuits where the possibility of overcurrents that exceed I_{DM} are minimized. However, it is impossible to eliminate this possibility entirely. Another step that can be taken is to slow down the IGBT at turn-off so that the rate of growth of the depletion region into the drift region is slowed down and the holes present in the drift region have a longer time to recombine, thus reducing the lateral current flow in the *p*-type body region during the turn-off. The increase in the turn-off time is easily accomplished by using a larger value of series gate resistance R_g, as will be explained in the next section.

The device manufacturer seeks to increase the latching current threshold I_{DM} by lowering the body-spreading resistance in the equivalent circuit of Fig. 25-4b. This is done in several ways. First the lateral width of the source regions, labeled L_s in Fig.

25-1*a*, is kept as small as possible consistent with other requirements. Second, the *p*-type body region is often partitioned into two separate regions of different levels of acceptor doping density, as is illustrated in Fig. 25-5*a*. The channel region where the inversion layer is formed is doped at a moderate level, on the order of 10^{16} cm^{-3}, and the depth of the *p*-region is not much deeper than the n^+ source region. The other portion of the body layer beneath the n^+ source regions is doped much more heavily, on the order of 10^{19} cm^{-3}, and is made much thicker (or equivalently deeper). This makes the lateral resistance much smaller both because of the larger cross-sectional area and because of the higher conductivity.

Another possible modification to the body layer is shown in Fig. 25-5*b*, where one of the source regions is eliminated from the basic IGBT cell. This allows the hole current

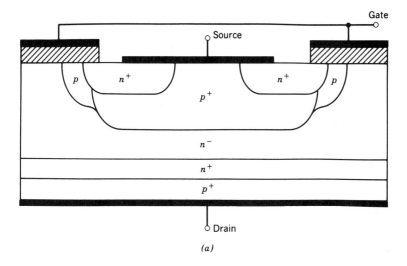

(a)

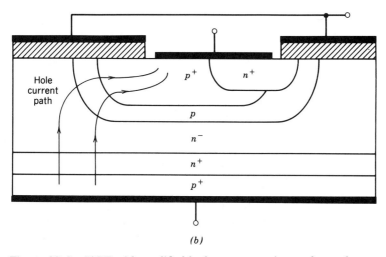

(b)

Figure 25-5 IGBT with modified body–source regions to lower the spreading resistance so that the drain current threshold for latchup is increased: (*a*) modification of the body region by heavier doping and greater depth to lower the spreading resistance; (*b*) modified IGBT with a hole current bypass structure to lower the spreading resistance.

to be collected by the entire side of the cell where the source has been removed. This so-called hole bypass structure in effect provides for an alternate path for the hole current component that does not have to flow laterally beneath a source region. This geometry is quite effective in raising the latchup threshold, but it does so at the expense of reducing the transconductance of the IGBT, since the effective width of the gate is reduced by the loss of the second source region in the basic cell.

By such means as these, the problem of latchup in IGBTs has been greatly minimized. Modern IGBTs are essentially latchup proof.

25-6 SWITCHING CHARACTERISTICS

25-6-1 TURN-ON TRANSIENT

The current and voltage waveforms for the turn-on of an IGBT embedded in a step-down converter circuit similar to that shown in Fig. 22-10 for the MOSFET are shown in Fig. 25-6. The turn-on portions of the waveforms appear similar to those of the power MOSFET shown in Fig. 22-11 of Chapter 22. The similarity is to be expected, since the IGBT is essentially acting as a MOSFET during most of the turn-on interval. The same equivalent circuits used in Chapter 22 for discussing the MOSFET turn-on waveforms can also be used for calculating the turn-on characteristics of the IGBT.

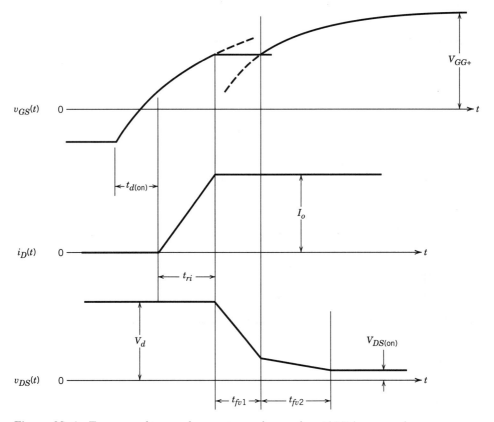

Figure 25-6 Turn-on voltage and current waveforms of an IGBT in a step-down converter circuit.

The t_{fv2} interval observed in the MOSFET drain–source voltage waveform in Fig. 22-11 is usually observed in the IGBT drain–source voltage waveform. Two factors will contributed to the t_{fv2} interval in the IGBT waveform. First the gate–drain capacitance C_{gd} will increase in the MOSFET portion of the IGBT at low drain–source voltages in a manner similar to that observed with power MOSFETs. Second, the *pnp* transistor portion of the IGBT traverses the active region to its on state (hard saturation) more slowly than the MOSFET portion of the IGBT. Until the *pnp* transistor is full on, the full benefit of conductivity modulation of the drain–drift region has not been achieved, and thus the voltage across the IGBT has not dropped to its final on-state value.

25-6-2 TURN-OFF TRANSIENT

The turn-off voltage and current waveforms for the IGBT in the step-down converter circuit are shown in Fig. 25-7. The observed sequence of a rise in the drain–source voltage to its blocking state value before any decrease in the drain current is identical to that observed in all devices used in a step-down converter circuit. The initial time intervals, the turn-off delay time $t_{d(off)}$, and the voltage rise time t_{rv} are governed by the MOSFET portion of the IGBT. The equivalent circuits used in Chapter 22 to describe these portions of the power MOSFET turn-off can also be applied to the IGBT.

The major difference between the IGBT turn-off and the power MOSFET turn-off is observed in the drain current waveform where there are two distinct time intervals. The rapid drop that occurs during the t_{fi1} interval corresponds to the turn-off of the MOSFET section of the IGBT. The "tailing" of the drain current during the second interval t_{fi2} is due to the stored charge in the n^- drift region. Since the MOSFET section is off and there

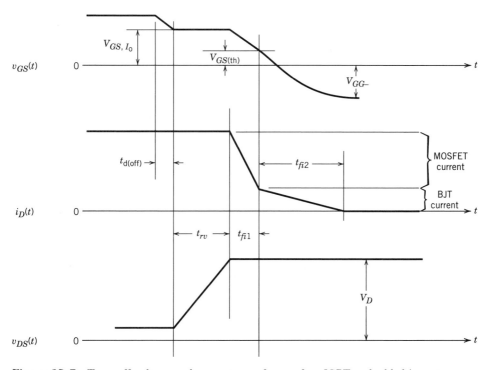

Figure 25-7 Turn-off voltage and current waveforms of an IGBT embedded in a step-down converter circuit.

is no reverse voltage applied to the IGBT terminals that could generate a negative drain current, there is no possibility for removing the stored charge by carrier sweep-out.

The only way that these excess carriers can be removed is by recombination within the IGBT. Since it is desirable that the excess-carrier lifetime in this region be large so that the on-state voltage drop is low, then the duration of the t_{fi2} interval at turn-off will be correspondingly long. However, a long t_{fi2} interval is undesirable because the power dissipation in this interval will be large since the drain–source voltage is at its off-state value. This time increases with temperature, as does the tailing time in a power BJT. Thus, a trade-off between on-state losses and faster turn-off times must be made in the IGBT, which are quite similar to those made with minority-carrier devices such as BJTs, thyristors, diodes, and the like. Electron irradiation of the IGBT is often used to set the carrier lifetime in the drift region to desired value.

Punch-through IGBTs attempt to minimize the current tailing problem by shortening the duration of the tailing time. This is done by means of the n^+ buffer layer described earlier and shown in Fig. 25-1. This layer is designed to have a much smaller excess-carrier lifetime than the n^- drift region and thus the buffer layer acts as a sink for excess holes. The greater recombination rate of the holes in the buffer layer sets up a hole density gradient in the n^- drift region during turn-off that causes a large flux of diffusing holes toward the buffer layer in a manner analogous to anode shorts in the GTO. This greatly enhances the removal rate of holes from the drift region and thus shortens the t_{fi2} interval. The buffer layer is relatively thin and heavily doped so that the ohmic losses in it are negligible in the on-state when large currents are flowing through the IGBT. The use of the buffer layer allows the drift region to be up to a factor of two smaller than for non-punch-through IGBTs. Hence, the on-state losses in the drift region of a PT device should be lower than those of a comparable (same voltage rating and same drift region carrier lifetime) NPT device.

Non-punch-through IGBTs attack the current tailing problem by minimizing the magnitude of the current during the tailing interval. This is done by designing the IGBT so that the MOSFET section carries as much of the total current as possible. When the IGBT turns off, the MOSFET section and the current carried by it turns off rapidly leaving only the small fraction of the total current that was flowing in the BJT section remaining as the tailing current. Newer NPT IGBTs designs have 90% or more of the total current carried by the MOSFET section of the device. This is accomplished by designing the PNP transistor section to have as low a value of beta as practical. The steps for doing this are described in the next section. The lifetime in the drift region is kept as large as possible in order to minimize drift region on-state losses.

25-6-3 NPT VERSUS PT STRUCTURES

Both IGBT geometries are effective in minimizing the effects of the tailing current. IGBTs are commercially available with voltage ratings as large as 1700 V, on-state currents of several 100 amperes, and turn-off times of 1 microsecond or less. Developmental efforts are in progress to realize IGBTs with voltage ratings of 2500 V.

While both geometries are being explored, there is currently the expectation that the non-punch-through geometry will have lower overall losses at higher voltage ratings. A wide base width for BJT section decreases its beta, and since the drift region of an IGBT is the base of the PNP transistor, higher voltage ratings requiring longer drift regions lead naturally to lower betas. Low emitter injection efficiency, which can be achieved by doping the emitter (the p-type anode of the IGBT) of the PNP transistor more lightly (10^{17} to 10^{18} cm^{-3}), will also reduce the beta or the PNP transistor. This step is also easier to implement at higher voltage ratings because the on-state losses contributed by the *p*-type anode become an increasingly smaller part of the overall on-state losses as the IGBT voltage rating increases. The on-state losses of high voltage IGBTs are dominated by the MOSFET channel losses and the drift region losses which completely mask any small increases in the ohmic resistance of the *p*-type anode due to smaller doping levels.

At the lower voltage ratings, 1000-1200 V and lower, the punch-through geometry appears to have lower overall losses compared to the non-punch-through structure for the reasons described previously. The steps necessary to keep the beta of the BJT section desirably small in NPT structures are not as effective as in higher voltage ratings. It is anticipated that the punch-through structure will have more severe problems with avalanche breakdown as the voltage rating increases beyond the 1000-1200 V range. Two factors support this expectation.

First, the BJT in a PT device has a larger beta because the drift region and thus the BJT base is smaller compared to a similar-rated NPT geometry. Larger betas lead to reduced breakdown voltage capabilities (recall the discussion in Chapter 21 regarding the reduction of BV_{CEO} compared to BV_{CBO} as the BJT beta increases). Secondly, detailed analysis (which is beyond the scope of this discussion) indicates that in the blocking state, the PT device has a larger electric field (by nearly a factor of two) at the blocking junction compared to a similar NPT geometry. As both device types are scaled to higher voltage operation, the electric field in the PT structure will reach breakdown magnitudes at lower blocking voltages than will the NPT structure.

These considerations lead to the expectation that both geometries will be used in future generations of IGBT designs. Some device manufacturers even question the expectation that PT structures are not the best choice for future higher voltage (greater than 1700 V) ratings and are planning to use PT geometries for these higher voltage ratings.

25-7 DEVICE LIMITS AND SOAs

The maximum drain current I_{DM} is set so that latchup is avoided and so problems with connecting wires from the chip to the case or in thin film metallizations are avoided. There is also a maximum permissible gate-source voltage $V_{GS(\max)}$. The value of this voltage is set by gate oxide breakdown considerations. The IGBT is designed so that when this gate-source voltage is applied, the maximum current that can flow under fault (short-circuit) conditions is approximately 4 to 10 times the nominal rated current. Under these conditions the IGBT will be in the active region with a drain-source voltage equal to the off-state voltage. Recent measurements indicate that the device can withstand such currents for 5-10 microseconds depending on the value of V_{DS} and can be turned off by V_{GS}.

The maximum drain-source voltage is set by the breakdown voltage of the *pnp* transistor. The beta of the transistor is quite small, so its breakdown voltage is essentially BV_{CBO}, the breakdown voltage of the drift-body junction (junction J_2 in Fig. 25-1*a*). Devices with blocking capabilities as large as 1700 V are commercially available and devices with larger voltage ratings are in development.

The maximum permissible junction temperature in commercially available IGBTs is 150°C. The IGBT can be designed to have an on-state voltage that changes little between room temperature and the maximum junction temperature. The reason for this is the combination of positive temperature coefficient of the MOSFET section and the negative temperature coefficient of the voltage drop across the drift region.

Individual IGBTs are available that have nominal current ratings as large as 200–400 amperes. IGBTs are easily paralleled because of the good control over the variation of

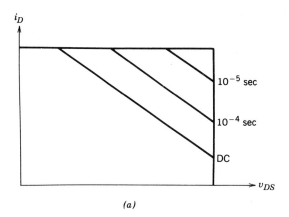

(a)

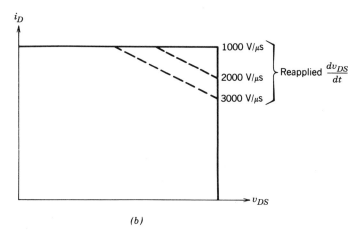

(b)

Figure 25-8 Safe operating areas of an IGBT: (a) forward-bias SOA; (b) reverse-bias SOA.

IGBT parameters from one device to another and also because of the small variation in on-state voltage with temperature. As many as four to six IGBTs connected in parallel are available as modules which have current ratings of 1000 to 1500 amperes.

The IGBT has robust SOAs both during turn-on and turn-off. The forward-bias safe operating area shown in Fig. 25-8a is square for short switching times, identical to the FBSOA of the power MOSFET shown in Fig. 22-19 for turn-on times shorter than 1 ms. For longer switching times the IGBT is thermally limited, as shown in the FBSOA, and this is also identical to the behavior FBSOA of the power MOSFET.

The reverse-bias safe operating area RBSOA is somewhat different than the FBSOA, as is illustrated in Fig. 25-8b. The upper-right-hand corner of the RBSOA is progressively cut out and the RBSOA becomes smaller as the rate of change of reapplied drain-to-source voltage dv_{DS}/dt becomes larger. The reason for this restriction on the RBSOA as a function of reapplied dv_{DS}/dt is to avoid latchup. Too large a value of dv_{DS}/dt during turn-off will cause latchup of the IGBT exactly as it can in thyristors and GTO. Fortunately, this value is quite large, comparing favorably with other power devices. In addition, the device user can easily control the reapplied dv_{DS}/dt by proper choice of V_{GG-} and gate drive resistance.

SUMMARY

This chapter has examined the structure and characteristics of a fairly new power device, the insulated gate transistor, or IGBT. The important conclusions are listed:

1. The IGBT is designed to operate as a MOSFET with an injecting region on its drain side to provide for conductivity modulation of the drain–drift region so that on-state losses are reduced.

2. The performance of the IGBT is thus midway between that of a MOSFET and a BJT. It is faster than a comparable BJT but slower than a MOSFET. Its on-state losses are much smaller than those of a MOSFET, and are comparable with those of a BJT.

3. The IGBT structure contains a parasitic thyristor that must not be allowed to turn on or else the gate will lose the ability to turn off the device.

4. Prevention of the turn-on of the parasitic thyristor involves special structural modifications of the IGBT structure by the device manufacturer and observance of maximum current and voltage ratings of the device by the user. New devices appear to be latchup proof.

5. The turn-on speed of the IGBT can be controlled by the rate of change of the gate–source voltage.

6. The IGBT has a rectangular SOA for switch-mode applications similar to the MOSFET and thus has minimal need for snubber circuits.

PROBLEMS

25-1 P-channel MOSFETs require about three times the area on a silicon wafer to achieve a performance comparable to an n-channel MOSFET. However, p-channel IGBTs have the same area as n-channel IGBTs. What are the reasons for the differences between the IGBT behavior and the MOSFET behavior?

25-2 During turn-off, the drain current in an IGBT will exhibit different behaviors depending on whether the carrier lifetime in the drift region is longer or shorter. Qualitatively sketch the drain current versus time during turn-off for a short-lifetime IGBT and for a long-lifetime IGBT and explain the reasons for the differences.

25-3 A punch-through IGBT will have a higher output resistance in the active region (flatter i_D-v_{DS} curves in the active region) than a non-punch-through IGBT. Explain why.

25-4 Estimate the forward and reverse breakdown voltages of the IGBT shown in Fig. 25-1. The doping levels are $p^+ = n^+ = 10^{19}$ cm^{-3}, $p = 10^{17}$ cm^{-3}, and $n^- = 10^{14}$ cm^{-3}. The length (dimension parallel to the current flow direction) of the drift region is 25 μm.

25-5 An IGBT and a MOSFET, both n-channel devices, are designed to block voltages as large as 750 volts in the off-state. The effective conducting area of both devices is 2 cm^2. If the on-state voltage is limited to 3 V or less, estimate the on-state current each device can conduct. Use $n_b = 10^{16}$ cm^{-3} as the excess carrier density at which the mobilities and carrier lifetime begin to decrease with increasing carrier density.

25-6 A punch-through (PT) IGBT and a non-punch-through (NPT) IGBT are each designed to block 1200 V and both have the same effective conducting area in the on-state. Estimate the relative on-state current capability of each device assuming that both devices have the same on-state voltage.

25-7 An IGBT can tolerate an overcurrent of considerable magnitude if its duration is not too long. Advantage can be taken of this characteristic in some circuit designs. Approximately estimate the overcurrent capability of an IGBT rated at $BV_{DSS} = 1000$V. Assume an effective conducting area of 0.25 cm^2, an overcurrent duration of 10 microseconds, and a maximum junction temperature of

300°C. Use $n_b = 10^{16}$ cm^{-3} as was suggested in Problem 25-5. *Hint:* Recall that $C_V dT = dQ$, where dQ = increase in heat energy, dT = temperature increase, and C_V = specific heat per unit volume.

25-8 Consider an IGBT and a MOSFET with the same BV_{DSS} rating and the same on-state current rating. Which device has the smaller values of C_{gs} and C_{gd} and why?

25-9 An IGBT circuit module complete with its own drive circuit has been made with the following performance specifications:

$$V_{DSM} = 800 \text{ V } I_{DM} = 150 \text{ A } \frac{dv_{DS}}{dt} < 800 \text{ V/}\mu\text{s, } R_{\theta j\text{-}a} = 0.5°\text{C/W}$$

$$t_{on} = t_{d(on)} + t_{ri} + t_{fv} = 0.3 \text{ }\mu\text{s } t_{off} = t_{d(off)} + t_{rv} + t_{fi} = 0.75 \text{ }\mu\text{s, } T_{j,max} = 150°\text{C}$$

This module is to be used in a step-down converter circuit with a diode-clamped inductive load. In this circuit the free-wheeling diode is ideal, the dc supply voltage is $V_d = 700$ V, the load current $I_o = 100$ A, and the switching frequency is 50 kHz with a 50% duty cycle. Determine if the IGBT module is overstressed.

REFERENCES

1. T. Rogne, N. A. Ringheim, J. Eskedal, B. Odegard, and T. M. Undeland, "Short Circuit Capability of IGBT (COMFET) Transistors," 1988 IEEE Industrial Applications Society Meeting, Pittsburg, PA, Oct. 1988.

2. B. Jayant Baliga, *Modern Power Devices*, Wiley, New York, 1987, Chapter 7.

3. B. Jayant Baliga, "The Insulated Gate Transistor (IGT)—A New Power Switching Device," *Power Transistors: Device Design and Applications*, B. Jayant Baliga and D. Y. Chen (Eds.), IEEE Press, Institute of Electrical and Electronic Engineers, New York, 1984, pp. 354–363.

4. H. Yilmaz, J. L. Benjamin, R. F. Dyer, Jr., Li. S. Chen, W. R. Van Dell, and G. C. Pifer, "Comparison of Punch-Through and Non-Punch-Through IGT Structures," *IEEE Transactions on Industrial Applications*, Vol. IA-22, No. 3, May/June 1986, pp. 466–470.

5. A. Nakagawa, Y. Yamaguchi, K. Watanabe, and H. Ohashi, "Safe Operating Area for 1200 V Nonlatchup Bipolar-Mode MOSFETs," *IEEE Transactions on Election Devices*, Vol. ED-34, No. 2, Feb. 1987, pp. 351–355.

6. M. S. Adler, K. W. Owyang, B. Jayant Baliga, and R. A. Kokosa, "The Evolution of Power Device Technology," *IEEE Transactions on Electron Devices*, Vol. ED-31, No. 11, Nov. 1984, pp. 1570–1591.

CHAPTER 26

EMERGING DEVICES AND CIRCUITS

26-1 INTRODUCTION

The number of semiconductor power devices available today is impressively large compared with just a few years ago. The list includes diodes, bipolar junction transistors, monolithic Darlingtons, MOSFETs, thyristors, GTOs, and IGBTs. Research efforts will continue to improve these devices, increasing their blocking voltage capabilities, lowering their on-state losses, and increasing their switching speeds.

Other device and integrated circuit concepts are also currently being explored that show significant potential for future power electronic applications. These concepts that have not yet found general commercial acceptance or are still in the laboratory prototype stage we term emerging devices and circuits. A list of such emerging devices would include power junction field effect transistors (also termed static induction transistors), field-controlled thyristors (bipolar static induction thyristors), MOS-controlled thyristors (MCTs), high-voltage integrated circuits and so-called smart power circuits and devices. Some of these devices may become widely used in the future, so it is important for designers of power electronic circuits to be aware of these potentially useful devices. In this chapter we briefly summarize the characteristics of these emerging devices and discuss their physical principles of operation and operational limitations. We also consider other semiconductor materials that may someday replace silicon for making power devices.

26-2 POWER JUNCTION FIELD EFFECT TRANSISTORS

26-2-1 BASIC STRUCTURE AND *I–V* CHARACTERISTICS

The vertical cross section of an *n*-channel power junction field effect transistor, JFET, (sometimes called a static induction transistor or SIT) is shown in Fig. 26-1*a*. This particular geometry is utilizing a so-called recessed gate structure, one of the more promising structures for the JFET. The gate and source regions are highly interdigitated in a manner similar to the interdigitation used in MOSFETs. Hundreds and even thousands of these basic gate–source cells are connected in parallel to make up a single power JFET. The doping levels indicated qualitatively in the figure are similar to those used in other power semiconductor devices and designated by the same qualitative symbols. The

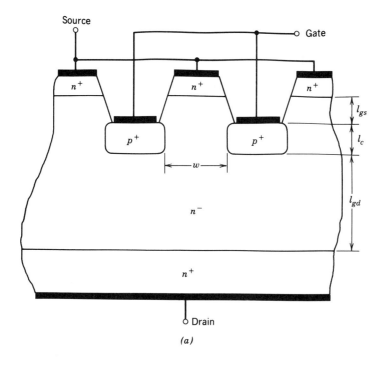

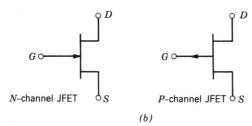

Figure 26-1 Structure and circuit symbols for JFETs:
(*a*) recessed gate JFET cross section; (*b*) JFET circuit symbols.

dimensions shown symbolically in the figure play a major role in determining the device characteristics, as will be discussed shortly.

In a power JFET the channel width w is fairly narrow, typically a few micrometers to a few tens of micrometers and the channel length l_c is made smaller than the width w. The device manufacturer also attempts to minimize the dimension l_{gs}. The length of the gate–drain region l_{gd} is dependent on the desired value of blocking voltage capability. This lightly doped region is essentially a drain–drift region analogous to that of the power MOSFET.

The circuit symbol for an *n*-channel JFET is shown in Fig. 26-1*b*. The arrow pointing into the gate indicates the direction of gate current that would flow if the gate–source junction becomes forward biased. The symbol for a *p*-channel JFET is also shown in this figure.

The *i–v* characteristics for an *n*-channel power JFET are shown in Fig. 26-2*a*. These characteristics are quite different from those of a MOSFET and are often referred to as triodelike characteristics because of their resemblance to the *i–v* characteristics of vacuum triodes. An approximate transfer curve of drain–source voltage versus gate–source volt-

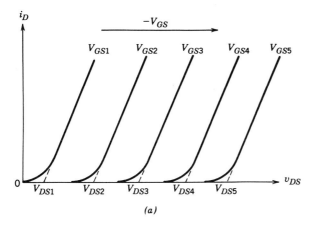

(a)

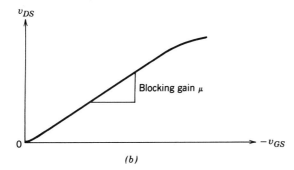

(b)

Figure 26-2 Current-voltage characteristics of a power JFET: (a) output characteristics; (b) transfer curve.

age is shown in Fig. 26-2b. The slope of this transfer curve is often termed the blocking gain μ because when the curves of i_D versus v_{DS} are extrapolated to zero drain current, as indicated in Fig. 26-2a, this slope represents the incremental increase in drain–source voltage that can be blocked by the device in the off state for a given incremental increase in the gate–source voltage.

The most important feature to note about the i–v characteristics is that the JFET is a normally-on device, meaning that when the gate is shorted to the source, the device is in the on state. In contrast, all other semiconductor power devices that we have discussed are normally-off devices. A normally-on characteristic is undesirable in most power electronic applications because the normally-on device may permit unacceptably large transient current flows at system power-up, whereas a normally-off device has a built-in safety feature of being off at system power-up. Because of this drawback, the power JFET has not found any widespread usage even though some are commercially available.

26-2-2 PHYSICS OF DEVICE OPERATION

As we indicated earlier, the JFET is in the on state when the gate–source voltage is zero. This occurs because there is no hindrance to the flow of current between drain and source in the simplified structure shown in Fig. 26-3a. The width of the channel region between the p^+ gate regions is large enough so that the depletion regions of the gate–source junction at $v_{GS} = 0$ do not meet in the center of the channel and thus pinch it off. In this on-state, the device designer seeks to minimize the ohmic resistance between the drain and source by making the channel length l_c short as well as the drain–drift region length l_{gd}. The lateral layout of the gate and source regions also impacts the on-state resistance of the device.

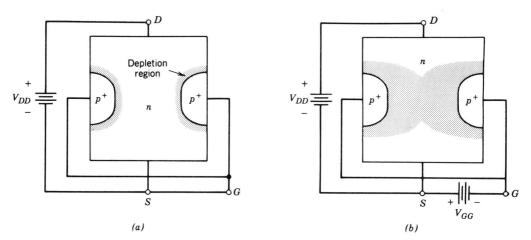

Figure 26-3 Gate depletion layers of a JFET: (a) in the on state; (b) in the blocking state.

When a reverse bias is applied to the gate–source pn junction, the depletion layers grow in width, and at the particular value of v_{GS}, termed the pinch-off voltage V_p, the depletion layers meet in the middle of the channel and pinch it off, as is indicated in Fig. 26-3b. In this circumstance there will be no flow of current between drain and source as long as the drain–source voltage is kept small. This is the off state of the JFET.

If the gate–source voltage is kept at a constant value $|V_{GG}| > |V_p|$, initially no drain current will flow as $v_{DS} = V_{DD}$ is increased from zero. The depletion layers that are blocking the channel have set up a potential barrier to the flow of drain current, as is illustrated in Fig. 26-4. The electric field E_{GS}, diagrammed in Fig. 26-4, which results from this potential barrier, opposes the electric field E_{DS}, which is set by the applied drain–source voltage. As V_{DD} increases, E_{DS} increases while E_{GS} remains fixed because of the fixed V_{GG} and the net electric field in the depletion region in the channel that is blocking the flow of current becomes progressively smaller. The result is that the potential barrier to the flow of drain current gets smaller and smaller, as diagrammed in Fig. 26-4.

When V_{DS} is large enough to suppress the potential barrier set up by V_{GG}, the depletion layer has been effectively eliminated from the center of the channel and current begins to flow. Increases in drain–source voltage beyond this, the threshold value of $\mu|V_{GG}| = V_{DD}$ will cause large increases in the drain current, since the incremental slope of the i–v characteristic in this operating region is essentially the on-state resistance. To maintain the JFET in the off state for some specified maximum value of $v_{DS} = V_{DSM}$, the gate–source voltage must be larger than V_{DSM}/μ. The length l_{gd} of the drain–drift region is set by the requirement that this region accommodate the depletion layer of the gate–drain junction at the maximum value of drain–source voltage V_{DSM} without avalanche breakdown occurring.

It is possible to operate the JFET in a so-called bipolar mode when it is in the on state. Instead of reducing v_{GS} to zero volts in order to turn on the JFET, the gate–source junction is forward biased. This leads to a significant reduction in the on-state resistance because the forward-biased junction injects minority carriers into the channel that conductivity modulates the on-state channel resistance. In this situation, significant gate currents must flow, and from a terminal viewpoint, the device appears similar to a BJT in the on state. In fact, the drain current that can flow in this mode of operation is proportional to the magnitude of the gate current, and the incremental current gain is typically 100 or more at small drain currents and falls as I_D is increased. If the JFET is designed

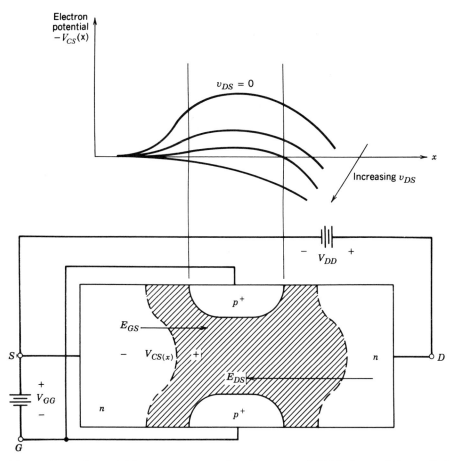

Figure 26-4 Potential barrier to current flow in a power JFET because of the applied gate–source voltage. As the applied drain-source voltage is increased, it gradually suppresses this barrier.

with narrow channel widths so that the channel is pinched off at zero gate–source bias, the only way to turn on the device is to forward bias the gate–source junction. This produces a JFET with a normally-off characteristic, and it has been called a bipolar static induction transistor (BSIT). The i–v characteristic of the BSIT looks similar to that of a BJT.

26-2-3 SWITCHING CHARACTERISTICS

A normally-on JFET is very similar to the MOSFET as far as its switching characteristics are concerned. The equivalent circuit for the normally-on JFET is identical to that of the MOSFET, and the switching waveforms and switching times will be the same as for a comparable MOSFET. The analysis presented in Chapter 22 for the MOSFET can be used with the JFET with only minimal changes. The major difference between the two devices is that the n-channel JFET requires a negative going gate–source voltage to turn off and a positive going v_{GS} to turn on, whereas just the opposite is required for the MOSFET.

If the JFET is operated in the bipolar mode or is fabricated as a normally-off device (BSIT), then it will behave in a switching application more as a bipolar junction transistor than as a MOSFET. This will include some of the phenomena associated with stored charge since the forward-biased gate–source junction injects minority carriers into the channel when the device is in the on state.

Two principal differences will be noted between the turn-off waveforms of a bipolar-mode JFET and the BJT. First, there will be only one current fall time interval with the bipolar-mode JFET because there is no quasi-saturation region such as is observed with the BJT. Second, the turn-off time of the JFET including both storage delay time and current fall time will be considerably shorter for the JFET than for a comparable BJT.

The reason is that the JFET has no *pn* junction in the path of the drain current that can interrupt the sweep-out of excess carriers as the device switches off. When an in-line *pn* junction, such as the collector–base junction or emitter–base junction becomes reverse biased, any stored charge in the device becomes trapped in the device and can be removed only by internal recombination. This considerably slows down the turn-off of the device. Special modifications to the device structure, such as anode shorts in GTO or special drive circuit arrangements such as emitter-open switching with the BJT, are often used to mitigate the effects of this trapped stored charge.

26-3 FIELD-CONTROLLED THYRISTOR

26-3-1 BASIC STRUCTURE AND *I*–*V* CHARACTERISTIC

If the drain of a power JFET structure is modified into an injecting contact by making it into a *pn* junction, then a new device is produced. This new device, which is variously termed a field-controlled thyristor (FCT), a field-controlled diode, and in Japan, a bipolar static induction thyristor (BSIThy), has the basic structure shown in Fig. 26-5. What would normally be the drain of an *n*-channel JFET is converted into a *pn* junction, as is shown, and it becomes the anode of the device. The source of the JFET portion of the new structure is now termed the cathode. The circuit symbol for the FCT is shown in Fig. 26-5*b* and is essentially a diode symbol with a gate terminal added. The arrow on the gate terminal indicates the direction of forward-bias current flowing into the gate–source *pn* junction.

The *i*–*v* characteristic of a normally-on FCT is shown in Fig. 26-6. In the forward-bias portion of the characteristic, the FCT appears similar to the power JFET. The difference in the forward-bias operation of the two devices is quantitative, with the FCT being able to conduct far larger currents than the JFET for the same on-state voltage. The FCT also blocks in the reverse direction because of the addition of the *pn* junction at the anode. This reverse blocking is independent of the voltage applied to the gate–source junction.

The FCT can also be made with a normally-off characteristic by using the same approach as was described for the normally-off JFET. The *i*–*v* characteristics of a normally-off FCT are similar to those of a BSIT except that the current levels are far larger because of the lower on-state resistance of the FCT.

26-3-2 PHYSICAL DESCRIPTION OF FCT OPERATION

As was explained in the preceding section, the FCT is basically a power JFET structure with an injecting contact at the anode (JFET drain). The injection of minority carriers from the anode into the anode drift region results in large conductivity modulation of the region. Consequently, there is a small value of on-state resistance compared with a JFET and a corresponding low value of on-state voltage, even at large values of currents.

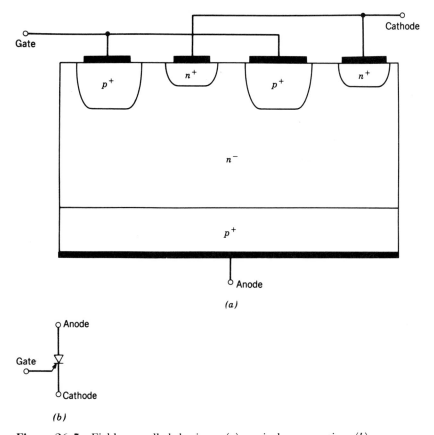

Figure 26-5 Field-controlled thyristor: (a) vertical cross-section; (b) circuit symbol.

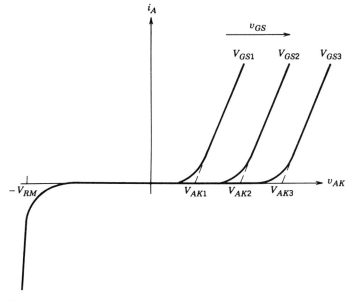

Figure 26-6 Current-voltage characteristic of a normally-on FCT.

The JFET-like gate structure gives the FCT a turn-off capability in addition to a turn-on capability. In the normally-on FCT, turn-off is accomplished by applying a large reverse bias to the gate–cathode terminals so that the gate–cathode junction is reverse biased. The depletion region of this junction then grows and pinches off the channel connecting the anode to the cathode, thus preventing any current flow. The negative bias on the gate draws the excess carriers out of the device as a large negative gate current, similar to the action in a GTO. If the FCT is fabricated as a normally-off device, then its turn-on and turn-off are similar to the turn-on and turn-off of a normally-off JFET.

Even though the FCT is termed a field-controlled thyristor, it is important to note that the device does not have any regenerative turn-on or turn-off as does a GTO. The FCT does not latch on or off. If the gate drive holding the normally-on FCT in the blocking state is removed, the FCT will turn on. Similarly, removal of the gate drive holding the normally-off FCT in the on-state will cause the device to turn off.

26-3-3 SWITCHING CHARACTERISTICS

The switching times of the FCT will be considerably slower than those of the normally-on JFET. This is because of the large amount of stored charge in the drift region and the channel region of the device. The switching waveforms will be qualitatively similar to those of a normally-off JFET, although the turn-off times will be significantly longer because of the large amount of drift region stored charge. As explained earlier, turn-off of the FCT requires a large negative gate current pulse, similar to a GTO. The normally-off JFET does not have nearly the amount of drift region stored charge, since it does not have the injection of charge from the anode.

The FCT would be expected to have large reapplied dv_{AK}/dt ratings, since it does not have any regenerative turn-on mechanism such as limits the conventional thyristor. This expectation has been verified in laboratory prototypes of FCT structures.

Laboratory prototypes of FCTs have also demonstrated that limits exist to the allowed values of di_A/dt, especially during device turn-on. Several factors lead to turn-on of the FCT in localized areas and then expansion of these localized areas to encompass the entire active region of the device as time proceeds. The localized turn-on leads to small regions of high-power dissipation, which will be too large if the rate of growth of the anode current di_A/dt exceeds some maximum rate. The excessive power dissipation in this small area can then cause device failure just as it can in comparable situations in BJTs or thyristors. Improvements in the di_A/dt rating have been made in recent years so that FCTs now have ratings in excess of 1000 A/μs.

26-4 JFET-BASED DEVICES VERSUS OTHER POWER DEVICES

The blocking voltage capability of the JFET that has been achieved to date compares reasonably well with the BJT or the MOSFET. The limiting factor in the JFET is not avalanche breakdown across the depletion region in the drift region but rather the achievable value of blocking gain. The maximum voltage that can be blocked between anode and cathode or drain and source is crudely modeled as the product of the blocking gain and the breakdown voltage of the gate–source junction.

The on-state losses in a JFET are larger than in a comparable MOSFET. The reasons are largely technological rather than fundamental, and further research will narrow the gap between the two types of devices. Intuitively one would expect that JFETs and MOSFETs made to block the same voltages in the off-state would have the same on-state losses. If the JFET is operated in the bipolar mode, then the on-state losses of the JFET lie between those of the MOSFET and the BJT.

The switching speeds of normally-on JFETs are presently somewhat slower than those of comparable MOSFETs. This is basically a technological limitation rather than a fundamental one. The normally-off JFET (BSIT) has switching speeds that are currently comparable or somewhat better than those of a comparable BJT. In principle a BSIT should have faster turn-off times than a BJT because of the lack of an in-line *pn* junction that can lead to an open-base turn-off problems such as can slow down the BJT. The FCT is somewhat faster than a GTO, the device against which it is normally compared.

Normally-on JFETs, like MOSFETs, are majority-carrier devices and have no serious propensities to second breakdown as do BJTs and other minority-carrier devices. Bipolar-mode JFETs and FCTs do have nonuniform turn-on of the active regions and, hence, some likelihood of second breakdown under the right conditions. This potential problem of current constrictions in bipolar-mode JFETs and FCTs also means that limits on *di/dt* and *dv/dt* may also exist. The seriousness of these potential problems is somewhat unclear, and further research will be needed to address this issue.

Probably more than any other reason, the normally-on characteristic of the JFET is most responsible for the lag in its use in switch-mode applications compared with other devices. The disadvantages of this normally-on characteristic have already been detailed. Although it is possible to fabricate normally-off JFET-based devices, they have not yet matched the capabilities of other normally-off devices. Further research and development with JFET-based structures will be needed to rectify this situation. It is likely that it will not happen on a broad front, since the other devices discussed in the previous chapters have already found wider acceptance and their capabilities are continuing to be developed. The JFET-based devices will most likely find niche applications where their unique properties offer advantages that other devices cannot match.

26-5 MOS-CONTROLLED THYRISTORS

26-5-1 BASIC STRUCTURE

The MOS-controlled thyristor (MCT) is a new device that has recently become commercially available (Harris Semiconductor). It is basically a thyristor with two MOSFETs built into the gate structure with one of the two MOSFETs, the ON-FET responsible for turning the MCT on, and the other MOSFET, the OFF-FET, responsible for turning the device off. There are two types of MCTs, the P-MCT and the N-MCT, and both combine the low on-state losses and large current capability of thyristors with the advantages of MOSFET-controlled turn-on and turn-off and relatively fast switching speeds.

A cross-sectional view of a single cell of a P-MCT is shown in Fig. 26-7a. A complete P-MCT is composed of many thousands of these cells fabricated integrally on the same silicon wafer and all the cells are connected electrically in parallel. The thyristor portion of the device has the same structure as a conventional thyristor. The *p*-type region nearest to the cathode, the base region of the *npn* transistor in the two-BJT model of the thyristor section shown in Fig. 26-7b, is the lightly doped region that must contain the depletion region of the blocking junction J_2 when the device is off. The ON-FET is a *p*-channel MOSFET and the OFF-FET is an *n*-channel MOSFET.

These MOSFETs are located around the anode of the MCT, as is shown in Fig. 26-7a, and thus the MOSFETs share the same side or surface of the silicon wafer as the anode. Every cell contains an OFF-FET, but most cells do not have an ON-FET. Typically about 1 out of 20 cells contains an ON-FET. Because of the close packing of the

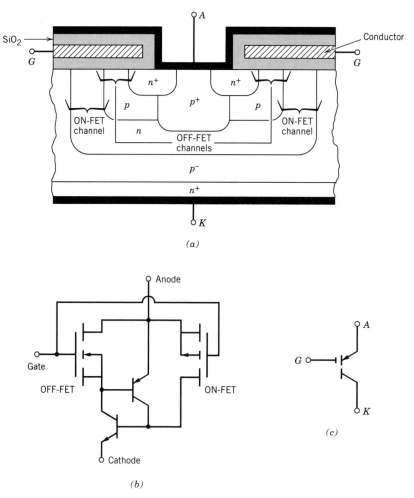

Figure 26-7 The P-MCT: (*a*) cross-sectional view; (*b*) equivalent circuit; (*c*) circuit symbol.

cells, if one cell turns on, it will cause adjacent cells to turn on because some of the excess carriers in the cell with the ON-FET will be able to diffuse to the adjacent cells and thus provide them with the excess carriers needed for turn-on. An equivalent circuit for the P-MCT is shown in Fig. 26-7*b*, and it includes not only the two-BJT model of the thyristor portion of the device but also the ON-FET and OFF-FET. The circuit symbol for the P-MCT is shown in Fig. 26-7*c*.

A cross-sectional view of a single cell of an N-MCT is shown in Fig. 26-8*a*. Like the P-MCT, a complete N-MCT is composed of many thousands of these cells fabricated integrally on the same silicon wafer, and all the cells are connected electrically in parallel. The thyristor portion of the device has the same *pnpn* structure as a conventional thyristor. The lightly doped region which must contain the depletion layer of the blocking junction J_2 is placed in the *n*-type region nearest to the anode. This *p*-type region also functions as the base of the *pnp* transistor in the two-BJT model of the thyristor. The ON-FET is a *n*-channel MOSFET and the OFF-FET is an *p*-channel MOSFET. These MOSFETs are located around the cathode, as shown in Fig. 26-8*a*, and share the same side of the silicon wafer as the cathode. Every cell contains an OFF-FET, but only about 1 in 20 cells has

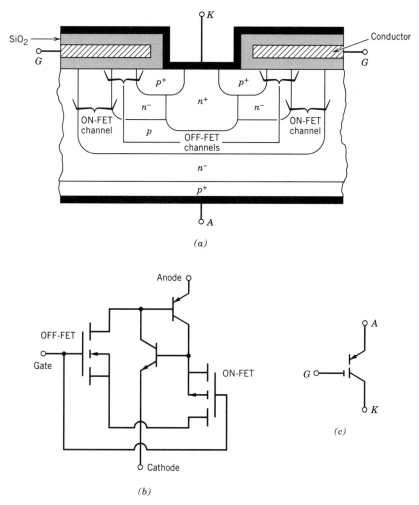

Figure 26-8 The N-MCT: (*a*) cross-sectional view; (*b*) equivalent circuit; (*c*) circuit symbol.

an ON-FET. An equivalent circuit for the N-MCT is shown in Fig. 26-8*b*, and it includes not only the two-BJT model of the thyristor portion of the device but also the ON-FET and OFF-FET. The circuit symbol for the N-MCT is shown in Fig. 26-8*c*.

The static *i–v* characteristic of both types of MCTs is essentially the same as for a GTO. Hence we will not repeat it here. Presently available MCTs are designed for asymmetrical blocking and little reverse-blocking capability, typically only about 25 V. Although the device is a latching switch, it is necessary to keep the gating signal appropriately activated in both the on and off states. Allowing the gating signal to go to zero opens up the possibility of unwanted turn-on or turn-off because of large *dv/dt* values applied to the anode–cathode terminals of the MCT.

26-5-2 MOSFET-CONTROLLED TURN-ON AND TURN-OFF

Assume that the N-MCT of Fig. 26-8 is on and it is to be turned off. Turn-off is accomplished by turning on the *p*-channel OFF-FET in the equivalent circuit of Fig. 26-8*b* by applying a negative gate–cathode voltage to the MCT. When the OFF-FET is conducting, it shorts out the base−emitter junction of the *npn* transistor in the thyristor pair, and the base current into the transistor is diverted through the OFF-FET. This causes the *npn* transistor to begin to turn off as the stored charge in the base of the transistor disappears and is not replenished. The *npn* BJT current gain falls to a low value where the latch-on condition of the thyristor is no longer satisfied. Once this occurs, the thyristor turns itself off by regenerative action. During the turn-off of the thyristor, the other MOSFET in the circuit, the *n*-channel ON-FET, is kept in the blocking state by the negative gate−cathode voltage.

The P-MCT is turned off in an analogous manner. However, because of the difference in structure, the gating signal must be a positive signal applied between the gate and the anode (gate positive and anode negative). The *n*-channel OFF-FET shorts out the base−emitter junction of the *pnp* transistor in the thyristor pair.

In turning off either type of MCT, it is essential that the drain−source voltage of the conducting OFF-FET be kept well below 0.7 V, the value of base−emitter voltage that causes the BJT to be in the active region. This requirement means that there is a maximum on-state current through the MCT that can be turned off by means of gate control. When the OFF-FET is activated to turn off the MCT, the on-state current must go through the OFF-FET. When this current is larger than a specified value, the voltage drop across the OFF-FET exceeds 0.7 V, and the BJT will not turn off.

If a P-MCT and an N-MCT are made to be as equivalent as possible, that is, the same size and relative doping levels, the P-MCT will be able to turn off a current approximately three times larger than an N-MCT. The reason for this is that the OFF-FET in a P-MCT is an *n*-channel MOSFET that has a value of on-state resistance that is three times smaller than the on-state resistance of a similar-sized *p*-channel device. The difference in the on-state resistances is caused by the mobility of electrons in silicon being about three times larger than the hole mobility.

Turn-on is accomplished by driving the ON-FET into the conducting state and simultaneously driving the OFF-FET into its blocking state. Turn-on of the ON-FET in a P-MCT by means of a negative gate−anode voltage permits the flow of base current into the *npn* transistor in the thyristor pair, which thus activates the *npn* transistor. The collector current into the *npn* transistor flows out of the base of the *pnp* transistor, thus turning it on. Once both transistors are on, the regenerative action of the connection will cause the thyristor to latch on. The better conduction characteristic of the *pnp* transistor ensures that it will carry most of the base current of the *npn* transistor in the on state rather than the ON-FET, which is in parallel with the *pnp* BJT. During the on state of the P-MCT, the OFF-FET is kept in the blocking state, which is ensured by maintaining a negative gate−anode voltage.

In an N-MCT the ON-FET is activated by means of a positive gate−cathode voltage that will simultaneously ensure that the OFF-FET is driven into its blocking state. Except for the changes in gating signal terminals and polarity of the gating signal, the turn-on process is analogous to that just described for the P-MCT.

26-5-3 RATIONALE OF OFF-FET PLACEMENT IN THE MCT STRUCTURE

In order to turn off a thyristor, the latching condition $\alpha_1 + \alpha_2 \approx 1$ must be broken by making $\alpha_1 + \alpha_2 \ll 1$. This is accomplished by turning off one of the two transistors in the two-transistor equivalent circuit. The turn-off of the selected BJT is done by shorting the base of the transistor to its emitter using the OFF-FET, which is placed in shunt with the base−emitter terminals. The transistor that should be selected for turn-off is the one with the larger gain, that is, the larger value of the base transport coefficient α because this will have the greatest effect in reducing the sum $\alpha_1 + \alpha_2$.

The transistor base that has the lightly doped region designed to contain the depletion layer of the blocking junction J_2 will have a low value of the base transport coefficient α because of the greater base thickness compared to the other transistor base. Hence the OFF-FET should be placed between the base–emitter terminals of the BJT, which has the narrower base width. In the case of the P-MCT, the higher gain transistor is the *pnp* BJT and thus the OFF-FET and ON-FET are placed around the anode terminal, as illustrated in Fig. 26-7a. For an N-MCT, the higher gain transistor is the *npn* BJT, and thus the OFF-FET and ON-FET must be placed around the cathode terminal as shown in Fig. 26-8a.

In most situations, the power electronics designer will want to drive the control terminals of the power switch from ground-referenced circuits in order to simplify the overall circuitry. Thus one of the two control terminals of the MCT should be grounded. This means that an N-MCT should have its cathode at ground and a P-MCT should have its anode at ground. Thus an N-MCT is used in circuits with positive dc voltages/power supplies such as the step-down converter shown in Fig. 26-9a. The same circuit powered with a negative dc supply would use a P-MCT as shown in Fig. 26-9b.

26-5-4 MCT SWITCHING BEHAVIOR

The MCTs can switch quite rapidly from off to on and from on to off with typical switching times being 1 μs. Figure 26-10 shows the approximate N-MCT switching waveforms for the step-down converter of Fig. 26-9a. The circuit contains no protective snubbers, so the MCT is exercised in a hard-switching mode of operation.

Two switching times are used to characterize the turn-on waveforms of the MCT. The turn-on delay time $t_{d,\mathrm{on}}$, which is shown in Fig. 26-10, is basically set by how fast substantial injection of excess carriers into the bases of the transistors can occur. Once there are substantial numbers of excess carriers in the base regions, the regenerative action of the thyristor can begin. Typical values of the turn-on delay time are about 0.5 μs. Toward the end of the turn-on delay time, the current begins to rise rapidly and the current rise time t_{ri} characterizes this interval. Typical values of t_{ri} are 0.5 μs and are governed by how fast the excess-carrier plasma can spread across the entire cross-sectional area of each cell. Since each cell has a relatively small area, this time is quite small.

During turn-on, the MCT does not appear to have the typical hard-switched waveforms but instead appears to be protected by a turn-on snubber circuit because the current

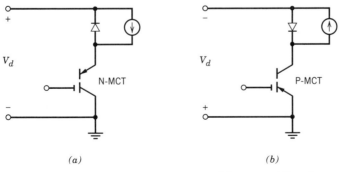

(a)   *(b)*

Figure 26-9 Step-down converter using *(a)* an N-MCT and *(b)* a P-MCT.

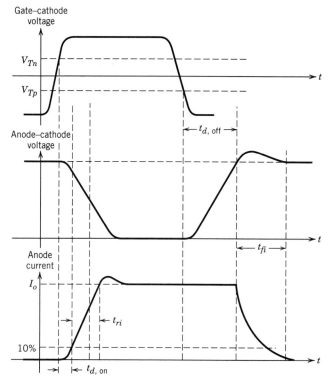

Figure 26-10 Switching waveforms in a step-down converter using an N-MCT.

is rising as the anode–cathode voltage is falling. This is due to the fact that the current rises very rapidly in the MCT due to the regenerative action of the thyristor structure. Values of di/dt in the range of 500–1000 A/μs are easily achieved, and hence even a small amount of stray inductance L_σ such as 0.3 μH can act as a turn-on snubber ($L_\sigma \, di/dt$ = $(3 \times 10^{-7}$ H)$(500$ V/μs) = 150 V).

During turn-on it is necessary that the gate–cathode voltage rise time to its final on-state value not be any slower than a specified upper limit, typically about 200 ns. This requirement stems from the need to ensure that all (typically 10^5 or more) paralleled MCT cells turn on at the same time so that current-crowding problems are minimized.

The turn-off of the MCT has the typical characteristics of a hard-switched device, and the waveforms are characterized by a turn-off delay time $t_{d,\text{off}}$ and a current fall time t_{fi}. The turn-off delay time is typically 1 μs or a little less and includes the time required for the anode–cathode voltage to reach its off-state value. The current fall time is controlled by how fast the excess carriers in the base regions of the thyristor structure recombine. Typical current fall times are 0.5–1 μs.

26-5-5 DEVICE LIMITS AND SAFE OPERATING AREA

The safe operating area (SOA) of an MCT is shown qualitatively in Fig. 26-11. At small anode–cathode voltages, the maximum controllable anode current is the limiting boundary of the SOA. By maximum controllable anode current, we mean the maximum on-state current that can be turned off under gate control. This value is determined by the char-

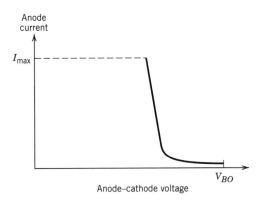

Figure 26-11 Qualitative sketch of the MCT safe operating area.

acteristics of the OFF-FET with P-MCTs having approximately three times greater capabilities than N-MCTs.

Above some anode–cathode voltage that is very technology dependent, the maximum current boundary must be decreased as the voltage is increased to larger values. This decrease is related to both power dissipation considerations and an internal device avalanche mechanism. At some significantly reduced current level, the allowable voltage values increase rapidly to zero-current breakover value. Since MCTs are new devices, the SOAs are very conservatively determined. They will become more ideally square as the technology improves.

Single MCT devices currently available have maximum current capabilities of approximately 100–200 A in the on state and can block 2000–3000 V in the off state. While the current ratings of single MCTs will increase as the technology improves, it will never match the single GTO capabilities. The reason is that MCTs have a very dense cell structure with very small feature sizes, approaching 1 μm. With a large number of cells per unit area, the greater the overall device area, the more likely will be the possibility of a cell being a short between anode and cathode. Considerations of reasonable fabrication yields limit the achievable device area in MCTs. Gate turn-off thyristors have a much lower cell density since each cell is much larger. Consequently the overall area of a GTO can be much larger and thus have a larger current rating than an MCT before fabrication yield considerations come into play.

Larger current ratings can be obtained by paralleling several MCTs. Modules of four to six MCTs connected in parallel have been demonstrated with current capabilities of the resultant module approaching 1000 A in the on state. These ratings will increase as the technology improves.

The MCTs are also subject to dv/dt and di/dt limitations. The dv/dt limitation of the MCT has the same physical origin as the dv/dt limitations in thyristors and GTO. In order to have reasonably large dv/dt limits, it is necessary to ensure that the OFF-FET is in the conducting state when the MCT is supposed to be off. This is accomplished by applying a continuous gating signal to the gate of the MCT so that the OFF-FET is held in the conducting state. If the gate signal is allowed to go to zero during the blocking state operation, then even moderate values of dv/dt will turn on the MCT. At present MCTs have dv/dt ratings of 500–1000 V/μs.

The MCTs have di/dt limits because of current-crowding problems. Since the ON-FET are not densely distributed (i.e., one per cell) across the entire die surface, some cells (those with ON-FETs) turn on before the rest. Hence some limits on di/dt are needed to prevent excessive currents from flowing in these cells. Increasing the density of the ON-FETs and other technology improvements will increase the di/dt ratings. Currently available MCTs have di/dt ratings of about 500 A/μs.

26-6 POWER INTEGRATED CIRCUITS

26-6-1 TYPES OF POWER INTEGRATED CIRCUITS

Modern semiconductor power control circuits have a considerable amount of control drive circuitry in addition to the power device itself. Several examples of this were presented in earlier chapters. The control circuitry often includes logic circuitry controlled by microprocessors. The inclusion of such control and drive circuitry on the same chip or wafer as the power device would greatly simplify the overall circuit design and broaden the range of potential applications. A cheaper and more reliable power control system would result from such integration. Overall, there would be a reduction in the complexity (fewer separate components) of circuits and systems using such power integrated circuits.

Such integration has already been demonstrated in many applications. There are three classes of power integrated circuits including so-called *smart* or *intelligent switches*, high voltage integrated circuits (HVICs), and discrete modules. The domain of power integrated circuits, particularly smart switches and HVICs, is considered to be current levels less than 50–100 A and voltages of approximately 1000 V or less. Discrete modules cover a much wider voltage–current range.

Smart switches are vertical power devices onto which additional components are added to the extent feasible without requiring major changes to the vertical power device process sequence. Features such as on-chip sensors for overcurrents and overtemperature as well as portions of drive circuits are examples of things that can be included. A somewhat hypothetical example of a smart switch is shown in Fig. 26-12. The *pn* junction

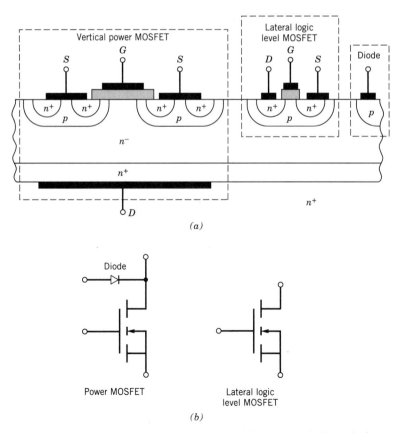

Figure 26-12 Simple hypothetical example of a smart switch vertical cross-section (*a*) and its equivalent circuit (*b*).

formed from the N⁻ drift region and the P-body region is always reverse biased if the drain of the vertical power MOSFET is positive with respect to the source and thus, this junction provides the electrical isolation between the lateral and vertical MOSFETs.

High-voltage integrated circuits (HVICs) are made using conventional logic-level device fabrication process but with some modifications so that lateral high-voltage devices can also be fabricated on the wafer compatibly with the low voltage devices. Two simple hypothetical examples of HVICs are shown in Fig. 26-13. These examples differ from each other in the manner in which electrical isolation between the various devices is realized. Actual HVICs have considerably more complexity.

Discrete modules are composed of multiple chips mounted on a common insulating substrate and hermetically sealed into a single package. The various chips may include vertical power devices, a drive circuit chip, and a control circuit chip (perhaps even a PWM controller), and possibly other functionality. Although this approach is not a completely integrated fabrication method, we include it because of its potential and its current widespread application compared to smart switches or HVICs.

26-6-2 CHALLENGES FACING PIC COMMERCIALIZATION

The use of power-integrated circuits in power electronics applications faces several challenges both technical and economic. The technical issues include:

1. Electrical isolation of high-voltage components from low-voltage components.
2. Thermal management – power devices usually operate at higher temperatures than low-voltage devices.
3. On-chip interconnections with high-voltage conductor runs over low-voltage devices or low-voltage regions.

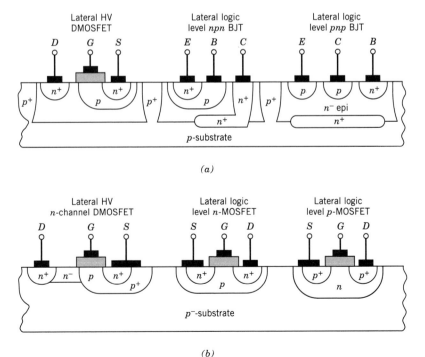

(a)

(b)

Figure 26-13 Simple hypothetical examples of high-voltage integrated circuits. In (a) electrical isolation is achieved by junction isolation while in (b) self-isolation is used.

4. Fabrication process must provide full range of devices and components—transistors (BJT, MOSFETs, IGBTs) diodes, resistors, capacitors, etc.

In addition, the use of power integrated circuits faces several economic issues. These include:

1. Large up-front development costs prior to any production runs.
2. Cost differentials between the three types of PICs.
3. Need for high volume applications to recover large development expenses.

26-6-3 PROGRESS IN RESOLVING CHALLENGES

Isolation of low-voltage devices from high-voltage elements, can be accomplished by either dielectric isolation, *pn* junction isolation, or self-isolation. Dielectric isolation can be implemented in two ways. In Fig. 26-14a, the isolation basically consists of etching a pocket in the chip or wafer and then growing a layer of silicon dioxide in it. Next, a layer of silicon is deposited over the SiO_2. After annealing the deposited silicon at a high temperature, it becomes recrystallized and can then be used for fabricating the low-voltage devices. Alternatively, the wafer-bonding technique is illustrated in Fig. 26-14b. Dielectric isolation is free of parasitic devices such as diodes that could become activated

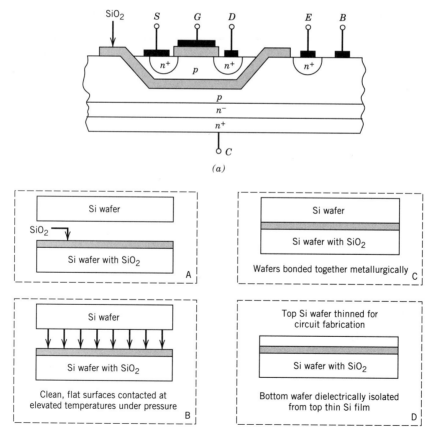

Figure 26-14 Examples of dielectric isolation with (*a*) oxide re-fill of an etched pocket followed by silicon thin film overgrowth and (*b*) wafer bonding followed by thinning of the bonded wafer.

in certain circumstances and negate the isolation. At present, dielectric isolation is relatively expensive to implement and results in lower yields compared to junction isolation or self-isolation, which are described below.

Junction isolation, on the other hand, is much cheaper and easier to implement. A *pn* junction is fabricated so as to completely surround the area to be isolated as is diagrammed in Fig. 26-13*a* for the HVIC. This junction is then reverse-biased at all times, thus achieving the desired isolation. A 1000-volt junction isolation has been demonstrated. The principal disadvantage of this isolation method is the parasitic diode that comes with it. Potential problems with this diode include possible turn-on and temperature dependent leakage currents.

If only MOSFET devices are used in the integrated circuit, then self-isolation can be used. In self-isolation, all the MOSFETs have the same doping type (*n*-type or *p*-type) for their body regions and the substrate also has this type of doping. This situation is illustrated in Fig. 26-13*b*. The drain and source dopings are contained in the body regions and the drain–body *pn* junction is reversed biased at all times, thus providing isolation between the drains of different MOSFETs. The sources of all the MOSFETs are connected together if body–source shorts are used.

A completely effective means of high-voltage interconnections on the chip or wafer has yet to be devised. The basic difficulty is that wherever an interconnect must pass over either an isolation region or some other heavily doped region that is at a significantly different potential than the interconnect, the equipotential lines between the low- and high-voltage regions have appreciable curvature. This results in a considerable amount of ''field crowding'' as is illustrated in Fig. 26-15 for the case of an isolation region. This field crowding will lead to premature breakdown of the isolation region and thus shorting of the high voltage to ground.

Two possible approaches to this problem include the use of thick insulators and the use of field shields. The use of thicker insulators will require further research and development. Questions such as what is the most appropriate insulator and how to prevent the inevitable strain (which increases in the deposited material as it gets thicker) from delaminating the film from the substrate remain to be answered. In the second approach, the field in the insulator can be made more uniform by the use of field shields, which are illustrated in Fig. 26-16. The rationale for the field shields is the same as for field plates,

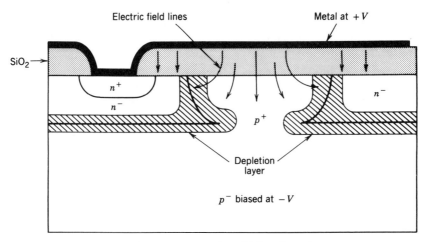

Figure 26-15 Electric field crowding where high-voltage interconnects cross over an isolation diffusion.

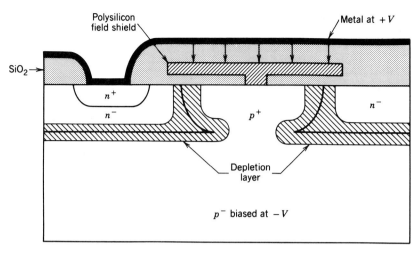

Figure 26-16   Use of a field shield to minimize field crowding in the case of an interconnect over an isolation region.

which were discussed in previous chapters. In addition the use of field shields would mean that the insulating film would not have to be as thick.

The problem of how to connect the low-voltage control circuits to the high-voltage power devices without leaving the chip has yet to be completely answered. For discrete circuits, special methods including the use of transformers or optocouplers are described in Chapter 28 for achieving this interconnection. The use of transformers is precluded if the interconnection is to be made on the chip or wafer. At the present time, it is not feasible to fabricate optocoupler circuits on the silicon wafer where the power device and low-voltage control circuitry reside. The basic difficulty is that the light-emitting portion of the optocoupler must be made from gallium arsenide, GaAs, and the integration of gallium arsenide devices on silicon is presently in an early research phase.

The most feasible method of connecting low-voltage circuits to high-voltage devices is with the use of level-shifting circuitry. Such circuitry maintains a relatively fixed voltage drop between two nodes while permitting the flow of any desired current. A zener diode in breakdown is an example of a device that could be used for level shifting. Other circuits are known that could be used including so-called *charge-pumps* (also known as *voltage multipliers*). The problem of how to maintain the voltage difference without excessive power dissipation is the basic issue that will require further research before a commonly accepted method of electrical connection will be available.

Several prototype fabrication processes have been recently demonstrated that include the full range of devices mentioned previously. But because of the problems mentioned in this section, none of these processes can yet be considered a "standard" process. The primary driving forces that will ultimately determine which, if any, process becomes "standard" are yield and cost. Based on what has already been demonstrated, it is clear that the high-voltage ICs have a promising future and will have wide applications. Some of the touted applications include automotive electronics, telecommunications, power supplies, office automation equipment, motor drives, and fluorescent lighting ballasts. Some have predicted that integrated power electronic devices and circuits could trigger a second electronic revolution that will surpass the present so-called *IC-based revolution*.

26-7 NEW SEMICONDUCTOR MATERIALS FOR POWER DEVICES

26-7-1 PROPERTIES OF CANDIDATE REPLACEMENT MATERIALS FOR SILICON

Silicon is presently the only semiconductor material used in making commercially available power devices. This is due to the fact that silicon can be grown in single crystal form with larger diameters and the greatest purity of any available semiconductor. There are, however, other materials that have superior properties compared to silicon for power device applications. Unfortunately, they are not available in high enough purity or large enough sizes to be considered for device manufacturing.

Gallium arsenide is a desirable material for device fabrication. It has a larger energy gap than silicon, which means that devices made in GaAs could be used to higher temperatures than silicon devices (larger energy gap translates into smaller intrinsic carrier densities in GaAs than in Si at the same temperature). The carrier mobilities in GaAs are larger than in silicon, which means that the on-state resistances, especially in majority carrier devices, would be smaller in gallium arsenide than in silicon. In addition it has a somewhat larger breakdown electric field strength. Table 26-1 contains a detailed listing of the important properties of GaAs.

Silicon carbide is a material of intense current interest for the fabrication power devices. It has a significantly larger energy gap than silicon, thus rendering it useful at higher temperatures than silicon. In addition, the thermal conductivity of SiC is several times larger than that of silicon. The most significant advantage of SiC compared to either silicon or GaAs is the order of magnitude larger breakdown electric field strength of silicon carbide. A summary of the properties of SiC is given in Table 26-1.

Diamond is the material with the greatest potential for power devices. It has the largest bandgap of any of the materials listed in Table 26-1. It also has the largest thermal conductivity and, most importantly, the largest breakdown electric field strength. Its carrier mobilities are larger than those of silicon.

Table 26-1 Properties of Semiconductor Materials with Potential for Power Devices

Property	Si	GaAs	3C–SiC	6H–SiC	Diamond
Bandgap at 300 K (eV)	1.12	1.43	2.2	2.9	5.5
Relative dielectric constant	11.8	12.8	9.7	10	5.5
Saturated drift velocity (cm/s)	1×10^7	2×10^7	2.5×10^7	2.5×10^7	2.7×10^7
Thermal conductivity (W/cm-°C)	1.5	0.5	5.0	5.0	20
Maximum operating temperature (K)	300	460	873	1240	1100
Melting temperature (°C)	1415	1238	Sublime ≫1800	Sublime ≫1800	Phase change
Electron mobility at 300 K (cm²/V-s)	1400	8500	1000	600	2200
Breakdown electric field (V/cm)	3×10^5	4×10^5	4×10^6	4×10^6	1×10^7

26-7-2 COMPARATIVE ESTIMATES OF POWER DEVICE PERFORMANCE USING OTHER MATERIALS

A casual glance of the material properties listed in Table 26-1 gives a rather incomplete view of how much power device performance could potentially be improved by fabricating the devices in these other materials. Consider the drift region ohmic resistance of a power device. This resistance is the dominant contribution to the on-state resistance of high-voltage MOSFETs and Schottky diodes. It can be shown that the specific on-state resistance, $R_{\text{on,sp}}$, can be expressed as (see Prob. 26-7):

$$R_{\text{on,sp}} = R_{\text{on}} A = \frac{4(BV_{BD})^2}{\epsilon \mu_n (E_{BD})^3} \qquad (26\text{-}1)$$

where A is the cross-sectional area of the drift region. A convenient way to compare the potential benefits of using other materials is to use Eq. 26-1 to compute the specific drift region resistance of devices having identical breakdown voltage ratings but which are made with other materials. If all devices are compared against silicon, a normalized specific resistance ratio would be

$$\frac{R_{\text{on}}(x)}{R_{\text{on}}(\text{Si})} = \text{resistance ratio} = \frac{\epsilon_{Si} \mu_{Si}}{\epsilon_x \mu_x} \left[\frac{E_{BD,\text{Si}}}{E_{BD,x}} \right]^3 \qquad (26\text{-}2)$$

where x is the material being compared against silicon. In Table 26-2 is shown a comparison of this resistance ratio for all of the materials listed in Table 26-1. The values for the parameters in Eq. 26-2 were taken from Table 26-1.

Another revealing way to compare the benefits of these other materials is to compute the drift region doping level and length required to support a specified value of breakdown voltage in a *pn* junction diode. Such a numerical comparison is made in Table 26-3 for a simple step junction having a breakdown voltage of 1000 V. Eq. 20-1 is used to estimate the doping level and Eq. 20-3 is used to estimate the drift region length. Numerical values for the equation parameters are taken from Table 26-1. The larger doping densities and shorter drift region lengths that are required as we go from silicon toward diamond clearly indicate the superior properties of the other materials compared to silicon.

A final telling comparison is to estimate the carrier lifetime required in a *pn* junction diode fabricated in each of the different materials. Such a comparison is shown in Table 26-4, which was compiled using Eq. 20-28 for the lifetime and expressing the drift region length W_d in Eq. 20-28 in terms of Eq. 20-3. The reader will recall that Eq. 20-28 is derived on the basis that the carrier diffusion length be approximately equal to the drift region length. A breakdown voltage of 1000 V was used for the comparison. The much shorter carrier lifetimes required in the other semiconductors compared to silicon mean

Table 26-2 Comparison of Specific Drift Region Ohmic Resistance of Devices Fabricated in Several Semiconductor Materials

Material	Resistance Ratio
Si	1
GaAs	6.4×10^{-2}
SiC	9.6×10^{-3}
Diamond	3.7×10^{-5}

Table 26-3 Comparison of Drift Region Doping Levels and Lengths Required for a 1000 V Step Junction Fabricated in Different Semiconductor Materials

Material	N_d (cm^{-3})	W_d (μm)
Si	1.3×10^{14}	100
GaAs	5.7×10^{14}	50
SiC	1.1×10^{16}	10
Diamond	1.5×10^{17}	2

Table 26-4 Carrier Lifetime Required in the Drift Region of a 1000 V Breakdown Voltage Rated pn Junction Diode Fabricated in Different Materials

Material	Lifetime
Si	1.2 μsec
GaAs	0.11 μsec
SiC	40 nsec
Diamond	7 nsec

that minority carrier devices fabricated in these materials will be significantly faster than the silicon counterpart.

26-7-3 CHALLENGES IN USING NEW SEMICONDUCTOR MATERIALS

Many problems remain to be resolved before these materials listed in Table 26-1 will be able to begin to replace silicon. Large amounts of time (upwards of 40 years) and funds (billions of dollars) have been expanded on research and development of silicon devices. The efforts expended in the other materials have been small by comparison. It is thus not surprising that silicon has such a dominant position in power device fabrication at present.

There has been a significant amount of research done in GaAs during the past 15–20 years, which has been motivated by its potential for high speed logic circuits and for microwave devices and its potential for light-emitting devices. However, this research has only indirectly benefited power devices. Single crystal wafers of GaAs are still too small (largest sizes available are three inches in diameter) for higher current devices. The control of unwanted impurities (which affect the breakdown voltage capabilities and carrier lifetimes) is rather poor. In addition, GaAs is handicapped by the lack of a native oxide that can be reliably used as an insulating layer or masking layer. Currently, achievable carrier lifetimes are too small for high-voltage minority carrier devices.

The state of silicon carbide technology is behind that of GaAs. The largest available wafers of silicon carbide are only one inch in diameter and the control of the impurities is behind that of GaAs. Significant improvements in ohmic contacts and the quality of the SiO_2–SiC interface are needed before reliable fabrication of power devices can be realized. Considerable improvement in the selectivity of etching methods is also needed.

The state of diamond device technology is primitive compared to that of other materials including SiC. There are no methods of fabricating single crystal wafers of diamond. The methods of producing thin films of diamond that have been developed within the past ten years produce polycrystalline films. Techniques for doing selective diffusion of impurities are poor and ohmic contacts to diamond require major research and development efforts. Selective etching methods for diamond are a major problem area.

26-7-4 FUTURE TRENDS

Significant improvements can be expected in the future in both the basic materials and device fabrication technology of all three alternative materials to silicon. The large research and development investment in GaAs over the past 20 years is already beginning to produce tangible results. During the first half of 1994, a 600 V GaAs Schottky diode will become commercially available from Motorola, Inc. However, only those GaAs devices which are already near commercial introduction are likely to find significant applications in power electronics. The comparisons given in Section 26-7-2 clearly indicate that GaAs power devices offer only incremental improvements over Si power devices compared to SiC and diamond power devices.

The rapid advances in SiC technology and the significantly better potential performance of SiC devices compared with GaAs will restrict the development of new GaAs power devices. As already indicated, single crystal wafers of SiC one inch in diameter are currently (early 1994) available and the expectation is that larger diameters will be available in the near future (1 to 3 years). An 1100 V (room temperature) SiC Schottky diode has been described in the literature, which operated at 400°C with a breakdown voltage of 460 V. This device had a specific on-state resistance an order of magnitude smaller than the theoretical value of a comparable silicon-based Schottky diode. Additionally, logic level MOSFETs and BJTs have been experimentally demonstrated in SiC. It is likely that there will be SiC power devices commercially available within 5–10 years.

The prospect for diamond-based power devices is much more long-term. A significant amount of basic research concentrating on improving the materials and fabrication technology will be needed. Methods of making reasonable-sized (1 to 3 inches in diameter) single crystal wafers of diamond is the first important benchmark to reach. The ancillary issues of ohmic contacts, dopants, etching, and the like, also will require significant research efforts. It is unlikely that any diamond-based power devices will become available in the foreseeable future (next 10–20 years).

SUMMARY

This chapter has examined the structure and characteristics of several devices and circuits that currently are in an early state of development (hence, our classification of emerging devices) but that appear to have potentially useful properties for power electronic applications. Power junction field effect transistors are included in this classification even though such devices are commercially available because they have not yet found widespread usage. The major conclusions are listed as follows:

1. Junction field effect transistors are a normally-on majority-carrier device with a triode-like $i-v$ characteristic. They are similar to MOSFETs in their switching characteristics and have somewhat higher on-state losses.

2. Field-controlled thyristors have a JFET structure with an injecting drain–channel pn junction that leads to heavy conductivity modulation of the drain–drift region and the channel region and, hence, lower on-state losses.

3. The FCT has both gate-controlled turn-on and turn-off and so is a potential supplement to the GTO. It appears to have faster switching speeds than the GTO.

4. The normally-on characteristics of the JFET and the FCT have limited their usage in power electronic applications. They can be made with normally-off characteristic, but this presently leads to higher on-state losses.

5. The MCT is essentially a GTO with integrated MOS-driven gates controlling both turn-on and turn-off that potentially will significantly simplify the design of circuits using GTO.

6. High-voltage integrated circuits offer the promise of drive circuits and logic-level control circuits even perhaps sensing elements for overcurrent and thermal protection

fabricated on the same chip as the power device. This would significantly lower costs and increase the reliability of power devices and systems.

7. Other semiconductor materials such as gallium arsenide have properties such as larger carrier mobilities than silicon that are highly desirable for the fabrication of power devices. However, the quality and availability of these other materials is currently inferior to silicon and, hence, cannot be used for commercial power device products.

PROBLEMS

26-1 Construct an equivalent circuit for the power JFET (normally-on SIT) similar to the equivalent circuit of the MOSFET that can be used for estimating switching times in circuit applications. Assume that the JFET is working in the triode mode.

26-2 Design a simple drive circuit for the normally-on FCT. Consider an arrangement of a MOSFET in series with the cathode of the FCT. Qualitatively describe the operation of the circuit and comment on the required characteristics of the MOSFET.

26-3 Consider a power diode such as is shown in Fig. 20-1. The diode is to be fabricated in gallium arsenide and silicon, and both diodes are to have the same on-state voltage drop and reverse-blocking capability. Which material can have the shorter carrier lifetime and by how much?

26-4 Approximately how thick does an insulating layer of silicon dioxide used in a high-voltage IC have to be in order to hold off 1000 V?

26-5 An N-MCT has 10^5 cells. The OFF-FET can conduct 15 mA before the drain-source voltage equals 0.7 V. What on-state current can be turned off in this MCT?

26-6 Assume the MCT in Problem 26-5 is a P-MCT but is otherwise identical. What is the maximum controllable MCT current?

26-7 Show that the specific on-state resistance of a Schottky diode is given by

$$R_{\text{on,sp}} = R_{\text{on}}A = \frac{4(BV_{BD})^2}{\epsilon \mu_n (E_{BD})^3}$$

Assume that the only significant resistance comes from the drift region of the diode.

26-8 Use the results of Problem 26-7 to compare the specific on-state resistance of Schottky diodes fabricated from Si, GaAs, SiC, and diamond. Assume $BV_{BD} = 500$ V in all cases.

26-9 Which of the semiconductor materials listed in Table 26-1 is most suited for devices which must operate at elevated temperatures. Explain on the basis of fundamental physical properties and not on the basis of current fabrication technology. Be specific and quantitative where possible.

26-10 Evaluate Eqs. 20-1 and 20-3 numerically for GaAs, SiC, and diamond.

26-11 Diodes having breakdown voltages of 200 V are to be fabricated with GaAs, SiC, and diamond. List, in tabular form, the doping level and drift region of the diodes fabricated with each material.

26-12 Diodes with identical cross-sectional areas are fabricated in Si, GaAs, SiC, and diamond. When used in the same circuit, each diode dissipates 200 watts. In the Si diode, the junction reaches 150°C. What is the junction temperature in the SiC, GaAs, and diamond diodes? You may assume that all of the diodes are mounted the same way and that the case temperature for each diode is 50°C.

26-13 Consider the JFET shown in Fig. 26-1. The dimensions shown in the figure have the values given below:

$$W = 10 \ \mu\text{m}; \ l_c = 15 \ \mu\text{m}; \ l_{gd} = 35 \ \mu\text{m}; \ l_{gs} = 10 \ \mu\text{m}$$

The doping levels are $N^- = 2 \times 10^{14} \ \text{cm}^{-3}$ and $P^+ = N^+ = 10^{19} \ \text{cm}^{-3}$. Estimate the pinch-off voltage, V_p, required to pinch off the channel.

26-14 For the JFET described in Problem 26-13, assume that there are 28 P^+ gate regions of depth $W = 700 \ \mu\text{m}$ (dimension perpendicular to the plane of the drawing). Further assume that the P^+ regions are each 10 μm wide (dimension parallel to the channel width W in Fig. 26-7a). Estimate the on-state resistance of the JFET (V_{GS} = 0 V).

26-15   Approximately estimate the blocking voltage capability of the JFET described in Problem 26-13.

REFERENCES

1.   B. Jayant Baliga, *Modern Power Devices*, Wiley, New York, 1987, Chapters 7–9.

2.   J. Nishizawa, "Junction Field-Effect Devices," *Semiconductor Devices for Power Conditioning*, Rolland Sittig and P. Roggwiller (Eds.), Plenum, New York, 1982, pp. 241–270.

3.   V. A. K. Temple, "MOS-Controlled Thyristors—A New Class of Power Devices," *IEEE Transactions on Electron Devices*, Vol. ED-33, No. 10, Oct. 1986, pp. 1609–1618.

4.   M. S. Adler, K. W. Owyang, B. Jayant Baliga, and R. A. Kokosa, "The Evolution of Power Device Technology," *IEEE Transactions on Electron Devices*, Vol. ED-31, No. 11, Nov. 1984, pp. 1570–1591.

5.   B. R. Pelly, "Power Semiconductor Devices—A Status Review," *1982 International Power Semiconductor Converter Conference Proceedings*, IEEE, New York, 1982.

6.   B. Jayant Baliga and Dan Y. Chen (Eds.), *Power Transistors: Device Design and Applications*, IEEE Institute of Electrical and Electronics Engineers, New York, 1984, Part IV, *Emerging Transistors Technology*, pp. 291–374.

7.   T. M. Jahns, R. W. A. A. De Dancker, J. W. A. Wilson, V. A. K. Temple, and D. L. Watrous, "Circuit Utilization Characteristics of MOS-controlled Thyristors," *IEEE Transactions on Industrial Applications*, Vol. 27, No. 3, May/June 1991, pp. 589–597.

8.   David L. Blackburn, "Status and Trends in Semiconductor Power Devices," *EPE '93, 5th European Conference on Power Electronics and Applications*, Conference Record Vol. 2, pp. 619–625.

9.   Tsunenobu Kimoto, Tatsuo Urushidani, Sota Kobayashi, and Hiroyuki Matsunami, "High-Voltage (> 1kV) SiC Schottky Barrier Diodes with Low On-Resistances," *IEEE Electron Device Letters*, Vol. 14, No. 12, Dec. 1993, pp. 548–550.

10.   Mohit Bhatnager and B. Jayant Baliga, "Comparison of 6H-SiC, 3C-SiC, and Si for Power Devices," *IEEE Trans. on Electron Devices*, Vol. 40, No. 3, March 1993, pp. 645–655.

PART 7

PRACTICAL CONVERTER DESIGN CONSIDERATIONS

CHAPTER 27

SNUBBER CIRCUITS

If a power electronic converter stresses a power semiconductor device beyond its ratings, there are two basic ways of relieving the problem. Either the device can be replaced with one whose ratings exceed the stresses or snubber circuits can be added to the basic converter to reduce the stresses to safe levels. The final choice will be a trade-off between cost and availability of semiconductor devices with the required electrical ratings compared with the cost and additional complexity of using snubber circuits. The power electronics circuit designer must be familiar with the design and operation of basic snubber circuits in order to make this comparison trade-off. In this chapter we discuss the fundamentals of snubber circuits commonly used in power electronics to reduce electrical stresses on power semiconductor devices.

27-1 FUNCTION AND TYPES OF SNUBBER CIRCUITS

The function of a snubber circuit is to reduce the electrical stresses placed on a device during switching by a power electronics converter to levels that are within the electrical ratings of the device. More explicitly a snubber circuit reduces the switching stresses to safe levels by:

1. Limiting voltages applied to devices during turn-off transients
2. Limiting device currents during turn-on transients
3. Limiting the rate of rise (di/dt) of currents through devices at device turn-on
4. Limiting the rate of rise (dv/dt) of voltages across devices during device turn-off or during reapplied forward blocking voltages (e.g., SCRs during the forward-blocking state)
5. Shaping of the switching trajectory of the device as it turns on and off

From the circuit topology perspective, there are three broad classes of snubber circuits. These classes include:

1. Unpolarized series R-C snubbers used to protect diodes and thyristors by limiting the maximum voltage and dv/dt at reverse recovery.
2. Polarized R-C snubbers. These snubbers are used to shape the turn-off portion of the switching trajectory of controllable switches, to clamp voltages applied to the devices to safe levels, or to limit dv/dt during device turn-off.
3. Polarized L-R snubbers. These snubbers are used to shape the turn-on switching trajectory of controllable switches and/or to limit di/dt during device turn-on.

Switching stresses are also controlled by utilizing a broad class of power electronic converter circuits termed *resonant* or *quasi-resonant* converters. Some of these converters are described in Chapter 9 of this book.

It must be emphasized that snubbers are not a fundamental part of a power electronic converter circuit. The snubber circuit is an addition to the basic converter, which is added to reduce the stresses on an electrical component, usually a power semiconductor device. Snubbers may be used singly or in combination depending on the requirements. As mentioned in the introduction to this chapter, the additional complexity and cost added to the converter circuit by the presence of the snubber must be balanced against the benefits of limiting the electrical stresses on critical circuit components.

27-2 DIODE SNUBBERS

Snubbers are needed in diode circuits to minimize overvoltages. These overvoltages occur in circuits such as the step-down converter shown in Fig. 27-1a due to the stray or leakage inductance in series with the diode and the snap-off of the diode reverse recovery current at the turn-on of switch T. The analysis of the snubber circuit that will protect the diode will be based on this step-down converter circuit where L_σ is the stray inductance. It is

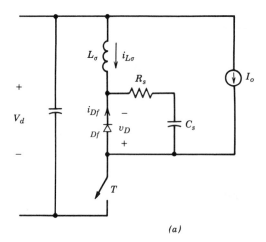

(a)

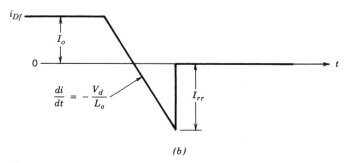

(b)

Figure 27-1 (*a*) A step-down converter circuit with stray inductance and a snubber circuit for the free-wheeling diode. (*b*) The diode reverse-recovery current.

shown later that for the purposes of snubber analysis, this circuit is an equivalent circuit for almost any converter where diodes are used. An R_s–C_s snubber is commonly used across the diode for overvoltage protection as is shown in Fig. 27-1a. To simplify the analysis, the diode reverse recovery current is assumed to snap off instantaneously as is shown in Fig. 27-1b. The load is inductive and it is assumed that the load current I_0 is constant during the switching transient.

27-2-1 CAPACITIVE SNUBBER

Although the capacitive snubber ($R_s = 0$) is not used in practice (for reasons that will become apparent), it provides an easily analyzed starting point for the analysis that illustrates the basic concepts. In obtaining an equivalent circuit, the switch in Fig. 27-2a is assumed to be ideal, which results in a worst-case analysis of this circuit. Treating the instant of diode snap-off at the peak reverse recovery current I_{rr} as $t = 0$, the initial inductor current in the equivalent circuit of Fig. 27-2 is I_{rr} and the initial snubber capacitor voltage is zero. To establish a baseline circuit, the snubber resistance R_s is assumed to be zero as is shown in Fig. 27-2b. The capacitor voltage (which is the negative of the diode voltage in this baseline circuit) can be obtained from the results of Chapter 9 as

$$v_{Cs} = V_d - V_d cos(\omega_0 t) + I_{rr}\sqrt{\frac{L_\sigma}{C_s}}\,sin(\omega_0 t) \qquad (27\text{-}1)$$

where

$$\omega_0 = \frac{1}{\sqrt{L_\sigma C_s}} \qquad (27\text{-}2)$$

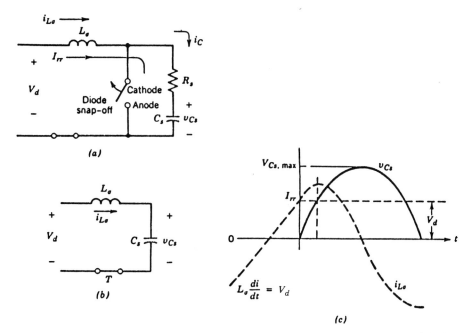

Figure 27-2 (a) Equivalent circuit of the step-down converter at the instant of diode reverse-recovery current snap-off and (b) the simplification that results when the snubber resistance is zero. (c) The voltage and current waveforms for $R_s = 0$ and $C_s = C_{base}$.

Introducing a baseline capacitance C_{base} given by

$$C_{base} = L_\sigma \left[\frac{I_{rr}}{V_d} \right]^2 \qquad (27\text{-}3)$$

it is possible to express Eq. 27-1 as

$$v_{Cs} = V_d \left[1 - \cos(\omega_0 t) + \sqrt{\frac{C_{base}}{C_s}} \sin(\omega_0 t) \right] \qquad (27\text{-}4)$$

Either by a time derivative or a phasor approach, the maximum value of v_{Cs} in Eq. 27-4 can be estimated as

$$V_{Cs,max} = V_d \left[1 + \sqrt{1 + \frac{C_{base}}{C_s}} \right] \qquad (27\text{-}5)$$

The waveforms for v_{Cs} and inductor current $i_{L\sigma}$ are shown in Fig. 27-2c for $C_s = C_{base}$. In this case the maximum reverse-diode voltage is the same as $V_{Cs,max}$ calculated from Eq. 27-5. For small values of C_s, the maximum diode voltage becomes excessive.

27-2-2   EFFECT OF ADDING A SNUBBER RESISTANCE

When the diode snubber resistance R_s is included, the equivalent circuit for the snubber becomes as shown in Fig. 27-2a. In analyzing this modified circuit the instant of diode snap-off is treated as $t = 0$, and the initial inductor current is I_{rr} and the initial capacitor voltage is zero. The differential equation governing the behavior of the diode voltage is

$$L_\sigma C_s \frac{d^2 v_{Df}}{dt^2} + R_s C_s \frac{dv_{Df}}{dt} + v_{Df} = -V_d \qquad (27\text{-}6)$$

The boundary conditions are $v_{Df}(0^+) = -I_{rr}R_s$ and

$$\frac{dv_{Df}(0^+)}{dt} = -\frac{I_{rr}}{C_s} - \frac{R_s V_d}{L_\sigma} - \frac{I_{rr}R_s^2}{L_\sigma}$$

The solution to Eq. 27-6 is given by

$$v_{Df}(t) = -V_d - \sqrt{\frac{L_\sigma}{C_s}} \frac{I_{rr}}{\cos(\phi)} e^{-\alpha t} \cos(\omega_a t - \phi - \gamma) \qquad (27\text{-}7)$$

where

$$\omega_a = \sqrt{1 - \frac{\alpha^2}{\omega_0^2}} \qquad \alpha = \frac{R_s}{2\omega_a} \qquad \phi = \tan^{-1}\left(\frac{V_d - I_{rr}R_s/2}{\omega_a L_\sigma I_{rr}} \right) \quad \text{and} \quad \gamma = \tan^{-1}\left(\frac{\omega_a}{\alpha} \right) \qquad (27\text{-}8)$$

The time $t = t_m$ at which the voltage given by Eq. 27-7 is a maximum can be found by setting the derivative dv_{Df}/dt equal to zero and solving for time. Doing this yields

$$t_m = \frac{\phi + \gamma - \pi/2}{\omega_a} \geq 0 \qquad (27\text{-}9)$$

Substituting $t = t_m$ into Eq. 27-7 yields the maximum reverse voltage across the diode as

$$\frac{V_{max}}{V_d} = 1 + \left\{ \sqrt{1 + \frac{C_{base}}{C_s} + \frac{R_s}{R_{base}} - 0.75 \left(\frac{R_s}{R_{base}} \right)^2} \right\} \exp\left(-\alpha t_m \right) \qquad (27\text{-}10)$$

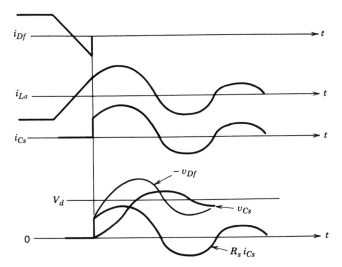

Figure 27-3 The current and voltage waveforms after the diode snaps off at $t = 0$.

In Eq. 27-10 the baseline capacitance C_{base} is given by Eq. 27-3 and the resistance R_{base} is given by

$$R_{\text{base}} = \frac{V_d}{I_{rr}} \qquad (27\text{-}11)$$

Typical circuit waveforms for $t > 0$ are shown in Fig. 27-3 for $C_s = C_{\text{base}}$. In these waveforms, the oscillations are damped out by R_s, and the maximum diode voltage depends on the values of R_s and C_s used. For a selected value of C_s, the maximum diode voltage varies with R_s. As an example, for $C_s = C_{\text{base}}$, the normalized maximum diode voltage is plotted in Fig. 27-4 as a function of R_s/R_{base}. It can be seen that for this value of C_s, there is an optimum value of $R_s = R_{\text{opt}} = 1.3 R_{\text{base}}$ that minimizes V_{max}. A snubber

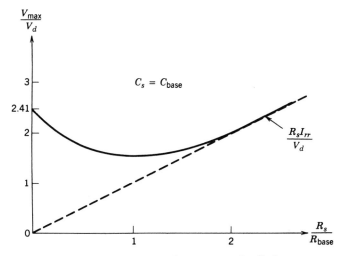

Figure 27-4 Maximum overvoltage across the diode as a function of snubber resistance for a fixed value of snubber capacitance.

design nomogram is shown in Fig. 27-5 where the optimum snubber resistance and corresponding V_{max} are plotted as a function of C_s. In this nomogram all quantities are normalized.

The energy loss in the resistor R_s is given by [1]

$$W_R = \tfrac{1}{2}L_\sigma I_{rr}^2 + \tfrac{1}{2}C_s V_d^2 \tag{27-12}$$

W_R, normalized with respect to $(\tfrac{1}{2}L_\sigma I_{rr}^2)$, the peak energy stored in the leakage inductance as the diode snaps off, is also plotted in Fig. 27-5. At the end of the current oscillations, the energy stored in C_s is equal to

$$W_{C_s} = \tfrac{1}{2}C_s V_d^2 \tag{27-13}$$

which is dissipated in the diode at the next turn-on of the diode. Assuming an instantaneous turn-on of the diode, the total energy dissipation in the diode and its snubber resistance is given by

$$W_{tot} = W_R + W_{C_s} = \tfrac{1}{2}L_\sigma I_{rr}^2 + C_s V_d^2 = \tfrac{1}{2}L_\sigma I_{rr}^2\left(1 + 2\frac{C_s}{C_{base}}\right) \tag{27-14}$$

and the normalized W_{tot} is also plotted in Fig. 27-5 as a function of the normalized C_s. It can be seen from Fig. 27-5 that the maximum voltage decreases only slightly by increasing C_s beyond C_{base}. However, the total energy dissipation increases linearly with C_s. Therefore, a snubber capacitor with C_s in a range close to C_{base} would be used. Once C_s has been selected, $R_s = R_{opt}$ can be obtained directly from Fig. 27-5.

In this analysis, it is assumed that the reverse-recovery current of the diode snaps off instantaneously. In practice, the diode reverse-recovery current can be assumed to decay

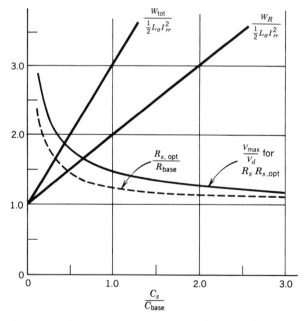

Figure 27-5 Snubber energy loss and the maximum diode voltage for the optimum value of snubber resistance R_s as a function of the snubber capacitance C_s.

exponentially. This can be accommodated in the equivalent circuit of Fig. 27-2 by adding a time-varying current source. The analysis can be carried out by computer simulation, and the results show that the snubber design remains essentially the same as before.

27-2-3 IMPLEMENTATION

The circuit of Fig. 27-2a has been used for the detailed analysis of the diode snubber. In the following paragraphs we consider how to derive similar equivalent circuits for diode snubber analysis in some commonly used converters. In the flyback converter of Fig. 27-6a operating in an incomplete demagnetization mode, when the switch is off, the diode is conducting. When the switch is turned on, the secondary side of the circuit can be represented by the circuit of Fig. 27-6b where the current in the leakage inductance and the diode decreases. Considering the instant at which the diode current I_{rr} snaps back as the time origin, the equivalent circuit of Fig. 27-6c is obtained that is identical to Fig. 27-2a, which was obtained for the step-down converter circuit of Fig. 27-1a.

Another example is that of a center-tapped secondary in a push-pull, half-bridge, or a full-bridge dc-dc converter circuit shown in Fig. 27-7a. Assuming that the converter is operating in a continuous-current mode, when all primary switches are off, one-half of the output current flows through each diode. At turn-on, if a positive voltage is applied across the primary, the secondary side of the circuit can be represented by Fig. 27-7b where the current through D_1 will increase and through D_2 will decrease. Considering the instant at which the peak reverse-recovery current I_{rr} snaps off in D_2 as the initial instant, the equivalent circuit of Fig. 27-7c, which is similar to that of Fig. 27-6c, results. There should be a snubber across each diode in this circuit.

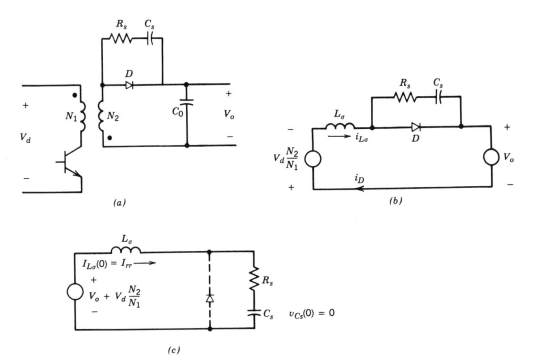

Figure 27-6 (a) Flyback converter circuit operating in an incomplete demagnetization mode. (b) Equivalent circuit on the secondary side and (c) the simplified equivalent circuit after the snap-off of the diode current. L_σ is the transformer leakage inductance.

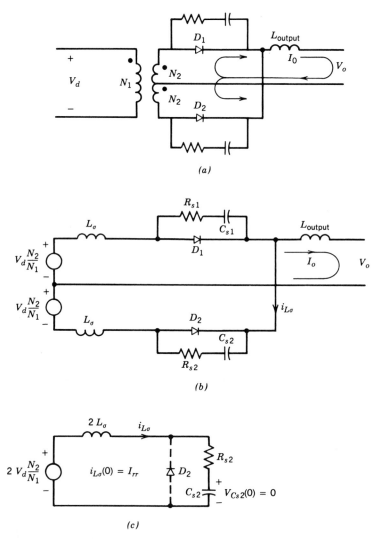

Figure 27-7 (*a*) Full-wave rectifier using a transformer with a center-tapped secondary. (*b*) Equivalent circuit on the secondary side and (*c*) the equivalent circuit at the instant of the snap-off of diode D_2.

In the preceding examples, converters consisted of an isolation transformer that introduced a substantial leakage inductance. Therefore diode snubbers were needed. However, the step-down converter circuit of Fig. 27-1*a*, which was used only to present the useful equivalent circuit for analyzing diode snubbers, in fact, may not require a diode snubber when BJTs and MOSFETs are used as the controlled switch and L_σ is minimized by proper circuit layout. The same comments also apply to half- and full-bridge switch-mode converters using BJTs and MOSFETs.

In single-phase line-frequency diode converters discussed in Chapter 5, if the diode bridge rectifier feeds a dc capacitor and operates in a discontinuous mode, then if filter inductors are used, they should be placed on the ac side, as shown in Fig. 27-8*a*. Such a configuration will provide overvoltage protection of the diodes against incoming line-voltage transients and against voltages induced by the *di/dt* at reverse recovery. There is no need for diode snubbers since, for example, if the reverse-recovery current of diode D_1

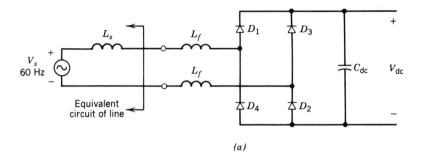

(a)

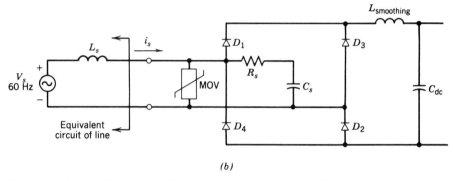

(b)

Figure 27-8 (a) Single-phase line-frequency diode rectifier. (b) Single-phase diode converter with an *RC* snubber to protect against overvoltages due to unknown inductive reactances on the ac side of the rectifier. An MOV is also shown for further overvoltage protection.

snaps off, then D_4 provides a path for the inductive current, and the reverse-voltage across D_1 is clamped to the dc capacitor voltage V_{dc}.

In the case of continuous conduction in a single-phase full-bridge rectifier, the filter inductor should be placed on the dc side as is shown in Fig. 27-8b. In practice, there is a finite inductance on the ac side, which is normally not known. For a worst-case analysis, the ac-side reactance $X_s(=\omega L_s)$ can be assumed to be 5%, which implies that

$$X_s = \omega L_s = 0.05 \frac{V_s}{I_{s1}} \qquad (27\text{-}15)$$

where V_s is the rms line voltage and I_{s1} is the rms value of the fundamental frequency component of the current at full load. In the circuit of Fig. 27-8b, one *RC* snubber may be used to protect all the diodes. In Fig. 27-8b, the waveforms due to the diode reverse-recovery snap-off are much faster than the variations in the 60-Hz line voltage input v_s. Therefore the value of v_s at the instant of diode snap-off can be treated as a constant dc, thus again allowing the use of the equivalent circuit of Fig. 27-2a. The detailed analysis of snubbers for such a circuit is given below for a thyristor rectifier where the diode bridge can be treated as a special case. The analysis for single-phase rectifiers also applies to three-phase diode bridge rectifiers.

It should be noted that in the converters connected to the line, the diode snubbers should also provide overvoltage protection against incoming line voltage transients. In fact this consideration may supersede the snubber design based on the reverse-recovery snap-off. Often metal-oxide varistors (MOVs) are used in addition for this transient overvoltage protection.

27-3 SNUBBER CIRCUITS FOR THYRISTORS

The reverse-recovery currents generated in thyristors when they are reverse biased may result in unacceptably large overvoltages because of series inductance if snubbers are not used. In the previous section, it was shown that an equivalent circuit for a step-down dc-dc converter could be used to analyze the diode overvoltage snubber in any converter. That equivalent circuit is used here for a three-phase line-frequency thyristor converter of the type discussed in Chapter 6. The ac-side inductances shown in Fig. 27-9a are due to line reactances plus any transformer leakage inductance. The dc side is represented by a current source where i_d is assumed to flow continuously.

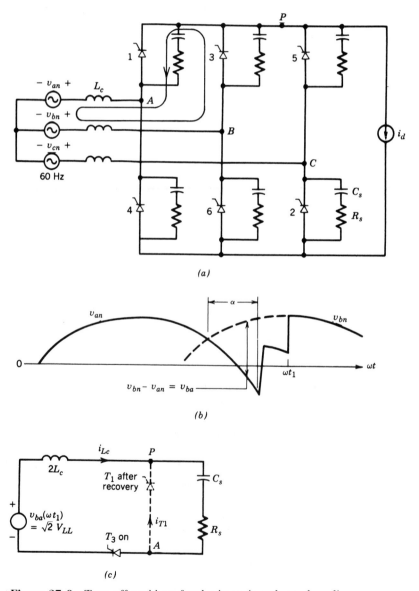

Figure 27-9 Turn-off snubbers for thyristors in a three-phase line-frequency converter circuit: (a) three-phase line-frequency converter; (b) trigger times; (c) equivalent circuit.

It is assumed that thyristors T_1 and T_2 have been conducting and that thyristor T_3 is gated on at a delay angle α, as is shown in Fig. 27-9b. The current i_d will commutate from thyristor T_1 (connected to phase a) to thyristor T_3 (connected to phase b). The voltage v_{ba} is responsible for the commutation of the current. The subcircuit consisting of T_1 and T_3 is shown in Fig. 27-9c with T_3 on and T_1 off and at its reverse recovery at ωt_1, with $i_\sigma = I_{rr}$. The voltage source in the circuit of Fig. 27-9c can be assumed to be a constant dc voltage with a value of v_{ba} at ωt_1 because of the slow variation of 60-Hz voltages compared to the fast voltage and current transients in this circuit. The snubber voltage and current waveforms will be identical to those described in Fig. 27-3.

To discuss the design of the snubber, a worst-case line impedance of 5% is used as explained in the previous section, and Eq. 27-15 becomes

$$x_c = \omega L_c = \frac{0.05 V_{LL}}{\sqrt{3} I_d} \tag{27-16}$$

where V_{LL} is the rms line-to-line voltage and I_d is the load current. For a worst-case design, the voltage source in Fig. 27-9c will have its maximum value of $\sqrt{2} V_{LL}$ which corresponds to $\alpha = 90°$. Here the reverse-recovery time is assumed to be 10 μs. Thus, during the current commutation, assuming that the commutation voltage has a constant value of $\sqrt{2} V_{LL}$, the di/dt through thyristor T_1 is

$$\frac{di}{dt} = \frac{\sqrt{2} V_{LL}}{2 L_c} \tag{27-17}$$

and therefore,

$$I_{rr} = \left(\frac{di}{dt}\right) t_{rr} = \frac{\sqrt{6} \omega V_{LL} t_{rr} I_d}{0.1 V_{LL}} = 0.09 I_d \tag{27-18}$$

where $t_{rr} = 10$ μs.

As was discussed in the previous section, $C_s = C_{\text{base}}$ is close to an optimum value. Relating Fig. 27-9c to Fig. 27-2a and Eq. 27-3 gives

$$C_{\text{base}} = L_c \left(\frac{I_{rr}}{V_{LL}}\right)^2 \tag{27-19}$$

Substituting L_c from Eq. 27-16 at $\omega = 377$ and I_{rr} from Eq. 27-18 into Eq. 27-19 yields

$$C_s = C_{\text{base}}(\mu F) = \frac{0.6 I_d}{V_{LL}} \tag{27-20}$$

$R_s = R_{\text{opt}}$ can be obtained from Fig. 27-5. Here, assuming the normalized $R_s = R_{\text{opt}} = 1.3 R_{\text{base}}$, and using the value $R_{\text{base}} = \sqrt{2} V_{LL}/I_{rr}$, we obtain, using Eq. 27-18,

$$R_s = R_{\text{opt}} = 1.3 \sqrt{2} \frac{V_{LL}}{I_{rr}} = 20 \frac{V_{LL}}{I_d} \tag{27-21}$$

In order to estimate the loss in each snubber, the voltage waveform across a thyristor having a worst-case trigger angle of $\alpha = 90°$ is shown in Fig. 27-10. It can be shown that the total energy loss in each snubber equals

$$W_{\text{snubber}} = 3 C_s V_{LL}^2 \tag{27-22}$$

or using Eq. 27-20

$$W_{\text{snubber}} = 1.8 \times 10^{-6} I_d V_{LL} \tag{27-23}$$

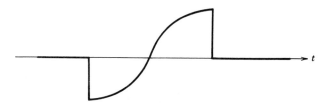

Figure 27-10 Voltage waveform across a thyristor triggered at a trigger angle of 90°.

If the three-phase converter kVA is S, then at 60 Hz, each snubber has a power loss equaling

$$P_{\text{snubber}}(\text{in watts}) = 10^{-4}S \qquad (27\text{-}24)$$

A similar procedure can be followed for any values of t_{rr} and the ac-line inductance. A conservative design may require C_s to be larger than C_{base}, and therefore R_s would be smaller than the value found above. In that case, the snubber losses would be higher since they are proportional to C_s.

27-4 NEED FOR SNUBBERS WITH TRANSISTORS

Snubber circuits are used to protect the transistors by improving their switching trajectory. There are three basic types of snubbers:

1. Turn-off snubbers
2. Turn-on snubbers
3. Overvoltage snubbers

To explain the need for these snubbers, a step-down converter without any snubbers is shown in Fig. 27-11a where the stray inductances in the various parts of the circuit are shown explicitly. For purposes of illustration, a bipolar junction power transistor is used for the controlled switch. However, the discussion that follows applies to all controlled switches including MOSFETS, IGBTs, GTOs, and newer devices such as MCTs. Initially, the transistor is conducting and $i_C = I_0$. During the turn-off switching, at $t = t_0$, the transistor voltage begins to rise, but the currents in the various parts of the circuit remain the same until t_1, when the freewheeling diode begins to conduct. Then the transistor current begins to decrease, and the rate at which it decreases is dictated by the transistor properties and its base drive. The transistor voltage can be expressed as

$$v_{CE} = V_d - L_\sigma di_c/dt \qquad (27\text{-}25)$$

where $L_\sigma = L_1 + L_2 + \ldots$. The presence of stray inductances results in an overvoltage since di_c/dt is negative. At t_3, at the end of the current fall time, the voltage comes down to V_d and stays at that value.

During the turn-on transition, the transistor current begins to rise at t_4 at a rate dictated by the transistor properties and the base drive circuit. Eq. 27-25 is still valid, but due to a positive di_c/dt the transistor voltage v_{CE} is slightly less than V_d. Due to the reverse-recovery current of the freewheeling diode, i_C exceeds I_o. The freewheeling diode

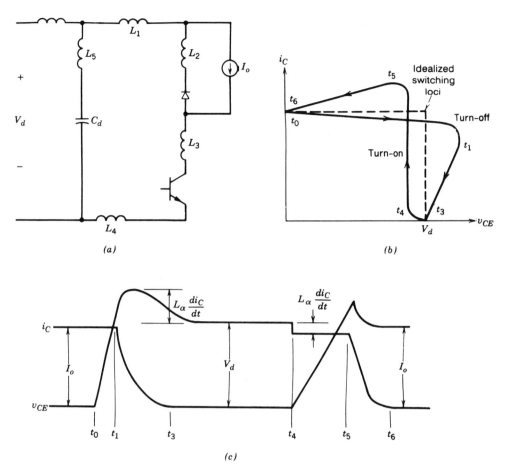

Figure 27-11 (a) Step-down converter circuit with stray inductance shown explicitly with (b) associated switching trajectory and (c) the current and voltage waveforms during turn-on and turn-off.

recovers at t_5 and the voltage across the BJT decreases to zero at t_6 at a rate dictated by the device properties.

These switching waveforms can be represented by switching loci as shown in Fig. 27-11b. The dotted lines represent idealized switching loci both for turn-on and turn-off, assuming zero stray inductances and no reverse-recovery current through the diode. They show that the transistor experiences high stresses at turn-on and turn-off when both its voltage and current are high simultaneously, thus causing a high instantaneous power dissipation. Moreover, the stray inductances result in overvoltage beyond V_d, and the diode reverse-recovery current causes overcurrent beyond I_o. If necessary, snubber circuits are used to reduce these stresses.

An important assumption that simplifies the snubber analysis is that the transistor current changes linearly in time with a constant di/dt, which is only dictated by the transistor and its base drive circuit. Therefore di/dt, which may be different at turn-on and turn-off, is assumed not to be affected by the addition of the snubber circuit. This assumption provides the basis for a simple design procedure for a laboratory prototype. The final design may be somewhat different depending on what is revealed by laboratory measurements on the prototype circuit.

27-5 TURN-OFF SNUBBER

To avoid the problems at turn-off, the goal of a turn-off snubber is to provide a zero voltage across the transistor while the current turns off. This can be approached by connecting a *RCD* network across the BJT as shown in Fig. 27-12*a*, where the stray inductances are ignored initially for ease of explanation. Prior to turn-off, the transistor current is I_o and the transistor voltage is essentially zero. At turn-off in the presence of this snubber, the transistor current i_C decreases with a constant di/dt and $(I_o - i_C)$ flows into

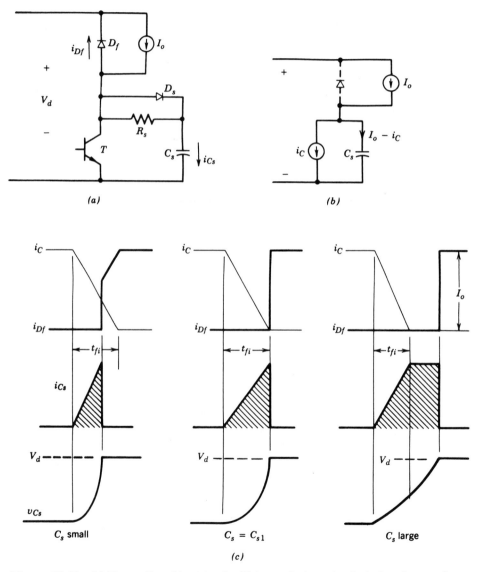

(a)

(b)

(c)

Figure 27-12 (*a*) Turn-off snubber circuit, (*b*) its equivalent circuit during the transient, and (*c*) current and voltage waveforms during the turn-off transient. The shaded areas represent the charge put on the snubber capacitance during turn-off that will be dissipated in the BJT at the next turn-on.

the capacitor through the snubber diode D_s. Therefore, for a current fall time of t_{fi}, the capacitor current can be written as

$$i_{Cs} = I_o t / t_{fi} \qquad 0 < t < t_{fi} \tag{27-26}$$

where i_{Cs} is zero prior to turn-off at $t = 0$. The capacitor voltage, which is the same as the voltage across the transistor when D_s is conducting, can be written as

$$v_{Cs} = v_{CE} = \frac{1}{C_s} \int_0^t i_{Cs} dt = \frac{I_o t^2}{2 C_s t_{fi}} \tag{27-27}$$

which is valid during the current fall time so long as the capacitor voltage is less than or equal to V_d. The equivalent circuit that represents this condition is shown in Fig. 27-12b.

The voltage and current waveforms are shown in Fig. 27-12c for three values of the snubber capacitance C_s. For a small value of capacitance, the capacitor voltage reaches V_d before the current fall time is over. At that time, the freewheeling diode D_f turns on and clamps the capacitor and the transistor to V_d, and i_{Cs} drops to zero due to dv_{Cs}/dt being equal to zero.

The next set of waveforms in Fig. 27-12c are drawn for a value of $C_s = C_{s1}$, which causes the capacitor voltage to reach V_d exactly at the current fall time t_{fi}; C_{s1} can be calculated by substituting $t = t_{fi}$ and $v_{Cs} = V_d$ in Eq. 27-27 and is given as

$$C_{s1} = \frac{I_o t_{fi}}{2 V_d} \tag{27-28}$$

For a large snubber capacitance with $C_s > C_{s1}$, the waveforms in Fig. 27-12c show that the transistor voltage rises slowly and takes longer than t_{fi} to reach V_d. Beyond t_{fi}, the capacitor current equals I_o and the capacitor and the transistor voltages rise linearly to V_d. The turn-off switching loci with the three values of C_s used in Fig. 27-12 are shown in Fig. 27-13.

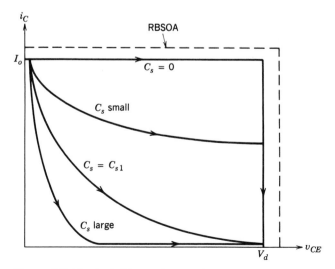

Figure 27-13 Switching trajectory during turn-off with various values of snubber capacitance C_s.

To optimize the snubber design it is necessary to consider the transistor turn-on in the presence of the turn-off snubber. To understand the transistor behavior at turn-on, initially it is assumed that the resistor is essentially zero, that is, a pure capacitor without R_s and D_s is used as the turn-off snubber, as is shown in Fig. 27-14a. The presence of C_s causes the turn-on current to increase beyond I_o and the freewheeling diode reverse-recovery current. We still assume a constant di_c/dt during turn-on. The shaded area in Fig. 27-14a represents the charge on the capacitor that is discharged into the transistor. This charge is equal to the area of one of the shaded areas in Fig. 27-12c depending on the value of C_s used. In the absence of the snubber capacitor C_s, the transistor voltage would have fallen almost instantaneously (since the voltage fall time is usually quite small) as shown by the dashed line in Fig. 27-14a, and hence the energy dissipated in the transistor during the voltage turn-on would have been small. The presence of C_s lengthens the voltage fall time

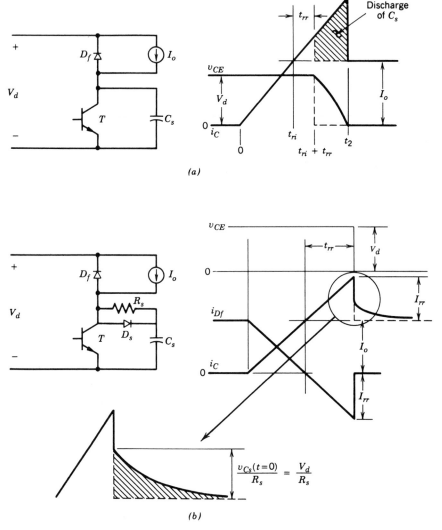

Figure 27-14 Effect of the turn-off snubber capacitance C_s on the (a) turn-off transient without a snubber resistance R_s and (b) with the resistance.

so that additional energy is dissipated in the transistor. The additional energy dissipated in the transistor during the capacitor discharge time can be expressed as

$$\Delta W_Q = \int_{t_{ri}+t_{rr}}^{t_2} i_C v_{CE} dt = \int_{t_{ri}+t_{rr}}^{t_2} i_{Cs} v_{CE} dt + \int_{t_{ri}+t_{rr}}^{t_2} I_o v_{CE} \, dt \qquad (27\text{-}29)$$

The first term in the right-hand side equals the energy stored in the capacitor, which is dissipated in the transistor at turn-on. However, there is additional energy dissipation in the transistor, as expressed by the second term in Eq. 27-29, which normally will be larger than the first term. This energy dissipation is due to the lengthening of the voltage fall time brought about by the presence of C_s.

The transistor turn-on waveforms in the presence of the snubber resistance R_s is shown in Fig. 27-14b. Here, unlike the pure capacitively snubbed transistor, the voltage can be assumed to fall almost instantaneously, and therefore no additional energy dissipation due to the snubber occurs in the transistor at turn-on. The capacitor energy, which is dissipated in the snubber resistor, is given by

$$W_R = \frac{C_s V_d^2}{2} \qquad (27\text{-}30)$$

In Fig. 27-14b, the snubber resistance should be chosen so that the peak current through it is less than the reverse-recovery current I_{rr} of the freewheeling diode, that is,

$$\frac{V_d}{R_s} < I_{rr} \qquad (27\text{-}31)$$

The circuit designer usually attempts to limit I_{rr} to $0.2I_o$ or less so that Eq. 27-31 becomes approximately

$$\frac{V_d}{R_s} = 0.2I_o \qquad (27\text{-}32)$$

Based on the above assumptions, the comparisons of Fig. 27-14a and 27-14b indicate that including the resistance R_s has the following beneficial effects during the transistor turn-on:

1. All the capacitor energy is dissipated in the resistor, which is easier to cool than the transistor.
2. No additional energy dissipation occurs in the transistor due to the turn-off snubber.
3. The peak current the transistor must conduct is not increased due to the turn-off snubber.

As an aid in choosing the appropriate value of C_s, the energy dissipated in the transistor during turn-off and the energy dissipated in the snubber resistance R_s during turn-on are plotted as functions of C_s in Fig. 27-15. Based on the previous assumptions, these plots are independent of R_s, and there is no additional energy dissipation in the transistor during turn-on due to the presence of the turn-off snubber. C_s should be chosen based on (1) keeping the turn-off switching locus within the reverse-bias safe operating area, (2) reducing the transistor losses based on its cooling considerations, and (3) keeping the sum (shown as the dashed line in Fig. 27-15) of transistor turn-off energy dissipation and snubber resistance energy dissipation low.

Having made initial selections of R_s based on Eq. 27-32 and C_s based on the design trade-offs just discussed, the designer must ensure that the capacitor has sufficient time to discharge down to a low voltage, say $0.1V_d$, during the minimum on-state time of the

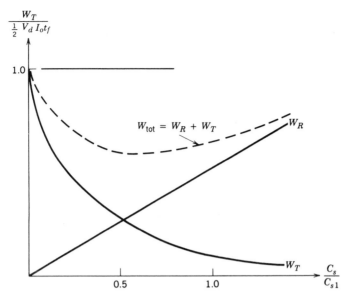

Figure 27-15   Turn-off energy dissipation in the BJT and the snubber resistance as a function of the snubber capacitance C_s.

transistor in order that the turn-off snubber be effective at the next turn-off interval. During the on state of the transistor, the capacitor discharges with a time constant $\tau_c = R_s C_s$ and

$$v_{Cs} = V_d e^{-t/\tau_c} \qquad (27\text{-}33)$$

and therefore, discharging v_{Cs} down to $0.1V_d$ requires a time interval of $2.3\tau_c$, and thus

$$t_{\text{on state}} > 2.3 R_s C_s \qquad (27\text{-}34)$$

As an example, if $C_s = C_{s1}$ (given in Eq. 27-28) and R_s is chosen using Eq. 27-32, then the minimum on-state time of the transistor must be six times the transistor current fall time t_{fi}.

27-6 OVERVOLTAGE SNUBBER

In describing the turn-off snubber, the stray inductances were neglected, and hence there was no overvoltage. The overvoltages at turn-off due to stray inductances, such as shown in Fig. 27-11a, can be minimized by means of the overvoltage snubber circuit shown in Fig. 27-16, assuming it is possible to lump all the stray inductances together as indicated in Eq. 27-25. The operation of the overvoltage snubber can be described as follows.

Initially the transistor is conducting and the voltage $v_{C,\text{ov}}$ across the overvoltage snubber capacitor equals V_d. At turn-off assuming the BJT current fall time to be small, the current through L_σ is essentially I_o when the transistor current decreases to zero, and the output current then freewheels through the free-wheeling diode D_f. At this stage the equivalent circuit is as shown in Fig. 27-16b where the D_f, I_o combination appears as a short circuit, and the transistor is an open circuit. Now the energy stored in the stray inductances gets transferred to the overvoltage capacitor through the diode D_{OV} and the

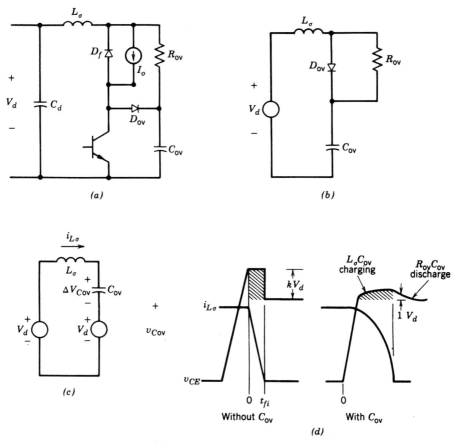

Figure 27-16 (a) Overvoltage snubber and (b, c) its equivalent circuit during transistor turn-off. (d) The collector-emitter voltage with and without the snubber.

overvoltage ΔV_{CE} across the transistor (noting that in this state, the capacitor C_{OV} and the transistor have the same voltage) can be obtained by replacing the precharged capacitor with its equivalent circuit as shown in Fig. 27-14c. Using energy considerations and noting that $\Delta V_{C,OV} = \Delta V_{CE}$, we obtain

$$\frac{C_{OV}\Delta V_{CE,\mathrm{max}}^2}{2} = \frac{L_\sigma I_o^2}{2} \qquad (27\text{-}35)$$

This equation shows that a large value of C_{OV} will minimize the overvoltage $\Delta V_{CE,\mathrm{max}}$. Once the current through L_σ has decreased to zero, it can reverse its direction due to the resistor R_{OV}, and the overvoltage on the capacitor decreases to V_d through the resistor R_{OV}. The capacitor discharge time constant $R_{OV}C_{OV}$ should be small enough so that the capacitor voltage has decayed approximately to V_d prior to the next turn-off of the transistor.

To aid in the estimation of the proper value of C_{OV}, the circuit waveforms with and without the overvoltage snubber are shown in Fig. 27-16d. The observed overvoltage of kV_d without the overvoltage snubber is used to estimate L_σ as

$$kV_d = \frac{L_\sigma I_o}{t_{fi}} \qquad (27\text{-}36)$$

If an overvoltage, for example $\Delta V_{CE,\max} = 0.1V_d$, is acceptable, then using this in Eq. 27-35 and substituting for L_σ from Eq. 27-36 yields

$$C_{ov} = \frac{100kI_o t_{fi}}{V_d} \qquad (\Delta V_{CE,\max} = 0.1V_d) \qquad (27\text{-}37)$$

In terms C_{s1} given by Eq. 27-37, C_{OV} from Eq. 27-37 can be rewritten as

$$C_{OV} = 200kC_{s1} \qquad (27\text{-}38)$$

which shows that substantially larger capacitance is needed for overvoltage protection compared to the values used in the turn-off snubber, which are on the order of C_{s1}. It can be shown that even with a large value of C_{OV}, the energy dissipated in R_{OV} is of the same order as the energy dissipated in the resistor of the turn-off snubber.

Both the turn-off and the overvoltage protection snubbers should be used simultaneously.

27-7 TURN-ON SNUBBER

Because of the large FBSOA of most controlled switches including BJTs, MOSFETS, GTOs, and IGBTs, turn-on snubbers are only used to reduce turn-on switching losses at high switching frequencies and for limiting the maximum diode reverse recovery current. Turn-on snubbers work by reducing the voltage across the switch (transistor) as the current builds up. A turn-on snubber can be in series with the transistor as in Fig. 27-17a or in series with the freewheeling diode as in Fig. 27-17b. In both circuits the turn-on and turn-off switching waveforms across the transistor and freewheeling diode are identical. The reduction in the voltage across the transistor during turn-on is due to the voltage drop across L_s. This reduction is given by

$$\Delta V_{CE} = \frac{L_s I_o}{t_{ri}} \qquad (27\text{-}39)$$

where t_{ri} is the current rise time as shown in Fig. 27-17c for small values of L_s. For such small values, di/dt is dictated only by the transistor and its base drive circuit and is assumed to be the same as without the turn-on snubber. Therefore, the diode peak reverse-recovery current is also the same as without the turn-on snubber.

If it is important to reduce the diode peak reverse-recovery current, it can be achieved with a large value of L_s as is shown by the waveforms in Fig. 27-17d. Here the current rate of rise is $di/dt = V_d/L_s$ and the voltage across the transistor is almost zero during the current rise time. Substituting this value of di/dt into Eq. 20-27 for I_{rr} yields

$$I_{rr} = \sqrt{\frac{2\tau I_F V_d}{L_s}} \qquad (27\text{-}40)$$

During the on-state of the transistor, L_s conducts I_o. When the transistor turns off, the energy stored in the snubber inductor, $L_s I_o^2/2$, will be dissipated in the snubber resistor R_{Ls}. The snubber time constant is $\tau_L = L_s/R_{Ls}$. In selecting R_{Ls} the following two factors must be considered. First, during transistor turn-off, this turn-on snubber will generate an overvoltage across the transistor given by

$$\Delta V_{CE,\max} = R_{Ls}I_o \qquad (27\text{-}41)$$

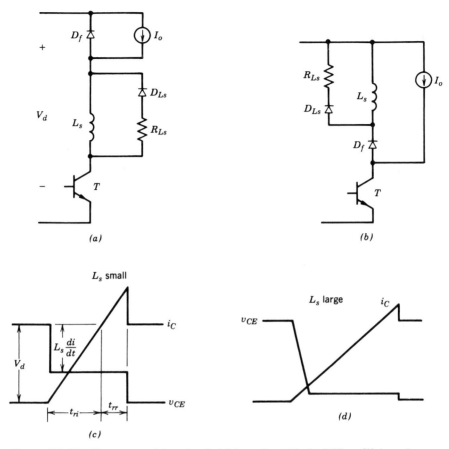

Figure 27-17 Turn-on snubber circuit (a) in series with the BJT or (b) in series with the free-wheeling diode. (c) The transistor voltage and current waveforms for small values of L_σ and (d) for large values of L_σ.

Second, during the off-state the inductor current must decay to a low value, for example, $0.1I_o$, so that the snubber can be effective during the next turn-on. Therefore the minimum interval for the off state of the BJT should be

$$t_{\text{off state}} > 2.3 \frac{L_s}{R_{Ls}} \tag{27-42}$$

Thus a large inductance will result in lower turn-on voltages and lower turn-on losses. But it will cause overvoltages during turn-off, lengthen the minimum required off-state interval, and result in higher losses in the snubber. Therefore L_s and R_s must be selected based on the above design trade-offs following a procedure similar to that described for the turn-off snubber. Since the turn-on snubber inductance must carry the load current, which makes this snubber expensive, it is seldom used alone. However, as will be shown in the next section, if turn-off snubbbers are to be used in transistor bridge configurations, then turn-on snubbers must be used.

It is possible to use all three snubbers simultaneously or in any other combination. A circuit configuration that includes all three snubbers but having a reduced component count (the Undeland snubber) is shown in Fig. 27-18.

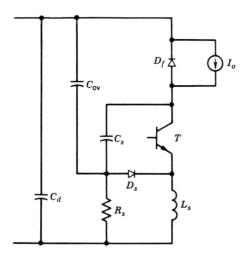

Figure 27-18 A modified circuit with an overvoltage snubber, a turn-on snubber, and a turn-off snubber; the Undeland snubber for step-down converters.

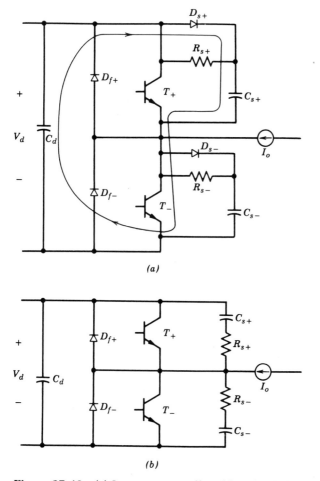

(a)

(b)

Figure 27-19 (a) Improper turn-off snubber circuit for BJTs used in bridge of half-bridge circuits. (b) The turn-off snubber circuit is satisfactory if the goal is to reduce dv/dt at turn-off to minimize EMI problems.

27-8 SNUBBERS FOR BRIDGE CIRCUIT CONFIGURATIONS

As discussed in Chapter 8, in pulse-width-modulated switch-mode converters with half-
or full-bridge configurations, such as in motor drives and uninterruptible power supplies,
the load current can be treated as a constant I_o over the switching cycle. In Fig. 27-19, I_o
is shown to be going into the converter leg, although it could be in the opposite direction
as well. The turn-off snubber shown in Fig. 27-19a, which was shown to be effective in
the step-down converter circuit, should *not* be used without including a turn-on snubber.
With the direction of I_o as indicated, the diode D_{F+} is conducting during the off-state of
T_- and the voltage across C_{s+} is zero. When T_- is turned on, there will be a capacitive
charging current through T_- at the recovery of D_{F+} as is shown by the current loop in Fig.
27-19a. This will result in additional turn-on losses in T_- as was explained in the
discussion of the step-down converter circuit of Fig. 27-14a with a turn-off snubber

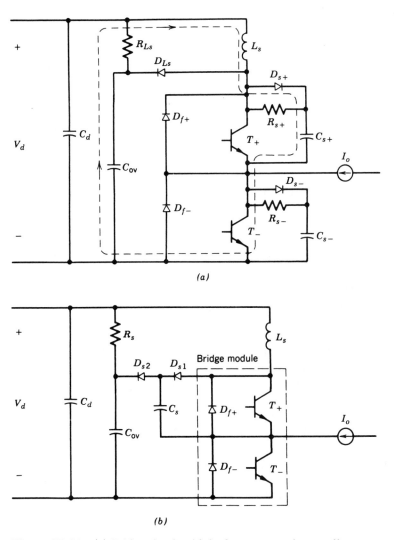

(a)

(b)

Figure 27-20 (a) Bridge circuit with both turn-on and turn-off
snubbers. (b) A modified arrangement; the Undeland snubber for
bridge configurations.

consisting of only C_s. T_+ will experience an identical problem at turn-on. Furthermore C_{s+} will not contribute to the reduction of the turn-off switching stresses of T_-.

An RC snubber of the type shown in Fig. 27-19b will also suffer from the same drawbacks as the snubber of Fig. 27-19a, if the same degree of improvement in the turn-off performance is required. However, the RC snubber of the type in Fig. 27-19b with small capacitance values is currently used in bridge configurations if the primary goal is to reduce dv/dt in order to reduce EMI problems.

The turn-on snubber of the type discussed earlier that will protect both the transistor and the freewheeling diode can be used in the bridge configuration as is shown in Fig. 27-20a. Here the RCD turn-off snubber can also be used. The reason why can be seen by looking at the current path (shown dashed) when T_- is turned off and D_{F+} recovers. This path includes the turn-on snubber inductor, thus reducing the problems compared to using such a turn-off snubber without a turn-on snubber. The two turn-off snubbers of Fig. 27-20a can be combined into one turn-off snubber as is shown in Fig. 27-20b [9], which will protect the two transistors and the freewheeling diodes at turn-on in the same manner as the separate snubbers of Fig. 27-20a.

In the circuit of Fig. 27-20b, it is easy to implement an overvoltage protection by connecting a capacitor C_{OV} as shown, and R_{Ls} also serves as R_{OV} of the overvoltage protection capacitor. This overvoltage snubber protects both the upper and lower transistors and the freewheeling diodes. Also the turn-off snubber capacitors for both the transistors are combined into a single capacitor, which will halve the losses at turn-on compared to the those in the circuit of Fig. 27-20a. All the snubber losses in the circuit of Fig. 27-20b occur in only one resistor, which can be replaced by a dc-dc converter for loss recovery. [10,11]

27-9 GTO SNUBBER CONSIDERATIONS

It was pointed out in the GTO chapter (Chapter 24) that snubber circuits are almost always required in GTO applications. While snubbers for GTOs have the same configurations as for other controlled switches, the large voltages and currents found in GTO circuits place additional requirements on the snubber circuits. Some of these additional considerations are discussed below.

A GTO is capable of turning off a significantly larger current compared to its rms or average current capability. The maximum controllable current for a given GTO in the circuit of Fig. 24-3 depends on the turn-off snubber capacitance C_s. This dependence arises because there is a maximum rate of change in the increase in the anode-cathode voltage at turn-off. Exceeding this maximum $dv_{AK}/dt|_{max}$ would cause retriggering of the GTO back into the on state due to large displacement currents. Now dv_{AK}/dt is inversely proportional to C_s according to Eq. 24-6, so for a given dv_{AK}/dt, the larger C_s is, the larger I_o can be. This, of course, assumes that the maximum controllable anode current given on the GTO specification sheet is not exceeded. A large C_s, however, results in higher overall switching losses and in a larger current through the GTO at turn-on. Therefore, the capacitance C_s should be just sufficient to turn-off the maximum current dictated by the particular application.

The capacitor C_s should have a low internal inductance and a large peak current rating. [12] In practice, this may require paralleling of many capacitors to achieve these required properties for C_s.

The turn-off snubber diode D_s needs to carry the entire load current for a short time. Its average current is very low, but since its dynamic forward voltage at turn-on must be low, often a diode with a larger average current rating is chosen.

The turn-off snubber resistance R_s must be selected based on trade-offs between maximum additional discharge current into the GTO at turn-on and the requirement on the minimum on-state time of the GTO to discharge C_s so it can properly operate during the next turn-off, as was described in the turn-off snubber section. There is a considerable power loss in R_s and therefore it may require mounting on a heat sink.

It has already been described why the stray inductance in the turn-off snubber current loop should be as small as possible. To achieve this, the snubber components should be mounted as close to the GTO as possible. Some of the layout and interconnection techniques that should be used are discussed in the next chapter. [12]

The design considerations for a turn-on snubber for the GTO are similar to those described in the turn-on snubber section.

For GTOs in a bridge configuration, the snubbers similar to those shown in Fig. 24-3 can be used with each GTO. Alternatively the turn-on snubbers can be combined as shown in Fig. 27-20b. The snubber losses in R_s can be significant in GTO applications because of the higher power level of operation and higher snubber losses due to a large C_s. It may be beneficial to replace R_s in Fig. 27-20b with a loss-recovery converter. [10,11]

SUMMARY

This chapter has discussed the design and operation of a variety of snubber circuits used to protect semiconductor power devices from electrical stresses that exceed the device ratings. The important conclusions follow.

1. Snubber circuits protect semiconductor devices during turn-on and turn-off transients by limiting voltage and current magnitudes and rates of rise and by shaping the switching trajectory of the device as it turns on and off.
2. There are three topological snubber configurations: unpolarized RC snubbers, polarized RC snubbers, and polarized RL snubbers.
3. Unpolarized RC snubbers protect diodes and thyristors from overvoltages generated by the reverse-recovery current snap-off in the presence of stray series inductance. There are optimum values of snubber resistance and capacitance that minimize the overvoltages.
4. A polarized RC circuit functions as turn-off snubber to protect all types of controllable switches (BJTs, MOSFETs, GTOs, IGBTs, MSCTs, etc.) by providing a low voltage across the switch while the current turns off. The turn-off switching trajectory traverses a low-power dissipation portion of the output $I-V$ plane and the rate of rise (dv/dt) of the voltage across the device during turn-off is controlled.
5. A polarized RC circuit with a different topology than a turn-off snubber can be used as an overvoltage snubber to limit the overvoltage experienced by a controlled switch when it turns off in the presence of series inductance.
6. A polarized RL snubber protects controlled switches by reducing the voltage across a device as the current through it builds up during the turn-on transient. These turn-on snubbers can also be designed to limit overcurrents in the device during the turn-on transient due to diode reverse recovery.
7. Snubber arrangements for bridge circuits differ in detail from the snubber configurations used to protect single devices.
8. The high voltages and currents commonly found in GTO circuits impose extra considerations in the design of snubber circuits for protecting the GTO.

PROBLEMS

27-1   Consider the step-down converter circuit shown in Fig. 24-3 without the turn-on snubber. The dc input voltage V_d is 500 V, the load current $I_o = 500$ A, and the switching frequency is 1 kHz. The free-wheeling diode has a reverse-recovery time $t_{rr} = 10$ μs. The GTO has a current fall time $t_{fi} = 1$ μs, a maximum reapplied voltage rate $dv/dt = 50$ V/μs, and a maximum controllable anode current $I_{AM} = 1000$ A.

   (a)   Find the appropriate values for resistance R_s and capacitance C_s for the turn-off snubber circuit.

   (b)   Estimate the power dissipated in the snubber resistance.

27-2   The GTO in the circuit of Problem 27-1 is to be protected by a turn-on snubber circuit such as is shown in Fig. 24-3. The maximum rate of rise of the anode current, di_A/dt, is 300 A/μs. Find appropriate values for the inductance and resistance.

27-3   Derive the equation (Eq. 27-5) for the maximum overvoltage across a purely capacitive snubber.

27-4   Consider the flyback converter circuit shown in Fig. 27-6. The input voltage is 100 V as is the dc output voltage. The transformer has a 1:1 turns ratio and a leakage inductance of 10 μH. The transistor, which can be considered as an ideal switch, is driven by a square wave with a 50% duty cycle. The snubber resistance is zero. The diode has a reverse-recovery time t_{rr} of 0.3 μs.

   (a)   Draw an equivalent circuit suitable for snubber design calculations.

   (b)   Find the value of snubber capacitance C_s that will limit the peak overvoltage to 2.5 times the dc output voltage.

27-5   Repeat Problem 27-4 with a resistance R_s included in the snubber circuit. Find both the value of snubber capacitance and optimum value of snubber resistance.

27-6   Estimate the power dissipated in the snubber resistance found in Problem 27-4 if the square-wave switching frequency is 20 kHz.

27-7   Consider the step-down converter circuit of Fig. 27-14 with a purely capacitive snubber having $C_s = C_{s1}$. Assume $I_o = 25$ A, an input voltage $V_d = 200$ V, and a reverse-recovery time $t_{rr} = 0.2$ μs for the free-wheeling diode.

   (a)   Calculate the reduction in the turn-off losses in the BJT due to the use of the snubber capacitor. Assume a switching frequency of 20 kHz and a current fall time t_{fi} of 0.4 μs.

   (b)   Calculate the increase in BJT losses during turn-on due to C_s. Assume that during the turn-on $di_C/dt = 50$ A/μs.

27-8   Repeat Problem 27-7 but now use a polarized R_s-C_s snubber such as is shown in Fig. 27-12.

27-9   Modify the single-phase line frequency diode rectifier in Fig. 27-8a by replacing the diodes with thyristors, replacing C_{dc} by a dc current source of value I_o, and consider the three inductors in series with the 60 Hz source V_s as a single stray inductance of value L_σ. Assume that the line voltage is 230 V rms, 60 Hz and that the stray line inductance L_σ has a magnitude given by $\omega L_\sigma = 0.05(V_s/I_{a1})$ where V_s is the rms phase voltage and I_{a1} is the rms value of the fundamental harmonic of the phase current. The thyristors have a reverse recovery time of 10 μs and $I_{a1} = 100$ A.

   (a)   Explain how a single series RC circuit connected across the line will serve as a turn-off snubber for all four thyristors.

   (b)   Derive equations for the proper values of snubber capacitance C_s and snubber resistance R_s as functions of V_s, I_{a1}, t_{rr}, and other circuit parameters given above.

   (c)   Find values for C_s and R_s that will limit the overvoltage to 1.3 times the peak line voltage.

27-10   The turn-off snubber for a thyristor does not include a diode as it does for the BJT and MOSFET. Explain why.

27-11   An IGBT circuit module complete with its own drive circuitry has been made with the following performance specifications: $V_{DSM} = 800$ V, $I_{DM} = 150$ A, $dV_{DS}/dt < 800$ V/μs, $t_{on} = t_{d(on)} + t_{fv} + t_{ri} = 0.3$ μs, $t_{off} = t_{d(off)} + t_{fi} + t_{rv} = 0.75$ μs, $R_{\theta ja} = 0.5$°C/W, $T_{jmax} = 150$°C. This module is to be used in a step-down converter circuit. In this circuit the free-wheeling diode is ideal, the dc supply voltage $V_d = 700$ V, the load current $I_o = 100$ A, and the switching frequency is 50 kHz with a 50% duty cycle.

(a) Show that a turn-off snubber is needed for the IGBT.

(b) Design a turn-off snubber that will provide a factor of safety of 2 to dv_{DS}/dt.

REFERENCES

1. W. McMurray, "Optimum Snubbers for Power Semiconductors," IEEE IAS 1971 Annual Meeting.

2. *SCR Manual*, 6th ed., General Electric Company, Syracuse, NY, 1979.

3. M. H. Rashid, *Power Electronics: Circuits, Devices, and Applications*, Prentice-Hall, Englewood Cliffs, NJ, 1988.

4. B. M. Bird and K. G. King, *An Introduction to Power Electronics*, Wiley, New York, 1983.

5. B. W. Williams, *Power Electronics, Devices, Drivers, and Applications*, Wiley, New York, 1987.

6. T. M. Undeland, A. Petterteic, G. Hauknes, A. K. Adnanes, and S. Garberg, "Diode and Thyristor Turn-off Snubber Simulation by KREAN and an Easy-to-Use Design Algorithm," IEEE IAS Proc., 1988, pp. 647–654.

7. E. S. Oxner, *Power FETs and Their Applications*, Prentice-Hall, Englewood Cliffs, NJ, 1982.

8. Thyristor Applications Notes, "Applying International Rectifier's Gate Turn-Off Thyristors," AN-315A, International Rectifier, El Segundo, CA, 1984.

9. T. M. Undeland, F. Jenset, A. Steinbakk, T. Rogne, and M. Hernes, "A Snubber Configuration for Both Power Transistors and GTO PWM Converters," *Proc. of 1984 Power Electronics Specialists Conference*, pp. 42–53.

10. Tore M. Undeland, "Switching Stress Reduction in Power Transistor Converters," *1976 IEEE Industrial Applications Society Conference Proceedings*, pp. 383–392.

11. "Loss Recovery," *IEEE Trans. on Power Electronics*, 1994.

12. H. Veffer, "High Current, Low Inductance GTO and IGBT Snubber Capacitors," Siemens Components, June 1990, pp. 81–85.

CHAPTER 28

GATE AND BASE DRIVE CIRCUITS

28-1 PRELIMINARY DESIGN CONSIDERATIONS

The primary function of a drive circuit is to switch a power semiconductor device from the off state to the on state and vice versa. In most situations the designer seeks a low cost drive circuit that minimizes the turn-on and turn-off times so that the power device spends little time in traversing the active region where the instantaneous power dissipation is large. In the on state the drive circuit must provide adequate drive power (e.g., base current to a BJT or gate-source voltage to a MOSFET) to keep the power switch in the on state where the conduction losses are low. Very often the drive circuit must provide reverse bias to the power switch control terminals to minimize turn-off times and to ensure that the device remains in the off-state and is not triggered on by stray transient signals generated by the switchings of other power devices.

The signal processing and control circuits that generate the logic-level control signals used to turn the power switch on and off are not considered part of the drive circuit. The drive circuit is the interface between the control circuit and the power switch. The drive circuit amplifies the control signals to levels required to drive the power switch and provides electrical isolation when required between the power switch and the logic-level signal processing/control circuits. Often the drive circuit has significant power capabilities compared to the logic-level control/signal processing circuits. For example, power BJTs have low values of beta, typically 5–10, so that the base current supplied by the drive circuit is often a significant fraction of the total load current.

The basic topology of the drive circuit is dictated by three functional considerations. First, is the output signal provided by the drive circuit unipolar or bipolar? Unipolar signals lead to simpler drive circuits, but bipolar signals are needed for rapid turn-on and turn-off of the power switch. Second, can the drive signals be directly coupled to the power switch, or is electrical isolation required between the logic-level control circuits and the power device? Most electrically isolated drive circuits will require isolated dc power supplies. Third, is the output of the drive circuit connected in parallel with the power switch (the usual situation) or in series with the switch (cascode connection)?

Additional functionality may be required of the drive circuit, which will further influence the topological details of the circuit. Provisions may be included in the drive circuit design for protection of the power switch from overcurrents. Then communication between the drive circuit and the control circuit is needed. In bridge circuits, the drive circuit must often provide blanking times for the power switch. Incorporation of these types of functionality requires design inputs to both the drive circuit and the logic-level control circuit. Waveshaping of the drive circuit output may also be included to improve the power switch performance.

The specific details of component values to be used in a drive circuit will vary depending on the characteristics of the power switch being driven. For example, BJT drive circuits must provide a relatively large output current (the base current of the power BJT) for the duration of the BJT on-state time interval, whereas MOSFET drive circuits need only provide an initial large current as the device turns on and for the rest of the on-state time interval merely provide a large gate-source voltage at low current levels.

It is a good idea to consider how the drive circuit will be configured on a circuit board even at the earliest stages in the design process. The placement of components to minimize stray inductance and to minimize susceptibility to switching noise may affect the choice of topology for the drive circuit.

28-2   dc-COUPLED DRIVE CIRCUITS

28-2-1   dc-COUPLED DRIVE CIRCUITS WITH UNIPOLAR OUTPUT

A very simple base drive circuit suitable for converters with a single-switch topology is shown in Fig. 28-1. At turn-on, the *pnp* driver transistor is turned on by saturating one of the internal transistors in the comparator (type 311, for example). This provides a base current for the main power BJT that can be calculated by noting in the circuit of Fig. 28-1a

$$V_{BB} = V_{CE(\text{sat})}(T_B) + R_1 I_1 + V_{BE(\text{on})} \tag{28-1}$$

and

$$I_{B(\text{on})} = I_1 - \frac{V_{BE(\text{on})}}{R_2} \tag{28-2}$$

For the specified maximum collector current I_C that the application demands of the transistor, the necessary base current $I_{B(\text{on})}$ and corresponding $V_{BE(\text{on})}$ can be found from the power transistor data sheets. Similarly, the $V_{CE(\text{sat})}$ for the *pnp* transistor in the base drive circuit can be obtained from its data sheets. In selecting R_1, R_2, and V_{BB}, it should be recognized that a small R_2 will allow a faster turn-off but will also cause the power dissipation in the drive circuit to be large. The approximate turn-off waveforms are shown in Fig. 28-1b where v_{BE} is shown as larger during the on state compared to the storage interval.

A step-by-step design procedure is shown below.

1. Based on the turn-off speed required, the negative base current, $I_{B,\text{storage}}$ during the storage time is estimated. From this, R_2 in Fig. 28-1a can be calculated as

$$R_2 = \frac{V_{BE,\text{storage}}}{I_{B,\text{storage}}} \tag{28-3}$$

2. Knowing the required on-state base current $I_{B(\text{on})}$ and the corresponding $V_{BE(\text{on})}$ and R_2 from the previous step, I_1 becomes

$$I_1 = I_{B(\text{on})} + \frac{V_{BE(\text{on})}}{R_2} \tag{28-4}$$

3. Two unknowns remain in Eq. 28-1, V_{BB} and R_1. The on-state losses in the drive circuit are approximately equal to $V_{BB}I_1$, which suggests that V_{BB} should be small. On the other hand, to reduce the influence of variations in $V_{BE(\text{on})}$, V_{BB} should be

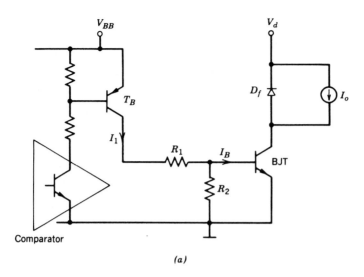

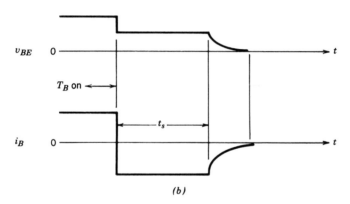

Figure 28-1 (*a*) Simple base current drive circuit for a power BJT and (*b*) the associated current and voltage waveforms at turn-off.

large. In practice, a V_{BB} of about 8 V is optimum. With $V_{BB} = 8$ V, R_1 can then be estimated using Eq. 28-1.

This base drive circuit should not be used in pulse-width-modulated bridge converter circuits for reasons that will be discussed shortly.

A simple MOSFET gate drive circuit with only one switch to control the gate current is shown in Fig. 28-2, where the output transistor of a comparator (e.g. LM311) controls the MOSFET. When the output transistor is off, the MOSFET is on and vice versa. When the comparator is on, it must sink a current V_{GG}/R_1, and to avoid large losses in the drive circuit, R_1 should be large. This will slow down the MOSFET turn-on time. This means that the drive circuit is only suitable for low switching speed applications.

The inadequacy of this circuit can be overcome by the MOSFET gate drive circuit shown in Fig. 28-3 where two switches are used in a totem-pole arrangement with the comparator (type 311) controlling the *npn–pnp* totem-pole stack. Here, to turn the MOS-FET on, the output transistor of the comparator turns off, thus turning the *npn* BJT on, which provides a positive gate voltage to the MOSFET. At the turn-off of the MOSFET,

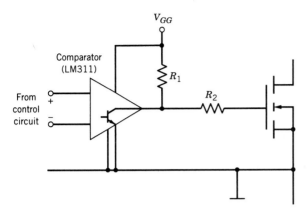

Figure 28-2 Simple MOSFET gate drive circuit suitable for low-speed and low-switching-frequency applications.

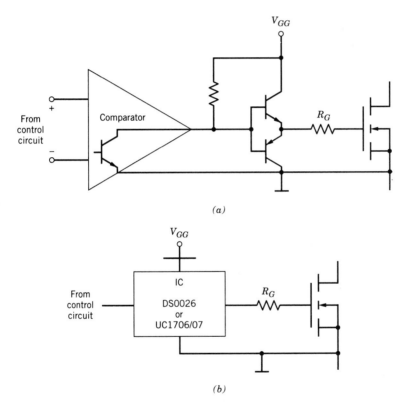

(a)

(b)

Figure 28-3 A MOSFET gate drive circuit with a totem-pole configuration for faster turn-off times: (a) discrete totem-pole gate drive circuit; (b) integrated circuit totem pole gate drive circuit.

the gate is shorted to the source through R_G and the *pnp* transistor. Since no steady-state current flows through R_G in contrast to R_1 described in the previous paragraph, R_G can be chosen to be much smaller in value, which results in much faster turn-on and turn-off times. Very often, instead of using discrete components, similar performance can be obtained, as is shown in Fig. 28-3b by using buffer ICs such as CMOS 4049 or 4050 if a low gate current is needed or a DS0026 or UC 1707, which can source or sink currents in excess of 1 A.

28-2-2 dc-COUPLED DRIVE CIRCUITS WITH BIPOLAR OUTPUT

In order to operate power semiconductor devices at high switching frequencies, drive circuits must be designed to turn-off the devices as rapidly as they turn on. The descriptions of the switching characteristics of BJTs, MOSFETs, IGBTs, and other devices clearly illustrated the need for a reverse bias to be applied to the control terminals of the power switch in order to affect a rapid turn-off. Drive circuits with unipolar outputs are unable to provide the needed reverse bias and thus are incapable of providing fast turn-off of power devices. In order to provide a reverse bias to the control terminals of the power device, the drive circuit must have a bipolar output (an output that can be either positive or negative). This in turn requires that the drive circuit be biased by a negative power supply as well as a positive power supply.

The BJT base drive circuit shown in Fig. 28-4 where both a positive and negative voltage supply with respect to the emitter are used provides a fast turn-off. For the turn-on

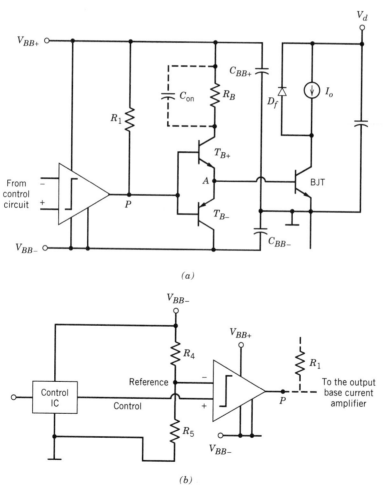

(a)

(b)

Figure 28-4 (a) A BJT base current drive circuit with both positive and negative voltages with respect to the BJT emitter for faster turn-off of the power device. (b) A pre-converter circuit should be used if the input from the control circuit has signal only between V_{BB+} and ground.

interval, the output transistor of the comparator turns off, thus turning the transistor T_{B+} on. The on-state base current is

$$I_{B(\text{on})} = \frac{V_{BB+} - V_{CE(\text{sat})}(T_{B+}) - V_{BE(\text{on})}}{R_B} \tag{28-5}$$

Arguments similar to step 3 of the previous BJT drive circuit design apply in the selection of V_{BB+} and R_B. The optional capacitor C_{on} shown as dashed, acts as a speed-up capacitor by providing a large transient base current to the power transistor at the instant of turn-on to speed up the turn-on sequence.

For turning the BJT off, the internal output transistor of the comparator turns on, thus turning the *pnp* transistor T_{B-} on (and automatically turns the *npn* transistor T_{B+} off). For a fast turn-off, no external resistance is used in series with T_{B-}. The magnitude of the negative voltage must be less than the *BE* breakdown voltage of the BJT, which is given on the data sheets and is normally in the $5-7$ V range. The switching waveforms will be similar to those described in Section 21-5 if the BJT is used in a similar circuit. If the BJT has a tendency for collector current tailing due to a too rapid turn-off of the *BE* junction compared to the *CB* junction, as described in Section 21-5, then a resistor or, if necessary, an inductor can be added in the turn-off base drive between points A and the emitter of T_{B-} in Fig. 28-4a.

If the control signal is supplied by a logic circuit that is connected between V_{BB+} and the emitter of the BJT, then the reference input to the comparator should be at the mid-potential between V_{BB+} and the BJT emitter terminal, as is shown in Fig. 28-4b where $R_4 = R_5$.

The modifications shown in Fig. 28-5a further enhance the BJT turn-off performance of the drive circuit of Fig. 28-4. An antisaturation diode D_{as} is added to keep the BJT voltage v_{CE} slightly above its saturation value $V_{CE(\text{sat})}$. This can be seen in Fig. 28-5a where

$$v_{AE} = V_{BE(\text{on})} + V_{D1} = V_{CE(\text{on})} + V_{Das} \tag{28-6}$$

and therefore,

$$V_{BE(\text{on})} = V_{CE(\text{on})} \tag{28-7}$$

since $V_{D1} = V_{Das}$. Since $V_{BE(\text{on})}$ is in general larger than $V_{CE(\text{sat})}$, the presence of the antisaturation diode keeps the transistor slightly out of saturation, thus reducing the storage time at the expense of increased on-state losses in the BJT. Therefore the antisaturation diode should only be used if the capability to use the BJT in a high switching frequency application is required. If still faster turn-off switching is needed, the on-state voltage $V_{CE(\text{sat})}$ can be adjusted by putting one or more diodes in series with D_1.

In the circuit of Fig. 28-5a, the diode D_2 is needed to provide a path for the negative base current. D_{as} should be a fast recovery diode with a reverse recovery time smaller than the storage time of the BJT. Moreover, its reverse voltage rating must be similar to the off-state voltage rating of the power transistor.

An improved version of the circuit of Fig. 28-5a is shown in 28-5b where the power loss in the positive portion of the base drive circuit is reduced compared to the original circuit. Here the antisaturation diode adjusts the base current of the drive transistor T_{B+} such that T_{B+} operates in the active mode and the current drawn from V_{BB+} now is only equal to the actual I_B needed to barely saturate the BJT. Moreover, the current rating required of D_{as} is reduced. A small resistance in series with the antisaturation diode can

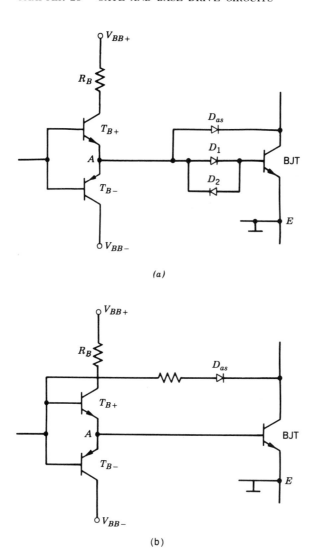

Figure 28-5 (a) Base drive circuit with antisaturation to minimize the storage time of the BJT and thus the turn-off time. The modifications in (b) permit the antisaturation diode to have a lower current rating compared to the situation in (a).

significantly help reduce oscillations at turn-on. Since T_{B+} operates in the active region, it must be mounted on a small heat sink.

A drive circuit for MOSFETs that provides positive gate voltages at turn on and negative gate voltages at turn off by means of a split power supply with respect to the MOSFET source is shown in Figs. 28-6a through 28-6c. If the control signal is supplied by logic circuit that is connected between V_{GG+} and the source of the MOSFET, then the reference input to the comparator should be shifted to be at the mid-potential between V_{GG+} and the source of the MOSFET using a pre-converter circuit similar to that in Fig. 28-4b.

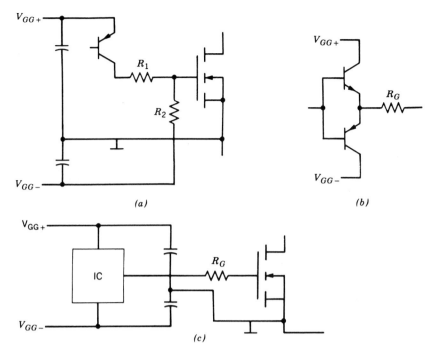

Figure 28-6 Various gate drive circuits using split dc power supplies for providing an n-channel MOSFET with positive gate-source voltages at turn-on and negative gate drive at turn-off.

28-3 ELECTRICALLY ISOLATED DRIVE CIRCUITS

28-3-1 NEED FOR AND TYPES OF ELECTRICAL ISOLATION

Very often, there is a need for electrical isolation between the logic-level control signals and the drive circuits. This is illustrated in Fig. 28-7 for the case of a power BJT half-bridge converter having a single-phase ac supply as its input where one of the power terminals is a grounded neutral wire. Now the positive dc bus is close to the ground potential during the negative half cycle of v_s, and the negative dc bus is near ground potential during the positive half cycle of v_s. Under these conditions the emitter terminals of both BJTs must be treated as "hot" with respect to power neutral. The logic-level control signals are normally referenced with respect to logic ground, which is at the same potential as the power neutral since the logic circuits are connected to the neutral by means of a safety ground wire.

The basic ways to provide electrical isolation are either by optocouplers, fiber optics, or by transformers. The optocoupler shown in Fig. 28-8 consists of a light-emitting diode (LED), the output transistor, and a built-in Schmitt trigger. A positive signal from the control logic causes the LED to emit light that is focused on the optically sensitive base region of a photo transistor. The light falling on the base region generates a significant number of electron-hole pairs in the base region that causes the photo transistor to turn on. The resulting drop in voltage at the photo transistor collector causes the Schmitt trigger to change state. The output of the Schmitt trigger is the optocoupler output and can be used as the control input to the isolated drive circuit. The capacitance between the LED and the

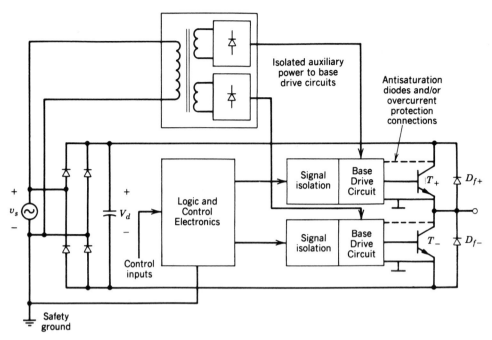

Figure 28-7 Power BJT base drive system showing the need for electrical isolation between the base drive circuitry and the logic level control circuitry.

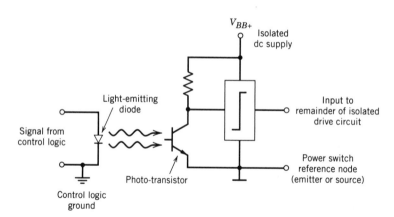

Figure 28-8 Schematic of an optocoupler used to couple signals to a floating (electrically isolated) drive circuit from a control circuit referenced with respect to the control logic ground (and power system neutral).

base of the receiving transistor within the optocoupler should be as small as possible to avoid retriggering at both turn-on and turn-off of the power transistor due to the jump in the potential between the power transistor emitter reference point and the ground of the control electronics. To reduce this problem, optocouplers with electrical shields between the LED and the receiver transistor should be used.

As an alternative, fiber optic cables can be used to completely eliminate this retriggering problem and to provide very high electrical isolation and creepage distance. When using fiber optic cables, the LED is kept on the printed circuit board of the control electronics, and the optical fiber transmits the signal to the receiver transistor, which is put on the drive circuit printed circuit board.

Instead of using optocouplers or fiber optic cables, the control signal can be coupled to the electrically isolated drive circuit by means of a transformer as is shown in Fig. 28-9a. If the switching frequency is high (several tens of kilohertz or more) and the duty ratio D varies only slightly around 0.5, a baseband control signal of appropriate magnitude can be applied directly to the primary of a relatively small and lightweight pulse transformer as implied in Fig. 28-9a, and the secondary output can be used to either directly drive the power switch or used as the input to an isolated drive circuit. As the switching frequency is decreased below the tens of kilohertz range, a baseband control

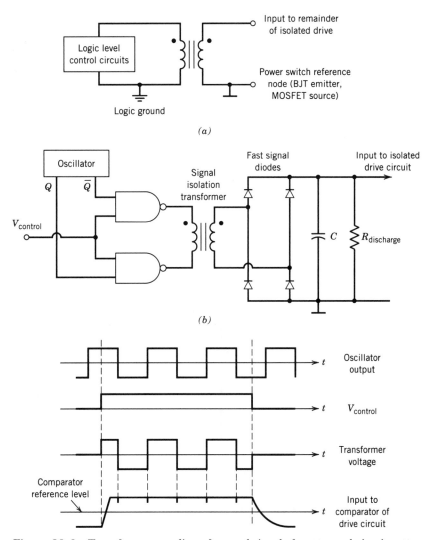

Figure 28-9 Transformer coupling of control signals from control circuits to electrically isolated drive circuits. In (a) the baseband control signal is directly connected to the transformer primary. In (b) the control signal modulates a high-frequency carrier that is then applied to the primary of a small high-frequency signal transformer. The waveforms associated with (b) are shown in (c).

signal directly applied to the transformer primary becomes impractical because the size and weight of the transformer becomes increasing larger.

Modulation of a high-frequency carrier by a low-frequency control signal enables a small high-frequency pulse transformer to be used for even low-frequency control signals. In Fig. 28-9b the control signal modulates a high-frequency (e.g., 1 MHz) oscillator output before being applied to the primary of a high-frequency signal transformer. Since a high-frequency transformer can be made quite small, it is easy to avoid stray capacitances between the input and the output windings, and the transformer will be inexpensive. The transformer secondary output is rectified and filtered and then applied to the comparator and the rest of the isolated drive circuit. The waveforms for this modulation scheme are shown in Fig. 28-9c.

28-3-2 OPTOCOUPLER ISOLATED DRIVE CIRCUITS

In optocoupler-isolated drive circuits, the optocoupler itself is the interface between the output of the control circuit and the input of the isolated drive circuit. The input side of the optocoupler is directly coupled to the control circuit and the output side of the optocoupler is directly connected to the isolated drive circuit. The topology of the isolated drive circuit between the output of the optocoupler and the control terminal of the power switch can take many different forms.

An optocoupler-isolated drive circuit for a power BJT is shown in Fig. 28-10. The drive circuit has a bipolar output so that rapid turn-on and turn-off of the BJT can be achieved. An *npn–pnp* totem-pole circuit couples the appropriate dc voltage to the base of the power BJT to turn it on or off as required. The isolated split dc power supplies are implemented by the circuit segment in the lower left side of Fig. 28-10.

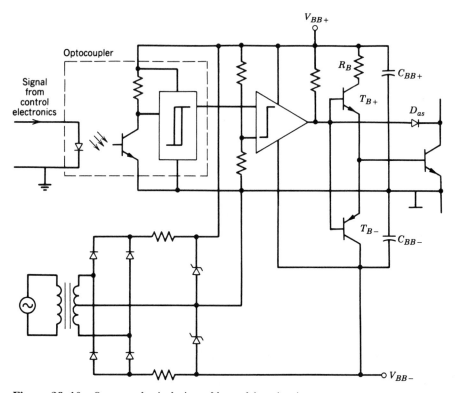

Figure 28-10   Optocoupler isolation of base drive circuits.

Optocoupler-isolated drive circuits can also be used with power MOSFETs and IGBTs. The circuit shown in Fig. 28-11 uses a high common-mode noise immunity optocoupler (HPCL-4503) and a high-speed driver (IXLD4425) with a 3-A output capability. The drive circuit uses a single-ended floating 15-V supply and provides a ±15-V output voltage for high noise immunity and fast switching to drive the gate of a power MOSFET or IGBT. The integrated high-speed driver circuit connects the gate of the power device to the 15-V bus bar while it simultaneously connects the source to the negative side of the bias supply in order to turn the power device on. To turn the power device off, the drive circuit connects the gate to the negative side of the single-ended supply while it connects the source to the +15-V bus bar.

28-3-3 TRANSFORMER-ISOLATED DRIVE CIRCUITS PROVIDING BOTH SIGNAL AND POWER

The use of transformers for electrically isolating the drive circuit from the control circuit introduces a great deal of flexibility into the design of the drive circuit. If floating dc power supplies are available, the transformer coupling scheme with a modulated carrier shown in Fig. 28-9b can be used as a replacement for the optocoupler in the circuits shown in Figs. 28-10 and 28-11.

However, the same transformer used to transfer the control signal from the control circuits to the isolated drive circuit can also be used to provide the isolated dc bias power as well and do away with a separate transformer for the isolated dc supplies. Consider the BJT base drive circuit shown in Fig. 28-12. This is an isolated base drive in which the base current is made to be proportional to the collector current. Here the need for an auxiliary dc power supply with respect to the emitter terminal is avoided. The transformer is a combination of a flyback converter transformer and a current transformer. When the drive transistor T_1 is on, the BJT is off and vice versa. When T_1 is conducting and the BJT is off, the transformer core is magnetized to the limit of saturation with $i_p = V_{BB+}/R_p$. Due to the stored energy in the slightly gapped transformer core, turning T_1 off forces a current to flow in the second winding as in a flyback converter, thus resulting in a positive base current to the BJT. This causes the BJT to start conducting, and its base current is mainly provided by the transformer action between windings 2 and 3, causing $i_B = N_3 i_C/N_2$. During the off interval of T_1, the voltage across the capacitor C_p discharges to zero due to the resistance R_p. Therefore, when T_1 is turned on in order to turn the BJT off, a voltage essentially equal to V_{BB+} is applied across winding 1 causing i_p to be large.

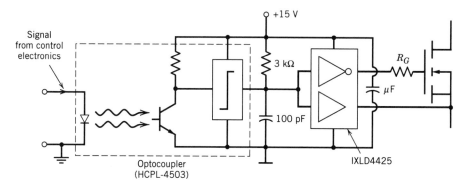

Figure 28-11 An optocoupler isolated drive circuit suitable for driving MOSFETs and IGBTs. The circuit providing the isolated single-ended 15 V bias supply is not shown for simplicity.

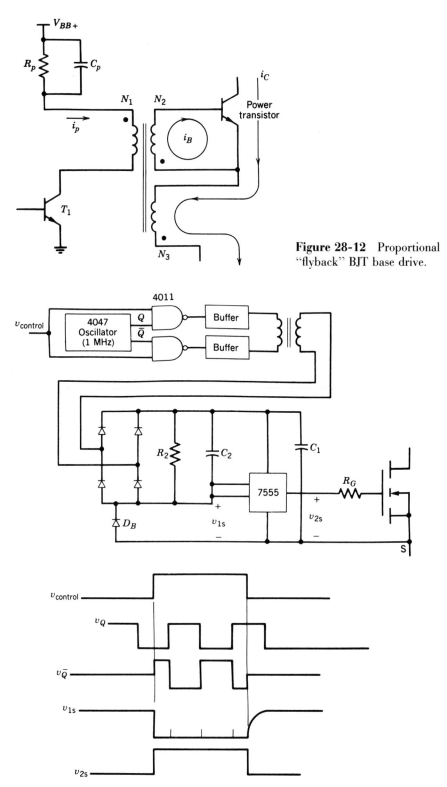

Figure 28-12 Proportional "flyback" BJT base drive.

Figure 28-13 Transformer-isolated MOSFET gate drive circuit using a high-frequency modulated carrier so that the MOSFET can be held on for long periods. No auxiliary dc power supplies are needed since both the control signal and bias power come through the transformer.

During the turn-off of the BJT, its base current is given as

$$i_B = N_3 i_C / N_2 - N_1 i_p N_2 \qquad (28\text{-}8)$$

The drive circuit must be designed so that the base current during turn-off is negative and of adequate magnitude and duration. This drive configuration is best suited for high-frequency applications where the variations in the duty cycle are limited.

If in a given application, the MOSFET to be controlled is to be on for a long time, the circuit shown in Fig. 28-13 can be used. In this circuit the control voltage is modulated by a high-frequency oscillator output before being applied to the buffer circuits. Now a high-frequency ac signal appears across the transformer primary when the control voltage is high, thus charging the energy storage capacitance C_1 and the capacitance C_2 at the input to the 7555 IC, which is used here as a buffer and a Schmitt trigger because of its low power consumption. With the input to the 7555 low, it provides a positive voltage to the MOSFET gate, thus turning it on as is shown in Fig. 28-13. At turn off, the control voltage goes low and the voltage across the transformer primary goes to zero. Now C_2 discharges through R_2 and the input voltage to the 7555 goes high, which causes its output voltage to go low, thus turning the MOSFET off. The diode D_B is used to prevent the energy stored in the capacitance C_1 from discharging into the resistance R_2.

In pulse-width-modulated inverters, such as motor drives and UPS, there is a need for a smooth transition in the duty ratio D from a finite value to either zero or one. The use of a phase shift resonant controller, UC3875, combined with two transformers, a demodulator, and some buffers will make a good gate drive circuit for a MOSFET or an IGBT as is shown in Fig. 28-14. As in the drive circuit of Fig. 28-13, the transformers may have a relatively large leakage inductance. This makes it easy to produce noise immune transformers with a high isolation test voltage.

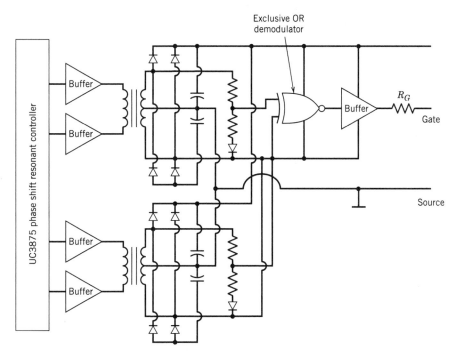

Figure 28-14 A transformer-coupled MOSFET or IGBT gate drive circuit, which has a smooth transition in the duty ratio D from zero to one.

28-4 CASCODE-CONNECTED DRIVE CIRCUITS

All of the drive circuits discussed so far can be characterized as shunt-connected to the power switch, meaning that the drive circuits are connected in shunt (parallel) with power switch control terminals. In this configuration the drive circuit conducts only a fraction (usually a small fraction) of the current carried by the power switch in the on state. However, there are some situations where it is advantageous to place the output of the drive circuit in series (a so-called cascode connection) with power switch so that the drive circuit must conduct the same current as the power switch.

28-4-1 OPEN-EMITTER BJT DRIVE CIRCUIT

An attractive alternative to conventional base-emitter drive circuits for switching the power BJT is the so-called open-emitter, or cascode switching, circuit shown in Fig. 28-15a. The switch in series with the BJT is chosen to be a MOSFET since it can switch very fast, is easy to control, and provides a very low on-state resistance during conduction since its breakdown voltage, which is required for this application, is very low, on the order of a few tens of volts. To turn the BJT on, the MOSFET is switched on, which causes the BJT base current to flow, thus turning the transistor on. When conducting, the main current flow is through the BJT and the MOSFET. To turn the BJT off, the MOSFET is quickly turned off, which causes the collector current to flow out of the base terminal through the capacitor, thus making the negative base current equal to the collector current. This negative base current quickly turns off the BJT and the problem of premature cutoff of the BE junction discussed in Section 21-4 that can occur in conventional base drive circuits and lead to collector current tailing is avoided. The potential of the base terminal is clamped by the zener diode, and therefore the MOSFET breakdown voltage is limited.

The safe operating area under emitter-open turn-off, which is shown in Fig. 28-15b, is much greater than the conventional RBSOA because the v_{CE} limit is BV_{CBO}, which can be as much as twice as large as BV_{CEO}. The larger voltage limit is a consequence of the BJT being used during turn-off as a simple diode composed of the CB junction. The degradation of the breakdown voltage to BV_{CEO} because of emitter current flow that was

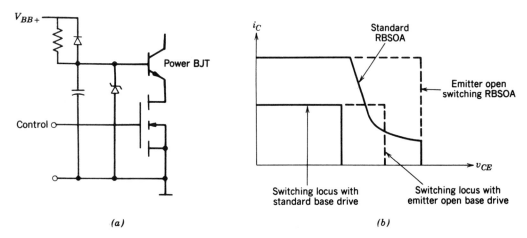

Figure 28-15   The emitter-open or cascode switching circuit (a) that takes advantage of a larger safe operating area (b).

discussed in Section 21-6 is absent in this case because the external circuit does not permit any emitter current to flow in the off state. Thus the BJT used in this circuit is chosen on the basis of its BV_{CBO} rating and not its BV_{CEO} rating. This provides for a combined switch that may have lower on-state conduction losses compared to a conventional BJT switch since the conventional switch would have a larger breakdown rating (the conventional BJT switch would be chosen on the basis of BV_{CEO} exceeding the off-state applied voltage) and consequently a much larger on-state voltage drop $V_{CE(\text{sat})}$.

The emitter-open switching circuit can be modified to provide the base current proportional to the collector current by means of a two-winding transformer. This eliminates the need for a dc voltage supply in a base drive circuit. The simple version of the open-emitter drive circuit shown in Fig. 28-15 cannot be used in bridge circuit configurations when one transistor is turned on while the opposite free-wheeling diode is conducting. The reasons will be discussed in a later section.

28-4-2 CASCODE DRIVE CIRCUITS FOR NORMALLY ON POWER DEVICES

Cascode-configured drive circuits are particularly attractive for normally on devices such as JFET-based power devices. The normally on characteristic of JFET-based devices, which has severely limited their application, is a natural fit to the requirements of a cascode drive circuit topology. Consider the cascode drive circuit for an FCT shown in Fig. 28-16. When the MOSFET Q_s is held off, the voltage at the cathode of the FCT (T_{sw}) rises up to a voltage $V_d R_{G2}/(R_{G1} + R_{G2})$, while the gate of the FCT is at the reference potential. If the cathode voltage is large enough, the FCT will be in the off state. The required negative gate-cathode voltage is given by

$$\frac{V_d R_{G2}}{R_{G1} + R_{G2}} \geq \frac{V_d}{\mu} \tag{28-9}$$

where μ is the blocking gain of the FCT. When the MOSFET is gated on, the gate-cathode voltage of the FCT drops to a very low value (the on-state voltage of the MOSFET), and the FCT turns on with the FCT current also passing through the MOSFET.

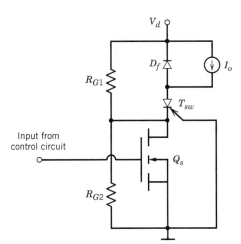

Figure 28-16 A concept cascode drive circuit for an FCT.

The blocking gain of an FCT typically can range from 10 to 100. For V_d values of 2000 V or less, this range of blocking gain would require negative gate-cathode voltages from 200 V (blocking gain equal to 10) to 20 V (blocking gain equal to 100) in order to hold the FCT off. MOSFETs are readily available that have blocking voltage capabilities that exceed these voltages and that simultaneously have on-state current capabilities of 100 A or larger. If the current capability of a single MOSFET is not commensurate with that of the FCT, several identical MOSFETs can be paralleled to reach the desired current capability.

The cascode drive circuit shown in Fig. 28-16 can equally be applied to other normally on devices as the power JFET or static induction transistor. Cascode circuits have also been applied to normally off devices including GTOs. Some attempts have been made to integrate the cascode output switch (e.g., the MOSFET Q_s in Fig. 28-16) onto the same silicon wafer as the power device. One such device is the so-called emitter-switched thyristor.

28-5   THYRISTOR DRIVE CIRCUITS

28-5-1   GATE CURRENT PULSE REQUIREMENTS

A pulse of gate current is needed to turn the thyristor on, and once triggered on, the thyristor continues to conduct without any continuous gate current because of the regenerative action of the device. In estimating how large the pulse of gate current should be to ensure device turn-on, the gate current-voltage characteristic shown in Fig. 28-17a must be used. For a given thyristor type, the range of variation possible in the gate characteristic is specified by maximum and minimum curves in the device data sheets, which are similar to the maximum and minimum curves shown in Fig. 28-17a. The minimum gate current and corresponding gate voltage needed to ensure that the thyristor will be triggered at various operating temperatures is also specified in the device data sheets and is shown in Fig. 28-17a as the dashed line. This minimum gate current curve is sometimes called the locus of minimum firing points.

The load line of the gate pulse amplifier should result in a gate current that is greater than that specified in the minimum gate current curve. An equivalent circuit for a gate pulse amplifier is shown in Fig. 28-17b, which consists of an open circuit output voltage of V_{GG} and an output resistance of R_G. This circuit will produce a gate current in the range of I_{G1} at low temperatures to I_{G2} at high temperatures as is shown in the load line construction of Fig. 28-17a. By proper selection of the load line parameters (V_{GG} and R_G), a gate current well in excess of the minimum required current is obtained. The minimum time duration of the current pulse, usually a few tens of microseconds, during which the gate current must flow, is specified on the device data sheets.

The data sheets also specify the maximum allowable gate current and gate power dissipation. A maximum gate power dissipation hyperbola is shown in Fig. 28-17a. These quantities are normally very large in relation to the gate trigger current needed and do not present any design constraint.

In order to allow a large di/dt during the turn-on of a thyristor, a large gate current pulse is supplied during the initial turn-on phase with a large di_G/dt. After the thyristor turns on, the gate current is then reduced and kept on for some time at a lower value in order to avoid unwanted turn-off of the device. Such a shaped gate current pulse is shown in Fig. 23-9.

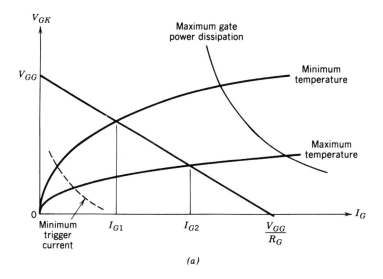

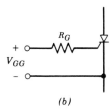

Figure 28-17 Design of thyristor gate drive circuits. (a) Gate-cathode $I-V$ characteristics used for designing gate trigger circuits; (b) equivalent circuit of gate-pulse amplifier.

Thyristors are often used in line-frequency converters where the devices are naturally turned off by the line-frequency voltages. In order to control the dc output voltage of the converter and the magnitude and direction of the power flow through the converter, the thyristors must be turned on at a proper delay angle relative to the zero crossing of the ac line voltages. In the case of load voltage commutated thyristor converters, such as those used in very large power synchronous motor drives and induction heating inverters, the gate trigger times of the thyristors are synchronized with the ac voltages of the load.

A general block diagram of a gate trigger circuit, for example, in a single-phase converter, is shown in Fig. 28-18. The thyristors are at line potential, and the trigger circuit must be referenced with respect to a logic ground associated with the control input. Therefore the zero crossing detection of the line-voltage synchronization and the gate pulse generated within the gate trigger circuit must be isolated from the line potential by means of transformers as is shown in Fig. 28-18. The gate trigger circuit also requires a dc power supply referenced with respect to the logic ground potential. This dc voltage can be supplied by rectifying the output of the line-voltage synchronization transformer as is shown in Fig. 28-18.

In the delay angle block, the ac synchronization voltage is converted into a ramp voltage that gets synchronized to the zero crossing of the line voltage as is shown by

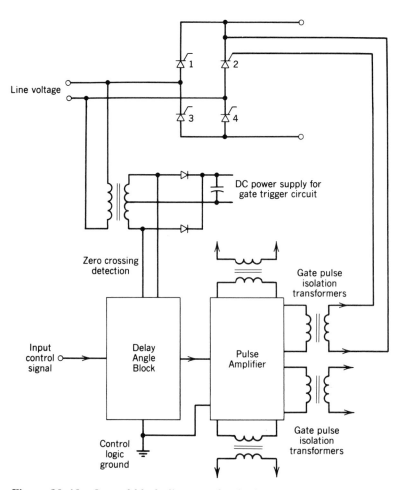

Figure 28-18 General block diagram of a thyristor gate trigger circuit.

waveforms in Fig. 28-19. This ramp voltage, which has a constant peak-to-peak amplitude, is compared with a control voltage. During alternate half-cycles when the ramp voltage equals the control voltage, a pulse signal of controllable duration is generated as is shown in Fig. 28-19. In this manner the delay angle can be varied over nearly the full range between 0° and 180° and the delay angle is proportional to the control voltage. Normally an integrated circuit such as one of the TCA780 family is used to implement this control function. Such integrated circuits also incorporate additional features for start-up, shut down, and so forth.

28-5-2 GATE PULSE AMPLIFIERS

In low-power thyristors used in consumer applications, the trigger current needed by the thyristor is small enough that it can be supplied by ICs without the need for an external pulse amplifier. In high-power thyristors, the trigger current requirement is purposely kept high in order to provide noise immunity. In such thyristors the initial peak gate current at turn-on, such as is shown in Fig. 23-9, may be as large as 3 A and then may drop to about 0.5 A for the duration of the pulse.

A pulse amplifier is shown in Fig. 28-20 where the pulse output from the delay angle block turns on the MOSFET, which supplies an amplified pulse of gate current to the

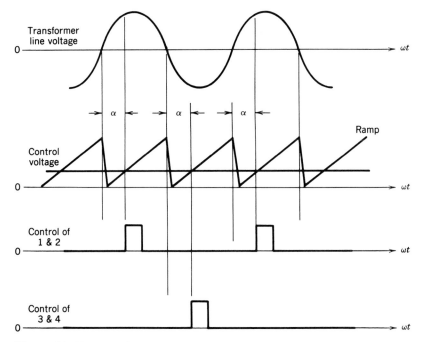

Figure 28-19 Waveforms in the gate trigger circuit of Fig. 28-18.

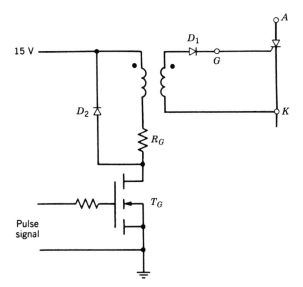

Figure 28-20 Pulse amplifier for a thyristor gating circuit.

thyristor through the pulse transformer. The diode D_1 in the secondary side is used to prevent a negative gate current due to the transformer magnetizing current when the MOSFET T_G turns off. The diode D_2 is used to provide a path for the transformer magnetizing current so that the energy in the magnetic core gets dissipated in R_G. Wave-shaping to produce a waveform similar to the waveform of Fig. 23-9 can be provided by an RC network in parallel with R_G.

A similar gate trigger circuit can be built for three-phase full-bridge thyristor converters where the six thyristors are triggered in sequence at 60° intervals. However, to get started and with discontinuous load currents, it is necessary to trigger a thyristor pair, one from the top group and one from the bottom group, in sequence at a 60° interval. Therefore a thyristor will receive gate pulses as is shown in Fig. 28-21 at turn-on.

28-5-3 COMMUTATION CIRCUITS

In the line-frequency and load-commutated converters, the thyristor current is naturally commutated and the device turns off when the next thyristor in the sequence is gated on. However, in a switch-mode converter, a commutation circuit such as is shown in Fig. 28-22 is needed to turn off the thyristor. Because of the cost, complexity, and losses associated with the commutation circuits and more importantly because of the evolution of the power-handling capabilities of BJTs, IGBTs, and GTOs, the switch-mode thyristor converter is not used in new designs, even at multimegawatt power ratings for dc and ac motor drives.

These commutation circuits circulate a current through a conducting thyristor in the reverse direction and thus force the total thyristor current to go to zero, thus turning it off. These circuits often consist of some form of an *LC* resonant circuit, similar to those

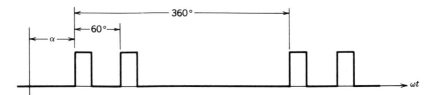

Figure 28-21   Gate trigger pulse waveform for a thyristor in a three-phase full-bridge converter.

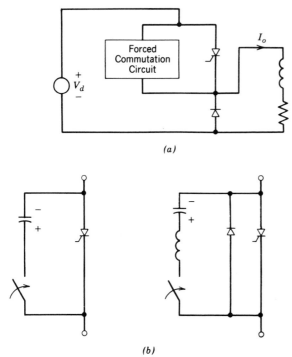

Figure 28-22   Turn-off of a thyristor by forced commutation: (*a*) thyristor switch-mode converter circuit illustrating the need for a commutating circuit; (*b*) idealized examples of commutation circuits. The plus and minus signs represent the polarity of the precharge voltage on the capacitors before the closure of the switches to turn off the thyristors.

discussed in Chapter 9. Such circuits are thoroughly discussed in the literature and so are not considered in this book.

28-6 POWER DEVICE PROTECTION IN DRIVE CIRCUITS

28-6-1 OVERCURRENT PROTECTION

In some applications the potential may exist for currents to flow through a power device that exceed the device's capabilities. If the device is not somehow protected against these overcurrents, it may be destroyed. Power devices cannot be protected against the over-currents by fuses because they cannot act fast enough. Overcurrents can be detected by measuring the device current and comparing it against a limit. At currents above this limit, the power device is turned off by a protection network in the drive circuit.

A cheaper and normally better way of providing overcurrent protection is to monitor the instantaneous output voltage of the device, for example, the collector-emitter on-state voltage of a BJT or the drain-source voltage of a MOSFET. Fig. 28-23a shows a simple

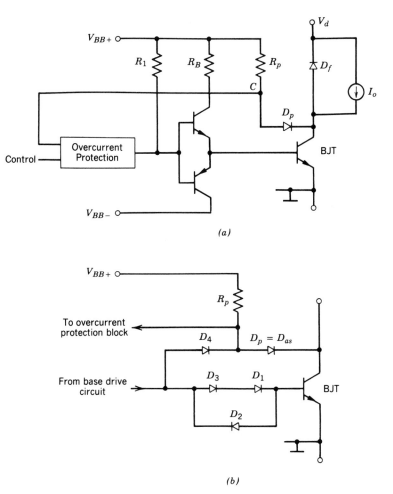

(a)

(b)

Figure 28-23 (a) Overcurrent protection by measuring the instantaneous on-state collector-emitter voltage of the power transistors. (b) This protection circuit can be used with an antisaturation network.

circuit to provide overcurrent protection to a BJT based on this principle. The voltage during the on-state at point C will be one forward-bias diode drop above $V_{CE,\text{sat}}$. This voltage signal is one of the inputs to the overcurrent protection block that requires the control signal as another input. When the transistor is supposed to be on, if the voltage at point C with some delay is above some predetermined threshold, the overcurrent is detected, and the protection block causes the base drive to turn the BJT off. Depending on the design philosophy, the overall system may be shut down after such an overcurrent detection and may have to be manually reset. The overcurrent detection network can be combined with the antisaturation network as is shown in the subcircuit of Fig. 28-23b.

The overcurrent protection should be combined with design measures that limit the maximum instantaneous current through the device. For example, consider the worst-case scenario shown in Fig. 28-24 where the output of a step-down converter circuit using a BJT is accidentally shorted. The instantaneous short-circuit current through the BJT can be estimated from the static $I-V$ characteristics shown in Fig. 28-24b where v_{CE} equals V_d under the short-circuit conditions. If the instantaneous short-circuit current is to be limited to a safe value, for example, twice the continuous current rating of the transistor, then the corresponding base current $I_{B,\text{max}}$ can be obtained from Fig. 28-24b. If the base current provided by the drive circuit remains less than $I_{B,\text{max}}$, the instantaneous short-circuit current would also be limited to the required safe value. The overcurrent protection circuit must act within a few microseconds to turn off the BJT, otherwise it will be destroyed. This design approach can be used with all power semiconductor devices that do not have a latching characteristic including BJTs, MOSFETs, IGBTs, FCTs, and power JFETs.

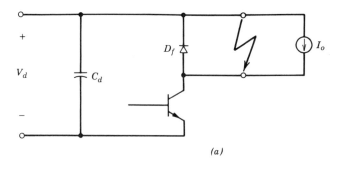

(a)

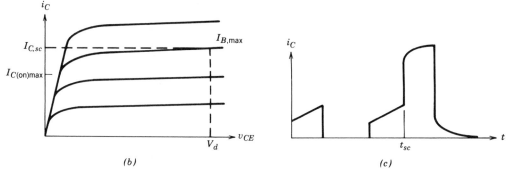

(b)   (c)

Figure 28-24   (a) Step-down converter circuit with an accidental short circuit. (b) The short-circuit current can be estimated from the transistor $I-V$ characteristics. (c) The current waveform at the onset of the short.

28-6-2 BLANKING TIMES FOR BRIDGE CIRCUITS

In the half-bridge and full-bridge circuits, where two transistors are connected in series in one converter leg, it is important to provide a blanking time so that the turn-on control input to one transistor is delayed with respect to the turn-off control input of the other transistor in the inverter leg. This blanking time should be chosen conservatively to be greater than the worst-case maximum storage time of the transistors being used to avoid cross conduction. Under normal operation, such a conservatively chosen blanking time will cause a dead time equal to the blanking time minus the actual delay time to occur in which both the transistors in the inverter leg are off. This dead time introduces an unwanted nonlinearity in the converter transfer characteristic as was discussed in Chapter 8. This dead time can be minimized by the use of design enhancements to drive circuits, which minimize turn-on and turn-off delay times in power semiconductor devices being used as the power switches. These design enhancements include the use of antisaturation diodes with BJTs, drive circuits with bipolar outputs, speed-up capacitors, and so forth.

This blanking time in the control inputs can be introduced by means of the circuit shown in Fig. 28-25a where the control signal is common to both BJTs of the converter leg. When the control signal is high, the upper transistor T_+ should be on and vice versa. The polarized RC network and the Schmitt trigger introduce a significant time delay in the turn-on of the BJT and almost no time delay in the turn-off of the transistor. The difference of these two time delays is the blanking time needed. The waveforms are shown in Fig. 28-25b where when the bridge control input goes low, a significant time delay occurs in the control signal to turn on the bottom transistor T_- and almost no time delay occurs in turning off the upper transistor T_+. The blanking time and the dead time are also shown in Fig. 28-25b.

As was mentioned previously, the simple base drive circuit of Fig. 28-1 cannot be used in a pulse-width-modulated bridge configuration where one of the free-wheeling diodes may conduct when the opposite transistor turns on. The waveforms for such a situation are shown in Fig. 28-26. Assuming I_o to be positive, the free-wheeling diode D_{F-} will conduct when T_+ is off, and then T_- is controlled to be on. Since T_{B2} is conducting during this interval, there will be a base current i_{B1} for T_-. This will make T_- conduct in its reverse direction, thus acting as a transistor in the reverse active region with the roles of collector and emitter being interchanged. In addition, due to the on-state voltage of D_{F-}, T_- will be provided with a base current i_{B2} supplied through R_2 and flowing through the base-collector diode of T_-. Assuming a D_{F-} forward voltage drop of 2 V and another 0.7 V across the forward-biased base-collector junction of T_-, then i_{B2} equals $1.3/R_2$.

During the blanking time, both T_{B1} and T_{B2} are off. However, this does not prevent the reverse conduction of T_- since i_{B2} will flow as long as D_{F-} is conducting. Instead of providing a negative base current, R_2 is a source of positive base current in this situation. The reverse current of T_- will now decay slowly to a steady-state value dictated by i_{B2}. Turning T_{B1} and thus T_+ on will cause a large forward current to flow through the barely saturated T_- in addition to the reverse-recovery current of D_{F-}. This current will normally destroy T_- and make it a short circuit, and in turn T_+ will be destroyed. Base drive circuits for bridge configurations must therefore be similar to that shown in Fig. 28-4 where the negative base current is supplied from a negative auxiliary power supply. A further improvement is to use an antisaturation diode. Then there will be no base current in the power transistor when its parallel free-wheeling diode is conducting, even if the transistor is controlled to be on.

It may seem odd to describe in such detail what should not be done. However, in some monolithic Darlingtons, the R_2 resistor of Fig. 28-1 is provided by overlapping the

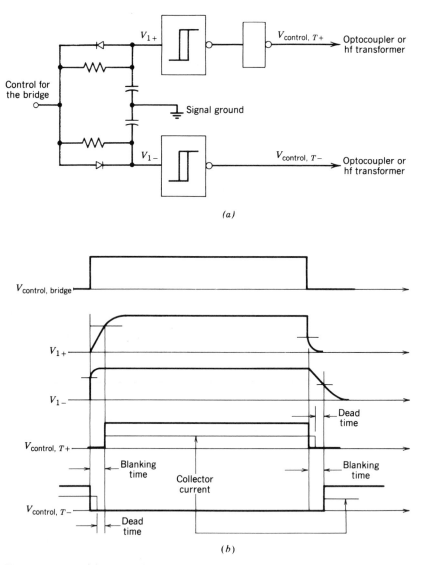

Figure 28-25 (a) Circuit for providing blanking times to the base drives of BJTs in a bridge configuration so as to avoid cross conduction of the BJTs. The dead times are the result of the BJT storage times, which are shown on (b) the collector current waveforms.

emitter metallization onto the base region. Such MDs will conduct reverse currents during the blanking time even with the base connected to V_{BB-}. Hence for the reasons explained in the preceding paragraphs, MDs constructed in this manner cannot be used in bridge configurations when one transistor is turned on while the opposite free-wheeling diode is conducting.

28-6-3 "SMART" DRIVE CIRCUITS FOR SNUBBERLESS SWITCHING

Since snubber circuits can add significant cost and complexity to a power electronics converter, the designer seeks ways to avoid their use. The availability of the newer power

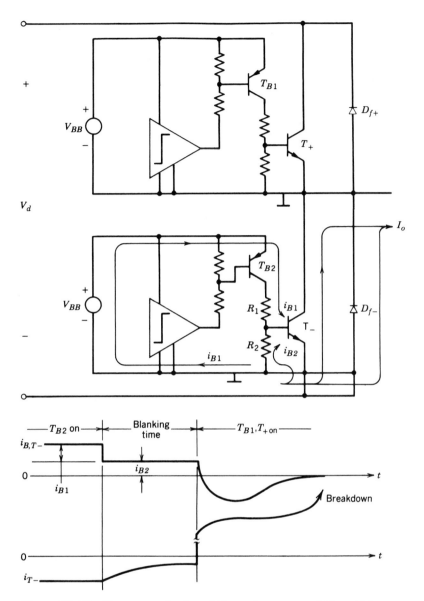

Figure 28-26   Reverse conduction of T_- at the turn-on of T_+, which causes destructive breakdown when the drive circuit of Fig. 28-1 is used.

semiconductor devices, such as IGBTs, which have robust and square safe operating areas, has prompted efforts to use these devices without snubbers. Snubberless operation is possible if two general requirements are met. First, the switching frequency must be relatively low (less than 10 kHz) so that switching losses are not a major concern. Second, the switching times of the power semiconductor device being used must be relatively easy to control. Devices that satisfy this second condition include MOSFETs, IGBTs, and to a lesser degree BJTs. Thyristors, GTOs, and related devices such as MCTs do not have switching times that are easily controllable.

The basic approach is to control the switching times so that the power device is not stressed beyond its specifications during switching transients. The control of switching times prevents dv/dt limits, di/dt limits, and overvoltage limits due to stray inductance from being exceeded. The controlled switching times will thus be significantly longer than the fastest speed at which the device is capable of being switched, which in turn means that each switching will dissipate a significant amount of energy in the device. In order to keep the power dissipation in the device due to these switching losses under control, the switching frequency must be kept relatively low, as was mentioned in the previous paragraph.

Several different drive circuit topologies can be envisioned that will perform satisfactorily as a so-called smart drive circuit. The basic requirements on any candidate drive circuit are that it have a bipolar output signal and that there be a separate current path for the turn-on and turn-off drive currents. The separate current paths needed for the independent control of the turn-off and turn-on transients dictate that the drive circuit topology use some type of totem-pole output circuit (either BJT based or MOSFET based) or CMOS inverter. Drive circuits such as those shown in Figs. 28-4 and 28-10 to 28-11 fulfill these requirements while those shown in Figs. 28-1, 28-2, 28-6, and 28-12 would be unsatisfactory for smart drive circuit applications. Detailed examples of smart drive circuits are given in the recent literature [11].

28-7 CIRCUIT LAYOUT CONSIDERATIONS

28-7-1 MINIMIZING STRAY INDUCTANCE IN DRIVE CIRCUITS

There are several practical considerations in the design and fabrication of drive circuits that are crucial to the successful operation of the circuits. The schematic shown in Fig. 28-27a serves as the focus of these considerations, and although it features a BJT the discussion based on this figure applies equally well to all power semiconductor devices. First, the length of the conductor that connects the base drive circuit to the emitter of the power BJT should be as small as possible to minimize the stray inductance illustrated in Fig. 28-27b. Otherwise the turn-off will be slowed down and possibly unwanted oscillation may occur. Consider a positive base current i_B that turns the BJT on, which in turn causes the collector current i_C to increase rapidly. The stray inductance illustrated in Fig. 28-27b will induce a voltage that will tend to reduce the base current. If this then causes a reduction in the collector current, there will be a subsequent negative di_C/dt and a voltage induced that will cause an increase in i_B. This then represents the start of unwanted oscillations.

In minimizing the stray inductance, all power devices including BJTs, MOSFETs, thyristors, GTOs, IGBTs, and so forth, should be treated as four terminal devices having two control terminals and two power terminals (as is illustrated for the BJT in Fig. 28-27c). To facilitate the reduction of this stray inductance, in high-power transistor modules, manufacturers provide a separate emitter terminal for the connection of the drive circuit as is shown in Fig. 28-27c. Such separate additional terminals are also found on thyristors, GTOs, and IGBTs.

28-7-2 SHIELDING AND PARTITIONING OF DRIVE CIRCUITS

Stray inductance must also be minimized in the high current power loop to which the output terminals of the power device are connected. An example of such a loop is the step-down converter circuit loop of Fig. 21-10 consisting of the power transistor, the free-wheeling diode, and the dc link capacitor C. At turn-off of the BJT, large overvoltages may develop across it due to large values of di_C/dt if the stray inductance is not

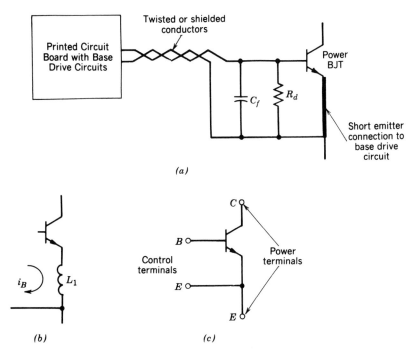

Figure 28-27 (*a*) Circuit layout and the interconnection considerations in
connecting base drive circuits to power BJTs in order to minimize stray
inductance (*b*) and other potential problems. Some BJTs have an extra
emitter connection as shown in (*c*) to help minimize such potential problems.

minimized. Even with careful layout, it may be necessary to further reduce the overvoltages at turn-off by means of snubber circuits or control of the turn-off times. It should be
kept in mind that one centimeter of unshielded lead has about 5 nH of series inductance.
Thus the lengths of all unshielded leads should be kept to an absolute minimum.

In many designs, the basic drive circuit may be on a printed circuit board at some
distance away from the power transistor, which is mounted on a heat sink. A twisted pair
of wires or even a shielded cable where the shield is connected to the emitter terminal
should be used to minimize the stray inductance and the inductive pick up of noise in the
base drive circuit. A small filter capacitor C_f and damping resistor R_D can be added across
the base and emitter terminals as shown in Fig. 28-27*a* to avoid oscillations and the
problem of retriggering at the turn-off of the BJT.

The output stage of base drive circuits such as shown in Fig. 28-4 should be put close
together in a corner of the base drive printed circuit board close to the terminals connecting it to the power transistor. This includes the components R_B, T_{B+}, T_{B-}, C_{BB+}, and
C_{BB-}. This will minimize the noise generated in the base drive circuit as well as minimizing the leakage inductances in both the positive and negative base current loops so that
the transistor can switch as rapidly as possible.

If more than one base drive circuit is put on the same printed circuit board, they must
be put on separately dedicated areas of the board with a minimum distance of at least 1
cm between the areas. This is especially important on double-sided or multiple-layer
circuit boards. There must never be an intermixing of the printed wires of the different
isolated base drive circuits on any area of the card.

In the rack of printed circuits of a converter, the logic and control electronic circuits
should be put on one side and the base drive circuit on the other side. A dummy aluminum

"card" used as a shield may be necessary on each side of the base drive circuit card in the rack for further noise reduction.

28-7-3 REDUCTION OF STRAY INDUCTANCE IN BUS BARS

Leakage or stray inductance in power supply leads (bus bars) can be a problem in circuits that experience large values of *di/dt*. Fortunately, means exist to reduce stray inductance in bus bars, and they have the added benefit of reducing stray magnetic flux and hence EMI.

Copper strips, with a thin insulator sandwiched between them as shown in Fig. 28-28, comprises a transmission line and provides an excellent means of reducing the stray

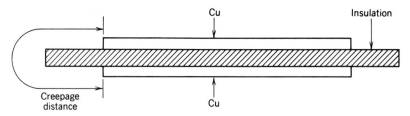

Figure 28-28 Transmission-line-like structure composed of copper strips for a low-inductance electrical interconnection.

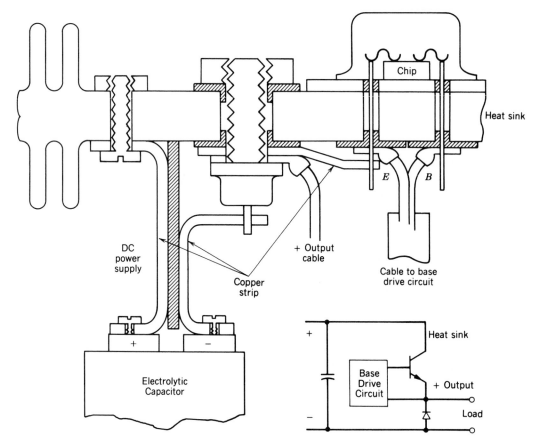

Figure 28-29 Use of low-inductance copper strips in the interconnection of an electrolytic capacitor, power transistor, and a free-wheeling diode.

inductance. The creepage distance can be increased as is shown in the figure. To use such strips may be clumbersome in practice, but such a design establishes the upper limit in reducing stray inductance. The actual design then can be adapted to meet the manufacturing and cost constraints. As an example, a step-down dc-dc converter layout is presented in Fig. 28-29 where a transistor in a TO-3 case and a stud-mounted free-wheeling diode are used.

These copper strips can also be used to make contact with high-power semiconductor modules as shown in Fig. 28-30. For a three-phase inverter, a parallel combination of electrolytic capacitors can be connected in a similar manner. [13]

28-7-4 CURRENT MEASUREMENTS

In many applications, electrical isolation is not needed between the current measurement circuit and the control electronics circuit. In such cases, the current can be measured by means of a shunt resistor and an op-amp (see Fig. 28-31). This circuit avoids common-mode voltage jumps in the measured output. Moreover, the slew rate of the op-amp provides excellent filtering of noise since the current to be measured often has a step waveform. The power loss in the resistor can be minimized by amplifying the voltage by the op-amp. Normally, the voltage across the resistor can be as low as 50–500 mV without unacceptable noise problems.

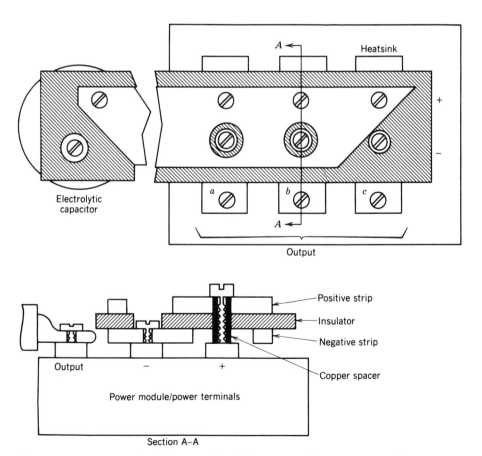

Figure 28-30 Example of the use of low-inductance to interconnect transistor modules of a three-phase inverter and an electrolytic capacitor. Several electrolytic capacitors may be connected in parallel.

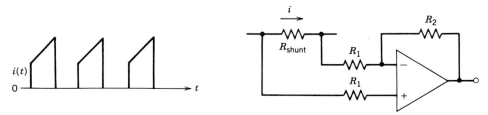

Figure 28-31 Current measurement with a resistive shunt and an op-amp.

Figure 28-32 shows two alternatives for current measurement in conjunction with control and driver ICs. In Fig. 28-32a, the voltage across R_{shunt} must be fairly large, and this may cause noise problems. Furthermore the inductance in the source-to-ground wire can result in oscillations in the gate signal due to the di/dt of the current through the MOSFET. In the circuit of Fig. 28-32b, these problems are reduced.

Often in high-power applications, electrical isolation is required between the current to be measured and the control electronics. For ac currents, current transformers can be used. If there is a dc bias in the current, the current transformer will not work. Often the instantaneous current must be measured that includes the dc bias. In such cases, the current can be measured by a Hall-effect current sensor such as that shown in Fig. 28-33. In this current sensor, a compensating current is applied to the secondary winding such that the field in the toroid is kept at zero. In this way, current from dc up to 100 kHz bandwidth can be measured.

28-7-5 CAPACITOR SELECTION

Capacitors must be selected based on the required capacitance, operating voltage, rms current, and frequency. In power electronics, there are basically three types of capacitors used: electrolytic, metallized polypropylene, and ceramic.

28-7-5-1 Aluminum Electrolytic Capacitors

Electrolytic capacitors offer a large capacitance per unit volume and are polarized. The large capacitance is due to the fact that the aluminum foil connected to the positive

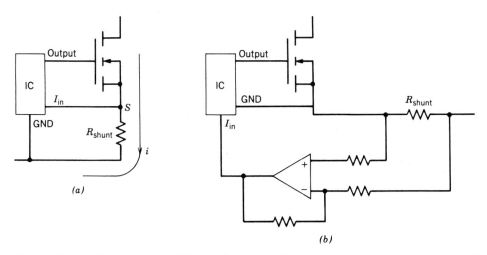

Figure 28-32 Use of a control IC in conjunction with (a) a current-measuring resistor and (b) an improved current-measuring method.

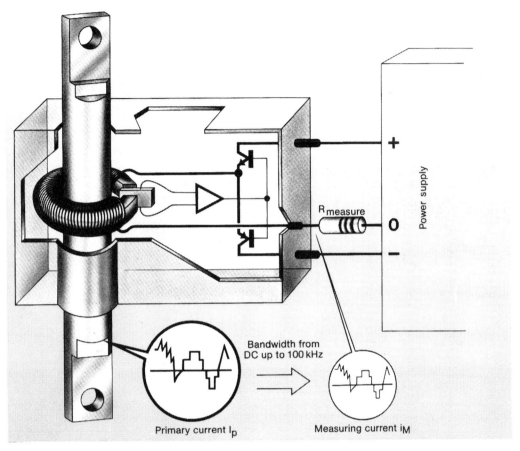

Figure 28-33 Hall-effect compensated current sensor with electrical isolation. (Courtesy of LEM, Geneva, Switzerland.)

terminal is etched so that its surface is porous like a sponge. This results in an increase in its surface area by as much as a factor of 100 compared to its original unetched area. On this etched foil, an insulating layer of aluminum oxide is formed electrochemically. The negative terminal of the capacitor is connected to another aluminum foil that is in electrical contact with the liquid electrolyte that is an electrically conducting material.

Because of the resistance of the electrolyte, these capacitors have a significant equivalent series resistance (ESR). These capacitors should not be used at temperatures below the specified minimum temperature since the tendency of the electrolyte to crystallize results in a larger resistance. As discussed in previous chapters, ESR in the output filter capacitors must be low to minimize the ripple in the output voltage.

The capacitor package or can is sealed at the top with an insulating layer that surrounds the electrical terminals. The rate of evaporation of the electrolyte through the seal increases significantly with temperature. Therefore the capacitor lifetime decreases significantly with temperature. It should be noted that the electrolytic capacitors have by far the shortest lifetime of any element, active or passive, used in power electronic converters. The temperature within the capacitor depends on the power loss. This loss increases with the rms current because of ohmic losses in the capacitor. For a given current, the ripple voltage across the dielectric decreases with increasing frequency. Therefore the dielectric power loss decreases with increasing frequency. For a given lifetime of the capacitor, its current-carrying capacity increases with increasing frequency and decreasing ambient temperature. Information about these factors are specified in the data sheets.

28-7-5-2 METALLIZED POLYPROPYLENE CAPACITORS AND CERAMIC CAPACITORS

In snubbers and thyristor commutation circuits, capacitors must handle large currents, but the capacitance value required is small. Metallized polypropylene capacitors are a good choice for such applications due to a very small loss coefficient of the polypropylene dielectric material. The dielectric losses are proportional to the square of the voltage and frequency. Since the voltage across the dielectric is proportional to the current and inversely proportional to the frequency, the dielectric power loss is proportional to the square of the current and inversely proportional to the frequency. Therefore, for a specified temperature, the current-handling capability increases slightly with frequency.

Ceramic capacitors have extremely low series inductance. They are used as filters, for example, on printed circuit boards to reduce ripple in the supply voltage.

SUMMARY

This chapter has discussed the general considerations that influence the design of drive circuits for power semiconductor devices. Several different topologies for drive circuits have been presented as well as ways of including power device protective measures into the design of the drive circuit. The important points follow.

1. The choice of drive circuit topology is influenced by several factors, including whether the output signal must be unipolar or bipolar, whether or not electrical isolation is required, whether the drive circuit output is in series or shunt with the power device, and whether or not additional functionality beyond simple on-off control is to be incorporated in the drive circuit.

2. DC-coupled drive circuits with unipolar outputs are the simplest drive circuits but are only suitable for driving grounded power devices at relatively low switching frequencies.

3. DC-coupled drives with bipolar outputs require split dc power supplies and are capable of driving power devices on and off rapidly. They are suitable for high switching frequencies but are restricted to driving grounded power switches.

4. Electrical isolation of drive circuits can be achieved by means of transformers, optocouplers, or fiber optic cables.

5. Optocoupler and fiber optic cable isolation require isolated dc bias supplies for the floating portion of the drive circuit.

6. Transformer-isolated drive circuits include topologies that require isolated dc supplies and some that do not.

7. Isolated dc supplies, both single ended or split outputs, use transformer coupling, and the ac input to the primary can be at power line frequencies, but high frequencies are preferred for minimizing the size of the isolation transformer.

8. Most drive circuit topologies are connected in shunt with the power device, but series-connected (cascode) drive circuits are useful in some circumstances such as driving normally on devices.

9. Provisions for overcurrent protection, blanking times in bridge circuits, and snubberless operation of the power switch can be incorporated in the design of the drive circuit.

10. Careful attention to the layout of the drive circuit components on the printed circuit board in order to minimize stray inductance is important to satisfactory operation of the drive circuit.

PROBLEMS

28-1 A MOSFET is to be used in a step-down converter circuit. The power supply voltage V_d for the circuit is 100 V, and the load current $I_o = 100$ A. Totem-pole drive circuit of Fig. 28-6b is used to drive the MOSFET. The BJTs can be considered as ideal switches. Complete the design of the drive circuit by specifying the values or a range of values for V_{GG+}, V_{GG-}, and R_G. The rate of rise and fall of v_{DS}, dv_{DS}/dt must be limited to 500 V/µs or smaller. The MOSFET parameters are

$$C_{gs} = 1000 \text{ pF} \quad C_{gd} = 400 \text{ pF} \quad I_{DM} = 200 \text{ A} \quad V_{GS(max)} = \pm 20 \text{ V}$$
$$BV_{DSS} = 200 \text{ V} \quad V_{GS(th)} = 4 \text{ V} \quad I_D = 60 \text{ A} \quad \text{at} \quad V_{GS} = 7 \text{ V}$$

28-2 The step-down converter of Fig. 28-16 employs an FCT with a blocking gain, µ, of 40. The load current $I_o = 200$ A and the dc input voltage $V_d = 1000$ V.

(a) What should be the values of R_{G1} and R_{G2} in order to ensure proper operation of the FCT? Assume $R_{G1} + R_{G2} = 1$ MΩ and include a 25% factor of safety in the blocking voltage capability of the circuit.

(b) Describe the characteristics the MOSFET in this circuit should have, including breakdown voltage and maximum average current capability.

REFERENCES

1. M. H. Rashid, *Power Electronics: Circuits, Devices, and Applications*, Prentice-Hall, Englewood Cliffs, NJ, 1988.
2. *Power Transistor in Its Environment*, Thompson-CSF, Semiconductor Division, 1978.
3. B. W. Williams, *Power Electronics, Devices, Drivers, and Applications*, Wiley, New York, 1987.
4. B. Jayant Baliga and Dan Y. Chen (Eds.), *Power Transistors, Device Design and Applications*, IEEE Press, Institute of Electrical and Electronic Engineers, New York, 1984.
5. *SCR Manual*, 6th ed., General Electric Company, Syracuse, NY, 1979.
6. B. M. Bird and K. G. King, *An Introduction to Power Electronics*, Wiley, New York, 1983.
7. P. Aloisi, *Power Switch*, Motorola Inc., 1986
8. E. S. Oxner, *Power FETs and Their Applications*, Prentice-Hall, Englewood Cliffs, NJ, 1982.
9. T. Rogne, N. A. Ringheim, J. Eskedal, B. Odegard, H. Seljeseth, and T. M. Undeland, "Short Circuit Capability of IGBT (COMFET) Transistors," 1988 IEEE Industrial Applications Society Meeting, Pittsburgh, PA, Oct. 1988.
10. J. G. Kassakian, M. F. G. Schlecht, Vergassian, *Principles of Power Electronics*, Addison-Wesley, Boston, 1991.
11. F. Blaabjerg and J. K. Pedersen, "An Optimum Drive and Clamp Circuit Design with Controlled Switching for a Snubberless PWM-VSI-IGBT Inverterleg," IEEE Power Electronics Specialists Conference, Madrid, Spain, June 29–July 3, 1992.
12. Harald Vetter, "GTO Snubber Capacitors for Low Inductance Current Loops," *Siemens Components*, Vol. 3, 1990, pp. 81–85.
13. Eric Motto, "Power Circuit Design for Third Generation IGBT Modules," *PCIM*, Jan. 1994, pp. 8–18.

CHAPTER 29

COMPONENT TEMPERATURE CONTROL AND HEAT SINKS

This chapter discusses the need to control the internal temperature of power electronic components and the factors to be considered in selecting passive components including resistors, capacitors, and heat sinks. Since excessive internal temperatures are deterimental to all power electronic components, especially power semiconductor devices, the bulk of the chapter will review heat transfer mechanisms including conduction, radiation, and convection. A fundamental understanding of heat transfer is needed not only for the design and specification of heat sinks but it is also needed in the design of inductors and transformers where thermal considerations are a major part of the design.

29-1 CONTROL OF SEMICONDUCTOR DEVICE TEMPERATURES

The theoretical upper limit on the internal temperature of a semiconductor device is the so-called intrinsic temperature, T_i, which is the temperature at which the intrinsic carrier density in the most lightly doped region of the semiconductor device equals the majority carrier doping density in that region. For example, in a lightly doped drift region of a silicon diode where the donor density is 10^{14} cm^{-3}, the intrinsic temperature is about 280°C. If this temperature is exceeded, the rectifying characteristics of the junction are lost because the intrinsic carrier density greatly exceeds the doping density, and the depletion region that gives rise to the potential barrier is shorted out by the intrinsic carriers.

However, the maximum internal temperatures specified on data sheets are much less than this limit. The power dissipation in power semiconductors normally increases with the internal temperature, and the losses become excessively high even at temperatures of 200°C. Device manufacturers typically will guarantee the maximum values of device parameters such as on-state conduction voltages, switching times, and switching losses at a specified maximum temperature, which varies from one type of device to another and is often at 125°C.

In a design process, one of the design inputs is the worst-case junction temperature. A system intended to have high reliability would be designed for a worst-case junction temperature in the semiconductor devices of 20–40°C below 125°C. Otherwise a value of 125°C is commonly used in the worst-case design input. An exception to this is the maximum junction temperature of thyristors, which should be kept below 125°C. Thyristors may retrigger or fail if their junction temperatures rise above 125°C at the same time as the maximum allowable dv/dt is applied to the device.

Some power semiconductor devices and signal level transistors and ICs can operate at temperatures even slightly above 200°C. However their reliability (expected operating lifetime) is low, and the performance characteristics may be poor compared to operation at 125°C, for instance. Moreover the manufacturer will not guarantee the parameters above the maximum temperature specified on the data sheet. If the designer or manufacturer of a power electronic converter decides to operate semiconductor devices above the data sheet maximum temperature, then the designer/manufacturer must screen (measure the high-temperature characteristics) a large number of devices for the application. Unless every component in all converters made to that particular design are screened, one cannot be sure that some components may not have such poor characteristics at the elevated temperatures that they will cause the converter to fail. Such comprehensive screening is time consuming and expensive.

Some special applications will require operating in extremely high ambient temperatures. In such cases, comprehensive screening is the only option. All such equipment should be tested for a day to a week at the factory in so-called burn-in tests conducted at full operating power and maximum ambient temperatures.

In designing power electronic equipment, especially for high ambient temperatures, the thermal layout must be considered at an early stage. The heat sink size and weight, its location in the equipment cabinet, and surrounding temperature should be considered at the beginning of the design process. It is important to be able to mount the heat sinks with their fins in a vertical position with ample room for natural convection of the air without a fan. The possibility of heating by the sun must be considered as part of a worst-case set of design inputs.

A bad thermal design will make the equipment much less reliable than intended. A rule-of-thumb to keep in mind is that the failure rate for semiconductor devices doubles for each 10–15°C temperature rise above 50°C.

The choice of the correct (most economical or cheapest in production) heat sink is only a part of the thermal design process for a power electronic system. At an early stage of the design, the designer should be free to consider a large or small heat sink that may be cooled by natural convection, by a fan (ac motor fans controlled by a small power electronic inverter are much more reliable than a dc motor fan), or even by the use of liquid cooling.

29-2 HEAT TRANSFER BY CONDUCTION

29-2-1 THERMAL RESISTANCE

When a section of material such as is shown in Fig. 29-1 has a temperature difference across it, there is a net flow of energy from the higher temperature end to the lower temperature end. The energy flow per unit time, that is, power, is given by

$$P_{\text{cond}} = \frac{\lambda A \, \Delta T}{d} \tag{29-1}$$

where $\Delta T = T_2 - T_1$ in °C, A is cross-sectional area in m^2, d is the length in m, and λ is the thermal conductivity in W-m^{-1} °C^{-1}. For 90% pure aluminum, which is typically used for heat sinks, the thermal conductivity is 220 W-m^{-1} °C^{-1}. Values of λ for other materials can be found in the literature.

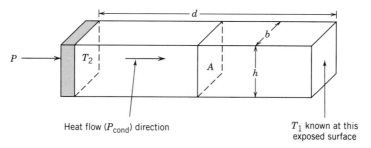

Figure 29-1   An isolated rectangular rod that conducts P watts of heat energy per unit time.

■ *Example 29-1*   Consider an aluminum rod such as shown in Fig. 29-1 with $h = b = 1$ cm and $d = 20$ cm. The rate of heat energy entering at the left end (where the temperature is T_2) is 3 W and the temperature at the right surface, $T_1 = 40°C$. Find T_2.

$$T_2 = \frac{P_{\text{cond}}\, d}{\lambda hb} + T_1 = \frac{(3)(0.2)}{(220)(0.01)(0.01)} + 40 = 67.3°C.$$ ■

■ *Example 29-2*   A transistor module is mounted on an aluminum plate having dimensions $h = 3$ cm, $b = 4$ cm, and $d = 2$ mm (refer to Fig. 29-1). A temperature drop of 3°C is allowed from one 3×4 cm^2 surface to the other. Find the maximum power that can be generated in the module. Ignore any heat losses to the surrounding air.

$$P = \frac{\lambda A(T_2 - T_1)}{d} = \frac{(220)\,(0.03)\,(0.04)\,(3)}{(.002)} = 396\ \text{W}$$ ■

The thermal resistance $R_{\theta,\text{cond}}$ is defined as

$$R_{\theta,\text{cond}} = \frac{\Delta T}{P_{\text{cond}}} \tag{29-2}$$

Referring to Fig. 29-1, we note that $\Delta T = T_2 - T_1$. Using this in Eq. 29-2 gives

$$R_{\theta,\text{cond}} = \frac{d}{\lambda A} \tag{29-3}$$

The thermal resistance has units of degrees centigrade per watt.

Often the heat must flow through several different materials, each having different thermal conductivity and perhaps different areas and thickness. A multilayer example, which models the heat conduction path from a region in the silicon device to the ambient, is shown in Fig. 29-2. The total thermal resistance from the junction to the ambient (ja) is given by

$$R_{\theta ja} = R_{\theta jc} + R_{\theta cs} + R_{\theta sa} \tag{29-4}$$

Each contribution to the total thermal resistance is computed using Eq. 29-4 with the proper values of λ, A, and d. The resulting junction temperature, assuming a power dissipation of P_d, is

$$T_j = P_d(R_{\theta jc} + R_{\theta cs} + R_{\theta sa}) + T_a \tag{29-5}$$

in analogy with electric circuits. If there are parallel paths for heat flow, then the thermal resistances are combined in exactly the same manner as electrical resistors in parallel.

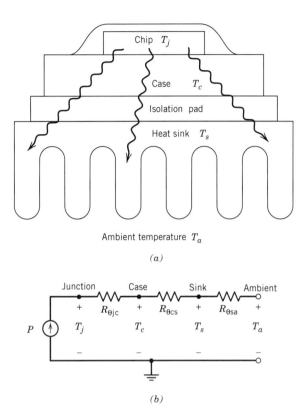

Figure 29-2 Steady-state heat flow and thermal resistance in a multiple layer structure including a (a) heat sink and (b) an equivalent circuit based on thermal resistances.

Manufacturers of power devices put great emphasis on keeping the thermal resistance as economically low as possible. This means keeping the length d, of all heat flow paths as short as possible, consistent with the requirements of breakdown voltage, mechanical ruggedness, and other requirements. It also means that the cross-sectional area A should be as large as possible consistent with other design requirements such as minimizing parasitic capacitance. In some high-power devices, such as thyristors, a hockey-puck-style package is used to maximize the heat conduction by using both surfaces of the silicon wafer for heat transfer rather than just one as is commonly used in low-power devices and standard power modules.

The package should be made of material with a high thermal conductivity. In high-power devices, the package may be mounted on a heat sink that is either air-cooled or even water-cooled. With these types of efforts, it is possible to achieve junction-to-case thermal resistances, $R_{\theta jc}$, of less than one degree centigrade per watt.

29-2-2 TRANSIENT THERMAL IMPEDANCE

In some situations, the user of power devices must be concerned with the transient thermal response of the device being used. For example, during transient overloads or at power-up or power-down of a system containing power devices, the instantaneous dissipation in the devices may greatly exceed the average power rating of the device. Whether or not these power surges cause the junction temperature to exceed the maximum permissible value depends on the magnitude and duration of the surge and the thermal properties of the device. If the duration of the power surge is short, we have the intuitive expectation that the temperature excursion should not be excessive.

In the transient case, the heat capacity, C_s, of the sample must be considered along with the thermal resistance. Heat capacity per unit volume of a material is defined as the rate of change of the heat energy density Q with respect to the material temperature T. Thus,

$$dQ/dT = C_v \tag{29-6}$$

where C_v is the heat capacity per unit volume and has dimensions of joules per unit volume per degree Kelvin. For a rectangular block of material of cross-sectional area A and thickness (in the direction of heat flow) d, the heat capacity of the block, C_s, is given by

$$C_s = C_v A d \tag{29-7}$$

The transient behavior of the junction temperature is governed by the time-dependent heat diffusion equation. A detailed solution of this equation is beyond the scope of this book. An approximate solution can be obtained by the use of the electric circuit analog shown in Fig. 29-3, which is suggested by the steady-state analog shown in Fig. 29-2. If the power input, $P(t)$, is a step function, the rise in temperature, $T_j(t)$, for short times is given by

$$T_j(t) = P_0 \left[4t/(\pi R_\theta C_s)\right]^{1/2} + T_a \tag{29-8}$$

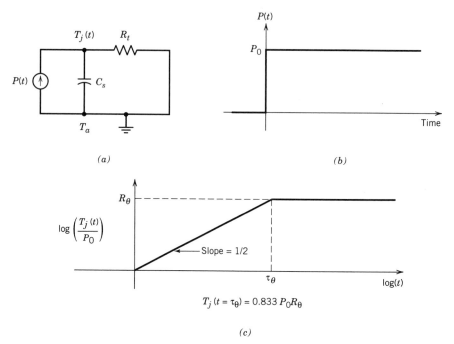

Figure 29-3 Equivalent circuit of (a) the transient thermal impedance (b) a step input power input, and (c) the transient response of the junction temperature illustrated as the transient thermal impedance.

where P_0 is the magnitude of the step and it is assumed that t is less than the thermal time constant

$$\tau_\theta = \pi R_\theta C_s / 4 \qquad (29\text{-}9)$$

For times large compared with τ_θ, T_j approaches the steady-state value $P_0 R_\theta + T_a$. The exact solution for $T_j(t)$ and the asymptotic solution are compared in Fig. 29-3c. The vertical axis of the plot, $T_j(t)/P_0$, is the transient thermal impedance, $Z_\theta(t)$.

Even though the terminology, thermal time constant, is used in this discussion and an electrical analogy of a resistor-capacitor circuit is used to characterize transient thermal effects, the reader should realize that transient heat transfer behaviors are not characterized by simple exponential time functions. The solution to the time-dependent heat diffusion equation involves error functions and complementary error functions. A power series expansion of an error function for times small compared to τ_θ yields the $t^{1/2}$ dependence shown in Eq. 29-8 rather than a linear time dependence that a simple exponential time dependence would give.

In real devices where the heat must flow through several different layers (Fig. 29-4a), the equivalent circuit is more complicated as shown in Fig. 29-4b. If the thermal time constants of each layer are widely different in value, then the total transient thermal impedance will be the sum of the individual contributions of each layer as is shown in the plot of the total $Z_\theta(t)$ versus time shown in Fig. 29-4c. Curves such as these are often given on the specification sheets of power devices.

Plots of $Z_\theta(t)$ can be used to estimate $T_j(t)$ if the power input $P(t)$ is known as a function of time since

$$T_j(t) = P(t) Z_\theta(t) + T_a \qquad (29\text{-}10)$$

For example, if the power being dissipated is a rectangular pulse starting at $t = 0$ and ending at $t = t_2$, then $T_j(t)$ is formally given by

$$T_j(t) = P_0 [Z_\theta(t) - Z_\theta(t - t_2)] + T_a \qquad (29\text{-}11)$$

Curves of $Z_\theta(t)$ such as those shown in Fig. 29-4c are used to graphically evaluate the thermal impedance and $T_j(t)$ in Eq. 29-11. If $P(t)$ is not a rectangular pulse, the approach just outlined can still be used if an equivalent rectangular pulse can be fitted to the actual $P(t)$. Consider the half-sine power pulse shown in Fig. 29-5. The rectangular power pulse in the figure has the same peak power and the same energy content as can be verified by direct calculation. The approximate junction temperature response to this rectangular power pulse is

$$T_j(t) = P_0 [Z_\theta(t - T/8) - Z_\theta(t - 3T/8)] + T_a \qquad (29\text{-}12)$$

It is apparent from this discussion that the way to increase the transient power capability of a device is to increase its thermal time constant $R_\theta C_s$. Unfortunately, this approach is not viable because the time constant is, using Eqs. 29-3 and 29-7, given by

$$\frac{\pi}{4} R_\theta C_s = C_v \lambda^{-1} d^2 \frac{\pi}{4} \qquad (29\text{-}13)$$

Unfortunately, materials normally used for packaging a device all have about the same value of heat capacity C_v. Moreover the thermal conductivity should be large, not small so that the thermal resistance is low. Finally, the length d of the heat flow path should be small in order to minimize R_θ. The trade-off between small values of R_θ and larger values of thermal time constant will always be made in favor of small values of R_θ. This is because the device is likely to be operated in a steady-state mode much more often than in a transient overload mode.

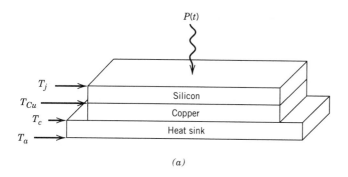

(a)

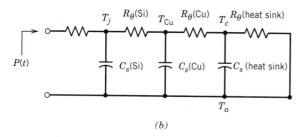

(b)

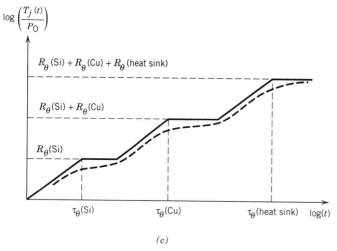

(c)

Figure 29-4 (a) Multiple-layer thermal structure (b) thermal equivalent circuit and (c) transient thermal impedance assuming widely separated thermal time constants.

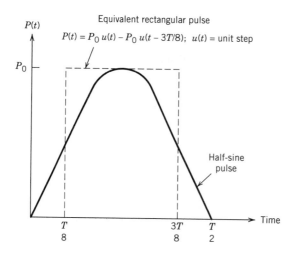

Figure 29-5 Example of modeling a power pulse transient by an equivalent (same total energy) rectangular pulse.

In spite of our inability to significantly modify the transient overload rating to any substantial degree, most power devices have overload capabilities that greatly exceed their average power ratings, with an order of magnitude or greater being common. In fact it is not so much the instantaneous power rating that is important as it is the total amount of energy deposited into the device during the transient. The overload ratings contain not only a specification on the instantaneous power rating but also its duration. Indeed there are often several different overload ratings given for different time durations. The user is well advised to carefully read the appropriate specifications and understand them thoroughly before using a power device.

29-3 HEAT SINKS

Keeping the junction temperature of a power device within reasonable bounds is the joint responsibility of the device manufacturer and the device user. The manufacturer minimizes the thermal resistance $R_{\theta jc}$ between the interior of the device where the power is dissipated and the outside of the case enclosing the device. The device user must provide a heat conduction path between the case of the device and the ambient so that thermal resistance $R_{\theta ca}$ between the case and the ambient (where the heat generated by device operation will ultimately be dissipated) is minimized in a cost-effective manner.

The user's responsibility is made easier by the wide availability of extruded aluminum heat sinks of various shapes that are used for cooling of the power semiconductor devices. If the heat sinks are cooled by natural convection, the distance between each fin, such as is shown in Fig. 29-2a, should be at least 10–15 mm. A coating of black oxide results in a reduction of the thermal resistance by 25%, but the cost may be higher by almost the same factor [1]. Thermal time constants of natural convection-cooled heat sinks are in the range of 4–15 min. If a fan is added, the thermal resistance, R_{θ}, goes down, and the heat sink can be made smaller and lighter, which also reduces the heat capacity C_s. The thermal time constants for force-cooled heat sinks are much smaller than for natural convection-cooled heat sinks. Typical values of τ_{θ} for forced-cooled heat sinks may be less than 1 min. Heat sinks that utilize forced cooling should have spacings between the cooling fins of not more than a few millimeters. In higher power ratings, water or oil cooling is used to further improve the thermal conduction.

The choice of the proper heat sink depends on the allowable junction temperature the device can tolerate. For a worst-case design, the maximum junction temperature $T_{j,\max}$, the maximum ambient temperature $T_{a,\max}$, the maximum operating voltage, and maximum on-state current are specified. The maximum on-state losses in the power device can be calculated if the maximum duty ratio, maximum on-state current, and maximum on-state resistance (obtainable from the data sheets corresponding to $T_{j,\max}$ and the maximum current) are known. The switching losses can be obtained by integrating the instantaneous power loss with respect to time and averaging it over the switching time period. Therefore, P_{Loss}, which is the sum of the on-state losses and the average switching losses can be estimated.

From this information the maximum allowable junction-to-ambient thermal resistance $R_{\theta ja}$ (Eq. 29-4) can be estimated as

$$R_{\theta ja} = (T_{j,\max} - T_{a,\max})/P_{\text{Loss}} \tag{29-14}$$

The junction-to-case thermal resistance $R_{\theta jc}$ can be obtained from the semiconductor device data sheets, and the case-to-sink thermal resistance $R_{\theta cs}$ depends on the thermal compound and the insulator (if any) used. Thermal resistance of insulators are found in handbooks such as in reference 3 and in data sheets from suppliers of such devices. As an example, a 75-μm thick mica insulator used for a TO-3 transistor package has a R_θ value of about 1.3 °C/W when used dry and a value of 0.4 °C/W when used with a thermal grease or heat sink compound. It is important to understand that the thermal grease is only to be used to remove the air from the between the microscopic high points of the mating surfaces (mica-heat sink and mica-transistor) and thus effectively utilize the entire surface area for heat conduction. If too much thermal compound is used, the layer will be excessively thick and will increase the thermal resistance [4, p.701]. Knowing $R_{\theta cs}$ and $R_{\theta jc}$, the thermal resistance of the heat sink-to-ambient thermal resistance $R_{\theta sa}$ can be calculated from Eqs. 29-4 and 29-14. A proper heat sink can then be selected based on the information provided by the heat sink manufacturer's data sheets such as shown in Fig. 29-6.

When using any of these heat sinks, it is imperative that the manufacturer's instructions be followed closely. Improper mounting of the power device on the heat sink could result in $R_{\theta ca}$ being much larger than anticipated and thus intolerably high values of junction temperature of the device during normal operation. For example, a small amount of thermal grease should be used to increase the contact area between the device and the heat sink. Application of the proper torque to the mounting bolts and nuts will also help ensure good contact between the device and the heat sink.

■ *Example 29-3*   For a junction temperature of 125°C, a TO-3 transistor has a power dissipation of 26 W. The transistor manufacturer specifies a value of 0.9°C/W for $R_{\theta jc}$. A 75-μm thick mica insulator is used with thermal grease, and the thermal resistance

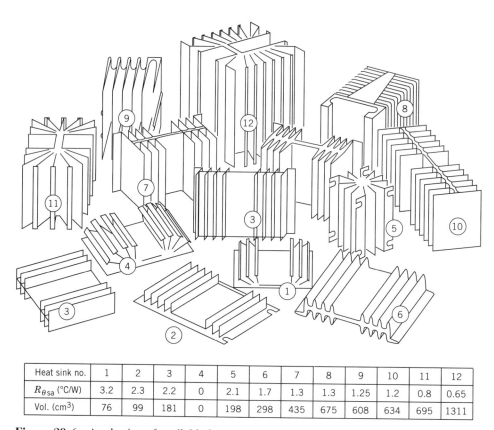

Heat sink no.	1	2	3	4	5	6	7	8	9	10	11	12
$R_{\theta sa}$ (°C/W)	3.2	2.3	2.2	0	2.1	1.7	1.3	1.3	1.25	1.2	0.8	0.65
Vol. (cm³)	76	99	181	0	198	298	435	675	608	634	695	1311

Figure 29-6   A selection of available heat sinks.

of the combination is 0.4 °C/W. The worst-case ambient temperature in the cabinet where the heat sink is to be used is 55°C. Hence the sink-to-ambient thermal resistance should be

$$R_{\theta sa} = \frac{125 - 55}{26} - (0.9 + 0.4) = 1.39 \ °C/W.$$

Heat sink number 7 in Fig. 29-6 has a thermal resistance of 1.3 °C/W, which is acceptable for this application. In fact using this heat sink will lower the junction temperature to 122.6°C, which is slightly cooler than assumed. The power dissipation in the transistor will then be somewhat smaller as a result and thus will lower the real junction temperature somewhat as well, perhaps to less than 120°C. If the converter that uses this heat sink is to be mass produced, it might make economic sense to look for a heat sink with $R_{\theta sa} = 1.39$ °C/W since it will be lighter and smaller than number 7 in Fig. 29-6. ∎

29-4   HEAT TRANSFER BY RADIATION AND CONVECTION

Heat sink–ambient thermal resistances, $R_{\theta sa}$, are provided by manufacturers of heat sinks as indicated by Fig. 29-6. These data are only valid for a specific set of heat sink and ambient temperature conditions. To understand how the heat sink will perform under different conditions, a fundamental understanding of the mechanisms of heat transfer from a body to its surroundings via convection and radiation is essential. In this section a brief introduction to convective and radiative heat transfer will be given. The equations developed will provide a basis for calculating the parallel effects of convective and radiative heat transfer on the heat sink–ambient thermal resistance.

29-4-1   THERMAL RESISTANCE DUE TO RADIATIVE HEAT TRANSFER

Heat transfer via radiation is given by the Stefan–Boltzmann law:

$$P_{\text{rad}} = 5.7 \times 10^{-8} \, EA(T_s^4 - T_a^4) \tag{29-15}$$

where P_{rad} is the radiated power in watts, E is the emissivity of the surface, T_s is the surface temperature in K, T_a is the ambient temperature or the temperature of surrounding objects in K, and A is the outer surface area (including the fins) of the heat sink in square meters. E is given in the literature [2] for various surfaces. For dark objects such as black oxidized aluminum heat sinks, $E = 0.9$. For polished aluminum, E may be as small as 0.05.

For black oxidized aluminum heat sinks, Eq. 29-15 can be rewritten as

$$P_{\text{rad}} = 5.1A\left[\left(\frac{T_s}{100}\right)^4 - \left(\frac{T_a}{100}\right)^4\right] \tag{29-16}$$

Combining Eqs. 29-2 and 29-16 yields

$$R_{\theta,\text{rad}} = \frac{\Delta T}{5.1A\left[\left(\dfrac{T_s}{100}\right)^4 - \left(\dfrac{T_a}{100}\right)^4\right]} \tag{29-17}$$

If $T_s = 120°C = 393$ K and $T_a = 20°C = 293$ K, then $R_{\theta,\text{rad}}$ is given by

$$R_{\theta,\text{rad}} = \frac{0.12}{A} \tag{29-18}$$

■ *Example 29-4* Find $R_{\theta,\text{rad}}$ for a cube of black oxidized aluminum 10 cm on a side. Assume $T_s = 120\ °C$ and $T_a = 20\ °C$. Using Eq. (29-18) yields

$$R_{\theta,\text{rad}} = \frac{0.12}{(6)(.1)^2} = 2\,°C/W$$

■

29-4-2 THERMAL RESISTANCE DUE TO CONVECTIVE HEAT TRANSFER

If a vertical surface has a vertical height d_{vert} less than about 1 m, it loses heat energy via convection to the surrounding air (at sea level) at a rate given by [2]

$$P_{\text{conv}} = 1.34A\,\frac{(\Delta T)^{1.25}}{(d_{\text{vert}})^{0.25}} \tag{29-19}$$

In Eq. 29-19 P_{conv} is the heat power lost via convection in watts, ΔT is the temperature difference between the surface of the body and the surrounding air in degrees centigrade or kelvin, A is the area of the vertical surface (or the total surface area of the body) in square meters, and d_{vert} is the vertical height of the body in meters. Combining Eqs. 29-2 and 29-19 gives

$$R_{\theta,\text{conv}} = \frac{1}{1.34A}\left(\frac{d_{\text{vert}}}{\Delta T}\right)^{1/4} \tag{29-20}$$

For $d_{\text{vert}} = 10$ cm and $\Delta T = 100°C$

$$R_{\theta,\text{conv}} = \frac{0.13}{A} \quad (°C/W) \tag{29-21}$$

■ *Example 29-5* A thin plate has a surface temperature of 120°C when the surrounding air temperature is 20°C. The plate is 10 cm high (vertical dimension) and 30 cm wide. Find $R_{\theta,\text{conv}}$.

$$R_{\theta,\text{conv}} = \frac{1}{(1.34)(2)(0.1)(0.3)}\left(\frac{0.1}{100}\right)^{1/4} = 2.2\ °C/W$$

■

If it is assumed that the cube in Example 29-4 has the same convection thermal resistance $R_{\theta,\text{conv}}$ as the plate in Example 29-5 (the surface area of the plate and the cube are the same), then the effects of radiation and convection can be combined to yield $R_{\theta sa}$ as

$$R_{\theta sa} = \frac{R_{\theta,\text{rad}}\,R_{\theta,\text{conv}}}{R_{\theta,\text{rad}} + R_{\theta,\text{conv}}} = 1\ °C/W \tag{29-22}$$

The result given in Eq. 29-22 may be somewhat optimistically low. The heat removed from a horizontal surface facing upward is 15–25% more than for a vertical surface. When facing downward, there is a reduction of approximately 33% relative to a vertical surface and could be even more if the area is larger. This means that one of the six sides of the cube in Example 29-4 should have 10–20% more convected thermal resistance than calculated in Example 29-5, which would give an increase in $R_{\theta,\text{conv}}$ of about 4%. If this correction is used in Eq. 29-22, the value of $R_{\theta sa}$ would increase by about 2%.

From the results discussed in the preceding paragraphs, it is seen that the thermal resistance depends on T_s, T_a, and ΔT. Usually a heat sink is never mounted with sufficient air flow around it, so for approximate calculations, Eqs. 29-19 and 29-20 can be used with the value of the total surface area A being used rather than accounting for the differences between upward and downward facing horizontal areas.

29-4-3 EXAMPLE HEAT SINK–AMBIENT CALCULATION

As an example of the use of the principles just discussed, the heat sink–ambient thermal resistance $R_{\theta sa}$ will be estimated for the heat sink shown in Fig. 29-7. Temperatures of $T_s = 120°C$ and $T_a = 20°C$ will be used. To estimate the $R_{\theta,rad}$ component of the thermal resistance, Eq. 29-18 is used with an effective area

$$A_{rad} = (2)(0.115)(0.075) + (2)(0.063)(0.075) = 0.0267 \text{ m}^2$$

which, using Eq. 29-18 gives

$$R_{\theta,rad} = \frac{0.12}{0.0267} = 4.5 °C/W$$

The spacing between the cooling fins is approximately 9 mm. This significantly reduces the effect of natural convective cooling of the heat sink. Equation 29-20 should be modified by including a reduction factor such that

$$R_{\theta,conv} = \frac{1}{1.34AF_{red}} \left(\frac{d_{vert}}{\Delta T}\right)^{1/4} \tag{29-23}$$

The factor F_{red} is graphed in Fig. 29-8 [1]. From Fig. 29-7, an approximate value of the area exposed to convective cooling is

$$A = 2A_2 + 16A_1 = ((2)(0.075)(0.092) + (16)(0.075)(0.063) = 0.089 \text{ m}^2$$

Using Fig. 29-8, a value of $F_{red} = 0.78$ is found when the fin spacing is 9 mm. Using these results in Eq. 29-23 gives

$$R_{\theta,conv} = \frac{1}{(1.34)(0.089)(0.78)} \left(\frac{0.075}{100}\right)^{1/4} = 1.8 °C/W$$

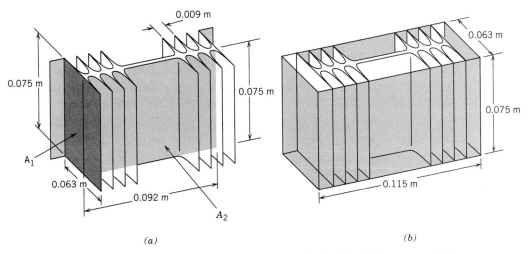

(a)   *(b)*

Figure 29-7 Dimensional details of heat sink number 7 in Fig. 29-6: (a) areas used for calculating the convective heat losses and (b) areas used in estimating radiative heat losses.

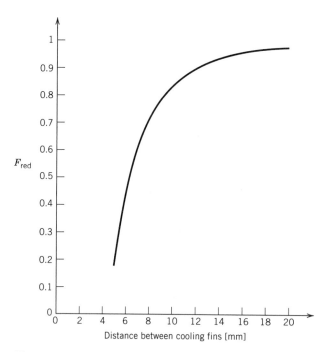

Figure 29-8 Reduction factor for the convection area to a naturally cooled heat sink with cooling fin separations less than 25 mm.

Combining the radiative and convective components of thermal resistance per Eq. 29-22 gives

$$R_{\theta sa} = \frac{(4.5)(1.8)}{(4.5 + 1.8)} = 1.3 \,^{\circ}C/W$$

which is the measured value shown in Fig. 29-6 for heat sink number 7.

SUMMARY

This chapter has explored the various mechanisms of heat transfer that are utilized to keep the temperatures of various power electronic components, especially power semiconductor devices, within manufacturers limits. The important conclusions follow.

1. The maximum junction temperatures of semiconductor devices given on data sheets should not be exceeded during device operation. Temperature excursions beyond these limits significantly reduce the reliability of the device.

2. Heat energy flows from the interior of a power electronic component mounted on external heat sink by means of conduction. In the steady state, the process is modeled by a series connection of thermal resistances.

3. Power dissipation of short duration (short compared to the thermal time constant) causes the internal temperature to rise by smaller amounts than is predicted by the thermal resistance alone. This is due to the finite thermal capacitance (or specific heat) of the component. As the term implies, it is modeled as the analog to an electrical capacitor.

4. The combined effects of thermal resistance and capacitance are modeled as the transient thermal impedance.

5. A large variety of heat sinks are commercially available for controlling the internal temperature of power electronic components.

6. The transfer of heat from the heat sink to the ambient is controlled by two heat transfer mechanisms, convection and radiation.

7. Heat transfer via radiation is proportional to the difference between the fourth power of the surface temperature of the component and the fourth power of the ambient temperature.

8. Heat transfer via convection is proportional to the fourth root of the ratio of the vertical height of the heat sink to the temperature difference between the surface temperature and the ambient temperature.

PROBLEMS

29-1 Calculate the thermal resistance of cubes with the same volume as the heat sinks given in Fig. 29-6. Compare the calculated $R_{\theta,\text{cube}}$ with the R_θ given in the figure. Explain why heat sink number 9 is only a little better than a cube, while heat sink number 1 is much better.

29-2 Find the value of $R_{\theta,\text{conv}}$, assuming $\Delta T = 100°C$ and $A = 10\text{ cm}^2$, for $d_{\text{vert}} = 1, 5$, and 20 cm. Plot a graph of the results.

29-3 Find the value of $R_{\theta,\text{conv}}$, assuming $d_{\text{vert}} = 5$ cm and $A = 10\text{ cm}^2$, for $\Delta T = 60, 80$, and $120°C$. Plot a graph of the results.

29-4 Find the value of $R_{\theta,\text{rad}}$, assuming $T_s = 120°C$ and $A = 10\text{ cm}^2$, for $T_a = 10, 20$, and $40°C$. Plot a graph of the results.

29-5 Find the value of $R_{\theta,\text{rad}}$, assuming $T_a = 40°C$ and $A = 10\text{ cm}^2$, for $T_s = 80, 100$, and $140°C$. Plot a graph of the results.

29-6 A MOSFET used in a step-down converter has an on-state loss of 50 W and a switching loss given by $10^{-3} f_s$ (in watts) where f_s is the switching frequency in hertz. The junction-to-case thermal resistance $R_{\theta,jc}$ is 1°C/W and the maximum junction temperature $T_{j,\text{max}}$ is 150°C. Assuming the case temperature is 50°C, estimate the maximum allowable switching frequency.

29-7 The MOSFET of Problem 29-6 is mounted on a heat sink and the ambient temperature $T_a = 35°C$. If the switching frequency is 25 kHz, what is the maximum allowable value of the case-to-ambient thermal resistance $R_{\theta,ca}$ of the heat sink. Assume all other parameters given in Problem 29-6 remain the same except the case temperature which can change.

REFERENCES

1. U. Fabricus, "Heat Sinks," ECR-45, Danish Research Center for Applied Electronics, 1974 (in Danish).

2. E. C. Snelling, *Soft Ferrites—Properties and Applications*, Butterworths, London, 1988.

3. Keith Billings, *Switchmode Power Supply Handbook*, McGraw-Hill, New York, 1989.

4. John G. Kassakian, Martin F. Schlecht, G. Vergassian, *Principles of Power Electronics*, Addison-Wesley, Boston, 1991.

CHAPTER 30

DESIGN OF MAGNETIC COMPONENTS

Magnetic components, inductors and transformers, are an indispensible part of most power electronic converters. However, they are not commercially available in a wide range of properties but are usually designed and constructed for the particular application. In this situation the power electronic equipment designer/user must be knowledgeable about the design and fabrication of these components in order to specify and use them properly in a given application. In this chapter the basic aspects of the design of inductors and transformers are presented with emphasis on high-frequency (tens of kHz to MHz) power electronic applications. The design procedures show that the size and rating of an inductor or transformer are dominated by the electrical loss in the component.

30-1 MAGNETIC MATERIALS AND CORES

The review of magnetic circuits and devices in Chapter 3 assumed that ideal materials were used to make the inductors and transformers. In particular it was assumed that loss-free magnetic materials were used for the cores. These assumptions are not satisfied in real materials, and the loss that occurs in them has a significant effect on the design and fabrication of inductors and transformers. Any inductor or transformer design procedure must take these losses into account, and the designer must have a good understanding of the material properties. This section discusses these material properties.

30-1-1 MAGNETIC CORE MATERIALS

Two broad classes of materials are used for magnetic cores for inductors and transformers. One class of materials are comprised of alloys principally of iron and small amounts of other elements including chrome and silicon. These alloys have large electrical conductivity compared with ferrites and large values of saturation flux density, near 1.8 tesla (T) (one $T = 1Wb/m^2$). Two types of loss are found in iron alloy materials, hysteresis loss and eddy current loss. Iron alloy core materials (often termed magnetic steels) are usually used only in low-frequency (2 kHz or less for transformers) applications because of eddy current loss. Iron alloy magnetic materials must be laminated to reduce eddy current loss even at modest frequencies such as 60 Hz. Cores are also made from powdered iron and powdered iron alloys. Powdered iron cores consist of small (less than a skin depth in their largest dimension even at moderately high frequencies) iron particles electrically isolated from each other and thus have significantly greater resistivity than laminated cores. Thus powered iron cores have lower eddy current loss than laminated cores and can be used to higher frequencies.

Various amorphous alloys of iron and other transition metals such as cobalt and nickel in combination with boron, silicon, and other glass-forming elements also offer interesting properties for inductor and transformer applications. These alloys, often as a group labeled by the trade name METGLAS, are roughly 70–80 atomic percent iron and other transition metal elements and approximately 20 atomic percent boron and other glass-forming elements. Alloy compositions containing cobalt such as METGLAS alloy 2705M appear particularly suitable for high-frequency applications. This alloy has a saturation induction of 0.75 T at room temperature and 0.65 T at 150°C, which is more than a factor of two larger than the saturation induction of ferrites at this elevated temperature. The electrical resistivity of METGLAS alloys is typically somewhat larger than the most magnetic steels. The METGLAS alloys are formed by rapid quenching techniques so that they have no crystalline order in their structure. The rapid quenching fabrication techniques also mean that the alloys are fabricated as long ribbons of material which are quite thin, typically 10–50 microns in thickness. This small thickness, together with larger resistivities than most magnetic steels, make amorphous alloys obvious candidate core materials for high frequency applications.

The second broad class of materials used for cores are ferrites. Ferrite materials are basically oxide mixtures of iron and other magnetic elements. They have quite large electrical resistivity but rather low saturation flux densities, typically about 0.3 T. Ferrites have only hysteresis loss. No significant eddy current loss occurs because of the high electrical resistivity. Ferrites are the material of choice for cores that operate at high frequencies (greater than 10 kHz) because of the low eddy current loss.

30-1-2 HYSTERESIS LOSS

All magnetic cores exhibit some degree of hysteresis in their B–H characteristic. A typical B–H characteristic (B–H loop) is shown in Fig. 3-20a. The details of the physical mechanisms that cause hysteresis are beyond the scope of this discussion but can be found in the literature. The area inside the B–H loop represents work done on the material by the applied field. The work (energy) is dissipated in the material, and the heat caused by the dissipation raises the temperature of the material.

The hysteresis loss increases in all core materials increases with increases in ac flux density, B_{ac}, and operating or switching frequency, f. The general form of the loss per unit volume (sometimes termed the specific loss), $P_{m,sp}$, is

$$P_{m,sp} = kf^a(B_{ac})^d \tag{30-1}$$

where k, a, and d are constants that vary from one material to another. This equation applies over a limited range of frequency and flux density with the range of validity being dependent on the specific material. The flux density B_{ac} in Eq. 30-1 is the peak value of the ac waveform as shown in Fig. 30-1a if the flux density waveform has no time average. When the flux density waveform has a time-average B_{avg} as shown in Fig. 30-1b, then the appropriate value to use in Eq. 30-1 is $B_{ac} = \hat{B} - B_{avg}$. Core manufacturers provide detailed information about core loss usually in the form of graphs of specific loss $P_{m,sp}$ as a function of flux density B_{ac} with frequency as a parameter. An example of such a graph is shown in Fig. 30-2a for the ferrite material 3F3, and Eq. 30-1 for this material is

$$P_{m,sp} = 1.5 \times 10^{-6} f^{1.3} (B_{ac})^{2.5} \tag{30-2a}$$

with $P_{m,sp}$ in mW/cm^3 when f is in kHz and B_{ac} is in mT. In selected METGLAS alloys, the core losses may be comparable to ferrites, in spite of the fact that the amorphous alloys

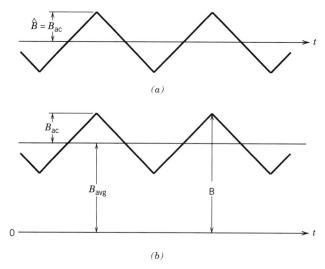

Figure 30-1   Magnetic flux density waveforms having (a) no time average and (b) with a time average.

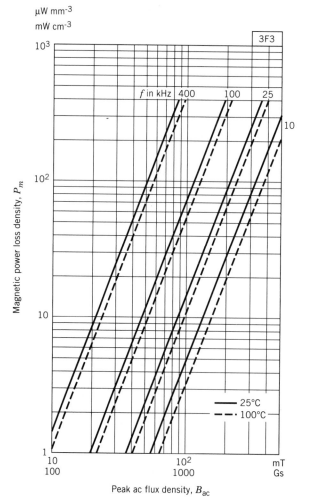

Figure 30-2   Core losses versus magnetic flux density for 3F3 ferrite cores. Note that the losses are approximately independent of waveform, but depend on the peak flux density.

have much lower resistivity than ferrites and thus will have eddy current losses. For the METGLAS alloy 2705M, the core losses are given by

$$P_{m,sp} = 3.2 \times 10^{-6} f^{1.8} (B_{ac})^2 \tag{30-2b}$$

The units in Eq. 30-2b are the same as in 30-2a. At a frequency of 100 kHz and a flux density B_{ac} of 100 mT, the 3F3 ferrite characterized by Eq. 28-2a would have $P_{m,sp} = 60$ mW/cm^3 while for the 2705M alloy, $P_{m,sp} = 127$ mW/cm^3

In a later section it will be shown that the volt-amp (V-A) rating of a transformer is proportional to the product $f B_{ac}$. For a constant specific core loss, it is convenient to define an empirical performance factor PF $= f B_{ac}$ for various ferrite materials that might be used for a transformer core. Using information from data sheets provided by the manufacturers of the different materials, plots of the performance factor as a function of frequency for several different ferrite materials are shown in Figure 30-3. As can be seen from the plots, a given material has the best performance factor only within a specific frequency range. In particular for the materials surveyed, 3C85 is the best below 40 kHz, 3F3 is the best from 40 to 420 kHz, and 3F4 is the best above 420 kHz. In addition it can be easily seen how much reduction in the performance factor will occur at a specific frequency if some material other than the optimum is chosen, for example, choosing 3B8 instead of 3F3 at a frequency of 100 kHz.

The reduction of the performance factors of all materials at high frequencies implies that the simple relationship between specific core loss and frequency and flux density given by Eq. 30-1 is not valid at high frequencies. The physical details for this behavior are beyond the scope of this discussion. The interested reader is referred to the literature.

Ultimately $P_{m,sp}$ is limited by the maximum temperature that can be allowed in the material. A commonly used maximum temperature for many applications is 100°C. At this temperature the maximum $P_{m,sp}$ in a typical design is in the low hundreds of mW/cm^3. The exact value of $P_{m,sp}$ will depend on how efficiently the heat dissipated is removed, that is, on the thermal resistance between the core and the ambient. In cores made of laminated magnetic steels, the maximum $P_{m,sp}$ is even smaller because

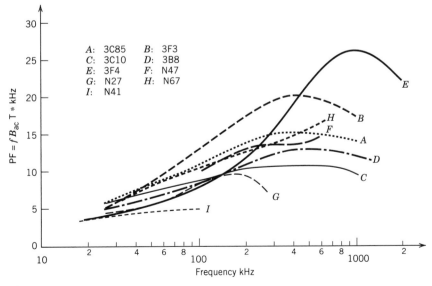

Figure 30-3 Empirical performance factor PF $= f B_{ac}$ versus frequency for various ferrite core materials. Measurements are made at a power density $P_{core} = 100$ mW/cm^3.

some allowance must be made for heat produced by eddy current loss, a subject to be discussed in a later section.

30-1-3 SKIN EFFECT LIMITATIONS

When a magnetic core is made from conducting materials such as magnetic steels, time-varying magnetic fields applied to the core will generate circulating currents as is diagramed in Fig. 30-4a. Using the right-hand rule, it can be seen that these currents, generically termed eddy currents, flow in directions such that secondary magnetic fields are produced that oppose the applied (primary) magnetic field. These opposing fields tend to screen the interior of the core from the applied field, and the total magnetic field in the core decays exponentially with distance into the core as is shown in Fig. 30-4b.

The characteristic decay length in the exponential is termed the skin depth and is given by

$$\delta = \sqrt{\frac{2}{\omega\mu\sigma}} \tag{30-3}$$

where $f = \omega/2\pi$ is the frequency (in hertz) of the applied magnetic field, μ is the magnetic permeability of the core material, and σ is the conductivity of the magnetic material. If the cross-sectional dimensions of the core are large compared to the skin depth, then the interior of the core carries little or none of the applied magnetic flux as is diagrammed in Fig. 30-4b and the core is ineffective in its intended role of providing a low reluctance return path for the applied magnetic field. Typical values of the skin depth are quite small even at low frequencies (typically 1 mm at 60 Hz) because of the large permeability of the materials and the skin depth becomes more of a problem as the applied frequency increases.

Thus magnetic cores for inductors and transformers that utilize conducting magnetic materials are made from stacks of many thin laminations as is shown in Fig. 30-5. Each

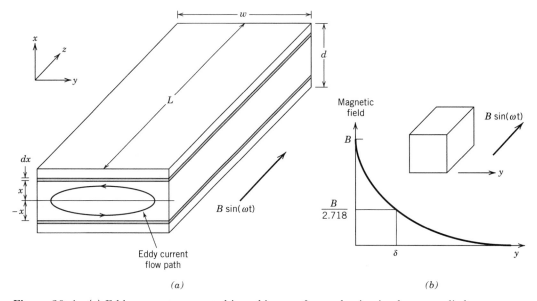

(a)   (b)

Figure 30-4 (a) Eddy currents generated in a thin transformer lamination by an applied time-varying magnetic field and (b) decay of the magnetic field versus depth y into the interior of a thick bar of magnetic material.

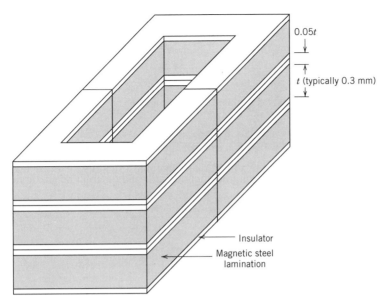

Figure 30-5 Magnetic core for a transformer or inductor made from a stack of magnetic steel laminations separated by insulators.

lamination is electrically isolated from the other by means of a thin insulating coating on each lamination. The core stacking factor is defined as the ratio of the cross sectional area of the magnetic material to the total cross-sectional area of the core. The stacking factor will be less than 1 (typical values are 0.9 to 0.95) because part of the total area of the core is occupied by the insulating layers.

Most magnetic steels have a small percentage of silicon added to the iron to increase the resistivity of the material and thus increase the skin depth. Addition of more than a few percent of silicon, however, reduces the magnetic properties such as saturation flux density more than it increases the resistivity. Hence a reasonable compromise for transformers for 50/60 Hz applications is an iron alloy of 97% iron–3% silicon and a lamination thickness approximately of 0.3 mm.

30-1-4 EDDY CURRENT LOSS IN LAMINATED CORES

The eddy currents generated in the conductive core dissipate power, generically termed eddy current loss, in the core and raise its temperature. Consider the magnetic conductor shown in Fig. 30-4a which is immersed in a uniform time-varying magnetic field having a flux density $B(t) = B \sin(\omega t)$. It is assumed that the thickness d is less than the skin depth δ so that the induced eddy currents do not reduce the magnetic field in the interior of the material. The conductor, which could represent a lamination used in a transformer core, has a conductivity σ. If a thin loop of thickness dx located at x and $-x$ is drawn in the xy plane of the conductor as shown in Fig. 30-4a, it intercepts a total flux given by

$$\phi(t) = 2xwB(t) \tag{30-4}$$

Using Faraday's law, (Eq. 3-66), this flux generates a voltage, $v(t)$, in the thin loop which is given by

$$v(t) = 2xw\,\frac{\partial B(t)}{\partial t} = 2wx\omega B \cos(\omega t) \tag{30-5}$$

The resistance r of this thin loop, using a cross-sectional area of width L and thickness dx and a length of $2w$ (the contribution of $2d$ to the overall length is neglected), is given by

$$r = \frac{2w\rho_{core}}{L\,dx} \qquad (30\text{-}6)$$

The instantaneous power $\delta p(t)$ dissipated in the thin loop is given by

$$\delta p(t) = \frac{v^2(t)}{r} \qquad (30\text{-}7)$$

Integrating over the volume of the lamination to obtain the total time-average eddy current power P_{ec} dissipated in the lamination gives

$$P_{ec} = \left\langle \int \delta p(t)\,dV \right\rangle = \left\langle \int_0^{d/2} \frac{[2wx\omega B\cos(\omega t)]^2 L\,dx}{2w\rho_{core}} \right\rangle = \frac{wLd^3\omega^2 B^2}{24\rho_{core}} \qquad (30\text{-}8a)$$

The brackets ($<\,>$) in Eq. 30-8a indicate time average. The specific eddy current loss, $P_{ec,sp}$ (loss per unit volume) are given by

$$P_{ec,sp} = \frac{d^2\omega^2 B^2}{24\rho_{core}} \qquad (30\text{-}8b)$$

Note that $P_{ec,sp}$ varies with the square of the lamination thickness d. This $P_{ec,sp}$ estimate represents an optimistic minimum in the eddy current loss. If the magnetic flux were inclined at some angle to the plane of the lamination (the yz plane), the loss would be considerably larger [1].

The problems of skin effect and eddy current loss that the finite conductivity of magnetic steels causes is largely avoided in ferrite materials. The very large resistivity found in ferrites minimizes these problems.

30-1-5   CORE SHAPES AND OPTIMUM DIMENSIONS

Cores are available in a wide variety of shapes and sizes to suit the given application. This is particularly true of ferrite cores, which are available as toroids, pot cores with an airgap, and in U, E, and I shapes. Laminated materials are available as tape wound toroids and C-cores. A double E-core is shown in Fig. 30-6a as an example.

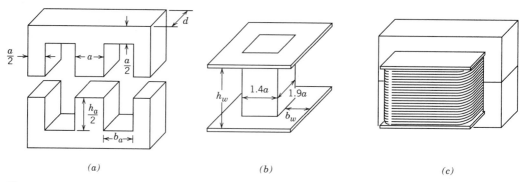

Figure 30-6   Dimensioned diagram of (a) a double-E core (b) bobbin, and (c) assembled core with winding.

A bobbin or a coil former is provided with most cores, and the effective cross-sectional area $A_w = h_w b_w$ available for the copper windings on the bobbin is given, as shown in Fig. 30-6b. These bobbins are also available in a wide variety of sizes and shapes.

Once a core such as shown in Fig. 30-6a is selected, the combination of dimensions d, h_a, and b_a should be optimized. The bobbin dimensions h_w and b_w shown in Fig. 30-6b are made to be as close as possible to the core dimensions h_a and b_a. The optimal combination of the core dimensions will often be found by the use of a computer optimization program. The criteria for the optimization can be the lowest total volume or weight for a given rating or the lowest cost. The lowest cost will depend on the relative cost of the copper wire winding compared to that of the core material. An optimum design may vary as this relation changes over time.

A producer of power electronic equipment will normally buy cores and bobbins from a manufacturer of these components. In doing this, the buyer assumes that the core manufacturer has done the optimization correctly. For large-scale mass production, it may be more economical to buy custom-made magnetic components because they can be optimized using the criteria of the producer of the equipment and furthermore a core size between two standard sizes can be used.

For the double-E core of Fig. 30-6a, experience has shown that an optimal set of dimensions could be $b_a = a$, $d = 1.5a$, $h_a = 2.5a$, $b_w = 0.7a$, and $h_w = 2a$. This combination will be used in examples later in this chapter. Using this optimal set of relative dimensions the core of Fig. 30-6a will have the sizes shown in Table 30-1. In addition to showing the relative size (scaled by the parameter a), absolute values are also presented corresponding to $a = 1$ cm. A database containing this type of information for all the cores that a particular organization commonly uses is an invaluable tool for simplifying the inductor/transformer design process. Such data can be obtained from vendor data sheets or can be compiled by the user of the core.

Table 30-1 Geometric Characteristics of a Near Optimum Core for Inductor/Transformer Design

Characteristic	Relative Size	Absolute Size for $a = 1$ cm
Core area A_{core}	$1.5a^2$	1.5 cm^2
Winding area A_w	$1.4a^2$	1.4 cm^2
Area product AP $= A_w A_c$	$2.1a^4$	2.1 cm^4
Core volume V_{core}	$13.5a^3$	13.5 cm^3
Winding volume V_w[a]	$12.3a^3$	12.3 cm^3
Total surface area of assembled inductor/transformer[b]	$59.6a^2$	59.6 cm^2

Notes: [a]Total volume is estimated as the volume in the winding windows $2A_w(d + 0.4a)$ plus the two rectangular volumes $A_w(a + 0.4a)$, one on either side of the core, and four-quarter circle cylinders (radius b_w and height h_w). The $0.4a$ factors are included to allow for the finite thickness of the bobbin.

[b]The total surface area is assumed to be composed of the outer area of the core ($50.5a^2$) plus the area of the top and bottom flanges ($5.9a^2$) of the windings plus the area of the rounded quarter-cylinder corners of radius $0.7a$ and height $2a$ (total area of four quarter-cylinders is $8.8a^2$) minus the core area covered by the four quarter-cylinder corners [total area $= 4(2a)(0.7a) = 5.6a^2$].

30-2 COPPER WINDINGS

The conductor windings in an inductor or transformer are made from copper because of its high conductivity. The high ductility of the copper makes it easy to bend the conductors into tight windings around a magnetic core and thus minimize the amount of copper and volume needed for the windings. High conductivity contributes to minimizing the amount of copper needed for the windings and thus to the volume and weight of the windings. At the current densities used in inductors and transformers, electrical loss is a significant source of heat even though the conductivity of copper is large. The heat generated raises the temperature of both the windings and the magnetic core. The amount of dissipation allowable in the windings will be limited by maximum temperature considerations just as was described for the core loss.

30-2-1 COPPER FILL FACTOR

A typical cross-sectional view of a multiturn winding on a magnetic core, in this case one winding window of the double-E core, is shown in Fig. 30-7. A perspective view of the winding on the core is shown in Fig. 30-6c. The copper conductor from which the winding is made has a cross-sectional area A_{Cu}. The conductor may be composed of a single round wire or it may be a special multistranded conductor such as Litz wire in which each strand has a diameter on the order of a few hundred microns or less. Litz wire is used where otherwise the skin effect would present problems.

The total number of turns N in the winding window of the core times the conductor area A_{Cu} gives the total copper area in the winding window. The total copper area will be less than the area A_w of the winding window for several reasons. First, the geometric shape of the conductors, usually circular, and the winding process prevents the N conductors from completely filling the window. Second, the conductor must be covered by an electrical insulator so that adjacent turns are not shorted to each other. This insulation takes up some of the winding window area. The ratio of the total copper area to the winding window area is termed the copper fill factor k_{Cu} and is given by

$$k_{Cu} = \frac{NA_{Cu}}{A_w} \tag{30-9}$$

Practical values of fill factor range from 0.3 for Litz wire to 0.5–0.6 for round conductors.

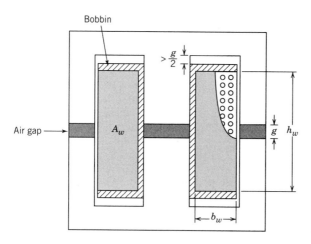

Figure 30-7 Cross-sectional view of a double-E core and bobbin assembled as an inductor. The air-gap spacing is supported by an insulating material (shaded area).

30-2-2 WINDING LOSS DUE TO dc RESISTANCE OF WINDINGS

The power $P_{\mathrm{Cu,sp}}$ dissipated per unit of copper volume in a copper winding due to its dc resistance is given by

$$P_{\mathrm{Cu,sp}} = \rho_{\mathrm{Cu}} \, (J_{\mathrm{rms}})^2 \tag{30-10}$$

where $J_{\mathrm{rms}} = I_{\mathrm{rms}}/A_{\mathrm{cu}}$ is the current density in the conductor and I_{rms} is the rms current in the winding. However, it is more convenient to express $P_{\mathrm{Cu,sp}}$ as power dissipated per unit of winding volume, $P_{w,\mathrm{sp}}$. The total volume V_{Cu} of the copper is given by $V_{\mathrm{Cu}} = k_{\mathrm{Cu}} V_w$, where V_w is the total winding volume. Using this result to express $P_{w,\mathrm{sp}}$ yields

$$P_{w,\mathrm{sp}} = k_{\mathrm{Cu}} \, \rho_{\mathrm{Cu}} \, (J_{\mathrm{rms}})^2 \tag{30-11}$$

If the resistivity of copper at 100°C (2.2×10^{-8} Ω-m) is used in Eq. 30-11 and J_{rms} is expressed in A/mm², the value of $P_{w,\mathrm{sp}}$ becomes

$$P_{w,\mathrm{sp}} = 22 k_{\mathrm{Cu}} \, (J_{\mathrm{rms}})^2 \qquad (\mathrm{mW/cm^3}) \tag{30-12a}$$

30-2-3 SKIN EFFECT IN COPPER WINDINGS

The skin effect occurs in the copper conductors used in inductor and transformer windings in exactly the same manner as described for the magnetic core. Consider the single copper conductor shown in Fig. 30-8a, which is carrying a time-varying current $i(t)$. This current generates the magnetic fields shown in Fig. 30-8a, and they in turn generate the eddy currents illustrated in Fig. 30-8b. These eddy currents flow in the opposite direction to the applied current $i(t)$ in the interior of the wire and thus tend to shield the interior of the conductor from the applied current and resulting magnetic field. As a result the total current density is largest at the surface of the conductor, and it decays exponentially with distance into the interior of the conductor as shown in Fig. 30-8c. The characteristic decay length is the skin depth given by Eq. 30-3. Table 30-2 shows the skin depth in copper at several different frequencies at a temperature of 100°C.

  If the cross-sectional dimensions of the conductor used in the winding are significantly larger than the skin depth, most of the current carried by the conductor will be constricted to a relatively thin layer at the surface approximately one skin depth in

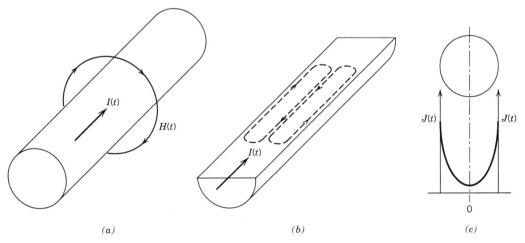

(a)   *(b)*   *(c)*

Figure 30-8 Isolated copper conductor carrying (*a*) a current $i(t)$, (*b*) eddy currents generated by the resulting magnetic field, and (*c*) the consequences of the skin effect on the current distribution.

Table 30-2 Skin Depth in Copper at 100°C for Several Different Frequencies

Frequency	50 Hz	5 kHz	20 kHz	500 kHz
δ	10.6 mm	1.06 mm	0.53 mm	0.106 mm

thickness as is illustrated in Fig. 30-8c. The net result of this is that the effective resistance of the conductor will be far larger than the dc resistance because the effective cross-sectional area for current flow is small compared to the geometric cross section of the conductor. This will cause the winding losses to be much larger than if it were a dc current.

The solution to this problem is to use conductors with cross-sectional dimensions on the order of the skin depth in size. If d is the diameter of a round conductor or the thickness of a rectangular conductor, calculations have shown that if $d \leq 2\delta$ the consequences of the skin effect can be neglected. Such considerations have led to the development of special conductor arrangements for high-frequency applications. These conductor arrangements include Litz wire, which was described earlier, and the use of thin foil windings. Eddy current loss in windings are treated in greater detail in later sections of this chapter. The net effect of these losses is to increase the effective resistance of the winding to a value R_{ac}. This modifies Eq. 30-12a to

$$P_{w,sp} = 22 \ k_{Cu} \frac{R_{ac}}{R_{dc}} (J_{rms})^2 \tag{30-12b}$$

where R_{dc} is the dc resistance of the winding.

30-3 THERMAL CONSIDERATIONS

Increases in the temperature of the core and winding materials degrade the performance of these materials in several respects. The resistivity of the copper windings increases with temperature, and so the winding loss increases with temperature, assuming a constant current density. In the magnetic materials, the core loss increases with increasing temperature above approximately 100°C, assuming the frequency and flux density remain constant. The value of the saturation flux density becomes smaller with increases in temperature.

To keep the performance degradation within bounds, the temperature of the core and windings must be kept at or below some maximum value. Winding and core loss cause the temperature increase, hence the loss must be kept below some maximum value. Thus two fundamental questions need to be addressed. First, what is the maximum allowable core/winding temperature? Second, what is the quantitative relationship between the loss (core and winding) and the temperature in the materials?

In practice the maximum temperature is usually limited to 100–125°C by several considerations. The reliability of the insulation on the copper windings, usually a thin layer of varnish, decreases rapidly with temperature increases much above 100°C. In many magnetic materials, especially ferrites, the core loss is a minimum around 100°C. Other components such as power semiconductor devices operated in close proximity to the inductors and transformers will have elevated temperatures due to internal power dissipation. Since heat will be transferred between components in close proximity via radiation and convection, the steady-state temperature of all components will equilibrate to a common value. In such circumstances, the overall maximum temperature of the power electronic system will be dictated by the component with the lowest maximum allowable temperature. The 100–125°C limit for magnetic components is about the same as for power semiconductor components and other passive components.

The internal temperature and surface temperature of the inductor or transformer is normally assumed to be nearly the same. This assumption is made because the power dissipation is approximately uniformly distributed throughout the volume of the core and the windings. This leads to a large cross-sectional area for heat conduction to the surface and for relatively short path lengths. Moreover, the thermal conductivities of the materials are large. Hence the thermal resistance of importance in determining the temperature of the inductor or transformer is the surface-to-ambient resistance, $R_{\theta sa}$, which is defined in Chapter 29, Eq. 29-22, which is called the sink-to-ambient thermal resistance. The radiative, $R_{\theta,rad}$, and convective, $R_{\theta,conv}$, components of $R_{\theta sa}$ are functions of the temperature difference and the surface area A of the component. To obtain a quantitative value of $R_{\theta sa}$, a particular core and winding with specific dimensions would have to be analyzed. This will be done as a specific example in a later section in this chapter. However, useful and interesting trends can be obtained without resorting to specific numerical dimensions.

Both $R_{\theta,rad}$ which is given by Eq. 29-17, and $R_{\theta,conv}$ [assuming that $(d_{vert})^{0.25} \approx$ constant for a limited variation in d_{vert}, which is given by Eq. 29-20] are inversely proportional to the surface area A of the inductor/transformer, assuming a fixed temperature difference ΔT. Hence $R_{\theta sa}$, by Eq. 29-22, is also inversely proportional to the surface area A. The surface area of any core, for example, the double-E core shown in Fig. 30-7a, is proportional to the square of a characteristic dimension, a, (see Table 30-1 for the surface area in terms of this characteristic dimension for double-E core of Fig. 30-7a). Thus, in general, we can write for any inductor or transformer with a specified value of ΔT

$$R_{\theta sa} = \frac{k_1}{a^2} \tag{30-13}$$

where k_1 is a constant. Since ΔT is fixed and known, the equation $\Delta T = R_{\theta sa} (P_{core} + P_w)$, where $P_{core} = P_{c,sp} V_c$ and $P_w = P_{w,sp} V_w$ (V_c = core volume and V_w = winding volume, see Table 30-1 for an example), can be inverted to find $(P_{core} + P_w)$ as

$$P_{core} + P_w = k_2 a^2 \tag{30-14}$$

where k_2 is a constant. In an optimal design, $P_{c,sp} \approx P_{w,sp} = P_{sp}$, and using this information along with the observation that the core and winding volumes are both proportional to the cube of the characteristic length a, we can use Eq. 30-14 to express P_{sp} as

$$P_{sp} = \frac{k_3}{a} \tag{30-15}$$

where k_3 is a constant. Using Eq. 30-2 in Eq. 30-15 to express P_{sp} in terms of the flux B_{ac} and frequency f yields

$$B_{ac} = \frac{k_4}{f^{0.52} a^{0.4}} \tag{30-16}$$

where k_4 is a constant. If the core loss is somewhat different than that given by Eq. 30-2, then Eq. 30-16 would be modified accordingly. When Eq. 30-12 is inverted to solve for the current density J_{rms} and Eq. 30-15 is used for the specific winding loss, then we find

$$J_{rms} = \frac{k_5}{\sqrt{k_{Cu} a}} \tag{30-17}$$

where k_5 is a constant.

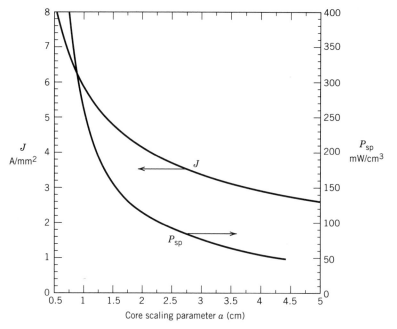

Figure 30-9   Maximum current density J and specific power dissipation (power density) P_{sp} as functions of the double-E core scaling parameter a. A maximum core surface temperature of 100°C and a maximum ambient temperature of 40°C is assumed. Litz wire with $k_{Cu} = 0.3$ is used for the current density versus core scaling parameter (Eq. 30-17).

Equations 30-15 to 30-17 are particularly useful because they show how the allowable specific loss, P_{sp}, in the core and windings, allowable flux density, B_{ac}, and allowable current density, J, for a given temperature difference $\Delta T = T_s - T_a$ scale with the physical size of the inductor/transformer. These equations can be plotted as a function of a.

If desired, the constants of proportionality, k_3, k_4, and k_5 can be evaluated quantitatively for specific core sizes such as the double-E core of Fig. 30-6 and Table 30-1. Then Eqs. 30-15 and 30-17 can be plotted as a function of a as is done in Fig. 30-9. If a database of such information, as is indicated in Fig. 30-9, is maintained for a variety of commonly used core shapes and sizes, it will significantly simplify the overall design process for inductors and transformers as will be demonstrated in a later section.

30-4   ANALYSIS OF A SPECIFIC INDUCTOR DESIGN

The purpose of the discussion in Sections 30-1 through 30-3 is to provide the basis for a detailed procedure for designing inductors and transformers. However, these sections demonstrate that many different and sometimes tightly coupled variables will be involved in the ultimate design procedure. Since analysis is almost always simpler to understand than design, we will analyze the performance of a specific inductor design. This analysis will serve to further illustrate the many factors that influence inductor and transformer design and will make it easier to describe the design procedure.

30-4-1 INDUCTOR PARAMETERS

In this design example, an inductor wound on the double-E core of Fig. 30-6a is used. The core width $a = 1$ cm and many of the important core dimensions are listed in Table 30-1. The inductor is used in a resonant circuit at 100 kHz with a worst-case or rated sine wave current of 4 A rms. Because of the high frequency and the rated current magnitude $I_{rms} = 4$ A, a Litz wire with a copper area of $A_{Cu} = 0.64$ mm^2 is used. The number of turns $N = 66$ and the bobbin is completely filled. In the inductor the total air gap (see Figs. 30-7 and 30-8), Σg, is 3 mm and composed of four air gaps (two in series in each leg) in the flux or H-field loop paths. The inductor is black with an emissivity $E = 0.9$. The ambient temperature $T_a = 40°C$ or less. The inductance L and its hot-spot temperature are to be estimated. To keep the analysis simple, eddy current losses due to the proximity effect will be ignored.

30-4-2 CHARACTERISTICS OF THE INDUCTOR

30-4-2-1 Copper Fill Factor k_{Cu}

Using Eq. 30-9 with $N = 66$, $A_{Cu} = 0.64$ mm^2, and $A_w = 140$ mm^2 from Table 30-1 yields

$$k_{Cu} = \frac{(66)\,(0.64)}{140} = 0.3$$

30-4-2-2 Current Density J and Winding Losses P_w

The current density in the winding at the maximum current density $I_{rms} = 4$ A is

$$J_{rms} = \frac{I_{rms}}{A_{Cu}} = \frac{4\ A}{0.64\ mm^2} = 6.25\ A/mm^2$$

Using Eq. 30-12a and using $V_w = 12.3$ cm^3 from Table 30-1, the total winding loss at the rated current is

$$P_w = P_{w,sp}\,V_w = 22k_{Cu}\,(J_{rms})^2\,V_w = (22)(0.3)(6.25)^2(12.3) = 3.17\ W$$

30-4-2-3 Flux Densities and Core Losses

The worst-case peak inductor current is $\hat{I} = \sqrt{2}\,I_{rms} = 5.66$ A, and thus $\hat{I}N = (5.66)(66) = 374$ At. From Eq. 3-54, assuming $H_{core} = 0$, the field in the gap, H_g, is

$$\hat{H}_g = \frac{\hat{I}N}{\Sigma_g} = \frac{374}{0.003} = 1.25 \times 10^5\ At/m$$

Using Eqs. 3-55 and 3-56, the flux density $\hat{B}_g = 4\pi \times 10^{-7}\,\hat{H}_g = 157$ mT. At every air gap in the magnetic core, there will be fringing flux as is shown in Fig. 30-10a. The total flux in the air gap must be the same as the total flux in the core, but since the total flux in the air gap is spread out over a larger cross-sectional area, the flux density in the gap is significantly lower than the flux density in the core. This decrease in air gap flux density is shown schematically in Fig. 30-10c as a decreased number of flux lines per unit area in the gap compared to the same number of lines being confined to the smaller cross-sectional area of the core. In this simplified analysis, we shall model the effect of the air

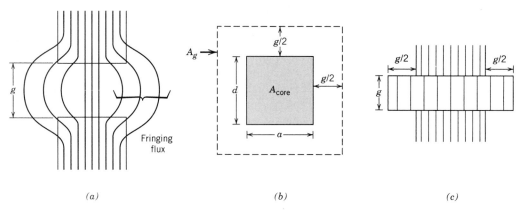

(a)   *(b)*   *(c)*

Figure 30-10   Fringing fields in the air gap of (*a*) an inductor (*b*) effective cross-sectional area of the gap and (*c*) an equivalent representation of the air gap.

gap as the rectangular block shown in Fig. 30-10*c* having a height or gap length *g* and cross-sectional area A_g given by

$$A_g = (a + g)(d + g) \qquad (30\text{-}18)$$

Then using Eq. 3-59 we obtain for $\hat{B}_{core}$

$$B_{core} = \frac{A_g}{A_c} B_g \qquad (30\text{-}19)$$

Using this equation along with the area information from Table 30-1 and $\hat{B}_g = 157$ mT, we obtain $\hat{B}_{core} = (1.69)(157)/1.5 = 177$ mT. This is also the value of B_{ac} since no dc current is flowing in the inductor.

The total maximum flux is calculated as

$$\hat{\phi}_{max} = \hat{B}_{core} A_{core} = (0.177)(1.5 \times 10^{-4}) = 2.6 \times 10^{-5} \text{ Wb} \qquad (30\text{-}20)$$

In Fig. 30-2, a nomogram of the specific core losses of the 3F3 core ferrite material is presented. From this figure, at a core temperature of 100°C and a frequency of 100 kHz, the power loss is 245 mW/cm^3, and thus the maximum total core loss $P_{core} = 3.3$ W (core volume = 13.5 cm^3 from Table 30-1).

30-4-3   INDUCTANCE *L*

Assuming a linear relationship between the flux and the current, Eqs. 3-57 and 3-67 can be used to find the inductance *L* as

$$L = \frac{N \hat{\phi}}{\hat{I}} = \frac{(66)(2.65 \times 10^{-5})}{(5.66)} = 309 \ \mu H \qquad (30\text{-}21)$$

In the above approach it is assumed that the leakage flux between the upper and lower yokes of the core is negligible, an assumption that is not well satisfied. This results in a measured value of inductance *L* that is significantly larger than the 0.31 mH value given above, which is based on an air gap $\Sigma g = 3$ mm. Further consequences of this neglected leakage flux are core flux densities greater than the calculated value of 177 mT and larger core loss. If the design goal is to keep the core losses at 3.3 W, then the total air gap should be increased until *L* = 0.31 mH. At this value of *L*, $\hat{B}_{core}$ will equal 177 mT.

30-4-4 TEMPERATURES IN THE INDUCTOR

Both radiative and convective heat transfer will contribute to the cooling of an inductor or transformer. The thermal resistance due to radiative heat transfer, $R_{\theta,\text{rad}}$, is given by Eq. 29-17. Using this equation and a total surface area for the inductor of 0.006 m^2 (see Table 30-1) along with $T_s = T_{\text{body}} = 100°C$ and $T_a = 40°C$, $R_{\theta,\text{rad}}$ becomes

$$R_{\theta,\text{rad}} = \frac{60}{(5.1)\,(0.006)\left[\left(\dfrac{373}{100}\right)^4 - \left(\dfrac{313}{100}\right)^4\right]} = 20.1°C/W$$

The thermal resistance due to convection, $R_{\theta,\text{conv}}$, is given by Eq. 29-20. For the inductor under consideration the appropriate numerical values for the parameters in Eq. 29-20 are $d_{\text{vert}} = 3.5a = 3.5$ cm, $\Delta T = 60°C$, and surface area $A = 0.006$ m^2. Using these values $R_{\theta,\text{conv}}$ is

$$R_{\theta,\text{conv}} = \frac{1}{(1.34)(0.006)} \sqrt[4]{\frac{0.035}{60}} = 19.3°C/W$$

These two thermal resistances are in parallel, and the total thermal resistance, inductor body to ambient when $T_{\text{body}} = 100°C$ and $T_a = 40°C$ is $R_\theta = 9.8°C/W$. The maximum body temperature T_{body} using this value of R_θ in Eq. 29-5 is

$$T_{\text{body,max}} = (9.8)\,(3.17 + 3.3) + 40 = 104\ °C$$

The so-called hot-spot temperature in the center of the middle leg of the inductor may be 5–10°C above the 104°C surface body temperature. The thermal conductivity of the ferrite is quite large, and thus the temperature of the center of the two outer legs may only be approximately 2°C above the surface temperature. The average core temperature can be estimated to be 106°C. The specific core loss is a minimum at 100°C so the core loss calculated at 100°C should be valid at 106°C.

A prototype inductor should be fabricated and put into the cabinet of the resonant converter. It should be operated at the rated conditions (4 A, 100 kHz, $T_a = 40°C$), and the surface temperature should be measured. If surface temperature is too high or too low, the design can be changed accordingly.

30-4-5 EFFECT OF AN OVERCURRENT ON THE HOT SPOT TEMPERATURE

The consequences of an overcurrent on the performance of the inductor, especially on the maximum hot spot temperature should be examined before we complete our examination of this specific example. For purposes of illustration, an overcurrent of 25%, that is, $I_{\text{rms}} = 5$ A, will be assumed, corresponding to $\hat{B}_{\text{core}} = 0.221$ Wb. This overcurrent will not change the inductance since the core will still be in its linear magnetic region. However, the loss in the windings will increase by a factor of $(\frac{5}{4})^2 = 56\%$ (the resistivity of the winding is assumed to stay constant). B_{ac} increases by the same factor of 1.25 as the current and using the graph of core loss in Fig. 30-2, the specific core loss at $B_{\text{ac}} = 0.221$ Wb is 440 mW/cm^3. Equation 30-2b with the appropriate values for a 3F3 core material predicts that an increase in B_{ac} by a factor of 1.25 increases $P_{m,\text{sp}}$ by a factor of 1.77, well within the accuracy of the nomogram of Fig. 30-2, which gives an increase of 1.8.

The total losses in the inductor when the current is 5 A and the frequency is 100 kHz is $P = P_w + P_{\text{core}} = (3.17)(1.56) + (3.31)(1.8) = 10.9$ W. If it is assumed that $R_\theta = 9.8°C/W$ (calculated at $T_{\text{body}} = 100°C$ and $T_a = 40°C$) can still be used, the body-to-ambient temperature difference increases from 56°C for a current of 4 A to 107°C for a current of 5 A. This large temperature increase is unacceptable in most designs. To keep the ΔT limited to 60°C in the presence of the overcurrent, R_θ must be reduced to 5.5°C/W. This could be achieved by either forced-air cooling with a fan or by using a heat sink bonded by a thermal conductive epoxy to the ferrite core. Laboratory tests should be done to confirm this expectation.

30-5   INDUCTOR DESIGN PROCEDURES

In the previous section a detailed analysis of a specific inductor design was done in order to highlight the important aspects of inductor/transformer design. This analysis was a straightforward application of the concepts of magnetostatic circuits from Chapter 3 and of thermal resistance from Chapter 29. However, in a power electronic converter design situation, the job facing the designer is not the analysis of a specific inductor or transformer where all the dimensions and magnetic properties of the core are known. Rather the job is just the inverse of the analysis example. In the typical design scenario, the dimensions and magnetic properties of the core as well as the number of turns and the type of copper winding must be specified so that the resulting inductor or transformer will meet a given set of electrical and thermal specifications. In order to avoid a haphazard cut-and-try approach, a methodical design procedure is needed that will lead to a correct design in a minimum number of confusing iterations and design trade-off decisions Such a design procedure is given and illustrated in this section. Since the design of inductors is usually limited by thermal considerations, the design procedure incorporates heat dissipation limitations as a basic part of the design process. In order to simplify the initial presentation of the design procedure, eddy current loss is assumed to be negligible. Corrections for eddy current loss will be discussed in a later section.

30-5-1   INDUCTOR DESIGN FOUNDATION: THE STORED ENERGY RELATION

An equation commonly termed the stored energy relation is the starting point for an inductor design. We begin its derivation by using results from Chapter 3 (Eq. 3-67) to express the inductance-current product as

$$L\hat{I} = N\hat{\phi} \tag{30-22}$$

The number of turns N is given by Eq. 30-9 ($N = k_{Cu}A_w/A_{Cu}$). The flux in the inductor is, using Eq. 3-57, $\phi = A_{core}B$. The required copper area A_{Cu} is obtained from $J_{rms} = I_{rms}/A_{Cu}$. Using these results in Eq. 30-22 yields

$$L\hat{I}I_{rms} = k_{Cu}J_{rms}\hat{B}A_wA_{core} \tag{30-23}$$

where $\hat{B}$ is in Wb and A_{core} is in m². If J_{rms} is in A/mm²,then A_w must be in mm². The area-product, AP, is given by

$$AP = A_wA_{core} \tag{30-24}$$

Equations 30-23 and 30-24 are the basis for an inductor design procedure because they relate design inputs (L, $\hat{I}$, and I_{rms}) to the product of material parameters (J_{rms} and $\hat{B}$) and geometric parameters (k_{Cu}, A_w, and A_{core}) of the core and winding.

Most inductor designs make use of standard commercially available cores. In this situation, the use of Eqs. 30-23 and 30-24 for design can be made even more explicit.

Equation 30-17 establishes the quantitative relationship between J_{rms} and the geometric size of the core while Eq. 30-16 does the same for flux density $\hat{B}$. Using Eqs. 30-16 and 30-17 to substitute for J_{rms} and $\hat{B}$ in Eq. 30-23 and noting that the area product is proportional to $(a)^4$, we can write

$$L\hat{I}I_{rms} = \frac{\text{const. } a^{3.1}}{f^{0.52}} \sqrt{k_{Cu}} \qquad (30\text{-}25)$$

The assumptions used in arriving at this equation are (1) specifying a given temperature difference between the surface of the inductor and the ambient, (2) all the geometric parameters of the core are available in terms of a scaling dimension a, and (3) ignoring eddy current loss due to the proximity effect, a topic to be discussed later.

30-5-2 SINGLE-PASS INDUCTOR DESIGN PROCEDURE OUTLINE

The step-by-step outline of a single-pass inductor design procedure is given in this section. For simplicity we will assume that the peak instantaneous flux density $\hat{B} < B_{sat}$. If magnetic saturation and not core loss limits $\hat{B}$, then modifications to the procedure are required. These modifications are discussed in a later section. The single-pass noniterative nature of the procedure hinges upon a complete database of core characteristics being available to the designer. The detailed nature of this database will be described shortly. A flowchart diagram shown in Fig. 30-11 summarizes the procedure. An example of the application of the procedure will be given to illustrate the details of the procedure.

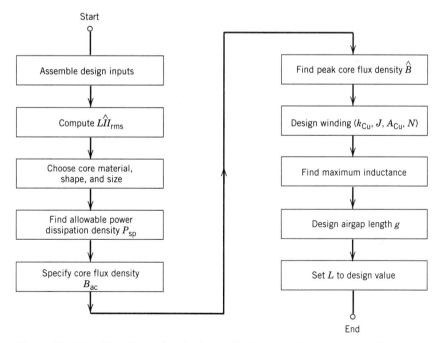

Figure 30-11 Flowchart of a single-pass inductor design procedure. The procedure assumes the existence of an extensive database of characteristics of all cores available to the inductor designer. The characteristics include the allowable power dissipation density for the design temperature range.

Step 1   Assemble design inputs

The design inputs consist of the following parameters:

a. Inductance value L

b. Rated peak current $\hat{I}$

c. Rated dc current I_{dc}

d. Rated rms current I_{rms}

e. Operating frequency f

f. Maximum surface temperature of the inductor T_s and the maximum ambient temperature T_a

The values for the first five inputs are found via the design calculations for the specific power electronic converter circuit in which the inductor is to be used. The maximum temperatures are determined from considerations of other temperature-limited components used in the same circuit, the temperature limitations of the inductor materials, and the environment in which the inductor must operate. A commonly used value of T_s is 100°C.

Step 2   Compute stored energy value $L\hat{I}I_{rms}$

The second step is to use the design inputs to compute $L\hat{I}I_{rms}$. $\hat{I}$ is the maximum peak value of the current and I_{rms} is the maximum rms value of the current.

Step 3   Choose core material, shape, and size

The core material, core shape, and size are chosen next. The choice of material will be influenced by the operating frequency. The material performance chart shown in Fig. 30-3 will be useful if ferrites are the material of choice. At lower frequencies, magnetic steels, powdered iron cores, and amorphous metallic glasses could also be considered. The choice of core shape, that is, E-core, U-core, toroid, and so forth, will depend on cost, availability, and ease of making the windings on the chosen core shape.

The core size is chosen based on the value of $L\hat{I}I_{rms}$ computed in Step 2. The designer looks in the core database for the core with the nearest value of $k_{Cu}J_{rms}\hat{B}A_wA_{core}$ (see Eq. 30-23), which is larger than $L\hat{I}I_{rms}$. This lookup step assumes that the characteristics of all cores of interest have been tabulated in a database such as is illustrated in Table 30-3. Such a database can be constructed independent of any particular inductor design using the methods illustrated in Sections 30-1 and 30-2. The database entries of J_{rms}, B_{ac}, and $k_{Cu}J_{rms}\hat{B}A_wA_{core}$ are calculated assuming that the power dissipated in the inductor is uniformly distributed throughout the inductor, that is, power dissipation density in the winding $P_{w,sp}$ equals the power dissipation density in the core, $P_{core,sp}$. It is shown in the problems (Problem 30-21) that uniform power dissipation density yields the highest value of the product $k_{Cu}J_{rms}\hat{B}A_wA_{core}$, assuming a maximum surface temperature T_s and thus a maximum total power dissipation $P_T = [T_s - T_a]/R_{\theta sa}$.

Table 30-3 Database of Core Characteristics Needed for Inductor Design

Core No.	Material	$AP = A_wA_{core}$	R_θ at $\Delta T = 60°C$	P_{sp} at $\Delta T = 60°C$	J_{rms} at $\Delta T = 60°C$ and P_{sp}	B_{ac} at $\Delta T = 60°C$ and 100 kHz	$k_{Cu}J_{rms}\hat{B}A_wA_{core}$
•	•	•	•	•	•	•	•
8^a	3F3	2.1 cm^4	9.8°C/W	237 mW/cm^3	$3.3/\sqrt{k_{Cu}}$	170 mT	$0.0125\sqrt{k_{Cu}}$
•	•	•	•	•	•	•	•

^{a}Core number 8 is the same as listed in Table 30-1 with $a = 1$ cm.

Note that the value of the stored energy given by Eq. 30-23 or equivalently the database entry, $k_{Cu}J_{rms}\hat{B}A_wA_{core}$, depends on the copper fill factor k_{Cu}. Since these factors range from 0.3 for Litz wire to approximately 0.6 for round conductors or even larger for foil windings, the value $k_{Cu}J_{rms}\hat{B}A_wA_{core}$ for a given core could vary by almost a factor of 2. Depending on the increments in the sizes of core ratings available to the designer, this factor of nearly 2 in core rating could mean that using Litz wire might require a larger core size than using solid round conductors. Thus the designer must choose the type of wire to be used for the winding in order to select the core size. If the subsequent design steps described below produce an acceptable design, then these steps will be a single-pass procedure. If an acceptable design is not produced, then it will be necessary to repeat the design steps after either choosing a larger core or choosing a wire type having a larger copper fill factor.

Step 4 Find R_θ and P_{sp}

The thermal resistance $R_{\theta sa}$ (surface-to-ambient), of the combined core and winding is the next item to be found. It can either be looked up in the core database as is shown in Table 30-3 or calculated using the approach illustrated in Section 30-3. In either case the value of $R_{\theta sa}$ must conform to the specified temperatures T_s and T_a. The allowable specific power density, P_{sp}, which can be dissipated in the core and the winding, can either be found in the core database or calculated using

$$P_{sp} = P_{core,sp} = P_{w,sp} = \frac{(T_s - T_a)}{R_{\theta sa}(V_{core} + V_w)} \qquad (30\text{-}26)$$

Step 5 Specification of the ac core flux density

The ac core flux density B_{ac} is now found using the core database, Table 30-3. Otherwise B_{ac} can be found using the allowable value of specific power density dissipation, P_{sp}, estimated in Step 5 in conjunction with data such as Fig. 30-2 or Eq. 30-2, which are provided by the core manufacturer.

Step 6 Calculation of the peak core flux density $\hat{B}$

The flux density in the inductor core is proportional to the current in the inductor. The application dictates the peak current $\hat{I}$, and the inductor core must be chosen so that the peak current $\hat{I}$ generates a peak flux density $\hat{B}_{core} < B_{sat}$. If the peak current has both a dc value I_{dc} and an ac value $\hat{I} - I_{dc}$, then the ratio

$$\frac{\hat{I} - I_{dc}}{\hat{I}} = \frac{B_{ac}}{\hat{B}_{core}} \quad \text{or} \quad \hat{B}_{core} = B_{ac}\frac{\hat{I}}{\hat{I} - I_{dc}} \qquad (30\text{-}27)$$

If Eq. 30-27 gives $\hat{B}_{core} > B_{sat}$, then B_{ac} must be reduced until $\hat{B}_{core} < B_{sat}$. In this case, the core flux is not loss limited.

Step 7 Specification of winding parameters (J_{rms}, A_{Cu}, N)

The type of winding conductor, round wire, Litz wire, and so forth to be used was chosen in Step 3 so the copper fill factor, k_{Cu}, is known. Selection of the conductor type will depend on the operating frequency and the importance of eddy current loss in the windings. The allowable current density can now be estimated using the core database, Table 30-3, or by inverting Eq. 30-12. The required area of the copper conductor A_{Cu} is then given as $A_{Cu} = I_{rms}/J_{rms}$. The required number of turns N is then found from Eq. 30-9.

Step 8 Calculation of L_{max} of selected core

The maximum inductance achievable with the specified core is given by

$$L_{max} = \frac{N A_{core} \hat{B}_{core}}{\hat{I}} \tag{30-28}$$

L_{max} should be greater than the design value of inductance if the design procedure has been followed properly, including accurate entries into the core database. The remaining steps of the design procedure will lower L_{max} and save on copper winding weight costs. If L_{max} is significantly greater than the design value of L, the next smaller core size should be used and the design sequence repeated.

Step 9 Specification of the air-gap length

The air-gap length Σg is the last dimension to be found in an inductor design. The air gap must be tailored to give the value of $\hat{B}_{core}$ when the inductor current is $\hat{I}$. The total reluctance, R_m, of the magnetic flux path is given by

$$R_m = \frac{N\hat{I}}{A_{core}\hat{B}_{core}} = R_{m,core} + R_{m,gap} = \frac{l_c}{\mu A_c} + \frac{\Sigma g}{\mu_0 A_g} \tag{30-29}$$

where l_c is the magnetic flux path length in the core material and the air-gap area A_g is given by Eq. 30-18 for a double-E core. In most situations

$$R_{m,gap} = \frac{\Sigma g}{\mu_0 A_g} \gg R_{m,core} = \frac{l_c}{\mu A_c}$$

so that the gap length Σg is given by

$$\Sigma g = \mu_0 A_g R_{m,gap} = \mu_0 A_g \frac{N\hat{I}}{A_{core}\hat{B}_{core}} \tag{30-30}$$

This equation cannot be used by itself to find Σg because the gap area A_g is a function of the gap length (see Eq. 30-18). Additionally the gap length parameter (g) in Eq. 30-18 is usually much smaller than g_{rated} because several smaller air gaps are uniformly distributed around the magnetic flux path. If there are N_g such distributed air gaps each having a length g, then

$$\Sigma g = N_g g \tag{30-31}$$

Distributed air gaps are used to keep the magnetic flux, which fringes into the copper windings to a minimum. This fringing flux causes additional eddy current loss in the windings. A single large air gap would have fringing magnetic flux paths that penetrate much farther into the copper windings and thus produce much larger eddy current loss than a distributed air gap. Substituting Eqs. 30-18 and 30-31 into Eq. 30-30 yields

$$\Sigma g = \mu_0 \frac{N\hat{I}}{A_{core}\hat{B}_{core}} \left(a + \frac{\Sigma g}{N_g} \right) \left(d + \frac{\Sigma g}{N_g} \right) \tag{30-32}$$

Expansion of Eq. 30-32 yields a quadratic equation in the desired parameter Σg. In practical design situations involving distributed gaps, the individual gap length $g \ll a$ or d or equivalently $\Sigma g/N_g \ll a$ or d. Thus the quadratic terms in Eq. 30-32 can be neglected in comparison to the linear terms in Σg and manipulated to yield

$$\Sigma g \approx \frac{A_{core}}{\dfrac{A_{core}\hat{B}_{core}}{\mu_0 N\hat{I}} - \dfrac{(a+d)}{N_g}} \tag{30-33}$$

Although Eq. 30-33 only applies to a double-E core, similar expressions can be developed for other core shapes.

Step 10   Adjustment of inductance value L_{max}

If the desired inductance value $L < L_{max}$, then L_{max} can be reduced with attendant savings in material size, weight, and cost. The reduction could be obtained by increasing the gap length beyond the value given by Eq. 30-33. However, this would make $\hat{B}_{core}$ less than the maximum allowable flux density, which in turn would lead to lower core loss and lower body temperature. This would be a poor design because it would not reduce the loss in the copper winding. A better approach would be to reduce the number of turns N until the desired value of L is reached. The copper weight and volume would be less and $P_{w,sp}$ could be increased and A_{Cu} reduced. This will result in cost saving as well. In doing this, the flux density is kept constant by adjusting the air-gap length Σg.

30-5-3   ITERATIVE INDUCTOR DESIGN PROCEDURE

If a complete database of available core characteristics similar to that shown in Table 30-3 is not available to the designer, then an iterative design procedure must be used. In this procedure, an initial estimate of the stored energy and thus the core size is made using Eq. 30-23 and typical values of J_{rms} and $\hat{B}$. For the initial estimate a reasonable value J_{rms} could be 2–4 A/mm^2. An initial estimate of B_{ac} can be found using an assumed value of $P_{core,sp}$ and either a graph of core loss versus B_{ac} such as Fig. 30-2 or Eq. 30-2. Maximum values of $P_{core,sp}$ in most design situations are usually 10–100 mW/cm^3, which for a 3F3 ferrite material at a frequency of 100 kHz gives a value of B_{ac} between 47 mT (10 mW/cm^3) and 105 mT (100 mW/cm^3).

Using the initial estimate of core size, corrected values of J_{rms} and $\hat{B}$, which are consistent with this core size, are then calculated and used to obtain a corrected value of the product $k_{Cu}J_{rms}\hat{B}A_wA_{core}$. If this corrected value is greater than $L\hat{I}_rI_{rms}$, then the correct size of core has been found. If not, then a larger core size must be selected and the procedure repeated. The iterations continue until the correct size is found. The flow-chart shown in Fig. 30-12 summarizes the procedure.

30-5-4   INDUCTOR DESIGN EXAMPLE

As an example of the design procedure, the inductor analyzed in Section 30-4 will be "designed" using the single-pass method.

Step 1   Design inputs

$L = 300$ μH.
Sine wave current $\hat{I} = 5.6$ A.
$I_{rms} = 4$ A.
Frequency $= 100$ kHz.
Maximum temperatures $T_s = 100°C$ and $T_a = 40°C$.

Step 2   Stored energy

$L\hat{I}I_{rms} = 0.0068$ H-A^2

Step 3   Core material, shape, and size

Since the operating frequency is at 100 kHz, a ferrite material will be used for the core. The specific material chosen is 3F3 because, from Fig. 30-3, it has the best performance factor. A double-E core is chosen for the core shape. According to the core database, Table 30-3, core number 8 with $a = 1$ cm has $k_{Cu}J_{rms}\hat{B}A_wA_{core}$ equal to

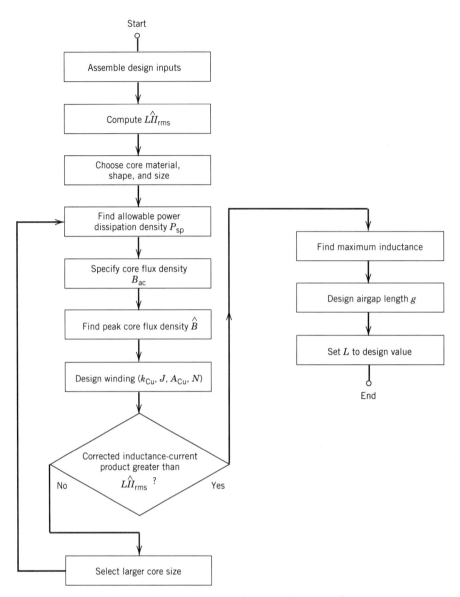

Figure 30-12 Flowchart for an iterative inductor design procedure.

$0.0125 \sqrt{k_{Cu}}$. This value is greater or equal to 0.0068 for all k_{Cu} values greater than 0.3. Litz wire, $k_{Cu} = 0.3$, is selected for the winding because it has the lowest eddy current loss of any conductor type at high frequencies like 100 kHz.

Step 4 R_θ and P_{sp}

From the core database $R_\theta = 9.8°C/W$ and $P_{sp} = 237 \text{ mW/cm}^3$.

Step 5   ac core flux density

Using the core database, $B_{ac} = 170 \text{ mT}$.

Step 6   Peak core flux density

There is no dc current in the inductor and thus $B_{ac} = \hat{B}$.

Step 7   Winding parameters (k_{Cu}, J_{rms}, A_{Cu}, N)

Since Litz wire is used, $k_{Cu} = 0.3$.
From the core database, $J_{rms} = 3.3/\sqrt{0.3} = 6$ A/mm^2.
The required conductor area $A_{Cu} = (4 \text{ A})/(6 \text{ A/mm}^2) = 0.67$ mm^2.
The number of turns

$$N = \frac{(140 \text{ mm}^2)\,(0.3)}{(0.67 \text{ mm}^2)} = 63 \text{ turns}$$

Step 8 L_{max}

Using Eq. 30-28

$$L_{max} = \frac{(63)(170 \text{ mT})(1.5 \times 0^{-4} \text{ m}^2)}{5.6 \text{ A}} = 287 \ \mu\text{H}$$

Step 9   Air-gap length

Using Eq. 30-33,

$$\Sigma g = \frac{1.5 \times 10^{-4} \text{ m}^2}{\dfrac{(1.5 \times 10^{-4} \text{ m}^2)(0.17 \text{ T})}{(4\pi \times 10^{-7} \text{ H/m})(63)(5.6 \text{ A})} - \dfrac{(0.025 \text{ m})}{4}} = 2.92 \text{ mm}$$

The length of each of the four distributed gaps is $g = (2.92)/4 \approx 0.73$ mm.

Step 10   Adjustment of L_{max}

The design value of L is about the same as L_{max} so no adjustments are needed.

30-6   ANALYSIS OF A SPECIFIC TRANSFORMER DESIGN

A transformer requires two or more conductor windings on the magnetic core compared to the single winding used for the inductor. These additional windings and the unique function of the transformer make the design of a transformer more complicated than for an inductor. To illustrate these complications, we will analyze a specific transformer design before undertaking the description of the transformer design procedure.

30-6-1   TRANSFORMER PARAMETERS

The transformer is wound on a double-E core shown in Fig. 30-6a. The core width $a = 1$ cm, the same sized core as used in the inductor example, and many of the important core dimensions are listed in Table 30-1. The primary current $I_{pri} = 4$ A rms is sinusoidal at a frequency of 100 kHz. The primary voltage is 300 V rms. The turns ratio $n = N_{pri}/N_{sec} = 4$ and $N_{pri} = 32$. Because of the high frequency and large primary and secondary currents, the transformer primary and secondary are wound with Litz wire. The bobbin is filled and the winding in the winding window is split between the primary and secondary windings as illustrated in Fig. 30-13. The transformer is black with an emissivity $E = 0.9$. The ambient temperature $T_a = 40°C$ or less. The transformer core flux density, leakage inductance, and hot-spot temperature are to be estimated.

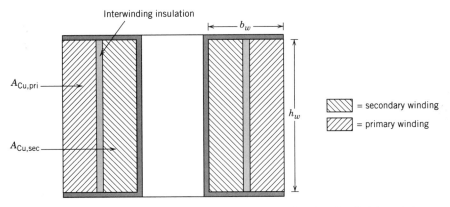

Figure 30-13 Winding window on a tranformer core showing the division of the windings in the window between primary and secondary.

30-6-2 TRANSFORMER ELECTRICAL CHARACTERISTICS

30-6-2-1 Areas of Primary and Secondary Conductors, A_{pri} and A_{sec}

The total area A_w is assumed to be completely occupied by the windings, which are split between the area occupied by the primary and that occupied by the secondary. The area occupied by the primary is

$$A_{w,pri} = \frac{N_{pri}A_{Cu,pri}}{k_{Cu,pri}} \tag{30-34}$$

and the area occupied by the secondary is

$$A_{w,sec} = \frac{N_{sec}A_{Cu,sec}}{k_{Cu,sec}} \tag{30-35}$$

The total winding area can be expressed as

$$A_w = A_{pri} + A_{sec} = \frac{N_{pri}A_{Cu,pri}}{k_{Cu}} + \frac{N_{sec}A_{Cu,sec}}{k_{Cu}} \tag{30-36}$$

where we assume that $k_{Cu,pri} = k_{Cu,sec} = k_{Cu}$ assuming the primary and secondary are wound with the same type of conductor. It is desirable to have the power dissipation density in the primary and secondary be the same so that the heat generated and thus the temperature rise, be uniformly distributed thoughout the windings. Using Eq. 30-12, this implies

$$k_{Cu}(J_{pri})^2 = k_{Cu}\left(\frac{I_{pri}}{A_{Cu,pri}}\right)^2 = k_{Cu}(J_{sec})^2 = k_{Cu}\left(\frac{I_{sec}}{A_{Cu,sec}}\right)^2 \tag{30-37a}$$

or

$$\frac{I_{pri}}{I_{sec}} = \frac{A_{Cu,pri}}{A_{Cu,sec}} = \frac{N_{sec}}{N_{pri}} \tag{30-37b}$$

Equations. 30-36 and 30-37 can be solved simultaneously to yield

$$A_{Cu,pri} = \frac{k_{Cu}A_w}{2N_{pri}} \tag{30-38}$$

and

$$A_{Cu,sec} = \frac{k_{Cu}A_w}{2N_{sec}} \tag{30-39}$$

For the transformer being analyzed, $k_{Cu} = 0.3$, $A_w = 140$ mm^2, $N_{pri} = 32$, and $N_{sec} = 8$. Using this information in Eqs. 30-38 and 30-39 yields

$$A_{Cu,pri} = \frac{(0.3)(140)}{(2)(32)} = 0.64 \text{ mm}^2 \quad \text{and} \quad A_{Cu,sec} = \frac{(0.3)(140)}{(2)(8)} = 2.6 \text{ mm}^2$$

30-6-2-2 Winding Loss P_w

The current densities in the primary and secondary windings are the same and are given by Eq. 30-37a. Evaluation of Eq. 30-37a yields

$$J_{rms} = \frac{4 \text{ A}}{0.64 \text{ mm}^2} = \frac{16 \text{ A}}{2.6 \text{ mm}^2} = 6.2 \text{ A/mm}^2$$

Since the current densities are the same in the primary and the secondary, Eq. 30-12 can be used to estimate the total winding losses $P_w = P_{w,sp}V_w$. The winding volume V_w, from Table 30-1, is 12.3 cm^3 so that the total winding loss is

$$P_w = (22)(0.3)(6.2)^2(12.3) = 3.1 \text{ W}$$

30-6-2-3 Flux Density and Core Loss

The peak flux density in the core, $\hat{B}_{core}$, can be estimated from the primary voltage since

$$\hat{V}_{pri} = N_{pri}A_c \left| \frac{d\hat{B}_{core}\sin(\omega t)}{dt} \right|_{max} = N_{pri}A_c \omega \hat{B}_{core} \tag{30-40}$$

Solving this equation for the flux density and putting in the numerical values for the parameters, including $\hat{V}_{pri} = \sqrt{2}\,(300) = 425$ V, yields

$$\hat{B}_{core} = \frac{425 \text{ V}}{(32 \text{ turns})(1.5 \times 10^{-4}\text{m}^2)(2\pi)(10^5 \text{ Hz})} = 0.141 \text{ T}$$

Using Eq. 30-2 for the specific core loss of the 3F3 ferrite, at a core temperature of 100°C and 100 kHz, and $\hat{B}_{core} = 0.14$ T yields $P_{sp,core} = 140$ mW/cm^3. The total core volume is 13.5 cm^3 (see Table 30-1) so the total core losses $P_{core} = 1.9$ W.

30-6-2-4 Leakage Inductance

It is shown in a later section that the leakage inductance L_{leak} of a transformer wound on a rectangular-shaped core such as the double-E core of Fig. 30-6 is given by

$$L_{leak} \approx \frac{\mu_0(N_{pri})^2 l_w b_w}{3h_w} \tag{30-41}$$

The mean length l_w of a single conductor turn, assuming that the winding volume is completely filled, is obtained from the top view of the double-E core shown in Fig. 30-14. The mean turn length is

$$l_w \approx 9a \tag{30-42}$$

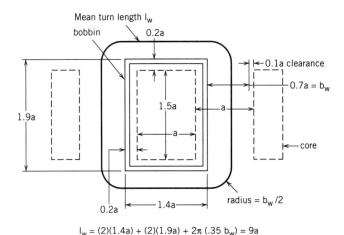

Mean turn length l_w

bobbin

0.2a

0.1a clearance

$0.7a = b_w$

1.5a

a

1.9a

a

core

0.2a

1.4a

radius = b_w/2

$l_w = (2)(1.4a) + (2)(1.9a) + 2\pi(.35\, b_w) = 9a$

Figure 30-14   Top view of the double-E core bobbin showing the mean length l_w of a single conductor turn (winding) assuming the winding window is completely filled.

The dimensions of the winding window are found in Section 30-1-5 to be $h_w = 2a$ and $b_w = 0.7a$. Using these dimensions in Eq. 30-41 along with $a = 1$ cm and $N_{\text{pri}} = 32$ yields

$$L_{\text{leak}} \approx \frac{(4\pi \times 10^{-9})(32)^2(9)(0.7)}{(3)(2)} = 14\ \mu\text{H}$$

30-6-3 TEMPERATURE IN THE TRANSFORMER

Both radiative and convective heat transfer will contribute to the cooling of the transformer. The surface-to-ambient thermal resistance due to radiation is given by Eq. 29-17. The surface-to-ambient thermal resistance due to convection is given by Eq. 29-20. The total surface-to-ambient thermal resistance for this transformer is the same as for the inductor example described in Section 30-4 because the core sizes are identical, the winding volume and surface area are identical, and the maximum surface temperatures, T_s, and ambient temperatures, T_a, are the same. Thus the total surface-to-ambient thermal resistance for the transformer is $R_{\theta sa} = 9.8°\text{C/W}$. The power dissipation in the winding is $P_w = 3.1$ W and in the core is $P_{\text{core}} = 1.9$ W. Hence the maximum surface temperature on the transformer is

$$T_{s,\text{max}} = (9.8)\,(3.1 + 1.9) + 40 = 89°C$$

30-6-4 EFFECT OF OVERCURRENTS ON TRANSFORMER TEMPERATURES

An overcurrent in the transformer primary or secondary will increase the temperature of the transformer exactly as an overcurrent increased the temperature of the inductor examined in Section 30-4. For purposes of illustration, we will assume an overcurrent of 25%, that is, $I_{\text{pri,max}} = 5$ A rms. The losses in the winding will increase by a factor of $(1.25)^2 = 1.56$, which will result in winding losses $P_w = (1.56)(3.1$ W$) = 4.8$ W. If it is assumed that $R_{\theta sa} = 9.8°\text{C/W}$ can still be used, the surface temperature T_s will increase to

$$T_{s,\text{max}} = (9.8)\,(4.8 + 1.9) + 40 = 106°C$$

This increase in temperature (17°C) is still within design parameters. If desired, the thermal resistance could be reduced by either forced-air cooling or by using a heat sink bonded to the core.

The reason the overcurrent produces a smaller temperature rise in the transformer compared to previous inductor example is because the flux in the core does not change significantly even with a 25% increase in the current. The transformer primary is driven by a nearly ideal voltage source whose magnitude is essentially unaffected by the current increase. Since the core flux is controlled by the applied voltage (see Eq. 30-40), it cannot change unless the voltage changes. In an inductor an increase in current leads to increases in flux density and thus core loss as well as winding loss.

30-7 EDDY CURRENTS

Until now, our discussion of inductors and transformers have assumed that eddy current loss in the windings could be neglected compared to the normal ohmic loss due to the dc resistance of the windings. The only concessions to possible eddy current loss has been in using Litz wire or thin foil windings to minimize skin effect problems. However, as operating frequencies continue to increase, the neglect of the proximity effect and attendant eddy current loss becomes increasingly difficult to justify. Inductors and transformers designed by procedures that do not take eddy currents into account have winding loss that become unacceptably large as the operating frequency increases. In this section we describe the phenomena of eddy current loss and how to optimize the design of windings so that the total loss, dc plus eddy current, are a minimum.

30-7-1 PROXIMITY EFFECT

Consider the cross-sectional view of an inductor winding carrying a current I shown in the winding window of a ferrite core in Fig. 30-15a. To simplify our initial considerations, the diameter of the winding conductor or the frequency are assumed to be small enough so that the skin effect can be neglected. The application of Ampere's law, Eq. 3-53, along path A in Fig. 30-15a encloses several ampere-turns of magnetomotive force (mmf), thus showing that a magnetic field is present in the winding window. If path B, which is located at a greater distance x into the winding window, is taken, a greater amount of mmf is enclosed and the magnetic field is even larger. The approximate distribution of the mmf and thus the magnitude of the magnetic field is indicated in Fig. 30-15a.

The magnetic field generates eddy currents in the conductor windings in exactly the same manner as was described earlier for the conductive magnetic core. In Fig. 30-15, the magnetic flux is contained in the plane of the winding window and thus is perpendicular to the longitudinal direction (direction of applied current flow) of the conductor windings. Hence the eddy currents flow either parallel or antiparallel to the applied current as diagrammed in Fig. 30-15b. This generation of eddy currents is termed the proximity effect because the eddy currents in a specific conductor or winding layer are caused by the magnetic fields of the other current-carrying conductors in proximity to the given conductor.

These eddy currents will dissipate power, P_{ec}, and thus contribute to the electrical loss in the winding in addition to those caused by the normal ohmic loss, P_{dc}, due to the dc resistance of the windings. Our earlier discussion of eddy currents in conductive magnetic cores indicated that the power dissipation due to eddy currents was proportional to the square of the local magnetic field intensity (see Eq. 30-8). Thus the eddy current

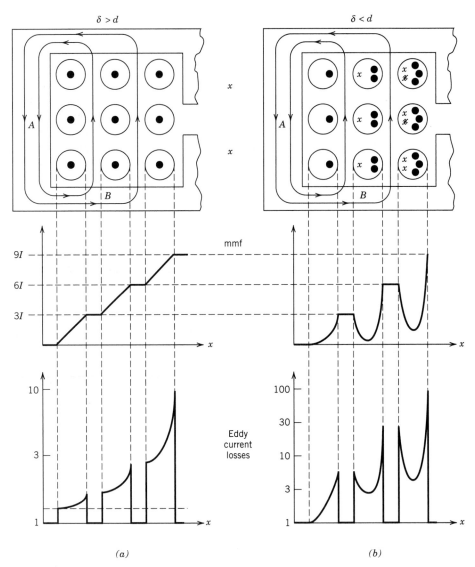

Figure 30-15   Cross-sectional view of an inductor winding in the winding window of a ferrite core and the associated spatial distribution of the mmf and eddy current loss density for the case where the skin depth is about the same as the conductor diameter (a) and for the case where the skin depth is much less than the conductor diameter. The dots (·) and x's in the conductor cross sections indicate the relative magnitude, spatial distribution of the current density, and current direction.

loss per unit length of conductor winding in Fig. 30-15 increases dramatically as the position x in the winding window increases, and thus the number of layers contributing to the local magnetic field increases. An approximate distribution of the eddy current loss with position is shown in Fig. 30-15a. In general the total eddy current loss increases dramatically with the number of layers in a given winding.

The total power dissipated in a winding is

$$P_w = P_{dc} + P_{ec} = (I_{rms})^2 R_{dc} + (I_{rms})^2 R_{ec} = (I_{rms})^2 R_{ac} \qquad (30\text{-}43)$$

where R_{ec} is the effective eddy current resistance and R_{dc} is the low frequency or dc resistance of the winding. The net resistance of the winding, R_{ac}, is given by

$$R_{ac} = F_R R_{dc} = \left(1 + \frac{R_{ec}}{R_{dc}}\right) R_{dc} \qquad (30\text{-}44)$$

where F_R is termed the resistance factor. For the situation illustrated in Fig. 30-15a where the diameter of the conductor is less than or about equal to the skin depth, the resistance factor will be somewhat greater than unity.

If the diameter of the winding conductor is significantly greater than a skin depth, then the eddy currents will flow only near the surface of the conductor. Because of the skin effect, no currents will flow in the interior as is sketched in Fig. 30-8c. Magnetic flux will be excluded from the interior of the conductor, and the mmf diagram will be modified as shown in Fig. 30-15b, which shows very little mmf in the interior of the conductor. The confinement of the total current to a thin area on the outside portion of the conductor means that current density will be much larger at the surface of the conductor than in the situation illustrated in Fig. 30-15a for the low-frequency case. As a consequence, for high frequencies the resistance factor will be much greater than what would be predicted on the basis of just the proximity effect at low frequencies. The resistance factor may be as large as one or two orders of magnitude larger than the dc resistance and hence so will the total loss.

30-7-2 OPTIMUM CONDUCTOR SIZE AND MINIMUM WINDING LOSS

The discussion in the previous section makes it clear that to minimize the eddy current loss in a winding, the diameter of the winding conductor (if a round conductor is used) or the thickness of a foil conductor should be made less than or at most equal to the skin depth. The difficulty is that the skin depth decreases as the operating frequency increases. Hence the dc losses would appear to become large at high operating frequencies. At any given frequency there is thus an optimum conductor diameter or thickness that minimizes the total loss in the conductor. The optimum size will be approximately the skin depth dimension but will vary to some degree with frequency and the number of layers in the winding. A general method to estimate the optimum conductor size will be presented shortly.

When the optimum diameter or thickness is used for the conductor winding, the resistance factor has the value [1]

$$F_R = 1.5 \qquad (30\text{-}45)$$

This means that the power dissipation due to eddy currents is equal to

$$P_{ec} = 0.5 P_{dc} \qquad (30\text{-}46)$$

and the total winding loss is

$$P_w = 1.5 P_{dc} \qquad (30\text{-}47)$$

30-7-3 REDUCTION OF LOSS IN THE INDUCTOR WINDING

At this point in our discussion of eddy current loss, we seem to be in a dilemma. In order to keep eddy current loss under control as the frequency increases, the diameter of round conductors used for the windings of inductors must decrease. This would appear to lead to large dc resistances and thus unacceptably large winding loss if the applied currents have any appreciable magnitude. What is required is a way to reduce the dc resistance without increasing the eddy current loss.

This can be done by connecting several of the small (significantly smaller than a skin depth) diameter wires in parallel. The paralleled wires must then be twisted or woven into a ropelike assembly in which each individual wire periodically moves from the interior to the exterior of the conductor assembly and back again. This periodic twisting or transposition of each individual wire causes the induced voltage in a given wire in one half-twist section to be opposite the induced voltage in the next half-twist section of the same wire, as is shown in Fig. 30-16. This causes eddy currents induced in one part of a given wire to be opposite those induced in the next half-twist section, as is illustrated in Fig. 30-16. The net result is that little if any eddy current will be generated in the twisted pair. Each wire must be electrically insulated from the other wires in the bundle, and all the wires are connected in parallel only at the terminals of the inductor, which are located outside the winding volume.

Wire bundles made in such a manner are called Litz wires. The diameter of a Litz wire bundle can be built up to many times larger than the diameter of an individual wire strand in order achieve the desired reduction in dc resistance without any significant increase in eddy current losses. The disadvantage of this approach is that Litz wires are much more expensive than solid wire conductors and the fill factor k_{Cu} is only about 0.3.

If the currents flowing in the inductor are large and the number of turns is relatively small, an alternative conductor arrangement consisting of thin wide rectangular-shaped conductors may be preferable to Litz wire. The width of the conductor can be almost as large as the height h_w of the winding window of the bobbin (see Fig. 30-6). The thickness, h, of this so-called foil conductor must be on the order of a skin depth or less in size, and there is an optimum value as was discussed previously. The fill factor k_{Cu} for foil conductors is approximately 0.6.

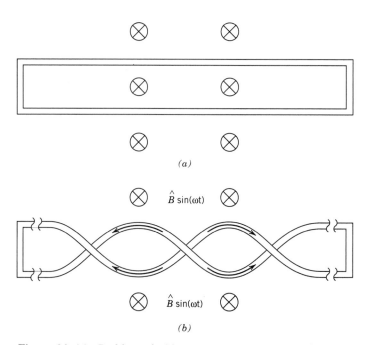

(a)

(b)

Figure 30-16 Problem of eddy currents in windings. In (a) two parallel wires shorted together at each end will intercept flux which will generate eddy currents around the loop. However in (b) the wires are twisted so that the effects of intercepted flux are largely canceled out.

30-7-4 SECTIONING TRANSFORMER WINDINGS TO REDUCE EDDY CURRENT LOSS

The presence of the secondary winding in a transformer enables the eddy current loss in a transformer to be minimized. Consider the mmf distribution in the transformer winding window shown in Fig. 30-17b. The mmf in the secondary section of the transformer has a negative slope and goes back to zero because the induced current in the secondary is opposite that in the primary.

Now consider sectioning the primary winding into two separate parts as shown in Fig. 30-18a and sandwiching the secondary between the two halves of the primary. The total number of primary and secondary turns remains unchanged as does the total volume taken up by each winding. The resulting mmf distribution is also shown in Fig. 30-18a. The peak value of the mmf with this sandwiched winding is now approximately one-half the peak value of that in Fig. 30-17 for the simple transformer winding since the number of ampere-turns in each primary half-section is one-half of the value shown in Fig. 30-17. Since the peak mmf is reduced by a factor of 2, so is the maximum magnetic flux in the

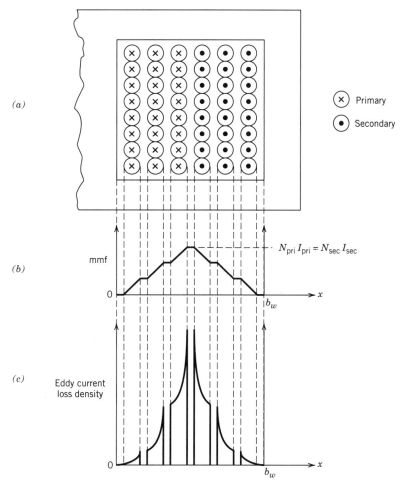

Figure 30-17 Winding window in a transformer containing (a) a simple winding arrangement, (b) mmf distribution versus position, and (c) eddy current loss density versus position.

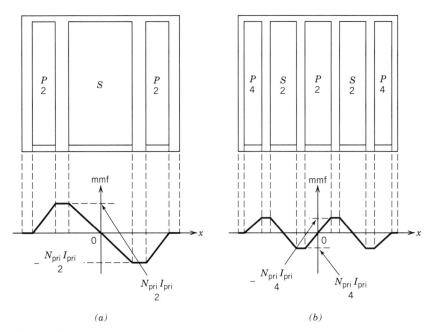

(a)   (b)

Figure 30-18 Partitioning of the primary and secondary windings into multiple sections to reduce eddy current loss. The resulting mmf distribution is indicated below the sectionalized winding diagram. In (a) there is a split primary and a single secondary and in (b) the primary is split into three sections and the secondary into two sections.

winding window. Eddy current loss is proportional to the square of the magnetic flux so the eddy current loss in the transformer of Fig. 30-18a should be approximately one-fourth that of the transformer of Fig. 30-17.

This approach can be extended by dividing both the primary and secondary windings into more sections as is indicated in Fig. 30-18b. Now the peak mmf is one-fourth that of Fig. 30-17, and so the eddy current loss should be one-sixteenth those of the simple transformer. In principle this subdividing of the windings into more sections could be continued until each section consists of one or two winding layers. This approach is not without its disadvantages. Winding a transformer in this manner is complex. The inter-winding capacitance is increased in proportion to the number of sections, and the amount of safety insulation between the primary and secondary windings increases, thus decreasing the reliability of the insulation and the copper fill factor.

30-7-5 OPTIMIZATION OF SOLID CONDUCTOR WINDINGS

The discussion in the previous sections (30-7-2 through 30-7-4) has provided several qualitative approaches for reducing the eddy current loss in inductor and transformer windings. However, optimizing the design of a winding requires a quantitative procedure for implementing these suggestions and assessing their benefits. Such a procedure has been developed that relates the power dissipated in the winding or section of winding to geometry of the winding (conductor cross-sectional dimensions, number of turns, and number of layers) and the skin depth of the winding conductor. [3,4] The procedure is based on a fairly general analysis that includes nonuniform magnetic fields across the cross-sectional area of the conductor, skin effect, and eddy current screening. The goal of the procedure is to find the combination of optimum conductor diameter or thickness and number of layers so that the total winding loss, dc and eddy current, is minimized.

The procedure is based on the set of curves shown in Fig. 30-19 of normalized power dissipation in the winding as a function of a normalized variable ϕ with the number of layers m in the winding section as a parameter. The normalized power dissipation is defined as

$$\frac{P_w}{R_{\mathrm{dc},h=\delta}\,(I_{\mathrm{rms}})^2} = \frac{R_{\mathrm{ac}}}{R_{\mathrm{dc},h=\delta}} = \frac{F_R R_{\mathrm{dc}}}{R_{\mathrm{dc},h=\delta}} \tag{30-48}$$

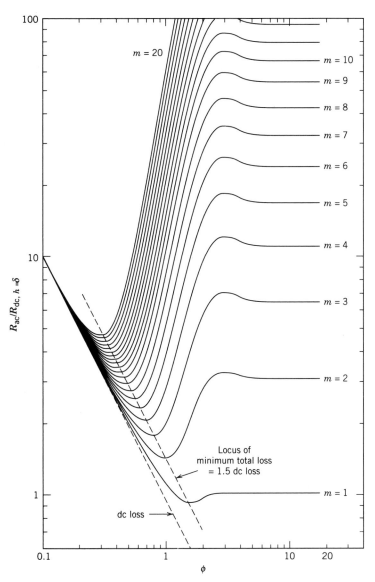

Figure 30-19 Normalized power dissipation in a winding or winding section as a function of $\phi = \sqrt{F_l}\,h/\delta$ with the number of layers m as a parameter. The normalizing value of power is $R_{\mathrm{dc},h=\delta}(I_{\mathrm{rms}})^2$.

where $R_{dc,h=\delta}$ is the dc resistance of the winding when the conductor diameter or thickness is equal to the skin depth. The parameter ϕ is given by

$$\phi = \frac{\sqrt{F_l}h}{\delta} \qquad (30\text{-}49)$$

In this equation, h is the effective conductor height, δ is the skin depth given by Eq. 30-3, and F_l is copper layer factor. For a rectangular conductor such as shown in Fig. 30-20a, the effective conductor height is the actual height h. For round conductors the effective conductor height is $[\sqrt{\pi/4}]d$ where d is the conductor diameter. The parameter m in Fig. 30-19 is the number of layers in the winding section under consideration. The copper layer factor is the fraction of the layer width (or equivalently the winding window height, see Fig. 30-6) h_w that is occupied by copper. For the winding made with rectangular conductor shown in Fig. 30-20a, the copper layer factor is b/b_0 and for the round conductors shown in Fig. 30-20b, the $F_l = d/d_0$. The dimensions b_0 and d_0 in Fig. 30-20 include the insulation on the conductor. For a layer composed of a single turn of foil conductor, the layer factor would equal unity.

A winding or winding portion of m layers has a low-frequency mmf distribution that varies linearly from zero on one side of the section to a maximum at the other side. For the split primary in Fig. 30-18a, each primary section has $M_{pri}/2$ layers where M_{pri} is the total number of layers in the primary. The secondary must also be considered to have two portions each having $M_{sec}/2$ layers. For the winding arrangement shown in Fig. 30-18b, the two outside primary sections have $M_{pri}/4$ layers while the central primary section is considered to be two separate portions each also having $M_{pri}/4$ layers. The two secondary sections should each be considered to have two portions each with $M_{sec}/4$ layers. The number of layers in winding sections for other divisions of the primary and secondary windings would be found in the same manner.

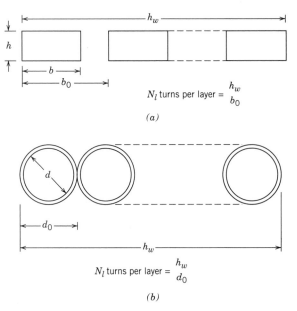

N_l turns per layer $= \dfrac{h_w}{b_0}$

(a)

N_l turns per layer $= \dfrac{h_w}{d_0}$

(b)

Figure 30-20 Evaluation of the winding parameters for (a) a rectangular (foil) conductor and (b) a round conductor.

The value of ϕ in the graph that corresponds to the minimum in the selected curve (selected by the number of layers in the winding section) corresponds to the optimum value of conductor diameter or thickness. Setting the conductor thickness or diameter to this optimum value results in the resistance factor F_R having the optimum value of 1.5. The availability of the set of curves in Fig. 30-19 permits several different winding designs (number of layers per section, number of sections into which the primary and/or the secondary are partitioned, and the type of winding conductor, round or rectangular) to be rapidly evaluated. A detailed explanation of how to apply these curves will be given in the transformer design procedure section (30-9).

30-8 TRANSFORMER LEAKAGE INDUCTANCE

Transformers used in power electronics in most cases should be designed for minimum leakage inductance because such inductance may be detrimental to the proper operation of the power electronic circuits. The leakage inductance can cause overvoltages in power switches at switch turn-off requiring the inclusion of a snubber circuit. The operation of some circuit topologies are sensitive to stray or leakage inductance.

The leakage inductance arises from magnetic flux that does not completely link the primary and secondary windings or does not completely link all the turns in the winding that generates the flux. An example of leakage flux is shown in Fig. 3-18. The magnetic flux shown in the winding window of the inductor in Fig. 30-15 is another example of leakage flux. Similarly the flux in the winding window of the simple transformer shown in Fig. 30-17 is a leakage flux which results in a leakage inductance.

The leakage inductance L_{leak} is defined as

$$\frac{1}{2} L_{\text{leak}} (I_{\text{pri}})^2 = \frac{1}{2} \int_{V_w} \mu_0 H^2 \, dV \tag{30-50}$$

The volume integration is taken over the winding volume V_w. Consider the simple transformer shown in Fig. 30-17. The mmf in the winding window is shown in Fig. 30-17b, and from this it is clear that the magnetic field in the window can be written approximately as

$$
\begin{aligned}
H_{\text{leak}} &= \frac{2N_{\text{pri}}I_{\text{pri}}x}{h_w b_w} \qquad 0 < x < \frac{b_w}{2} \\
&= \frac{2N_{\text{sec}}I_{\text{sec}}}{h_w} \left(1 - \frac{x}{b_w}\right) \qquad \frac{b_w}{2} < x < b_w
\end{aligned}
\tag{30-51}
$$

In Eq. 30-51, $N_{\text{pri}}I_{\text{pri}} = N_{\text{sec}}I_{\text{sec}}$. A volume element for the integral in Eq. 30-51 is $dV = l_w h_w \, dx$ where l_w is the length of turn located at position x in the winding window. The length l_w increases with position x. However, for simplicity we will consider it to be a constant that is equal to the mean turn length in the winding volume. For the double-E core used for the transformer in Fig. 30-17, $l_w \approx 9a$ (see Eq. 30-42 and Fig. 30-15). This approximation overestimates the turn length for $x < b_w/2$ but compensates for it by underestimating the turn length for $x > b_w/2$. Inserting the expressions for H_{leak} and volume element into Eq. 30-51 yields

$$
\begin{aligned}
\frac{1}{2} L_{\text{leak}} (I_{\text{pri}})^2 &= \frac{1}{2} 2 \int_0^{b_w/2} \mu_0 \left(\frac{2N_{\text{pri}}I_{\text{pri}}x}{h_w b_w}\right)^2 l_w h_w \, dx \\
&= \frac{\mu_0 (N_{\text{pri}})^2 l_w b_w (I_{\text{pri}})^2}{6 h_w}
\end{aligned}
\tag{30-52}
$$

Dividing both sides of the equation by $\frac{1}{2}(I_{\text{pri}})^2$ yields the expression for L_{leak} given in Eq. 30-41.

If the windings are split or partitioned as shown in Fig. 30-18 in order to reduce the eddy current loss, there will also be a reduction in the leakage inductance because of the reduction in stored magnetostatic energy resulting from the smaller peak magnetic fields. The winding arrangement shown in Fig. 30-18a has one-fourth the leakage inductance of the simple winding arrangement of Fig. 30-17 while the reduction in Fig. 30-18b is a factor of 16. A general expression for the leakage of a split winding arrangement is

$$L_{\text{leak}} \approx \frac{\mu_0(N_{\text{pri}})^2 l_w b_w}{3p^2 h_w}$$ (30-53a)

In this equation p is the number of interfaces between winding sections. A more exact treatment that takes into account the insulation between adjacent conductors in the same layer yields

$$L_{\text{leak}} \approx \frac{\mu_0(N_{\text{pri}})^2 l_w}{p^2 h_w}\left(\frac{b_{\text{Cu}}}{3} + b_i\right)$$ (30-53b)

where b_i is the interwinding insulation thickness and b_{Cu} is the total width of the copper in the winding window For the winding arrangement of Fig. 30-18a, $p = 2$ and in Fig. 30-18b, $p = 4$.

30-9 TRANSFORMER DESIGN PROCEDURE

A methodical step-by-step procedure is outlined for the design of small naturally cooled (convection and radiation) transformers used in power electronic applications. A single-pass procedure is presented that assumes access to a complete database of core characteristics. An iterative procedure, similar to that discussed for the inductor, can also be derived for the situation where a complete core database is not available. Heat dissipation considerations form an integral part of the design procedure. Eddy current loss is also taken into account in the design procedure. In most situations the same type of conductor is used for both primary and secondary windings, so we shall assume that $k_{\text{Cu,pri}} = k_{\text{Cu,sec}} = k_{\text{Cu}}$.

30-9-1 TRANSFORMER DESIGN FOUNDATION: THE VOLT-AMPERE RATING

The volt-ampere rating, S, of a transformer is defined as $S = V_{\text{pri}}I_{\text{pri}}$ where V_{pri} and I_{pri} are the rated rms values of voltage and current. Equation 30-40 expresses the voltage V_{pri} in terms of transformer design parameters such as frequency, flux density, core area, and number of turns in the primary winding. If the operating frequency is low so that the skin effect can be neglected or if Litz wire is used for the winding, Eq. 30-37 expresses the primary current in terms of current density and primary conductor cross-sectional area. Using these two equations to construct the volt-ampere product yields

$$S = V_{\text{pri}}I_{\text{pri}} = \frac{N_{\text{pri}}A_{\text{core}}\omega\hat{B}}{\sqrt{2}}J_{\text{rms}}A_{\text{Cu,pri}}$$ (30-54)

Equation 30-38 gives the the primary conductor cross-sectional area $A_{\text{Cu,pri}}$ in terms of the winding area A_w, number of primary turns N_{pri}, and the copper fill factors k_{Cu}. Using this equation yields

$$S = V_{\text{pri}}I_{\text{pri}} = 2.22k_{\text{Cu}}fA_{\text{core}}A_w J_{\text{rms}}\hat{B}$$ (30-55)

If eddy current loss must be taken into account, then the effective resistance of the winding R_{ac} can be found from Fig. 30-19. Equation 30-12 must then be modified to

$$P_{w,sp} = 22 \frac{R_{ac}}{R_{dc}} k_{Cu}(J_{rms})^2 \qquad (30\text{-}56)$$

Equation 30-55 is the starting point of the transformer design procedure because it relates design requirements (V_{pri} and I_{pri}) from the circuit application to transformer design variables such as core area, conductor area, flux density, and current density. This equation plays the same role in transformer design as the stored energy value (Eq. 30-23) does in inductor design.

30-9-2 SINGLE-PASS TRANSFORMER DESIGN PROCEDURE

In this section we present a step-by-step outline of a single-pass (no iterations) transformer design procedure. The procedure relies on the availability of a complete database of core characteristics as was required for the single-pass inductor design procedure. A flowchart of the design steps is given in Fig. 30-21. An example of the use of the design procedure will be given to illustrate the details of the procedure.

Step 1   Assemble design inputs

The design inputs consist of the following parameters.

(a) The rated rms primary voltage V_{pri}

(b) The rated rms primary current I_{pri}

(c) The transformer turns ration $n = N_{pri}/N_{sec}$

(d) The operating frequency f

(e) The maximum body temperature T_s of the transformer and the maximum ambient temperature T_a

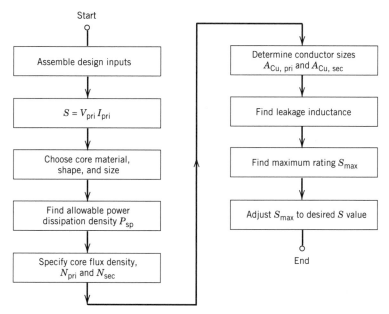

Figure 30-21   Flowchart of a single-pass transformer design procedure. The procedure assumes the existence of an extensive database of all cores available to the transformer designer. The characteristics include the allowable power dissipation density for the design temperature range.

The values for the first four parameters are found via design calculations for the specific power electronic converter circuit in which the transformer is to be used. The maximum temperatures are determined from considerations of other temperature-limited components that are used in the same circuit, the temperature limitations of the transformer materials, and the environment in which the transformer must operate. A commonly used value of T_s is 100°C.

Step 2   Compute the volt-amp rating S

The second step is to use the design inputs to compute the volt-amp rating of the transformer. I_{pri} is the maximum rms value of the primary current and V_{pri} is the maximum rms value of the primary voltage.

Step 3   Choose core material, shape, and size

The third step is to choose the core material and core shape. The considerations are the same as for the inductor design procedure described in Section 30-5-3. In particular, if ferrite cores are to be used, then the material performance factor shown in Fig. 30-3 should be an integral part of the selection process.

The core size is chosen based on the value of S computed in Step 2. The designer looks in the core database for the core with the nearest value of $2.22k_{\text{Cu}}fJ_{\text{rms}}\hat{B}A_{\text{core}}A_w$ that is larger than S. This lookup step assumes that the characteristics of all cores of interest have been tabulated in a database such as is illustrated in Table 30-4 . Such a database can be constructed independent of any particular transformer design using the methods illustrated in previous sections of this chapter.

Similar to the inductor design procedure, it is necessary to select a winding conductor type (Litz wire, round conductor, etc.) so that the copper fill factor is known in order to complete the selection of the core size. Selection of the conductor type will depend on the operating frequency and the importance of eddy current losses in the windings. For round wire, $k_{\text{Cu}} = 0.55$ and for Litz wire, $k_{\text{Cu}} = 0.3$. If the subsequent design steps described below produce an acceptable design, then these steps will be a single-pass procedure. If an acceptable design is not produced, then it will be necessary to repeat the design steps after either choosing a larger core or choosing a wire type having a larger copper fill factor.

Step 4   Find $R_{\theta\text{sa}}$ and P_{sp}

The thermal resistance, $R_{\theta\text{sa}}$ (surface-to-ambient) of the transformer is the next item to be found. It can either be looked up in the core database as is shown in Table 30-4 or calculated using the approach illustrated in Section 30-4-4. In either case the value of $R_{\theta\text{sa}}$

Table 30-4 Database of Core Characteristics Needed for Transformer Design

Core No.	Material	$AP =$ A_wA_c	R_θ $\Delta T = 60°C$	P_{sp} at $T_s = 100°C$	J_{rms} at $T_s = 100°C$ and P_{sp}	$\hat{B}_{rated}$ at $T_s = 100°C$ and 100 kHz	$2.22k_{Cu}f$ $J_{rms}\hat{B}A_wA_{core}$ $(f = 100 \text{ kHz})$
•	•	•	•	•	•	•	•
8^a	3F3	2.1 cm⁴	9.8°C/W	237 mW/cm³	$3.3\sqrt{\frac{R_{dc}}{k_{Cu}R_{ac}}}$ A/mm²	170 mT	$2.6\times10^3\sqrt{\frac{k_{Cu}R_{dc}}{R_{ac}}}$ V-A
•	•	•	•	•	•	•	•

^{a}Core number 8 is the same as listed in Table 30-3.

must conform to the specified temperatures T_s and T_a. For the same reasons as detailed in the inductor design procedure, we assume uniform power dissipation in the transformer and thus the allowable specific power densities in the winding, $P_{w,sp}$, equals that in the core, $P_{core,sp} = P_{sp}$. The value of P_{sp} that can be dissipated in the core and windings can either be found in the core database or calculated using Eq. 30-26.

Step 5 Specify core flux density and number of primary and secondary turns

The core flux density $\hat{B}$ is found using the core database (Table 30-4). Otherwise $\hat{B}$ can be found using the allowable value of specific power density dissipation, P_{sp}, estimated in Step 5 in conjunction with data such as Fig. 30-2 or Eq. 30-2, which are provided by the core manufacturer. The number of turns, N_{pri}, needed in the primary winding are found using Eq. 30-40 using the flux density value $\hat{B}$ just found in this step. The number of turns in the secondary, N_{sec}, are found using the specified turns ratio $n = N_{pri}/N_{sec}$.

Step 6a Determine primary and secondary conductor sizes: eddy currents neglected

The type of winding conductor, round wire, Litz wire, and so forth has already been tentatively chosen in Step 3, thus the copper fill factor is known. The required areas, $A_{Cu,pri}$ and $A_{Cu,sec}$, of the primary and secondary conductors are then found using Eqs. 30-38 and 30-39 and the values of N_{pri} and N_{sec} found in Step 5.

There is an alternative way to estimate the conductor areas. If the operating frequency is low so that eddy currents can be neglected or if Litz wire is used, the allowable current density J_{rms} can be estimated using the core database (Table 30-4), or by inverting Eq. 30-12. The areas $A_{Cu,pri}$ and $A_{cu,sec}$ of the winding conductors can then be estimated from Eq. 30-37a.

The two seemingly separate ways of calculating the required winding conductor cross-sectional areas are not in conflict. They both yield the same result. This is demonstrated in the problems at the end of the chapter.

Step 6b Determine primary and secondary conductor sizes: eddy currents included

When eddy current losses are included in the design procedure, only solid conductors of either round or rectangular cross section will be considered for use in the primary and secondary windings. We assume that if Litz wire is used because of eddy current loss concerns, the diameter of the strands that make up the Litz wire are much smaller than a skin depth so that the low-frequency design procedure is appropriate. Both round and rectangular conductors initially will be considered in the winding design before a final choice is made. For round wire $k_{Cu} = 0.55$ while for rectangular wire or foils $k_{Cu} = 0.6$.

The conductor cross-sectional areas are found by one of two equivalent methods. Since the number of primary and secondary turns, N_{pri} and N_{sec} were found in Step 5, Eqs. 30-38 and 30-39 can be used to find the areas. Alternatively, the allowable current density J_{rms} is found using the J_{rms} column in the core database with $\sqrt{R_{ac}/R_{dc}} = \sqrt{1.5}$ and this value of J_{rms} is used in Eq. 30-37a to find the required areas.

To complete the sizing of the conductors, it is necessary to choose the dimensions of the conductor so that the cross-sectional area requirements are satisfied simultaneously within the constraints of Fig. 30-19. This figure indicates that many different conductor dimensions may satisfy the condition $\sqrt{R_{ac}/R_{dc}} = \sqrt{1.5}$ depending on how many layers there are in each transformer winding section. An iterative procedure will be needed in order to find an acceptable combination of conductor dimensions and number of layers per section.

To begin the iterative process, it is convenient to assume a single turn per layer and to have the winding split into only two sections, a primary section with N_{pri} layers and a secondary section with N_{sec} layers. The required thickness h of the primary conductor is

$$A_{Cu,pri} = F_l h_w h \quad \text{or} \quad h = \frac{A_{Cu,pri}}{F_l h_w} \tag{30-57}$$

where F_l is the copper layer factor (or layer utilization factor) and h_w is the bobbin height. A typical value of F_l is 0.9 for a rectangular conductor or foil. The next step is to compute the normalized conductor thickness, which is given by Eq. 30-49. Figure 30-19 is then used to find the optimum value of ϕ for N_{pri} layers in the primary winding section. If the optimum value is the same or somewhat (10–20%) larger than the value found in Eq. 30-49, then the initial design is usable. If the optimum value is substantially smaller than the value found using Eq. 30-49, then the primary winding will have to be sectionalized so that the number of layers per section is reduced.

In the next (second) iteration, we will split the primary into two sections, and the secondary is sandwiched between the two primary sections as illustrated in Fig. 30-18a. For purposes of using Fig. 30-19, the secondary section must be also considered to be split into two sections as well as the primary, with one-half of the secondary being associated with one-half of the original primary and the number of layers in each primary and secondary section being a factor of 2 smaller than the first iteration. The same procedure as described in the preceding paragraph is repeated, and the optimum value of ϕ is again checked against the value calculated using Eq. 30-49. The process is repeated until the optimum value of ϕ is equal to or somewhat larger than the value estimated using Eq. 30-49.

If the optimum value of ϕ is substantially (approaching a factor of 2) larger than the value estimated by Eq. 30-49, then it may be possible to have two or more turns per layer in the winding. If two turns per layer are used, then the number of layers in the section will be reduced by a factor of 2. The iterative process should be repeated assuming two or more turns per layer and continued until the optimum value of ϕ equals or moderately exceeds the prediction of Eq. 30-49.

Once the primary winding has been satisfactorily designed, it will be necessary to repeat the same process for the secondary winding. Depending on the turns ratio, it may be necessary to adjust the number of primary and secondary turns either up or down from the values estimated in Step 5 so that the required sectioning of the windings can be done. Additionally it may be necessary to return to the primary winding design and adjust it so that the primary and secondary windings are compatible.

The iterative process needed to find the conductor dimensions seems complex. However, the fact that the number of turns must be an integer and the number of turns per layer must be an integer limits the number of feasible winding arrangements and thus keeps the number of iterations to a small number. A detailed application of the procedure is given in an example in a later section of this chapter. The complication of the design procedure by the use of solid conductors in the presence of eddy current loss is justified in most situations because Litz wire is considerably more expensive than solid conductors and has a much poorer copper fill factor.

Step 7   Estimation of leakage inductance

The leakage inductance is found using Eq. 30-53.

Step 8   Estimate of maximum V–I rating, S_{max}, of selected core

The maximum V–I rating S_{max} of the selected core is given by

$$S_{max} = 2.2 k_{Cu} f A_{core} A_w J_{rms} \hat{B} \tag{30-58}$$

The transformer $V-I$ rating $S = V_{pri}I_{pri}$ should be less than S_{max} if the design procedure has been properly followed but greater than the S_{max} of the next available smaller core. If S is significantly less than S_{max} (say $S < 0.8\ S_{max}$), then the next step in the design procedure will lower S_{max}, and savings in copper winding cost and weight will be realized. The occurrence of S being significantly less than S_{max} is likely when only a relative few numbers of cores are available in the database and thus relatively large differences in their attainable volt-ampere ratings. If S is smaller than the S_{max} of the next smaller core size, then the design procedure repeated with the smaller core size.

Step 9 Adjustment of S_{max}

The value of S_{max} can be reduced by decreasing the number of turns in the primary and secondary windings if the voltage is too large. If the current is too large, the copper area can be reduced. Either step will reduce the copper weight and volume and thus the copper costs.

In the unlikely event that S is significantly greater than S_{max} (say $S > 1.2\ S_{max}$) but still less than the S_{max} of the next available larger core, then S_{max} should be increased. This could most easily be done by increasing the specifc power density in the winding $P_{sp,w}$. This will increase the temperature of the transformer by a small amount above the design limit, but if it is not too large, it may be preferrable to going to a larger core. If S is greater than the S_{max} of the next available core size, the design procedure should be repeated with the larger core size.

30-9-3 TRANSFORMER DESIGN EXAMPLE

As an example of the design procedure, the transformer analyzed in Section 30-6 will be "designed" using the step-by-step sequence just described.

Step 1 Design inputs

I_{pri} = 4 A rms sine wave current.
V_{pri} = 300 V rms sine wave voltage.
Frequency f = 100 kHz.
Turns ration N_{pri}/N_{sec} = n = 4.
Maximum temperatures T_s = 100°C and T_a = 40°C.

Step 2 Transformer volt-ampere rating S

$$S = V_{pri}I_{pri} = (300\ \text{V})\ (4\ \text{A}) = 1200\ \text{V-A}$$

Step 3 Choose core material, shape, and size

The relatively high operating frequency suggests that a ferrite core be used. The performance factor curves for ferrite materials in Fig. 30-3 indicates that at this frequency, the 3F3 ferrite is the best choice. A double-E core is chosen for the core shape.

From the core database (Table 30-4), core 8 with a = 1 cm has $2.22k_{Cu}fJ_{rms}\hat{B}A_{core}A_w$ equal to $(2.6 \times 10^3)\ \sqrt{k_{Cu}}\ \sqrt{R_{dc}/R_{ac}}$ V-A. This value is greater than S = 1200 V-A for all values of copper fill factor larger than k_{Cu} = 0.21 if $\sqrt{R_{dc}/R_{ac}}$ = 1 or greater than S = 1200 V-A for k_{Cu} > 0.32 if $\sqrt{R_{dc}/R_{ac}} = \dfrac{1}{\sqrt{1.5}}$.

Step 4 Find $R_{\theta sa}$ and P_{sp}

Using the core database (Table 30-4), $R_{\theta sa}$ = 9.8°C/W and P_{sp} = 237 mW/cm³.

Step 5   Specify core flux density and number of primary and secondary turns

Using the core database (Table 30-4), the maximum core flux density $\hat{B}$ = 170 mT. Using this value in Eq. 30-40 along with the design inputs $\hat{V}_{pri}$, = 300 $\sqrt{2}$ = 424 V and ω = $(2\pi)(100$ kHz) and using A_{core} = 1.5 cm^2 yields

$$N_{pri} = \frac{\hat{V}_{pri}}{A_{core}\omega\hat{B}_{core}} = \frac{424}{(1.5 \times 10^{-4} \text{ m}^2)(2\pi)(10^5 \text{ Hz})(0.17 \text{ T})} = 26.5 = 24 \text{ turns}$$

The number of primary turns is rounded down to 24 because of the requirement that the number of turns be an integer divisible by 4. The number of secondary turns is N_{sec} = N_{pri}/n = 24/4 = 6. We could have chosen N_{pri} = 28 but that would have made N_{sec} = 7, which meant that the secondary would probably either be a single section of seven layers or seven sections of 1 layer. This would have greatly reduced the flexibilty in designing a sectionalized transformer winding.

The number of primary and secondary turns estimated here are somewhat less than the example of Section 30-6 because the winding window in that example was assumed to be completely filled. In this design procedure it was found unnecessary to completely fill the window. The reason for this difference is that the core loss in this example was set at the maximum of 237 mW/cm^3 whereas in the example of Section 30-6, the core losses were only 140 mW/cm^3. The lower losses in Section 30-6 meant lower values of flux density and thus more turns were needed to achieve the desired primary and secondary voltages.

Step 6a   Specify primary and secondary conductor sizes: foil conductors

The frequency is high enough to make it necessary to consider eddy current losses in the design procedure if solid copper conductors are to be used or else Litz wire must be used. Initially we shall try to design with solid rectangular (foil) copper conductors instead of Litz wire in order to minimize copper costs. The allowable current density J_{rms} is, using the core database and assuming that rectangular conductors have a copper fill factor k_{Cu} = 0.6,

$$J_{rms} = \frac{3.3}{\sqrt{.6}\sqrt{1.5}} = 3.5 \text{ A/mm}^2$$

The area of the primary conductor $A_{Cu,pri}$ = (4 A)/(3.5 A/mm^2) = 1.15 mm^2. The area of the conductor for the secondary winding is $A_{Cu,sec}$ = $nA_{Cu,pri}$ = (4)(1.15 mm^2) = 4.6 mm^2.

To minimize the number of iterations needed to reach a practical design, we will use Fig. 30-19 to estimate the number of layers needed per section to obtain the required conductor areas. We assume that the rectangular conductors have a layer factor F_l = 0.9 and that we use a single turn per layer. The skin depth δ = 0.24 mm in copper at 100 kHz and 100°C. The required thickness h of the primary conductor is

$$A_{Cu,pri} = 1.15 \text{ mm}^2 = (0.90)\ (20 \text{ mm})\ (h)\quad \text{or}\quad h = 0.064 \text{ mm}$$

The value of the normalized primary conductor thickness is

$$\sqrt{0.9}\ \frac{h}{\delta} = (0.95)\frac{0.064}{0.24} = 0.25$$

Examination of Fig. 30-19 indicates a single primary section composed of 24 layers, one turn per layer has an optimum normalized thickness value of approximately $\sqrt{0.9}\ h/\delta$ = 0.3, thus meeting the requirement.

The required thickness of the secondary conductor is

$$A_{Cu,sec} = 4.6 \text{ mm}^2 = (0.90)(20 \text{ mm})(h) \quad \text{or} \quad h = 0.256 \text{ mm}$$

The value of the normalized secondary conductor thickness is

$$\sqrt{0.9}\,\frac{h}{\delta} = (0.95)\,\frac{0.256}{0.24} = 1.01$$

Examination of Fig. 30-19 indicates that this requirement can be met if there are two layers per section of secondary winding. This means that the secondary winding will be composed of three sections, each with two layers and each layer having a single turn.

To be compatible with this, the primary winding will have four sections as shown in Fig. 30-22. The two center primary sections would have eight layers and the two outer sections would have four layers, assuming a single turn per layer. The optimum value of normalized conductor thickness for eight layers is $\sqrt{0.9}\,h/\delta = 0.5$. This value is nearly twice as large as needed, and it suggests that we could have more than one turn per layer. Based on previous experience, we will try using four turns per layer of primary winding which will result in the outer primary sections each with one layer of four turns and the two inner primary sections having two layers of four turns each. The required thickness h for four per layer is $A_{Cu,pri} = 1.15 \text{ mm}^2 = (0.90)(5 \text{ mm})(h)$ or $h = 0.256 \text{ mm}$ and the normalized thickness is $(0.95)\,0.256/0.24 = 1.01$. The optimum value of $\sqrt{0.9}\,h/\delta$ for two layer is $\sqrt{0.9}\,h/\delta = 1.0$. Hence the two inner primary sections are optimally designed. The two outer sections, having only one layer, have a slightly increased value of R_{ac}/R_{dc} of about 1.7 (estimated from Fig. 30-19), which is only about 20% larger than the optimum value of 1.5. Overall the primary winding has a nearly optimum value of R_{ac}/R_{dc}.

The final design for the transformer winding is shown in Fig. 30-22 along with the corresponding mmf diagram. The primary is divided into four sections, with the central two sections having two layers and the outer two sections having one layer. Each layer has

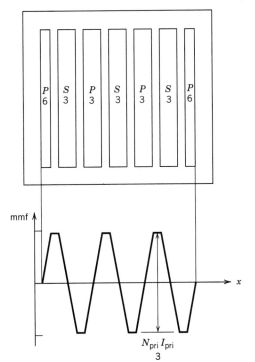

Figure 30-22 Sandwich winding arrangement required for the transformer design example wound with solid copper conductors. Rectangular or foil conductors are used on both secondary and primary windings, and the resulting mmf distribution is also shown.

four turns, and the thickness of the primary conductor is 0.26 mm and the width is 5 mm. The secondary is divided into three sections, two layers per section with one turn per layer. The secondary conductor is approximately 20 mm wide and has a thickness $h = 0.26$ mm.

This arrangement should only be considered as a starting point of the final design. Practical fabrication considerations may require design modifications such as narrower primary winding widths, and so forth.

Step 6b   Determine primary and secondary conductor sizes: Litz wire

The complicated winding geometry needed for a solid conductor winding behooves us to consider a transformer wound with Litz wire. In this case the design procedure for the winding is simpler. The allowable current density $J_{rms,rated}$, using the core database and assuming a copper fill factor for Litz wire of $k_{Cu} = 0.3$, is

$$J_{rms,rated} = \frac{3.3}{\sqrt{.3}} = 6 \text{ A/mm}^2$$

The area of the primary conductor $A_{Cu,pri} = (4 \text{ A})/(6 \text{ A/mm}^2) = 0.67 \text{ mm}^2$. The area of the conductor for the secondary winding is $A_{Cu,sec} = nA_{Cu,pri} = (4)(0.67 \text{ mm}^2) = 2.7 \text{ mm}^2$.

The approximate diameter of a Litz wire bundle having a copper area of 0.67 mm^2 is approximately

$$d_{Litz} \approx \sqrt{\frac{4A_{Cu}}{\pi k_{Cu}}} = \sqrt{\frac{(4)(0.67 \text{ mm}^2)}{(\pi)(0.3)}} = 1.69 \text{ mm}$$

while that of the secondary conductor is 3.37 mm. The diameter of the secondary is somewhat larger than conventionally available Litz wires and thus would either be a special order or would have to be made by hand by the person making the transformer. There is also some question as to whether such a large diameter wire bundle can be made to fit into the limited winding window area. The next larger core size would ease the problem of fitting the wire bundles into the winding window.

Step 7   Estimate leakage inductance

For the winding geometry shown in Fig. 30-22a, the leakage inductance, using Eq. 30-53a with all dimensions in cm and $\mu_0 = 4\pi \times 10^{-9}$ H/cm, is

$$L_{leak} = \frac{(4\pi \times 10^{-9})(24)^2(9)(0.7)(1)}{(3)(6^2)(2)} = 0.2 \text{ } \mu\text{H}$$

If Litz wire is used in simple winding arrangement of Fig. 30-17, there is only one interface between windings so $P = 1$ in Eq. 30-53a and the leakage inductance is $L_{leak} = 8.1$ μH.

Step 8   Maximum V–I rating S_{max} for selected core

The maximum V–I rating for the core selected for the transformer when solid rectangular conductors are used for the winding is given by Eq. 30-55 or the right-most entry in Table 30-4. The value S_{max} using the entry from Table 30-4 is

$$S_{max} = 2.6 \times 10^3 \sqrt{\frac{k_{Cu}R_{dc}}{R_{ac}}} = 2.6 \times 10^3 \sqrt{\frac{0.6}{1.5}} = 1644 \text{ V-A}$$

For the Litz wire winding, S_{max} is given by Eq. 30-58 with $k_{Cu} = 0.3$ and has the value $S_{max} = 1,424$ V-A. The better copper fill factor of the solid conductor accounts for the larger S_{max} value.

Step 9   Adjustment of S_{max}

The $V{-}I$ rating needed for the transformer is 1200 V-A, which is somewhat smaller than the S_{max} of the selected core. In principle, this would permit a reduction in the number of turns or in the copper conductor areas and thus a savings in copper cost and weight. However, in this case S is only marginally smaller than S_{max} (about 25% smaller in the case of the solid conductor windings). It is questionable whether there is much to be gained in reducing S_{max} especially if only a few such transformers are to be fabricated.

30-10 COMPARISON OF TRANSFORMER AND INDUCTOR SIZES

For a specific core, a useful comparison of the inductor size (value of inductance) to the transformer size ($V{-}I$ rating) can be made. Assuming the same operating frequency so that the maximum flux density is the same in both the inductor and transformer, the inductance-current product Eq. 30-23 is equal to the transformer $V{-}I$ rating S, Eq. 30-55 divided by the parameter $2.2f$ where f is the frequency of operation. Thus

$$\frac{S}{2.2f} = L\hat{I}I_{rms} = k_{Cu}A_cA_wJ_{rms}B \tag{30-59}$$

or

$$S = 2.2fL\hat{I}I_{rms} \tag{30-60}$$

Given the inductance and inductor currents, it is possible to equate the inductor size to that of a transformer at a frequency f whose volt-ampere rating S can be calculated using Eq. 30-55. If the comparison is made for sinewave currents in the inductor and transformer, then $\hat{I} = \sqrt{2}I_{rms}$ and Eq. 30-60 becomes

$$S = \pi fLI_{rms}^2 \tag{30-61}$$

SUMMARY

This chapter has discussed the design and fabrication of inductors and transformers intended for high-frequency (tens of kilohertz to megahertz) operation in power electronic circuits. The important conclusions follow.

1. Magnetic materials used for the cores of inductors and transformers have two types of electrical losses, eddy current losses due to finite electrical conductivity and hysteresis (magnetic) losses. High-frequency operation mandates the use of ferrites, which have large electrical resistivity and thus have only magnetic losses.
2. Magnetic cores are available in a wide variety of shapes and sizes and materials to suit almost any application.
3. Windings for inductors and transformers are made from copper wire, which is available in a wide range of sizes and geometric shapes in order to minimize electrical losses. The copper losses include not only dc resistance losses but additional ohmic losses caused by nonuniform current density concentrations arising from the proximity effect and skin effect.
4. The maximum permissible temperature of an inductor or transformer is approximately 100°C and is limited by both magnetic material and winding insulation material considerations. This temperature limit along with the surface-to-ambient thermal resistance of the component limit the average power dissipation density (W/cm^3) in the component.

5. The power dissipation density limit translates into a maximum current density limit in the copper windings and a maximum peak ac flux density in the core material.

6. A single-pass inductor design procedure can be developed that is based on the inductor stored energy value and the existence of a complete database of properties of available cores. This database incudes thermal resistance, current density limits, flux density limits, and so forth. In the absence of such a complete database, the procedure becomes an iterative design method.

7. Minimizing copper winding losses at high frequencies requires special efforts including the use of Litz wire and sectionalizing the primary and secondary windings in a transformer. A procedure is described for the optimum manner in which to sectionalize a transformer winding.

8. A single-pass transformer design procedure is developed that is based on the volt-amp rating of the transformer and the existence of a complete database of properties of available cores. In the absence of such a database, the procedure becomes an iterative design method.

PROBLEMS

30-1 A core such as shown in Fig. 30-5 is made from magnetic steel laminations whose outer dimensions are 4 cm on a side and whose inner dimensions (the winding window) are 2 cm on a side. The laminations are 0.25 mm thick and the stacking factor is 0.95. Forty such laminations are used to make the core. The resistivity of the core ρ_{core} = 30 $\mu\Omega$-cm and the relative permeability is 900. An inductor winding wound on the core produces a sinusoidal flux density $\hat{B}$ = 0.5 T at a frequency of 100 Hz.

(a) What is the skin depth in the core?

(b) What are the total average core losses due to the eddy currents?

30-2 Assume that the maximum surface temperature T_s of the core of Problem 30-1 cannot exceed 100°C and that the ambient T_a never exceeds 40°C. Model the core as a solid rectangular parallelpiped whose outer dimensions are those given in the previous problem and assume an emissivity E = 0.9.

(a) What are the maximum allowable core losses per cubic centimeter?

(b) What is the allowable $\hat{B}$ at a frequency of 800 Hz?

30-3 What is the ratio of energy stored in the air gap to the energy stored in the core of the example inductor analyzed Section 30-4? Assume a relative permeability μ_r = 200 for the ferrite.

30-4 Design an iterative transformer design procedure for the situation where a complete core database is not available. Show the design procedure flowchart and state reasonable values for any initial values of parameters to get the iteration launched. Use the inductor iterative design procedure as a model.

30-5 An inductance of 750 μH is needed for a power electronic converter operating at 100 kHz. A sinusoidal current of 5 A rms maximum flows through the inductor. The only core available is a double-E core having a dimension a = 1.5 cm and made from 3F3 ferrite material. The maximum surface temperature $T_s \leq$ 125°C and the ambient $T_a \leq$ 35°C. A core database is shown below. Litz wire is used for the winding.

a (cm)	A_w (cm^2)	A_{core} (cm^2)	V_w (cm^3)	V_{core} (cm^3)	$R_{\theta sa}$ (°C/W)
1.5	3.15	3.38	34.1	45.6	3.4

(a) Determine the maximum inductance L_{max} that can be wound on the core.

(b) Determine the required air-gap length Σg that will result in the maximum core flux density when the current in the inductor is maximum (5 A rms). Assume four distributed gaps.

30-6 Verify Eq. (30-12a) for copper at 100°C. Assume ρ_{Cu} (100°C) = 2.2 × 10^{-8} Ω-m.

30-7 Show that δ_{Cu} (100°C) = 75/$\sqrt{f}$ (mm) where f is in hertz.

30-8   An inductor that has winding loss and core loss can be modeled with an equivalent circuit consisting of an inductance L in series with a resistance R. For the example inductor examined in Section 30-4, find the value of resistor R in the equivalent circuit. Assume that the current in the inductor is at the maximum value of 4 A rms.

30-9   Estimate the quality factor Q of the inductor of Problem 30-8. Assume a frequency of 100 kHz, the same specified in Section 30-4.

30-10 What is the ac voltage across the inductor of Problems 30-8 and 30-9 at a frequency of 100 kHz?

30-11 An inductor is used in a circuit that causes the core flux to have the waveshape shown in Fig. 30-1 with B_{avg} = 200 mT and $\hat{B}$ = 300 mT. The ripple frequency is 400 kHz and the core material is 3F3 whose loss characteristic is shown in Fig. 30-2. Find the specific power loss in the inductor core.

30-12 Assume that the inductor of Problem 30-11 can be modeled as a cube that is 2 cm on each side. The inductor is black and has an emissivity of 0.9. Find the surface temperature T_s of the inductor if the ambient temperature T_a = 40°C. (Several iterations of assuming a trial value of T_s, calculating $R_{\theta sa}$, and then calculating a corrected value of T_s may be necessary.) Assume $P_{c,sp} = P_{w,sp}$.

30-13 The inductor of Problem 30-11 is used in a different circuit that causes a flux density $B(t) = B$ $\sin(\omega t)$ with B = 300 mT and the frequency f = 100 kHz. The inductor can still be modeled as a cube as was done in Problem 30-12.

   (a) Find the specific core loss $P_{sp,core}$.

   (b) The surface temperature T_s is to be held at 90°C when the ambient temperature T_a = 30°C. The inductor can be mounted on a heat sink, if necessary, to provide an additional heat flow path to keep the inductor temperature at 90°C. Determine if the heat sink is needed and if so, what the thermal resistance of the heat sink should be.

30-14 An inductor is used in a circuit that causes the maximum inductor current to be I_{rms} = 8 A (sine wave). The inductor is identical to that discussed in Section 30-4 except that the number of turns N = 33 and the copper area A_{Cu} = 1.28 mm². What is the value of the inductance L? Assume that the same type of conductor is used in both inductors.

30-15 An inductor is to be designed to have a value L = 150 μH. The current through the inductor is to be I_{rms} = 4 A (sine wave) at a frequency of 100 kHz. The inductor is wound on the same core (a = 1 cm) as was used in the example of Section 30-4. Assume that the air-gap length Σg remains constant at the value Σg = 3 mm found in Section 30-4. Find the required number of turns N.

30-16 Assume that the inductor of Problem 30-15 has the same surface area as the inductor discussed in Section 30-4, the same maximum surface temperature of 100°C, and that the average length of each turn is the same as the inductor of Section 30-4. If the power dissipated in the winding P_w = 3.17 W for the inductor of Problem 30-15, find the following:

   (a) The current density in the winding of the new inductor.

   (b) The ratio of the copper weight in the new inductor to that of the inductor of Section 30-4.

30-17 The inductor of Problem 30-15 dissipates a total power P_{tot} = 6.3 W. The current I_{rms} = 4 A (sine wave) at a frequency of 100 kHz. Find the following:

   (a) The current density in the winding of the new inductor.

   (b) The ratio of the copper weight in the new inductor to that of the inductor of Section 30-4.

30-18 The air-gap length Σg for the inductor of Problem 30-15 is reduced to keep the core flux $\hat{B}_{core}$ constant at 177 mT. Find the number of turns N now needed to realize an inductance L = 150 μH.

30-19 Assume that the inductor of Problem 30-18 has the same surface area as the inductor of Section 30-4 and the same temperature difference $T_s - T_a$. Find the following:

   (a) The current density in the winding of the new inductor.

   (b) The ratio of the copper weight in the new inductor to that of the inductor of Section 30-4.

30-20 An inductor is to be made using an double-E core similiar to that in Fig. 30-6 where d = 1.5a. However, the window dimensions h_a = 2.5b_a are independent of a. Assume NA_{Cu} = 0.2A_a where A_a = 2.5b_a^2 is the window area. The maximum current density J_{rms} = 6.25 A/mm², the peak flux density $B_{ac} = 0.2$ Wb/m^2, the inductance $L = 0.3$ mH, and the maximum current in the inductor is 4 A rms (sine wave).

(a) Find b_a and h_a as functions of the number of turns N.

(b) Find a and d as functions of the number of turns N.

(c) Find V_w and V_{core} as functions of N.

(d) Plot $V_w + V_{core}$ as a functions of N. For what value of N is the total volume a minimum?

(e) Assume that the cost of the (Litz) winding copper material per unit volume is twice the cost of the core material per unit volume. For what value of N is the cost of the inductor material (core plus winding) a minimum?

30-21 The objective of this problem is to show that the volt-ampere rating of a transformer, given by Eq. 30-55, is approximately a maximum when the power dissipation density in the transformer is uniform, that is, when $P_{w,sp}$ equals $P_{core,sp}$. This result also applies to inductors.

(a) Show that $J_{rms} = \sqrt{P_T(1 - \alpha)/V_w C_w k_{cu}}$ where V_w is the winding volume and C_w is a numerical constant.

(b) Show that $B_{ac} = [P_T \alpha/V_{core} C_c f^{1.3}]^{0.4}$ where V_{core} is the core volume, C_c is a numerical constant, $\alpha = \dfrac{P_c}{P_T}$, and P_c is the power dissipated in the core.

(c) Use the results of parts (a) and (b) to show that S is a maximum when $\alpha = 0.44$.

(d) Graph $S(\alpha)/S_{max}$ for $0.1 < \alpha < 0.9$. For what range of α is the ratio > 0.9, that is, for what range of α is $S(\alpha) > 0.9\,S_{max}$?

(e) What is α and $P_{w,sp}/P_{c,sp}$ for the transformer designed in Section 30-9-3?

30-22 In the discussion of the transformer design procedure, two seemingly different ways of finding the required cross-sectional area, $A_{Cu,pri}$, of the winding conductors were presented (Eqs. 30-37a and 30-38). Demonstrate that these two ways are equivalent by showing that each method yields the same values for the areas. (Hint: recall that the transformer is designed subject to the constraint of the volt-ampere product.)

30-23 An equivalent circuit for a transformer is given in Fig. 3-21b. Find numerical values for the components of this circuit using the transformer designed in Section 30-9-3, which uses solid rectangular conductors. Split the leakage inductance into two equal values and assume $\mu_r = 200$ for the ferrite core.

30-24 Repeat Problem 30-23 for the transformer wound with Litz wire.

30-25 The transformer designed in Section 30-9-3 is to be used at a frequency of 300 kHz. Otherwise all other input electrical parameters remain the same. What will be the temperature T_s of the transformer? Assume Litz wire is used for the windings.

REFERENCES

1. E. C. Snelling, *Soft Ferrites—Properties and Applications*, Butterworths, London, 1988.

2. John G. Kassakian, Martin F. Schlecht, and G. Vergassian, *Principles of Power Electronics*, Addison-Wesley, Boston, 1991.

3. P. I. Dowell, "Effects of Eddy Current in Transformer Windings," *Proc. of IEEE*, Vol. 113, No. 8, August, 1966.

4. Bruce Carsten, "High Frequency Conductors in Switchmode Magnetics," *HFPC Proceedings*, May, 1986, pp. 155-176.

INDEX

ac–dc converter, 10
ac voltage waveform, phase-controlled
    converters, 138, 150
  distortion, 99, 153
  line notching, 150
  See also Harmonic, voltage distortion
Acceptor impurity, 509
Accumulation layer resistance, 589
Active current shaping, 488, 498
Active filters, 480
Active harmonic filtering, 480
Active region, BJT, 550
Active region, MOSFET, 576
Adjustable-speed drives, 391, 399
Aerospace applications, 8
Air conditioning, 451
Air-gap in inductors, 764
Aluminum electrolytic capacitor, 726
Ampere's law, 46
Amplitude modulation ratio, m_a, 203
Anode shorting structure, GTO, 615
Anode tail current (GTO), 622
Antisaturation diode, 701
Antisymmetric IGBT, 629
Apparent power, S, 35
Applications, 7
Arc welding, 457
Area product, inductor/transformer, 751
Armature current, 278
  discontinuous, 393
  form factor, 382
  ripple, 388
Armature winding, 377
Asymmetrical silicon-controlled rectifier
    (ASCR), 20
Asynchronous PWM, 208
ATP, 72
Auger recombination, 511, 533
Avalanche breakdown, 520
Average inductor voltage, 44
Average on-state power loss, 23

Average power, 34, 35, 42
Average switching power loss, 23

Back-emf, 378
Backporch current, GTO, 617, 619
Back-to-back connected converters, 122, 393
Ballast, fluorescent lighting, 452
Base width (thickness), 551
Base width (thickness) modulation, 563
Basic rectifier concepts, 80
Battery, 359
  constant charging current, 360
  lead-acid, 359
  trickle charged, 359
Beta, 547, 552
  fall-off at large currents, 552
Bipolar junction transistor (BJT), 546
  npn, 24, 546
  pnp, 546
Bipolar static induction thyristor (BSI Thy),
    646
Bipolar static induction transistor (BSIT),
    645
Bipolar voltage switching PWM, 190, 212,
    388
BJT. *See* Bipolar junction transistor
Blanking time, 189, 387, 719
  effect of, 236
  nonlinearity, 389
Blocking gain, 643, 648
Bobbin, 750, 751
Body-source junction, 571, 574, 590, 627
Body-spreading resistance, 632
Body-to-source short, 572, 574, 590
Bond (bonding), covalent, 508, 521
Boost converter, 172
Braking, 413, 421, 426
Breakdown:
  avalanche, 520
  primary, 550
  second(ary), 550, 563

Breakover current, 598
Bridge converter, twelve-pulse, 153, 461
Brushless-dc motors, 435
Buck converter, 164
Buck–boost converter, 178

Capacitance, space charge, 536
Capacitive snubber, 671
Capacitor discharge time constant, 686
Capacitor lifetime, 727
Carbon brush, 378
Carrier, charge, 507
Carrier frequency, 203
Carrier injection, 519
Carrier mobility, 512, 533
Carrier sweep-out, 537
Cascode switching circuit, 710
Cathode shorts, 609
C-cores, 750
Center-tapped transformer winding, 221
Centrifugal pump, 399
Ceramic capacitor, 728
Channel, 571, 578
Channel resistance, 589
Channel-to-source voltage, 578
Circuit-commutated recovery time, t_q, 19, 137, 149
Circuit layout, 722
Class-E converters, 253, 271
    nonoptimum mode, 272
    optimum mode, 271
Collector drift region, 546, 554
Commercial applications, 8
Common-mode noise, 357, 500
Commutation circuits, 716
Commutation failure, 137
Commutation interval, u, 131, 144
Commutator arcing, 383
Commutator segment, 378
Conducted noise, 500
    common mode, 500
    differential mode, 500
    line impedance stabilization network (LISN), 501
Conductivity modulation, 528, 531, 553, 602, 631
Contact potential, 516
Controllable switch, 16, 20
    comparison, 29
    desired characteristics, 20
    switching times, 22, 23
Convection, 740
Convention, symbol, 14
Converter, 9
    classification, 9
Copper strips, bus-bar, 724

Core:
    area, 751
    database, 762
    loss, 320, 745
    materials, 744
    shapes, 750
    sizes, 750
Coupling mechanism, 369
Coupling ratio, 369
Covalent bond, 508, 521
Creepage distance, 724
Crest factor, 42
Crossover frequency, 333
Crowbarring technique, 624
Cúk converter, 184, 195
Current control, 241
    constant frequency, 242, 339, 492
    constant off-time control, 338
    discontinuous, 241, 338, 492
    tolerance-band, 147
    variable tolerance-band, 492
Current crowding, 547, 553, 565
Current fall time, t_{fi}, 22
Current filament, 564
Current gain, 24
Current harmonics, 461, 484
Current limiting, 343, 373, 374, 423
Current-mode control, 337, 491, 497
    See also Current-regulated
Current ratio, 485
Current-regulated:
    modulation, 241, 373
    voltage source converter, 373, 440
    voltage source inverter (CR-VSI), 424, 442
    See also Current-mode control
Current rise time, t_{ri}, 22
Current sensor, 726
Current-source, 456
    dc–dc converter, 319
    inverters (CSI), 201, 418, 425, 456
    parallel-resonant converter, 253
    parallel-resonant dc-to-ac inverter, 269, 456
Current tailing. *See* Tailing current
Cycloconverter, 445
Cylindrical junction, 528

d/dt limiter, 394
Darlington connection, 24, 547
    See also Monolithic Darlington (MD)
dc blocking capacitor, 211
dc–dc converters, 10, 161
    boost, 172
    buck, 164
    buck–boost, 178
    comparison, 195
    control, 162

Cúk, 184
current-source, 319
electrical isolation, 304
full-bridge, 188
step-down, 164
step-down/up, 178
step-up, 172
dc motor drives, 377
adjustable speed, 391
dc motors, 377
discontinuous current, effect of, 393
field weakening, 394
nonlinearity due to blanking time, 389
permanent-magnet motors, 380
power electronic converter, 386
power factor, 395
separately excited field winding, 381
servo drives, 383
torque constant, k_t, k_T, 378, 380
torque pulsations, 383
transfer function model, 383
voltage constant k_e, k_E, 378, 380
dc servo drive, 383
dc-side current i_d in switch-mode inverters, 213, 218, 234
dc-to-ac inverters. *See* Switch-mode dc-to-ac inverters
dc-to-dc switch-mode converters. *See* dc–dc converters
Dead zone, 390
Delay angle, 125
Demagnetizing winding, 312
Density:
acceptor, 510
donor, 510
excess carrier, 511
free-carrier, 508
free-election, 508
minority carrier, 511, 520
Depletion
capacitance, 536, 583
layer/region, 514, 516, 526
width (thickness), 516, 526
di/dt rating, 24
Dielectric constant, 517
Dielectric isolation, 658
Diffusion
constant, 513, 519
current, 513
length, 519, 532, 539
Diode, 16, 524
fast recovery diodes, 17, 701
forward biased, 17
idealized characteristic, 17
leakage current, 17
line frequency, 17

reverse biased, 17
reverse breakdown, 17
reverse recovery current, 17, 524
reverse recovery time, t_{rr}, 17, 535, 538
Schottky, 17, 539
snap-off, 670
snappiness factor, 535, 538
Diode rectifiers, 79
basic concepts, 80
comparison, 112
inrush current, 112
line-frequency diode, 79
single-phase diode bridge, 82
three-phase diode full-bridge, 103
voltage-doubler, 100
Displacement power factor (DPF), 43
Displacement power factor angle, 43
Distortion, 40
total harmonic (THD), 42
Disturbance:
chopped voltage waveform, 354
EMI, 355
outage (blackout), 354
overvoltage, 354
powerline, 354
sources of, 355
tolerance, 356
undervoltage (SAG or brownout), 354
voltage spike, 354
Dithering technique, 477
Donor impurity, 510
Doping:
density, 521
profile, 513, 521
Double injection, 532
Drain:
body junction, 571
drift region resistance, 589
MOSFET, 571
Drift, 512
current, 512
region, 524, 526
region length, 526, 539
region MOSFET, 571, 583, 589
velocity, 512
Drive circuit, 30
dv/dt rating, 24
Dynamic current limit, 386
Dynamic latchup mode in IGBT, 632
Dynamic performance, small-signal, 383

Ebers–Moll equations, 600
Eddy current losses, 749, 771
Effective base width (thickness), 556, 600
Einstein relation, 513
Electric utility applications, 8, 460

Electric utility interface, 483
Electrical isolation, 304, 344
    of drive circuits, 696, 703
    need for, 696, 703
Electrical shields, 723
Electrolytic capacitor, 726
Electromagnetic interference (EMI), 80, 249,
        348, 500
    filter, 502
    reduction, 501
    standards, 501
Electron irradiation, 511
Electronically-commutated motors. *See*
        Synchronous-motor drives
Emitter current crowding, 553
Emitter-open
    switching circuit, 710
    turn-off, 710
EMTP, 70
Energy conservation, 7
Energy gap, 509, 521
Energy storage systems, 475
Enhancement mode field effect transistor, 578
Equivalent series resistance (ESR) of
        capacitor, 348, 727
Excess carrier, 511
    density, 511, 533
    injection, 519, 532
    lifetime, 511, 533
Extinction angle, 136, 149, 465

Faraday's voltage law, 50
Fast recovery diode, 17, 701
Feed-forward PWM control, 336
Feed-screw, 369
Ferrite material, 745
    performance factor, 747
Ferroresonant transformer, 357
Fiber optic cable, 704
Field-controlled thyristor (FCT), 646
Field crowding, 528, 659
Field effect, 576
Field-oriented space-vector-based control, 424
Field plate, 529
Field shields, 659
Field weakening in dc motors, 381
Field-weakening region, 394
Fill factor, 752
Filter:
    electromagnetic interference (EMI), 502
    shielding, 502
Fluorescent lighting, 452
Flyback converter, 308
    paralleling, 310
    two-transistor, 310
Foldback current limiting, 343

Forced-air cooled, 737
Forced commutation, 598, 602, 716
Forced-commutated thyristor, 29
Form factor, 382
Forward bias, 516
    safe operating area (FBSOA), 567
    voltage, 516, 519
Forward converter, 311
    paralleling, 314
    paralleling, two-switch, 314
Forward recovery current, 607
Fourier analysis, 39
Four-quadrant inverter, 202
Four-quadrant operation, 386
Free carrier, 507, 508
    density, 508
    electron, 507, 508
    electron density, 508
    hole, 508, 509
Frequency modulation ratio, m_f, 204
Fuel-cells, 478
Full-bridge:
    converter, 188, 317
    inverter, 211

Gallium arsenide, 661
Gate pulse amplifier, 714
Gate region of MOSFET, 576
Gate width-to-length ratio, 573
Gate-turn-off thyristor (GTO), 26, 613
Gate-assisted turn-off thyristor (GATT), 20,
        609
Generic switch, 20
GTO. *See* Gate-turn-off thyristor
Guard ring, 530

h_{FE}, 24
Half-bridge converter, 316
Half-bridge inverter, 211
Half-cycle controllers, 458
Hall-effect current sensor, 726
Hard saturation, 556
Harmonic
    current limits, 486
    elimination, 240, 243
    filters, HVDC, 468
    sidebands, 206
    spectrum, 205
    standards, 485
    voltage distortion limits, utility, 486
H-bridge, 387
Heat transfer:
    via conduction, 731
    via convection, 740
    via radiation, 739
Heatsinks, 452, 713, 737

High-frequency fluorescent lighting, 452
High-frequency noise, 355
High-frequency-link integral-half-cycle converters, 253, 289
High-voltage dc (HVDC) transmission, 460
High-voltage integrated circuits, 656, 657
Highly interdigitated gate cathode structure, GTO, 613, 616
Holding current, thyristor, 598
Holding torque, 440
Holding voltage, thyristor, 598
Hold-up time, 347
Hole bypass structure in IGBT, 633
Hybrid resonant dc–dc converter, 253, 268
Hydro power (small) interconnection, 477
Hysteresis loss, 745

Ideal switch, 12, 16
Idealized device characteristics, 31
Impact ionization, 521
Impurity (dopant), 509
Incremental position encoders, 373
Induction cooking, 455
Induction heating, 269, 455
Induction motor drives, 399
    adjustable speed-control (PWM-VSI), 422
    braking, 413, 421, 426
    capabilities, 411
    comparison of drives, 427
    constant power region, 411
    constant slip frequency region, 413
    constant torque region, 411
    current-limiting, 423
    current-source inverter (CSI), 418, 426
    field-oriented control, 424
    harmonics, impact, 420
    important relationships, 404
    induction motor, basic principle, 400
    See also induction motors
    Kramer drive, 431
    line-frequency variable voltage drives, 428
    power factor (input), 421, 426
    PWM voltage source inverter (PWM-VSI), 418, 419
    reduced voltage starting, 430
    regenerating, 421
    Scherbius, 431
    servo drives, 424
    slip compensation, 424
    soft-start, 430
    speed control by static slip-power recovery, 431
    speed control by varying stator frequency and voltage, 406
    speed ripple, 417
    square-wave voltage source inverter (square-wave VSI), 418, 425
    start-up considerations, 408
    torque pulsations, 417
    torque ripple, 417
    torque-speed characteristic, 407
    variable-frequency, 406
    variable-frequency converter classifications, 418
    voltage boost at low speed, 409, 424
Induction motors:
    characteristics at rated V and f, 405
    harmonic currents, 415
    harmonic losses, 416
    nonsinusoidal excitation, 415
    per-phase equivalent circuit, 401
    squirrel-cage rotors, 400
    torque pulsations, 417
    torque ripple, 417
    torque-speed characteristic, 407
Inductive current switching, 249
Inductive switching circuit, 21
Inductor:
    definition, 758
    design, 760
    stored energy relationship, 760
Industrial applications, 8, 451, 455
Injection of excess carrier, 519, 532
Input filter, 346
Inrush current, 112, 347
Instantaneous var control, 474
Insulated gate bipolar transistor (IGBT), 27, 626
Integral-half-cycle controllers, 458
Integral-half-cycle converters, 289
Intrinsic, 509
    carrier density, 509
    temperature, 730
Inversion layer, 576
Inverter, 10
Inverter-grade thyristor, 20
Ionization, thermal, 508
Ionized acceptor density, 509
Ionized donor density, 509
Iron laminations, 744, 749
Iron powder core, 744

Junction isolation, 657, 659
Junction temperature, 730

Kramer drive, 431

Latching action in thyristors, 600, 601
Latching current threshold in IGBTs, 632
Lattice, 508
Law of mass action, 510

Leakage inductance, 769, 779
Level shifting circuitry, 636
Lifetime, 511
    control of, 511
    excess carrier, 511
    reduction, 511, 512
Light-activated thyristor, 20
Limiting inrush (surge) current, 347
Line notching, 150
Linear:
    electronics, 4
    modulation, 208, 226
    power supplies, 301
Line-frequency converters, 121
    See also Thyristor converters;
        Phase-controlled; Twelve-pulse
Line-frequency diode, 17
Line-frequency noise, 150
Litz wire, 774
Load-resonant converters, 252, 258

Magnetic:
    circuits, 46
    core, 744
    core, amorphous, 745
    core loss curves, 745, 746
    core material, 744
Majority carrier, 509
    device, 534
MATLAB, 72
Matrix converter, 11
Maximum controllable anode current, 622
Metallized polypropylene capacitor, 728
Metal–oxide–semiconductor field effect
        transistor (MOSFET), 25, 571
Metal–oxide varistor (MOV), 355
Metal–semiconductor interface, 541
Minimum off-state time of GTO, 622
Minimum on-state time of GTO, 622
Minnesota interface, 478
Minnesota rectifier, 500
Minority carrier, 509
    density, 509, 519
    distribution, 517, 532
    lifetime, 511, 519
    power device, 534
Monolithic Darlington (MD), 24, 549, 560
MOS-controlled thyristor (MCT), 29, 649
MOS field effect transistor (MOSFET). See
        Metal–oxide–semiconductor
Motor drives, 367
    adjustable-speed, 368
    coupling mechanism, 369
    speed, position sensors, 374
    thermal considerations, 370
    See also Motors

Motors:
    current rating, 373
    power loss, 372
    thermal considerations, 370

Natural convection, 737, 740
n-channel MOSFET, 25, 571
Negative gate current, GTO, 615
Neutral currents, 101
Normally-off device, 643
Normally-on device, 643

Ohmic contacts, 541
Ohmic region, 575
On-state:
    losses, 23, 511, 531, 565, 588
    resistance, $r_{DS(on)}$, 26, 581, 582, 588
Optocoupler, 703
Overcurrent, 717
    detection, 717
    inductors, in, 759
    protection, 717
    protection of GTO, 623
    transformers, in, 770
Overmodulation, 208, 228
Overvoltage snubber, 686
    capacitor, 688
Oxide capacitor, 578

Parallel-loaded resonant (PLR) dc–dc
        converter, 253, 264
Parallel-resonant circuits, 257
Parasitic:
    BJT in MOSFET, 572, 574, 590
    diode in MOSFET, 572, 574
    thyristor in IGBT, 626, 632
Passive filter, 489
p-channel MOSFET, 571
Penetration depth, skin effect, 748
Performance factor, ferrites, 747
Permanent-magnet dc motor, 380
Perturb-and-adjust method, 476
Phase margin (PM), 334
Phase-controlled:
    converter, 121
    inverter, 121
    rectifier, 121
    thyristor, 121
Phasor representation, 34
Photovoltaic interconnection, 475
Pilot thyristor, 608
Pinch-off voltage, 644
Plasma spreading time, 604
pn junction, 513
Polypropylene capacitor, 728
Position:

encoder, 374
   sensor, 374
Potential barrier, 514, 541, 644
Power bipolar junction transistor (BJT), 24, 546
Power conditioners, 354
Power converter, 9
Power diodes, 524
Power electronic converter classifications, 9
Power electronic applications, 7
Power factor (PF), 35, 42
Power factor displacement (DPF), 43
Power factor correction capacitors, 468
Power junction field effect transistor (JFET), 641
Power processor, 3, 9
Power semiconductor switches, 16
Powerline disturbances, 354
   See also Disturbances
Primary breakdown, 550
Protection, power supply, 341
Proximity effect, 771
Pulse-width-modulated (PWM) inverter, 201
   See also Switch-mode dc-to-ac inverters
Pulse-width modulation (PWM) control, 162, 202
   asynchronous, 208
   digital, 341
   linear modulation, 209, 226
   overmodulation, 228
   programmed harmonic elimination, 240
   sinusoidal, 206
   synchronous, 208, 226
Pulse-width-modulated voltage-source inverter (PWM-VSI), 418, 419
Punch-through:
   BJT base, 563
   diode, 526
Push-pull converter, 220
Push-pull inverter, 315

Quadrant operation:
   four-, 122, 393
   single-, 392
   two-, 122, 392
Quasi-resonant converter, 253
Quasi-saturation, 553

Radiated noise, 450
Ramp-limiter, 394, 423
R-C snubber, 669, 671
Reach-through, 526, 563, 588
Reactive power, Q_1, 35
Real power, 34
Recombination, 511
   center, 511

Rectifier mode of operation, 122, 243
Rectifiers. *See* Diode rectifiers, Switch-mode rectifiers
Regenerative action in thyristors, 600
Regenerative braking, 205, 494
Regulated power supplies, 301
Renewable energy source interconnection, 475
Residential applications, 8, 451
Resonant circuits, basic concepts, 253
Resonant converters, 249
   class-E, 271
   classifications, 252
   current source, parallel-resonant, 269
   high-frequency-link integral-half-cycle, 289
   hybrid resonant, 253, 268
   load resonant, 252, 258
   parallel-loaded resonant (PLR), 264
   parallel-resonant circuits, 257
   pseudo-resonant, 280
   resonant dc-link, 287
   resonant switch, 273
   resonant transition, 280
   series-loaded resonant (SLR), 258
   series-resonant circuits, 255
   voltage cancellation, 283
   zero-voltage-switching, clamped-voltage, 280
Reverse bias safe operating area (RBSOA), 567
Reverse blocking state, thyristor, 599
Reverse recovery:
   charge, Q_{rr}, 17, 535, 538
   current, 535, 537, 538
   time, t_{rr}, 17, 538
Reverse saturation current, 518
Reverse-conducting thyristor (RCT), 20
Ridethrough, 356
Ripple, armature current, 388
Ripple, inverter output, 220, 231
Rise time, thyristor, 603

Safe operating area (SOA), 569
   BJT, 569
   IGBT, 637
   MCT, 654
   MOSFET, 591
Saturation flux density, 744
Scherbius drive, 431
Schottky diode, 17, 539
Second breakdown, 550, 563
Sectionalized transformer windings, 775
Semiconductor, 508
Semiconductor controlled rectifier (SCR), 596
   See also Thyristors
Short-circuit capacity, 487
Short-circuit current, 487

Silicon, 508
Simulation, 61
  ATP, 72
  circuit-oriented simulators, 64
  EMTP, 70
  equation solvers, 65, 72
  linear differential equations, 66
  MATLAB, 72
  nonlinear differential equations, 68
  PSPICE, 69
  SIMULINK, 73
  solution techniques, 65
  SPICE, 69
  trapezoidal method of integration, 67
Sinusoidal PWM, 205
Six-step inverter, 418
Skin depth, 748, 753
Skin effect, 748, 753
Slip compensation, 424
Slip, s, 402
Small signal transfer function, 323
Smart switches, 656
Snappiness factor, 535, 538
Snubber circuit, 30
  diode, 670
    baseline capacitance, 672
    baseline resistance, 673
    capacitive, 671
  GTO, 692
  overvoltage, 686
  recovery times, 686, 687, 689
  thyristor, 678
  turn-off, 682
  turn-on, 688
Soft start, 342, 430
Solar cells, 475
Source, MOSFET, 571
Space charge, 514
  capacitance, 536
  layer/region, 514
Specific core loss, 745
Static slip power recovery, 431
Speed sensor, 374
Speed-up capacitor, 701
Spreading time, thyristor, 604
Square-wave inverters, 201
  operation, 218, 229
  switching scheme, 210
  voltage-source inverter (square-wave VSI), 418, 425
  VSI drives, 425
Square-wave pulse switching, 239
Squeezing velocity, GTO, 620
Standards, harmonics, 485
Standby power supply, 362
Start-up of induction motor drives, 408

State-space averaging, 323
Static induction transistor, 641
Static latchup mode of IGBT, 632
Static transfer switch, 362
Static var control (SVC), 471
Step-down converter, 164
Step-down/up converter, 178
Step junction, 518
Step-up converter, 172
Stored charge distribution, 520, 532, 551
Stray inductance, 670, 680
Stress-reduction snubber, 30
Superconductive energy storage inductors, 478
Surge arrestor, 355
Switching dc power supplies, 301
  bulk capacitor, 347
  compensation (feedback control), 333
  control, 322
  current limiting, 343
  current-mode control, 337
  current-source, 319
  design specifications, 346
  digital pulse-width modulation, 341
  direct duty ratio PWM, 333
  electrical isolation, 304, 344
  EMI, 348
  equivalent series resistance, 348
  flyback, 308
  forward, 311
  full bridge, 317
  half bridge, 316
  hold-up time, 347
  inrush current, 347
  K-factor approach, 335
  linear, 301
  linearization, 323
  multiple outputs, 303, 348
  overview, 302
  protection, 341
  push-pull, 315
  soft-start, 342
  state-space averaging, 323
  synchronous rectifiers, 348
  transformer core, 304, 319
  voltage feed-forward, 336
Switching frequency, f_s, 20, 163
Switching power loss, P_s, 23
Switch-mode converter, rectifier mode of operation, 243
Switch-mode dc power supplies. *See* Switching dc power supplies
Switch-mode dc-to-ac inverters, 205
  amplitude-modulation ratio, m_a, 203
  basic concepts, 202
  blanking time, 236

current-regulated (current-mode) modulation, 241
current source inverters, 201
dc-side current, i_d, 213, 218, 234
fixed-frequency current control, 242
frequency modulation ratio, m_f, 204
full-bridge inverter, 211
half-bridge inverter, 211
harmonics, 206, 228
linear modulation, 208, 226
overmodulation, 208, 228
programmed harmonic elimination switching, 240
pulse-width modulated inverter, 203
pulse-width-modulated switching scheme, 203
push-pull inverters, 221
PWM with bipolar switching, 212
PWM with unipolar voltage switching, 215
rectifier mode of operation, 243
ripple in inverter output, 220, 231
single-phase inverter, 211
square-wave operation, 218, 229
square-wave pulse switching, 239
square-wave switching scheme, 210
switch-mode rectifier, 243
switch-utilization, 220, 223, 230
three-phase inverters, 225
tolerance-band control, 241
voltage cancellation for output control, 218
voltage-source inverters, 201
Switch-mode inverters, 200
Switch-mode rectifiers, 200, 243
Symmetrical IGBT, 629
Synchronous PWM, 208, 226
Synchronous rectifier, 348
Synchronous motor drives, 435
brushless dc motor drives, 435
current-regulated voltage-source inverter, 440
cycloconverters, 445
electronically-commutated motors, 435
load-commutated inverter (LCI), 445
sinusoidal waveforms, 439
synchronous motor, 435
torque constant, k_T, 438
trapezoidal waveforms, 440
trapezoidal waveform, synchronous motor, 440
voltage constant k_E, 440
Synchronous speed, 401, 436

Tailing current:
BJT, 560
GTO, 622
IGBT, 635

Tap changing scheme, 357
Telecommunications applications, 8
Thermal conductivity, 731
Thermal resistance, 732
due to conduction, 732
due to convection, 740
due to radiation, 739
Thermal equilibrium carrier density, 510
Thermal ionization, 508
Thermal runaway, 564, 605
Thermal stabilization effect, 589
Thermal time constant, 735
Three-phase inverters, 225
Three-phase pulse-width-modulated (PWM) inverter, 226
Threshold voltage, 25, 575, 578
Thyristor-controlled inductor (TCI), 472
Thyristor converters:
back-to-back connected, 122, 393
discontinuous operation, 134, 148
inverter operation, 135, 148
other three-phase, 153
three-phase, 138
See also Phase-controlled converter
Thyristors, 18, 596
circuit-commutated recovery time, t_q, 19
current commutation circuit, 273
forward blocking state, 18
gate current, 18
gate trigger circuit, 124
inverter-grade, 20
latched on, 18
light-activated, 20
phase-controlled, 20
reverse bias, 19
turn-off time inverval, t_q, 19
Thyristor-switched capacitor (TSC), 474
Tolerance-band control, 241, 338
Total harmonic distortion (THD), 42, 486
Transformers, 52
Transformer core, 305
selection, 782
Transformer design, 780
Transformer, volt-second imbalance, 320
Transient thermal impedance, 733
Transportation applications, 8
Triac, 358
Turn-off gain, GTO, 614
Turn-off snubber, 682
capacitance, 682
resistance, 685
Turn-on delay time, thyristor, 603
Turn-on snubber, 688
inductance, 688
resistance, 688
Twelve-pulse line-frequency converters, 461

Uninterruptible power supply (UPS), 354
   inverter, 360
   rectifier, 358
   static transfer switch, 362
Utility interface, 483
   bidirectional power flow, 494, 499
   improved single-phase, 488
   improved three-phase, 498
Utility-load leveling, 478

Var control, 471, 474
VDMOS, 572
Velocity saturation (electrons), 580

Volt-ampere rating of transformers, 780
Voltage, breakdown, 521
Voltage, forward overshoot, 535
Voltage source inverters (VSI), 201
Voltage-doubler rectifier, 100

Welding, 457
Wind system interconnection, 477
Winding area of inductors and transformers,
   751

Zero-current switchings, 249, 251
Zero-voltage switchings, 249, 251